Illuminating Engineering
From Edison's Lamp To The LED

Second Edition

Illuminating Engineering
From Edison's Lamp To The LED

Second Edition

Joseph B. Murdoch
University of New Hampshire

VISIONS COMMUNICATIONS

Illuminating Engineering: From Edison's Lamp to the LED, 2nd Edition
Joseph B. Murdoch, 1927–

First Edition, published by Macmillan, 1985, ISBN 1-885750-05-6
Reprinted in 1994 by Visions Communications.

Second Edition, 2003, Published by Visions Communications

Printed in the United States of America.

ISBN: 1-885750-05-6
LCCN: 2002116417

To our wonderful grandchildren:
Andrew, Sierra, Katie, Corey, Ben, Winslow,
Max, Ian, Molly, Sarah, Ned, Jessie, and Sydney.
May your feet make happy tracks,
may you always dream
and may rainbows touch your shoulders.

Contents

Preface to the First Edition

When I was writing a book on electrical network theory several years ago, the publisher's representative told me "Joe, one does not complete a book, one abandons it." This statement has a certain applicability here, because, as I wrote this book, it became clear to me that I could not possibly cover all of the myriad of topics in the field of lighting. Choices had to be made, and were.

The book is a scientific and technical treatment of lighting, rather than an architectural and design treatment. It presumes that the reader has a working knowledge of algebra, trigonometry, and introductory calculus. It further presumes a background in physics, such as would be obtained from a one-year college-level experience. It is intended both for students in college courses and for self-study by practitioners in the field.

The book evolved from notes for a course in illumination engineering which I have taught regularly at the University of New Hampshire since the 1950s. Typically, juniors and seniors in electrical and mechanical engineering elect the course, but I have had students from other engineering disciplines, and from physics, chemistry, mathematics, and psychology as well. It seems to me that students in architecture also could take the course, but we do not have an architecture program at UNH.

The purposes of the book (and of the course) are several fold:

1. To provide a foundation in the science and mathematics of light, sight, and lighting.
2. To develop the analytical tools and skills required for quantitative lighting calculations.
3. To present lighting measurement techniques and instruments.
4. To give an historical perspective—the scientific and engineering achievements of the past relevant to lighting.
5. To present the current state of knowledge in the field of illumination.

6. To provide a reference and updating base for those whose knowledge of lighting is rusty or incomplete.
7. To include sufficient detail, examples, problems (with answers), and references that self-study is possible without formal instruction.
8. To point out that, although the book is largely technical and quantitative, there is another important aspect to lighting, namely considerations of quality and aesthetics. A good lighting design is a blend of both.

The book contains too much material to be covered in a one-semester course. Ideally, two semesters should be allocated, with the first devoted to the science of lighting and the second to lighting calculations and applications. If this is not possible, then a choice must be made as to which aspect to feature. If the science of lighting is to be paramount, it is suggested that Chaps. 1, 2, 3, 4, 5, 6, and 10 be stressed. If lighting calculations and applications are of prime importance, then Chaps. 1, 2, 5, 7, 8, 9, and 11 should receive emphasis. Whichever emphasis is chosen, there is an extensive set of problems for each chapter, with answers to selected problems, at the back of the book.

I think it has been my desire to write a book on lighting ever since my first exposure to the subject while I was attending Case Institute of Technology in the late 1940s. Along with many other Case graduates, I owe a special debt of gratitude to Professor Russell Putnam for getting me started in lighting and for being my friend and mentor in those earlier days.

In recent years, it has been my good fortune that UNH is only 60 miles away from Dr. Robert Levin of GTE Lighting Products. Bob has given most generously of his time in helping me with the lighting program here. Also, as the prime reviewer for the book, he has read and critiqued each chapter in detail, for which I am most grateful. The book is much better than it would have been because of Bob's painstaking review and comments.

I would also like to thank Mr. Calvin Gungle of GTE Lighting Products for his careful review of the electrical discharge material for Chaps. 4 and 6, Mr. John Kaufman, Technical Director of the Illuminating Engineering Society, for his reading of the first several chapters and his encouragement, Mr. William Pierpoint of the Naval Construction Battalion Center at Point Hueneme, California, for his comments on the daylighting chapter, and Dr. Gary Yonemura of the National Bureau of Standards for his review of the material on vision. Finally, I am very much indebted to Mrs. Alice Greenleaf of our Word Processing Center for her patience and diligence in typing the manuscript. It will seem strange not to be taking something to type down the hall to Alice each day.

Joseph B. Murdoch

Preface to the Second Edition

In the 18 years since the first edition appeared, there has been a myriad of changes and advances in lighting. Thus, although the revision is labeled as a second edition, it is in many respects a new book on lighting. The major changes are:

1. The material on lighting basics, and in particular the coverage of the six lighting metrics, has been expanded, presented at several levels, and applied through numerous examples.
2. The chapter on vision and color has been separated into two chapters. The vision chapter includes new material on visual perception and the aging eye, plus a detailed presentation of the major methods of estimating visual performance. The chapter on color includes new sections on the LAB and LUV systems, the linking of the CIE and Munsell systems, and the ratio of scotopic to photopic lumens.
3. The chapters on radiation as waves and radiation as particles have only minor revisions, but a new section on electroluminescence and LEDs has been added.
4. The lamp chapter has extensive additions—halogen IR lamps, MR lamps, compact fluorescent lamps, triphosphors, ballast metrics, electronic ballasts, ceramic and pulse start metal halide lamps, and cold cathode tubing.
5. The first edition gave a cursory treatment of luminaires; the second edition contains a chapter on them. New topics included are: Luminaire Efficacy Rating, optical controls for luminaires (diffusers, baffles, louvers, reflectors), luminaire photometry, unified glare rating, and lighting control methods and sensors.
6. The chapter on interior lighting design has been expanded to include qualitative as well as quantitative considerations. The work of Dr. John Flynn on the psychological aspects of lighting is presented. The evolution of illuminance selection procedures is covered, ending with the current IESNA procedure.

An expanded treatment of the Lumen Method is given, including partitioned spaces, non-rectangular rooms, and the latest modifications in determining light loss.

7. Additions to the coverage of optics include consideration of non-point sources and flashing of reflectors, and an expanded treatment of fiber optics and light guides.
8. There have been major changes in the field of exterior lighting and many of these are discussed, including the new Coefficient of Variation metric (CV), the revised method of floodlighting calculations, changes in sports lighting design, and the new small target visibility procedure for roadway lighting analysis.
9. The daylighting chapter has been extensively revised. It includes an expanded and updated coverage of daylight availability, skylight and window treatments, and glazing materials. Daylight models are discussed and the three methods of determining interior illuminance are updated.
10. The material on advanced lighting calculations from Chapters 2, 7, and 8 in the first edition has been reorganized and placed in a new chapter near the end of the book. The coverage of flux transfer is expanded considerably and includes a treatment of one-bounce, two-bounce, and multiple-bounce flux transfer analysis.
11. The first edition did not treat lighting economics. The final chapter of the second edition does, including both simple payback and life cycle cost benefit analysis. A final section covers environmental concerns and costs.
12. Many additional problems, with answers, have been added at the back of the book.

The goals of the book, as set forth in the Preface to the First Edition, still apply. The book continues to be a scientific and technical treatment of lighting, rather than an architectural and design treatment, but many non-engineers and scientists have told me that the first edition was quite readable, and I hope that will be true of the second edition. I have indeed added a number of design-oriented topics to the second edition, such as the psychological aspects of lighting.

I continue to be the beneficiary of Dr. Robert Levin's (OSRAM/SYLVANIA) assistance in the UNH Lighting Program. Dr. Levin reviewed several chapters of the second edition. I also want to thank Bo Budinger of OSRAM/SYLVANIA for reviewing the lamp chapter, Cheryl English, Richard Heinisch, and Brent Nedearis, all of Lithonia, for reviewing the chapter on luminaries, and Morad Atif, National Research Council of Canada, for reading the daylighting chapter. Special thanks to Beth Bay of Visions Communications for guiding the project capably from concept to completion.

Finally I am very much indebted to Nan Collins in the UNH Word Processing Center for typing the manuscript. Not only is Nan extremely competent, but she is always cheerful, making my endless trips from my office to hers a most pleasant experience. Thank you, Nan.

or represent that its use will not infringe any privately-owned rights, including but not limited to patents, trademarks or copyrights. References to downloading the 2001 Edition should be referred to the New Buildings Institute web site at http://www.newbuildings.org to allow the Institute to track distribution and notify users of future editions.

Joseph B. Murdoch

Chapter 1

Introducing Light and Seeing

1.1 PRE-ELECTRIC LIGHTING

It is likely that interest in lighting had its beginnings when one of man's prehistoric ancestors carefully gathered up the blazing embers of a forest fire and brought them into his cave to provide warmth and light. Much later, in medieval times, metal fire baskets called *cressets* were developed. Holding glowing pine knots, these containers were suspended from ceilings or hung on wall brackets. Pine wood was also used for light in colonial times. It was called *candlewood*, and bundles of slivers were placed in metal stands near the hearths in homes.

It could be said that the flaming torch was the first "portable" lamp. American Indians used pine knots as torches, whereas early colonial settlers found that certain reeds or rushes, when dried and dipped in grease or fat, could be burned and carried about to provide illumination. These were called *rushlights*.

For a variety of obvious reasons, wood did not turn out to be satisfactory as a source of light. Through recorded history four other basic light sources have emerged. These are, in chronological order: oil lamps; candles; gas lighting; and electric lamps. The rise and fall of the first three of these will be considered in the remainder of this section, and the early development of electric lighting will be addressed in the section to follow.

The ancient civilizations of Babylonia and Egypt (around 3000 B.C.) had crude oil lamps. These were in the form of small open stone or sea shell bowls containing animal fat, fish oil, or vegetable oil. Later, the bowls were made of clay and, still later, of metal, particularly bronze, pewter, and tin. Around 500 B.C., the Greeks and Romans developed the oil reservoir lamp, which had a wick projecting from a channel and a cover to keep the fuel in the reservoir clean.

Early colonists used virtually the same type of oil lamp as did the Romans. It is true that their lamps were made of metal and were constructed better, particularly with regard to the wick, but they were basically the same lamp that had been developed 1000 years earlier. The most widely used was the Betty lamp, brought over from Germany. Then, in the mid 1700s, the rise of the whaling industry provided a better fuel. Whale oil lamps produced steadier and brighter light, but the lamps were still smoky and rather dangerous. About this time Benjamin Franklin discovered that placing two wicks side by side and slightly apart produced more light than two single-wick lamps. Lamps employing the Franklin principle were used in the United States until after the Civil War.

The discovery of oil by Colonel Edwin Drake in 1858 provided a new liquid fuel for lighting. Kerosene (coal oil) lamps were safer, gave better light, and were less expensive to operate. Three other developments aided their acceptance. One was the invention of the friction match, which made it much easier to light the lamp. Second was the development of the flat wick and the associated mechanical button for turning it in and out of the lamp. Last was the creation of the lamp chimney. In 1783, a Swiss chemist named Ami Argand discovered that placing a glass cylinder over a lamp flame caused the flame to stop smoking and become steadier and brighter. Thomas Jefferson, writing from Europe in 1784, stated "there has been a lamp called a cylinder lamp lately invented here. It gives a light equal to six or eight candles. This improvement is produced by forcing the wick into a hollow cylinder so that there is a passage of air through it." When it is realized that lamps of that time generally provided the equivalent light of one candle per wick, the significance of Argand's discovery can be appreciated.

The development of the candle, from about 400 A.D., is closely associated with the development of the Christian church. Few candles were used in nonchurch settings until the fourteenth century. Church candles were made of beeswax (the bee was the symbol of purity) and were quite expensive. Inexpensive, smelly, and smoky tallow candles were the lot of the common people, if they were able to obtain them.

Most early American colonists fashioned candles from tallow also, using crude wicks of flax. The wealthier people could afford wax from the bayberry bush, which was cleaner and more fragrant than tallow.

The rise of the whaling industry produced not only a better fuel for oil lamps, but also a better wax for candles. Spermaceti, a translucent and slightly crystalline wax that vaporizes readily, is derived from oil from the head cavity of the sperm whale. It was superior to all other candle materials; consequently, spermaceti candles were used in the past as standards for basic lighting measurements.

In the 1850s it was discovered that much of the smoke and odor of tallow could be eliminated by removing the glycerine from animal fats. The result was the stearine candle. At nearly the same time, paraffin, a petroleum product, was introduced. Most inexpensive candles today are made of molded paraffin.

The Chinese are believed to be the first people to use natural gas for lighting, piping it from salt mines through bamboo tubes. Around 1800, several people discovered how to distill coal to obtain natural gas, and gas lighting was introduced. By 1823, London had 40,000 gas lamps. However, it was the mid 1800s before the gaslight era began in the United States, with fuel for both light and heat distributed from central station facilities to residential and commercial users.

The first gas burners were simply iron pipes with holes in the end. These clogged, and a lava tip for gas burners was developed. Later the fishtail burner, with two jets at 90° forming a flat flame, was introduced.

Throughout most of the nineteenth century, gas lighting was provided by the light from an open gas flame, rather than by the burning gas heating an object to incandescence. Then, in 1886, C.A. Von Welsbach, in an effort to improve gas lighting to compete with the infant electric lighting industry, invented the Welsbach burner. This "mantle" consisted of a cone of knit cotton fabric soaked in nitrates (later oxides). When heated for the first time, the cotton was burned, leaving a thin fish net of ash that became incandescent when placed over a gas flame. Refinements of the Welsbach mantle are still in use today, especially in camping lanterns.

The invention of the incandescent lamp by Thomas Edison in 1879 triggered a 30–year battle between the approximately 500 gas companies existing in the United States at that time, with their central stations and distribution networks, and the fledgling electric industry. Obviously, the latter won the battle, for gas lighting devices were severely limited. They had to be burned upright; nothing inflammable could be near them; they required air, but could not tolerate drafts; they were difficult to start; and, above all, they were dirty. Despite the merits of electric lighting, nearly one-quarter of the people in the United States in the mid 1900s still depended on flame sources for light.

1.2 EARLY DEVELOPMENTS IN ELECTRIC LIGHTING

The first recorded man-made electric light was by Otto Von Guericke, a German scientist, in 1650. Von Guericke's apparatus consisted of a chain attached to a spinning sulfur ball. When he held his hand against the ball, sparks jumped from the chain.

In 1710, Sir Francis Hauksbee, an Englishman, noted a glow inside a glass globe from which air had been evacuated, and mercury added. He called the glow "electric light" and claimed that his experiment proved that electricity could produce light.

Benjamin Franklin, with his famous kite experiment in 1752, gave further proof that electricity could produce light, by collecting electric charge from clouds into Leyden jars during a thunderstorm.

As the nineteenth century began, the development of electric lighting took three rather distinct paths, two of which were initiated by the work of Sir Humphrey Davy, a British scientist. In 1802, Davy showed that if thin strips of metal were raised to high temperatures by passing electricity through them, they would emit light. This observation ultimately led to the development of the incandescent lamp.

Then, in 1808, before members of the Royal Institution, Davy demonstrated the first carbon arc, a luminous discharge about 4 in long between pieces of charcoal connected to a battery of 2000 cells. Davy's source was the forerunner of the carbon arc lamp used extensively in outdoor lighting in the late 1800s.

Davy's observation of incandescence led to a crude incandescent lamp by Warren De la Rue in England in 1820. De la Rue's lamp consisted of a coiled filament of platinum wire mounted in a glass tube. The ends of the coil were connected to brass caps at the ends of the tube.

Early experiments had shown that a metal filament has a very short life if heated in air—it literally burned up. Thus, it is likely that De la Rue's lamp was evacuated. However, it is also likely that the vacuum was not very good, resulting in a very short lamp life. The mercury vacuum pump, which was so important in later incandescent lamp development, was not invented until 1865, by Herman Sprengel.

From De la Rue's lamp in 1820 until Edison lit his first successful lamp on October 21, 1879, many people worked on incandescent lamp development. However, the lamps they created did not prove practical, largely because of unreliability, short life, and excessive operating expense. William Sawyer, Albon Mann, Moses Farmer, and Hiram Maxim in the United States and Frederick DeMoleyns, St. George Lane-Fox, and Sir Joseph Swan in England all developed lamps during this period. Farmer, a native of New Hampshire, used the platinum filament incandescent lamp he developed to light the parlor in his home in Salem, Massachusetts, in 1859.

All of these experimenters used thick filaments, of either platinum or carbon, which had low electrical resistance and required large currents to heat them to incandescence. The controversy over who truly invented the incandescent lamp continues to this day, particularly with regard to Swan and Edison. Swan showed that carbon should be the filament material, but it remained for Edison to show that a fine hairlike filament, which had high resistance and thus required a small current to cause it to emit light, was the answer for a commercially-practical incandescent lamp. Edison was the only one who understood that lamps should be operated from constant voltage, thus permitting additional lamps to be added in parallel. Distinguished scientists, such as Lord Kelvin, rejected the idea that light could be "subdivided".

Concurrent with early incandescent lamp development was the development of the arc light, again stemming from Davy's early work. The first commercial arc lamp was installed in the Dungeness Lighthouse in England in 1862. The earliest arc lamp systems in the United States were developed by Edward Weston, Elihu Thomson, William Wallace, and Charles Brush in the 1870s and 1880s. These were open carbon arc systems, as contrasted to the enclosed gas discharge arc lamps that exist today. Weston made the first carbon arc street lighting installation in Philadelphia in 1879, whereas Brush was responsible for providing the first all-night street lighting system in Cleveland, also in 1879. Two sets of 8-h carbons were installed in each Brush arc lamp, the second set automatically going into operation after the first set was consumed.

The third path of light source development in the 1800s was that of enclosed gaseous discharge lamps. The German scientist, Herman Geissler, had observed in 1856 that a tube containing a small amount of air would glow if a high voltage were placed across its ends. In 1860, John Way, an Englishman, created the first mercury arc by breaking a stream of mercury connected to a battery and observing its characteristic blue-green color. Arons developed a mercury vapor arc in 1892, which led Peter Cooper Hewitt to produce a commercial tubular mercury lamp with a mercury pool cathode and an iron anode in 1901. When the lamp was tilted, a column of mercury bridged the gap between the anode and cathode. As the lamp was righted, the mercury column broke, and the electric arc discharge began. The Cooper Hewitt lamp was widely used in the early 1900s and led to the development of high-pressure mercury lamps in the 1930s and metal halide and sodium lamps in the 1950s and 1960s.

None of the lamps discussed thus far makes use of fluorescence. In 1852, Sir George Stokes discovered the basic principle of transforming ultraviolet radiation into light through irradiation of a quinine sulfate solution. Seven years later, A.E. Becquerel discovered that certain powdered materials when placed in an evacuated tube to which a high voltage was applied would fluoresce. These early discoveries led to extensive research on fluorescent materials in the early 1900s and to the development of the fluorescent lamp in the 1930s.

1.3 THE WAVE NATURE OF LIGHT

Light is defined as visually evaluated radiant energy. This definition suggests that light is first of all energy; secondly, that it is transmitted by radiation; and, lastly, that it is a form of radiant energy to which the eye is sensitive. All of these facets of light need to be explored.

Until the middle of the seventeenth century, most scientists, particularly in the field of optics, accepted the corpuscular theory of light as enunciated by Sir Isaac Newton (1642–1727). According to Newton, all hot bodies emitted elastic corpuscles, each having the same, very high velocity and each having a size dependent upon its color. These corpuscles were considered to be minute particles that traveled in straight lines and could be reflected and refracted. Reflection was explained much as one would explain the bouncing of a ball against a wall. Refraction, the bending inward of light as it enters a denser medium, was justified on the basis of increased gravitational attraction of the more dense medium, causing a particle entering it to bend toward the perpendicular to the surface of the medium. Because the bending indicated an acceleration toward the perpendicular, it was postulated that the velocity of the particles would be greater in the denser medium. Newton explained diffraction, the bending of light around a sharp edge, in the same way, but he could not explain the phenomena of polarization and interference of light with his corpuscular theory.

In 1670, Christian Huygens (1629–1695), Dutch astronomer, physicist, and mathematician, developed a wave theory of light, in contrast to the corpuscular theory of Newton. Huygens described the bending of light as a fundamental principle of wave motion, Huygens' principle, which states that every point on an advancing wavefront behaves as a source of secondary wavelets that are sent out radially. Thus, a wavefront encountering an obstruction will send out a secondary wave in a direction different from that of the incident wave. The similarity to the action of ripples on a lake encountering an obstruction should immediately come to mind. Huygens showed that the laws of reflection and refraction could be explained by his wave theory. He also proposed that color was a function of wavelength, again in contradiction of Newton, and conducted many experiments involving polarization of light.

The English physicist Thomas Young (1773–1829) reinforced Huygens' wave theory with his famous experiment on interference in 1801. Young showed that a narrow beam of light passing through a pinhole, and then through an additional pair of pinholes to a screen, created alternating light and dark bands on the screen. The light bands occurred when the secondary waves from the pair of pinholes reinforced each other by arriving in time phase at the screen; the dark bands occurred when the waves arrived out of time phase and thus interfered with each other.

It remained until 1862 for the controversy between the Newton corpuscular theory and the Huygens-Young wave theory to be resolved. In that year, Jean Bernard Leon Foucault (1819–1868), a French physicist known for the Foucault pendulum, conducted an experiment in which light was passed through air and a tube of water and its velocity measured by a rotating mirror apparatus. Foucault showed that the speed of light was less in the denser medium (water), conclusively disproving the corpuscular theory of Newton.

Thus, in the mid 1800s light was considered as wave motion by most scientists. The waves vibrated at right angles to the direction of propagation of the wavefront (transverse waves, as opposed to longitudinal waves that vibrate along the direction of propagation), and light "rays" were nothing more than lines showing the paths followed by different portions of the wavefronts. Reflection, refraction, interference, diffraction, and polarization could all be explained in terms of wave theory.

James Clerk Maxwell (1831–1879), the noted Scottish physicist, unified these concepts in his Treatise on Electricity and Magnetism, published in May 1873. Essentially, Maxwell showed that electromagnetic waves consist of electric and magnetic fields, vibrating at right angles to each other in space and at right angles to the direction of propagation. These fields are in phase with each other in time and have the same frequency. His theory made it unnecessary to have a medium, such as the fictitious "ether" of the early nineteenth century, for transmission of a wave. He further theorized that the velocity of electromagnetic waves could be computed from purely electric and magnetic measurements. He calculated this velocity as very nearly the measured velocity of light and, thus, was able to infer that light consisted of electromagnetic waves of short wavelength.

Maxwell's work paved the way for Heinrich Hertz (1857–1894), a German physicist, to produce electromagnetic waves experimentally in the microwave region of the spectrum in 1887, to show that they had the properties of light waves, and to verify that Maxwell's theories were indeed valid and universally applicable to all electromagnetic radiation. Hertz's work proved Maxwell's inference that the common denominator in all electromagnetic radiation is the velocity of light.

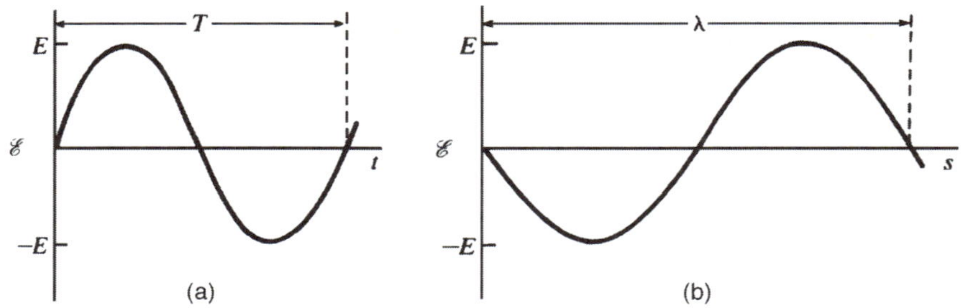

Figure 1.1 Electric field intensity in time and space.

It is useful to examine Maxwell's electromagnetic waves in further detail. The diagrams in Fig. 1.1 depict the electric field density $\mathscr{E}$ of an electromagnetic wave propagating through space. Figure 1.1a shows the variation in $\mathscr{E}$ with time at a fixed point in space (s = 0 is assumed). An expression for this variation can be written as:

$$\mathscr{E} = \text{E} \sin 2\pi \text{ft} \tag{1.1}$$

What this equation says is that the electric field intensity at s = 0 is a sinusoidal function of time with period T = 1/f seconds, where f is frequency in hertz, and amplitude E volts/meter. Once the wave is established, the waveshape at any other point in space will be the same except that it will be shifted in time, and perhaps attenuated. If an observer stands a fixed distance from a radio antenna, light source, or some other kind of electromagnetic wave generator and holds an "electric field intensity detector" attached to some kind of graphical read-out device, such as a chart recorder or an oscilloscope, he sees the waveform in Fig. 1.1a after the wave first reaches him.

Now suppose the observer could move speedily outward from the source with time fixed. Assuming no attenuation, he would see a pattern such as that in Fig. 1.1b after the wave is established, where s is the distance from the source. When he reaches the point at which the wave begins to repeat itself, he has traversed one wavelength, λ, in space, as labeled in Fig. 1.1b.

An equation for this variation of the electric field intensity with distance at a fixed point in time after the wave is established can be written as:

$$\mathscr{E} = -\text{E} \sin 2\pi \frac{\text{s}}{\lambda} \tag{1.2}$$

This is also a sinusoidal variation, but as a function of s rather than t, with wavelength λ playing the same role that period did in Eq. (1.1).

The time and space variations in Eqs. (1.1) and (1.2) can be combined to obtain an overall expression for the electric field intensity, given by

$$\mathcal{E} = E \sin 2\pi \left(ft - \frac{s}{\lambda} \right) \tag{1.3}$$

At $s = 0$, Eq. (1.3) becomes Eq. (1.1); at $t = 0$, it becomes Eq. (1.2).

What is the velocity of propagation of the wave? This will be the velocity of a fixed point on the wave and is obtained by requiring that the quantity $(ft - s/\lambda)$ in Eq. (1.3) remain constant as time passes. Thus, as t changes from t_1 to t_2, s must change from s_1 to s_2 such that

$$ft_1 - \frac{s_1}{\lambda} = ft_2 - \frac{s_2}{\lambda} \tag{1.4}$$

Rearranging this expression gives

$$\frac{s_2 - s_1}{t_2 - t_1} = \lambda f \tag{1.5}$$

The left side of Eq. (1.5) is indeed the velocity of the wave, the rate of change of position of a point on the wave with time. The minus sign in Eq. (1.3) can now be explained by noting that if a plus sign had been chosen, the velocity in Eq. (1.5) would have been negative and the wave would have been traveling in the wrong direction (from observer to source).

To summarize what has been shown: The electric field intensity of an electromagnetic wave is sinusoidal in both time and space. At a fixed point in space, it is sinusoidal in time with a period T. At a fixed point in time, it is sinusoidal in space with a wavelength λ. The wave propagates with a velocity given as the product of wavelength and frequency.

The magnetic field intensity H of an electromagnetic wave is similar to the electric field intensity, except that the two waves lie at right angles to each other in space. Both lie at right angles to the direction of propagation. The fields are in time phase in that, at a given point in space, $\mathcal{E}$ reaches a positive maximum at the same instant of time as does H. The ratio of the two fields intensities is always the same and can be shown to be given by:

$$\frac{\mathcal{E}}{H} = \frac{\mu_0}{\varepsilon_0} \tag{1.6}$$

where μ_0 is the permeability of free space and ε_0 is its permittivity. These two quantities occur throughout the study of electromagnetic theory. Maxwell's calculation in 1873 of the velocity of electromagnetic waves radiating from an oscillating electric circuit depended on them in the form of the following simple equation

$$v = \sqrt{\frac{1}{\mu_0 \varepsilon_0}} \tag{1.7}$$

where

$$\mu_0 = 4\pi \times 10^{-7}\ \mathrm{N \cdot s^2/C^2} \quad \text{and} \quad \varepsilon_0 = \frac{1}{36\pi \times 10^9}\ \mathrm{C^2/N \cdot m^2} \tag{1.8}$$

where C denotes "coulomb."

Inserting these values in Eq. (1.7) gives $v = 3 \times 10^8$ m/s, which is the speed of light. This led Maxwell to conclude that light consisted of electromagnetic waves and allows the revision of Eq. (1.5) to show that, for all electromagnetic radiation, frequency and wavelength are related by

$$c = \lambda f \tag{1.9}$$

where c is the velocity of light. If c is in meters per second, then λ is in meters and f in cycles per second or hertz. The full gamut of the electromagnetic spectrum is shown in Fig. 1.2. Note the immense range of wavelengths present, from the very short cosmic rays (10^{-14} m) to the very long 60-Hz electric power waves (3100 mi).

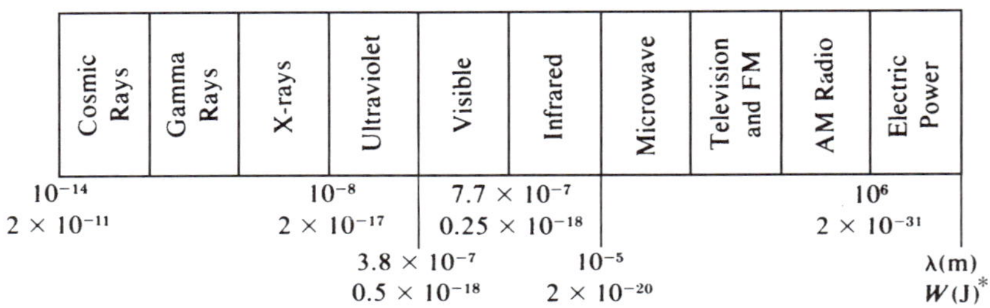

Figure 1.2 The electromagnetic spectrum: wavelengths and photon energies.

The portions of the electromagnetic spectrum with which we will be concerned are the ultraviolet, visible, and infrared, as shown in Fig. 1.3, with emphasis on the visible. It is seen that light waves occupy a very small portion of the spectrum, extending roughly from 380 to 780 nm. A nanometer (nm) is 10^{-9} m, about 50 billionths of an inch.**

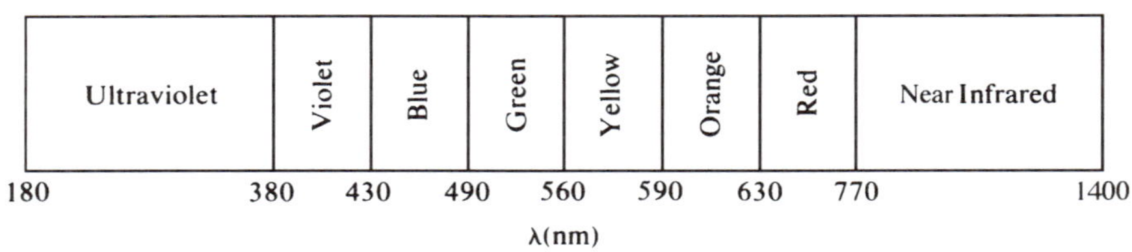

Figure 1.3 The visible and near-visible spectra.

Six basic colors are generally associated with the visible spectrum, as indicated in Fig. 1.3. These colors are not distinct bands, as implied by Fig. 1.3, but rather blend

* The energies (W) associated with Fig. 1.2 will be discussed in Sect. 1.5.

** Light wavelengths in the past were also measured in Angstroms and microns. An Angstrom (Å) is 10^{-10} m and a micron (μ) is 10^{-6} m. The Angstrom unit is named for A.J. Angstrom (1814–1874), a Swedish physicist noted for his atlas of the solar spectrum.

together, as anyone who has looked at a rainbow or examined the light emerging from a prism realizes.

Why is this narrow band of wavelengths of such importance? The answer, of course, is that human eyes respond to these wavelengths, and to these wavelengths only. This concept of sensitivity of a receptor to certain wavelengths and not to others is an important one, and is not unique to the eye. For example, skin sensitivity for tanning centers at about 300 nm in the ultraviolet. Bacteria are critically sensitive to radiation centered at 265 nm, and germicidal lamps provide radiation near this wavelength (at 253.7 nm) for product sanitation and to kill air-borne bacteria in sanitariums and nurseries. In a different field, resonant circuits in radios and television sets are "tuned" to frequencies in the 300-m and 3-m regions, respectively, of the spectrum.

While the various spectral colors are of interest, the primary interest is in the sensation of whiteness. Equal-energy white light can be defined as the simultaneous presence, in equal energy amounts, of all the wavelengths within the visible spectrum. Incandescent, fluorescent, and high-intensity discharge lamps do not provide equal-energy white light. Incandescent lamps are much stronger in the reds and yellows than they are in the blues and greens. Standard fluorescent lamps are somewhat weak in blues and reds. Clear mercury lamps have only blues, greens, and yellow-greens, and sodium lamps are strongest in the yellows. Sunlight provides a nearly uniform spectral distribution, but is somewhat deficient in blues and violets. However, each of these sources, by itself, appears to the eye to produce white light, even though one white light may seem "warmer" or "cooler" than another. Thus, it is not necessary to have equal-energy white to create the sensation of whiteness. It is only when these lamps are used to illuminate colored objects that the differences in spectral content between them can be detected.

1.4 WORK, ENERGY AND POWER

Because so much of the subject of illuminating engineering has to do with power and energy, it is important to understand at the outset what these two quantities are, how they relate to each other, and how they are connected to the concept of work. The concept of work will be considered first. Let a constant force F be applied to a body such that the body moves a distance s in a direction making an angle θ with the line of application of the force. The work done by the force F is given by $F \cos \theta \times s$, the product of the component of the force in the direction of motion and the displacement. For work to be done, there must be motion, and there must be a force component in the direction of the motion (work is not done if you hold a baby still in your arms, but work is done if you rock the baby to and fro).

In Systeme Internationale (SI) units, the unit of force is the Newton, and the unit of displacement is the meter. Thus, the unit of work is the newton-meter and is called a joule. One joule of work is done when a force of one newton in the direction of motion moves a body one meter.

Now it will be determined how work done on a body relates to the potential and kinetic energy gained by that body. The energy exchanges that occur within gaseous discharge light sources will be used as an example. Consider the situation shown in Fig. 1.4 where two parallel plates are separated by a distance d, with an applied voltage V, positive at the left plate. The region between the plates is assumed to be filled with gas atoms and a single electron sits just outside the positive plate, held there by

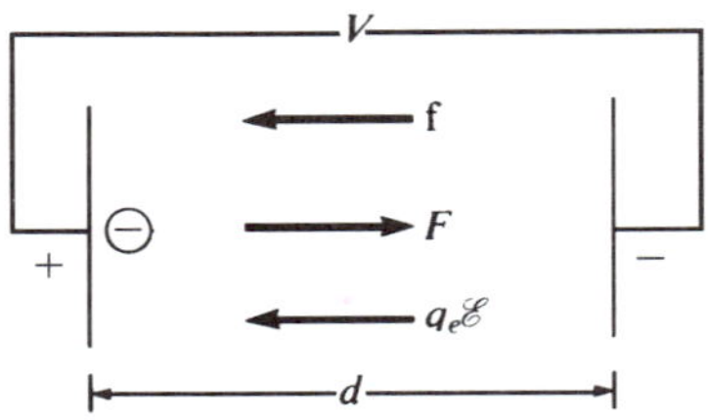

Figure 1.4 Relating work and energy.

the positive voltage. A force is applied so as to move the electron toward the negative plate and it is desired to relate the work done by this force to the energy gains of the electron.

There are three forces in Fig. 1.4. One is the applied force F which must somehow be exerted on the electron to move it toward the negative plate. The second is the force of the applied electric field, which resists efforts to move the electron. The electric field is given by V/d and its force by $q_e V/d$, where q_e is the charge on the electron (=1.60 × 10^{-19}C). The third force is a "friction" force f_r imposed by the medium between the plates. In a gaseous discharge device this is the effect of the collisions of the electron with the many gas atoms in the space. Most of these are collisions in which the kinetic energy of the two colliding bodies is conserved, called *elastic collisions*, and result in the exchange of a small amount of kinetic energy between the electron and the gas atom.

With the forces defined, Newton's second law of motion for the situation in Fig. 1.4 can be written as

$$F - f_r - \frac{q_e V}{d} = m_e a \tag{1.10}$$

where m_e is the mass of the electron (9.11 × 10^{-31} kg) and a is the acceleration.

Equation (1.10) is a force equation. To convert it to a work equation, each side is multiplied by d.

$$Fd - f_r d - q_e V = m_e ad \tag{1.11}$$

Since the forces in Fig. 1.4 are assumed constant, the acceleration is constant and is given by

$$a = \frac{v^2}{2d} \tag{1.12}$$

where v is the velocity of the electron as it reaches the negative plate.

Inserting Eq. (1.12) into Eq. (1.11) and rearranging yields

$$Fd = f_r d + q_e V + \frac{1}{2} m_e v^2 \tag{1.13}$$

Each term in Eq. (1.13) should be carefully examined. The term on the left side is the work done by the external force applied to the electron. This is the "input" to the system. The "output" consists of the three terms on the right side of the equation. The leftmost of these is the work done against the "friction" between the electron and the medium. This will cause the gas atoms to increase in temperature, so this work ultimately manifests itself as heat energy imparted to the system (mechanical energy converted to thermal energy). There are, of course, other ways of imparting heat to a body. For example, passing an electric current through a wire, as in an incandescent lamp, causes the wire to become hot. This is the conversion of electric energy to thermal energy.

The middle term on the right side of Eq. (1.13) is the increase in potential energy of the electron. This energy is called potential energy because it is determined by the position of the electron, rather than by its motion. Note that the increase in potential energy is independent of the path taken in going from one plate to the other. It depends only on the difference between the final and initial positions. Note also that the electron at the negative plate has a greater potential to do work than it did at the positive plate. Thus the work going into potential energy is recoverable, unlike the friction work which is lost as heat to the gas atoms.

The rightmost term in Eq. (1.13) is the increase in the kinetic energy of the electron and is governed by the increase in its velocity in going from rest at the left plate to a velocity v at the right plate.

To sum up, the work done by the external force in Fig. 1.4 manifests itself as an increase in temperature due to friction, an increase in the potential energy of the electron due to a change in its position, and an increase in the kinetic energy of the electron due to an increase in its velocity. For Eq. (1.13) to balance dimensionally, the unit of energy must be the same as that of work, namely the joule in the SI system. When work is done on a body it gains energy; when work is done by a body, it loses energy. The results in Eq. (1.13) will be referred to in subsequent chapters during the discussion of the mechanisms by which light sources process energy and produce visible radiation.

Thus far, the time element has been excluded from discussion. If a force of one newton is exerted on a body and moved a distance of one meter, one joule of work has been done, whether the task takes a second, an hour, or some other period of time to perform. An additional concept to describe the rate at which work is done or energy is produced or consumed is required. This concept is power, defined as the time rate of doing work. Energy and power are related by

$$W = \int_{t_1}^{t_2} P\,dt$$

$$P_{av} = \frac{W}{t_2 - t_1} \left(\text{Average power} = \frac{\text{work done or energy produced}}{\text{time interval}} \right) \tag{1.14}$$

In the SI system, the unit of power is the watt, which is a joule per second. Thus the joule is a watt-second.

Electric energy is generally measured in kilowatthours, where one kilowatthour = 3.6×10^6 J. Another common unit of energy measurement is the British thermal unit

(Btu), which is the amount of heat to raise a pound of water one degree Fahrenheit. The unit of power in this system is the Btu per hour, which is equivalent to 0.293 W. Thus each watt represents 3.413 Btu/h.

In these days of worldwide concern about energy, there is often public confusion about what the terms power and energy mean. There is a wish to conserve energy, but there is no particular virtue in saving power.* For example, if reducing the light level (lighting power) in a large office by 30% reduces the amount of work accomplished to the extent that an extra shift is required to maintain work output level, more lighting energy will be used than before. It is not only that the lights are on that is important, it is how long they are on.

1.5 THE PARTICLE NATURE OF LIGHT

It might appear from the foregoing that Maxwell's treatise provided a final and unified theory of light. As a matter of fact, most scientists at the close of the nineteenth century felt that little new knowledge concerning the nature of light would be forthcoming in the future. However, they were quite wrong.

Despite its great accomplishments, electromagnetic wave theory could not explain the radiation spectrum from a blackbody or photoelectric emission from the surfaces of metals. These phenomena will be discussed in detail in Chaps. 4 and 5, covering the work of Max Planck, Albert Einstein, and the other physicists who made such notable contributions during the first part of the twentieth century. At this point, it should be noted that Planck postulated in 1900 that the atoms in the walls of a blackbody radiator oscillated with only discrete (quantized) energies given by

$$W = nhf \tag{1.15}$$

where n is an integer and h is Planck's universal constant (6.63×10^{-34} J · s). Then in 1905, Einstein, drawing on Planck's work, postulated that the energy in a light beam was concentrated in small bundles, later named photons**, instead of being distributed throughout space in its electric and magnetic fields, with each photon having an energy in joules given by $W = hf$.

Plank believed that light, although emitted from its source in discrete quanta, traveled outward as an electromagnetic wave. Einstein concluded that light propagated in bundles of particles.

A parallel development with which the reader who has taken a chemistry course is presumably familiar should be noted. In 1911, the New Zealand physicist Earnest Rutherford (1871–1937), while working in England, verified the existence of the atomic nucleus. He proposed a model for the atom consisting of a nucleus, positively charged, surrounded by orbiting electrons, negatively charged, very similar in concept to the sun and planetary system.

Specifically the Rutherford model had these characteristics:

1. Positively charged nucleus of relatively large mass surrounded by negatively charged electrons of much lesser mass.

---

* Except for the possibility of a demand charge to users who have a large kilowatt connected load.

** By G.N. Lewis in 1926.

2. Electrically neutral atom: the number of protons in the nucleus equals the number of orbiting electrons.
3. Diameter of atom is 10^5 times diameter of nucleus.
4. Electrons attracted toward the nucleus by Coulomb's law:

$$F = \frac{q_e^2}{4\pi\varepsilon_0 r^2} \tag{1.16}$$

where r is the distance of the electron from the nucleus.
5. Electrons revolving in fixed orbits about the nucleus, requiring that each electron have a centripetal acceleration toward the nucleus and thus a centripetal force given by Newton's second law as

$$F = \frac{m_e v^2}{r} \tag{1.17}$$

This model came into immediate conflict with Maxwell's electromagnetic theory, which asserts that an accelerating electron radiates energy thus causing its total energy to decrease. Since an electron orbiting a nucleus is accelerated continually toward that nucleus, it should continually radiate electromagnetic energy until its energy is gone and it collapses into the nucleus. Furthermore, as the electron spirals in to rest, its angular velocity ($\omega = v/r$) should continually increase. This predicts a continuous spectrum, in contradiction to the observed line spectra of hydrogen, mercury, etc.

Niels Bohr (1885–1962), the Danish physicist who had studied atomic physics under Rutherford in England, attempted to resolve this difficulty. In a series of papers from 1913 to 1915, he developed a theory of atomic structure that came to be known as the Bohr atom and earned for him the Nobel prize in physics in 1922. His theory consists of three fundamental postulates:

1. An electron in an atom can exist only in certain specified circular orbits. When in one of these orbits, the electron is in a stable state and does not emit radiation.
2. An electron may jump suddenly from one stable state to another of lower energy. When it does so, a single photon is emitted, whose energy is given by

$$W_i - W_f = hf = \frac{hc}{\lambda} \tag{1.18}$$

where W_i is the energy of the initial orbit and W_f is the energy of the final orbit.
3. Only those orbits are permitted in which the angular momentum of the electron is an integral multiple of $h/2\pi$. Stated mathematically

$$m_e vr = n\frac{h}{2\pi}\ \mathrm{J \cdot s} \tag{1.19}$$

where n is an integer.

The spectral lines of hydrogen, the simplest element, were very well known at the time Bohr set forth his theories. Thus his postulates were tested thoroughly on hydrogen and were found to predict the frequencies of its spectral lines within one part per million.

The radii of the orbits, the permitted levels and the resulting spectral lines of hydrogen can be calculated. From Rutherford's third and fourth postulates, the centripetal force causing the electron to travel in a circle is provided by Coulomb's law of attraction between the electron and proton. Combining Eqs. (1.16) and (1.17) gives

$$\frac{q_e^2}{4\pi\varepsilon_0 r^2} = \frac{m_e v^2}{r} \tag{1.20}$$

Equations (1.19) and (1.20) may be solved simultaneously for r and v, since all other quantities in these equations are presumed known. The results are:

$$r = \frac{n^2 h^2 \varepsilon_0}{\pi m_e q_e^2}, \qquad v = \frac{q_e^2}{2nh\varepsilon_0} \tag{1.21}$$

The only variable in these two equations is n, the integer which quantizes the orbits.

Unlike the situation in Fig. 1.4 and Eq. (1.13), there is no friction force associated with an electron rotating about a nucleus. Thus the energy of an orbiting electron consists of just two parts. First is its potential energy, which is the work done in bringing it from infinity to the orbit in question. This is given by

$$PE = \int_{\infty}^{r} F dr = \int_{\infty}^{r} \frac{q_e^2}{4\pi\varepsilon_0 r^2} dr = -\frac{q_e^2}{4\pi\varepsilon_0 r} \tag{1.22}$$

The second energy of the electron is kinetic and is obtained from Eq. (1.20) as

$$KE = \frac{1}{2} m_e v^2 = \frac{q_e^2}{8\pi\varepsilon_0 r} \tag{1.23}$$

Thus the total energy of the electron in an orbit is

$$TE = KE + PE = -\frac{q_e^2}{8\pi\varepsilon_0 r} \tag{1.24}$$

Equations (1.21) through (1.24) show that as n decreases, r decreases, PE decreases, v increases, KE increases, and TE decreases. Thus an electron closest to the nucleus (n = 1) has the least potential energy, the greatest kinetic energy, and the least total energy. Also, all energies are zero at $r = \infty$.

A tabulation of radii and energies for the single hydrogen electron appears in Table 1.1. The radii vary directly as n^2; the total energies inversely as n^2.

TABLE 1.1 HYDROGEN: RADII AND ENERGIES

n	r(nm)	TE ($J \times 10^{-19}$)
1	0.053	−21.65
2	0.21	−5.41
3	0.48	−2.40
4	0.85	−1.36
5	1.33	−0.86
∞	∞	0

Equation (1.18) gives the energy of a photon resulting from a transition from one energy level in an atom to a lower one. Assume the electron in a hydrogen atom has been lifted from its normal state (n = 1) to a higher energy state (n = 3) due to a collision of the atom with a fast-moving external electron. Assume further that, following the collision, the atom's electron returns to the n = 2 state. A photon will be emitted whose wavelength is given by

$$\lambda = \frac{hc}{W_i - W_f} = \frac{6.63 \times 10^{-34} \times 3 \times 10^{8}}{(-2.40 + 5.41) \times 10^{-19}}$$

$$= 6.61 \times 10^{-7} = 661 \text{ nm}$$

This is indeed one of the observed spectral lines of hydrogen, which truly confirmed Bohr's work.

The work of Planck, Einstein, Rutherford, and Bohr led to many years of controversy in physics, culminating in the reluctant acceptance of the fact that light appears to be dualistic in nature. At a source or at a receiver, the concepts of quantum theory appear to explain better what is going on, whereas in the medium between source and receiver, electromagnetic wave theory seems to provide the better explanation. This dualistic nature of electromagnetic radiation is illustrated in Fig. 1.2, where the photon energies in joules throughout the spectrum, obtained from Eq. (1.15) have been included. Thus, Fig. 1.2 may be considered as either an energy spectrum or a wave spectrum.

1.6 THE EYE AND SEEING

In this section, elements of seeing and the structure of the eye and its various parts will be described. Then in Chap. 3, the process of seeing, the internal and external factors affecting that process and the ultimate goal of visual performance will be explored in detail. Figure 1.5 depicts the elements of seeing. The first prerequisite is a source of visible radiant energy—the sun, a flame, or one of the many sources of electric light available. The second requirement is some kind of a light modifier—a natural or manufactured object that reflects or transmits light to the eye (it may also absorb light). This may be a window, the moon, the page of a book, a gravel pit, etc. We call such an object a modifier because in most instances it will alter the spectral character of the radiation

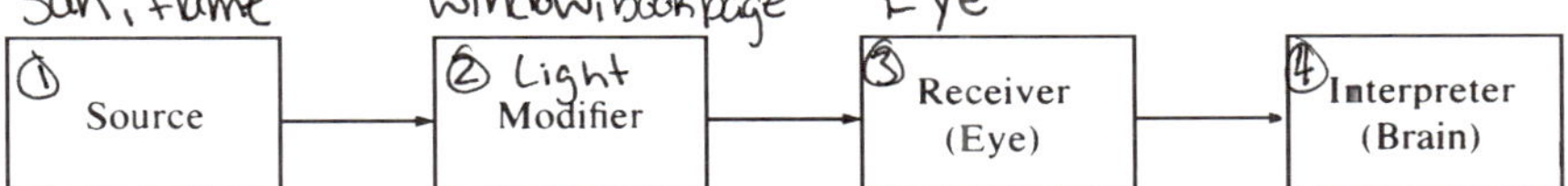

Figure 1.5 The elements of seeing.

from the source, as well as its intensity and directionality. This spectral modification is illustrated in the extreme in Fig. 1.6. In Fig. 1.6a, white light containing all spectral colors, impinges on red, green, and blue surfaces. Three light waves are reflected to the eye, a red wave from the red surface, and green and blue waves from the green and blue surfaces, respectively. The red surface appears red to the viewer, the green surface green, and the blue surface blue, assuming adaptation to a neutral (gray) background.

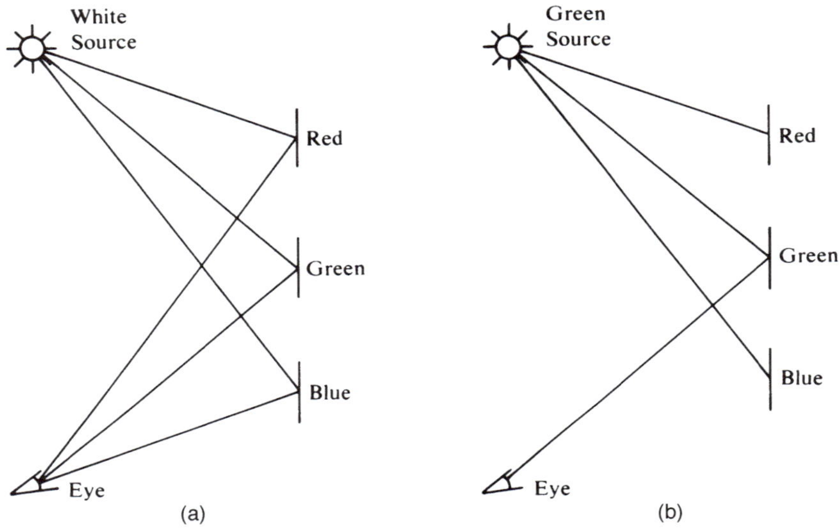

Figure 1.6 Spectral modification of light.

In Fig. 1.6b, the source has only green wavelengths. Green light is reflected to the eye from the green surface, which then appears green to the viewer. The other two surfaces appear dark, because no light is reflected to the eye from them. Thus, it is the spectral character of both the source and the modifier that determines what the viewer perceives.

Returning to Fig. 1.5, we note that the third ingredient of seeing is the eye, which serves as both the receiver of light "messages" and also the processor of them for transmission to the brain. The brain is the fourth element, the deprocessor and interpreter link in the seeing process. It receives the signals from the optic nerve, decodes them, and thus provides the viewer with perception and understanding of the object being viewed.

The eye is shown diagrammatically in Fig. 1.7. Each of its parts will be considered in some detail.

1. *Sclera and choroid.* These are the outer jacket of the eye and its lining, about one mm in total thickness. The sclera maintains the eye in nearly spherical shape and about 2.5 cm in diameter. It is translucent (the white of the eye), except at the cornea, where it is clear and bulges outward. The choroid is a 0.2 mm thick black coating that lines the sclera and contains much of the eye's blood supply. It absorbs stray light inside the eye and thus makes vision more distinct, serving the same function as the black interior of a camera.
2. *Cornea.* This is the first of the three major parts of the optical system of the eye, occupying about 1/5 of the eyeball at its front. It is a bulging transparent membrane consisting of five layers, including an outer thin liquid film that optically

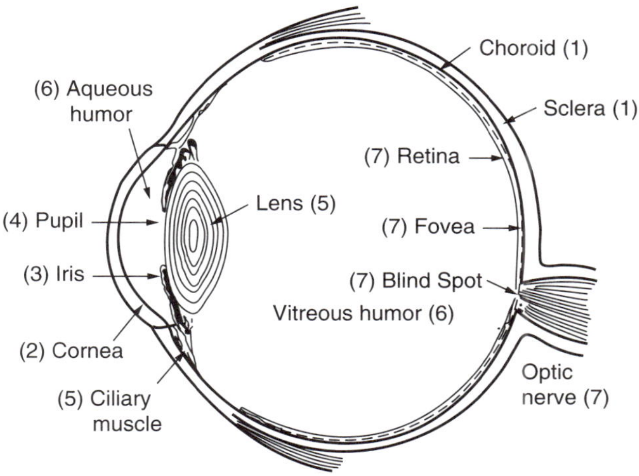

Figure 1.7 The eye (numbers in parentheses refer to text discussion). *Source*: *IESNA Lighting Handbook*, 9th edition, 2000. Courtesy of the Illuminating Engineering Society of North America.

smoothes irregularities in the cornea's surface. The cornea has a radius of curvature of about 7.5 mm, a thickness of 0.4 mm, and a refractive index of 1.37. It is essentially a fixed-focus lens and supplies 70% of the converging power of the eye. It absorbs most wavelengths less than 295 nm.

3. *Iris.* This is the continuation of the choroid in the front of the eye. It is the colored portion of the eye (blue, brown, etc.) and it serves the same role as the diaphragm in a camera by controlling the size of the aperture (pupil) through which light is admitted, thus protecting the eye against excessive light (the eye is well-shielded from excess light from above by the eyebrow, eyelash, and eyelid, but it is not at all well-shielded from below).
4. *Pupil.* This is the small circular opening in the iris. It is the second part of the eye's optical system, controlling the amount of light entering the eye and confining the light to the central area of the lens to minimize distortion of whatever is being viewed. Under normal illumination levels, pupil diameter is 3 to 4 mm and ranges from 8 mm under dim light to 2 mm under bright light. Thus, the optical adaptation range of the eye is about 16 to 1; but, every day illumination levels vary over a 10^7 to 1 range and the overall range of eye response is 10^{12} to 1. Thus, the eye must also adapt neurologically and photochemically to changes in illumination level, since its optical system is not designed to do anywhere near the entire job of adaptation. It has been shown that 10^{-15} W of green light, equivalent to a candle at a distance of 2 mi, can produce the sensation of sight, so the eye is indeed a very sensitive instrument.
5. *Lens.* This is the third part of the eye's optical system. It is crystalline, located just behind the pupil, and suspended in place by the ciliary muscle in back of the iris. In a camera, accommodation (focusing) is accomplished by moving the lens; in the eye, the ciliary muscle alters the shape of the lens to change focus.

   The outer surface of the lens has a radius of curvature of 10 mm; the inner surface 6 mm. The lens has an average index of refraction of 1.44, having a higher index than this at its center and lower at its periphery. It is the outer surface of the lens which changes most in shape during accommodation. When the eye is viewing a distant object, the ciliary muscle relaxes, allowing the ligaments attached to the lens to pull it into a more flattened shape. The reverse occurs for viewing near objects. The lens absorbs nearly all wavelengths less than 350 nm.

6. *Aqueous and vitreous humors.* These are the transparent liquids filling the two inner chambers of the eye. The aqueous humor lies between the cornea and the lens and is a waterlike substance having an index of refraction of 1.34. The vitreous humor, located behind the lens, is more dense and jelly-like and also has an index of refraction of 1.34. Both these media serve to bend and focus, as well as transmit, the incident light.
7. *Retina.* This is the inner of the three linings at the back of the eye and is the "film of the camera." Contained in it are the two types of photoreceptors: rods and cones. Also present are two small areas: (a) the macula, a depression about 1.5 mm in diameter, in the center of which is the fovea, a small pit 0.5 mm in diameter subtending an angle of 2° at the pupil, where most of the distinct three-dimensional seeing takes place and where color is distinguished, and (b) the blind spot on the nasal side of the fovea, which is the connection of the eye to the optic nerve and contains no photoreceptors. The overall visual field is much larger than the foveal field, subtending about 180° in the horizontal direction and 130° in the vertical direction.

   The retina is less than 0.25 mm thick and is composed of 10 layers, with the rods and cones residing in the second layer in from the vitreous humor. The radius of curvature of the retina is 12.0 mm, and the distance from the lens to the retina is 14.6 mm.
8. *Extraocular muscles.* These enable the eye to focus on an object and also to look in any forward direction. Each eye is provided with six extraocular muscles, which are coordinated by the brain to work in harmony with each other so that in the majority of persons neither set dominates.

   These muscles provide motions of two types: gross contractions to change fixation from one object to another, and small, involuntary contractions that occur even when the eye is fixated on an object.

   There are three major types of gross ocular movements, denoted as saccades, pursuit, and vergence. Saccades are rapid eye movements to reorient the line of sight, such as when reading across a line on a page or when scanning different icons on a computer screen. Approximately five line-of-sight changes per second are the maximum. Pursuits are smooth eye movements to track a mostly laterally-moving object after focusing on it. During such tracking, the angle between the eyes and the object remains nearly constant. Vergences occur in tracking an object moving toward or away from an observer. In this case the angle between the eyes and the object does change. Many visual tasks are combinations of the three types of motions, involving jumps, fixations, and smooth transitions.

   In addition to the three gross movements, our eyes constantly undergo small, vibratory motions, even when they are focused on an object. These are called microsaccades and are essential to vision. If they did not occur, focusing intently on an object would cause that object to fade and possibly disappear from view. This is because the receptors at the point of focus on the retina respond to changes in signal, rather than a steady signal.
9. *Rods and cones.* These are the two types of photoreceptors in the retina, which transform the light entering the eye ultimately into electric impulses that are sent through the optic nerve to the brain. Their properties are summarized in Table 1.2.

TABLE 1.2

	Rods	Cones
Number	130×10^6	7×10^6
Shape	Cylindrical. 0.07 mm long and 0.002 mm in diameter.	Somewhat cylindrical in the fovea, conical in the periphery, 0.005 mm base diameter and 0.07 mm high.
Distribution	Located outside fovea. Number of rods increases rapidly to 160,000 per square millimeter at 20° from fovea then decreases gradually.	150,000 per square millimeter in fovea, dropping precipitously to 5000 per square millimeter outside the fovea.
Adaptation time*	Several minutes to several hours.	200 ms to 10 minutes
Sensitivity**	Scotopic, <0.01 cd/m², high luminance sensitivity, nighttime vision.	Photopic, >3 cd/m², low luminance sensitivity, daytime vision.
Connection to optic nerve	Connected in large groups in periphery, providing poor resolution of detail but high sensitivity.	Connected in fovea, individually or in small groups, providing fine resolution of detail but low sensitivity.
Characteristics of vision	Achromatic, all objects appear gray. More sensitive to shorter wavelengths. Provide peripheral vision with poor depth perception.	Chromatic, with a peak sensitivity at 555 nm. Cones dominant in fovea, providing binocular vision and good depth perception.

* Dark adaptation is always slower than light adaptation.

** There is a third category called mesopic vision, which overlaps the photopic and scotopic categories and covers the luminance range in between. In this region, both cones and rods are active.

It was mentioned that the overall adaptation range of the eye is 10^{12} to 1, but that the optical system of the eye accounts for only a 16 to 1 range. Thus additional mechanisms must account for the major part of the eye's ability to adapt to changing light levels. One of these is neural adaptation, which occurs generally in less than a second at most ordinary light levels and makes it possible for people to adapt quickly as interior spaces are scanned. The range of this adaptation is up to 1000 to 1. The second method of adaptation is photochemical. Research indicates that light interacts with the photopigment molecules in the retina, changing their shape and breaking them apart into a protein called opsin and an aldehyde called retinene, a derivative of Vitamin A. The entire process takes less than 1/1000 second and includes the release of energy in the form of electrical currents, which are processed and transmitted through various cell layers in the retina and thence through the optic nerve to the brain. In Chap. 3, these cell layers will be described, along with how they accept the signals from the photoreceptors and convert them into meaningful information for transmission to the brain. This allows for the explanation of many of the characteristics of seeing, such as spectral sensitivity, light sensitivity, spatial resolution, contrast sensitivity, depth perception, light and dark adaptation, temporal resolution, visual tracking, and response to color.

1.7 THE SIX LIGHTING METRICS

The purpose of this section is to introduce, in a brief idealized way, the six major lighting terms and their units of measure and to show how these terms are interrelated. Then, in Chap. 2, more rigorous definitions and descriptions of these quantities will be derived and procedures for calculating them will be developed.

The reader may wonder why it is necessary to visit the lighting terms twice. It has been the author's experience, from many years of teaching, that these terms and units are difficult to master, and that they are best understood when presented in an idealized way first.

Every discipline has its own special terminology, and lighting is no exception. Fortunately, in lighting there are only six basic metrics that need to be learned; unfortunately, they have similar names and unusual units and thus are a challenge for the beginner in lighting to master.

Luminous Flux

An incandescent lamp* is shown in Fig. 1.8a. Diagrammatically, the arrowed line into the base shows that power (P), in watts, flows into the lamp. This causes the lamp to emit luminous flux, shown by arrows leaving the lamp in nearly all directions. The symbol for luminous flux is Φ, and its unit of measure is the lumen.

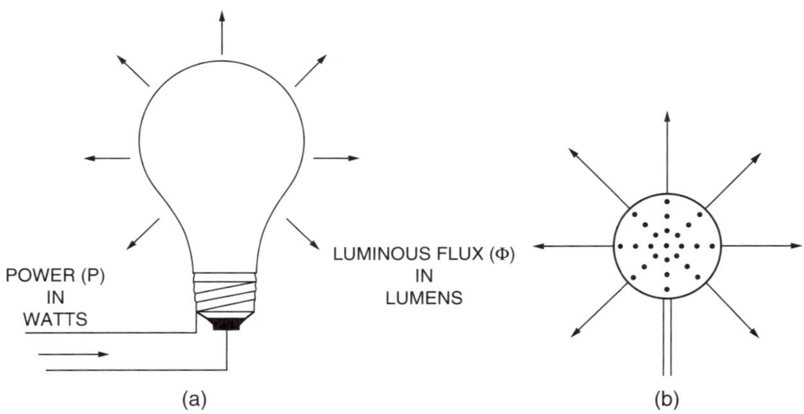

Figure 1.8 Luminous flux. ***Source*: *Lighting Metrics* (TM-1), J. Murdoch, 1999. Courtesy of the Illuminating Engineering Society of North America.**

Luminous flux is the time rate of flow of light. The common analogy is fluid flux, which is the time rate of flow of water. This analogy is developed in Fig. 1.8b, where a small sphere perforated with holes is shown connected to a garden hose. When the valve is opened, fine streams of water emerge from the holes and travel radially outward. Were it not for gravity, the analogy would be complete, but the water streams, of course, eventually bend toward the ground. A second analogy is the electrical one,

* In illumination parlance, a light bulb is called a "lamp" and a lighting fixture is called a "luminaire".

in which luminous flux is compared to electric current, in amperes, the time rate of flow of electric charge in a wire.

A lamp receives watts and emits lumens. The measure of its success in doing this is called "efficacy" and is measured in lumens per watt. Lumen outputs and efficacies of a few common lamps are given in Table 1.3.

TABLE 1.3 LAMP EFFICACIES

LAMP	LUMENS	LUMENS/WATT
100-watt incandescent	1690	17
100-watt tungsten halogen	1800	18
26-watt compact fluorescent	1800	70
34-watt fluorescent	2800	82
100-watt metal halide	9000	90
250-watt high pressure sodium	26000	104

Illuminance

As luminous flux travels outward from a source, it ultimately impinges on surfaces, where it is reflected, transmitted, and/or absorbed. The illuminance on a surface, symbol E, is the density of luminous flux incident on that surface.

$$E(\text{illuminance}) = \frac{\Phi(\text{incident luminous flux})}{A(\text{area})} \tag{1.25}$$

where A is the surface area upon which the luminous flux lands.

The units of illuminance are lumens per square meter in the metric system and lumens per square foot in the English system. A lumen per square meter is called a lux; a lumen per square foot a footcandle. One footcandle is approximately equal to 11 lux (because there are 10.76 square feet in a square meter). For example, if 500 lumens fall on a desk that is 2.5 feet by 5 feet, the illuminance is

$$E = \frac{500}{2.5 \times 5} = 40 \text{ footcandles (approximately 440 lux)}$$

Typical illuminance levels recommended by the IESNA for several common visual activities are shown in Table 1.4.

TABLE 1.4 RECOMMENDED ILLUMINANCE LEVELS

Visual Activity	Footcandles	(Lux)
Lighting for Safety	2	(22)
Stairways	10	(110)
Lobbies	20	(220)
Reading #2 Pencil	30	(330)
Reading #3 Pencil	50	(550)
Product Inspection	100	(1100)
Store Fitting Rooms	100	(1100)
Fine Machine Work	300	(3300)

Luminous Intensity

Luminous intensity is the force generating the luminous flux. It is analogous to the force due to pressure in a fluid system and voltage in an electrical system.

The previously–considered incandescent lamp is shown again in Fig. 1.9a, only this time the radial arrows are used to indicate the intensities in various directions. The lamp has its greatest intensity horizontally, somewhat less intensity vertically down, and no intensity vertically up (because of its base). The lengths of the radial arrows portray these differences.

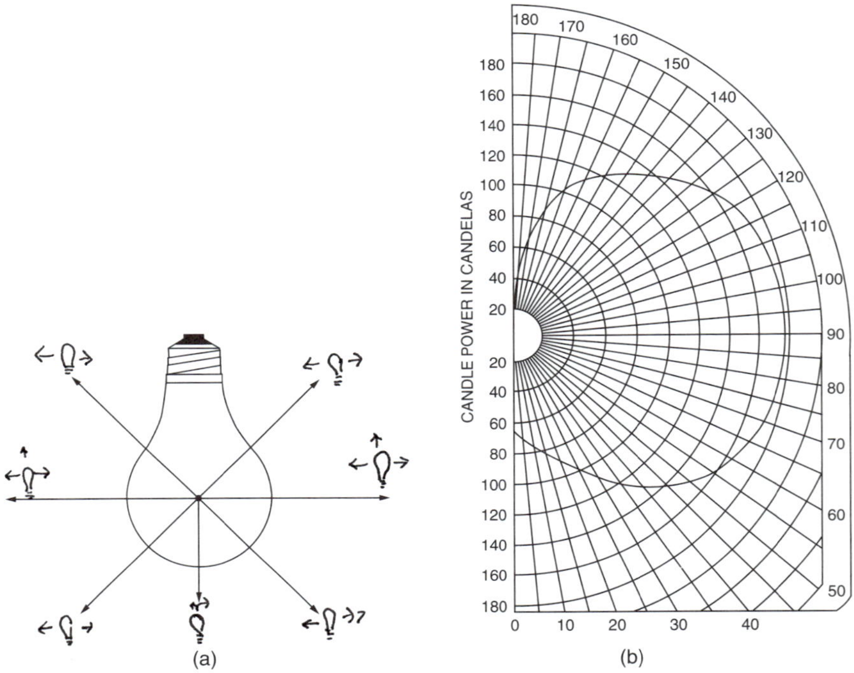

Figure 1.9 Intensity. ***Source*: *Lighting Metrics* (TM-1), J. Murdoch, 1999. Courtesy of the Illuminating Engineering Society of North America.**

The unit of intensity is the candela, symbol I. A lamp is described as having an intensity of I candelas in a particular direction or, alternatively, an intensity of I candlepower in that direction.

Rather than show the intensity profile of a lamp as a collection of radial arrows, it is more common to connect their ends to form an intensity distribution diagram, such as the one shown in Fig 1.9b. The angles shown are called polar angles, with 0° down (nadir), 90° horizontal, and 180° up (zenith) in the diagram. The plot is for a standard A-line 100-watt incandescent lamp. Note that the intensity is indeed 0 at 180° (base region), about 175 candelas at 90° (horizontally), and about 65 candelas at 0° (opposite the base). The lamp intensity is symmetrical about a vertical axis. Thus, the intensity at 90° is the same all around the lamp. The same is true for other polar angles. The intensity distribution curve can be considered as the cross section of a three-dimensional doughnut-shaped solid, obtained by rotating the distribution curve about its vertical axis.

It is not surprising that a name sounding like candle emerged as the unit of luminous intensity. In the early days of lighting in the United States, the whale oil lamp and the spermaceti candle were the major sources of light. In 1860, a unit of luminous intensity known as the "candle" was established using a spermaceti candle of specific weight burning at a specified rate as the basic standard. Later gas flames were calibrated in terms of this definition. The usual gas flame had a luminous intensity of 16 candles. The first incandescent lamps were burned at such voltage that they also had a luminous intensity of 16 candles. This voltage was approximately 110 V, but varied over a rather wide range depending on the length and diameter of the lamp filament, which was not as closely controlled during manufacture as it is today. From 1948 to 1979, the "standard" candle was defined in terms of the luminance of a blackbody radiator. Since then, it is defined radiometrically (see Sect. 4.6.).

To relate Φ, E, and I, in Fig. 1.10a a uniform point source of 1 cd is placed at the center of a 1-m radius non-reflecting (all incident flux is transmitted or absorbed) sphere. By uniform is meant that the source has the same intensity in every direction. By point is meant that the source is much smaller than any other dimension present. This source emits luminous flux (Φ) uniformally in all directions and, by definition, will provide 1 lm of luminous flux to each square meter on the inside surface of the sphere.

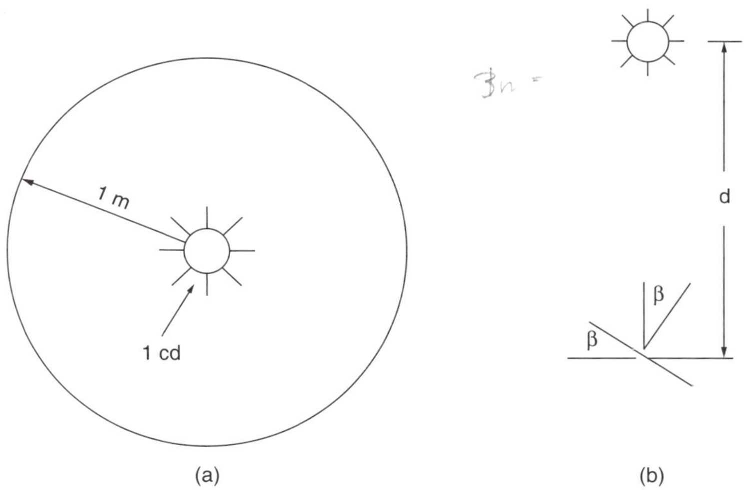

Figure 1.10 Inverse Square Law. ***Source*: *Lighting Metrics*** **(TM-1), J. Murdoch, 1999. Courtesy of the Illuminating Engineering Society of North America.**

Since the surface area of a sphere is $4\pi r^2$, the total luminous flux emitted by the source in Fig. 1.10a is 4π lm, and the illuminance on the inside of the sphere is

$$\frac{4\pi\,\text{lm}}{4\pi\,\text{m}^2} = 1\ \text{lx} = 0.093\ \text{fc}.$$

This latter result can be generalized to the Inverse Square Law of illumination (ISL)

$$E = \frac{I}{r^2} \tag{1.26}$$

This equation states that the illuminance on a surface perpendicular to the line from the point source to the surface varies directly with the intensity of the source and inversely with the square of the distance from the source to the surface. Thus, if the sphere in Fig. 1.10a were 2 m in radius, with the intensity still 1 cd, the illuminance would be 0.25 lx. For 100 cd at 2 m, it would be 25 lx.

In Fig. 1.10b, we show the more general case in which the surface is tilted at an angle β with respect to the line from the source to the surface. For this situation, Eq. (1.26) is modified to read

$$E = \frac{I\cos\beta}{d^2} \tag{1.27}$$

where "d" is substituted for distance, instead of "r" for radius, to make the result more general (not limited to a sphere). The rationale for cos β is that if a surface is tilted at an angle β with respect to the direction of the incoming flux, that flux will be spread over a larger area $(\frac{A}{\cos\beta})$, resulting in reduced illuminance.

Luminance

The fourth and fifth metrics pertain to luminous flux emitted from sources or reflected or transmitted from surfaces. An observer cannot see illuminance because it involves luminous flux hitting a surface. The observer can see the luminance of a surface.

The luminance of a source or surface is defined as the intensity of the source or surface in the direction of an observer divided by the projected area of the source or surface seen by the observer.

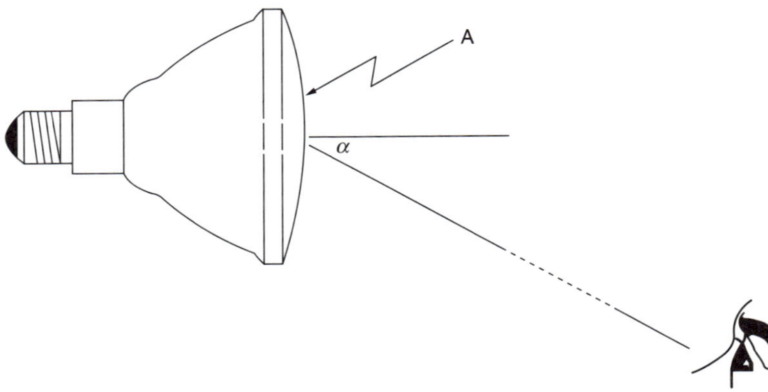

Figure 1.11 Luminance. ***Source*: *Lighting Metrics* (TM-1), J. Murdoch, 1999. Courtesy of the Illuminating Engineering Society of North America.**

The situation is illustrated in Fig. 1.11, where a pressed aluminized reflector (PAR) lamp is shown. The line from the observer to the lamp makes an angle α* with the perpendicular to the lamp face. The observer does not see the lamp face as circular. Rather, it appears elliptical (shrunken in one dimension) and, thus, reduced in size. This shrunken area is called the "projected area."

* Throughout the text, whenever source and receiver are considered together and there might be confusion, α will be used to represent the angle with the perpendicular at the source and β for the angle with the perpendicular at the receiver surface. Otherwise, θ will be used to represent angles in a vertical plane

The luminance of the lamp in the direction of the observer is the intensity in that direction divided by the projected area. Thus,

$$L(\text{luminance}) = \frac{I(\text{intensity})}{A\cos\alpha(\text{projected area})} \tag{1.28}$$

Note that the projected area is the actual area of the lamp face, A, reduced by $\cos\alpha$ because the lamp is not viewed end-on.

The units of luminance are candelas per square meter in the metric system and candelas per square inch or per square foot in the English system. A candela per square inch is equivalent to 1550 candelas per square meter; a candela per square foot to 11 candelas per square meter. Typical luminance values are given in Table 1.5.

TABLE 1.5 TYPICAL LUMINANCE VALUES

Source of Surface	L (cd/in^2)	L cd/m^2
Sun	1,000,000	1,550,000,000
Moon	1.6	2,500
Candle	6.5	10,000
60-watt incandescent lamp	77	12,000
40-watt fluorescent lamp	5.3	8,200
zenith sky	0.7	1,100
snow in sunlight	20	31,000
500-watt tungsten filament	7,500	11,600,000
high pressure mercury arc	1,000	1,600,000

As mentioned previously, luminance is a directional quantity. To illustrate, let the lamp intensity in Fig. 1.11 be 5000 candelas perpendicular to the lamp (at 0°) and 3000 candelas at an angle $\alpha = 30°$. Assume that the circular lamp face has an area of 20 square inches. The luminance values are

$$L = \frac{5000}{20\cos 0}$$

$$= 250 \text{ candelas per square inch}$$

$$= 388{,}000 \text{ candelas per square meter}$$

perpendicular to the lamp and

$$L = \frac{3000}{20\cos 30}$$

$$= 173 \text{ candelas per square inch}$$

$$= 258{,}000 \text{ candelas per square meter}$$

at the angle of 30°.

Luminous Exitance

The fifth metric also describes luminous flux leaving a surface, but in this case it is not directionally dependent. In Fig. 1.12 luminous flux is shown reflecting from a surface [Fig. 1.12a] and transmitting through a surface [Fig. 1.12b]. In each case the concern now is not with luminance in a particular direction, but rather with the total amount of luminous flux reflected or transmitted.

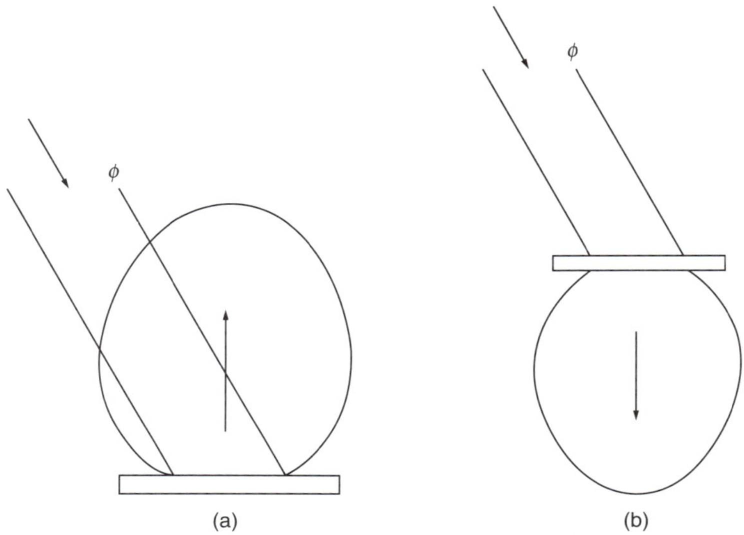

Figure 1.12 Luminous Exitance. ***Source*: *Lighting Metrics* (TM-1), J. Murdoch, 1999. Courtesy of the Illuminating Engineering Society of North America.**

Three new parameters need to be defined. Reflectance, denoted by (ρ), is a number less than one that is the ratio of the luminous flux reflected from a surface to the luminous flux incident on that same surface. Transmittance, denoted by (τ), is the ratio of the luminous flux transmitted through a surface to that incident on the same surface. Finally, luminous exitance, symbol M, is the density of luminous flux leaving a surface. Thus,

$$\text{M (luminous excitance)} = \frac{\Phi\ \text{(departing luminous flux)}}{\text{A (area)}} \tag{1.29}$$

The units of luminous exitance are the same as those for illuminance, namely, lumens per square foot and lumens per square meter. However, luminous exitance describes departing luminous flux, whereas illuminance refers to incident luminous flux. There are no special names for the units of luminous exitance as there are for those of illuminance (footcandle and lux).

It is now possible to relate E, M, ρ, and τ. The departing luminous flux from a reflecting (transmitting) surface is simply the reflectance (transmittance) times the incident flux. Thus,

$$\Phi(\text{reflected}) = \rho\Phi(\text{incident}) \tag{1.30}$$

$$\Phi(\text{transmitted}) = \tau\Phi(\text{incident}) \tag{1.31}$$

Then, using Eqs. (1.25), (1.30), and (1.31),

$$M = \frac{\Phi\,(\text{departing})}{A} = \frac{\rho\Phi\,(\text{incident})}{A} = \rho E \tag{1.32}$$

for a reflecting surface and

$$M = \frac{\Phi\,(\text{departing})}{A} = \frac{\tau\Phi\,(\text{incident})}{A} = \tau E \tag{1.33}$$

for a transmitting surface.

As an example, suppose a book on a desk is lighted to 100 footcandles. Assume that the reflectance of the white pages is $\rho = 0.85$, and that of the black print is $\rho = 0.08$. This means that the white background reflects 85 percent of the lumens hitting it, whereas the black print reflects only 8 percent. The luminous exitances are

$$M = \rho E = 0.85 \times 100$$
$$= 85 \text{ lumens per square foot for the white background and}$$
$$M = \rho E = 0.8 \times 100 = 8 \text{ lumens per square foot for the black print}$$

There is one particular kind of surface for which luminance and luminous exitance have a simple relation. This is a perfectly diffusing surface, which by definition has constant luminance for any angle of viewing (see Fig. 1.13). Such a surface is said to be Lambertain.*

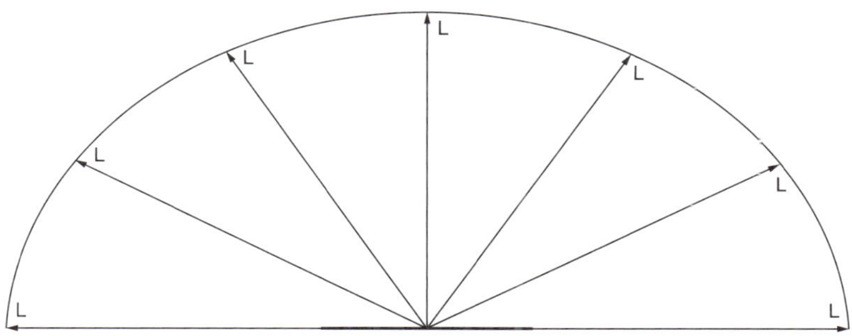

Figure 1.13 Perfectly diffusing surface. *Source*: *Lighting Metrics* (TM-1), J. Murdoch, 1999. Courtesy of the Illuminating Engineering Society of North America.

Many common surfaces, especially room surfaces, are nearly perfectly diffusing. Such surfaces are often called "matte finished," in contrast to more directionally reflecting surfaces, which are called "specular" (such as a mirror) or "glossy" (such as white enamel). For perfectly diffusing surfaces,

$$\pi L \text{ (luminance)} = M \text{ (luminous exitance)} \tag{1.34}$$

* Named for J.H. Lambert, who made many early contributions to photometry.

The area units must be the same on each side of Eq. (1.34). Thus, if L is given in candelas per square foot, M must be given in lumens per square foot. For the previous book example, if the white pages are perfectly diffusing, their luminance, constant from any angle of viewing, is

$$L = \frac{M}{\pi} = \frac{85}{\pi} = 27 \text{ candelas per square foot}$$

As a second example of perfect diffusion, consider that the sphere in Fig. 1.10a is made of translucent glass or plastic and that it transmits 80% of the luminous flux it receives (reflecting none back to the inside surface and absorbing the remaining 20%). Thus, $0.8 \times 4\pi = 3.2\pi$ lm leave the sphere, and, recalling that the sphere surface area is 4π m^2, the density of luminous flux leaving the sphere (its luminous existance) is 0.8 lm/m^2.

What is the sphere luminance? To find out, one should stand far away from the sphere and look toward it. What is seen is no longer a luminous sphere, but rather a luminous circular plane, much as the full moon is seen as a flat luminous disc when gazed at from Earth. Because $\tau = 0.8$, the sphere appears from outside as though it has an intensity of 0.8 cd. Its projected area is π m^2. Thus its luminance is $0.8/\pi$ cd/m^2. Once again $\pi L = M$ because the sphere is perfectly diffusing.

The final term to be considered is quantity of light (Q). Quantity of light is luminous energy and is related to luminous flux, which is luminous power, through the parameter of time. The lumen is used as the measure of luminous flux and the lumen-second or lumen-hour as the measure of light. The lumen-second is useful when the concern is with bursts of lumens lasting for short periods of time, as in flashtube photography. The lumen-hour is used, much as the kilowatthour is used, in evaluating the consumption of energy, in this case luminous energy over extended periods of time.

The six lighting metrics, with their symbols and units are listed in Table 1.6.

TABLE 1.6 ENTITIES IN THE ILLUMINATION SYSTEM

Symbol	Concept	English Unit	Metric Unit
I	Luminous intensity	Candela (cd) candlepower	Candela (cd)
Φ	Luminous flux*	Lumen (lm)	Lumen (lm)
E	Illuminance	Lumen per square foot [footcandle (fc)]	Lumen per square meter [lux (lx)]
M	Luminous exitance	Lumen per square foot (lm/ft^2)	Lumen per square meter (lm/m^2)
L	Luminance†	Candela per square foot (cd/ft^2)	Candela per square meter (cd/m^2)
Q	Quantity of light*	Lumen-second (lm – s)	Lumen-second (lm – s)

*Luminous flux corresponds to power in the radiation system and quantity of light corresponds to energy. Thus the lumen and the watt are the same dimensionally, as are the lumen-second and the joule (watt-second).

†Luminance is also expressed in candelas per square inch and candelas per square centimeter. The footlambert is a unit of luminance equal to $1/\pi$ candelas per square foot. Its use is deprecated, but it does still appear in much of the older daylighting and vision literature.

EXAMPLE 1.1

A glass globe is shown in Fig. 1.14. The upper hemisphere is silvered on the inside so that 90% of the lumens hitting it are reflected to the lower hemisphere. The latter is translucent glass and has a transmittance of 75% and an absorptance of 25%. The sphere has a radius of 0.5 m with a 100-cd lamp at its center. The lamp emits lumens uniformly in all directions. The lower hemisphere is perfectly diffusing. Find the illuminance on the inside of the lower hemisphere, the luminous exitance and luminance of the outside of the lower hemisphere, and the intensity of the lower hemisphere, as seen directly below it from afar.

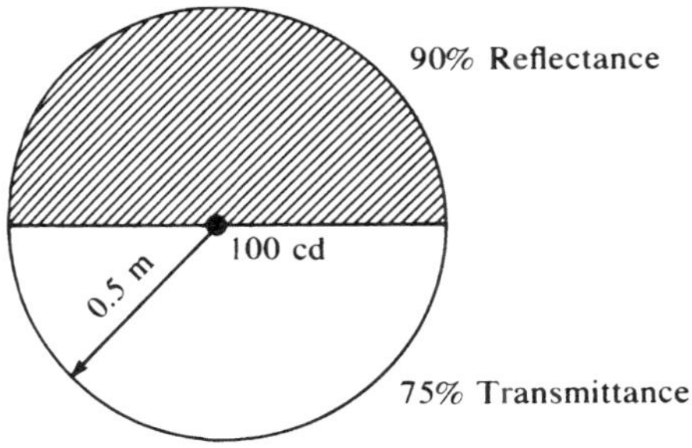

Figure 1.14

Solution. A 1-cd lamp emits 4π lm, so a 100-cd lamp emits 400π lm. Half of these, 200π lm go upward. Of these upward lumens, 90% or 180π lm are reflected downward. In addition, 200π lm go directly downward from the lamp. This gives a total of 380π lm incident on the inside of the lower hemisphere. The surface area of the sphere is π m^2; the area of the lower hemisphere is 0.5π m^2. Thus, the illuminance on the inside of the lower hemisphere is $380\pi/0.5\pi = 760$ lx.

The lower hemisphere has a transmittance of 75%. Therefore 285π lm leave the lower hemisphere, and its luminous exitance is $285\pi/0.5\pi = 570$ lm/m^2. Since the lower hemisphere is perfectly diffusing, its luminance is $570/\pi = 181$ cd/m^2. The projected area of the hemisphere from afar is 0.25π m^2. Multiplying luminance by projected area gives the intensity of the bottom hemisphere directly below from afar as 142 cd.

EXAMPLE 1.2

In Fig. 1.15, B is a perfectly diffusing translucent panel with a transmittance of 45%. It is lighted from the left, resulting in a luminance viewed from the right at A of 0.5 cd/in^2.

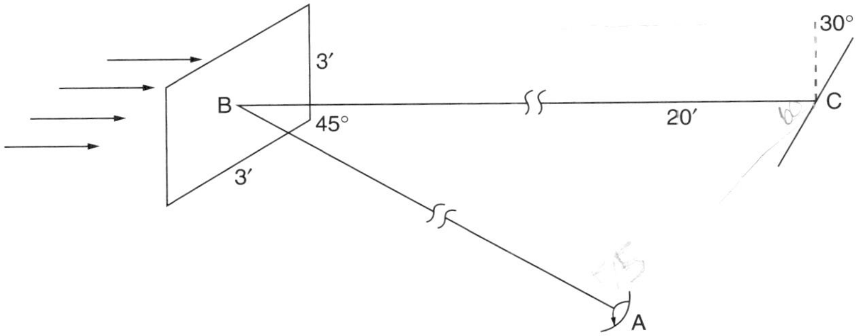

Figure 1.15

Find the following:

a) The intensity of the panel viewed from afar at A.
b) The illuminance on the left panel face.
c) The illuminance on the plane at C.
d) The lumens hitting and leaving the panel.

Solution.

a) The projected area seen from A is

$$A_p = A \cos \alpha = 9 \times 144 \times 0.707 = 916 \text{ in}^2$$

Then the intensity toward A is

$$I_A = LA_p = 0.5 \times 916 = 458 \text{ cd}$$

b) First the luminous exitance of the right side from Eq. (1.34) is found.

$$M = \pi L = 0.5 \times 144 \times \pi = 226 \text{ lm/ft}^2$$

then, from Eq. (1.33),

$$E = \frac{M}{\tau} = \frac{226}{0.45} = 502 \text{ fc}$$

c) To obtain the illuminance at C requires the ISL from Eq. (1.27). First, the intensity toward C is given by

$$I_C = LA_p = 0.5 \times 9 \times 144 = 648 \text{ cd}$$

In this case the projected area is the actual area because the line from B to C is perpendicular to the plane at B. Then

$$E = \frac{I \cos 30}{20^2} = \frac{648 \times 0.866}{400} = 1.4 \text{ fc}$$

d) From Eq. (1.25), the luminous flux hitting B is

$$\varphi = 503 \times 9 = 4530 \text{ lm}$$

and from Eq. (1.33), the luminous flux leaving B is

$$\varphi = 226 \times 9 = 2030 \text{ lm}$$

REFERENCES

American National Standards Institute: *Nomenclature and Definitions for Illuminating Engineering,* ANSI/IES RP-16, New York, 1996.

Cox, J.A.: *A Century of Light*, The Benjamin Co., New York, 1979.

IESNA: *Intermediate Level Lighting Course*, IESNA, New York, 1993, Chap 1.

General Electric Co.: *Light and Color*, TP119, Nela Park, Cleveland, Ohio, 1968.

General Electric Co.: *Light Measurement and Control*, TP118, Nela Park, Cleveland, Ohio, 1971.

Halliday, D., and R. Resnick: *Physics, Third Edition*, John Wiley and Sons, New York, 1978.

Levin, R.E.: "Luminance—A Tutorial Paper," *Journal of the Society of Motion Picture and Television Engineers*, 77(10): 1005, 1968.

Luckiesh, M.: *Light, Vision and Seeing*, D. Van Nostrand, New York, 1944.

Millman, J., and C.C. Halkias: *Electronic Devices and Circuits*, McGraw-Hill, New York, 1967.

Murdoch, J.B.: *Lighting Metrics (TM-1)*, IESNA, New York, 2001.

Oetting, R.L.: "Electric Lighting in the First Century of Engineering," *Transactions of AIEE*, November 1952.

Rea, M.S., editor: *IESNA Lighting Handbook, Ninth Edition*, IESNA, New York, 2000, Chaps. 1 and 3.

Rinalducci, E.J.: *Models of Visual Performance, Interim Report*, Orlando, Fl, 1995.

Sears, F.W., M.W. Zemansky, and H.D. Young: *University Physics, Sixth Edition*, Addison-Wesley, Cambridge, Massachusetts. 1983.

Sekuler, R., and R. Blake: *Perception, Third Edition*, McGraw-Hill, 1994.

Thwing, L.: *Flickering Flames*, Charles E. Tuttle, Rutland, Vermont, 1958.

Chapter 2

Lighting Metrics, Calculations and Measurements

2.1 SI UNITS

In most areas of human endeavor, it is necessary to have standardized units of measure. For each area of the physical sciences and engineering, four basic units are required. Three of these are common to all areas; the fourth is a chosen unit unique to the particular area being examined.

In Systeme International (SI) units, the three common base units are mass in kilograms, length in meters, and time in seconds. In the caloric system, the fourth basic unit is temperature in kelvin's; in electromagnetics, it is electric current in amperes. For the science of light, it is luminous intensity in candelas.

Once the fourth basic unit is identified for a particular area, all other units in that area may be defined. Thus, with the candela defined, the other basic lighting units, such as the lumen and the lux, can be defined, as was done in a rudimentary way in Sect. 1.7. In this chapter, the lighting terms are defined more rigorously and then utilized in calculations and measurements relating to point and area sources of light.

For those not familiar with the SI system, a few comments about it may be useful before proceeding. Until relatively recently there have been two systems of units throughout the world, namely the inch-pound system, used by the United States and the British Commonwealth, and the old metric system, used by the rest of the world. Over the past several years, the world has been moving slowly toward a new third system of units, the international system called SI that evolved from the metric system.

The SI system is based on seven fundamental entities, whose units, definitions, and symbols have been agreed to internationally. These are length in meters, mass in kilograms, time in seconds, electric current in amperes, temperature in kelvins, the amount of a substance in moles, and luminous intensity in candelas. In addition, there are two supplementary units, namely the radian as the measure of a plane angle and the steradian as the measure of a solid angle. There are 17 derived units, obtainable from the seven fundamental and two supplementary units. Among these are force in newtons, energy in joules, and power in watts.

There are three major advantages of the SI system over its predecessors. First, SI is a coherent system in that the factors relating the various units are all unity. For example, as noted in Sect. 1.4, a force of 1 N moving a mass of 1 kg a distance of 1 m does 1 J of work. Second, the SI system is absolute. It does not have the same units for different quantities as, for example, in the inch-pound system, where force and mass can each be expressed in pounds. Third, the SI system is unique in that each quantity has only one unit. For example, energy is always in joules whether we are talking about an electrical system, a mechanical system, or a thermal system.

There are a few exceptions to SI which, through custom and conditioning, are not likely to change. The major of these are the use of liter, rather than cubic meter to represent liquid volume; the more-prevalent use of degree, rather than radian; the use of minutes, hours, days, and years to represent time increments; and the use of degrees Celsius to represent temperature.

2.2 THE LIGHTING TERMS REVISITED

The introduction of the six lighting terms in Sect. 1.7 was based on the notion of a 1-cd point source of light radiating uniformly at the center of a 1-m translucent sphere. That description was, by intention, oversimplified. In practice, most light sources are not point sources nor do they have the same intensity in all directions, as was seen in connection with Fig. 1.9.

Nonuniform and/or nonpoint light sources can be treated in one of two ways. They can be treated as composed of small differential elements, each of which may be considered as a point source, or the entire source can be observed at a distance that is large compared with the maximum source dimension, so that its luminous flux may be considered as radiating from a point. With either approach it is useful to consider that light is emitted in conical tubes of flux radiating outward from an area whose dimensions are small compared with the distance to the point of observation. This requires the conceptual development of a solid angle.

Refer to Fig. 2.1a, in which the radian measure of a two-dimensional angle is presented. When s, the arc length on the circumference of the circle, is equal to r, the radius of the circle, the angle θ is 1 rad. Since the circumference of the circle is $2\pi r$, there are 2π rad in a plane around a point.

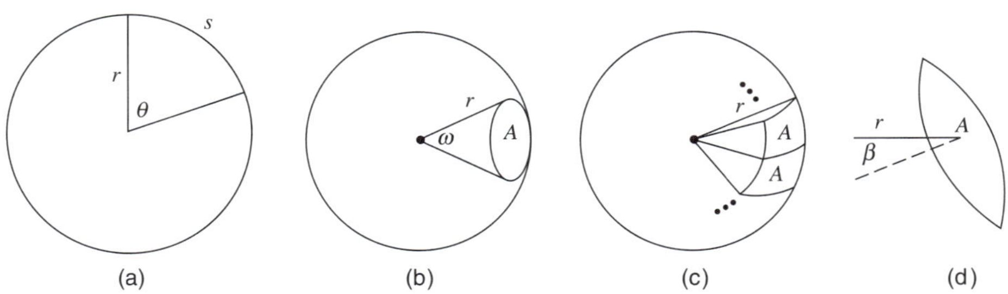

Figure 2.1 Illustrating the radian and steradian.

The extension of the two-dimensional case to three dimensions is shown in Fig. 2.1b. When A, the intercepted area on the sphere, equals r^2, where r is now the radius of the sphere, the angle ω, which is referred to as a solid angle, is 1 steradian (sr). It is said that the area A subtends a solid angle of 1 sr at the center of the sphere. Since the surface area of the sphere is $4\pi r^2$, there are 4π sr surrounding a point in space, i.e. 12.6 one-steradian cones, although it would be necessary to distort the sides of the cones to make them fit, as indicated in Fig. 2.1c. If the area A is tilted at an angle β with respect to a radius vector through its center, as shown in Fig. 2.1d, it is the projected area of A, given by $A \cos \beta$, that determines the solid angle.

The simplest geometry for a solid angle is a cone whose apex is at the center of a sphere. This is illustrated in Fig. 2.2. In this case, it is easy to relate the solid angle ω to the cone angle ψ_c.

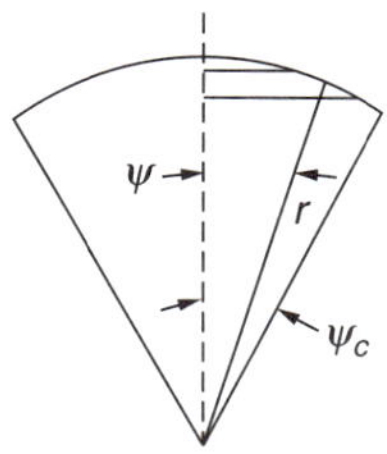

Figure 2.2 Comparison of solid angle and cone angle.

The differential area intercepted on the sphere surface is

$$dA = 2\pi r \sin\psi \, r \, d\psi \tag{2.1}$$

which, when integrated, yields

$$A = \int_0^{\psi_c} 2\pi r^2 \sin\psi \, d\psi$$

$$= 2\pi r^2 (1 - \cos\psi_c) \tag{2.2}$$

Then ω and ψ_c can be related by

$$\omega = \frac{A}{r^2} = 2\pi(1 - \cos\psi_c) \tag{2.3}$$

When $\omega = 1$ sr, solution of Eq. (2.3) yields $\psi_c = 32.8°$. For $\omega = 0.1$ sr, $\psi_c = 10.2°$, for $\omega = 0.01$ sr, $\psi_c = 3.2°$, and for $\omega = 0.001$ sr, $\psi_c = 1.0°$.

The development in Sect. 1.7 leading to the ISL was oversimplified; first in order to present the lighting terms initially in a simple way; and, second, because the concept of a solid angle had not yet been developed. What can now be stated is that luminous flux and intensity are related directly through solid angle by

$$I = \phi/\omega \tag{2.4}$$

In Eq. (2.4), it is seen that intensity is solid angular flux density (lm/sr), and, because the steradian is dimensionless, that the lumen and the candela are dimensionally equivalent.

Now, using Eqs. (1.25), (2.3), and (2.4)

$$E = \frac{\phi}{A} = \frac{I\omega}{\omega r^2} = \frac{I}{r^2} \tag{2.5}$$

which shows the two interpretations of illuminance: luminous flux per unit area at the receiver surface and intensity at the source divided by the separation distance squared.

2.3 INVERSE SQUARE LAW EXAMPLES

At this point, sources whose intensities vary with angle and whose dimensions are small compared with distances to receiver points will be considered. Several examples will be shown to illustrate the ISL and the lighting metrics under these conditions. Then in Sect. 2.5, the scope will be expanded to include large area sources, such as the sky.

EXAMPLE 2.1

A globe street light has a uniform intensity of 2000 cd. Find the illuminance on the vertical stop sign shown in Fig. 2.3.

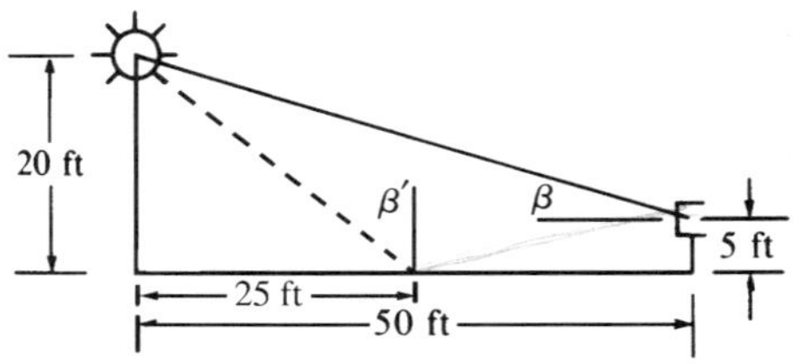

Figure 2.3

Solution. From the geometry, the angle of incidence (angle with the perpendicular) at the sign is $\beta = 16.7°$. Then Eq. (1.27) gives

$$E = \frac{2000\cos 16.7}{15^2 + 50^2} = 0.7 \text{ fc}$$

EXAMPLE 2.2

If the pavement in Fig. 2.3 has a reflectance of 35%, what is the luminous exitance at a point on the pavement halfway between the pole and the sign?

Solution. The angle of incidence is $\beta' = \tan^{-1}\frac{25}{20} = 51.3°$. Then the illuminance at the designated point, from Eq. (1.27), is

$$E = \frac{2000\cos 51.3}{20^2 + 25^2} = 1.2 \text{ fc}$$

The luminous exitance, from Eq. (1.32), is

$$M = 0.35 \times 1.2 = 0.4 \text{ lm/ft}^2$$

EXAMPLE 2.3

A flat circular luminaire 15 cm in diameter is mounted on a ceiling in a recreation room as shown in Fig. 2.4. Its luminance in the direction of point P on a table is 40 cd/cm^2. Find its intensity in the direction of P and the horizontal illuminance on the table at P.

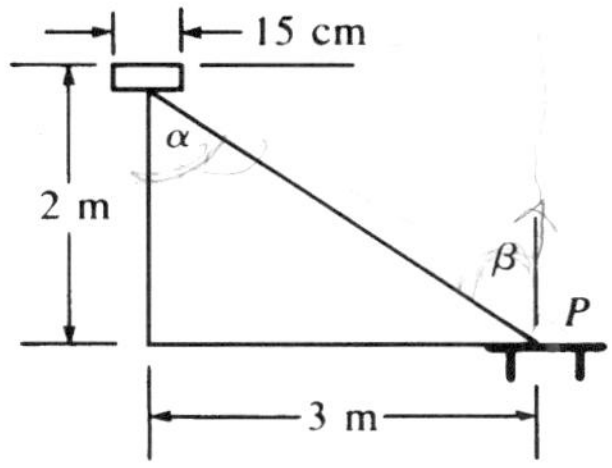

Figure 2.4

Solution. The luminous surface area of the fixture, ignoring its edges, is $56.25\,\pi\ \text{cm}^2$. Angle α is 56.3°. Thus the intensity of the source toward P is, from Eq. (1.28) (source size is small compared with source to receiver distance).

$$I = 40 \times 56.25\pi \times \cos 56.3 = 3920\ \text{cd*}$$

Then, noting that $\beta = \alpha$, the illuminance at P is given by Eq. (1.27) as

$$E = \frac{3920}{13} \times 0.555 = 167\ \text{lx}$$

EXAMPLE 2.4

A recessed downlight in a ceiling has the intensity distribution tabulated in Fig. 2.5a. Find the illuminances at points O, P and Q.

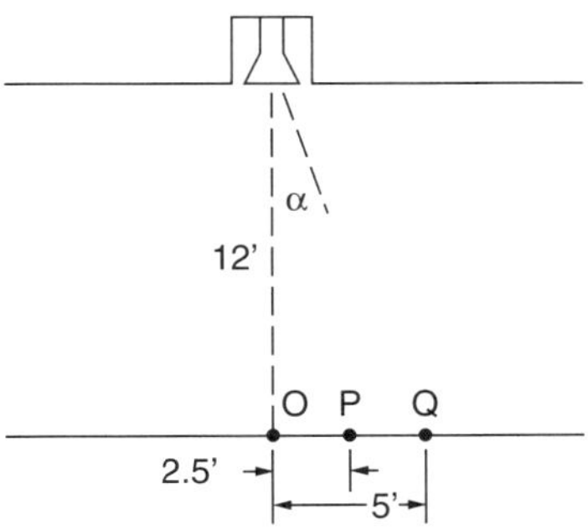

(a)

I_{cd}	α (deg)
5000	0
4500	5
4000	10
3000	15
2000	20
500	25
0	30

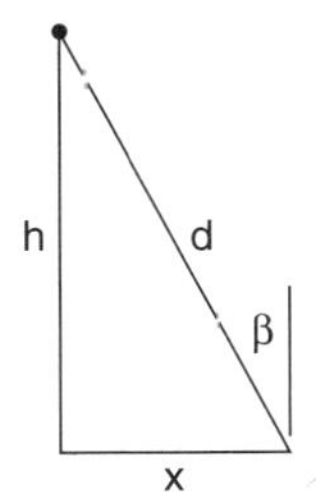

(b)

Figure 2.5

Solution. When the illuminances at several locations on a horizontal surface are desired, an alternative form to Eq. (1.27) is useful. Using the notation in Fig. 2.5b, Eq. (1.27) can be rewritten as

$$E = \frac{I\cos\beta}{d^2} = \frac{I\cos\beta}{(h/\cos\beta)^2} = \frac{I\cos^3\beta}{h^2} \tag{2.6}$$

* In general, no more than three significant figures will appear in final answers. More than three digits may be carried internally to avoid round-off errors.

Equation (2.6) is often preferable to Eq. (1.27) for calculations at multiple horizontal locations for sources of fixed height because it has one less variable.

At Point O

$$h = 12', \quad \beta = \alpha = 0 \quad \text{and}$$

$$E = \frac{5000}{144} = 35 \text{ fc}$$

At Point P

$$h = 12' \quad \text{and} \quad \beta = \alpha = 11.8°$$

To obtain I, an interpolation of the intensity data needs to be done. Generally a linear interpolation is sufficiently accurate.

$$I = 4000 - \frac{1.8}{5} \times 1000 = 3640 \text{ cd}$$

The second term is obtained by noting that 11.8° is $\frac{1.8}{5}$ of the angular change from 10° to 15° and that the ΔI from 10° to 15° is 1000 cd. Now

$$E = \frac{3640 \cos^3 11.8}{144} = 24 \text{ fc}$$

At Point Q

$$h = 12' \quad \text{and} \quad \beta = \alpha = 22.6°$$

$$I = 500 + \frac{2.4}{5} \times 1500 = 1220 \text{ cd}$$

$$E = \frac{1220 \cos^3 22.6}{144} = 7 \text{ fc}$$

These results are very typical of the fall-off in illuminance that occurs as one moves outward from directly under a downlight.

EXAMPLE 2.5

In Fig. 2.6, QR is the width of a badminton court. It is lighted by a flood lamp on a 30-ft high pole. Intensity data for the lamp are given. At what point P should the lamp be aimed to provide equal horizontal illuminances at Q and R? What is this value of illuminance? What is the ratio I_R/I_Q?

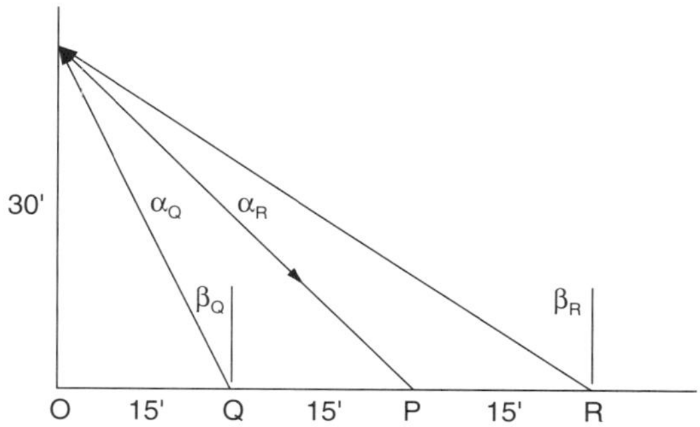

I(cd)	α(deg)
5000	0
4600	5
4000	10
3000	15
2000	20
500	25
0	30

Figure 2.6

Solution. The illuminances at Q and R must be equated. The β angles are given by

$$\beta_Q = \tan^{-1}\frac{15}{30} = 26.6^\circ$$

$$\beta_R = \tan^{-1}\frac{45}{30} = 56.3^\circ$$

Equating illuminances gives

$$\frac{I_Q \cos 26.6}{900+225} = \frac{I_R \cos 56.3}{900+2025}$$

This gives $I_R/I_Q = 4.2$

The next step is trial and error. Assume P is 30 ft from O, midway across the field. Then

$$\alpha_Q = 45 - \tan^{-1}\frac{15}{30} = 18.4^\circ$$

$$\alpha_R = \tan^{-1}\frac{45}{30} - 45 = 11.3^\circ$$

Interpolating the intensities gives

$$I_Q = 2000 + \frac{1.6}{5}\times 1000 = 2320 \text{ cd}$$

$$I_R = 4000 - \frac{1.3}{5}\times 1000 = 3740 \text{ cd}$$

$$\frac{I_R}{I_Q} = \frac{3740}{2320} = 1.6 \neq 4.2$$

Clearly mid-point aiming does not provide sufficient illuminance at R.

Next try aiming at a point 2/3 of the distance across the field. Thus, P is now 35 ft from the pole. The parameters become

$$\alpha_Q = 22.8°, \ \alpha_R = 6.9°$$

$$I_Q = 1160\,\text{cd}, \ I_R = 4370\,\text{cd}$$

$$\frac{I_R}{I_Q} = \frac{4370}{1160} = 3.8$$

This value is close to the desired ratio of 4.2. To find the illuminances for this case,

$$E_Q = \frac{1160\cos 26.6}{1125} = 0.92\ \text{fc}$$

$$E_R = \frac{4370\cos 56.3}{2925} = 0.83\ \text{fc}$$

The results from this example state a rough rule-of-thumb: when lighting a field, as in Fig. 2.6, in order to obtain nearly equal illuminances on the near and far sides, aim the lamps at a point $^2/_3$ to $^3/_4$ of the distance across the field.

2.4 GENERAL INVERSE SQUARE LAW CASE

The general case of a flood lamp on a pole or building lighting a point on the ground (Q) while being aimed at second point on the ground (P) is illustrated in Fig. 2.7.

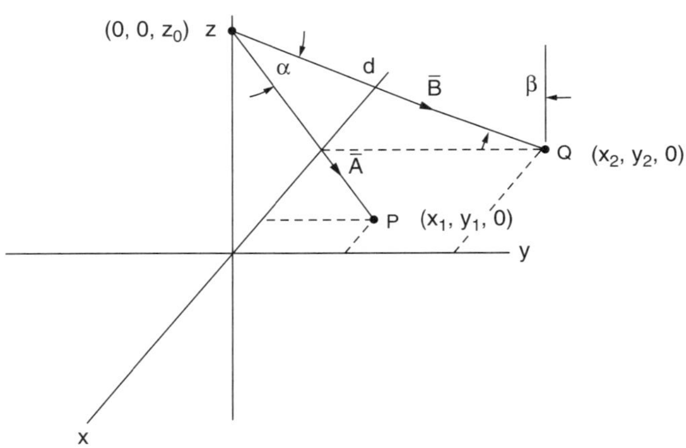

Figure 2.7 General ISL Case.

In Fig 2.7, a right hand coordinate system (x, y, z) is established and the lamp is located at $(0, 0, z_0)$ on the z axis. I, β and d must be found to use Eq. (1.27) to find E.

$$d = \sqrt{x_2^2 + y_2^2 + z_o^2} \tag{2.7}$$

$$\beta = \cos^{-1}\frac{z_o}{d} \tag{2.8}$$

To find I, α must be found.
To obtain α, vectors $\overline{A}$ and $\overline{B}$ are defined in the directions of their arrows by

$$\begin{aligned}\overline{A} &= -x_1\overline{a}_x + y_1\overline{a}_y - z_o\overline{a}_z \\ B &= -x_2\,\overline{a}_x + y_2\,\overline{a}_y - z_o\overline{a}_z\end{aligned} \tag{2.9}$$

Then, the dot product is formed

$$\overline{A}\cdot\overline{B} = x_1x_2 + y_1y_2 + z_0^2 = |A||B|\cos\alpha \tag{2.10}$$

where |A| and |B| are the magnitudes of the vectors $\overline{A}$ and $\overline{B}$. Solving for $\cos\alpha$ gives

$$\cos\alpha = \frac{x_1x_2 + y_1y_2 + z_0^2}{\sqrt{x_1^2 + y_1^2 + z_0^2}\,\sqrt{x_2^2 + y_2^2 + z_0^2}} \tag{2.11}$$

EXAMPLE 2.6

The flood lamp whose intensity distribution is presented graphically, rather than in tabular form, in Fig. 2.8, is aimed at point P. Find the horizontal illuminance at point Q.

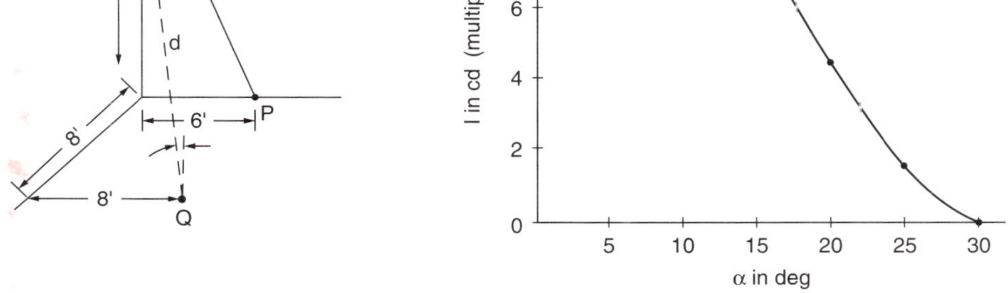

Figure 2.8

Solution. The coordinates in feet are:

$$\text{Lamp:}\quad (0, 0, 20)$$

$$\text{P:}\quad (6, 0, 0)$$

$$\text{Q:}\quad (8, 8, 0)$$

From Eq. (2.7), $d = \sqrt{8^2 + 8^2 + 20^2} = \sqrt{528} = 23.0'$

From Eq. (2.8), $\cos\beta = \dfrac{20}{\sqrt{528}} = 0.870\ (29.5°)$

and from Eq. (2.11),

$$\cos\alpha = \frac{0\times 8 + 6\times 8 + 20^2}{\sqrt{6^2 + 20^2}\ \sqrt{8^2 + 8^2 + 20^2}}$$

$$= \frac{448}{\sqrt{436}\ \sqrt{528}} = 0.934\ (21.0°)$$

From the graph, I = 4000 cd. Then

$$E_Q = \frac{4000 \times 0.870}{528} = 6.6\ \text{fc}$$

2.5 LUMINANCE AS THE FUNDAMENTAL LIGHTING PARAMETER

Thus far, the lighting terms have been presented with luminous intensity as the fundamental entity, in accord with the SI system in which the candela is one of the seven base units. This is satisfactory for point and near-point sources, but for several reasons, it is preferable to use luminance, rather than intensity, as the fundamental quantity (from which the other lighting terms are derived) when considering non-point (area) sources. Luminance relates better to fundamental concepts in optics; it is the parameter of a blackbody radiator that is specified in developing a lighting standard; it is a quantity that relates to a ray of light, and thus it can be used to describe both source and receiver surfaces, as well as the transmitting medium in between; it is a density quantity and, thus, is much more suitable for describing differential source areas; it is much more applicable than intensity for integrations over area sources; it does not depend on source-to-receiver distance.

Consider the situation in Fig. 2.9a. Luminous flux is traveling from left to right and encounters two apertures, one of which is tilted at an angle θ with respect to the direction of propagation. An amount of flux ΔΦ diverges from the right aperture.

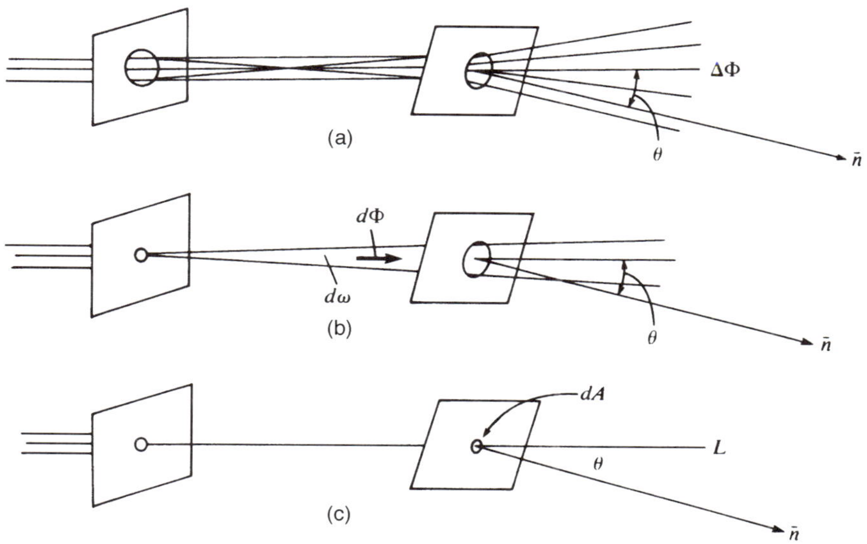

Figure 2.9 The definition of luminance.

Shrink the left aperture so that it nearly closes, as shown in Fig. 2.9b. Now a differential amount of flux dΦ emerges from the left aperture within a differential solid angle dω. This small bundle of rays diverges toward the right aperture. Next shrink the right aperture so that it nearly closes, as shown in Fig. 2.9c. This permits just one ray of the bundle to pass. The projected area of the right aperture in the direction of the ray is

$$dA_p = dA \cos\theta \tag{2.12}$$

and the luminance of the single emerging ray is defined by

$$L = \frac{d^2\Phi}{d\omega\, dA \cos\theta} \tag{2.13}$$

Thus, luminance (at a point and in a direction) is the quotient of the differential luminous flux at a differential element of the surface surrounding the given point and propagated in a differential cone containing the given direction, by the product of the differential solid angle of the cone and the projection of the differential area on a plane perpendicular to the given direction. The differential luminous flux may be arriving, leaving, or passing through the differential surface element. Luminance is thus an inherent characteristic of a ray, meaning that the luminous flux associated with a ray possessing a certain luminance is invariant in a loss-free medium.

With the fundamental definition of luminance established, Eq. (2.13) can be used to derive the remaining lighting terms and a more general form of the ISL. There are three situations to consider. First, recast Eq. (2.13) in the form

$$L = \frac{d\left(\frac{d\phi}{d\omega}\right)}{dA\cos\theta} \tag{2.14}$$

This form defines intensity as the ratio of differential luminous flux to its differential solid angle.

$$I = \frac{d\phi}{d\omega} \tag{2.15}$$

Then

$$L = \frac{dI}{dA\cos\theta} \tag{2.16}$$

which is illustrated in Fig. 2.10a. Equation (2.16) ties intensity and luminance together by showing that the former is the latter multiplied by projected differential source area. This interpretation of Eq. (2.13) is very useful in obtaining the total luminous flux from a light source or luminaire, as will be discussed in Sect. 2.9.

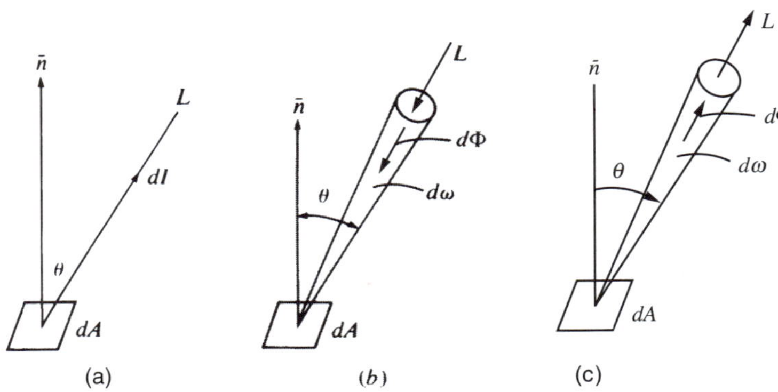

Figure 2.10 Interpretations of Eq. (2.13).

Intensity is the property of a point source in a given direction. This is confirmed in Eq. (2.15), because a solid angle must have a point as its apex. In practice, as was noted previously, finite source size is accommodated by either observing a source at a distance that is large compared with the largest source dimension so that the luminous flux may be treated as though it were diverging from a point, or by breaking the source into small elements, each of which may be considered as a point source with its own intensity.

In either case, the applicability of Eq. (2.15) is determined by

1. Whether the average luminance of the projected source area remains constant as distance is varied;
2. Whether the solid angle subtended by the *source* at the *receiving point* varies inversely with the square of the distance.

This latter constraint can be justified through Fig. 2.11. The solid angle subtended by the differential source element dA at the receiver point is

$$d\omega = \frac{dA\cos\theta}{d^2} \tag{2.17}$$

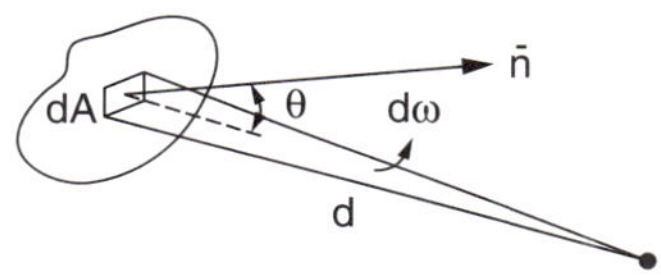

Figure 2.11 Test of Eq. (2.15).

If source to receiver distance is large compared with source area, cos θ and d^2 will remain essentially constant as source area is traversed. This gives

$$\omega = \int_s \frac{\cos\theta}{d^2} dA = \frac{\cos\theta}{d^2}\int_s dA = \frac{A\cos\theta}{d^2} \tag{2.18}$$

Thus ω is seen to vary inversely as distance squared, the source may be considered as a point source, and the concept of intensity in Eq. (2.15) is valid.

The second interpretation of Eq. (2.13) is obtained by rewriting it in the form

$$L = \frac{d\left(\frac{d\Phi}{dA}\right)}{d\omega\cos\theta} = \frac{dE}{d\omega\cos\theta} \tag{2.19}$$

Here illuminance appears as a differential area density of luminous flux, as depicted in Fig. 2.10b, where the luminous flux is shown approaching dA and the luminance is from an external source.

The flux in Fig. 2.10b impinges on the receiver surface at an angle. This is usually the case. If dA were rotated so as to be perpendicular to the direction of L, more flux would be intercepted, and E would increase. In general,

$$E = E_n \cos\theta \tag{2.20}$$

where E_n is the illuminance when the receiver surface is normal to the incident flux. This result is known as Lambert's cosine law, which was introduced in Sect. 1.7 and Eq. (1.27).

Equation (2.19) is useful in calculating the illuminance at a point from a large area source whose distance from the receiver surface is undefined. To this end,

$$dE = L\cos\theta\, d\omega$$
$$E = \int_s L\cos\theta\, d\omega \tag{2.21}$$

where the integral is taken over the entire source area. This interpretation of Eq. (2.13) will be used in obtaining the illuminance from the sky in Sect. 2.8.

The final interpretation of Eq. (2.13) concerns luminous exitance. Illuminance was defined in differential form through Eq. (2.19) when the flux dΦ approached the area dA in Fig. 2.10b. Similarly, luminous exitance is defined as

$$M = \frac{d\Phi}{dA} \tag{2.22}$$

where it is understood now that the differential flux is leaving the differential area, as in Fig. 2.10c. This interpretation of Eq. (2.13) will be used in Sect. 2.7 to develop the properties of a perfectly diffusing surface.

Finally, quantity of light can be defined more rigorously by

$$Q = \int_0^T \Phi(t)\, dt \tag{2.23}$$

which gives the luminous energy emitted or received during a period of time T from a source whose luminous flux output varies with time.

2.6 INVERSE-SQUARE LAW—DIFFERENTIAL ELEMENTS

The general situation of a finite, nonuniform source radiating luminous flux to a point on a receiver surface is shown in Fig. 2.12. The source element dA is assumed to have a luminance L in the direction of point P on the receiver element dA′. The source

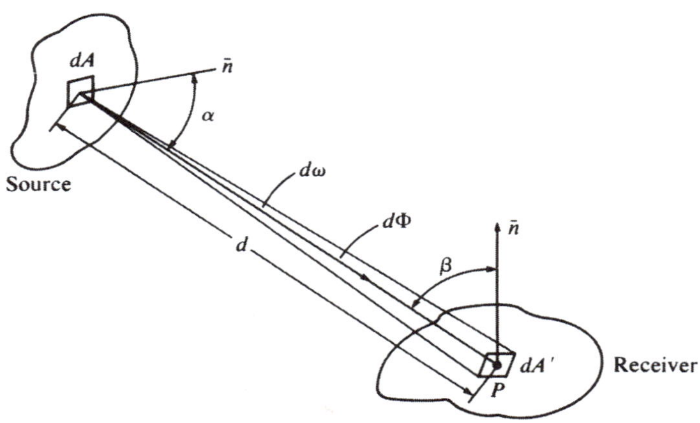

Figure 2.12 General inverse-square law situation.

emits $d\Phi$ lumens in the solid angle $d\omega$ that dA' subtends at the source. The line from source to receiver makes angles α with the normal to dA and β with the normal to dA'.

From Eqs. (2.15) and (2.17),

$$d\Phi = I d\omega = \frac{I dA' \cos\beta}{d^2} \tag{2.24}$$

Thus the illuminance at P is

$$E = \frac{d\Phi}{dA'} = \frac{I \cos\beta}{d^2} \tag{2.25}$$

This is the form of the inverse-square law for a point source. If dA is now considered to be part of a finite-area source, Eq. (2.25) is modified to read

$$dE = \frac{dI \cos\beta}{d^2} \tag{2.26}$$

From Eq. (2.16),

$$dI = L \cos\alpha \, dA \tag{2.27}$$

which, when inserted into Eq. (2.26), yields

$$dE = \frac{L \cos\alpha \cos\beta \, dA}{d^2} \tag{2.28}$$

Equation (2.28) is the inverse-square-law formulation for an element of an area source.

To find the intensity of a finite-area source in the direction of P and the illuminance it provides at P, Eqs. (2.27) and (2.28) are integrated over the source area and the following is obtained

$$I = \int_s L \cos\alpha \, dA \tag{2.29}$$

$$E = \int_s \frac{L \cos\alpha \cos\beta \, dA}{d^2} \tag{2.30}$$

Equation (2.30) is perfectly general and may be used to calculate the illuminance provided by sources of any shape and luminance distribution, subject only to one's ability to perform the integration, either in closed form or by numerical approximation.

Equation (2.29) is not perfectly general. Intensity is a property of a point source in a given direction. If the distance d in Fig. 2.12 is large compared with the maximum dimension of the source (far-field case), Eq. (2.29) will yield the intensity of the source. In this case, the receiver is far enough away so that the lines drawn from each source element to the point may be assumed parallel. If the distance d is not large compared with the maximum source dimension (the near-field case), Eq. (2.29) fails to yield the correct source intensity. Thus, Eq. (2.29) can be used with nonpoint sources, but only if the receiver point is relatively distant from the source. This will be illustrated quantitatively with an example in Sect. 2.12 when inverse square law approximations for area sources are considered.

2.7 PERFECT DIFFUSION

The constraint of luminance being independent of angle is illustrated in Fig. 2.13a. What does this require of luminous intensity? Luminous intensity and luminance are related through projected area, which varies as the cosine of the angle with the normal, as shown in Fig. 2.13b. Thus, for luminance to be constant with angle, intensity must also follow a cosine law, as shown in Fig. 2.13c, and is given by

$$I(\theta) = I_m \cos\theta \tag{2.31}$$

where I_m is the maximum intensity of the surface, occurring when viewed normally.

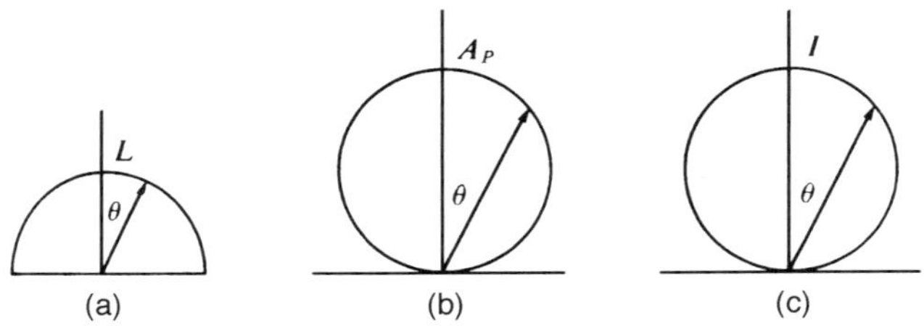

Figure 2.13 Interpretation of perfectly diffusing surfaces.

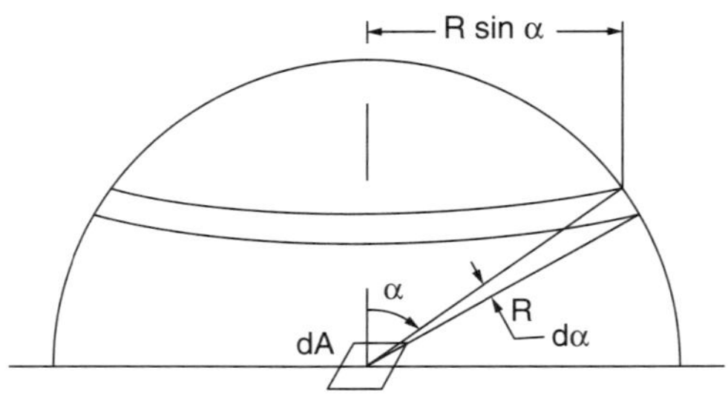

Figure 2.14 Perfectly diffusing source element.

It was stated in Sect. 1.7 that luminous exitance and luminance are related through a π factor for a perfectly diffusing surface. To derive this relationship, consider Fig. 2.14, which shows a source element dA radiating from the center of a hemisphere, drawing on the interpretation in Fig. 2.10c. The illuminance on the hemispherical ring shown is, from Eq. (2.28) ($\cos\beta = 1$),

$$dE_r = \frac{L \cos\alpha \, dA}{R^2} \tag{2.32}$$

where r denotes *ring*. The differential area of the ring is

$$dA = (2\pi R \sin\alpha)(R\, d\alpha) \tag{2.33}$$

Multiplying dE_r by this area gives the luminous flux that the ring receives

$$d\Phi_r = \frac{L\cos\alpha\, dA}{R^2}(2\pi R^2)\sin\alpha\, d\alpha$$
$$= 2\pi L\, dA \sin\alpha \cos\alpha\, d\alpha \tag{2.34}$$

For a perfectly diffusing surface, L is independent of α and the total luminous flux emitted by dA is

$$d\Phi = 2\pi L\, dA \int_0^{\pi/2} \sin\alpha \cos\alpha\, d\alpha$$
$$= \pi L\, dA \tag{2.35}$$

Thus the luminous exitance of dA is, from Eq. (2.22)

$$M = \frac{d\Phi}{dA} = \pi L \tag{2.36}$$

If L is in candelas per square meter, M is in lumens per square meter; if L is in candelas per square foot, M is in lumens per square foot.

The key to the development leading to Eq. (2.36) is being able to move L outside the integral in Eq. (2.35). For nonperfectly diffusing surfaces, the procedure just described could still be used to find M, but the integration would be more difficult, and the result in Eq. (2.36) would not be valid.

EXAMPLE 2.7

A spherical glass globe 1 ft in diameter is perfectly diffusing. At its center is a uniform-intensity point source of 50 cd. The luminance of the globe is 0.3 cd/in^2. Find the transmittance of the globe (the percent of the lumens arriving on the inside of the hemisphere that depart from the outside).

Solution. The problem will be worked in two ways:

1. *Lumen procedure.* The lamp emits $50 \times 4\pi = 200\pi$ lm. The surface area of the sphere is π ft^2. Thus, the illuminance on the inside of the sphere is $200\pi/\pi = 200$ fc. The luminance of the sphere is $0.3 \times 144 = 43.2$ cd/ft^2 and, thus, its luminous exitance is $43.2\pi = 135.7$ lm/ft^2. The transmittance is the ratio of M to E, giving

$$\tau = \frac{135.7}{200} = 0.68$$

2. *Candela procedure.* The projected area of the globe is $\pi r^2 = 36\pi$ in^2. Thus, the apparent intensity of the globe is $0.3 \times 36\pi = 33.9$ cd. The ratio of the apparent intensity of the globe to the intensity of the lamp inside (for the perfect spherical diffuser) is also the transmittance. The result is

$$\tau = \frac{33.9}{50} = 0.68$$

EXAMPLE 2.8

The luminance of a small flat differential source element dA is $L = 10\cos\theta$ cd/m^2, where θ is the angle with the normal to dA. Find the luminous exitance of dA.

Solution. For this situation, Eq. (2.35) becomes

$$d\Phi = 2\pi\, dA \int_0^{\pi/2} 10 \sin\theta \cos^2\theta\, d\theta$$

i.e. L cannot be moved outside the integral. The result of the integration is $d\Phi = (20\pi/3)dA$. Thus, the luminance exitance is $M = 20\pi/3$ lm/m^2 and is less than it would be if dA were perfectly diffusing with a luminance of 10 cd/m^2. ($M = 10\pi$ lm/m^2).

EXAMPLE 2.9

An 8-in square perfectly diffusing flat lighting fixture has a luminance of 5 cd/in^2. Find its luminous exitance, emitted luminous flux, and intensity.

Solution. Because of perfect diffusion,

$$M = \pi L = 15.7 \text{ lm/in}^2$$

Emitted luminous flux is given by

$$\varphi = MA = 5\pi \times 64 = 1005 \text{ lm}$$

Maximum intensity is

$$I_m = 5 \times 64 = 320 \text{ cd}$$

Thus

$$I = 320 \cos\alpha \text{ cd}$$

2.8 THE SKY AS A SOURCE

Suppose it is desired to determine the illuminance at a point on the ground from the sky. Attempts to use the ISL [Eq. (2.25) or (2.28)] fail because there are no values available for the intensity of the sky and its distance.

Consider Fig. 2.14 again, but this time let the ring on the hemisphere be a portion of the sky (the source) and dA be the receiver, drawing on the interpretation in Fig. 2.10b.

Assume the luminance of the sky hemisphere in the direction of dA can vary with the polar angle, customarily denoted by θ in daylighting parlance.

It is desired to find the horizontal illuminance on dA due to the entire hemisphere. The differential illuminance produced by the ring is, from Eq. (2.28),*

$$dE = \frac{L(\theta)\cos\theta}{R^2}(2\pi R^2)\sin\theta\, d\theta \tag{2.37}$$

Assume $L(\theta)$ to be constant between the limits θ_1 and θ_2. Then integration of Eq. (2.37) yields

$$\begin{aligned} E_{\theta_1\to\theta_2} &= 2\pi L_{\theta_1\to\theta_2}\int_{\theta_1}^{\theta_2}\sin\theta\cos\theta\, d\theta \\ &= \pi L_{\theta_1\to\theta_2}(\sin^2\theta_2 - \sin^2\theta_1) \end{aligned} \tag{2.38}$$

If $L(\theta)$ is constant over all the hemisphere, Eq. (2.38) becomes

$$E_{0\to 90} = \pi L \tag{2.39}$$

Thus, the horizontal illuminance produced by a sky of constant luminance is π times that luminance. Note that the distance of the sky from the observer does not appear in either Eq. (2.38) or (2.39).

EXAMPLE 2.10

Assume an overcast sky has the following luminances:

θ, degrees	L, cd/in^2
0–40	4.0
40–60	3.0
60–90	2.0

Find the horizontal illuminance produced at a point on the ground.

Solution. From Eq. (2.38),

$$\begin{aligned} E &= 144\pi[4(\sin^2 40 - \sin^2 0) + 3(\sin^2 60 - \sin^2 40) + 2(\sin^2 90 - \sin^2 60)] \\ &= 144\pi(1.65 + 1.01 + 0.50) = 1430 \text{ fc} \end{aligned}$$

The 144 is necessary to convert cd/in^2 to cd/ft^2 so that the illuminance will be in lm/ft^2 (fc).

* Eq. (2.21) could also be used with the note that $d\omega = 2\pi\sin\theta\, d\theta$.

EXAMPLE 2.11

The standard CIE* overcast sky is given by the following expression:

$$L = \frac{L_z}{3}(1 + 2\cos\theta)\,\text{cd/m}^2 \tag{2.40}$$

where L_z is the zenith luminance, given by

$$L_z = 123 + 8600 \sin a_t \text{ cd/m}^2 \tag{2.41}$$

where a_t is the solar altitude (angle above the horizon) in degrees. Find the horizontal illuminance produced by this sky.

Solution. Using Eqs. (2.21) and (2.40),

$$E = \int_0^{\pi/2} L\cos\theta\, d\omega = \int_0^{\pi/2} \frac{L_z}{3}(1 + 2\cos\theta)\cos\theta\, 2\pi\sin\theta\, d\theta$$

$$= \frac{2\pi L_z}{3}\left(\frac{\sin^2\theta}{2} - \frac{2\cos^3\theta}{3}\right)_0^{\pi/2} = \frac{7\pi L_z}{9}$$

Then, inserting Eq. (2.41), we have

$$E = \frac{7\pi}{9}(123 + 8500\sin a_t) = 300 + 21{,}000\sin a_t \text{ lx}$$

$$= 28 + 1950\sin a_t \text{ fc}$$

When the solar altitude is 30°, for example,

$$E = 28 + 975 = 1003 \text{ fc}$$

The property of a perfectly diffusing surface that luminance is constant from any angle of observation is of considerable help in dealing with otherwise difficult architectural lighting geometries. Consider the three ceiling surfaces in Fig. 2.15. The first is a portion of a sphere; the second of a cylinder; the third of a cone. Each subtends the same solid angle at point O. Assuming the surfaces are perfectly diffusing and have the same luminance, the observer at O cannot tell them apart. However, the geometry for the portion of the sphere is much easier to deal with mathematically than the other two geometries, as has been seen in the development of Eq. (2.37). This concept will be illustrated with an example.

---

* International Commission on Illumination. The letters CIE derive from the French label, Commission Internationale de L′ Eclairage.

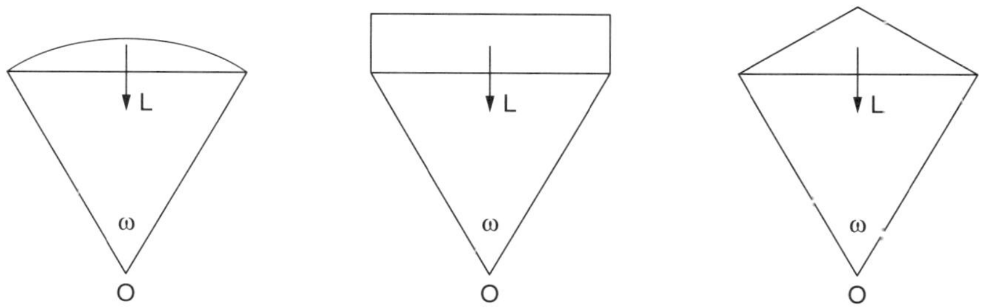

Figure 2.15 Three surfaces of revolution with the same solid angle.

EXAMPLE 2.12

A is a circular translucent perfectly diffusing skylight with $\tau = 55\%$. B is a skylight well with perfectly diffusing walls having a luminance of 0.5 cd/in^2. The illuminance on the outside of the skylight is 1250 fc. Find the horizontal illuminance at P.

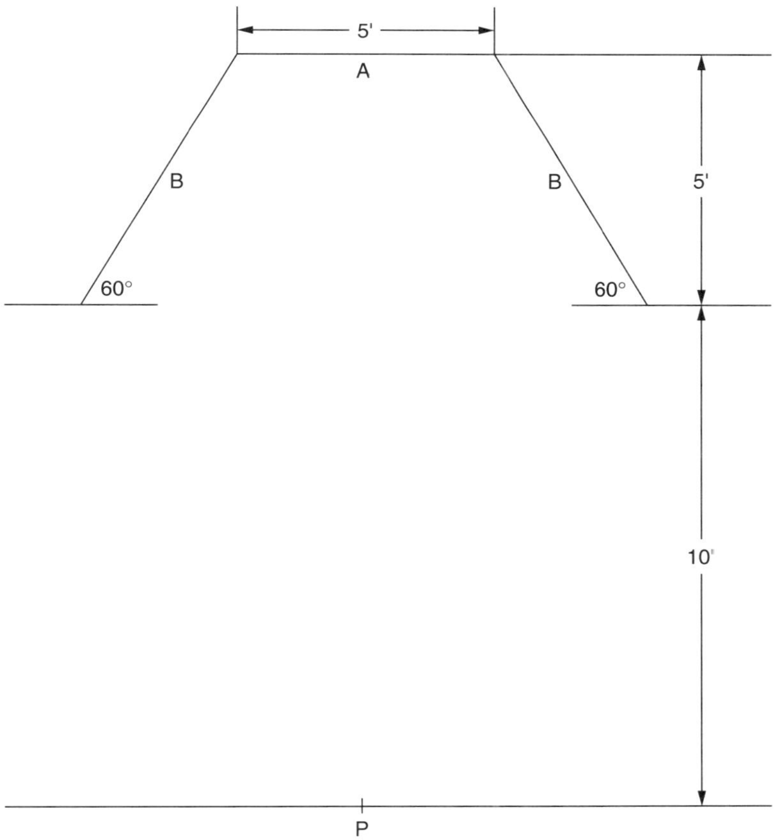

Figure 2.16

Solution. Since both the skylight and the well are perfectly diffusing, their geometries can be replaced with the portion of the sphere shown in Fig. 2.17. From the

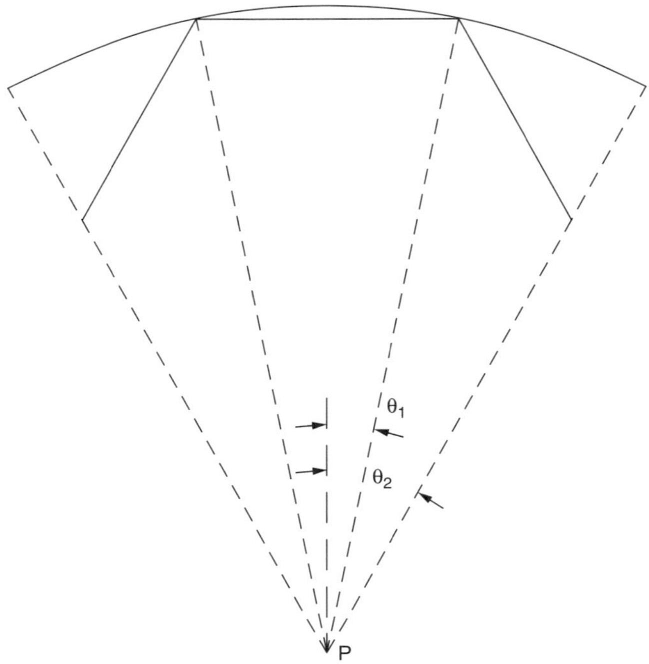

Figure 2.17

geometry in Fig. 2.16, θ_1, and θ_2 can be calculated as $\theta_1 = 9.5°$, $\theta_2 = 28.3°$. The luminance on the inside of the skylight is

$$L = \frac{E\tau}{\pi} = \frac{1250 \times 0.55}{\pi} = 218.8 \text{ cd/ft}^2$$

The luminance of the well is

$$L = 144 \times 0.5 = 72 \text{ cd/ft}^2$$

Then, from Eq. (2.21)

$$dE = L\cos\theta \, 2\pi \sin\theta \, d\theta$$

$$E = 2\pi \times 218.8 \int_0^{9.5} \sin\theta\cos\theta \, d\theta$$

$$+ 2\pi \times 72 \int_{9.5}^{28.3} \sin\theta\cos\theta \, d\theta$$

$$E = 1375 \left.\frac{\sin^2\theta}{2}\right|_0^{9.5} + 452 \left.\frac{\sin^2\theta}{2}\right|_{9.5}^{28.3}$$

$$= 18.7 + 44.6 = 63 \text{ fc}$$

2.9 MEASURING LUMINOUS FLUX

An incandescent lamp is shown mounted horizontally on a stand in Fig. 2.18. A person walks slowly around the lamp, keeping the distance d constant and following the path shown. Every 10°* the person stops, points an illuminance meter at the lamp, and takes a reading. After all readings from 0 to 180° are taken, Eq. (2.25) (with $\beta = 0$) is used to find the intensity of the lamp at each viewing position. These intensity values are then plotted versus viewing angle θ on a polar diagram. The result is shown in Fig. 2.19 and is the intensity distribution curve for the given source. The semicircles labeled 500, 1000, and 1500 are intensity loci in candelas. To find the intensity at any angle, one interpolates visually between these loci. For example, the intensity at 30° is approximately 1250 candelas.

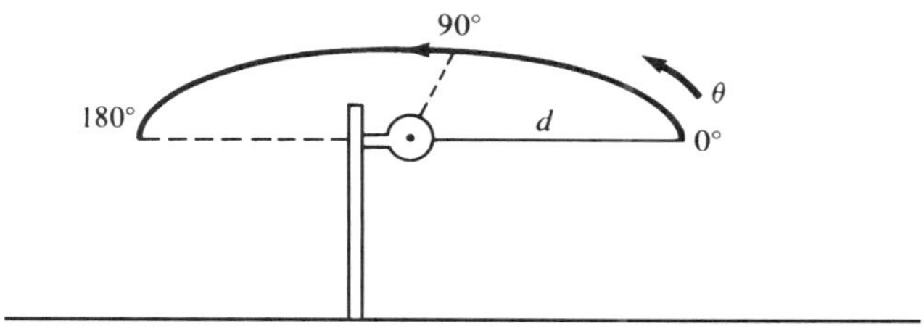

Figure 2.18 Laboratory setup for measuring lumens.

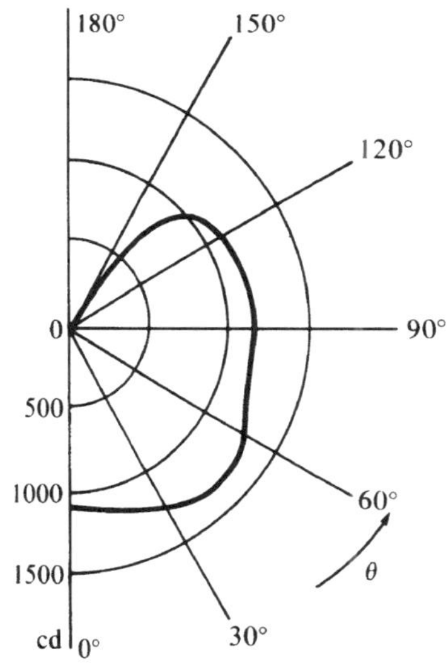

Figure 2.19 Intensity distribution curve of an incandescent lamp.

It should be noted that the graph contains values from 0 to 180° only. This is because the lamp is presumed symmetrical and, thus, its intensity at 330°, for example, is the same as that at 30°. If symmetry is not assumed, the incandescent lamp may be rotated about the horizontal in Fig. 2.18 to obtain an average intensity in each angular direction.

* In modern goniophotometers, readings are taken at least every 5° and often at steps less than 1°, especially for lamps with rapidly changing intensity with angle.

Lamp and luminaire manufacturers obtain intensity distribution curves using devices called goniophotometers, where the prefix "gonio" means directional. These devices are very similar in principle to that shown in Fig. 2.18. Each consists of a calibrated light-sensitive cell for measuring illuminance, mounted at the end of an arm that rotates about a source at a fixed distance from the source. A considerable amount of space is required for such an instrument because, for the ISL to hold, the distance between source and meter must be large compared with the maximum dimension of the source. Generally this distance is kept greater than five times this dimension, which gives an accuracy of the order ±1% for many sources and luminaires. For an 8-ft-long fluorescent lamp, the longest commercially available source, the minimum distance is 40 ft, requiring an 80-ft-diameter circle through which the meter rotates and, thus, an 80-ft-high room. This distance can be reduced by two means:

1. Mounting the meter near the source and using a mirror to reflect the source rays to the meter. Such an arrangement can provide a 40-ft light path in a 20-ft space.
2. Raising the lamp to the ceiling to obtain the 0 to 90° portion of the candlepower distribution curve and then lowering the lamp to the floor for the 90 to 180° portion of the curve. Instead of an 80-ft-high room, one only 40-ft high is needed.

A much greater than 5:1 ratio is required for certain luminaires, such as spotlights. In such cases, the minimum distance for accuracy is a function of the reflector and lens characteristics of the luminaire.

Goniophotometers are calibrated with incandescent lamps which have, in turn, been calibrated against standards from the National Institute of Standards and Technology (NIST) in the USA or the National Research Council (NRC) in Canada. Modern goniophotometers are completely automated and computerized so that the movement of the meter arm, the recording of data, and the plotting of the candlepower distribution curve are all done automatically. Some models place a light cell at each required angle, rather than using a single movable cell.

The intensity distributions of many lamps and luminaires, particularly fluorescent, are nonsymmetrical, and, thus, a family of intensity distribution curves is required. Generally candlepower data are taken in at least three vertical planes, oriented at 0°,

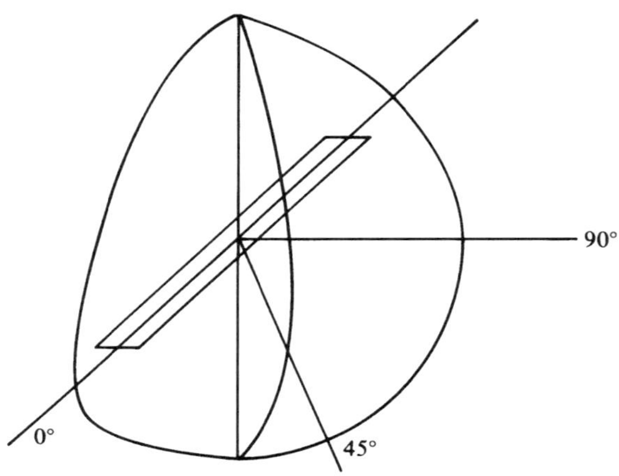

Figure 2.20 Fluorescent photometry.

45°, and 90° with respect to the lamp or luminaire axis, as shown in Fig. 2.20. In more complete photometry, additional planes are also used.

The lamp manufacturer must be able to tell the customer how many lumens a given lamp emits. For an ideal source, which has uniform intensity in all directions, the answer is easily found by multiplying the intensity by 4π. When the intensity is nonuniform, which is usually the case, the problem is more difficult and calculations from the candlepower distribution curve must be made.

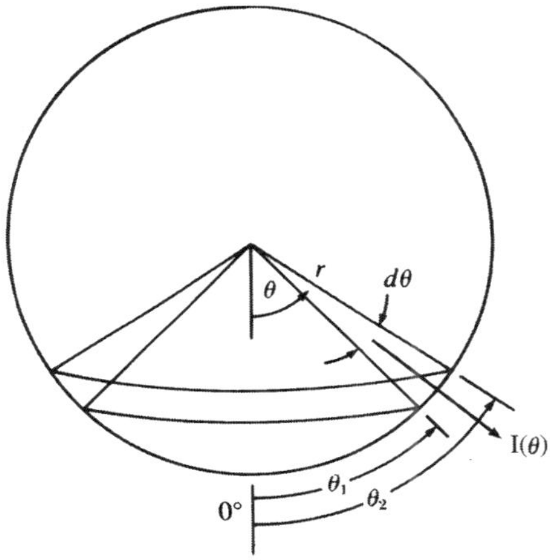

Figure 2.21 Zonal lumens.

Let a source have an intensity $I(\theta)$ in a small angular band $d\theta$, as shown in Fig. 2.21. Place a sphere of radius r around the source. Then, from Eq. (2.15), the luminous flux intercepted by the $d\theta$ band on the sphere is

$$d\phi = I(\theta)\, d\omega = I(\theta) \times 2\pi \sin\theta\, d\theta \tag{2.42}$$

If $I(\theta)$ is assumed constant in an angular band from θ_1 to θ_2, integration of Eq. (2.42) yields

$$\Phi_{\theta_1 \to \theta_2} = 2\pi I(\theta)(\cos\theta_1 - \cos\theta_2) \tag{2.43}$$

where $I(\theta)$ is taken as the intensity at the midpoint of the zone between θ_1 and θ_2. If $I(\theta)$ has a constant value I for all θ, then Eq. (2.42) gives $\Phi = 4\pi I$, as would be expected.

The quantity $2\pi(\cos\theta_1 - \cos\theta_2)$ is the solid angle $\Delta\omega$ of a ring lying between θ_1 and θ_2 on the sphere and is called a zonal constant. The zonal constants for the 10° zones appear in Table 2.1.

Note that the zonal constants for angles greater than 90° are the same as for angles less than 90°; for example, the 30°–40° zone has the same zonal constant (same area) as the 140°–150° zone.

With the concept of $\Delta\omega$s and zonal constants established, Eq. (2.43) can be rewritten to read

$$\Phi_{\theta_1 \to \theta_2} = I(\theta)\Delta\omega_{\theta_1 \to \theta_2} = I(\theta) ZC_{\theta_1 \to \theta_2} \tag{2.44}$$

TABLE 2.1

Zone (degrees)	Zonal Constant	Zone (degrees)
0–10	0.095	170–180
10–20	0.283	160–170
20–30	0.463	150–160
30–40	0.628	140–150
40–50	0.774	130–140
50–60	0.897	120–130
60–70	0.993	110–120
70–80	1.058	100–110
80–90	1.091	90–100

where ZC means zonal constant. Equation (2.44) says that the lumens in the zone bounded by θ_1 and θ_2 are obtained by multiplying the intensity in that zone (presumed constant throughout the zone) by the zonal constant of the zone.

To get the total lumens emitted by the lamp, the contributions from all zones must be summed. Using Eq. (2.44), this sum can be written as

$$\Phi = \sum_{0^\circ}^{180^\circ} I(\theta) ZC_{\theta_1 \to \theta_2} \tag{2.45}$$

EXAMPLE 2.13

Find the total lumens given off by the lamp whose candlepower distribution curve is given in Fig. 2.19.

Solution. This problem will be worked using the zonal constants and assuming that the intensity in each 30° zone is taken at the midpoint of that zone, and may be considered constant in that zone. This assumption is fairly good, except for the 120 to 150° zone, where it is an obvious approximation.

Zone	Average/(candelas)	ZC	Zonal Lumens
0–30	1150	0.841	967
30–60	1300	2.299	2989
60–90	1150	3.142	3613
90–120	1100	3.142	3456
120–150	850	2.299	1954
150–180	0	0.841	0
		total lumens = 12,979 ≃ 13,000	

EXAMPLE 2.14

A lamp having a uniform candlepower of 1000 cd in all directions is placed in a reflector, as shown in Fig. 2.22. The reflector intercepts all of the lumens in the 60 to 180° zone and reflects 80% of these. All of these reflected lumens enter the 0 to 60° zone. Find the total lumens in the 0 to 60° zone and the average intensity of the beam.

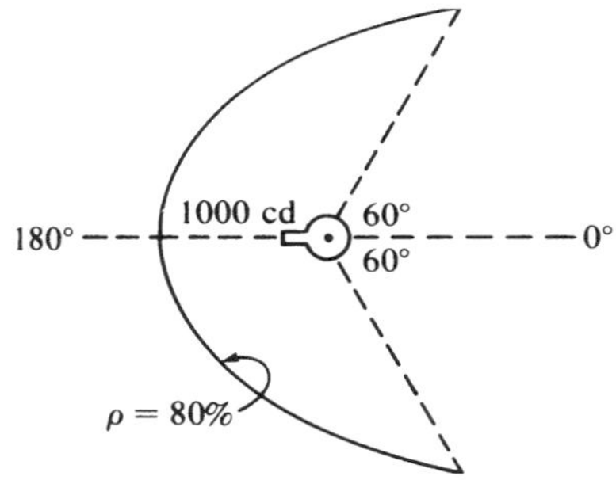

Figure 2.22

Solution. Since I is constant, the region from 60 to 180° is considered as a single zone, and Eq. (2.43) is used to obtain

$$\Phi_{60-180} = 2\pi 1000(\cos 60 - \cos 180)$$
$$= 2000\pi(0.5 + 1) = 3000\pi \text{ lm}$$

Of these lumens, 80% are reflected into the 0 to 60° zone. Thus

$$\Phi_{\text{refl.}} = 0.8(3000\pi) = 2400\pi \text{ lm}$$

The lumens going directly from the lamp to the 0 to 60° zone are given by

$$\Phi_{0-60} = 2\pi 1000(\cos 0 - \cos 60)$$
$$= 2000\pi(1 - 0.5) = 1000\pi \text{ lm}$$

Thus the total lumens in the 0 to 60° beam are

$$\Phi_{\text{beam}} = 2400\pi + 1000\pi = 3400\pi \text{ lm}$$

To find the average beam intensity, Φ_{beam} is inserted into Eq. (2.43) and $I_{\text{beam avg}}$ is solved for:

$$3400\pi = 2\pi I_{\text{beam avg}}(\cos 0 - \cos 60)$$

$$I_{\text{beam avg}} = 3400 \text{ cd}$$

It is noted that the effect of the reflector is to convert a 1000-cd source, uniform in all directions, into a directed beam of 3400-cd intensity and 120° spread (60° each side of the center line), with no intensity in any other direction. This, of course, is the principle behind all spot and flood lamps, auto headlights, searchlights, and projection lamps.

EXAMPLE 2.15

The lamp whose intensity distribution is given, is mounted in a recessed ceiling cavity. What is the average illuminance in a 7-ft diameter circle directly under the luminaire?

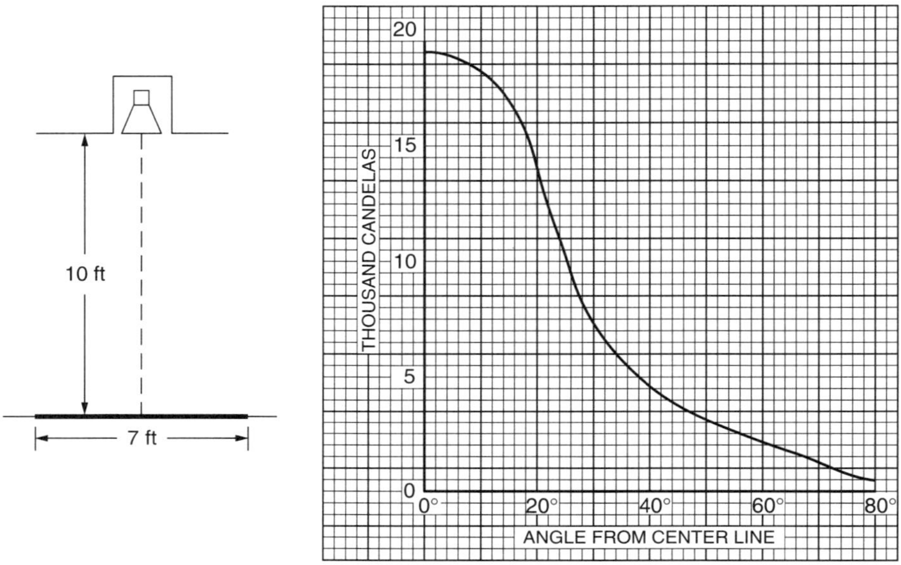

Figure 2.23

Solution. Figure 2.21 is replicated in Fig. 2.24 by drawing a portion of a 10-ft radius sphere enclosing the 7-ft diameter circle. All lumens incident on the circle will also be incident on the spherical saucer below.

At the edge of circle,

$$\tan\alpha = \frac{3.5}{10} = 0.35, \quad \alpha = 19.3° \approx 20°$$

Then, with 5° zones, Eq. (2.45) gives

Zone (degrees)	I_{mid} (cd)	ω (sr)	Φ lumens
0–5	18,900	0.0239	451.7
5–10	18,400	0.0715	1315.6
10–15	17,500	0.1186	2075.5
15–20	15,400	0.1648	2537.9
			6381

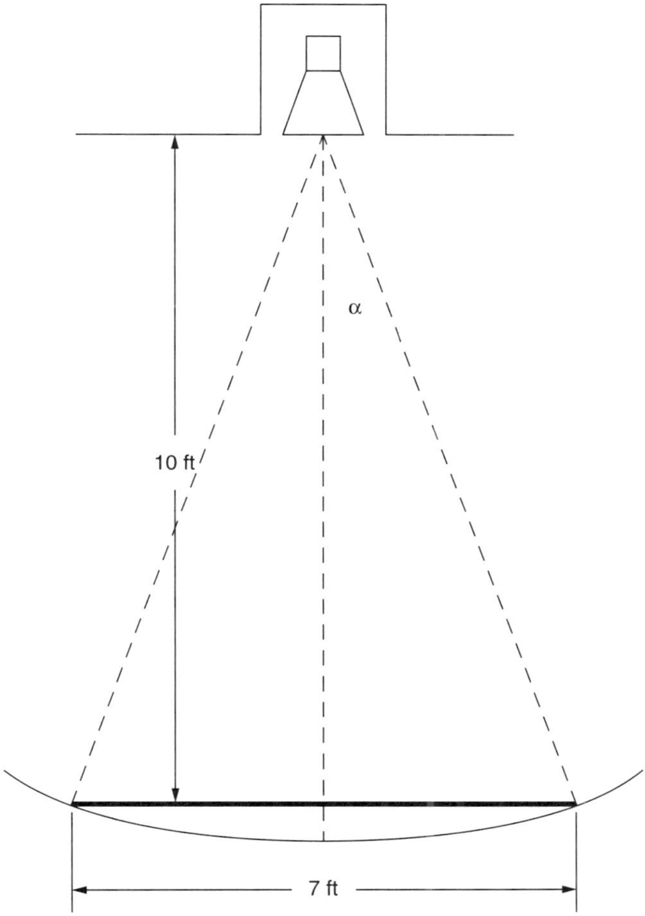

Figure 2.24

The average illuminance is

$$E_{AVE} = \frac{6381}{\pi \times 3.5^2} = 166 \text{ fc}$$

EXAMPLE 2.16

If the intensity of the lamp in Ex. 2.15 is now changed to

$$I = 1000 \cos \alpha \text{ cd } (0°–90°)$$

$$I = 0 \text{ cd } (90°–180°)$$

where α is the angle from the centerline, find the average illumnance in the 7-ft. diameter circle and the total lumens emitted by the lamp.

Solution. From Eq. (2.15)

$$d\phi = I\,d\omega = 1000\cos\alpha \times 2\pi \sin\alpha \, d\alpha$$

$$\phi = 2000\pi \int_0^{20} \sin\alpha \cos\alpha \, d\alpha$$

$$= 2000\pi \left.\frac{\sin^2\alpha}{2}\right|_0^{20} = 367 \text{ lm}$$

$$A = \pi r^2 = 38.5 \text{ ft}^2$$

$$E = \frac{367}{38.5} = 9.5 \text{ fc}$$

To obtain the total luminous flux,

$$\phi = 2000\pi \left.\frac{\sin^2\alpha}{2}\right|_0^{90} = 1000\pi = 3140 \text{ lm}$$

2.10 ALTERNATIVE METHODS OF DISPLAYING INTENSITY DISTRIBUTIONS AND OBTAINING LUMINOUS FLUXES

Intensity data are not always displayed on a polar diagram, as in Fig. 2.19. For lamps with a directed beam, a rectangular plot is often more useful, as was illustrated earlier in Fig. 2.8. Such a presentation is repeated in Fig. 2.25 for four 150-W PAR and R lamps.

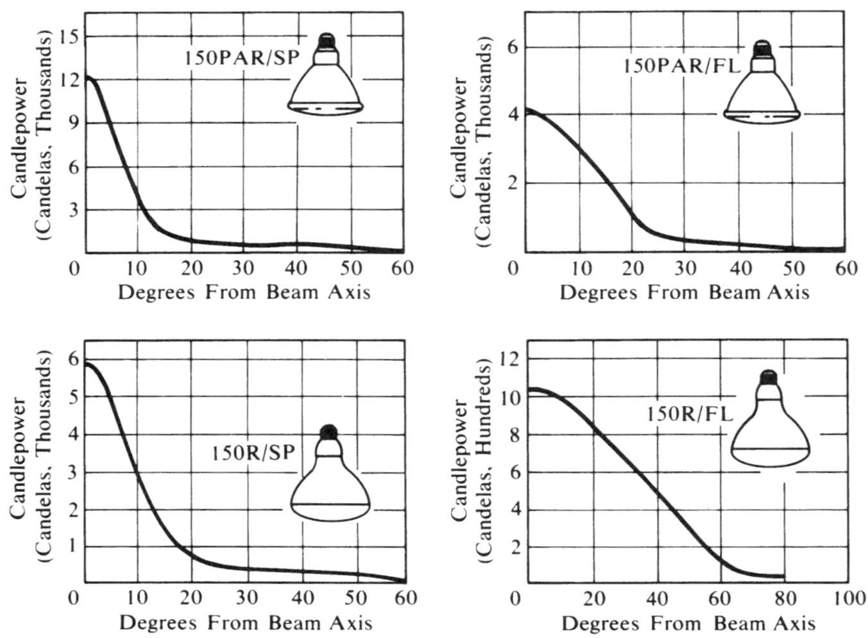

Figure 2.25 Intensity data on a rectangular plot. *Source*: "Incandescent Lamps," Bulletin TP110R1, 1980. Courtesy of General Electric Co.

The notation PAR, as was noted in Sect. 1.7, means *pressed aluminized reflector.* These lamps are of two-piece heat-resistant glass construction: one piece for the pressed reflector, which is similar to parabolic or elliptical; and the other for the lens. R (reflector) lamps consist of a one-piece blown glass bulb, again with the reflector similar to a parabola or an ellipse. More recently, two-piece, pressed glass multifaceted reflector (MR) lamps have appeared, adapted from similar lamps used in projection systems. All of these lamps will be discussed in detail in Chap. 6.

When dealing with floodlighting luminaires and reflectorized lamps, it is useful to define two additional metrics. The first is center beam candlepower (CBCP), expressed in candelas, which is the intensity of the source on the center line of the beam. This is usually, but not always, the maximum intensity of the beam. The second metric is beam angle, which is the angular beam width between the two directions at which the intensity is 50% of its maximum value.*

Lighting calculations with reflectorized lamps usually require that the ISL be inverted, i.e., it is not the illuminance for a given intensity that is required but rather what intensity is needed to produce a desired illuminance. Thus, for example, to produce an illuminance E on the table directly under a downlight d feet away requires, from Eq. (1.27), a CBCP of Ed^2 cd. This procedure is illustrated with an example.

EXAMPLE 2.17

Reconsider the recessed downlight in Ex. 2.4. Its CBCP is 5000 cd and one-half of its beam angle is 17.5°, the angle at which its intensity is 2500 cd. Now specify a new recessed downlight that provides an illuminance of 50 fc directly beneath its center at point 0 and, in conjunction with another downlight of the same size recessed in the ceiling 5 ft to the right of the first one, produces 50 fc at point P.

Solution. The required CBCP of the new downlight is

$$I = \frac{Ed^2}{\cos\beta} = \frac{50 \times 144}{1} = 7200 \text{ cd}$$

At point P, $\beta = \alpha = 11.8°$ and $d^2 = 144 + 6 = 150$. Thus, the required intensity is

$$I = \frac{Ed^2}{\cos\beta} = \frac{25 \times 150}{0.979} = 3830 \text{ cd}$$

The one-half beam angle of the new downlight is the angle at which I = 3600 cd and, thus, is an angle that is somewhat greater than 11.8°, perhaps 14°. A down-lighting catalog would be used to find a luminaire with a CBCP of 7200 cd and a beam angle of 28°.

Initially reflectorized lamps were available in just spot and flood distributions, with the latter having about twice the beam angle of the former (about 60° compared with about 30°). Now, PAR and MR lamps have intensity distributions ranging from very narrow spots (beam angle of 8°) to very wide floods (beam angle of 60°).

Returning now to the measurement of luminous flux, there is an alternative method of obtaining the total lumens emitted by a light source that involves the use of an integrating or Ulbricht sphere. Consider Fig. 2.26, which shows an incandescent lamp at

---

* A second beam metric is field angle, which is the angular beam width between the two directions at which the intensity is 10% of its maximum value.

the center of a hollow sphere of radius r. The inside of the sphere is coated with a matte-finish white paint to make it perfectly diffusing and very highly reflecting. It is desired to show that the total luminous flux emitted by the lamp is directly related to the illuminance at a small opening in the sphere wall.

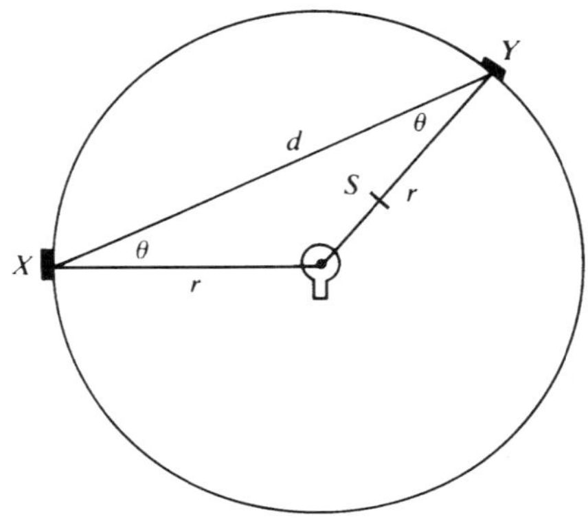

Figure 2.26 Ulbricht Sphere.

The illuminance at any point on the inside of the sphere is composed of two components: a direct component due to luminous flux traveling directly from the source to the point and an indirect component due to the luminous flux reaching the point by reflection from all other points within the sphere. Let the sphere in Fig. 2.26 have a uniform reflectance ρ. Further, let the luminance at point X be L. Then, from Eq. (2.28), the illuminance at a point Y on the inside of the sphere due to a small element of sphere area dA at point X is

$$dE_Y = \frac{L\,dA\cos^2\theta}{d^2} \tag{2.46}$$

Noting that $d = 2r\cos\theta$. Equation (2.46) becomes

$$dE_Y = \frac{L\,dA}{4r^2} \tag{2.47}$$

It is observed that the illuminance at point Y due to a small element at point X is independent of the position of point Y (independent of θ). Thus, Eq. (2.47) gives the illuminance at any point on the inside of the sphere due to the element at point X. Also, the illuminance at point Y due to any other element on the inside of the sphere is given by Eq. (2.47), again because that equation does not depend on θ. It is concluded that every element on the inside of the sphere illuminates every other element equally, and that the indirect illuminance is uniform throughout the sphere.

Now assume that the source at the center of the sphere emits Φ lumens. Of these, ρΦ are reflected. When these reflected lumens hit the inside of the sphere, $\rho^2\Phi$ are reflected, and so on. The total reflected lumens are

$$\Phi_{refl.} = \rho\Phi(1+\rho+\rho^2+\cdots+\rho^{n-1}) \tag{2.48}$$

Since ρ is a number less than 1, the sum in the parentheses has a finite value of 1/(1 – ρ) and thus Eq. (2.48) can be rewritten as

$$\Phi_{\text{refl.}} = \frac{\rho}{1-\rho}\Phi \tag{2.49}$$

Noting that the sphere surface area is $4\pi r^2$, the illuminance due to reflected lumens (the indirect illuminance) is

$$E = \frac{\rho}{4\pi r^2(1-\rho)}\Phi \tag{2.50}$$

This is the relation that is sought. It shows that if an illuminance meter is placed in the sphere wall (at point Y in Fig. 2.26) and a small shield is placed between the source and the meter to block out direct illuminance (shown at S in Fig. 2.26), the reading of the meter will be proportional to the total lumens emitted by the lamp.*

In practice, the sphere is calibrated by measuring a standard source whose total lumen output is known, and whose spatial and spectral properties are similar to lamps being measured. This enables the calibration constant of the sphere to be established as $\rho/[4\pi r^2(1-\rho)]$. Calling this constant k, the luminous flux of any source placed within the sphere is obtained from

$$\Phi = \frac{1}{k}E \tag{2.51}$$

where E is the reading of the illuminance meter placed in the sphere wall.

For example, if an incandescent lamp is placed inside a 2-ft-radius integrating sphere whose reflectance is 0.95 and the meter in the sphere wall reads 300 fc, the number of lumens emitted by the lamp is, from Eq. (2.50),

$$\Phi = \left[\frac{4\pi(4)(0.05)}{0.95}\right]300 = 794 \text{ lm}$$

2.11 IDEAL INTENSITY DISTRIBUTIONS

There are four ideal intensity distributions that will be considered, namely uniform, cosine, torroidal, and narrow beam. Although these are indeed ideal sources, they have real-world counterparts whose intensity distributions are close matches to the ideal cases.

Uniform

From Eq. (2.15), a source with a uniform intensity distribution I, such as in Fig. 2.27a, gives

$$\begin{aligned} d\phi &= I\,d\omega = I \times 2\pi \sin\alpha\, d\alpha \\ \phi &= 2\pi I(-\cos\alpha)_0^\pi = 4\pi I \text{ lm} \end{aligned} \tag{2.52}$$

* Unfortunately, practical spheres are not perfect integrators, See, for example, IESNA LM 9-99 and LM45-00.

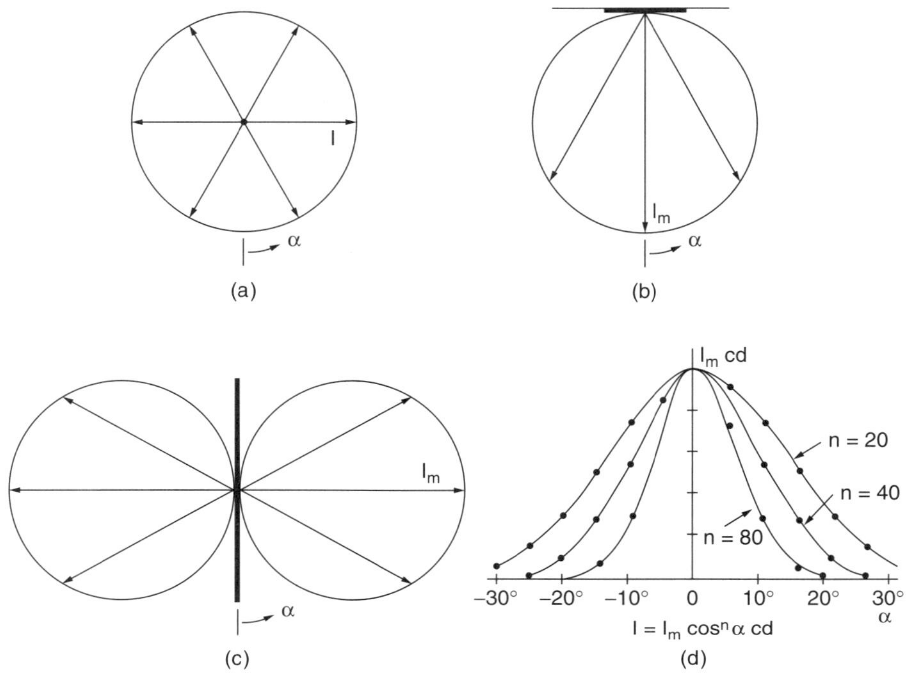

Figure 2.27 Ideal Intensity Distributions.

A 100-watt, inside-frosted incandescent lamp emits 1750 lm. If its intensity were uniform, its value would be $I = \frac{1750}{4\pi} = 139\,\text{cd}$. The actual intensities of the lamp at $\alpha = 45°$, 90°, and 135° are 134, 153, and 140 cd, respectively, very close to the uniform intensity value. The near-zero intensity in the base region ($\alpha = 0$) and the reduced intensity at the top of the bulb ($\alpha = 180°$) account for the differences.

Cosine

If a source, such as a flat circular disc, has a cosine intensity distribution, as in Eq. (2.31) and Fig. 2.27b, its luminous flux output is given by Eq. (2.15) as

$$d\phi = I\,d\omega = I_m \cos\alpha \times 2\pi \sin\alpha\, d\alpha$$

$$\phi = 2\pi I_m \left.\frac{\sin^2\alpha}{2}\right|_0^{\pi/2} = \pi I_m \text{ lm} \tag{2.53}$$

An example of this type of source is a shallow translucent cylindrical luminaire. Its luminance is nearly constant with viewing angle, and, thus, its intensity nearly obeys the cosine law.

Toroidal

A cylindrical source that is a Lambertian emitter has the toroidal intensity distribution shown in Fig. 2.27c with $I = I_m \sin\alpha$ cd. Again, using Eq. (2.15), and assuming the viewing distance is large compared with source length,

$$\begin{aligned} d\phi &= I\,d\omega = I_m \sin\alpha \times 2\pi \sin\alpha\, d\alpha \\ &= 2\pi I_m \sin^2\alpha\, d\alpha = 2\pi I_m \frac{1-\cos 2\alpha}{2} \\ \phi &= \pi I_m \int_0^{\pi} (1-\cos 2\alpha)\, d\alpha \\ &= \pi I_m \left(\alpha - \frac{\sin 2\alpha}{2}\right)_0^{\pi} = \pi^2 I_m \text{ lm} \end{aligned} \tag{2.54}$$

giving a lumen/maximum candela ratio of $\pi^2 = 9.87$.

A 32-watt T8 fluorescent lamp emits 2850 lm. If its intensity distribution were perfectly toroidal, its maximum intensity would be $\frac{2850}{9.87} = 289$ cd. The actual value of the ratio is 9.15, rather than 9.87, because of attenuation by the bulb wall at large angles from a normal to it. Thus, Eq. (2.54) is modified to $\phi = 9.15\, I_m$ for this case, giving $I_m = 311$ cd.

Narrow Beam

Floodlight and spotlight intensity distributions can often be approximated by a cosine function raised to a power.

$$I = I_m \cos^n \alpha \tag{2.55}$$

Values of I for three values of n are plotted in Fig. 2.27d.

In Sect. 2.10, beam angle was defined as the angular beam width between the two directions at which beam intensity was 50% of maximum. Calling this angle α_B,

$$\left.\begin{aligned} &\cos^n \frac{\alpha_B}{2} = 0.5 \\ &n \log\left(\cos\frac{\alpha_B}{2}\right) = -0.30 \\ &n = \frac{-0.30}{\log\left(\cos\frac{\alpha_B}{2}\right)} \end{aligned}\right\} \tag{2.56}$$

Values of n from Eq. (2.56) for several values of α_B are listed in Table 2.2.

TABLE 2.2

α_B°	n	α_B°	n
5	730	25	29
10	180	30	20
15	81	35	15
20	45	40	11

EXAMPLE 2.18

A 150-watt 2800 lm incandescent lamp is mounted in a shallow 1-ft diameter circular white reflector with $\rho = 80\%$, as shown in Fig. 2.28. Assume the following:

a) The lamp is a uniform point source.
b) The reflector intercepts 50% of the lamp lumens.
c) The reflector is flat and perfectly diffusing.

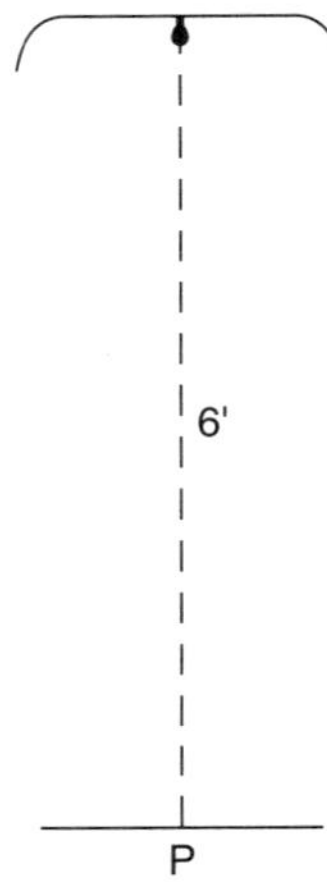

Figure 2.28

Find the horizontal illuminance at P, 6 ft away from the lamp.

Solution. The uniform lamp intensity is, from Eq. (2.52),

$$I = \frac{2800}{4\pi} = 223\,\text{cd}$$

The reflector reflects

$$2800 \times 0.5 \times 0.8 = 1120\ \text{lm}$$

For a flat, perfectly diffusing disc, Eq. (2.53) applies and

$$I_m = \frac{1120}{\pi} = 357\,\text{cd}$$

Then, since the overall source size is small compared with 6 ft,

$$E = \frac{223 + 357}{36} = 16\,\text{fc}$$

2.12 INVERSE-SQUARE-LAW APPROXIMATIONS

In Sect. 2.9, it was noted that there is no way of directly measuring luminous intensity or luminous flux. Rather illuminance is measured and then the ISL is used to obtain these quantities. This leads to further inquiry about the accuracy of ISL calculations.

Recall that the ISL is based on a point source. For a nonpoint source, it was reasoned that the ISL could still be used if the source size were small compared with the distance from the source to the receiver. It was noted that the ratio of this distance to the maximum source dimension should be at least 5:1 for the ISL to hold within the order of ±1%. This constraint can be investigated more fully by considering the flat disc (circular) source, viewed on edge in Fig. 2.29. Assume the source has a radius R and a constant luminance of L in the direction of P from all points on its surface. Assume that D is the distance from source to receiver and is not necessarily large compared with 2R, the maximum source dimension.

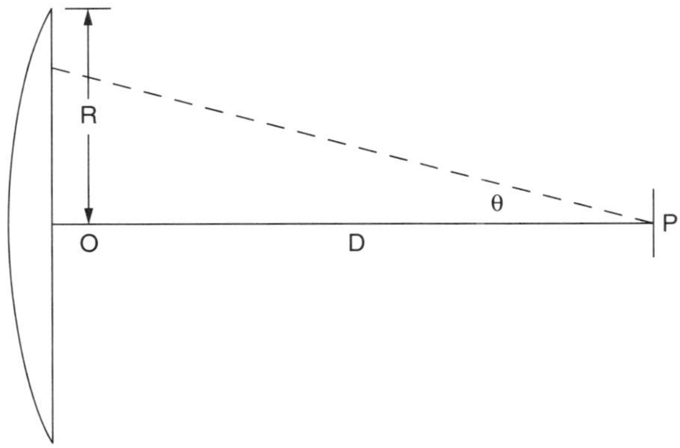

Figure 2.29 Ascertaining ISL Limitations.

Since L is constant toward P the source appears perfectly diffusing from P, and thus the disc can be replaced with a saucer-shaped portion of a sphere centered at P, to simplify integration. Then, much as was done with the sky as a source in Sect. 2.8, Eq. (2.37) is used to obtain.

$$dE_p = L\cos\theta\, d\theta = L\cos\theta \times 2\pi \sin\theta\, d\theta \tag{2.57}$$

$$E_p = 2\pi L \left.\frac{\sin^2\theta}{2}\right|_0^{\theta_m} = \pi L \sin^2\theta_m = \pi L \frac{R^2}{R^2 + D^2} \tag{2.58}$$

As an example, let R = 1 ft and L = 500 cd/ft^2. Substituting into Eq. (2.58) yields

$$E_p = \frac{1570}{1 + D^2} \tag{2.59a}$$

This is the actual illuminance at any distance D ft from the source, and is tabulated in the second column of Table 2.3 for values of D from 0.5 ft to 40 ft.

TABLE 2.3

D (ft)	Actual E [Eq. (2.59a)]	ISL E [Eq. (2.59b)]	% Error [Eq. (2.60)]	Apparent I [(Actual E)D^2]
0.5	1256	6280	400	314
1	785	1570	100	785
2	314	392.5	25	1256
4	92.35	98.1	6.25	1478
4.5	73.38	77.53	4.94	1486
5	60.38	62.80	4.00	1510
6	42.43	43.61	2.78	1528
8	24.15	24.53	1.56	1546
10	15.54	15.70	1.00	1554
15	6.95	6.98	0.44	1563
20	3.92	3.93	0.25	1566
30	1.74	1.74	0.11	1568
40	0.981	0.981	0.06	1569

Assume now that the ISL holds for all values of D. The intensity of the source becomes

$$I = LA = \pi R^2 L = 1570 \text{ cd}$$

Then, from the ISL,

$$E_p = \frac{1570}{D^2} \tag{2.59b}$$

These values of E_p appear in the third column in Table 2.3. The percent error between the actual and ISL values of E_p appears in the fourth column of Table 2.3. It is given by

$$\%\,\text{Error} = 100\frac{\dfrac{1570}{D^2} - \dfrac{1570}{D^2+1}}{\dfrac{1570}{D^2+1}} = 100 \times \frac{1}{D^2} \tag{2.60}$$

The fifth column in Table 2.3 is labeled "apparent I". This is the intensity that is obtained by multiplying the actual illuminance by D^2, as though the ISL were valid.

From Table 2.3 it is seen that for the ISL illuminance to be within 1% of the actual illuminance, the ratio of D to 2R (the maximum source dimension) must be 5 to 1 or greater. This is the criterion that has been used up to now. However, if a 3% error can be tolerated, a ratio of only 3 to 1 is required. These numbers become 2.5 to 1 for a 4% error and 2.25 to 1 for a 5% error. Although it cannot be generalized directly to other source shapes from this example, the result is generally similar if the source luminance distribution toward the point of interest from all points on its surface is nearly Lambertian. The 4% criterion will be used quite often in problems in the remainder of the text. For example, this means that a 2-ft diameter fluorescent luminaire 5 ft from a desk top can be treated as a point source and the ISL applied.

If the source in Fig. 2.29 were viewed in the laboratory and its intensity from far away measured using an illuminance meter, the values of "Actual E" in Table 2.3 would be obtained. Then the "Apparent I" column in Table 2.3 would be found by multiplying each actual E by D^2, as though the ISL held. As D increases, it is seen that the apparent intensity approaches the actual intensity of the whole source from afar. In limit notation,

$$I = \lim_{D \to \infty} ED^2 \tag{2.61}$$

This result can be obtained from Eq. (2.58) as well. For D >> R,

$$E_p = \frac{\pi LR^2}{D^2} = \frac{LA}{D^2} = \frac{I}{D^2} \tag{2.62}$$

For D << R, Eq. (2.58) becomes

$$E_p = \pi L \tag{2.63}$$

This is indeed what one would expect, for as one moves very close to the source, it becomes a uniform-luminance, infinite sky, giving the result of Eq. (2.39). Thus, the disc source in Fig. 2.29 appears as a point source from far away and as a uniform infinite source from close by.

2.13 MEASURING ILLUMINANCE AND LUMINANCE

Most modern instruments for measuring illuminance and related quantities make use of the photoelectric effect, first discovered by Hertz in 1887. There are four basic forms of photoelectric phenomena, namely photoemissive, photoconductive junction, photoconductive bulk, and photovoltaic.

The photoemissive effect is the liberation of electrons from a photosensitive surface when light impinges on that surface. As will be discussed in Chap. 5, it was this effect that could not be explained by electromagnetic wave theory and which led Planck, Einstein, and others to postulate the photon. Vacuum and gas phototubes and multiplier phototubes operate in the photoemissive mode.

The photoconductive effect is manifested by certain semiconductor devices. The basic phenomenon is that the conductivity of a piece of semiconductor material is altered drastically by incident radiation. The radiant energy causes covalent bonds to be broken within the semiconductor, thus creating many hole-electron pairs in excess of those created normally by thermal generation. The resulting increase in current carriers decreases the resistance of the material.

There are two types of photoconductive devices. Photoconductive junction devices, such as the photodiode and phototransistor, are made of germanium or silicon. These devices exhibit very fast response, but relatively low sensitivity (mA/lx). Photoconductive bulk devices utilize crystals, such as cadmium sulfide, which are doped with impurities of silver, antimony, or indium. These devices have very high sensitivity; their resistances may vary from 2 MΩ in absolute darkness to 10 Ω under strong light.

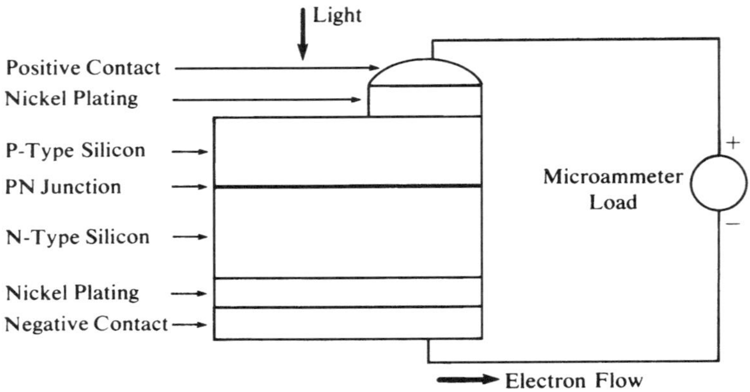

Figure 2.30 Silicon cell light meter.

The fourth type of photoelectric phenomena, which is utilized in illuminance measuring devices, is the photovoltaic effect. A typical photovoltaic cell is shown in Fig. 2.30. The P-type silicon is doped with boron; the N-type silicon with arsenic. When light strikes the P-type material, excess hole-electron pairs are created. The holes migrate across the junction into the P-type material; the electrons move to the N-type material. The resulting unbalance of charge carriers creates a voltage at the diode terminals. This is the photovolatic voltage and is approximately 0.5 V for a silicon photodiode.

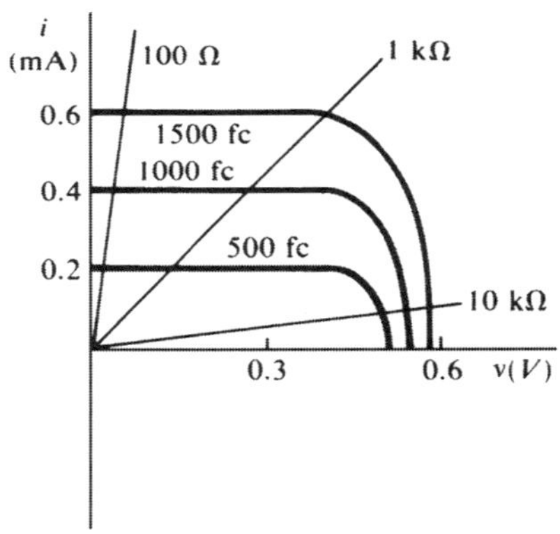

Figure 2.31 Silicon cell volt-ampere characteristic.

The volt-ampere characteristic of a typical silicon photocell used to measure illuminance is shown in Fig. 2.31.

Note that the short-circuit current ($v = 0$) varies linearly with illuminance level, and the current for low-resistance (100-Ω) loads nearly so. This condition is utilized in an illuminance meter, that is, the output current is directly proportional to the illuminance level.

The first photovoltaic cell light meter, using a selenium cell, was developed by the Weston Instrument Company in 1931 and led to the lighting industry's first portable illuminance meters and photographic exposure meters. Modern illuminance meters are more sophisticated than their predecessors, but embody the same basic principles. Selenium has given way to silicon. The latter is much more linear and less temperature dependent than the former.

Most contemporary meters are color and cosine corrected. Figure 2.32 shows spectral sensitivity curves of an uncorrected silicon cell, a corrected cell, and the human eye. The problem, of course, is that it is desired that an illuminance meter evaluate luminous flux in the same way that the eye does. This means that the light cell's spectral sensitivity curve should match that of the eye. As can be seen in Fig. 2.32, an uncorrected cell has a much greater spectral sensitivity to wavelengths in the violet-blue and red-orange regions of the spectrum than does the eye. Also the cell responds to ultraviolet radiation whereas the eye does not. What the meter manufacturers have done is to integrate a carefully designed color filter with the cell so that the filter-cell spectral sensitivity curve nearly matches that of the human eye, as shown in Fig. 2.32, within 1% of the CIE photopic spectral luminous efficiency curve.

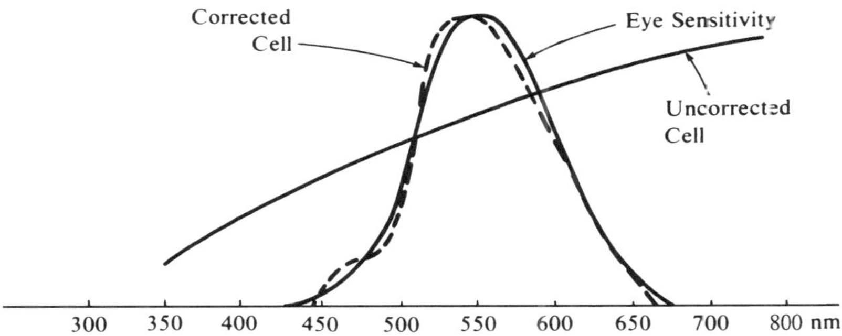

Figure 2.32 Color correction.

Cosine correction is also necessary for accurate illuminance measurement. When luminous flux strikes the meter at an angle, the meter should evaluate this flux so that its reading is the product of the reading that would be obtained with the cell held perpendicular to the incoming light times the cosine of the angle of incidence.

A meter which is not cosine corrected generally has a glass plate covering the cell. Light rays striking the glass plate at a grazing angle are for the most part reflected and, thus, do not enter the cell and contribute to the meter reading. Cosine correction is achieved by replacing the glass plate with a slightly convex diffusing cover of white plastic, designed to intercept wide-angle rays and direct them to the cell.

The effect of cosine correction is shown in Fig. 2.33. The plot of a cosine is shown dotted. This is how the cell should respond. The response of the uncorrected cell with a glass cover is the dashed curve, and this results in illuminance readings that are lower than they should be. The solid curve is for the cell with cosine correction and matches closely the cosine curve. Contemporary precision light meters often provide less than ±1% cosine error up to an incident angle of 85°. Cosine direction at larger angles is very difficult to achieve.

Another issue that should be considered in choosing a light meter is its accuracy. The calibration of light meters is often "traceable to the National Institute of Standards and Technology (USA) or the National Research Council (Canada)," but the number

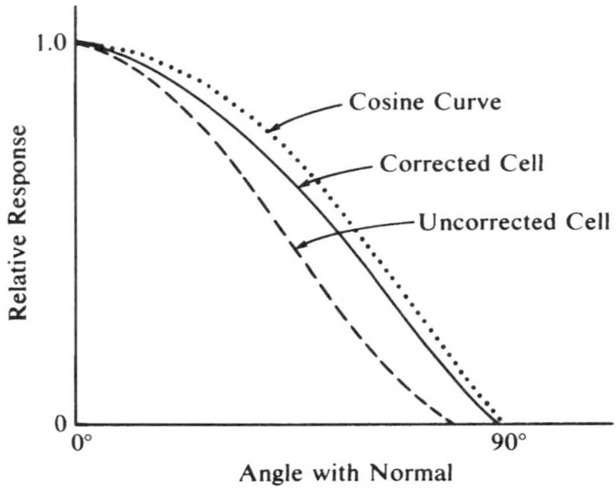

Figure 2.33 Cosine correction.

of links in the chain of calibration between the meter and NIST or NRC is not always clear. Documentation of accuracy has become even more important in recent years with the advent of the ISO 9000* standards, to which equipment and parts suppliers must adhere. Laboratory meters should be accurate within less than ±5% and small portable meters within ±15%. An accuracy of about ±2% is the current state-of-the-art.

It is possible to measure crudely luminous exitance, reflectance, and transmittance with an illuminance meter. To estimate luminous exitance, the luminous flux density leaving a surface, the cell is held close to the surface and facing it and then withdrawn a short distance until the meter reading becomes constant. The reading of the meter is now in lumens per square meter or lumens per square foot of luminous exitance. Care must be taken that the meter is withdrawn far enough so that the meter (or the measurer) does not create a shadow on the surface being measured.

To estimate the reflectance of a surface, the cell is held against the surface facing outward, and the illuminance reading is noted. Then the luminous exitance of the surface is measured. The ratio of the two is the reflectance. To estimate the transmittance of a piece of glass or plastic, the sample is placed over the cell for one reading, and then a second reading is taken with the sample removed. The ratio of the two readings is the transmittance.

Frequently, the light cell of an illuminance meter is in the form of a separate light paddle, connected to the display by an extension cord. This enables the user to read the meter without imposing a shadow on the cell face. In most contemporary illuminance meters, the analog microammeter has been replaced by solid-state amplification, an analog-to-digital converter, logic circuitry, and a liquid crystal display with light-emitting diode readout, which consists usually of three 10 mm-high numbers, each composed of seven LED bars. The illuminance range for a digital meter greatly exceeds that of its analog ancestor, often accommodating inputs from 10^{-2} to 10^{6} lx.

Early luminance meters, such as the Luckiesh-Taylor brightness meter, required the user to make a visual photometric match between the luminance of the object being measured and that produced by a small calibrated incandescent lamp within the instrument.

* International Standards Organization.

Such matches were difficult to make, largely because of color temperature differences between the object and calibrated lamp fields, and several measurements of a given luminance by the same or different observers were likely to show quite a wide spread.

Contemporary luminance meters do not require visual matching. Their optics permit luminance measurements of fields from $^1/_3°$ to 3°. One simply aims the device at the area of interest by viewing through a sight tube and then pushes a button to display the luminance digitally. The meters are color corrected and have ranges from 10^{-3} to 10^6 cd/m^2.

A luminance meter is susceptible to all of the errors of an illuminance meter and, in addition, its optics are very dust sensitive. Luminances outside the intended field of view will be included in the luminance reading if the lenses are dirty. Some meters have a rubber lens hood to minimize the admission of light outside the target zone. Also an objective lens focuses accurately the entering light so that only the rays from the target area are sent through to the photo cell.

REFERENCES

American National Standards Institute: *Nomenclature and Definitions for Illuminating Engineering*, ANSI/IESNA RP 16-96, New York, 1996.

Boast, W.B.: *Illumination Engineering*, McGraw-Hill, New York, 1953.

Coaton, J.B., and A.M. Marsden: *Lamps and Lighting*, Fourth Edition, Arnold, London, 1997.

Collins, R.G.: "A Sphere Photometry System for the Nineties," *IESNA Journal*, Volume 26, Number 1, 1997.

IESNA Lighting Design Practice Committee: "The Determination of Illumination at a Point in Interior Spaces," *Journal of IES*, January 1974.

Johnston, N.B.: "Introducing the Modern Metric System into Engineering Education," *Engineering Education*, April 1977.

Levin, R.E.: "Luminance-A Tutorial Paper," Journal of the Society of Motion Picture and Television Engineers, 77(10): 1005, 1968.

Levin, R.E.: "The Photometric Connection—Parts 1–4," *Lighting Design and Application*, September–December 1982.

Lindsey, J.L.: *Applied Illumination Engineering*, Fairmont Press, Lilburn, GA, 1991.

Millman, J., and C.C. Halkias: *Integrated Electronics*, McGraw-Hill, New York, 1972.

Murdoch, J.B.: "Inverse Square Law Approximations of Illuminance," *IES Journal*, 1981.

Murdoch, J.B.: *Lighting Metrics, TM–1-99*, IESNA, New York, 1999.

Rea, M.S. editor: *IESNA Lighting Handbook*, Ninth Edition, IESNA, New York, 2000.

Roberts, J., editor: *Advanced Lighting Guidelines*, New Buildings Institute, White Salmon, WA, 2001.

Chapter 3

Vision

3.1 RECEPTIVE FIELDS

In Sect. 1.6, the eye was examined, and it was established that light is absorbed by the pigments in the rods and cones, changing the electric current flow around the photoreceptors and causing a chemical substance to be secreted. This is the first step in a lengthy process leading to vision.

In Fig. 3.1, a model of the retina is shown. Light, after passing through the vitreous humor, penetrates the largely-transparent cell regions of the retina and is absorbed by the photoreceptors, causing the release of neural signals. These are received and organized by the collector cells, and sent on to the ganglion cells for interpretation and recasting as nerve impulses, which are then forwarded via the axons and the optic nerve to the brain. The complexity of the connections from photoreceptors to collector cells to ganglion cells is portrayed by the straight lines in the model.

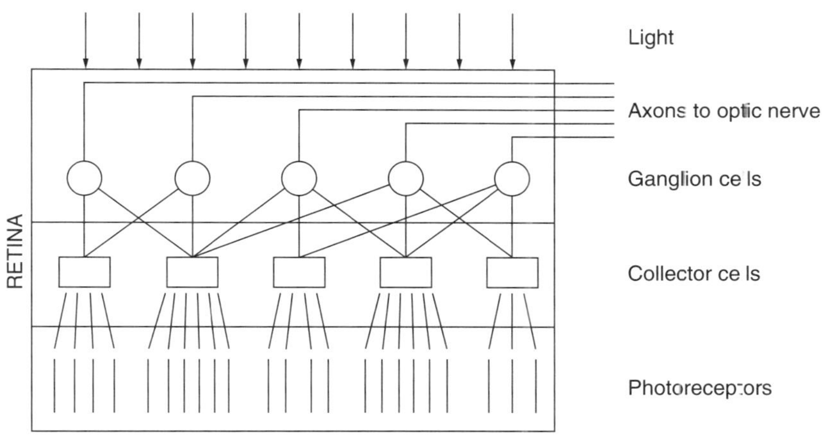

Figure 3.1 Model of the cross section of the retina.

Each eye contains about 130 million photoreceptors. Thus, at any instant, there are 130 million light signals received by each eye, and it is the job of each photoreceptor to sense the amount of light received and pass this information on.

Each eye contains only about 1 million ganglion cells, whose job it is to condense and interpret the 130 million photoreceptor signals. This 130 to 1 relationship between receptors and ganglion cells is called convergence. Convergence varies throughout the retina. It is nearly 1 to 1 in the fovea, but increases with distance from the fovea to several hundred to 1.

In order to see objects, the eye must be able to define boundaries, which, in turn, means that it must be able to distinguish light-level differences. There are three major factors that govern the retina's ability to accomplish this.

Receptive Field Geometry and Excitation

Fig. 3.2 depicts the organization of receptive fields in the retina. A receptive field is a small area of the retina whose photoreceptors feed a single ganglion cell. Research shows that these patches are circular, with a central circle and a surrounding ring. When the photoreceptors in the central circle are excited by light, the ganglion cell emits a sequence of neural signals. When the light is removed, it stops emitting. When photoreceptors in the ring portion are stimulated by light, the cell output drops below the normal (background noise) level. When the light ceases, the cell emits a signal burst.

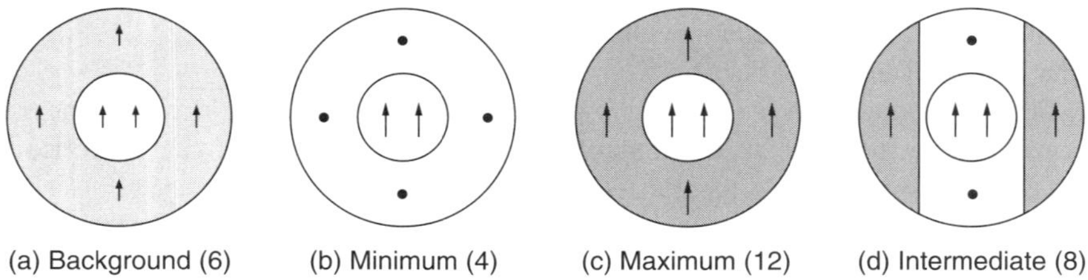

Figure 3.2 Receptive field geometry and excitation.

The receptive fields in Fig. 3.2 represent half of the total. In the other half, the role of the center and the ring is reversed, i.e., the receptors in the center turn "off" in response to light and those in the ring turn "on".

How is it that this circle-ring receptive field geometry provides edge detection of objects? In Fig. 3.2a, a receptive field is shown in its normal, unexcited condition. The six small arrows indicate background signal strength (assuming each arrow is 1 unit long, total signal strength is 6 units). Now apply a light signal over the entire receptive field in Fig. 3.2b. The signals from the receptors in the center will increase while those in the ring will decrease, resulting in minimal change (with the center arrows twice as long and the ring arrows zero, the total signal strength is now 4 units).

Next assume a lighted center disc and a dark ring, as in Fig. 3.2c. The center of the receptive field receives an increase in light, and the ring receives a decrease in light. The arrow lengths tell the story, an increased signal strength everywhere (a total of 12 units). Finally, an intermediate case is shown in Fig. 3.2d, where about one-half of the total receptive field is lighted and the other one-half is not lighted (a total of 8 units).

It now should be clear that the center-ring receptive field geometry has enabled the ganglion cells to perform the important task of condensing the information from the photo receptors so as to permit detection of light-dark boundaries (contrasts) and edges of objects.

Receptive Field Size

The size of a receptive field increases as its distance radially from the fovea increases. Within the macula, receptive fields are as small as 0.01 mm in diameter, whereas in the periphery they may be as large as 0.5 mm in diameter. Also, at a given distance from the fovea, receptive field diameters vary statistically about the mean. What is the influence of receptive field size? A smaller receptive field will have fewer photo receptors feeding it, which leads to the conclusion that ganglion cells with small receptive fields are best for imaging small objects, since larger objects would overlap adjacent receptive fields.

Types of Ganglion Cells

Ganglion cells are of two types: large magnocellular (M cells) and small parvocellular (P cells), named for the distinct parallel channels from the cells to the cortex of the brain. The characteristics of these cells are presented in Table 3.1 and will be utilized in succeeding sections in explaining various characteristics of vision.

TABLE 3.1 GANGLION CELL CHARACTERISTICS

Parvocellular	Magnocellular
Predominant in fovea	More uniformly-distributed throughout retina
Small size	Large size
Respond better to small objects	Respond better to large objects
Respond to more sustained light signals—better for stationary objects	Respond to rapidly-changing light signals—better for moving objects.
Account for 80% of ganglion cells	Account for 20% of ganglion cells
Smaller receptive fields at a given distance from fovea	Larger receptive fields at a given distance from fovea
Respond to large differences in light level	Respond to small differences in light level
Low sensitivity to contrast	High sensitivity to contrast
Respond poorly to rapid on-off light signals	Respond well to rapid on-off light signals
Respond to wavelength (color) selectively; respond to one color and reject its opponent color—chromatic	Respond to light based on amount, independent of color—achromatic
Respond preferentially to high rates of change of object luminance in space	Respond preferentially to low rates of change of object luminance in space
Fine depth perception	Course depth perception
Process inputs from heterochromatic brightness matching	Process inputs from flicker photometry
Process recognition of objects	Process detection of objects

3.2 CHARACTERISTICS OF VISION

The information from Sect. 3.1 on how the eye deciphers luminance variations and object details permits several characteristics of vision to be analyzed.

Resolution

Resolution, specifically spatial resolution, is defined as the ability of the eye to distinguish between two adjacent luminances in a visual image. Resolution is inversely proportional to convergence—the more photo receptors feeding information to a single ganglion cell, the worse the resolution. Since convergence is about 1 to 1 in the fovea and receptive field size is smallest there, resolution is greatest. Visual acuity, which is a measure of resolution, will be discussed in a later section.

The small receptive fields and low convergences in the fovea are essential to accurately locating and replicating objects and delineating contrast boundaries. This is cone or photopic vision.

Light Sensitivity

Whereas high convergence is an anathema to resolution, it is a boon to sensitivity. In the periphery, receptive fields are large and are connected to a large number of photo receptors. This means that individually weak light signals can generate large overall responses in the ganglion cells. Each ganglion cell collects and sums the signal inputs from the large areas on the retina, without regard for which receptors contributed to the overall signal. The action of a retinal cell in processing the collection of weak signals is called spatial summation. The result is rod vision (scotopic vision).

Adaptation

It was stated earlier that the eye has an accommodation range of about 10^{12}, from 10^{-6} to 10^{6} cd/m^2, and that three adaptation mechanisms—optical, neural and photochemical provide this accommodation. Optical changes account for $10^{1.2}$ (16 to 1) of this range, with constriction occurring in 0.3s and dilation in 1.5s. Neural changes account for 10^2 to 10^3 of the range in normal room-lighting levels and take less than 200 ms. Photochemical adaptation represents up to 10^8 of the overall accommodation range and takes much longer. The cones adapt faster than the rods, 10 to 12 minutes compared with 60 minutes or more. Responses to higher levels of illuminance happen faster than to lower levels. This is evident when entering and leaving a movie theater. Upon entering the theater from a bright lobby, one cannot see clearly. It may take up to 30 minutes for a person's eyes to become thoroughly dark adapted. When leaving the theater, the time it takes to adapt to the higher-level lobby lighting is relatively brief.

The visual system is capable of adapting to three levels of vision: photopic, mesopic, and scotopic. Photopic vision occurs at luminances higher than 3 cd/m^2 and is cone vision. Scotopic vision lies below a luminance of 0.01 cd/m^2 and is rod vision. In between is a region called mesopic vision, where both rods and cones are energized.

Spectral Sensitivity

Rods and cones are selectively sensitive to wavelength, and each has a spectral sensitivity curve. The curve for cones is shown in Fig. 3.3a (solid curve) and the data is

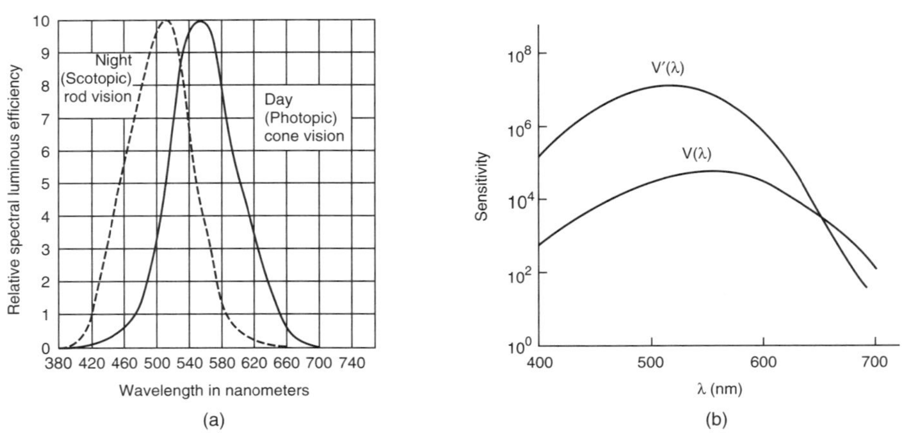

Figure 3.3 Spectral luminous efficiency curves. *Source*: *IESNA Lighting Handbook*, 9th edition, 2000. Courtesy of the Illuminating Engineering Society of North America.

TABLE 3.2 SPECTRAL LUMINOUS EFFICIENCY VALUES FOR PHOTOPIC VISION

Wavelength (nm)	Spectral Luminous Efficiency	Wavelength (nm)	Spectral Luminous Efficiency
380	0.00004	580	0.870
390	0.00012	590	0.757
400	0.0004	600	0.631
410	0.0012	610	0.503
420	0.0040	620	0.381
430	0.0116	630	0.265
440	0.023	640	0.175
450	0.038	650	0.107
460	0.060	660	0.061
470	0.091	670	0.032
480	0.139	680	0.017
490	0.208	690	0.0082
500	0.323	700	0.0041
510	0.503	710	0.0021
520	0.710	720	0.00105
530	0.862	730	0.00052
540	0.954	740	0.00025
550	0.995	750	0.00012
555	1.000	760	0.00006
560	0.995	770	0.00003
570	0.952		

given in Table 3.2. What this curve shows is that, at photopic levels, the eye is most sensitive to light at 555 nm (in the yellow-green portion of the spectrum). The spectral luminous efficiency function V(λ) is normalized to the value 1.0 at that wavelength. At all other wavelengths, the eye is less sensitive. For example, if an observer focuses on a one-watt source of blue light at 450 nm and a one-watt source of yellow light of the same size at 600 nm, the 600-nm light will be $\frac{0.631}{0.038} = 16.6$ times as luminous (values from Table 3.2).

The minimum spectral power density input to yield a prescribed constant luminance for photopic vision occurs at 555 nm. Thus, spectral luminous efficiency can be defined by

$$V(\lambda) = \frac{M_{555}}{M_{\lambda}} \tag{3.1}$$

where M_{555} is the spectral power density at 555 nm required for a luminance balance with spectral power density M_{λ} at wavelength λ.

The spectral luminous efficiency function $V'_{(\lambda)}$, adopted by the CIE in 1951 for scotopic vision, is also shown in Fig. 3.3a (dashed curve), and the data is given in Table 3.3. The curve peaks at 507 nm and $V'_{(\lambda)}$ is normalized to 1.0 at that wavelength. Note the

TABLE 3.3 SPECTRAL LUMINOUS EFICIENCY VALUES FOR SCOTOPIC VISION

Wavelength (nm)	Spectral Luminous Efficiency	Wavelength (nm)	Spectral Luminous Efficiency
380	0.0006	555	0.402
390	0.0022	560	0.329
400	0.0093	570	0.208
410	0.035	580	0.121
420	0.097	590	0.066
430	0.200	600	0.033
440	0.328	610	0.016
450	0.455	620	0.0074
460	0.567	630	0.0033
470	0.676	640	0.0015
480	0.793	650	0.0007
490	0.904	660	0.0003
500	0.982	670	0.0001
507	1.000	680	0.00007
510	0.997	690	0.00004
520	0.935	700	0.00002
530	0.811	710	0.00001
540	0.650		
550	0.481		

shift of 48 nm from V(λ) toward the shorter wavelengths. This is known as the Purkinje Effect, named for Johannes von Purkinje, a physiologist from Bohemia. The shift can be seen by observing red and violet flowers as dusk approaches. In daylight, the red flower will likely appear brighter, but as dusk approaches, the red flower will dim, and the violet flower will appear brighter. This is the transition from photopic, through mesopic, to scotopic vision. As the transition occurs, both cones and rods are operable, but cone sensitivity is decreasing while rod sensitivity is increasing. Also vision is shifting from the central fovea to the periphery, causing loss of resolution and color distinction.

The curves in Figs. 3.3a have been normalized to 1.0. Absolute curves are shown in Fig. 3.3b, which show that the rods are indeed significantly more sensitive than the cones, except at longer wavelengths. These curves are based on the luminance required for detection of a test spot target, in the periphery for $V'_{(\lambda)}$ and in the fovea for V(λ). The vertical axis in Fig. 3.3b is the reciprocal of the number of photons required to just see the test spot (threshold seeing).

3.3 PHOTOPIC SPECTRAL SENSITIVITY

The spectral luminous efficiency curve for photopic vision was developed from two studies, by Coblentz and Emerson in 1918 and Gibson and Tyndall in 1921. Gibson and Tyndall combined the two sets of results and recommended a single set of values at 10-nm intervals. These were adopted by the CIE in 1924 and by the CIPM in 1933.

There were two types of photometry used in the studies. Coblentz and Emerson, with 125 subjects, used flicker photometry. A reference field, consisting of white light, and a test field, composed of monochromatic light, were alternated rapidly. As the frequency of alternation (the flicker frequency) was increased, a point was reached at which the colors fused into one. Then the flicker frequency was increased further and the intensity of the test field adjusted until a condition known as "minimum perceptible flicker" was achieved, indicating a brightness match. The key to flicker photometry is that the colors always fuse before brightnesses. Thus, only the achromatic (black and white) channel of vision is involved in a flicker photometer match.

Gibson and Tyndall's study involved 52 observers and used heterochromatic brightness matching. In this method, two semicircular fields are used side-by-side, but they do not alternate. The fields are illuminated by two monochromatic light sources 10 nm apart. The test field intensity is adjusted until a brightness match is achieved. Then the test wave-length is moved 10 nm and the process repeated. In heterochromatic brightness matching, the match is a function of both the chromatic (color) and achromatic (black and white) channels of vision. The results differ from a flicker photometer match.

A comparison of the two methods of obtaining V(λ) is shown in Fig. 3.4. Brightness matching gives consistently higher values for V(λ), particularly at the shorter and longer wavelengths. The root of the discrepancy is that lights of equal perceived brightness are generally not equal in luminance. Methods of calculating luminance, meters that measure it, and flicker photometry assume that luminances are additive. Heterochromatic brightness matches are not. The ultimate V(λ) curve approved by the CIE in 1924 was based on 177 observers, 70% of whom obtained V(λ) from flicker photometry and 30% from heterochromatic brightness matching. Thus V(λ) is indeed a compromise standard. Also there is considerable spread in the data at any given wavelength. For example, at $\lambda = 490$ nm, where V(λ) is listed as 0.208, actual measured values varied from 0.10 to 0.38. At 590 nm (V(λ) = 0.757), values ranged from 0.62 to 0.95, and at 640 nm (V(λ) = 0.175), from 0.09 to 0.32.

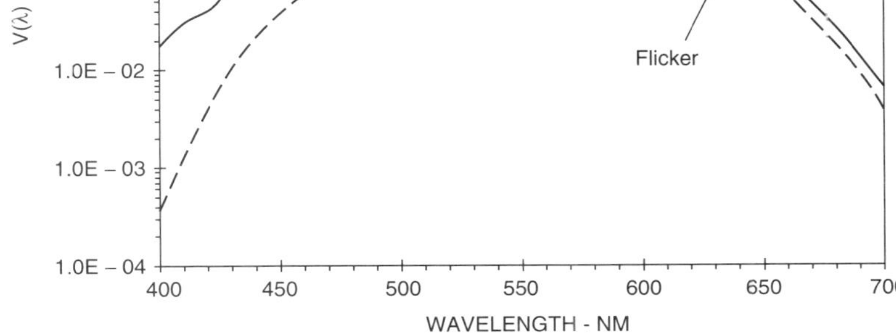

Figure 3.4 Comparing the two methods for obtaining V(λ). Courtesy of R. Levin.

Modifications to V(λ) have occurred over the years. The Judd correction, attributed to D.B. Judd, increased V(λ) at the shorter wavelengths (less than 450 nm) in an effort to counteract the fact that flicker photometry underestimated the effect of blue light. The Judd correction is indicated in Fig. 3.5.

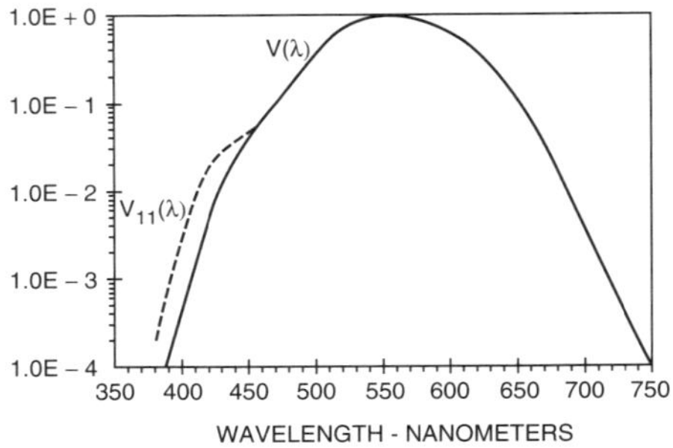

Figure 3.5 Judd correction to V(λ). Courtesy of R. Levin.

An additional correction was published by the CIE in 1964. The test fields in the earlier studies of Coblentz and Emerson and Gibson and Tyndall were 2°, largely restricted to foveal vision. In 1964, V(λ) values were added for a 10° field, extending the view beyond the macula (visual angle of 5°). The curve for a 10° field is shown in Fig. 3.6. Again, the underestimation of V(λ) for 2° at the shorter wavelengths is noted.

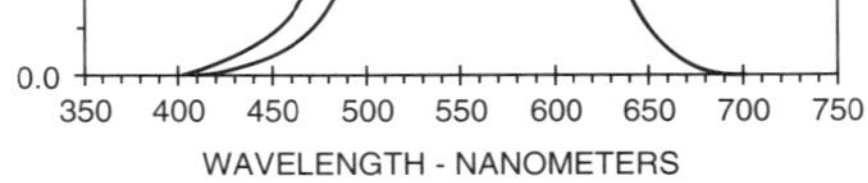

Figure 3.6 10° Field correction to V(λ). Courtesy of R. Levin.

3.4 THE EXTERNAL FACTORS OF SEEING

Up to now the process of seeing has been considered from the point of view of the functioning of the human eye. The eye's optical system; its two types of photoreceptors; the photochemical-electrical processing of light energy within the eye, and the phenomena of resolution, sensitivity, accommodation, adaptation, and selective response to wavelength have been discussed. In short, the physiological factors of seeing have been considered.

In this section, some of the physical factors of seeing will be discussed; those factors external to the eye, somewhat under a person's control, and of paramount importance in determining the ability to see a visual task and to do visual work. There are five major factors: namely, size, contrast, time, luminance, and color. The first four of these will be discussed together in detail. It will be shown how they are interrelated. They will be used to establish criteria for prescribing the amount and type of illumination needed to perform a visual task. The fifth factor, color, which not only affects seeing but also our perception of objects and the spaces that they occupy, will be treated in a later chapter.

Before discussing each of the first four factors in detail, it should be defined more carefully what is meant by a visual task. Visual potential and visual performance should be distinguished. Driving a car, threading a needle, and washing a dog are examples of human tasks. Each of these contains a visual component, which is called a visual task. Success in performing a total task depends in part on the ability to perform the visual portion of the task. This, in turn, depends on the visual characteristics of the task and on the physical characteristics of the observer, particularly those of his/her seeing system. A task may have excellent visibility, but may not be well performed because of a tired, aged, untrained, or distracted worker. The visual potential is present, but the visual performance is not. Thus, optimization of the four factors of seeing creates the potential for excellence in visual performance, but does not guarantee it.

To explore the relationship of the factors of seeing to the visual task an example from the field of electrical communications will be considered. To perform a visual task, a person must be able to receive clearly the "signal" being sent from the task, in the form of reflected or transmitted lumens. One's ability to detect the signal is influenced by what other visible radiation one is receiving from the areas surrounding the task. This is the lighting system "noise" and can be as strong as the signal itself. One's ability to detect the signal is also influenced by how fast the signal is changing, for one's visual receiver, like a radio or television receiver, has an upper limit on its capacity to process rapidly changing signals. The strength of the signal is certainly of significance, and this, in turn, is a function of both the density of luminous flux from the task and the surface area of the task. Last, the bandwidth of the system channel, which is 5 kHz for a radio channel, 6 MHz for a TV channel, and 320×10^6 MHz for the visual channel, the latter obtained from

$$\Delta f = c\left(\frac{1}{400} - \frac{1}{700}\right) \times 10^9 \text{ Hz} \tag{3.2}$$

and selectivity within that channel [Q for radio and TV and $V(\lambda)$ for the visible spectrum] determine the quality (sound, picture, or color fidelity) of the output signal. System noise, rapidly changing signals, signal intensity, source surface area, and system bandwith and selectivity are nothing more than other ways of identifying contrast, time, luminance, size, and color as the external factors of seeing.*

In predicting visual performance, not only is it required to assess the characteristics of a visual task, but it is also required to describe what level of response is desired from the observer. At the simplest level, an observer is asked to determine the presence or absence of an object. This is called *detection*—an object in the road or a crack

* No analogy should be stretched too far. In electronic systems, information is processed sequentially in time. In the visual system, there is simultaneous temporal and spatial processing of information.

in an industrial part. The concern is largely whether the object or crack is present or absent, not what it is or what its shape is.

The next higher level of visual response is *recognition*, being able to describe an object as a member of a group—a geometrical object, a letter, a person. This level requires the observer to distinguish shape and form, in a somewhat gross sense, but still sufficiently to classify the object generically.

The third level is *identification*. Now more than just shape and form are required; the observer must be able to identify surface details and features and discriminate edges and boundaries. From a geometrical object, a letter, and a person, the observer now sees, for example, a hexagon, the letter E, and Mr. Jones, i.e., the observer is now at the level of discrimination between members of the same group.

The three levels are usually described in terms of threshold requirements, having enough light to just accomplish the stated visual requirements. Threshold implies getting it right 50% of the time, i.e., if a person approaches a stranger on the street, 5 times out of 10 the person will be able to tell if the stranger is a friend or an assailant. This is certainly not a satisfactory lighting condition, but threshold experiments are often the easiest to conduct in the laboratory.

Just as one would not want a bridge to be able to just handle the weight of cars passing over it, so one would not choose to perform visual-tasks at threshold every day. Rather, suprathreshold lighting conditions at levels that provide easy assimilation of visual-task details are usually desired. Threshold and supra-threshold visual performance models are described later in the chapter.

3.5 SIZE

What is meant when size of an object is discussed? Consider Fig. 3.7a, where a road is shown disappearing into the distance with a row of telephone poles at its side. While all the telephone poles are the same height, those in the distance appear much smaller than those close by, and it is harder to see the details of the distant poles (their crossarms, for example). Evidently physical size of an object is not a good criterion for determining whether or not that object can be seen.

Consider now Fig. 3.7b, where an eye is shown observing two telephone poles, each of height h. The angles which these poles subtend at the eye are labeled θ_1 and θ_2; the distances from the eye to the poles are labeled d_1 and d_2. From the definition of radian measure,

$$\theta_1 = 2\tan^{-1}\frac{h}{2d_1} \cong \frac{h}{d_1} \quad \text{and} \quad \theta_2 = 2\tan^{-1}\frac{h}{2d_2} \cong \frac{h}{d_2}, \tag{3.3}$$

where we have used the approximation for small angles that an angle in radians and its tangent are equal. θ_1 and θ_2 are the visual sizes of the objects, measured in radians, or more commonly in minutes of arc*, where 1 radian = 3438 minutes of arc. Thus an object 10 ft high and 200 ft from an observer would have a visual size of $\frac{10}{200}$ = 0.05 rad = 172 minutes of arc (about 3°). It is observed that even though the pole heights are the same, their visual sizes are different. It is visual (angular) size, rather than physical size, that helps to determine how well an observer can see an object.

---

* A minute of arc is 1/60 of a degree.

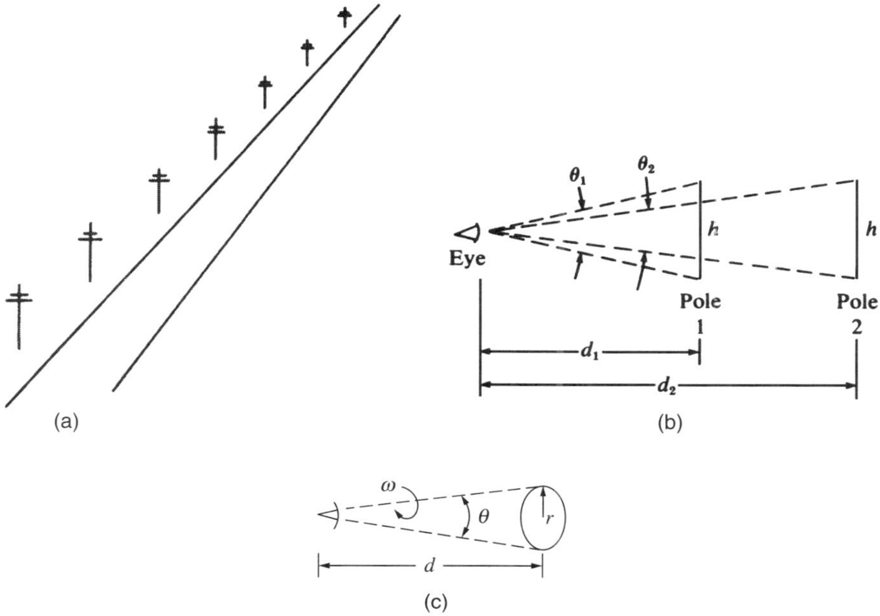

Figure 3.7 Representations of size.

Visual size of a two-dimensional circular object is expressed in terms of solid angle, as shown in Fig. 3.7c.

$$\omega = \frac{\pi r^2}{d^2} = \pi \tan^2 \frac{\theta}{2} \tag{3.4}$$

For r << d, Eq. (3.4) becomes

$$\omega = \pi \frac{\theta^2}{4} \tag{3.5}$$

For a visual angle of 3°, the solid angle is 2.15×10^{-3} sr = 2150 μsr.

The shape of the object being viewed determines whether visual angle or solid angle is used. When objects have reasonable symmetry, such as discs and rectangles, visual-angle is customary; for non-symmetrical objects, such as many letters, solid angle is preferred.

It is not always easy to decide what dimension of a visual task to use in determining its visual size. Basically the visual size of the smallest detail that needs to be seen in order to perform the visual task needs to be determined.

Consider Fig. 3.8. The first diagram is the standard parallel bar test object—a square with the center third removed. The bars are generally black and the space white. This test object originated in the General Electric Lighting Research Laboratory in the early 1900s and served as the fundamental test object for visibility studies until after

World War II. Assuming the visual task is simply to determine the presence or absence of the object in the field of view (detection), the dimension d is the one to use in determining visual angle. If, on the other hand, one is interested in resolution of detail so that the test object may be identified as parallel bars (identification), the separation s should be taken. The latter situation of identifying, rather than simply detecting, is usually the case.

Consider the other seeing tasks in Fig. 3.8. If concern is with the details of the objects, in each case the dimension s must be used. Thus all four objects in Fig. 3.8 have the same visual size from the point of view of identifying the object, not just detecting its presence.

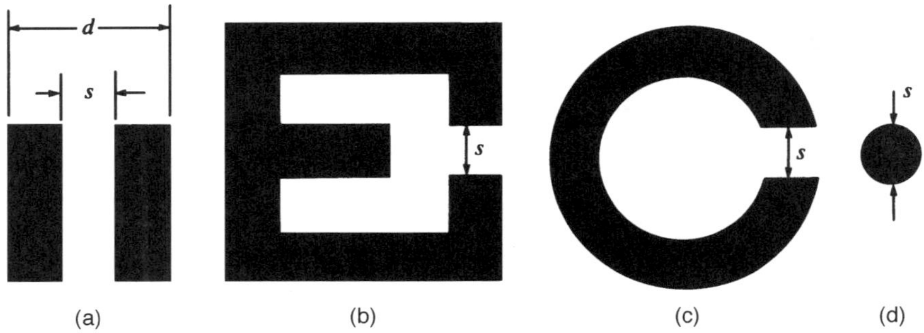

Figure 3.8 Visual test objects.

Two other terms concerning size require definition. One is threshold size, which is the visual size in minutes of arc of an object that can just be seen 50% of the time in a given seeing situation. At threshold, there is zero certainty of seeing. An observer scoring 5 correct identifications out of 10 in a forced-choice, yes or no, situation could just as easily have achieved this score by chance. Threshold size is not a fixed parameter. It varies with the contrast between the object and its background, with the time available for seeing, and with the luminance of the background.

The other term requiring definition is visual acuity, which is simply the reciprocal of threshold size. Thus, a visual threshold angle of 0.4 minutes of arc gives a visual acuity of 2.5. It should be understood that visual size is a physical quantity independent of the lighting environment, whereas threshold size and visual acuity are very dependent on the other factors of seeing.

A plot of visual acuity versus background luminance in footlamberts* for a circular black object on a white background appears in Fig. 3.9. It is noted that as background luminance decreases, object size must increase in order to continue to just see the object. Also, the maximum visual acuity possible for this object is 2.46, corresponding to a threshold size of 0.41 minutes of arc and a background luminance of 100,000 fL. Ninety percent of maximum visual acuity occurs at about 100 fL. Thus, the

* In many publications on vision, background luminance is expressed incorrectly in footlamberts (lumens per square foot). It should be either luminance exitance in lumens per square foot or luminance in candelas per square foot, preferably the latter. However, the tasks and backgrounds involved are generally assumed to be perfectly diffusing surfaces so that one can convert to candelas per square foot if necessary by dividing by π. The misnomer will be used in order to be consistent with the literature.

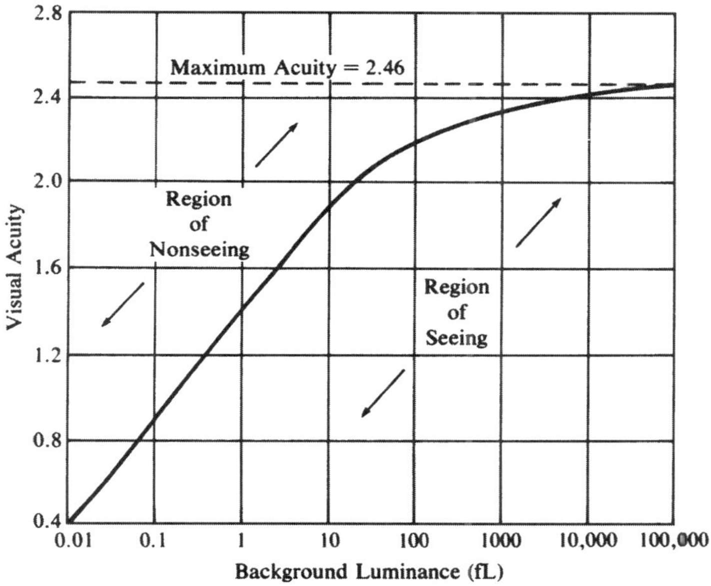

Figure 3.9 Visual acuity: Black object on white background. ***Source: IESNA Lighting Handbook*****, 5th edition, 1972. Courtesy of the Illuminating Engineering Society of North America.**

final 10% gain in visual acuity requires a 1000 to 1 increase in background luminance. At the very low background luminance of 0.01 fL, visual acuity is 0.4 and threshold size is 2.5 minutes.

It is useful to develop some "feel" for the visual size of real-world seeing tasks that are encountered every day. A fingernail at arms length subtends a visual angle of about 1°. The most common seeing task is the reading of bookprint. Customarily a book is held about 14 in from the eye. At that distance, the visual size of various type sizes in minutes of arc, measured using the distance between the dot and the body of the letter i, is given in Table 3.4a. One point of book type is approximately $\frac{1}{72}$ inch, measured from the top of the longest ascending letter to the bottom of the longest descending letter. Thus, 12 point type is 0.167 in high. Visual and solid angles of common wire sizes are given in Table 3.4b and of distant circular objects in Table 3.4c.

The smallest visual task normally encountered is 6 point type at 14 in (1.7 minutes of arc). The width of the eye of a small needle at 14 in subtends a visual angle of 2.5 minutes of arc at the eye. As an extreme, the letters in the third from the bottom line of an eye chart, which is the lowest line a person with 20/20 vision can just read at the specified test distance, subtend an angle of 1 minute of arc at the eye.

Referring to Fig. 3.9, and assuming black on white, it is found that 0.025 fL is required to just see (threshold) 6 point type, 0.01 fL to just see the needle eye, and 0.18 fL to just see the row on the eye chart. These are all quite low levels of luminance and show that, in general, people are well above threshold in contemporary seeing situations. The IESNA suggests a boundary of 4 microsteradians between "small" and "large" size targets. It is not clear why or how this distinction was made.

TABLE 3.4 VISUAL SIZES OF COMMON OBJECTS

(a) Book Type at 14 inches (35.6 cm)

Type Size	Visual Angle (minutes of arc)
6 point	1.7
8 point	3.9
10 point	4.9
12 point	6.4
14 point	7.8
18 point	17.0

(b) American Wire Gauge (AWG) at 15 inches (38.1 cm)

Wire Gauge	Wire Size (diameter in mm)	Visual Angle (minutes of arc)	Solid Angle (microsteradians)
30	0.25	2.4	0.39
24	0.51	4.9	1.61
20	0.81	7.8	4.05
16	1.29	12.4	10.3
12	2.05	19.8	26.0
8	3.28	31.6	66.4

(c) Circular Objects Viewed from 100 ft (30.5 m)

Object Diameter in (cm)	Visual Angle (minutes of arc)	Solid Angle (microsteradians)
3.0 (7.6)	8.6	4.88
6.0 (15.2)	17.1	19.5
9.0 (22.9)	25.8	44.3
12.0 (30.5)	34.4	78.5

3.6 CONTRAST

Contrast, or more properly luminance contrast, between an object and its adjacent background is defined by the following equation:

$$C = \frac{|L_b - L_t|}{L_b} \tag{3.6}$$

where L_t is the luminance of the target (task), L_b is the luminance of the background (also the adaptation luminance of the observer), and the vertical bars in the numerator indicate that the absolute value of the difference is taken. Thus, contrast is positive whether the object luminance is greater than that of its background or vice versa, and can range from 0 to 1 for $L_t \le L_b$ and from 0 to ∞ for $L_t \ge L_b$. For example, from Eq. (3.6) an object whose luminance is 30 fL has the same contrast against a background of 10 fL as it does against a background of 60 fL. In each case the contrast is 0.5, or 50%.

Eq. (3.6) is most often used when $L_b > L_t$. Other forms of Eq. (3.6) are

$$C = \frac{L_{max} - L_{min}}{L_{max}} \tag{3.7a}$$

and

$$C = \frac{L_{max} - L_{min}}{L_{max} + L_{min}} \tag{3.7b}$$

Both these equations give values of C between 0 and 1 for all task and background luminances. Eq. (3.7a) is useful when it is difficult to decide which portion of a scene is the target. Eq. (3.7b) is useful when repeated patterns, such as gratings, are being viewed.

If a visual task and its background are lighted to the same illuminance level and if both are perfectly diffusing, contrast becomes a function of only the reflectances of the two surfaces. Then Eq. (3.6) can be rewritten as:

$$C = \frac{|\rho_b - \rho_t|}{\rho_b} \tag{3.8}$$

If the task is black print and the background is good white paper, contrast can approach 1.0 (100%); but, for faded copy, it will be considerably less than 1.0. The diagram in Fig. 3.10 is intended to illustrate the wide differences in contrast that can exist. The black lines against the white background are relatively easy to see, but against the dark gray background are difficult to detect. However, the former situation is not always the case. A dark task on a dark background is not an uncommon seeing task, sewing black thread on black cloth being an example. It requires a high level of illumination to provide sufficient visibility for such a task.

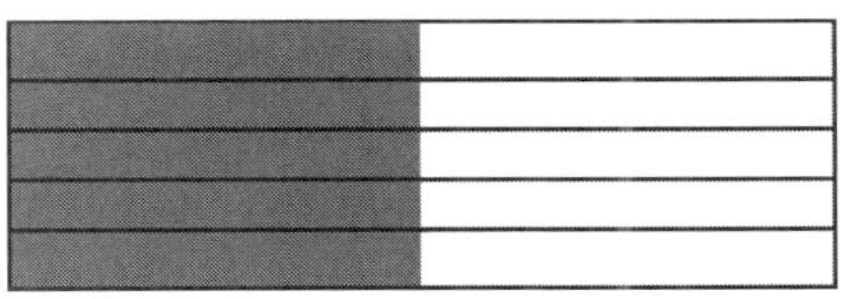

Figure 3.10 Contrast differences.

Just as with size, there are two additional terms concerning contrast at threshold that should be defined. The first is minimum perceptible contrast (MPC), which is the least contrast a task may have with its background and be seen 50% of the time in a given seeing situation, that is, with the other three factors specified. Its reciprocal is called contrast sensitivity (CS). Thus, a minimum perceptible contrast of 0.1 yields a contrast sensitivity of 10. Minimum perceptible contrast and contrast sensitivity are each very dependent on background luminance, visual size of the object, and the time available for seeing, and, in that sense, they are similar to threshold size and visual acuity.

Contrast sensitivity is plotted versus background luminance in Fig. 3.11 for a 4-minute test object with a time for seeing of 0.1 s. It would be expected that someone with high sensitivity to contrast would require very little contrast between an object and its background to see the object, and this is indeed the case. Note that, as with visual acuity, a saturation effect occurs as background luminance increases. The maximum contrast sensitivity for this size object is 45, corresponding to a minimum perceptible contrast of 0.022. This occurs when object and background reflectances are nearly the same, for example, a task of 88% reflectance against a background of 90% reflectance. The value of contrast sensitivity at 0.001 fL is 0.035, with a minimum perceptible contrast of 28.6. Such a contrast requires a very dark background, for example, a task with 60% relectance and a background with 2% reflectance. From these extremes, it can be seen that Fig. 3.11 covers the full range of contrast possibilities. The IESNA defines C = 0.3 as the boundary between "low" and "high" contrast, although again it is not clear why or how this distinction was made.

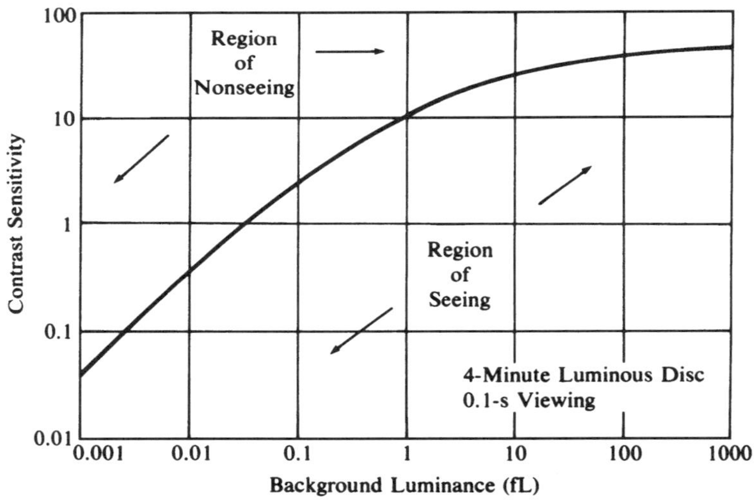

Figure 3.11 Contrast sensitivity. ***Source*: *IESNA Lighting Handbook*, 5th edition, 1972. Courtesy of the Illuminating Engineering Society of North America.**

Size and contrast are closely interrelated. As size increases, there is a gradual transition from size being the governing factor in determining visual threshold to contrast assuming this role. The transition occurs at 1.5 to 2.5 minutes of arc. For small objects (less than 1.5 minutes), a small percent change in visual size requires a larger percent change in contrast to return to threshold conditions with the same background luminance. For larger objects, (greater than 2.5 minutes), the reverse is true. Since, as was seen earlier, the concern is usually with objects whose visual size is greater than 2.5 minutes, visual threshold is in general a much more sensitive function of contrast than it is of size.

3.7 BRIGHTNESS AND LIGHTNESS, CONSTANCY AND CONTRAST

Brightness is a comparative subjective assessment of the light emitted from one surface as compared to that emitted from another surface. It is not luminance, but is related to it. Luminance is a measured, replicable metric; brightness is a perceived response to luminance.

Snow on a sunny day, neon signs at night, and movie screens are examples of objects that are usually considered bright; but brightness is a relative assessment. A full moon on a clear night can appear bright to an observer, but the same observer would not consider the moon bright in the daytime sky. Yet the moon's luminance (2500 cd/m^2) is the same in each case. The same is true of street lights and headlights—bright by night, but not by day. What has changed? It is the background luminance of the entire surround that is high during the day and low at night, creating significant contrast boundaries at night.

Brightness and luminance are related by a power relationship known as Stevens' Law:

$$B = kL^{0.3} \tag{3.9}$$

where L is in cd/m^2 and k is a proportionality constant. What Eq. (3.9) reveals is that equal changes in luminance do not result in equal changes in perceived brightness. Rather, the pattern is as shown in Table 3.5 for k normalized to 1. Note that a 10 to 1 change in L produces a 2 to 1 change in B. Also note that L and B bear a straight-line relationship on a log-log plot.

TABLE 3.5

L cd/m^2	1	2	3	5	7	10
B	1.0	1.2	1.4	1.6	1.8	2.0

There is a subtle distinction between brightness and lightness. The former is a judgment of the amount of light emitted from an object. The latter is a judgment of the amount of light being reflected from an object.

There are two attributes of lightness that will be considered. The first is lightness constancy, which is a person's perception that the lightness of an object remains nearly the same over wide ranges of luminance. This phenomenon allows us to continue to recognize objects when light levels change markedly. The second attribute is lightness contrast, often termed simultaneous contrast. This is the property by which two adjacent areas, for example two squares, that have the same luminance, have different lightnesses because of their different backgrounds. The situation is shown in Fig. 3.12. The central gray squares are equal in luminance. The right square appears darker than the left square because of its lighter background luminance.

To summarize, in light constancy the perception of brightness remains constant over a wide variation in light level. In lightness contrast, the perception of brightness does not remain constant, even though the object luminances are the same.

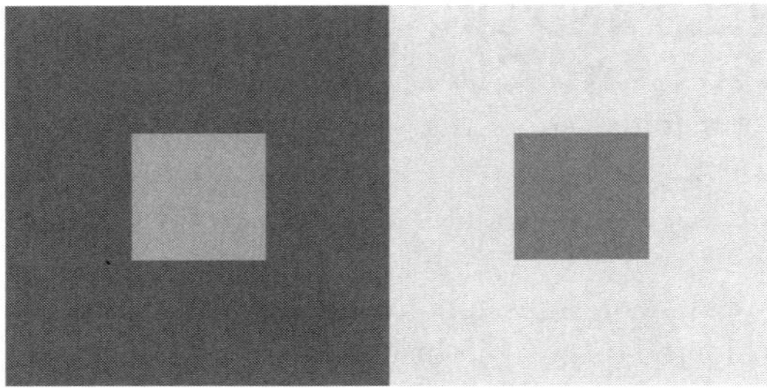

Figure 3.12 Simultaneous Contrast.

3.8 SPATIAL FREQUENCY AND CONTRAST SENSITIVITY

A grating is a pattern of alternating light and dark rectangular bars. A sinusoidal grating is one whose intensity of its light and dark bars changes sinusoidally as the grating is traversed. One characteristic of a grating is its spatial frequency. Spatial frequency is the number of cycles (light bar to dark bar to light bar) that occur within a visual angle of one degree. A grating of narrow bars has a higher spatial frequency than one with wide bars. Note that spatial frequency changes with distance from the eye. As this distance increases, spatial frequency increases.

A second characteristic of a grating is contrast—the luminance difference between the light and dark bars. Customarily Eq. (3.7b) is used to quantify grating contrast, and thus the contrast of a grating can vary from 0 to nearly 1.

A scene to be viewed can be considered as an assemblage of sinusoidal gratings of different spatial frequencies and contrasts. These can be analyzed mathematically by a technique known as Fourier Analysis, named for Jean Baptiste Joseph Fourier (1786–1830), a French mathematician. Fourier showed that any periodic waveform in time can be represented by a series of sinusoidal waves, one with the same frequency (the fundamental) as that of the waveform to be synthesized, plus as many harmonics (integer multiples) of the fundamental as are necessary to replicate the given wave. Using this same principle, a visual scene with a Fourier Series of grating waveforms in space can be replicated and analyzed, where spatial frequency in cycles per degree replaces temporal frequency in cycles per second (Hertz).

Visual acuity has been presented as the reciprocal of threshold size and contrast sensitivity as the reciprocal of minimum perceptible contrast. The contrast sensitivity function blends these two concepts. It is a plot of contrast threshold as a function of spatial frequency and is effectively the transfer function of the eye, as applied to all the grating elements in a visual scene. A typical contrast sensitivity function (CSF) curve is shown in Fig. 3.13.

The curve displays an area of visibility (light) within which contrasts and spatial frequencies are such that a pattern can be seen. In the surrounding (dark) areas, patterns cannot be seen. One should note that the CSF has both a low and high frequency cut off, i.e., it is only in the middle range of spatial frequencies that patterns can be readily seen (with lower contrast).*

* A much more detailed interpretation of CSF appears in Sekuler and Blake.

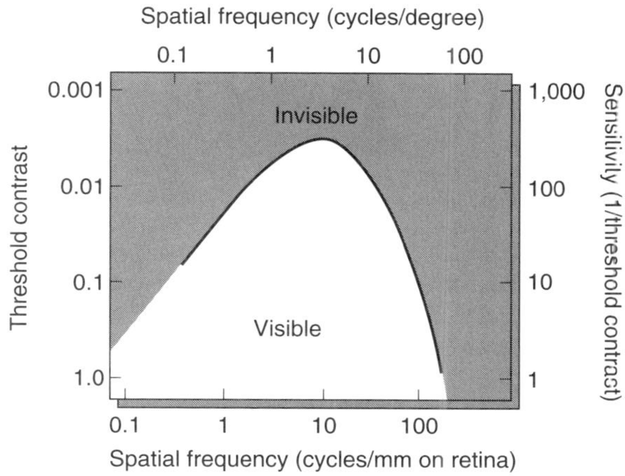

Figure 3.13 Contrast sensitivity function. *Source*: *Perception*, 3rd edition, R. Sekuler and R. Blake, 1994, McGraw-Hill. Courtesy of The McGraw-Hill Companies.

3.9 TIME

It takes time to see. Older people might recall the Burma Shave signs along the nation's highways during the 1930s and 1940s. These were spaced so that a motorist traveling at a speed of up to 45 mi/h had time to read the message on each sign while passing it (at interstate highway speeds, the signs would have to be placed much further apart). As another illustration, one cannot see quickly enough to catch a balloon in the act of popping. A person sees it fully inflated and fully deflated, but not in the process of deflating right after it is pricked with a pin. This is a job for high-speed photography, which is capable of catching an event that lasts for only a thousandth of a second or less. A third example is an assembly line, where integrated electronic circuit boards move rapidly past an inspector.

The time required for seeing decreases as background luminance increases, if contrast and size are held constant. For example, it takes about 1 s to see a 4-minute black test object having a contrast of 1.0 with its background when the background luminance is 0.015 fL. This becomes 0.1 s at 0.03 fL and 0.01 s at 0.54 fL. Thus, for normal lighting levels of 20 to 70 fc, and for most seeing tasks that are encountered, time is not a problem.

An interesting experiment is to prepare a 35 mm slide with several large black capital letters on a white background and flash this slide to a group of observers for about one second, asking them to identify as many of the letters as they can. Then repeat the experiment with a second slide of large black letters on a gray background. Finally, reduce the projector's intensity and ask them to repeat both parts.

Most observers will be able to identify 4 or 5 letters when background luminance and contrast are high, and a smaller number when they are lower. After the experiment, tell the observers that the average fixation time when reading a group of letters under good seeing conditions is 0.2 seconds. Thus, it does take time to see, and 4 to 5 letters identified in a second is acceptable performance. Also point out to them that it takes longer to perform a given visual task if either the contrast or the background luminance is reduced.

To summarize what has been learned about the external factors of seeing.

1. *Size.* Increasing visual size increases visibility.
2. *Time.* The more time given to see, the better the likelihood of seeing becomes.
3. *Contrast.* Increasing contrast improves visibility. Given a dark object on a light background, if the object is made darker or the background lighter, seeing improves.
4. *Luminance.* Visibility improves with an increase in luminance, i.e. deficiencies can be overcome in the other three factors by increasing luminance.

These conclusions are very rudimentary, but they do, at least, give a feeling for the effect of each factor on visibility and pave the way for consideration later in the chapter of various models for measuring visual performance.

3.10 THE AGING EYE

At birth, a baby's eyes are not connected completely to its brain. Seeing is fuzzy, but the eyes can follow light and shadow. At four months, the eyes can focus, and the baby can perceive spaces and distinguish color. At six months, the eyes can move together, and three-dimensional vision begins. By one year, close space is mastered and perspective develops—large, small, near, far. At 18 months, texture shape and hand-eye coordination are mastered. Visual development continues until completed by age 7 to 9.

From this time on, vision begins to deteriorate, slowly at first, and then more rapidly. The following changes occur.

1. "Near Point" advances (presbyopia)

   The "near point" is the nearest distance from the eye at which the eye is able to focus an image on the retina. As a person ages, the lens becomes less flexible, and there comes a time when the ciliary muscles are unable to alter its curvature sufficiently for it to focus on close work (the lens must become thicker for focusing on nearby objects). The accommodation of the eye is expressed in diopters, which is the reciprocal of the distance in meters from a plane 1.4 cm in front of the cornea (the spectacle plane) to the nearest point on which the eye can focus when using a lens that permits a distant object to be seen clearly. In a young person, 12 diopters (8 cm) is average. This reduces to 5 diopters (20 cm) at age 40, 2 diopters (50 cm) at age 50, and 1 diopter (100 cm) at age 65.

   It is certainly not uncommon to see people in their 40s holding a newspaper or a menu at arm's length to read it, not realizing that visual size decreases with distance making seeing more difficult. Fortunately, bifocals are available to correct the problem.
2. Pupil size decreases

   Under an adaptation exitance of one lm/ft^2, the diameter of the pupil is 5 mm for 20 to 30-year olds, but decreases to 4 mm for ages 40 to 50, and 3.6 mm for people 60 to 70 years old. Since the amount of light reaching the retina is a function of pupil area, the reduction from 5 mm to 3.6 mm represents roughly a 2 to 1 decrease in the amount of admitted light.

3. Lens yellows

As people age, the spectrum of the light reaching the retina is altered; notably the shorter wavelengths are absorbed reducing blue and violet responses and making those colors look gray. In addition, the overall illuminance on the retina is reduced.

The combination of pupil size reduction and lens yellowing results in about a 3 to 1 reduction in retinal illuminance from age 20 to age 60. It is very clear that older people need more illuminance than younger people for the same visual task, but for reasons unknown, the IESNA has chosen not to include higher levels for the elderly in its illuminance recommendations.

4. Adaptation time increases

As elderly people move from a well-lighted to a dimly-lighted space, it takes longer for them to readapt to the lower level. An example indoors is entering a dark movie theater. An example outdoors is driving away from under the canopy of a gas station-convenience store. A driver may experience a drop in illumination from 100 fc to 0.1 fc at the drive entrance to the roadway. Older eyes require a short period of time to adapt to this lower level and, thus, may temporarily not see cars on the roadway.

5. Sensitivity to glare increases

Glare is the sensation produced by luminances within the visual field that are sufficiently greater than the luminance to which the eyes are adapted to cause annoyance, discomfort, and/or decrease in visibility and visual performance. There are two types of glare. Discomfort glare is glare which produces discomfort, but does not necessarily impair visibility and visual performance.* Disability glare, on the other hand, is glare that reduces visibility and visual performance, but does not necessarily produce discomfort. It comes in two forms: reflected glare, which produces veiling reflections (to be discussed in Sect. 3.14) and direct glare, caused by luminances in the field of view that create stray light within the eye, producing a veil of light over the task image on the retina. Calling this veiling luminance L_v, the total luminances of task and background in the presence of direct disability glare become $L_t + L_v$ and $L_b + L_v$, respectively. Then from Eq. (3.6), contrast becomes

$$C = \frac{|L_b + L_v - L_t - L_v|}{L_b + L_v} = \frac{|L_b - L_t|}{L_b + L_v} \tag{3.10}$$

It is easily seen from Eq. (3.10) that L_v reduces contrast. How does the luminance L_v produced by a source relate to the illuminance E_o at the eye from the source? An empirical relation is

$$L_v = \frac{9.2 E_o}{\theta(\theta + 1.5)} \tag{3.11}$$

where L_v is in cd/m^2, E_o is the illuminance at the eye in lux and θ is the angle in degrees between the glare source and the line of sight. For multiple sources

* Procedures for calculating this type of glare will be developed in Chap. 8.

of veiling luminance, one would need to do a summation of L_v values in Eq. (3.11).

Elderly people are much more sensitive to glare than their younger counterparts. Scattering of light within the eye increases with age, caused by an increase in "floaters" in the vitreous humor and a decrease in the tear cover of the cornea, and these effects permit a veiling luminance that often produces glare in the elderly eye. This can be particularly serious in night time driving, not just approaching headlights but also, for example, the glare from drop-lens lighting units, again predominantly under gas station canopies.

Eq. (3.11) is for young, clear eyes. For older eyes, L_v is much greater. A person 65 can require 2.5 times the contrast of a 20-year old, to see equally well. This is caused by a thickening of the lens which reduces ocular transparency.

6. Visual acuity decreases

   Visual acuity (VA), the eye's ability to detect size, decreases with age. If VA is set at 100% for ages 14 to 19, it will decrease to 80% for ages 60 to 69.

7. Contrast sensitivity decreases

   For the same contrast, an older person is less able to distinguish an object from its background. This is due to the increased scattering of light in the older eye and the oveall reduction in retinal illuminance.

8. Speed of perception decreases

   It takes older people longer to see. Their average detection speed decreases with age. This can be especially detrimental for inspectors on assembly lines or for a driver trying to avoid a dog running into the road.

The eight effects listed *will happen* to everyone's eyes as they age. All of these effects can be somewhat ameliorated by increased light levels, but mere quantity is not enough. Directionality and optical control of light are also major players, and visual tasks should be increased in size and improved in contrast.

Optical disorders, with which many elderly people must cope, have not been addressed, particularly cataracts, macular degeneration, and glaucoma. A recent survey observed that 59% of those over 60 years of age can be classified as "partially-sighted". Partially-sighted people are helped only minimally by increased illumination. Of greater help is to simplify the person's environment and use high-contrast boundaries and large, single-colored areas wherever possible. Properly admitted and controlled daylight can also be beneficial.

3.11 MODELS OF VISUAL PERFORMANCE

Section 3.4 distinguished between visual potential and visual performance, noting that creating the former does not guarantee producing the latter. Before examining models of visual performance, an additional distinction should be made between visual performance and task performance. Visual performance is one of three components of task performance. The other two components are cognitive and motor. Consider the driver of a car as an example. A approaching a railroad grade crossing, and the cross arm warning lights come on. The driver must stop the car before reaching the crossing. First the driver must see the warning lights. This is the visual component of the driver's task, and may be made difficult by such things as weather conditions, distance from the crossing, a dirty windshield, glare, age, and the maintenance of the cross arm and its lighting.

Assuming that the driver "sees" the lights, the cognitive component of the task begins. For this illustration, it is called perception time and has a typical value of one second. Finally, the motor component occurs, in this case the brake reaction time of about one-half second, the time required for the driver to apply his foot to the brake. Then, of course, there is additional time for the vehicle to come to a stop, which again is a function of several variables.

The point of drawing out this scenario is to demonstrate that visual performance, though important, is only one part of task performance. However, it is the only part that can be affected through changing the lighting and viewing conditions.

Throughout the years since the 1930s many vision researchers have developed models of visual performance, usually with the ultimate goal of being able to specify illuminance levels for a myriad of visual tasks. For the most part, these models have failed to achieve this goal, and thus the procedure for prescribing illuminances today is largely consensus based, i.e., based on the collected wisdom of those who design lighting installations, and it is over-simplified. This is not to say that the research leading to these models is worthless. Rather, it has provided professionals with considerable insight as to the effect of the external factors of seeing on visibility.

In the sections to follow, five of these models will be reviewed, two of the earlier ones (1920s to 1940s) by P.W. Cobb and F.K. Moss and by H.C. Weston, each remarkably incisive with regard to the interrelationships between the external factors in determining visibility; the extensive threshold studies of H.R. Blackwell from the 1950s to the 1980s, and the more recent (1980s and 1990s) work on relative visual performance by M.S. Rea and M.J. Ouellette and on visibility level by R. Clear and S. Berman.

Before examining these models, it is noted that two of them (Cobb-Moss and Blackwell) are threshold models, two (Weston and Rea-Ouellette) are suprathreshold models, and one (Clear-Berman) is both. In a threshold experiment, observers are required to detect the presence or absence of an object or the discrimination of an object's detail under threshold lighting conditions, i.e. can "just be seen" (50% correct answers). In a suprathreshold experiment, the tasks and their details under consideration are above threshold, and the issues are speed and accuracy of seeing, rather than simply detecting or recognizing. Attempts have been made to relate the two types of conditions through multipliers, to raise threshold data to suprathreshold levels, but the validity of these efforts has been questioned, as will be noted.

3.12 COBB-MOSS MODEL

P.W. Cobb and F.K. Moss expressed the relationships between visual size, contrast, time and luminance in the 1920s in a series of experiments. Their work is reported in the literature (Cobb and Moss, 1928) and is described in detail in "The Science of Seeing". (Luckiesh and Moss, 1937).

Cobb and Moss utilized a square parallel bar test object, such as shown in Fig. 3.8, with the middle third of the square white and the outer bars variable from black to light gray. The experiment was a threshold test. Each observer was asked to detect the space between the two parallel bars. A wide range of contrasts from $C \approx 1$ to $C = 0.02$ were used. Gap size ranged from 0.5 to 15 minutes of arc. Exposure times varied from 0.075 to 0.3 seconds. Background luminances spanned a range from 1 to 100 fL.

The results were displayed in a three-dimensional plot, as shown in Fig. 3.14a, with contrast plotted against size in planes of constant luminance parallel to the plane of the page. Two curves, one dotted and one solid, showed the limits of exposure time.

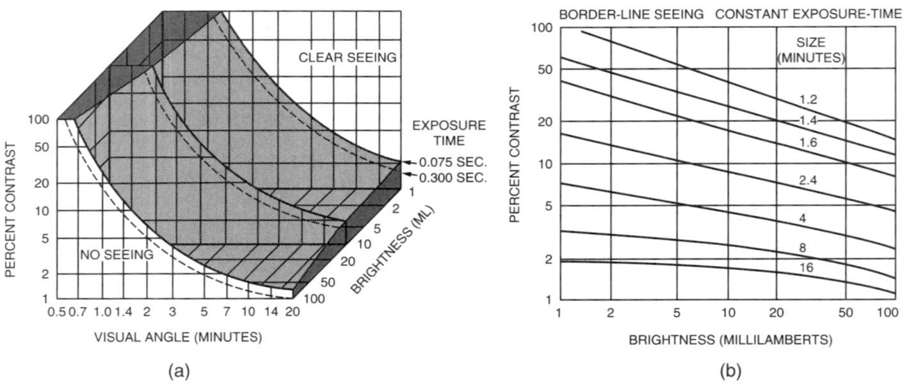

Figure 3.14 Results of Luckiesh-Moss visibility experiment. *Source*: *The Science of Seeing*, M. Luckiesh and F. Moss, 1937. Courtesy of Van Nostrand.

The curves are replotted in two dimensions in Fig. 3.14b, using the more normal coordinates of log contrast vs log background luminance, for various values of threshold size.

The curves are nearly linear on this log-log plot, except for the larger threshold sizes. Note that for constant size, very large increases in background luminance are required to compensate for small decreases in contrast. Also with contrast constant, small increases in size permit large decreases in background luminance. The trends of these results reappear constantly in later studies.

3.13 WESTON MODEL

The work of H.C. Weston occurred between 1935 and 1945 and is reported in the literature(Weston, 1945). Weston's experiment was at suprathreshold and involved detecting the gaps in variously-oriented Landolt C rings (see Fig. 3.15). Each test sheet had 256 rings, in 8 different orientations, corresponding to the 8 major compass directions. Subjects were asked to "cancel out" those rings having a certain orientation. Time and errors were tabulated. There were four contrasts, ranging from 0.97 to 0.28; five sizes of gap width, ranging from 1 to 6 minutes of visual arc, background reflectances from 38 to 90 percent; and eight values of illuminance, from 0.5 to 512 fc. Fifteen subjects performed the experiment, and each subject processed 65 sheets at each sitting.

Weston was sensitive to the fact that the physical action of canceling a ring took time (the motor component). He devised a separate test to determine what he called "action time" and used this in calculating visual performance. His calculation procedure was as follows:

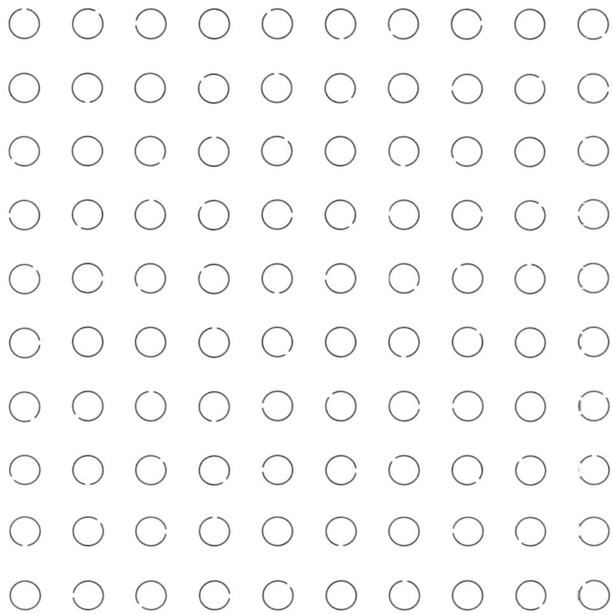

Figure 3.15 Landolt C rings used in Weston's experiment.

Define an accuracy factor by

$$\text{AF} = \frac{\text{number of rings correctly canceled in one minute}}{\text{number of rings that should have been canceled in one minute}} \tag{3.12}$$

Then define an index of gross performance by

$$\text{IP} = \text{number of rings correctly canceled in one minute} \times \text{AF} \tag{3.13}$$

Note that the index of gross performance includes both speed and accuracy parameters.

The next step is to modify IP to include action time (AT). This allows the separation of visual performance (VP) from IP.

$$\frac{1}{\text{IP}} - \frac{1}{\text{AT}} = \frac{1}{\text{VP}} \frac{\text{minutes}}{\text{ring}} \tag{3.14}$$

EXAMPLE 3.1

In a Weston test, assume 18 rings were canceled in a minute, 20 should have been canceled, and it takes 1/2s to cancel a ring. Find VP. Then assume 36 rings are canceled per minute, 40 should have been canceled, and it still takes 1/2s to cancel a ring. Recalculate VP.

Solution. For the first case

$$AF = 18/20 = 0.9$$

$$IP = 18 \times 0.9 = 16.2 \text{ rings/minute}$$

$$\frac{1}{IP} - \frac{1}{AT} = \frac{1}{16.2} - \frac{1}{120} = 0.053 \frac{\text{minutes}}{\text{ring}}$$

$$VP = \frac{1}{0.053} = 18.7 \frac{\text{rings}}{\text{minute}}$$

For the second case,

$$AF = \frac{36}{40} = 0.9$$

$$IP = 36 \times 0.9 = 32.4 \text{ rings/minute}$$

$$\frac{1}{IP} - \frac{1}{AT} = \frac{1}{32.4} - \frac{1}{120} = 0.0225 \frac{\text{minutes}}{\text{ring}}$$

$$VP = \frac{1}{0.0225} = 44.4 \text{ rings/minute}$$

Note that in the second case, the visual performance has more than doubled for a doubling of the index of performance. This, of course, is because the action time remained the same.

Weston prepared curves of relative visual performance (RVP) plotted against task illuminances for values of contrast and gap size. RVP is given as the ratio of actual visual performance at a given illuminance to the maximum performance obtainable for that gap size and contrast. One such set of curves appears in Fig. 3.16, for a background reflectance of 0.9. What these curves show is that, for a given size and contrast, RVP increases as task illuminance increases, but at an ever decreasing rate. Further, the steepness of the curves increases as contrast decreases, i.e., the visibility of tasks of high contrast is not increased as much by increases in illuminance as it is for tasks of low contrast. Finally, curves similar to Fig. 3.16 for small sizes show that illuminance alone cannot bring a task to the same level of performance as for large sizes, especially for low contrasts.

Westson's work was truly seminal for its time and the British used his RVP of 90% as the basis for their illuminance recommendations* for many years. However, Weston's model did have deficiencies. It included a limited number of visual skills and thus could not be applied to a variety of visual tasks; it required that a response be made to one task (canceling a ring) before proceeding to the next task (finding the next ring to be canceled); it had a limited range of contrasts and sizes and the data did not warrant extrapolation, particularly toward threshold.

* "IES Code for the Lighting of Building Interiors."

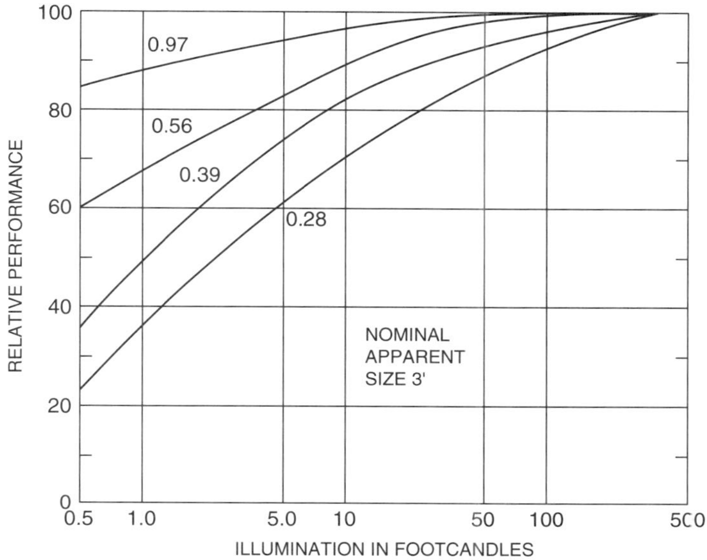

Figure 3.16 Relative visual performance in Landolt C ring test. ***Source: IESNA Lighting Handbook*, 5th edition, 1972. Courtesy of the Illuminating Engineering Society of North America.**

It has been somewhat unclear how Weston conducted his experiments and how he handled the scatter in his data. Thus, it has not been possible for current vision researchers to completely replicate and validate his work. Weston conducted two major experiments that yielded somewhat different results. Werner Adrian, in work for the Lighting Research Institute, examined Weston's work and developed equations that fit the data of Weston's second experiment, thus offering a contemporary model based on the Weston data (Adrian, 1994). The studies of other researchers have also produced results that support the trends shown by Weston.

3.14 BLACKWELL MODEL (CIE 19/2)

A comprehensive study of the four factors of seeing was undertaken by H. Richard Blackwell in the 1940s. Blackwell had been working for the army during World War II on a project to determine how well servicemen see, and he continued this work at the University of Michigan and Ohio State University after the war. His early work was reported in the literature (Blackwell, 1946 and 1952) and was later adopted and published (with some alterations and caveats) by the International Commission on Illumination as CIE 19/2.

Blackwell's basic experiment consisted of having observers detect the presence (or absence) of a translucent disc target lighted from behind at the center of a darker uniform background. The disc was located at the back of a 7 ft × 7 ft × 7 ft cubical box. The background was lit by shielded incandescent lamps, surrounding the periphery of the front of the box. Four target sizes were used from 1 to 64 minutes of arc, five target luminances from 0.001 to 800 fL, and seven time deviations from 0.001 to 1 s. The observers, knowing the location of the target (fixed foveal vision), were required to identify the time interval during which the target appeared from four possible 2.5-second time intervals. Each observer was given 250 target presentations at each two-hour viewing session, 50 of each of five target luminances. The observers were young, trained, and had 20/20 vision.

Important in Blackwell's work was his use of the "forced choice" technique. His studies were done at threshold, and his observers were required to choose one of the four time intervals. He corrected for guessing correctly by adjusting his data using the following equation:

$$N_a = \frac{N_o - 25}{75} \times 100 \tag{3.15}$$

where N_o is the percent correct time-interval identifications and N_a is the corrected percent. For example, if an observer had 90% correct identifications, the corrected score would be $\frac{90-25}{75} \times 100 = 87\%$. This procedure placed Blackwell's results on a sound statistical base.

The results of the experiment were plotted as a family of curves, one set of which appears in Fig. 3.17. Each set presents contrast plotted against background luminance on logarithmic scales for various values of target visual size and for one value of target time duration. Seven such sets appear in his research, one for each of the seven target time durations. The vertical axis of each set of curves has four scales, one for each value of performance level (99% performance means the target is found 99 out of 100 times; 50% performance is threshold). The threshold values are experimental. The values above threshold are not. They were obtained mathematically by extrapolation using a calibrated ogive curve.

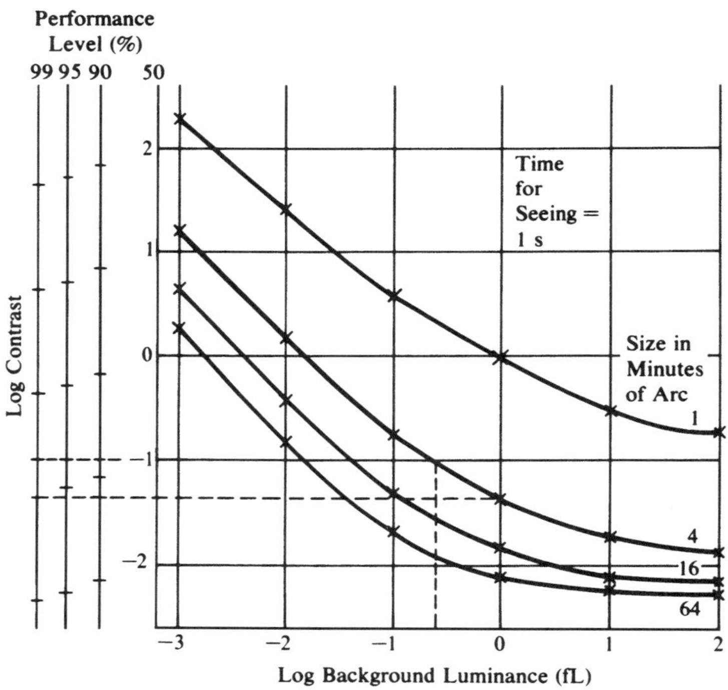

Figure 3.17 Relating the four factors of seeing. ***Source*****: "Brightness Discrimination Data for the Specification of Quantity of Illumination," H. Blackwell, *IESNA Transactions*, November 1952. Coutesy of the Illuminating Engineering Society of North America.**

EXAMPLE 3.2

An example of the use of these curves is finding the background luminance necessary to see with 50% accuracy (threshold seeing) a 4-minute disc target having a contrast with its background of 0.1 when the target is lighted for 1s.

Solution. To enter the curves in Fig. 3.17, move horizontally from the –l log contrast line on the 50% accuracy scale and intersect the 4-minute size curve at a log background luminance of –0.62. This gives a background luminance of 0.24 fL. Then task luminance from Eq. (3.6) can be found as

$$0.1 = \frac{|0.24 - L_t|}{0.24}$$

$$L_t = 0.26 \text{ fL}$$

Suppose that a 99% accuracy of discernment is desired. There are two choices. Size and contrast can be kept constant, and background luminance can be increased. From the curves in Fig. 3.17, L_b = 1.0 fL for this situation. Solving for task luminance gives L_t = 1.10 fL. Ninety-nine percent accuracy of discernment can also be achieved by increasing contrast while keeping size and background luminance fixed. Proceeding vertically from –0.62 on the log L_b axis to the 4-minute size curve and then horizontally to the 99% scale, log C = –0.65 can be found, giving C = 0.22. This gives a task luminance of 0.28 fL. The ratio of the contrasts required for 99% and 50% visual performance, with all other factors held constant, is 0.22/0.1 = 2.2, whereas the ratio of background luminances required, with all other factors held constant, is 1.0/0.24 = 4.2. Thus, task visibility is a more sensitive function of contrast in this instance than it is of background luminance.

In Blackwell's early experiments, observers knew when and where the visual task would appear. Also their viewing was static, not dynamic; that is, the observers were not required to scan with their eyes prior to focusing on the visual task. Last, they were required to detect only the presence or absence of a task, not its details.

Blackwell postulated that each of these effects could be represented by a contrast multiplier, a number greater than one by which the contrast values from his original study could be multiplied to take into account these additional constraints. From numerous studies, these multiplying factors were determined as:

1.5	When and where factor
2.78	Dynamic seeing factor
2.51	Detection of detail factor (also called "common-sense" factor)
1.9	Accuracy factor

The accuracy factor (99%) came from Blackwell's earlier experiments, as previously noted.

Blackwell's original curves (one set has been shown in Fig. 3.17) are a bit cumbersome to work with, especially if they have to be shifted by the multipliers just presented. What was needed was a standard visibility reference function to which all other seeing situations could be related. Threshold seeing is the easiest condition to measure and it was decided to relate all seeing to this level. It was also decided to include the common-sense factor in the threshold level, the rationale being that most of the time an object should be identified, not just its presence detected.

What standards for size and time should be used? Reading is the most common seeing task, and the average duration of fixation pauses in reading is 0.2s. Thus, it was decided to standardize on a $\frac{1}{5}$s exposure. Also, a 4-minute luminous circular disc was chosen as the standard test object, largely because so much data had been gathered on the visibility of such an object, and, because, 8 point type at 15 in subtends an angle of about 4 minutes at the eye.

Last, it was necessary to determine what characteristics of task, background, and surround luminance should be required. "Surround" means outside the immediate task background but still in the field of view. It was decided to stipulate a background luminance that is diffuse and to require a surround luminance that is uniform and equal in value to the background luminance. Such a lighting environment can be produced by placing the task at the center of a luminous diffusing hemisphere, and thus can be readily achieved in the laboratory.

In Fig. 3.18, the agreed-upon visibility reference function (VRF) is shown as VL1 (visibility level 1). To repeat, this is a common-sense visibility-threshold curve for a 4-minute circular task with $\frac{1}{5}$s time for seeing under diffuse illumination.

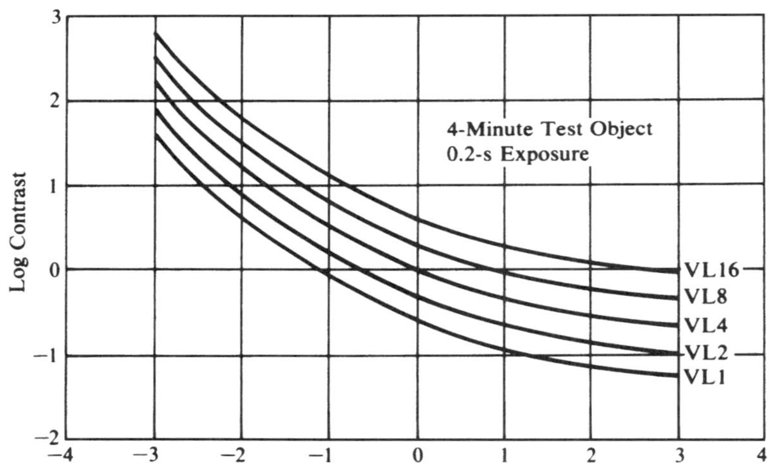

Figure 3.18 Visibility level curves.

The next step was to denote what visibility level is reasonable to expect in "normal" seeing tasks. It was decided that normal seeing should require 99% accuracy, that it should not rely on knowing when and where an object might appear, and that it should include normal eye movement. Using the contrast multipliers previously listed, this means an overall multiplier of $1.5(2.78)(1.9) = 7.92 = 8$. The second curve from the top in Fig. 3.18 is labeled VL8 (visibility level 8) and was named the visual performance criterion function (VPCF). For a given background and luminance, VL8 represents eight times the contrast of VL1. It is this visibility level that was aimed for prescribing task illuminance levels.

EXAMPLE 3.3

Assume the background of the 4-minute standard test object has a luminance of 10 fL. What luminance of the test object is required to achieve VL1? VL8?

Solution. From Fig. 3.18, to reach VL1 (threshold) requires a contrast of 0.11 at a background luminance of 10 fL. To reach VL8 requires a contrast of 0.91. Using Eq. (3.6), test object luminance for VL1 is 11.1 fL and for VL8 is 19.1 fL. Note that the ratio of the two contrasts is approximately 8, as it should be.

Assume the luminous disc in a lighted environment has a background luminance L_b, a task luminance L_t, and thus, from Eq. (3.6), a contrast C. It is desired to reduce the task to threshold by changing its contrast while keeping background luminance constant. This may be done by adding a veiling luminance L_v and simultaneously reducing both L_t and L_b by a factor $K < 1$ such that

$$L_v + KL_b = L_b \tag{3.16}$$

and

$$\overline{C} = \frac{|L_v + KL_b - L_v - KL_t|}{L_v + KL_b} = KC \tag{3.17}$$

where $\overline{C}$ is the contrast of the disc task at threshold in the lighted environment. The threshold condition is reached by adding the veiling luminance while reducing uniformly the output of all luminaires so as not to alter the distribution of luminous flux within the environment. From Eq. (3.12), the visibility level present in the actual lighted environment is simply

$$VL = \frac{C}{\overline{C}} = \frac{1}{K} \tag{3.18}$$

Several visibility level curves in addition to VL1 and VL8 are shown in Fig. 3.18.

An instrument called a Visual Task Evaluator (VTE) was devised to perform the measurement of $\overline{C}$ in a given environment without disturbing the lighting in that environment. The instrument places a veiling luminance over the task, while simultaneously, through neutral density filters, reducing the luminance of both the task and its background until threshold is reached.

Thus far only a 4-minute luminous disc task has been dealt with. These results need to be related to other, more realistic, visual tasks. Perhaps the most important factor affecting the visibility of a visual task in a given environment is veiling reflections. This effect results when rays from lighting units, usually on the ceiling of a room, reflect from the visual task directly into a person's eyes (as though from a mirror), rather than being absorbed or reflected diffusely. The net result is that the details of the visual task are partially or sometimes totally obscured.

Veiling reflections can be quantified as an effective reduction in contrast between the task and its background. To develop this concept, the equivalent contrast $\tilde{C}_0$ of an actual task is defined as the contrast of the 4-minute luminous reference task which makes the reference task equal in visibility to the actual task when both are viewed for $\frac{1}{5}$ s under threshold seeing conditions at the center of a diffusing luminous hemisphere.

The measurement procedure is as follows: the reference and actual tasks are placed in the sphere and viewed through the VTE. The background luminances of the two tasks, as seen in the VTE, are matched so as to insure that the adaptation of the operator is the same when observing each task. The contrast of the actual task is then reduced to threshold while keeping background luminance constant by using veiling luminance and neutral density filters, as described earlier. The fraction (K_a) of the actual task luminance remaining is measured. The process is repeated for the reference task and the fraction (K_r) of reference task luminance remaining is noted. Generally $K_a > K_r$. $\tilde{C}_0$ is obtained by multiplying the physical contrast of the reference task (C_r) by the ratio of the two K's.

$$\tilde{C}_0 = \frac{K_r}{K_a} C_r \tag{3.19}$$

With $\tilde{C}_0$ in hand, whatever VL curve is desired is entered and the necessary background luminance L_r is determined. In Fig. 3.19, VL8 is chosen. Dividing by the background reflectance ρ_b gives the amount of hemisphere illuminance E_r required. Thus E_r is the illuminance necessary to place the actual task at a visibility level of VL8 under reference lighting conditions in a hemisphere.

Can the same procedure be followed to determine required illuminance in the actual task in its real lighting environment? If the measurements of contrasts are repeated at threshold and at the same viewing angle with the VTE, this time with the actual and reference tasks in the real lighting environment, there will usually be a lower value for equivalent contrast. The main reason for this is the veiling reflections previously mentioned. Particularly troublesome in this regard are luminaire luminances slightly in front of desks, which cause light to reflect into a person's eyes from printed material on the desk tops.

The new value of equivalent contrast is labeled $\tilde{C}$ and with it the VL8 curve in Fig. 3.19 can be entered as had been done previously to find the required design values of task background luminance L_d and task illuminance E_d. E_d can then be compared with the actual illuminance present in the real lighting environment to see if the VL8 criterion is being met.

The effect of veiling reflections was to reduce the equivalent contrast of the task from $\tilde{C}_0$ to $\tilde{C}$. Contract rendition factor (CRF) is defined as the measure of this change in going from sphere conditions to the actual lighting environment. Thus

$$\text{CRF} = \frac{\tilde{C}}{\tilde{C}_0} \tag{3.20}$$

and is generally less than unity. It should be noted that a small percentage change in equivalent contrast produces a large percentage change in required illuminance if the same visibility level is to be maintained. In the example in Fig. 3.19, $\tilde{C}_0 = 1.3$ and $\tilde{C} = 0.7$. Thus, CRF = 0.54, quite a low value. This gives $L_r = 2.6$ fL and $L_d = 30.9$ fL. Their ratio is 11.9. Thus a reduction in contrast of 46% resulted in an increase in luminance, and thus illuminance, of 1090% for the same task visibility.

Measurements of $\tilde{C}_0$ and $\tilde{C}$ using the contrast-reducing visibility meter and the standard reference task are awkward to perform. Blackwell found that CRF may be obtained reasonably accurately in terms of *physical* measurements and use of Eq. (3.6). Thus the

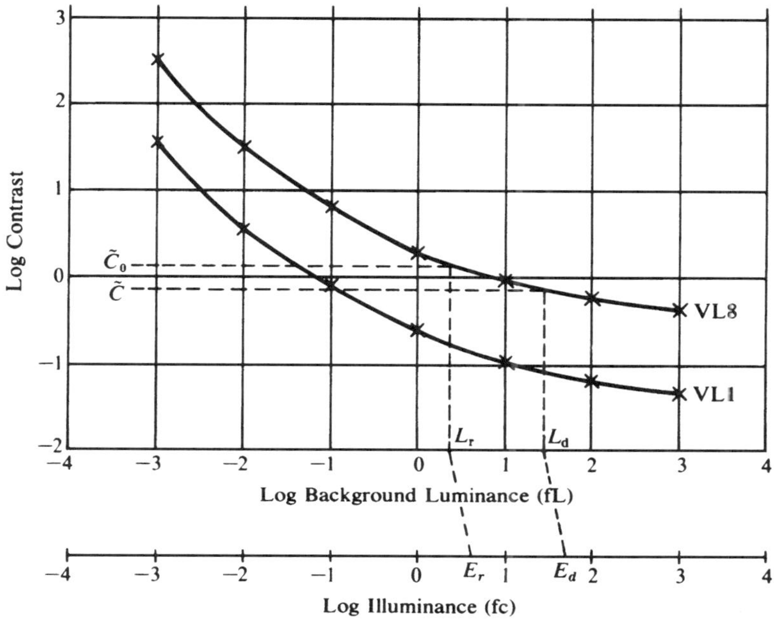

Figure 3.19 Equivalent contrasts.

task is placed in the hemisphere, its luminance measured and that of its background in the direction of viewing, and Eq. (3.6) is used to obtain $\tilde{C}_0$. The process is repeated in the actual lighting environment to obtain $\tilde{C}$. Then Eq. (3.20) gives the value of CRF.

It would appear that the curves in Figs. 3.17 and 3.18, as employed in Exs. 3.2 and 3.3, provide an effective, easy-to-use procedure for assessing the four factors, establishing visibility levels, and prescribing illuminances. Over the years, however, serious problems and shortcomings have arisen:

1. Blackwell's model assumes (Fig. 3.18) that suprathreshold visibility curves are parallel to the threshold visibility curves, thus allowing for a single contrast multiplier in going from one VL to another. Subsequent research has shown that this is not the case.
2. As noted following Eq. (3.20), small percentage changes in CRF often produce huge percentage changes in required illuminance for the same visibility level, not in keeping with real-world experience and practice. This is caused again by the assumption of parallel curves and their leveling off as background luminance increases.
3. Blackwell's VL curves were based on a 4-minute luminance disc task viewed for 0.2 seconds. He extrapolated this task, through equivalent contrast, to a variety of practical tasks. There were no guidelines advanced as to how far one could deviate from the luminous disc task before the results became invalid.
4. The assumption by Blackwell is that, along any VL curve, a constant visibility condition exists. Thus, one can trade off luminance for contrast, and vice-versa, to maintain the same visibility. The concept that luminance (illuminance) can be traded for contrast is valid, but not as quantified on the Blackwell VL curves.

Because of the problems cited above, the Blackwell, CIE 19/2 procedure has descended from its dominant role as the visual performance model of choice from 1960 to 1980. However, many of the ideas advanced by Blackwell are very valid today, as examples, the concept of contrast rendition, the tradeoff of luminance and contrast, and the forced-choice procedure for obtaining data. The Clear-Berman model (Sect. 3.16) is a visibility-level model that draws heavily upon Blackwell's work.

3.15 REA-OUELLETTE MODEL

The Rea-Ouellette model (Rea, 1981, 1986, 1987, and Rea-Ouellette, 1988, 1991) is one of two contemporary visual performance models that have achieved some acceptance. It is an on-axis (no visual search) model that is based on two experiments: a numerical verification task where observers recorded the differences in two nearly identical columns of numbers, and a reaction time task in which observers detected the presence of a square against its background on a computer monitor.

In the numerical verification test, observers were asked to compare two printed-number lists (see Fig. 3.20). The left hand column is a reference sheet, consisting of 20 five-digit numbers. There were 64 reference sheets, 32 with gloss black print on a matte white background and the other 32 with matte gray print on matte white. The right-hand column in Fig. 3.20 is a response sheet. The numbers in the response list were the same as those in the reference list, except for one digit being different in some numbers. There were 128 response sheets, four groups of 32 sheets each. All response sheets were high-contrast matte black print on a matte white background. The four groups of response sheets differed in frequency of errors, which varied from 0 to 6.

Eight young observers participated in the experiment. They were seated at a desk in a black room. Eight desk positions were used. Four illumination levels were provided by an overhead fluorescent luminaire mounted slightly in front of the desk. Background luminances of the reference sheets, at the observer's eyes, were held constant at 12, 22, 31, and 169 cd/m^2 (approximately 47 to 185 lux on the task). The middle digits of the reference list subtended angles of 13 minutes of arc horizontally and 19 minutes vertically at the eyes of the observer. Separation between digits was 7.1° of visual angle.

Response sheet contrasts varied from 0.670 to 0.826, as a function of desk orientation with respect to the overhead illumination. For the gloss black reference sheet, contrasts varied from 0.199 to 0.894; for the matte gray reference sheet from 0.092 to 0.156.

Observers were asked to compare the lists "quickly and accurately." Total elapsed time, number of discrepancies missed, and false positives were tabulated.

Results of the experiment are shown in Fig. 3.21, where the reciprocal of the total elapsed time to compare the reference and response sheets (1/s) is plotted against contrast of the numbers on the reference sheets for the four values of background luminance. The curves show the suprathreshold saturation (compression) effect inherent in all visual performance models.

From the results of this experiment, Rea derived a computational model of suprathreshold relative visual performance (RVP) that appears in the three-dimensional representation in Fig. 3.22. He altered the time measure from the total elapsed time per sheet, first to the reading time per sheet (by removing the time to note errors), and then to the reading time of a reference list, arguing that the time to read the response

69179	69179
27982	27982
15179	15179
39940	39940
60468	60468
18602	18602
71194	71124
94595	94595
57740	57740
38867	38867
56865	56865
18663	18663
36320	36320
67689	67689
47564	47562
60756	60756
55322	55322
18594	18594
83149	83149
76988	76988

Figure 3.20 Numerical verification test. *Source*: "Visual performance with realistic methods of changing contrast," in Rea, *JIES*, April 1981. Courtesy of the Illuminating Engineering Society of North America.

list for a given background luminance was constant because its contrast was constant. Next he used a suprathreshold sensory compression equation to calculate VP/VP_{max}, where VP is defined as the reciprocal of the time in seconds to read the reference sheet at each adaptation luminance, and VP_{max} is the maximum attainable visual performance at a given background luminance. This process required estimating threshold readability contrasts for the four luminance levels used and then employing a regression computer analysis to calculate parameters in the equation. The result appears in Fig. 3.22.

The second experiment upon which the Rea-Ouellette model is based was conducted not to create an additional model, but rather to validate and extend the results of the numerical verification test. This experiment measures detection threshold values and reaction times of observers viewing a square target of variable size and

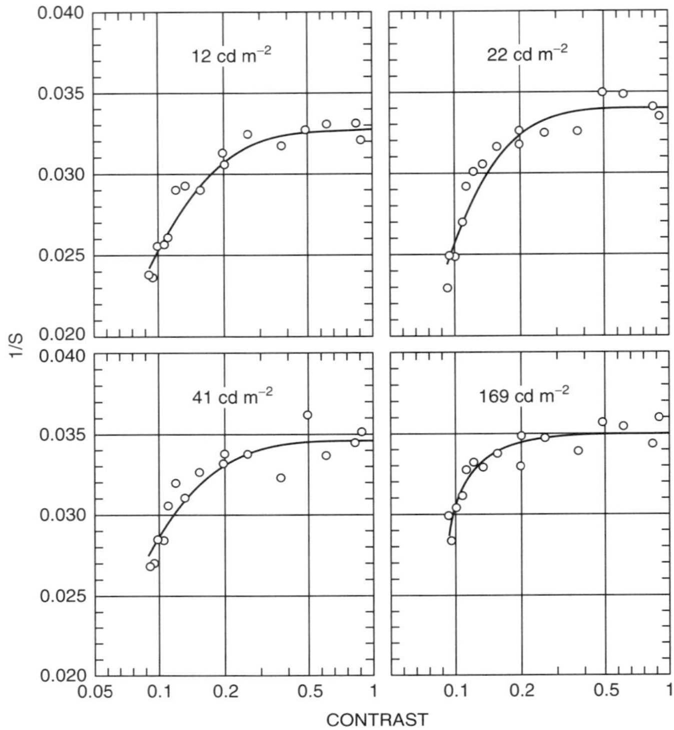

Figure 3.21 Visual performance as a function of contrast. ***Source*****: "Toward a model of Visual Performance: Foundations and data," M. Rea, *JIES*, Summer 1986. Courtesy of the Illuminating Engineering Society of North America.**

contrast against an adjustable luminance background, presented on a computer monitor screen. Both positive and negative contrasts were used. The observers were young adults.

The squares were flashed briefly on the monitor screen. The monitor was viewed through an opening 2 mm in diameter that controlled retinal illuminance, given in trolands. A troland (T) is, by definition, the product of the luminance in cd/m^2 of the object being viewed and the pupil aperture area of the eye in mm^2. One troland is the illuminance on the retina produced by 1 cd/m^2 viewed through 1 mm^2 of pupil area.

Reaction times and detection threshold values of contrast were obtained for retinal illuminances ranging from 0.53 to 80 T and for target sizes ranging from 2 to 2800 μsr in solid angle.

The calculation procedure used to develop the RVP model for the numerical verification data was modified for the detection and reaction time model, with similar equations being used to calculate response times from detection threshold to suprathreshold saturation. A parameter ΔT_{vis} was defined as the difference in visual response time for a particular set of values of retinal illuminance, target size and target contrast, as compared with set of suprathreshold reference conditions (high contrast, large size, high background luminance).

Thus, two performance metrics are available: RVP from the numerical verification experiment and ΔT_{vis} from the detection and reaction time model.

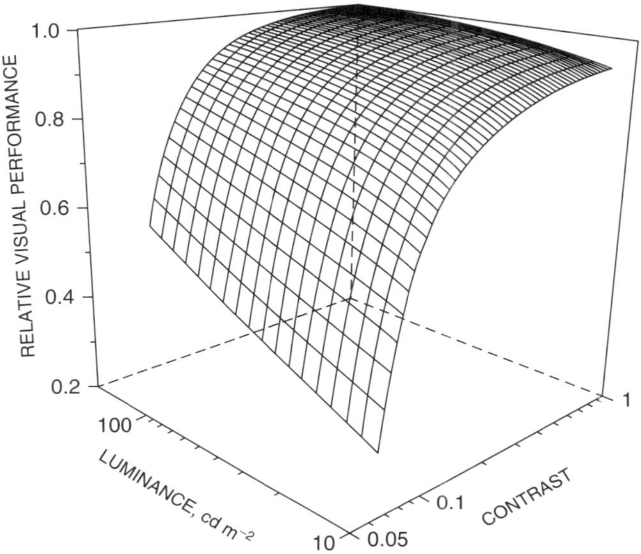

Figure 3.22 Relative visual performance based on numerical verification data. *Source*: "Toward a model of Visual Performance: Foundations and data," M. Rea, *JIES*, Summer 1986. Courtesy of the Illuminating Engineering Society of North America.

It remained to relate RVP to ΔT_{vis}. It was argued that if each experiment is scaling visual performance, it should be possible to relate the scaling parameters by a linear transformation. This was done and a modified relative visual performance was derived. The results are shown in Fig. 3.23 (the RVP's in Figs. 3.22 and 3.23 are somewhat different). Four three-dimensional diagrams are presented in Fig. 3.23 covering target sizes from 1.9 to 130 μsr (the diagram for 4.8 μsr is roughly equivalent to the diagram in Fig. 3.22).

Each diagram plots relative visual performance against target contrast and retinal illuminance (replacing background luminance). The plots clearly show the escarpment-plateau features of the Rea-Ouellette model.

Many questions have been raised about the Rea-Ouellette model and the combining of the two experiments (Mistrick, 1995, Rinalducci, 1995, Clear, 1996).

Among these are:

1. Size Measure

   Should size include the entire area occupied by a number (for example), the stroke width, the critical gap distance, or the area of the print? In the numerical verification test, size was the entire area—visual angles of 13 and 19 minutes of arc, which translated into a solid angle of 20.8 μsr. In the development of the combined model, the printed area was used, giving 4.8 μsr. Rea-Ouellette contend that the solid angle of the print is the relevant parameter, whether the target is a square or a number.
2. Non-Visual Performance Parameters

   It is not convincing that the non-visual performance factors remain constant and thus can be factored out in order to determine visual performance.

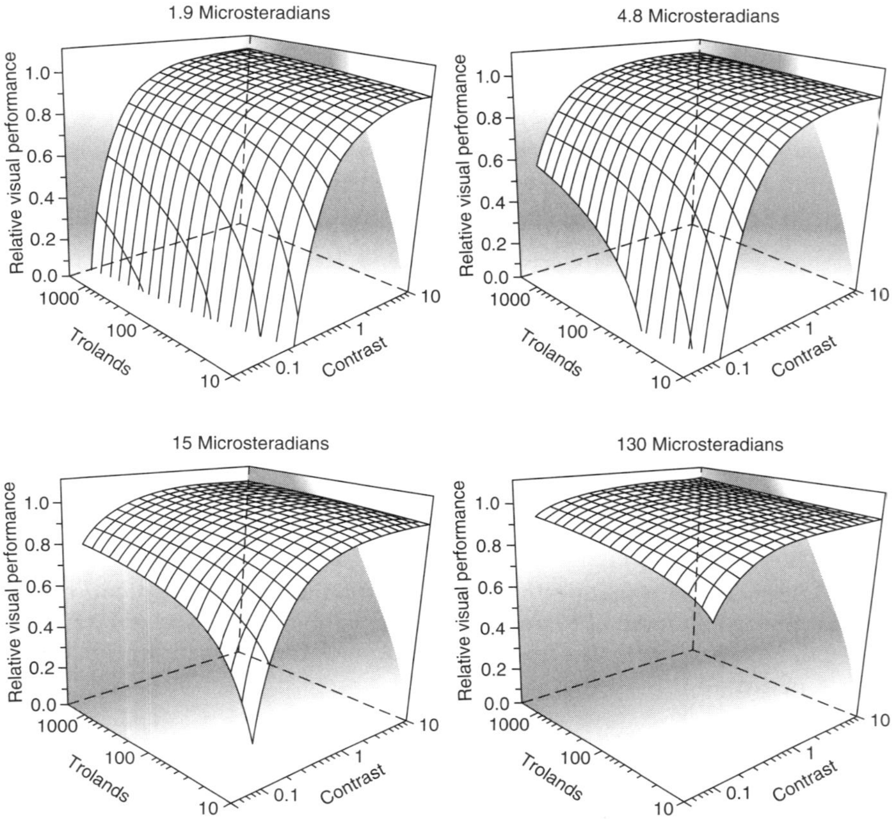

Figure 3.23 Relative visual performance based on reaction time data. *Source*: *IESNA Lighting Handbook*, 9th edition, 2000. Courtesy of the Illuminating Engineering Society of North America.

3. Speed and Accuracy

   The following equation was used to obtain a response measure:

$$\text{Score} = \frac{100(\text{T} - \text{E})}{\text{S} + 5} \tag{3.21}$$

where T = comparisons per trial (20)
E = misses and false positives per trial
S = time in seconds per trial.

   Based on this equation, Rea concluded that E played a minor role, and thus that accuracy was unimportant and relative visual performances depended on speed. Many feel that both speed and accuracy are needed in the real world.

4. ΔRVP and ΔT_{vis}

   In combining the two experiments, it was assumed that ΔRVP and ΔT_{vis} were linearly related. Since RVP is the inverse of time, there are inherent difficulties in this assumption.

5. Extrapolation of Size
   The model was developed for relatively large sized targets (4.8 μsr). This is equivalent to 10-point type at 19 in or #18 AWG wire from 15 in. Thus the model would not seem to apply to an inspection task such as detecting a small crack in a piece of metal.
6. Background Luminance
   The RVP model was based on a limited range of background luminances, representing a maximum task illuminance (for young adults) of around 300 lux. This is certainly on the low end for interior lighting.
7. Types of Tasks
   It is not clear for the Rea-Ouellette model, and for the Clear-Berman model to follow, which types of tasks are served—the orientation of a Landolt C, errors in a number list, detection of an object, industrial inspection, sewing, drawing, identifying words, reading to edit, reading for comprehension…the list goes on and on. This question about which tasks each model serves is perhaps the most important unanswered question about visual performance models.

3.16 CLEAR-BERMAN MODEL

This model (Clear, 1990, Bailey, Clear and Beaman, 1993) is an additive visibility-level (VL) model based on on-axis viewing (no visual search). It can handle both threshold and suprathreshold viewing situations and, thus, a wide variety of visual tasks. It predicts absolute, not relative, performance. The basic metric is the speed to perform a task, or its reciprocal, total time, the sum of the time needed for the non-visual components (T_{NV}), which includes saccade, cognition and motor response times, and the time needed for the strictly visual components (T_V), which includes detection, recognition and/or identification. The T_{NV} term can vary with both size and background luminance, but it can also be constant for some tasks. The T_V term depends upon size,* with VL as its principal component, but also includes accuracy—it yields speed for a given accuracy. Having speed and accuracy as co-dependent variables is a unique feature of the Clear-Berman model.

The basic Clear-Berman equation is

$$S = (T_{NV} + T_V)^{-1} \tag{3.22}$$

where S is speed in $(s)^{-1}$, and T_{NV} and T_V are the non-visual and visual times in seconds to perform the task. Clear and Berman postulated that the visual time could be given by

$$T_V = \frac{TC}{f(VL)} \tag{3.23}$$

where TC is a time constant which is independent of VL, but dependent on background luminance and angular size of the target. Its value has been determined by Adrian** to range from 0.12 to 0.35s, for a wide range of target sizes and background luminances. Clear and Berman found that a single value of TC = 0.21s provided a good data fit.

* Clear and Berman feel that size is more important than contrast in predicting reading speeds, for example.

** Adrian, W., "On the Contrast Sensitivity of the Human Eye," Public Works Canada, 1982.

The term f(VL) is a function of visibility level. Let VL(t) be the visibility level as a function of exposure time t. VL(t) can be expressed in terms of VL(∞), VL for infinite exposure time, by

$$VL(t) = VL(\infty)\frac{t}{t+TC} \tag{3.24}$$

Solving for t gives

$$t = TC \times \frac{VL(t)}{VL(\infty) - VL(t)} \tag{3.25}$$

Now let V be the visibility level required to obtain a certain accuracy. Then $t = T_V$ and

$$T_V = TC \times \frac{V}{VL(\infty) - V} \tag{3.26}$$

Thus f(VL) is

$$\frac{VL(\infty) - V}{V} \tag{3.27}$$

When Blackwell (Sect 3.14) developed his visibility level curves, he used an exposure time of 0.2s and assigned VL1 based on "common-sense" seeing (resolution of detail). This required a contrast multiplier of 2.51 to lift his VL curves above forced-choice detection (which would be VL0.4). Rewriting VL(∞) in Eq. (3.26) in terms of VL(0.2) yields an expression for T_V in terms of the forced-choice base.

$$T_V = T_c \times \frac{V}{VL(0.2) \times \frac{TC+0.2}{0.2} - V} \tag{3.28}$$

Then, from Eq. (3.22),

$$S = \left(T_{NV} + \frac{TC}{\frac{TC+0.2}{0.2} \times \frac{VL(0.2)}{V} - 1} \right)^{-1} \tag{3.29}$$

If T_c is constant at 0.21s, S becomes

$$S = \left(T_{NV} + \frac{0.21}{2.05 \times \frac{VL(0.2)}{V} - 1} \right)^{-1} \tag{3.30}$$

The result in Eq. (3.29) shows that S is a function of VL but can also be a function of background luminance and target size, through TC. Accuracy is a function of only VL, but, as noted earlier, speed and accuracy are codependent variables, i.e., V corresponds to a particular desired accuracy.

Clear-Berman and others have noted that detection, as used by Blackwell for luminous discs, does not work for small targets. Such targets require resolution. There is a certain visual size below which the number of photons per unit time required for detection is constant. The relationship is called Ricco's Law

$$I \times A = K \tag{3.31}$$

where I is the threshold luminous flux, in photons per unit area per unit time, A is target area, and K is a constant. Ricco's Law applies only for foveal vision. For fixed background luminance, the critical size is given by

$$S_R = \frac{4 + 2L_b}{1 + L_b} \tag{3.32}$$

where L_b is background luminance in cd/m^2, and S_R is the size in minutes of arc below which Ricco's Law applies. Blackwell's data follow Ricco's Law when target sizes are less than S_R.

Clear and Berman tested their mathematical model with both the Rea-Ouellette models, Blackwell's model, and a reading test devised by I. Bailey. They concluded that their model gave a better fit to numerical verification data than did the Rea-Ouellette model, and did it with fewer free parameters. When they examined the Rea-Ouellette reaction time model (Clear, 1996), they found considerable disagreement, particularly faulting the notion that it was applicable to a wide range of tasks. Also they pointed out that the Blackwell threshold contrast formula is much more sensitive to luminance and size than is the Rea threshold contrast formula, and thus suggested that the Blackwell data should be used. Rea and Ouellette disagreed with Clear and Berman on most of these matters.

Clear and Berman did not obtain a good fit with Blackwell's detection of contrast data, indicating that it may not be possible to extrapolate that data to more general viewing conditions, as was indicated earlier.

The final verification by Clear-Berman utilized a reading test, and set out to show that a sized-based performance model was preferable to a contrast-based model for predicting reading speed. The reading test had complex resolution and identification components, far beyond simple detection. The test was a speed test to read unrelated words on Bailey-Lovie word reading charts (see Fig. 3.24). Each chart consisted of 11 rows of words of decreasing letter size, ranging from 20 point to 2 point type and having six words per row, two each of four, seven, and ten letters.

The charts were black print on clear transparencies, lit from behind by a translucent light box containing incandescent lamps, which could be dimmed. Contrast was varied by superimposing a veiling luminance on the target from a second light box and mirrors.

Fifteen subjects, with 20/20 corrected vision, performed the experiment. Each testing session lasted $3^1/_2$ hours, during which time subjects viewed 40 charts under various values of contrast and background luminance, reading an average of 800 six-word rows of printed letters.

nations tank
liquidated oven
whip justify
receptacle idle recital
quay lubricates darning
measure ventilator fade
able yawning particular just management holiday
emotion jars separately butcher cork forbidding
disapoint salt drawing yourselves much zoology
neck locomotive weather ridiculous quartet king
volcano lefthanded inch undress only thoughtful
helicopter gradual numb to Sisle collection Ash

Figure 3.24 Word reading chart. *Source*: "Size as a Determinant of Reading Speed," I. Bailey, R. Clear and S. Berman, *JIES*, Summer 1993. Courtesy of the Illuminating Engineering Society of North America.

Performance was measured as the speed of reading. Reading time per row of words was obtained by recording each subject's eye movements—the shift left of a subject's eyes to begin fixation on a next row of words. In most cases, subjects required only one fixation per word (six fixations per row).

Clear and Berman fit the experimental data to their VL model. The end result is shown in Fig. 3.25, where reading speed in words/second is plotted against visibility level as a function of size. The plotted points are derived from the experimental data, whereas the solid line is obtained from an equation for reading time derived from the Clear-Berman model. The agreement between measured and calculated values is excellent and was shown by Clear and Berman to provide a better fit to the data than that provided by the reaction time, detection-based model of Rea and Ouellette. This lends further credence to the notion that no one model is suitable for all visual tasks.

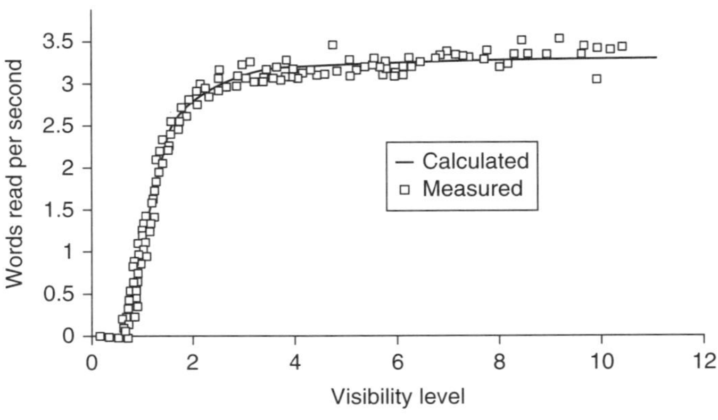

Figure 3.25 Reading speed vs. visibility level (size). *Source*: *IESNA Lighting Handbook*, 9th ed., 2000. Courtesy of the Illuminating Engineering Society of North America.

REFERENCES

Adrian, W., and R. Gibbons: "Visual Performance and its Metric," Lighting Research Institute, New York, 1994.

Bailey, I., R. Clear, and S. Berman: "Size as a Determinant of Reading Speed," *Journal of the IES*, 1993.

Blackwell, H.R.: "Contrast Threshold of the Human Eye," *Journal of the Optical Society of America*, 1946.

Blackwell, H.R.: "Brightness Discrimination Data for the Specification of Quantity of Illumination," *IES Transactions*, November, 1952.

Blackwell, H.R.: "A More Complete Quantitative Method for Specification of Interior Illumination Levels on the Basis of Performance Data," *Illuminating Engineering*, 1969.

Boyce, P.R.: "Illuminance Selection Based on Visual Performance—and Other Fairy Tales," *Journal of the IESNA*, Summer, 1966.

Clear, R., and S. Berman: "Speed, Accuracy and VL," *Journal of the IESNA*, 1990.

Clear, R.: "Relationships Between the VL and Reaction Time Models," *Journal of the IESNA*, 1996.

Cobb, P.W., and F.K. Moss: "The Four Variables of the Visual Threshold," *Journal of the Franklin Institute*, 1928.

Coblentz and Emerson: "Relative Sensibility of the Average Eye to Light of Different Colors," *National Bureau of Standards, Bulletin #14*, 1918.

Gibson, K.S., and E.P.T. Tyndall: "Visibility of Radiant Energy," *National Bureau of Standards, Bulletin #14*, 1918.

Hancock, S.: "Bright Eyes," *American Baby*, May, 1992.

Jensen, B.J.: *IES Lighting Handbook, Second Edition*, Illuminating Engineering Society, New York, 1952, pages 2–24 to 2–23.

Kaiser, P.K.: "Photometry of Brightness: Problems, Physiology, Models," *Journal of Light and Vision*, Illuminating Engineering Institute, Japan, 1984.

Luckiesh, M., and F.K. Moss: *The Science of Seeing*, Van Nostrand, New York, 1937.

Luckiesh, M.: "Footcandle Levels—Threshold, Ideal, Optimum and Recommended," *Illuminating Engineering*, 1948.

McSweeney, D.: "Report on the IESNA Visibility Metric Forum," New York, 1995.

Mistrick, R.G.: "An Overview of Visual Performance Models," IESNA Visibility Metric Forum, New York, 1995.

Mistrick, R.G.: "Outline of a Procedure for the Development of a Visual Performance Model," IESNA Visibility Metric Forum, New York, 1995.

Noel, Eunice, D.: "Lighting and the Visual Environment for Senior Living," *RP-28-98*, IESNA, New York, 1998.

Rea, M.S.: "Visual Performance with Realistic Methods of Changing Contrast," *Journal of the IESNA*, April, 1981.

Rea, M.S.: "Toward a Model of Visual Performance: Foundations and Data," *Journal of the IESNA*, Summer 1986.

Rea, M.S.: "Toward a Model of Visual Performance: A Review of Methodologies," *Journal of the IESNA*, 1987.

Rea, M.S., and M.J. Ouelette: "Visual Performance Using Reaction Times," *Lighting Research and Technology*, 1988.

Rea, M.S., and M.J. Ouelette: "Relative Visual Performance: A Basis for Application," *Lighting Research and Technology*, 1991.

Rea, M.S.: "An Overview of Visual Performance," *LD+A*, November, 1982.

Rea, M.S., editor: *IESNA Lighting Handbook, Ninth Edition*, IESNA, New York, 2002.

Rinalducci, E.J.: "Models of Visual Performance," IESNA Visibility Metric Forum, New York, 1995.

Sekuler, R., and B. Blake: *Perception, Third Edition*, McGraw Hill, New York, 1994.

Tinker, M.A.: "Trends in Illumination Standards," *Illuminating Engineering*, 1948.

Weale, R.A.: *The Senescence of Human Vision*, Oxford Press, New York, 1992.

Weston, H.C.: "The Relation Between Illumination and Industrial Efficiency: The Effect of Size of Work," London Industrial Health Research Board, 1935.

Weston, H.C.: "The Relation Between Illumination and Visual Efficiency: The Effect of Brightness Contrast," London Industrial Health Research Board, Report #87, 1945.

Chapter 4

Radiant Energy as Waves

4.1 INTRODUCTION

Sections 1.3 and 1.5 reviewed the theories that have been developed since Newton's time to explain the nature of light, reaching the somewhat discomforting conclusion that light is dualistic in nature, with the properties of both waves and particles.

This chapter and the next will be devoted to exploring this wave-particle duality of light, with concentration on the wave nature of light in this chapter and its particle nature in Chap. 5. The chapter begins with a discussion of power density and spectral power density, quantities used to describe the total energy content of a wave and the energy content in selected spectral bands. This leads to a description of blackbody radiation and the laws of Planck, Stefan, and Wien governing that radiation.

Then the evolution of luminous intensity standards based on the blackbody radiator is discussed, and it is shown how the lumen and the watt can be related through the current standard and the spectral luminous efficiency curve of the eye.

The chapter concludes with a discussion of selective radiators (of which the tungsten filament is an example), the concept of the color temperature of a light source, the design of a tungsten filament and methods of making spectral power measurements.

4.2 POWER DENSITY AND SPECTRAL POWER DENSITY

Much of the discussion on radiation to follow will be as much concerned with power density as with power, that is, it is important to know the watts per unit of surface area emitted by a radiator or received by a surface, as well as the total watts emitted or received. Radiant exitance (M) is defined as emitted, transmitted, or reflected radiant power density and irradiance (E) is defined as received radiant power density. Thus M is radiation leaving a surface, whereas E is radiation arriving at a surface.*

Also of interest is power density within a narrow wavelength band, called spectral power density. Spectral power density leaving a surface is called spectral radiant exitance, M_λ;** that arriving at a surface is called spectral irradiance, E_λ. To illustrate these new quantities, consider Fig. 4.1 in which the spectral radiant exitance of a fictitious source is shown. The graph is a plot of M_λ in microwatts per square centimeter of source surface area per 10 nm of wavelength. What this plot provides is the radiant exitance in any 10-nm band centered about the wavelength that is under consideration. For example, at 400 nm, the radiant exitance in a 10-nm band from 395 to 405 nanometers

* It is sometimes confusing that the symbols for radiant and luminous quantities are the same in the standards literature. The context in which a given symbol is used should clarify which entity is being discussed. Otherwise the subscript e for radiant quantities and v for luminous ones will be used.

** The subscript λ indicates the derivative of M with respect to λ, whereas parentheses indicate that the entity is a function of λ, for example $V(\lambda)$.

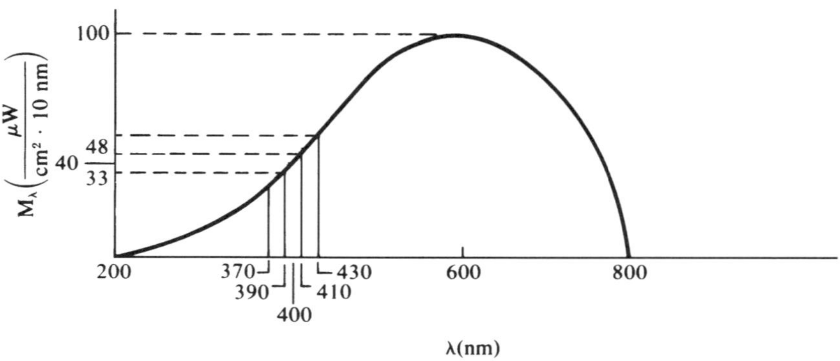

Figure 4.1 Emission from a radiant source.

is 40 $\mu W/cm^2$. In a 2-nm band centered at 400 nm (from 399 to 401 nm) it is $40(0.2) = 8\ \mu W/cm^2$. The factor 0.2 is necessary because the wavelength band is now 2 nm wide, whereas M_λ is based on a 10-nm width.

It is possible to find the total radiant exitance M emitted by the source in Fig 4.1. To do this, the region beneath the curve is blanketed with narrow rectangles, each of the same width. In the case illustrated 20-nm widths have been chosen, and three of these rectangles are shown. The height of each is taken as the value of M_λ at the midpoint wavelength of that rectangle. Thus for the leftmost of the three rectangles in Fig. 4.1, the height is the value of M_λ at 380 nm, which is 33 $\mu W/cm^2 \cdot 10$ nm). This rectangle contributes $33(2) = 66\ \mu W/cm^2$ to the total radiant exitance of the source, the factor 2 being necessary because of the 20-nm width. The other two rectangles contribute 80 and 96 $\mu W/cm^2$ to the total. Summing the values for all rectangles would give the total radiant exitance.

What has been done above is to find the approximate area beneath the curve in Fig. 4.1. If the equation for M_λ as a function of λ were known, the total emitted power density could be found exactly through integration as

$$M = \int_0^\infty M_\lambda \, d\lambda \tag{4.1}$$

To sum up, the vertical distance to a spectral power density curve at a specified wavelength gives the spectral power density at that wavelength and the area bounded by the curve and the horizontal axis gives the total power density.

EXAMPLE 4.1

To illustrate the concepts of a spectral distribution curve, consider Fig. 4.2, which is a graph of spectral population density (people per square mile per 2-year age interval) plotted against age. The total number of people per square mile is sought.

Solution. Break the horizontal axis up into 4-year intervals and read the spectral population density at the midpoint of each interval. For example, from 16 to 20 years, the midpoint is 18 years, and the value is 300 people per square mile per 2-year interval or 600 people per square mile per 4-year interval. The total population density is the sum of all such values. Starting with the 0- to 4-year age interval, $(240 + 380 + 390 + 330 + 300 + 390 + 410 + 410 + 400 + 370 + 300 + 240 + 170 + 120 + 80 + 50 + 30 + 10)2 = 9920$

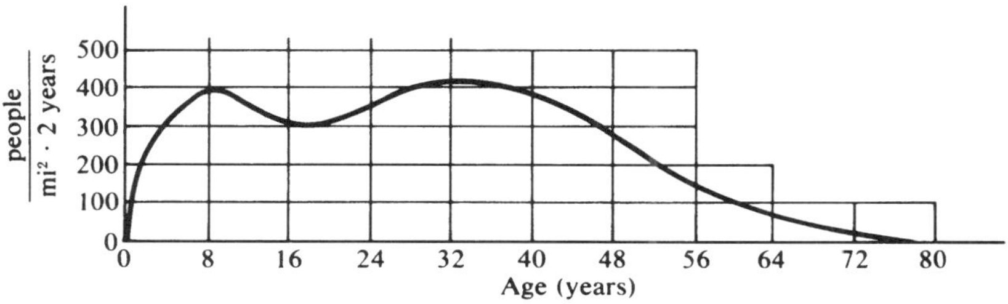

Figure 4.2

people per square mile, the "2" being necessary because a 4-year age interval has been used.

In Fig. 4.1, M_λ is expressed in $\mu W/cm^2 \cdot 10$ nm and the x axis is in nm. Suppose it is desired to have the x axis in μm. What must the values of M_λ on the y axis become? For any M_λ value,

$$M_\lambda \frac{\mu W}{cm^2 \cdot 10\,nm} \times \frac{100(10\,nm)}{1\,\mu m} = M_\lambda \frac{\mu W}{cm^2 \cdot \mu m} \tag{4.2}$$

since there are 100 10 nm bands in a 1 μm band. Thus, each value on the M_λ scale in Fig. 4.1 should be multiplied by 100 and the scale relabeled $\frac{\mu W}{cm^2 \cdot \mu m}$.

Fig. 4.1 displays a continuous spectrum source, typical of incandescent radiators. In Chap. 6, discharge sources will be considered, which have spectral lines, as shown diagrammatically for a mercury lamp in Fig. 4.3. Spectral lines are generally of irregular shape. It is customary to standardize them as rectangles of width 10 nm and height such that the area of the rectangle equals the power in the spectral line. Thus in Fig. 4.3, the power in the 405-nm line is $20 \frac{mw}{10\,nm}$ × one 10nm-band, giving 20 mw.

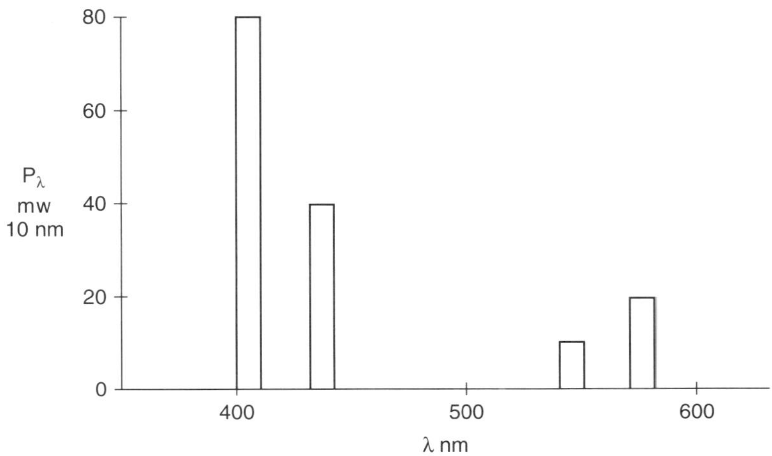

Figure 4.3 Line spectrum source.

EXAMPLE 4.2

A lamp with both a continuous and a line spectrum is shown in Fig. 4 4. The y axis gives the number of μW required per 10 nm to produce one lumen. Assuming that 25%

of the input watts appear in the visible portion of the spectrum (75% in the infrared and ultraviolet regions), find the efficacy of the lamp in lm/W.

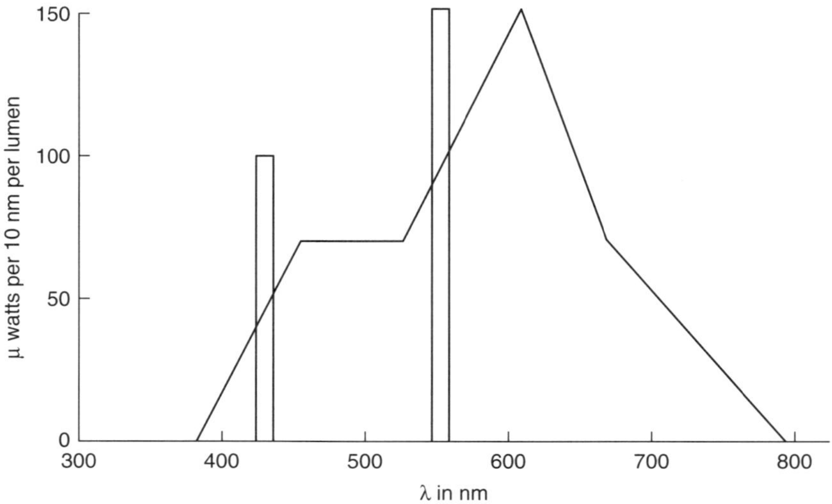

Figure 4.4

Solution. The problem involves finding the area under the curve and of the rectangles in Fig. 4.4, i.e. performing the integration in Eq. (4.1) graphically. The following is tabulated:

$\Delta\lambda$ (nm)	μW/10 nm·lm	no. of 10 nm bands	μW/lm
380–450	35	7	245
450–520	70	7	490
520–600	110	8	880
600–660	110	6	660
660–760	35	8	280
line	100	1	100
line	150	1	150
			2805

resulting in

$$\frac{1}{0.002805} = 357\,\text{lm/visible W}$$

and

$$\frac{357}{4} = 89\ \text{lm/W}$$

4.3 BLACKBODY RADIATION AND PLANCK'S EQUATION

There are three major ways in which heat energy is transferred from one body to another, namely conduction, convection, and radiation.

Assume a person is standing near a campfire. If he (or she) inadvertently touches one of the glowing embers, heat reaches his hand by conduction. If he holds his hands above the fire, heat reaches his hands by convection, the upward flow of air that has been heated. Now if he stands back from the fire, he still feels the sensation of heat. Energy is now reaching him by radiation.

Radiation is the emission of energy in the form of electromagnetic waves from all bodies. Such energy travels with the speed of light and does not require a medium for its passage; in fact, it travels best in a vacuum. When such radiation encounters a body, it is absorbed, transmitted, or reflected. That which is absorbed may be reradiated, increase the temperature of the body, and/or be converted to other energy forms.

Light is radiant energy in the narrow band of wavelengths from 380 to 780 nm. It is customary to divide the generation of light into two categories—incandescence and luminescence. Incandescence is visual radiation from the surface of a radiator that is directly related to the temperature of that radiator. The entire family of incandescent lamps falls into this category, each lamp containing a filament that is heated to incandescence by the passage of electric current through it. Gas lamps, candle flames, and the sun are other examples of incandescent sources. Luminescence is visual radiation not directly related to temperature. Glow lamps, "neon" signs, fluorescent lamps, high-intensity discharge lamps, electroluminescent panels, light-emitting diodes, and lasers are included in this grouping. In the remainder of this chapter incandescence will be considered. The various forms of luminescence will be discussed in later chapters.

In dealing with incandescent, or temperature, radiation, it is convenient to consider an ideal radiator known as a blackbody. A blackbody radiator absorbs all of the radiant energy reaching it, transmitting and reflecting none. Because it is a perfect absorber, it is also a perfect radiator. A blackbody radiates more power at each wavelength and more total power than any other incandescent radiator of the same surface area operating at the same temperature.

To demonstrate the concept of the blackbody, obtain a small closed cardboard box, cut a hole the size of a dime in one side, cover the hole with a piece of cardboard, taped on one edge to form a flap, and paint the outside of the box and the flap matte black. State rhetorically to the observers that the box certainly is very black. Now uncover the flap and the observers will note that the hole is much blacker than the surrounding side of the box. Explain that some incident light flux does reflect from the side of the box, but the light flux entering the hole is almost completely absorbed.

From 1897 to 1900, Lummer and Pringsheim at the National Bureau of Standards in Washington, D.C. obtained blackbody radiation curves experimentally for a wide range of temperatures, using the open end of a specially-constructed and uniformly-heated tube as their blackbody source. Then in 1900, Lord Rayleigh and J.H. Jeans attempted to derive an expression for the radiation from a blackbody radiator using classical means. These men proceeded directly from the assumptions of classical statistical and Newtonian mechanics and Maxwell's electromagnetic equations. They assumed that a blackbody cavity contains electromagnetic waves bouncing back and forth between and reflecting from the cavity walls and that the superposition of these waves creates stationary waves in space within the cavity. With respect to energy, each of these waves should behave like a mechanical harmonic oscillator to which the laws of statistical mechanics are applicable. These are:

1. The average kinetic energy of the collection of harmonic oscillators is $\frac{1}{2}$ kT times the number of degrees of freedom that they have inside the box, where k is the Boltzmann constant* and T is temperature in kelvins.
2. The average potential energy equals the average kinetic energy. Thus the average total energy of the oscillators is kT times the number of degrees of freedom.
3. The number of stationary waves (which is the number of degrees of freedom) per unit volume and per unit wavelength is $8\pi/\lambda^4$, which gives an average total energy per unit volume and per unit wavelength $(8\pi/\lambda^4)kT$.

From these basic relations, Rayleigh and Jeans derived the following expression for the spectral power density distribution of a blackbody radiator:

$$M_\lambda = \frac{2\pi kTc}{\lambda^4} \frac{W}{cm^2 \cdot cm} \tag{4.3}$$

where c (the velocity of light) is in centimeters per second, and λ is in centimeters. This expression is known as the Rayleigh-Jeans Law. The "cm^2" in the denominator of the units is surface area and the "cm" is wavelength.

It was found that Eq. (4.3) does not agree with the experimental results of Lummer and Pringsheim. At wavelengths in the ultraviolet and visible regions of the spectrum it is extremely in error**, and it leads to the obviously incorrect result that the total emitted power density of a blackbody radiator is infinite. It does agree with experimental values for λT greater than 1.5×10^5 K · µm (long-wavelength infrared radiation).

In the year 1901, Max Planck (1858–1947), the famous German physicist and originator of quantum theory, developed a purely empirical formula to represent the Lummer-Pringsheim results. The equation bears Planck's name and is

$$M_\lambda = c_1 \lambda^{-5} \left(e^{c_2/\lambda T} - 1\right)^{-1} \tag{4.4}$$

where M_λ is the spectral power density in watts per square centimeter radiated by the blackbody per micrometer of wavelength band, λ is in micrometers, T is in kelvins, and c_1 and c_2 are constants given by

$$c_1 = 2\pi hc^2 = 37{,}418\,(W \cdot \mu m^4)/cm^2 \qquad c_2 = \frac{hc}{k} = 14{,}388\ \mu m \cdot K \tag{4.5}$$

In these equations, c is again in centimeters per second, k is the Boltzmann constant, and h is a new constant, Planck's constant whose value (6.63×10^{-34} J · s) is chosen so that M_λ in Eq. (4.4) matches the experimental data.

Planck's constant, like the Boltzmann constant, is a universal constant of nature. In deriving (Eq. 4.4) for blackbody radiators, Planck discovered that the emitted spectral radiation could not be described in terms of harmonic oscillators possessing all possible energies, as Rayleigh and Jeans had done. Rather he found that it was necessary to consider the emitted energy as occurring in discrete packets, which he called quanta. As was

* The Boltzmann constant is a universal constant relating the average kinetic energy of an aggregate of molecules to their temperature. Its value is 1.38×10^{-23} joules per degree Kelvin (J/K). It is named after Ludwig Boltzmann (1844–1906), professor of physics at Vienna, who made many contributions to kinetic theory and thermodynamics.

** This error was described as the "violet catastrophe".

noted in Sect. 1.5, each quantum has an energy in joules given by*

$$W = hf = h\frac{c}{\lambda} \tag{4.6}$$

Thus, Planck's constant is the proportionality constant relating the frequency and energy of a quantum of radiation. It is the smallest unit of energy that cannot be subdivided.

Returning now to Eq. (4.4), the only change Planck made in what Rayleigh and Jeans had done was to assume that the stationary waves in the blackbody cavity could have only integrally related discrete energies given by 0, hf, 2hf, 3hf.... He further assumed that the probability that a given stationary wave would have a particular one of these energies could be calculated from classical statistical mechanics, as before. Thus, he showed the average total energy per oscillator to be

$$W = kT\frac{hc/\lambda kT}{e^{hc/\lambda kT} - 1}\ \text{joules} \tag{4.7}$$

instead of kT joules, as used by Rayleigh and Jeans. Applying this extra factor to the Rayleigh-Jeans Law gives Planck's equation,

$$M_\lambda = \frac{2\pi kTc}{\lambda^4}\frac{hc/\lambda kT}{e^{hc/\lambda kT} - 1} = \frac{2\pi hc^2\lambda^{-5}}{e^{hc/\lambda kT} - 1} \tag{4.8}$$

Planck's assumption that harmonic oscillators can have only discrete energies opened the door to quantum theory. His assumption implies that matter does not radiate energy continuously, as had been previously assumed, but rather in very small discrete amounts. For example, the energy of a visible quantum at 600 nm is

$$hf = \frac{(6.63\times10^{-34})(3\times10^{8})}{6\times10^{-7}} = 33\times10^{-20}\ \text{J}$$

which is indeed very small. In large-scale macroscopic mechanics the energies involved are so large that the tiny discrete changes that are occurring with time cannot be observed. Thus, for example, a sled sliding down a hill in a series of increasing energy increments is seen as having a smooth, continuous motion. This is not the case in atomic, microscopic mechanics, which will be demonstrated when particle radiation and luminescent sources are considered in later chapters.

4.4 STEFAN, BOLTZMANN, AND WIEN

In Fig. 4.5, blackbody radiation curves for several values of temperature are shown. The visible spectrum is labeled to show that portion of each curve to which our eyes are sensitive.

Two important characteristics of blackbody radiation can be seen in Fig. 4.5. First the total radiated power density from a blackbody increases dramatically as temperature increases. Second, as temperature goes up, the peak of the radiation curve shifts toward the shorter wavelengths. Each of these phenomena will be examined in detail.

---

* The preferred symbols for energy and power are Q and Φ, respectively. However, W for work and energy and P for power are well-established in the engineering and physics literature and will be used here. M and E will continue to be used to represent both radiant and luminous power densities.

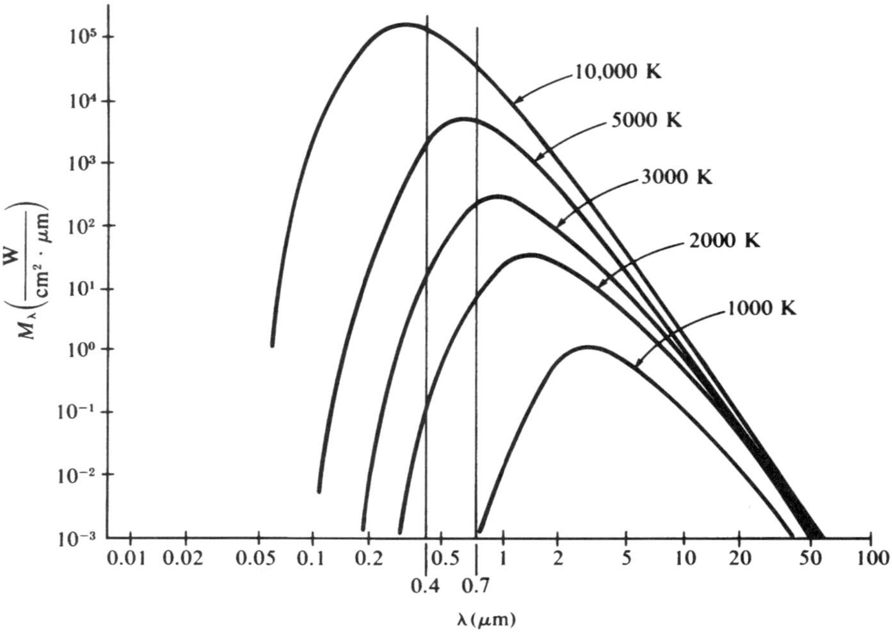

Figure 4.5 Blackbody radiation curves.

The total radiated power density from a blackbody radiator as a function of temperature is obtained by finding the areas under the curves in Fig. 4.5. This is done by integrating Planck's equation over all λ for each temperature.

$$M = \int_0^\infty c_1 \lambda^{-5} (e^{c_2/\lambda T} - 1)^{-1} \, d\lambda \tag{4.9}$$

To evaluate Eq. (4.9), the term in parentheses is expanded into a Taylor series:

$$\frac{1}{e^{c_2/\lambda T} - 1} = e^{-c_2/\lambda T} + e^{-2c_2/\lambda T} + e^{-3c_2/\lambda T} + \cdots \tag{4.10}$$

Then integration term by term and insertion of the limits yield

$$M = \frac{6c_1 T^4}{c_2^4} \left(\frac{1}{1^4} + \frac{1}{2^4} + \frac{1}{3^4} + \cdots \right) \tag{4.11}$$

The quantity in parentheses is an infinite series whose sum can be shown to be $\pi^4/90$. Thus

$$M = \frac{4c_1 T^4 \pi^4}{90 c_2^4} \equiv \sigma T^4 \tag{4.12}$$

which is the Stefan*-Boltzmann law.

* J. Stefan (1835–1893), Austrian physicist.

The value of σ, the Stefan-Boltzmann constant is $5.67032 \times (10^{-12}\ \text{W/cm}^2 \cdot \text{K}^4)$, which gives M in watts per square centimeter.

The result in Eq. (4.12) that the radiated power density of a blackbody radiator depends only on the temperature raised to the fourth power is remarkably simple. Thus, if M at one temperature, call it T_1, is known, M at the second temperature, call it T_2, can be found by writing

$$M_{\text{at } T_2} = \frac{T_2^4}{T_1^4}(M_{\text{at } T_1}) \tag{4.13}$$

EXAMPLE 4.3

The filament of an incandescent lamp is 0.006 cm in diameter and 13 cm long. It consumes 100 W. Assuming that the filament can be considered a blackbody radiator,* at what temperature is it operating? How many watts would it consume at a temperature of 3100 K?

Solution. The surface area of the filament is $2\pi rl = 2\pi(0.003)(13) = 0.245\ \text{cm}^2$. Thus the radiated power density is $M = 100/0.245 = 408\ \text{W/cm}^2$.

From Eq. (4.12),

$$408 = 5.67 \times 10^{-12}\ T^4$$

giving

$$T = 2910\ \text{K}$$

For a temperature of 3100 K, Eq. (4.13) yields

$$P = \frac{3100^4}{2910^4}(100) = 129\ \text{W}$$

Note that input power increases by 29% for a temperature rise of 6.5%.

It was noted that two things happen as the temperature of a blackbody radiator increases. The second is that the peaks of the blackbody radiation curves shift to shorter wavelengths. This result might be expected from personal experience. Suppose the surface unit of an electric stove is set on low heat. In addition to getting warm, it glows a deep red. On medium heat, the unit gets hotter and brighter, but also the color has more orange in it. On high heat the color is definitely orange. Thus the shift toward the visible spectrum is indeed noticeable.

A quantitative measure of this shift of the peak to shorter wavelengths can be derived by reconsidering Eq. (4.4). For temperatures (2000 to 4000 K) and wavelengths (0.30 to 0.80 μm) of interest in dealing with incandescent light sources and the visible and near-visible portions of the spectrum, this equation can be simplified to

$$M_\lambda = c_1\lambda^{-5}e^{-c_2/\lambda T} \tag{4.14}$$

* An incandescent lamp filament is actually a selective radiator, to be discussed in Sect. (4.7). Also a filament 60 cm long would require supports, which would produce temperature gradients along its length. Thus the example is idealized.

which is known as Wien's* Radiation law. What has been done to arrive at Eq. (4.14) is to note that, for the temperatures and wavelengths indicated, $e^{c_2/\lambda T} >> 1$ and thus the 1 may be neglected in Planck's equation. For example at T = 4000 K and $\lambda = 0.8$ μm, $e^{c_2/\lambda T} \approx 90$. For lower temperatures and shorter wavelengths, this quantity will be still larger.

To find the value of λT in Eq. (4.14) at which M_λ is a maximum, the derivtive of M_λ with respect to λ is set equal to zero. The result is

$$-5c_1\lambda^{-6}e^{-c_2/\lambda T} - \frac{c_2}{T}\left(\frac{-1}{\lambda^2}\right)c_1\lambda^{-5}e^{-c_2/\lambda T} = 0 \tag{4.15}$$

yielding

$$\lambda_{max}T = \frac{c_2}{5} = 2878\ \mu m \cdot K \tag{4.16}$$

Inserting this result into Eq. (4.14) gives

$$M_{\lambda_{max}} = \frac{c_1 T^5 e^{-5}}{(2878)^5} = 1.28 \times 10^{-15}\, T^5\ W/(cm^2 \cdot \mu m) \tag{4.17}$$

Equations (4.16) and (4.17) show that as T increases, the point at which M_λ is maximum occurs at shorter wavelengths and $M_{\lambda\,max}$ increases markedly. When T is 2900 K, which is, approximately, the temperature of an incandescent lamp filament, M_λ is a maximum at $\lambda = 1$ μm, well into the infrared region of the spectrum. This is why an incandescent lamp is rich in reds and yellows, but deficient in blues and greens. It is also why it is an inefficient light source. To cause the M_λ peak to occur in the red region of the spectrum say at 0.7 μm, requires that T = 2878/0.7 = 4110 K. But tungsten, the filament material in an incandescent lamp, melts at 3655 K, so operating at this temperature would cause the lamp to expire immediately. This is the light versus life tradeoff that always occurs with an incandescent lamp design. Normally at least a 1000-h life for general lighting lamps is wanted, so less light is accepted. Where maximum light output and more nearly balanced white light are necessary, as in the case of photo flood lamps for studio photography, life is sacrificed and the filament is operated at 3100 to 3400 K.

It should be noted that M_λ in Eq. (4.12) varies as T^4 and $M_{\lambda\,max}$ in Eq. (4.17) varies as T^5. Thus, each 100 K (≈3.5%) increase that a lamp manufacturer can coax out of a filament without impairing its life excessively represents a 14.5% increase in total radiant exitance and an 18.5% increase in the peak value of the spectral radiant exitance curve, as well as a shift of the peak of that curve approximately 33 nm toward the visible portion of the spectrum. Each of these factors works to increase the lm/W efficacy of an incandescent lamp.

EXAMPLE 4.4

A cylindrical filament, assumed to behave as a blackbody, operates on 120V and draws 1.4A. Its diameter is 50 μm, and its length is 20 cm. Find

---

* Wilhelm Wien (1864–1928), German physicist noted for his work in optics and radiation.

a) Radiant exitance
b) Operating temperature
c) λ at which M_λ is maximum
d) Value of $M_{\lambda\,max}$

Solution. The filament consumes $120 \times 1.4 = 168$W, and, because of the assumption of a blackbody, radiates the same amount. The surface area of the filament neglecting its ends, is

$$SA = 2\pi rl = 2\pi \times 25 \times 10^{-4} \times 20 = 0.314 \text{ cm}^2$$

Thus the radiant exitance is

$$M = \frac{168}{0.314} = 535 \text{ W/cm}^2$$

Inserting M into Eq. (4.12) yields

$$535 = 5.67 \times 10^{-12}\, T^4$$

$$T = 3120 \text{ K}$$

Then Eq. (4.16) gives

$$\lambda_{max} \times 3120 = 2878$$

$$\lambda_{max} = 0.92\ \mu\text{m} = 920 \text{ nm}$$

and $M_{\lambda\,max}$, from Eq. (4.17), is

$$M_{\lambda_{max}} = 1.28 \times 10^{-15} \times 3.12^5 \times 10^5$$

$$= 378 \text{ W/cm}^2 \cdot \mu\text{m}$$

4.5 LUMINOUS INTENSITY STANDARD

In Sect. 1.7, it was pointed out that, in the days of pre-electric lighting, a spermaceti candle and later a gas flame were used to define the candle as the unit of luminous intensity.

In 1909, the U.S. National Bureau of Standards (NBS) redefined the candle in terms of a group of 45 carbon filament incandescent lamps having precise filament dimensions and operating at a closely-controlled voltage. This standard was accepted as the “international candle” in both the United States and Europe. While it was an improvement over the candle and the gas flame, it was still not wholly satisfactory because of the difficulty in controlling the several variables involved.

In 1937, the International Committee on Weights and Measures (CIPM) adopted the following resolution: “From January 1, 1940, the unit of luminous intensity shall be such that the brightness of a blackbody radiator at the temperature of solidification of platinum is 60 candles per square centimeter.” Their choice of 60 was to make the “new candle” and the “old candle” nearly the same in brightness. The solidification temperature is 2042 K.

World War II intervened, and little was done to implement the new definition until 1948, when the International Commission on Illumination (CIE) recommended that the new unit of luminous intensity be named the candela. In that same year, the General Conference on Weights and Measures (CGPM) refined the prewar definition to read: "The candela is the unit of luminous intensity. The magnitude of the candela is such that the luminance of a blackbody radiator at the temperature of solidification of platinum is 60 candelas per square centimeter at a pressure of 101,325 Pa (one atmosphere)."

The Waidner and Burgess blackbody standard that was used from 1948 to 1979, is illustrated in Fig. 4.6. The thorium oxide test tube is surrounded by platinum in a thermally insulated housing. The platinum is maintained at a pressure of 101,325 Pa and is heated by an applied electromagnetic field until it melts (the melting temperature of thorium oxide is much higher than that of platinum). Then the platinum is allowed to cool, and the luminance of the thorium oxide is measured in the sight tube at the temperature (2042 K) at which the platinum returns to solid form. This luminance is assigned the value of 60 cd/cm^2, and thus each square centimeter has a luminous intensity of 60 cd.

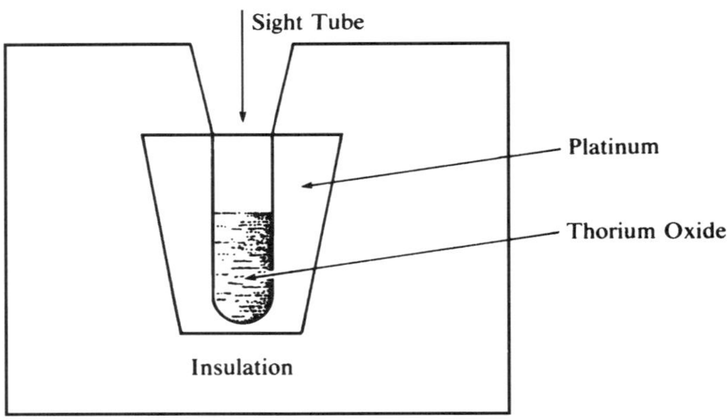

Figure 4.6 Waidner and Burgess standard.

Although the blackbody radiator was superior to the carbon filament lamp as a standard because its radiant power density could be calculated accurately and because the variables associated with its measurement could be more closely controlled, it still suffered from being based, in part, on a photometric measurement (luminance) done at a temperature that was difficult to attain. There was considerable interest worldwide in having an illumination standard that depended only on radiometric quantities (watts). In 1979, the CGPM adopted the following definition: "The candela is the luminous intensity in a given direction of a source which is emitting monochromatic radiation of frequency 540×10^{12} Hz (555 nm) and whose radiant intensity in that direction is $\frac{1}{683}$ W/sr."

What this definition does is relate the candela to a single spectral line of radiation at the peak of the eye's spectral luminous efficiency curve (555 nm). Thus it eliminates the need for a photometric measurement and defines the candela strictly in radiometric terms, which are independent of temperature and can be easily and accurately measured. It remains to show where the value $\frac{1}{683}$ comes from and how the earlier (1948) definition of the candela is consistent with the newer (1979) definition.

4.6 RELATING THE LUMEN AND THE WATT

The watt is the time rate of flow of radiant energy; the lumen is the time rate of flow of luminous energy. To link these two units, the V(λ) function and the blackbody standard radiator are needed. This linkage through the eye is illustrated in Fig. 4.7.

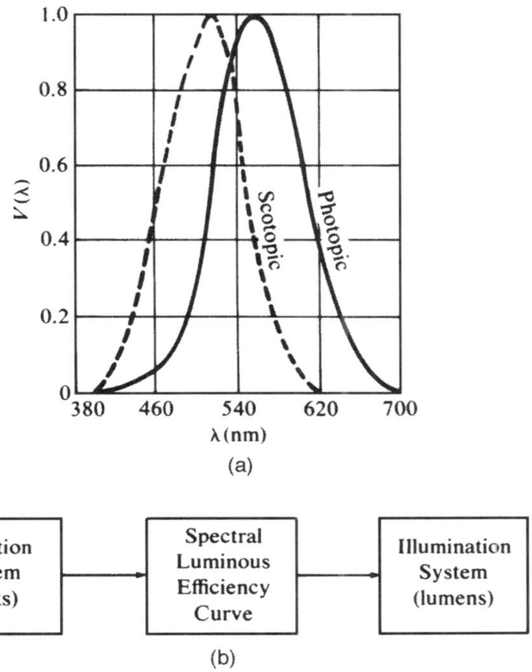

Figure 4.7 Relating watts to lumens.

The spectral luminous efficiency curve for photopic vision is the screen that enables the evaluation of the visual effect of radiant energy. For the purpose of illustrating how such a calculation is made, consider the hypothetical linearized spectral power density and spectral luminous efficiency curves in Fig. 4.8. To obtain the visual effect of this source, the spectral power density curve of the source (M_λ versus λ) is multiplied,

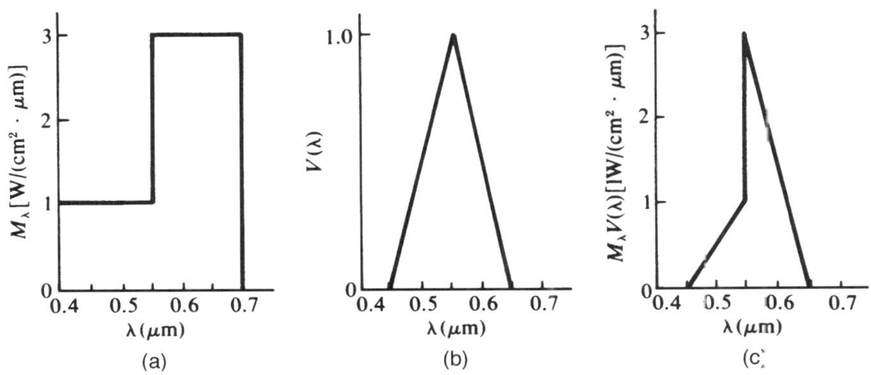

Figure 4.8 Illustrating the calculation of light watts.

point by point, by the spectral luminous efficiency curve of the eye [$V(\lambda)$ versus λ]. In Fig. 4.8 this multiplication is accomplished easily because straight-line representations for both the source and the eye curves have been used. The resulting product curve, $M_\lambda V(\lambda)$ versus λ, is shown in Fig. 4.8c. The quantity $M_\lambda V(\lambda)$ is called "spectral light power density", with units of light watts (lW) per square centimeter per micrometer. The area under this curve is the total light watts emitted by the source, and for the example is $\frac{1}{2}(1)(0.1)+\frac{1}{2}(3)(0.1)=0.20$ lW/cm^2.

Light power density [$M_\lambda V(\lambda)$] must now be related to lumen density. To do this, return to the blackbody standard and recall that it has a luminance of 60 cd/cm^2 at 2042 K. Since it is perfectly diffusing, its luminous exitance is 60π lm/cm^2 at that temperature. If the blackbody spectral power curve for 2042 K is plotted, and multiplied point-by-point by the spectral luminous efficiency curve, as was done in Fig. 4.8, the spectral light power density curve is obtained for the blackbody at that temperature. The area under this curve is then found to obtain the total blackbody light power density. This figure has been calculated accurately to be 0.27598 lW/cm^2.

60π lm/cm^2 is thus required to be equivalent to 0.27598 lW/cm^2. This means that each light watt must generate $60\pi/0.27598 = 683$ lm. Thus, for any source, it is necessary to multiply the number of light watts present by a constant $k = 683$ lm/lW* to get the number of lumens produced. The fictitious source in Fig. 4.8 provides $683(0.20) = 137$ lm/cm^2.

What k shows is that if radiant power could be concentrated at the single wavelength of 555 nm, each watt of such power would yield one light watt and thus 683 lm of yellow-green light. Because radiant power is customarily distributed over a wide band of wavelengths, much of it in the infrared, and because the eye has less response at other wavelengths than it has at 555 nm and has essentially zero response outside the visible spectrum, the luminous efficacy in lumens per watt (lm/W) is generally far below the theoretical maximum. The source in Fig. 4.8 emits 0.6 W/cm^2 (the area under the M_λ graph). Since it also emits 137 lm/cm^2, its luminous efficacy is 228 lm/W (quite high, because all of the watts were in the visible spectrum).

For a radiant power density such that M_λ is constant throughout the visible spectrum and zero at all wavelengths outside the visible spectrum (giving white light), the luminous efficacy is approximately 187 lm/W. To obtain this value, the visible spectrum from 380 to 780 nm is divided into forty 10-nm bands and $\frac{1}{40}$ W is assigned to each band. Then

$$\Phi = 683 \sum_{380}^{780} \frac{1}{40} V(\lambda)\Delta\lambda$$

$$\equiv 683\, V(\lambda)_{av} \tag{4.18}$$

The summation of $V(\lambda)$s is obtained from Table 3.2 and is 10.6858. Then $V(\lambda)_{av} = 10.6858/40 = 0.2671$ and $\Phi = 683(0.2671) = 182$ lm. Thus a single watt of power uniformly distributed between 380 and 780 nm yields 182 lm. In practice, even though M_λ may be reasonably constant throughout the visible spectrum, thus giving white light, M_λ is not zero in the infrared. Consequently, the luminous efficacy is considerably below the 182 lm/W value.

* A similar constant has been developed for the scotopic case with the value 1700 scotopic lumens per watt and was adopted by the CIE in 1951.

For general service incandescent lamps, so much of the radiation is in the infrared that only about 20 lm/W can be achieved in practice. For tungsten-halogen lamps, up to 35 lm/W can be obtained. Greater luminous efficacies can be realized with luminescent sources. Mercury lamps yield up to 65 lm/W, fluorescent lamps up to 100 lm/W, metal halide lamps up to 125 lm/W, and high-pressure sodium lamps up to 140 lm/W.*

The development leading to the constant k allows us to derive expressions for the illumination metrics in terms of their radiant power equivalents. Let M_λ be the average watts per unit of surface area per nanometer emitted by a radiator in the neighborhood of wavelength λ. Then $M_\lambda\, d\lambda$ is the watts per unit area emitted in a range of wavelengths $d\lambda$ wide, centered at λ, and $M_\lambda V(\lambda) d\lambda$ gives the light watts per unit area in this range of wavelengths. The total light watts per unit area are obtained by integration over the visible spectrum. Then, multiplying by k converts the result to lumens per unit area, which is luminous exitance M.

$$M = k \int_{380}^{780} M_\lambda V(\lambda)\, d\lambda \tag{4.19}$$

Similar expressions follow for luminous flux, illuminance, and quantity of light.

$$\Phi = k \int_{380}^{780} P_\lambda V(\lambda)\, d\lambda \tag{4.20}$$

$$E = k \int_{380}^{780} E_\lambda V(\lambda)\, d\lambda \tag{4.21}$$

$$Q = k \int_{380}^{780} W_\lambda V(\lambda)\, d\lambda \tag{4.22}$$

EXAMPLE 4.5

A 250-watt metal halide lamp has three spectral lines at 500 nm, 550 nm, and 600 nm. The watts radiated in these lines are in the ratio 5:3:2. If 20% of the input power to the lamp appears in the lines, find the lumens emitted by the lamp and its lumen per watt efficacy.

Solution. The power in the lines is $250 \times 0.2 = 50$ W. Thus 25 W are at 500 nm, 15 W at 550 nm and 10 W at 600 nm. The light-watt calculations are best presented in tabular form, using $V(\lambda)$ values from Table 3.2

λ nm	P_λ W	$V(\lambda)$	$P_\lambda V(\lambda)$ l-W
500	25	0.323	8.07
550	15	0.995	14.92
600	10	0.631	6.31
			29.30

* A luminescent source requires auxiliary equipment (ballast) that consumes power and thus reduces overall efficacy.

From Eq. (4.20),

$$\varphi = 683 \times 29.30 = 20{,}000 \text{ lm}$$

and

$$\frac{20{,}000}{250} = 80 \text{ lm/W}$$

4.7 SELECTIVE RADIATORS AND COLOR TEMPERATURE

No radiator of a practical sort is truly a blackbody. A graybody radiator is defined as one whose radiation curve is the same shape as that of a blackbody for a given temperature but has a lower radiated spectral power density at each wavelength. For a graybody radiator, Eq. (4.11) becomes

$$M = \varepsilon M_{BB} = \varepsilon \sigma T^4 \tag{4.23}$$

where M_{BB} is the spectral power density of a blackbody, and ε is called the emissivity of the graybody and is a number between 0 and 1. If the radiator is almost a blackbody, ε is almost 1. If the source is a poor radiator, ε is nearly 0. A graybody radiation curve is simply a scaled-down version of a blackbody radiation curve. Its total exitance is given by Eq. (4.12) multiplied by ε, and Eq. (4.16) continues to give λ_{max}. Carbon is an example of graybody radiator.

Consider a unit surface area of a graybody radiator of emissivity ε operating at a temperature T. Let its reflectance be ρ. The radiator emits εM_{BB}, where M_{BB} is the exitance of a blackbody at the same T. Now place the radiator in an opaque enclosure whose walls are at the same temperature T. The graybody is now part of a blackbody and thus its exitance becomes M_{BB}. It receives radiation from the walls (M_w) and reflects a portion of it (ρM_w). Thus,

$$M_{BB} = \rho M_w + \varepsilon M_{BB} \tag{4.24}$$

Because of thermal equilibrium,

$$(1 - \rho) M_w = \varepsilon M_{BB} \tag{4.25}$$

where the left side is absorbed radiation, which must equal emitted radiation on the right side. Solving Eqs. (4.24) and (4.25) gives

$$\varepsilon = 1 - \rho \equiv a \tag{4.26}$$

where a is absorbtion ($0 \le a \le 1$). Eq. (4.26) shows that the best emitter is the best absorber and also the worst reflector.

EXAMPLE 4.6

A cylindrical graybody filament has a diameter of 2.5 mm and a length of 3.0 cm. Its maximum emission is at a wavelength of 900 nm. It operates on 120 volts in a gas-filled incandescent lamp, consumes 600 W, and has 20% losses. Find its emissivity.

Solution. From Eq. (4.16),

$$T = \frac{2878}{0.9} = 3198\,K$$

The surface area of the filament is

$$A_s = \pi d\ell = \pi \times 3.0 \times 0.25 = 2.36\ cm^2$$

The filament radiates 480 W. Its radiant power density is

$$M = \frac{480}{2.36} = 203.4\ W/cm^2$$

Its radiant power density is also given by Eq. (4.12) times ε.

$$M = \varepsilon \times 5.67 \times 10^{-12} \times 3.198^4 \times 10^{12} = 593\varepsilon\ W/cm^2$$

Thus

$$\varepsilon = \frac{203.4}{593} = 0.34$$

A selective radiator is one whose radiation curves differ in shape from those of a blackbody. For example, tungsten is a selective radiator, as shown in Fig. 4.9. It matches the shape of a graybody radiator in the visible and near infrared regions

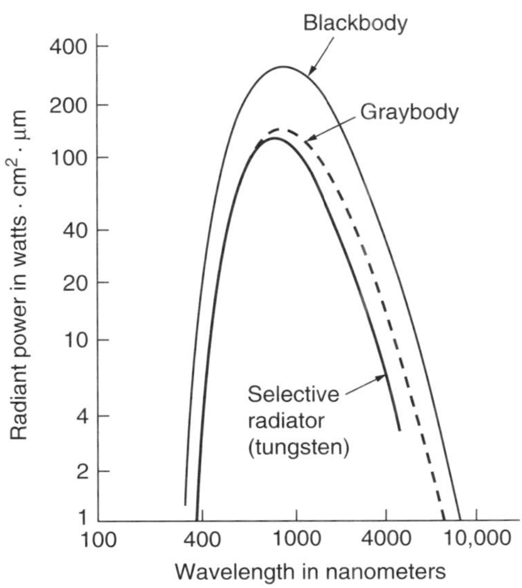

Figure 4.9 Comparison of a blackbody, graybody and selective radiator at 3000 K. *Source*: *IESNA Lighting Handbook*, 9th edition, 2000. Courtesy of the Illuminating Engineering Society of North America.

quite well, but falls considerably below the graybody in the far infrared. Equation (4.23) does not apply to a selective radiator. Rather Eq. (4.4) must be revisited and rewritten to read

$$M_\lambda = \varepsilon(\lambda)M_{\lambda_{BB}} = c_1\varepsilon(\lambda)\lambda^{-5}(e^{c_2/\lambda T} - 1)^{-1} \tag{4.27}$$

where $\varepsilon(\lambda)$ is the spectral emissivity of the selective radiator. Thus, a selective radiator has a different emissivity at each wavelength, whereas a graybody radiator has a single emissivity for all wavelengths.

It is possible to define a weighted-average emissivity for a selective radiator as the ratio of the total radiated power density of the selective radiator to the total radiated power density of a blackbody operating at the same temperature.

$$\bar{\varepsilon} = \frac{\int_0^\infty \varepsilon(\lambda)M_{\lambda_{BB}}\, d\lambda}{\int_0^\infty M_{\lambda_{BB}}\, d\lambda} \tag{4.28}$$

Then the total radiated power density of the selective radiator can be expressed in the form of Eq. (4.23) as

$$M = \bar{\varepsilon}\sigma T^4 \tag{4.29}$$

Values of $\varepsilon(\lambda)$ as a function of λ and of $\bar{\varepsilon}$ as a function of T for tungsten are given in Table 4.1. Note that $\varepsilon(\lambda)$ peaks at 0.4 μm (400 nm) and decreases steadily throughout the infrared as λ increases. This is typical of metals. Note also that $\bar{\varepsilon}$ increases as T increases.

TABLE 4.1 EMISSIVITY AS A FUNCTION OF WAVELENGTH AND TEMPERATURE FOR TUNGSTEN

$\lambda_{(um)}$	$\varepsilon(\lambda)$
0.3	0.43
0.4	0.46
0.5	0.45
0.6	0.43
0.7	0.41
0.8	0.39
1.0	0.36

T(K)	$\bar{\varepsilon}$
2400	0.308
2500	0.316
2600	0.328
2700	0.332
2800	0.336
2900	0.342
3000	0.348

As an example, for tungsten operating at 2900 K, $\bar{\varepsilon} = 0.342$, as indicated in Table 4.1. Inserting this value into Eq. (4.29) gives M = 137 W/cm^2 for tungsten at this temperature.

EXAMPLE 4.7

An idealized radiation curve of a selective radiator operating at 3000 K is shown in Fig. 4.10. Find the watts per square centimeter emitted by the radiator and its emissivity.

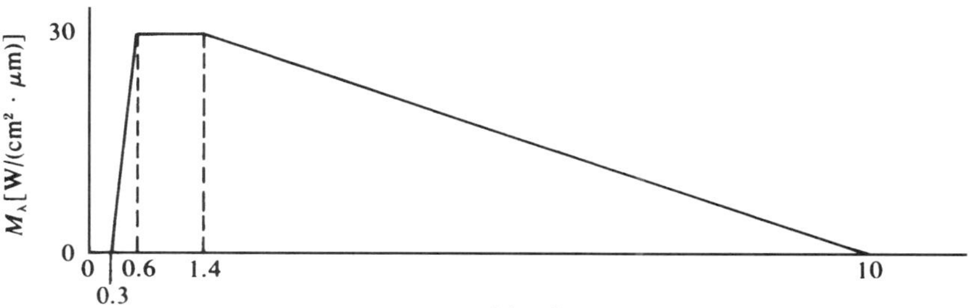

Figure 4.10 Idealized radiator spectral distribution.

Solution. To find M, the radiant exitance from the radiator, the area under the radiation curve is needed. This can be found, in this straight-line case, as the area of two triangles and one rectangle.

$$M = 30[(0.3)](0.5) + (0.8)(1) + (8.6)(0.5)] = 30(5.25) = 157.5 \text{ W/cm}^2$$

From Eq. (4.12), a blackbody at the given temperature would emit

$$M_{BB} = (5.67 \times 10^{-12})(81 \times 10^{12}) = 459 \text{ W/cm}^2$$

Thus, the weighted-average emissivity, from Eq. (4.28), is

$$\bar{\varepsilon} = \frac{157.5}{459} = 0.34$$

EXAMPLE 4.8

A coiled tungsten filament operates at 2800 K in a gas-filled incandescent lamp having 18% losses. The filament may be considered as a cylinder 2.0 mm in diameter and 2.1 cm long. Find its power input.

Solution. The surface area is:

$$A_s = \pi d\ell = \pi \times 0.2 \times 2.1 = 1.32 \text{ cm}^2$$

From Table 4.1, $\bar{\varepsilon} = 0.336$
Then Eq. (4.29) gives

$$M = 0.336 \times 5.67 \times 10^{-12} \times 2.8^4 \times 10^{12}$$
$$= 117 \text{ W/cm}^2$$

The radiated power is

$$P_r = 117 \times 1.32 = 154 \text{ W}$$

and the input power is

$$P_m = \frac{154}{0.82} = 188 \text{ W}$$

In describing the color of radiated power, it is convenient to use the concept of color temperature (CT). This quantity is defined as the absolute temperature of a blackbody radiator that provides a color match between a blackbody and a given radiator. A blackbody is black at room temperature (300 K), red at 800 K, yellowish-white at 3000 K, white at 5000 K, bluish-white at 8000 K, and blue at 60,000 K.

Generally, an exact color match between a blackbody and an arbitrary radiator is impossible. This is because, as will be shown in Chap. 7, two degrees of freedom are necessary to color match two radiators. In this case, there is only one variable, temperature. However, if the color to be matched does not differ appreciably from that of a blackbody, such as in the case of many selective radiators, the concept of correlated color temperature (CCT) can be used. CCT is defined as the absolute temperature of a blackbody at which the closest possible color match with the given radiator is obtained. For example, a 100-W incandescent lamp operating at a filament temperature of 2850 K has a CCT of 2870 K. This means that a blackbody must be operated at 2870 K to provide the closest possible color match to the 100-W lamp at 2850 K. The reason for the 20 K difference is that the spectral emissivity of tungsten is slightly greater in the blue-green region of the spectrum than it is in the yellow-red region.

Luminescent sources and daylight are not heated objects giving off light. Still the concept of CCT may be used. For example, northwest blue sky has a CCT of 25,000 K, which means that a blackbody would have to operate at this temperature to best match the color of this sky. In the area of luminescent sources, a warm white fluorescent lamp has a CCT of 3000 K, whereas the CCT of a cool white fluorescent lamp is 4200 K. The former gives yellowish-white light; the latter more bluish-white. Also, metal halide lamps (CCT = 3700 K) appear much cooler and bluish-white than high pressure sodium lamps (CCT = 2100 K), which have a warmer, yellowish-white appearance.

4.8 FILAMENT DESIGN PARAMETERS

In 1927, Jones and Langmuir developed a set of curves describing the properties of a straight tungsten filament in a vacuum as a function of filament temperature. These curves are shown in Fig. 4.11, where

1. γ is the evaporation rate of tungsten in a vacuum in grams of filament material per square centimeter of filament surface area per second. For gas-filled lamps, the evaporation rate is 2% of the values given in Fig. 4.11. Knowing γ allows the calculation of the life of a filament, which is defined from the arithmetic

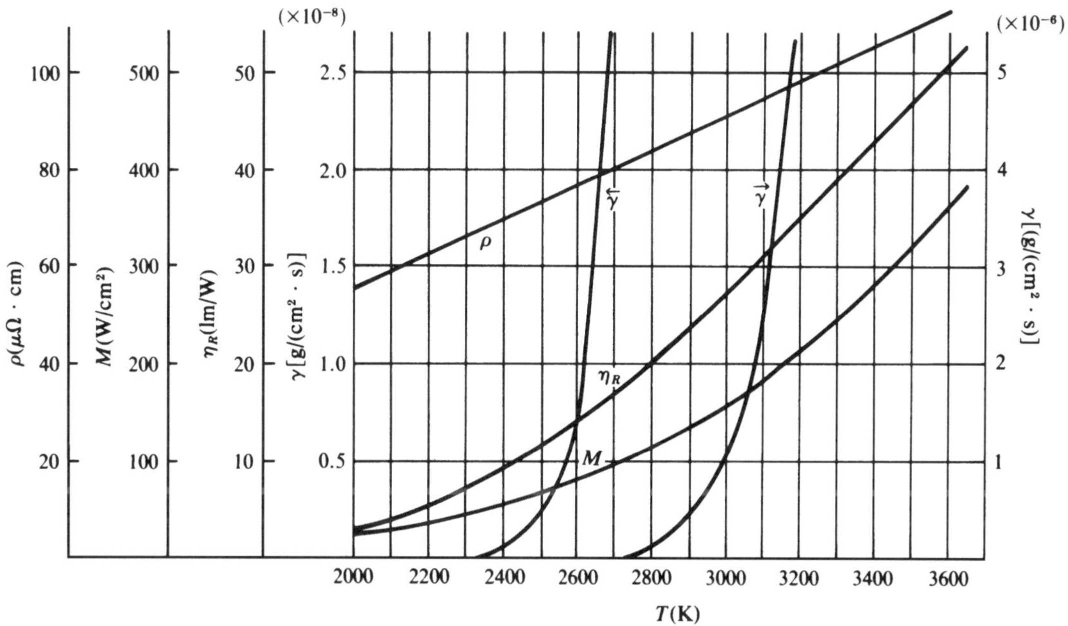

Figure 4.11 **Properties of a straight tungsten filament in a vacuum. *Source*: Jones and Langmuir, "The Characteristics of Tungsten Filaments as Functions of Temperature," GE Review, 30, 1927, pp. 310, 354, 408. Courtesy of General Electric.**

mean of the burning hours of a large group of lamps tested and is the number of hours for 25% of the filament to be evaporated.

2. ρ is the resistivity of tungsten in micro-ohm-centimeters. The resistance of the filament in ohms is given by

$$R = \frac{\rho \ell}{A_c} \times 10^{-6} \qquad (4.30)$$

where ℓ is filament length in centimeters and A_c is filament cross-sectional area in square centimeters.

3. M is the emitted power density in watts per square centimeter of filament surface area.
4. η_R is the luminous conversion efficiency in initial lumens emitted per radiated watt.

The use of these curves will be illustrated by working several examples.

EXAMPLE 4.9

Assume a vacuum incandescent lamp has an uncoiled filament operating at a temperature of 2500 K. Find the following: life, watts consumed, lumens emitted, luminous efficacy, and operating voltage.

The density of tungsten is 19.35 g/cm^3. The filament is 44 cm long and 0.03 mm in diameter at this temperature. Assume losses are 7%.

Solution. From the curves in Fig. 4.11, for T = 2500 K, the following values are obtained

$$M = 70 \text{ W/cm}^2 \qquad \rho = 73.8\ \mu\Omega \cdot \text{cm}$$

$$\gamma = 0.2 \times 10^{-8} \text{ g/(cm}^2 \cdot \text{s)} \qquad \eta_R = 11.5 \text{ lm/radiated watt}$$

The volume, mass, surface area, and cross-sectional area of the filament are

$$v = \pi \frac{d^2}{4} \ell = \pi \left(\frac{9 \times 10^{-6}}{4} \right)(44) = 3.11 \times 10^{-4} \text{ cm}^3$$

$$m = (3.11 \times 10^{-4})(19.35) = 6.02 \times 10^{-3} \text{ g}$$

$$A_s = \pi d \ell = \pi (3 \times 10^{-3})(44) = 0.415 \text{cm}^2$$

$$A_c = \frac{\pi d^2}{4} = 7.07 \times 10^{-6} \text{ cm}^2$$

The evaporation of the filament is

$$r = \gamma A_s\ (0.2 \times 10^{-8})(0.415)(3600) = 2.99 \times 10^{-6} \text{ g/h}$$

Thus, the life of the filament is

$$L = 0.25\frac{m}{r} = \frac{0.25(6.02\times10^{-3})}{2.99\times10^{-6}} = 503\text{ h}$$

The filament radiates

$$P_R = MA_s = 70(0.415) = 29.1\text{ W}$$

With losses included, the input power is

$$P = \frac{29.1}{0.93} = 31.3\text{ W}$$

The luminous efficacy is

$$\eta = 11.5\ (0.93) = 10.7\text{ lm/W}$$

giving a lumen output of

$$\Phi = \eta P = 10.7(31.3) = 335\text{ lm}$$

or

$$\Phi = \eta_R P_R = 11.5(29.1) = 335\text{ lm}$$

From Eq. (4.30), the filament resistance is

$$R = \frac{\rho\ell}{A_c} = \frac{(73.8\times10^{-6})(44)}{7.07\times10^{-6}} = 459\ \Omega$$

Thus

$$V = \sqrt{RP} = \sqrt{459(31.3)} = 120\text{ V}$$

EXAMPLE 4.10

Repeat Ex. 4.9 if the lamp is gas-filled. Assume the losses increase to 17% and that the filament has the same dimensions, but is operating at a new temperature of 2800 K.

Solution. From the curves, at 2800 K,

$M = 117\text{ W/cm}^2$ $\qquad$ $\rho = 84.5\ \mu\Omega\cdot\text{cm}$

$\gamma = 0.1\times10^{-6}\times0.02 = 0.2\times10^{-8}\text{ g/(cm}^2\cdot\text{s)}$ $\qquad$ $\eta_R = 20.3$ lm/radiated watt

Note that, in this case, the filament evaporation rate is the same as before because γ is unchanged, the effect of the increase in filament temperature being exactly offset by the introduction of filling gas. Thus, the life is unchanged at 503 h.

The filament radiates

$$P_R = 117(0.415) = 48.6 \text{ W}$$

giving an input power of

$$P = \frac{48.6}{0.83} = 58.6 \text{ W}$$

The lumen output and luminous efficacy are

$$\Phi = 20.3(48.6) = 987 \text{ lm}, \qquad \eta = 20.3(0.83) = 16.8 \text{ lm/W}$$

The resistance of the filament is

$$R = \frac{84.5(44)}{7.07} = 526\ \Omega$$

requiring an applied voltage of

$$V = \sqrt{526(58.6)} = 176 \text{ V}$$

Note that this lamp is much more efficient, for the same life, than the lamp in Ex 4.9, but requires a much higher voltage.

Examples 4.9 and 4.10 have dealt with the analysis of an already existing filament. An equally interesting problem is that of filament design. Assume the curves in Fig. 4.11 are available and a filament to operate at a voltage V, a power input P, a life L, and a lumen output ϕ is to be designed.

Two design equations are needed to find its length and diameter

$$\pi d\ell = \frac{P_R}{M} \tag{4.31}$$

$$\frac{V^2}{P} = \frac{4\rho\ell}{\pi d^2} \tag{4.32}$$

The first equation is simply two expressions for the surface area of the filament; the second, for the resistance of the filament.

d and ℓ in Eqs. (4.31) and (4.32) can be separated, giving

$$\frac{P}{V^2} \times \frac{P_R}{M} = \frac{\pi d^2}{4\rho\ell} \times \pi d\ell = \frac{\pi^2 d^3}{4\rho} \tag{4.33}$$

$$\frac{V^2}{P} \times \left(\frac{P_R}{M}\right)^2 = \frac{4\rho\ell}{\pi d^2} \times \pi^2 d^2 \ell = 4\pi\rho\ell^3 \tag{4.34}$$

There are four parameters of interest (V, P, L, ϕ), but only two of these can be independent variables, because there are only two equations. A design solution is obtained by trial and error. A temperature is chosen and values are read from the curves in Fig. 4.11. If P and V are then prescribed, φ is established. Then d and ℓ are calculated, and L is determined. The process can be iterated at a new temperature, if desired.

EXAMPLE 4.11

Design a tungsten filament to operate on 120 V ac, consume 500 W, have at least 1000 h life, and provide at least 9000 initial lumens when operated in a gas-filled incandescent lamp. Assume losses of 17.7%.

Solution. Try a temperature of 2800 K. From Fig. 4.11,

$$M = 117\ \text{W/cm}^2 \qquad \gamma = 0.2 \times 10^{-8}\ \text{gm/cm}^2 \cdot \text{s}$$

$$\eta_R = 20.3\ \text{lm/radiated watt} \qquad \rho = 84.5\ \mu\Omega \cdot \text{cm}$$

If P is 500 watts,

$$P_R = 500\ (0.823) = 411.5\ \text{W}$$

and thus

$$\Phi = 20.3 \times 411.5 = 8350\ \text{lm}$$

From Eq. (4.33)

$$\frac{500}{(120)^2} \times \frac{411.5}{117} = \frac{\pi^2 d^3}{4 \times 84.5 \times 10^{-6}}$$

giving

$$d = 0.0161\ \text{cm}$$

and from Eq. (4.34),

$$\frac{(120)^2}{500} \times \frac{(411.5)^2}{(117)^2} = 4\pi \times 84.5 \times 10^{-6}\ \ell^3$$

and thus

$$\ell = 69.5\ \text{cm}$$

Next it is found

$$v = 1.41 \times 10^{-1}\ \text{cm}^3, \qquad m = 0.274\ \text{g}, \qquad A_s = 3.52\ \text{cm}^2$$

The evaporation rate is

$$r = (0.2 \times 10^{-8})\ (3.52)\ (3600) = 2.53 \times 10^{-5}\ \text{g/h}$$

giving a life of

$$L = \frac{0.274(0.25)}{2.53 \times 10^{-5}} = 2700\ \text{h}$$

The first trial has given too long a life and too low a lumen output. Choosing a higher temperature will increase Φ and lower L.

The preceding three examples should be considered as simplifications of the very complex problem of filament design. In practical coiled or coiled-coil filaments, each part of the filament radiates to other parts. Some of this radiation is absorbed; the rest is multiply reflected. Also the filament temperature varies from coil to coil due to conduction to support wires. If the filament is coiled, the curves in Fig. 4.11 theoretically do not apply. But if coil diameter and coil length are used in calculating surface area, radiated power, and filament evaporation, and wire diameter and wire length are used in calculating volume, mass and resistance, the answers will still be reasonable approximations.

EXAMPLE 4.12

A gas-filled incandescent lamp with a coiled filament has the following data:

$$\ell_{fil} = 50\ \text{cm} \qquad d_{fil} = 0.08\ \text{mm}$$

$$\ell_{coil} = 1\ \text{cm} \qquad d_{coil} = 0.1\ \text{cm}$$

$$T = 2950\ \text{K} \qquad D = 19.35\ \text{gm/cm}^3$$

Find the life and lumen output of the filament.

Solution. From the curves in Fig. 4.11,

$$\gamma = 0.75 \times 10^{-6} \times 0.02 = 1.5 \times 10^{-8}\ \frac{\text{gm}}{\text{cm}^2 \cdot \text{sec}}$$

$$M = 150\ \text{W/cm}^2\ \text{radiated}$$

$$\eta_R = 25.5\ \text{lm/W}$$

The coil surface area is

$$A_{s_{coil}} = \pi d_{coil}\,\ell_{coil} = \pi \times 0.1 \times 1 = 0.314\ \text{cm}^2$$

The filament volume is

$$V_{fil} = \frac{\pi d_{fil}\ell_{fil}}{4} = \frac{\pi \times 64 \times 10^{-6} \times 50}{4} = 2.51 \times 10^{-3}\ \text{cm}^3$$

The mass of the filament is

$$m = 19.35 \times 2.51 \times 10^{-3} = 48.6 \times 10^{-3}\ \text{gm}$$

The evaporation rate is

$$r = 1.5 \times 10^{-8} \times 0.314 \times 3600 = 1.70 \times 10^{-5} \frac{\text{gm}}{\text{h}}$$

The filament life is

$$L = \frac{0.25 \times 48.6 \times 10^{-3}}{1.70 \times 10^{-5}} = 715\ \text{h}$$

Power radiated is given by

$$P_R = 150 \times 0.314 = 47.1\ \text{W}$$

giving a lumen output of

$$\Phi = 47.1 \times 25.5 = 1200\ \text{lm}$$

4.9 SPECTRORADIOMETRIC MEASUREMENTS

In previous sections of this chapter, it has been shown how to calculate the power density and spectral power density of a light source. Now consider how to measure these quantities.

An elementary radiometer for measuring power density is shown in Fig. 4.12. This device is a radiation thermocouple and consists of two junctions (J_1 and J_2) connected by fine wire and including an indicating instrument such as a galvanometer G. The junction J_2 is maintained at a constant temperature. The junction J_1 is heated by the incoming radiation, and its temperature is directly proportional to the power density of that radiation. The temperature difference between the two junctions causes a small voltage to be induced in the circuit, which then produces a current in the galvanometer causing its needle to deflect. The deflection is a measure of the power density, and the galvanometer scale can be labeled to give this density directly in microwatts per square centimeter.

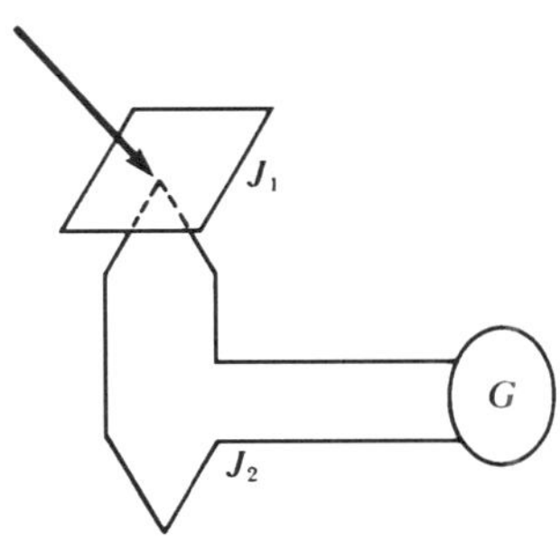

Figure 4.12 A radiation thermocouple.

The radiation thermocouple is an example of a radiometer whose detector is relatively nonselective in wavelength response. Many modern radiometer/photometer combinations use solid-state sensors, such as photodiodes and phototransistors, which are wavelength selective. Also, their overall response is often modified by optical filters, either to produce a specified response over a narrow range of wavelengths, such as duplicating the eye's spectral luminous efficiency curve, or to level the detector's response over a wide range of wavelengths. The overall ranges of spectral response of common detector types are shown in Table 4.2.

TABLE 4.2

Detector	Wavelength Range
Photomultiplier tube	125–1100 nm
Silicon photodiode	200–1200 nm
Intrinsic germanium	0.9–1.5 μm
Lead sulfide	1.0–4.0 μm
Indium arsenide	1.0–3.6 μm
Indium antinomide	2.0–5.4 μm
Doped germanium	2.0–40 μm
Mercury cadmium telluride	1–13 μm

Contemporary radiometer/photometers employ solid-state digital logic and operational amplifier technology throughout. This provides very high sensitivity, excellent linearity over a wide range, stability and reliability over extended periods of time, and good signal-to-noise ratio.

If it is desired to measure spectral power density, a spectroradiometer or spectrophotometer is required. The former measures radiant flux of light sources as a function of wavelength whereas the latter measures the reflectance and transmittance of surfaces and materials as a function of wavelength. The instruments are very similar, with the former designed to measure the spectral content of a source and the latter the spectral reflectance or transmittance of a sample in the presence of a known source.

All spectrophotometers are composed of four basic components:

1. An internal light source and integrating sphere;
2. A light-dispersing system, called a monochromator;
3. A detector system;
4. A sample-processing and readout system.

Item 1 is absent if the device is used as a spectroradiometer.

An elementary spectroradiometer is shown in Fig. 4.13. Radiation from a source enters the instrument at the left. It is imaged at the slit S_1 by a condensing lens system L_1.

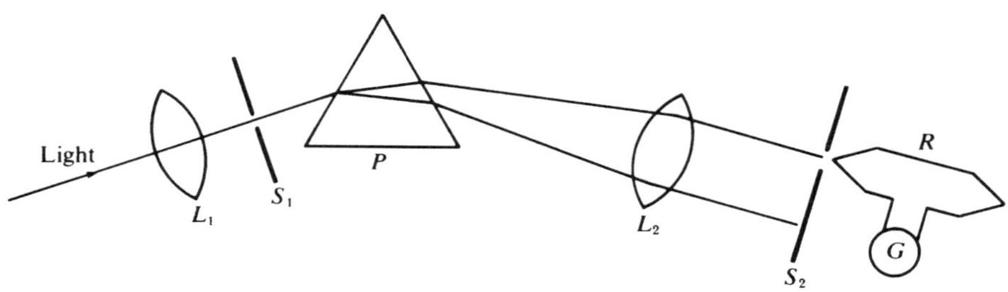

Figure 4.13 A spectroradiometer.

Then it enters a glass prism P, where two things take place. First, the entire beam is refracted from the path it previously followed. Second, the phenomenon of dispersion occurs, in which the radiation is separated into its spectral components. The reason for the dispersion lies, of course, in the fact that the prism glass has a different refractive index for each wavelength, bending red wavelengths the least and violet wavelengths the most. The rays emerging from the prism are somewhat diverging, so a focusing lens system L_2 converges them so that they are parallel. They then enter the slit S_2 and those that pass through impinge on a detecting device, such as a radiation thermocouple R. The slit is adjustable in width, but is generally set at 2 to 4 nm so that the incoming spectrum can be analyzed in reasonably fine detail. The width used depends on such factors as detector sensitivity, signal strength, scan speed, and required spectral resolution. The slit and sensor are movable through a fine gear control.

The galvanometer is calibrated to read in watts per square centimeter per nanometer or per 10 nm, regardless of the actual slit width. This is to provide easy comparison of data from one spectral curve to another.

The parts of a spectrophotometer are now discussed in greater detail. First, consider choosing a source. The most commonly used sources and their wavelength ranges are shown in Table 4.3. Xenon sources have a high source luminance and small size and when used with relatively narrow slits produce a high monochromator output. Mercury sources are similar in intensity to xenon sources, but because they produce line spectra, are not suitable for wavelength scanning applications. Tungsten-halogen lamps have a wider wavelength range than xenon lamps, but have much larger source size and less luminance. Deuterium sources are ideal for ultraviolet spectroradiometry since they have small source size and very little long-wavelength radiation. Silicon carbide resistors provide approximate blackbody radiation at 1000 to 1200 K and are useful for infrared spectroradiometry.

TABLE 4.3

Source	λ Range
Xenon	200–700 nm
Mercury	Line spectrum
Tungsten-halogen	300–2500 nm
Deuterium	180–400 nm
Silicon carbide	2–30 μm

The function of the monochromator (the light-dispersing system) is to separate the source radiation into intense narrow wavelength bands of 10 nm or less. A typical monochromator is shown in Fig. 4.14. The input lens-aperture system is designed to perform two main functions:

1. Send as much radiation as possible through the slit S_1.
2. Just fill the acceptance cone of the monochromator. This is the cone defined by the angle θ in Fig. 4.14.

Radiation emerging from S_1 is reflected to a concave mirror and then to a diffraction grating, which is used instead of a prism in most contemporary monochromators because it produces a linear dispersion. Following reflection from the concave mirror,

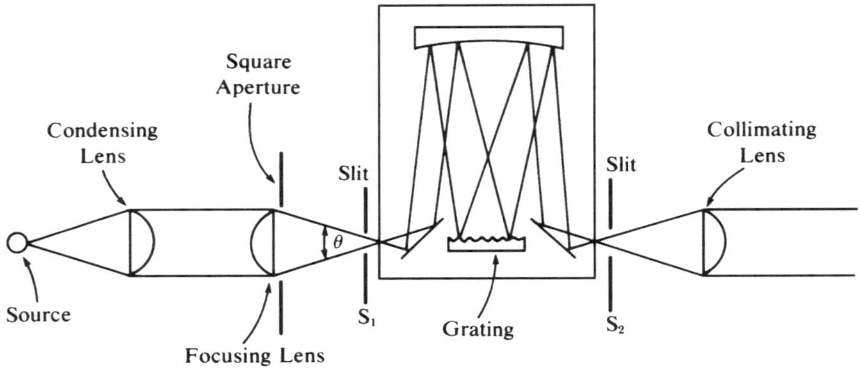

Figure 4.14 Monochromator detail.

the beam exits through S_2 and is collimated by a lens placed one focal length from S_2. A subsequent optical system directs it to the detector.

Detector types have been presented previously in Table 4.2. In contemporary spectradiometers, the detector system is entirely automated, with the spectral scanning and the detector output signal amplification and processing all under microcomputer control.

In a device called a spectrograph, the exit slit (S_2) and detector are replaced with photographic film. The exposed film presents the wavelengths in the source spectrum. This is particularly helpful in showing the structure of line spectra. In many contemporary spectrographs, the film is replaced by silicon photodiode arrays, scanned electronically to yield spectral power densities. Such arrays permit measurements to be taken much more rapidly than in slit-scanning systems.

REFERENCES

Bitter, F.: *Currents, Fields and Particles*, John Wiley, New York, 1957.

Boast, W.B.: *Illumination Engineering*, McGraw-Hill, New York, 1953.

Coaton, J.R., and A.M. Marsden: *Lamps and Lighting*, Fourth Edition, Arnold, London, 1997.

Dicke, R.H., and J.P. Wittke: *Introduction to Quantum Mechanics*, Addison-Wesley, Reading, MA, 1961.

"IES Guide to Spectroradiometric Measurements," *JIES*, April, 1983.

Gilbert, R.G.: *Optimizing Monochromator Performances*, Oriel Corporation, Stanford, Conn.

Halliday, D., and R. Resnick: *Physics, Third Edition*, John Wiley and Sons, New York, 1978.

Hawking, S.W.: *A Brief History of Time*, Bantam, New York, 1988.

Moon, P.: *The Scientific Basis for Illuminating Engineering*, McGraw-Hill, New York, 1936.

Rea, M.S., editor: *IESNA Lighting Handbook*, Ninth Edition, IESNA, New York, 2000.

Waymouth, J.F., and R.E. Levin: *Designers Handbook—Light Source Applications*, GTE Products Corp., Danvers, Mass., 1980.

Chapter 5

Radiant Energy as Particles

5.1 DISCREPANCIES IN WAVE THEORY

In Chap. 1 it was pointed out that as the year 1900 approached, the electromagnetic theory of Maxwell was widely accepted, and it was felt that very little additional information would be added in the future to the knowledge of light.

However, there were two major radiation phenomena (and a few phenomena in other areas of physics) that were not explained by Maxwell's wave theory. One of these was the radiation from a blackbody. As discussed in Chap. 4, Planck's work in deriving the equation for a blackbody radiator led him to the conclusion that light energy is contained in discrete packets called quanta, with each quantum having an energy in joules given by

$$W = hf = \frac{hc}{\lambda} \tag{5.1}$$

A second phenomenon involving electricity and light that was not explained by electromagnetic wave theory was the photoelectric effect, the emission of electrons from the surfaces of metals when illuminated by visible energy. The experimental evidence concerning the photoelectric effect was as follows:

1. When a beam of light falls on a photosensitive surface, electrons are emitted from the surface with velocities ranging from 0 to some value υ_{max}.
2. υ_{max} is independent of the intensity of the incident light, but is inversely proportional to its wavelength. Thus, a weak beam of short wavelength causes a greater emission velocity than a strong beam of long wavelength.
3. The photoelectric current increases as the intensity of the incident light increases.
4. For a given photoelectric surface, there is no emission of electrons if the wavelength of the incident light is greater than λ_c, where λ_c is defined as the critical wavelength for that surface.

This experimental evidence was contrary to electromagnetic theory, which states that an increase in the intensity of the light (an increase in the electric field) should increase the energy imparted to the electrons and thus increase their emission velocities from the surface. This is not what occurred.

In 1905, Albert Einstein (1879–1955), drawing on Planck's earlier work in 1900, published three papers which revolutionized physics and provided the foundation for the atomic age. One of these, which formed the cornerstone of what was soon to be called "quantum theory," was an explanation of the photoelectric effect. Using Eq. (5.1),

Einstein stated that a single quantum of radiant energy may collide with an electron in the surface of a metal and transfer its energy to the electron in an all or none exchange. If the energy exchange takes place, the photon ceases to exist and the electron is ejected from the metal with a velocity given by

$$\frac{1}{2}m_e\upsilon^2 \leq hf - W_m \tag{5.2}$$

In this inequality, the term on the left is the kinetic energy of the departing electron and υ is its ejection velocity. On the right side of the inequality, hf is the energy of the incident photon, and W_m is the "work function" of the metal, the least amount of energy required to eject an electron from its surface. The inclusion of the inequality sign in Eq. (5.2) is explained by noting that, at a given temperature, electrons in a metal exist at energies from 0 to some maximum value. If a photon collides with an electron having the maximum energy, the equal sign in Eq. (5.2) holds. Otherwise the inequality sign holds.

The four pieces of experimental evidence concerning the photoelectric effect can now be explained:

1. If a beam of monochromatic light with hf greater than W_m impinges on a metal surface, electrons will be emitted with velocities up to a maximum of

$$\upsilon_{max} = \sqrt{\frac{2(hc/\lambda - W_m)}{m_e}} \tag{5.3}$$

2. υ_{max} is independent of the intensity of the light, but does depend (inversely) on wavelength.
3. The photoelectric current is the number of electrons per second ejected from the surface, which is a function of the number of photons per second striking the surface (the intensity of the beam).
4. There is no ejection of electrons if the wavelength of the incident radiation is above a certain value, given by setting $\upsilon = 0$ in Eq. (5.2) to obtain

$$\lambda = \frac{hc}{W_m} \equiv \lambda_c \tag{5.4}$$

While it is tempting to suggest that knowledge of radiation has come full circle—from the corpuscular (particle-oriented) theory of Newton to the electromagnetic (wave-oriented) theory of Huygens and Maxwell, to the quantum (particle-oriented) theory of Planck and Einstein, this would be an oversimplification. The quanta of Planck and Einstein are quite different from the corpuscles of Newton and, in fact, do not even obey Newton's laws of motion.

As suggested in Sect. 1.5, it turns out that both electromagnetic wave theory and quantum theory are needed to adequately describe light and other radiant energy. In other words, light must be thought of as being dualistic in nature. Light propagation is best treated as waves, while the interaction of light and matter is best considered as particles.*

* Newton anticipated the dual nature of light by noting, "The interchangeability of bodies and light conforms with the essence of nature which seems delighted with transmutations." Much later Einstein stated," "Light is composed of quanta like electricity is composed of electrons."

5.2 ELECTRONVOLTS

In dealing with energies of electrons, a more convenient measure of energy is the electronvolt (eV), rather than the joule. In electrical terms, a joule is the energy gained by a *coulomb* of electric charge as it falls through a voltage of one volt. Similarly, an electronvolt is the energy gained by an *electron* in falling through a voltage of one volt. Since $q_e = -1.60 \times 10^{-19}$ C is the charge on the electron, the joule and the electronvolt are related by

$$1 \text{ eV} = 1.60 \times 10^{-19} \text{ J} \tag{5.5}$$

With the electronvolt defined, Eq. (5.1) can be rewritten as

$$W = \frac{hc}{\lambda} \text{ J} = \frac{hc}{q_e \lambda} \text{ eV} = \frac{1240}{\lambda} \text{ eV} \tag{5.6}$$

where λ is in nanometers.

EXAMPLE 5.1

A tungsten surface having a work function of 4.52 eV is irradiated by a quantum of radiation from mercury at 253.7 nm. Find the maximum velocity of the emitted electron.

Solution. Using Eq. (5.2), with the equal sign, and Eq. (5.6), we have

$$\frac{1}{2} m_e \upsilon^2 = \frac{1240}{\lambda} - W_m$$

where the energies are expressed in electronvolts. Inserting values gives

$$\frac{1}{2} m_e \upsilon^2 = \frac{1240}{253.7} - 4.52 = 0.36 \text{ eV} = 0.576 \times 10^{-19} \text{ J}$$

Then

$$\upsilon_{max} = \sqrt{\frac{2(0.576 \times 10^{-19})}{9.11 \times 10^{-31}}} = 3.56 \times 10^5 \text{ m/s}$$

EXAMPLE 5.2

One-tenth of one percent of the electric energy supplied to a 50-W mercury lamp appears in the spectral line at 253.7 nm. Find the number of photons per second emitted by the lamp at this wavelength.

Solution. 0.001(50) = 0.05 W emitted at 253.7 nm. From Eq. (5.1)

$$W = \frac{hc}{\lambda} = \frac{(6.63 \times 10^{-34})(3 \times 10^8)}{253.7 \times 10^{-9}} = 0.784 \times 10^{-18} \text{ J/photon}$$

With the number of joules per second (watts) emitted and the number of joules per photon now known, their ratio gives the number of photons per second.

$$\frac{0.05}{0.784 \times 10^{-18}} = 6.38 \times 10^{16} \text{ photons/s}$$

Note the immensity of this number.

In Sect. 1.5 the Rutherford-Bohr model of the atom was reviewed and Table 1.1 was developed to show the energy levels of hydrogen in joules. These levels are converted to electronvolts, and the 0 energy reference is shifted from n = ∞ to n = 1 in Table 5.1. This causes all energy levels to be positive, making it easier to calculate transitions from one orbit to another. Table 5.1 can be used to construct an energy-level diagram for hydrogen as shown in Fig. 5.1.

TABLE 5.1 ENERGY LEVELS OF HYDROGEN

n	TE (eV)	TE + 13.53 (eV)
1	−13.53	0
2	−3.38	10.15
3	−1.50	12.03
4	−0.85	12.68
5	−0.54	12.99
∞	0	13.53

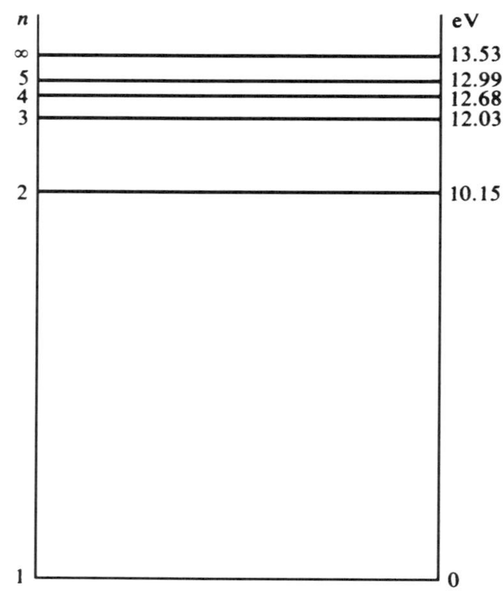

Figure 5.1 Energy-level-diagram of hydrogen.

Equation (1.18) gives the energy of a photon resulting from a transition from one energy level in an atom to a lower one. Equation (5.6) is the relationship between the energy of a photon in electronvolts and its wavelength in nanometers. Combining these two equations gives

$$\lambda = \frac{1240}{W_i - W_f} \tag{5.7}$$

as the wavelength in nanometers of a photon emitted when an electron in an orbit with energy W_i electronvolts falls toward the nucleus to an orbit of lesser energy W_f electronvolts.

EXAMPLE 5.3

An electron traveling at 2×10^6 m/s collides with a hydrogen atom. Find:

(a) The energy level to which the hydrogen electron is likely to be raised.
(b) The λ of the photon emitted.
(c) The velocity of the departing electron.

Solution.

(a) The KE of the arriving electron is

$$KE = 0.5(9.11 \times 10^{-31})\ (4 \times 10^{12}) = 1.822 \times 10^{-18}\ J = 11.39\ eV$$

From Fig. 5.1 this can raise the hydrogen electron to the n = 2 level (10.15 eV). (There is not enough energy present to raise the hydrogen electron to the n = 3 level).

(b) The hydrogen valence electron will return to its rest state (n = 1) emitting

$$\lambda = \frac{1240}{10.15} = 122\,\text{nm}$$

(c) The KE remaining after the collision is 11.39 – 10.15 = 1.24 eV = 1.984×10^{-9}J. Thus

$$1.984 \times 10^{-19} = 0.5(9.11 \times 10^{-31})\upsilon^2$$

$$\upsilon = 0.66 \times 10^6\ \text{m/s}$$

5.3 EXTENSIONS OF BOHR'S MODEL

The Bohr model was very successful in explaining the observed spectral lines of hydrogen and a few other elements, but it was unsuccessful in explaining the spectra of multi-electron atoms and of molecules. Moreover, even for hydrogen, it gave incorrect values for angular momentum.

At this point, research proceeded along two paths:

1. Extensions of the Bohr model by Somerfeld, Zeeman, Goudsmit, and Uhlenbeck.
2. Development of wave and quantum mechanics by DeBroglie, Schrodinger, Heisenberg, and Dirac.

The first of these developments will be reviewed in this section and the second in the following section.

It soon became obvious that a single quantum number n was not sufficient to describe all that goes on in each electron orbit. Thus, it was necessary to modify Bohr's work in the following ways:

1. Elliptical, rather than circular, orbits were assumed, with the nucleus at one of the foci of the ellipse.
2. A three-dimensional, rather than two-dimensional, model of the atom was allowed, in which the planes of the electron orbits were inclined at certain angles with respect to the earth's magnetic field.
3. The spin of an electron about its axis was taken into account.

Each of these modifications resulted in a separate quantum number, so that each electron ended up with four quantum numbers, one to describe its main energy level, one to describe its angular momentum and the shape of its orbit, one to describe the inclination of its orbit with respect to the earth's magnetic field, and the last to describe its spin and spin momentum. The measurable effect of these refinements in the single-quantum-number Bohr model was to split each main energy level into several levels of differing energy, thus splitting the spectral lines.

The first change in the Bohr model was to allow elliptical orbits, and was introduced by Arnold Somerfeld, a German physicist. In a circular orbit, the electron has no velocity or momentum in a radial direction. With an elliptic orbit, the electron has two kinds of momentum, one angular and one linear. Somerfeld postulated that each of these momenta must be quantized, requiring two quantum numbers. These are specified as:

1. n = principal quantum number = 1, 2, 3, ... This quantum number quantizes the main energy levels, similar to the Bohr model.
2. $$l = \text{orbital quantum number} = 0, 1, 2, \ldots, n-1 \tag{5.8}$$

   This quantum number subdivides a main energy level into two or more energy levels, defines the shape of the orbit, and quantizes the electron's angular momentum. Its maximum value is $n - 1$ because it does not include energy associated with linear radial momentum.

A second change was to allow tilted orbits. It had been observed that an atom in a magnetic field had a more complicated emission spectra than the same atom in the absence of a magnetic field. This splitting of energy levels in the presence of a magnetic field is called the Zeeman effect after its discoverer, Pieter Zeeman, a Dutch physicist.

Actually it is quite reasonable that an atom should be considered as three-dimensional with the electron orbits at various inclination angles with respect to some reference axis. But what reference axis should be used? All atoms on earth are subject to the earth's magnetic field. Zeeman postulated that these tilted orbits could assume only certain inclination angles with respect to this field. These angles are determined by the quantum number m, which quantizes the components of orbital angular momentum in the direction of the field. The only permissible values of this quantum number are

$$m = 0, 1, 2, \ldots, \pm l \tag{5.9}$$

Thus for each value of l, there are $2l + 1$ values for m, each representing a slightly different energy level.

This quantizing of angular momentum in the direction of the earth's magnetic field can be shown diagrammatically. Assume $l = 3$ and refer to Fig. 5.2, where each of the seven radial l arrows is 3 units long. The inclination of each of the arrows is such that its projection on the vertical axis, which is the direction of the earth's magnetic field is 0, ±1, ±2, or ±3. This agrees with Eq. (5.9) which says that when $l = 3$, m can have seven integer values only. To conclude, l quantizes angular momentum in general, while m quantizes that portion of angular momentum in the direction of the earth's magnetic field.

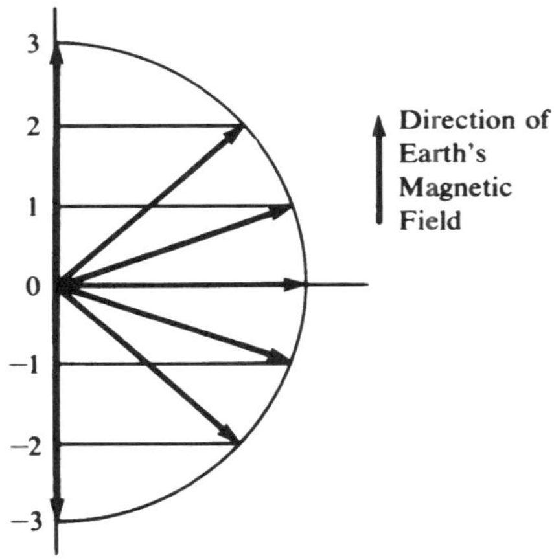

Figure 5.2

The third change was proposed by George Uhlenbeck and Samuel Goudsmit, Dutch physicists. They found that some of the earlier discrepancies in the elliptic-orbit model of Somerfeld could be resolved if the electrons in the orbits were allowed to spin about their axes. This spinning produces a small magnetic field, which is either in the same direction as the earth's magnetic field or exactly opposite to it, depending on the direction of the spin. Like orbital angular momentum, spin angular momentum is quantized. The spin quantum number is s and has the two values $\pm^1/_2$, corresponding to permitted angular momenta of $\pm^1/_2$ $(h/2\pi)$ joule-seconds.

The inclusion of spin splits each energy level and sublevel in an atom into two energy levels very close together. The energy level for $s = +^1/_2$ is slightly higher than that for $s = -^1/_2$.

There are now four quantum numbers, n, l, m, and s, with which to describe the state of a given electron within an atom. Before the quantum states of all electrons in an atom can be defined, a very important principle postulated by Wolfgang Pauli, Austrian physicist (1900–1958), in 1925 is needed. This is the Pauli Exclusion Principle, which states simply that no two electrons in an atom can have the same four quantum numbers.

The four quantum numbers, the restrictions of Eqs. (5.8) and (5.9), and the Pauli exclusion principle, can now be used to construct the various shells and subshells permitted in an atom.

1. *First shell ($n = 1$).* When $n = 1$, the only possible value for l is 0. Thus m is 0. The quantum number s can be $\pm^1/_2$ so there can be a maximum of two electrons in the first shell.

2. *Second shell (n = 2).* When n = 2, l can be 0 or 1.
   (a) For $l = 0$, m = 0, s = ±½. This gives a maximum of two electrons for this subshell, as in the first shell.
   (b) For $l = 1$, m can be −1, 0, +1, and s can be ±½. This gives a maximum of six electrons in this subshell.

   Thus, the n = 2 shell can have a maximum of eight electrons.
3. *Third shell (n = 3).* When n is 3, l can be 0, 1, or 2.
   (a) For $l = 0$, two energy states are permissible as before.
   (b) For $l = 1$, six energy states are permissible as before.
   (c) For $l = 2$, m = −2, −1, 0, 1, 2 and s = ±½. Thus, there are ten permitted energy states.

   For the entire n = 3 shell, there can be a maximum of 2 + 6 + 10 = 18 electrons. Additional shells are constructed similarly.

The elements that are of interest from a lighting standpoint are shown in Table 5.2, where the distributions of the electrons among the energy shells are summarized. An atom always seeks its lowest energy state, so the inner shells fill before electrons are permitted in the outer shells. Thus, most atoms have an inner core of tightly-bound electrons surrounded by an outer shell containing the remaining electrons. It is generally these outer-shell electrons that are affected by collisions of the atom with fast-moving electrons.

TABLE 5.2 DISTRIBUTION OF ELECTRONS IN THE LIGHTING ELEMENTS

	Atomic	Main Energy Level n					
Element	Number	1	2	3	4	5	6
Helium	2	2					
Carbon	6	2	4				
Neon	10	2	8				
Sodium	11	2	8	1			
Sulfur	16	2	8	6			
Argon	18	2	8	8			
Scandium	21	2	8	9	2		
Krypton	36	2	8	18	8		
Indium	49	2	8	18	18 + 3		
Xenon	54	2	8	18	18 + 8		
Neodymium	60	2	8	18	22	8	2
Dysprosium	66	2	8	18	28	8	2
Tungsten	74	2	8	18	32	12	2
Mercury	80	2	8	18	32	18 + 2	
Thallium	81	2	8	18	32	18 + 3	
Thorium	90	2	8	18	32	19	9 + 2

5.4 QUANTUM MECHANICS AND QUANTUM THEORY

Quantum theory occupied the first quarter of the twentieth century; quantum mechanics the second. The work of Planck, Einstein, Rutherford, Bohr, Sommerfeld, and others, which has been discussed previously, forms the basis for quantum theory. It is not the purpose in this section to engage in a lengthy discussion of quantum mechanics.

Rather, what it entails and how it differs from its simpler, more intuitive, and more familiar predecessor will be explained briefly.

Quantum mechanics brings together the wave mechanics of the German physicist, Erwin Schrodinger, the matrix mechanics and the uncertainty principle of another German physicist, Werner Heisenberg, and Einstein's relativity theory, as related to quantum mechanics by P.A.M. Dirac, the British physicist. Quantum mechanics differs from quantum theory in the following ways:

1. It is essentially mathematical in nature, describing atoms in terms of partial differential equations rather than solar system-like models.
2. It makes fewer arbitrary assumptions. Quantum theory contained a number of these that matched experimental evidence.
3. It deals with readily observable and measurable spectral lines rather than with less-accessible electron orbits.
4. It does not rely on Newtonian mechanics but instead is self-contained and consistent.
5. It explains the intensities of spectral lines (which quantum theory does not).

In 1925, the French physicist Louis DeBroglie postulated that, if light rays have particle properties, then particles should exhibit wave properties. He stated that what he called "wave-particle duality" was not confined to photons, but rather is a characteristic of all fundamental particles, such as electrons, atoms, protons, and molecules.

Reconsider the photon. Its energy is given by Eq. (5.1), and it travels at the speed of light. Following DeBroglie's postulate, a photon should behave like a particle and thus have mass. From Einstein's relativity theory (another of his famous papers in 1905), the equivalence of mass and energy can be simply stated as

$$W = mc^2 \tag{5.10}$$

The two energies for the photon should be equal:

$$mc^2 = \frac{hc}{\lambda}$$

Thus

$$\lambda = \frac{h}{mc} \tag{5.11}$$

Equation (5.11) was expressed for a photon. DeBroglie applied it more generally to any particle, obtaining

$$\lambda = \frac{h}{m\upsilon} \tag{5.12}$$

Thus, the wavelength of any moving particle is the ratio of Planck's constant to the particle's linear momentum.

In 1927, C. Davisson and L. Germer of the Bell Laboratories demonstrated the diffraction of electrons by a crystal of nickel, thus establishing their wavelike property.

EXAMPLE 5.4

What is the wavelength of an electron that is traveling with a velocity of 10^4 m/s? What is the mass of a photon having that wavelength? What is the wavelength of a 200-g baseball traveling at 30 m/s?

Solution. From Eq. (5.12),

$$\lambda = \frac{h}{m_e \upsilon} = \frac{6.63 \times 10^{-34}}{9.11 \times 10^{-31} \times 10^4} = 0.728 \times 10^{-7}\ \text{m} = 72.8\,\text{nm}$$

This radiation is in the far ultraviolet region of the spectrum.

From Eq. (5.11),

$$m = \frac{h}{\lambda c} = \frac{6.63 \times 10^{-34}}{(72.8 \times 10^{-9})(3 \times 10^8)} = 3.04 \times 10^{-35}\ \text{kg}$$

Note that the photon has a very tiny mass, much less than that of the electron.

For the baseball,

$$\lambda = \frac{6.63 \times 10^{-23}}{0.2(30)} \approx 10^{-34}\ \text{m}$$

This wavelength is too small for ordinary observation.

Recall that Bohr's third postulate stated that the angular momentum of an electron must be an integral multiple of $h/2\pi$. With DeBroglie's wave particle duality, it is possible to derive this relationship, previously given without proof in Eq. (1.19).

According to Eq. (5.12), a particle of mass m moving with a velocity υ has a distinct wavelength associated with it. If the particle is an electron in a circular orbit around the nucleus of an atom, it is reasonable to assume that the circumference of the orbit must be an integral number of these electron wavelengths, as shown in Fig. 5.3, where four wavelengths per circumference are chosen. If this were not the case, an electron's wave motion would tend to interfere with itself as the electron completed one circumference and started another. This quantization of the orbit may be stated mathematically as

$$2\pi r = n\lambda \tag{5.13}$$

Now combining Eqs. (5.12) and (5.13) yields the result in Eq. (1.19), which was anticipated by Bohr 12 years earlier.

DeBroglie's work was the first radical departure from classical mechanics. It was followed in 1926 by Schrodinger's development of a mathematical analysis of the atom, which he called wave mechanics. Schrodinger essentially replaced all of quantum theory with a single, rather complex, partial differential wave equation. Schrodinger's equation had certain discrete ("eigenvalue") solutions which he showed were identical with the quantum numbers of quantum theory.

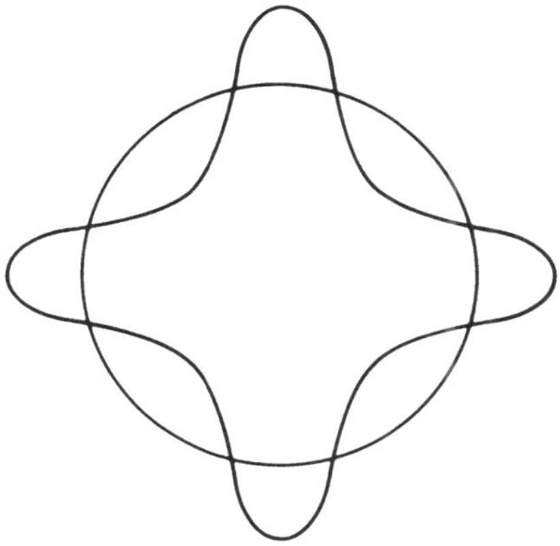

Figure 5.3

At about the same time, Heisenberg developed what he called matrix mechanics. Heisenberg's model of the atom was also mathematical, but was a matrix of rows and columns of quantities representing the locations and velocities of electrons within an atom. The rows were the quantum numbers of quantum theory. These matrices could be solved for the spectral lines of an atom.

Schrodiner subsequently showed that wave mechanics and matrix mechanics were different mathematical models of the same thing, and he blended the two theories into what has since become known as quantum mechanics.

Some physical interpretations of quantum mechanics can be given. In quantum mechanics, there is still the atomic nucleus. Surrounding it are groups of stationary waves, which have their maximum amplitudes at locations matching the orbits of quantum theory. An electron's location is determined probabilistically: there is a certain probability that an electron will occupy a given space at a given time. Each electron can be thought of as being somewhere between two spheres, with the mean radius between them the most probable distance of the electron from the nucleus.

5.5 EXCITATION, IONIZATION, AND RADIATION

Under normal conditions, an atom is electrically neutral and in its rest state. The number of orbiting electrons equals the number of protons in the nucleus and all orbiting electrons are at their lowest possible energy levels (in those orbits nearest the nucleus).

If sufficient energy is added to an atom, through collision with an electron, with another atom, or by some other means, one or more electrons in the atom may be raised to higher energy levels in orbits farther from the nucleus. It can then be said that the atom is in an *excited state* and the energy so applied is the *excitation energy*.

The outermost (valence) electrons in an atom are the most likely to be excited. This is true because:

1. The next higher permitted levels are empty and thus there are no electrons tending to repel the valence electron back to its original level.
2. The difference in energy between the valence electron level and the first vacant level is small compared with the difference in energy between lower levels and the first vacant level or between lower levels and the valence electron level.

Once excited, the electron normally remains in its elevated state for a brief instant (10^{-8} s) before returning to a lower energy state and releasing a photon, whose wavelength is given by Eq. (5.7). However, there are energy levels in atoms from which transitions cannot occur downward directly. These are called *metastable states*. An electron excited to or falling to a metastable state must sit there until it receives additional energy to lift it to a higher state, from which it can fall to a lower state. Residence in a metastable state generally lasts about 10^{-2} s.

If the energy supplied to the atom is sufficient to hoist an electron to the $n = \infty$ level, the electron will be separated from the atom, and a positive ion will remain. It is said that the atom has become *ionized* and the least amount of energy required to do this is denoted as the *ionization energy*.

The ionization and first excitation energies for several of the elements included in Table 5.2 appear in Table 5.3. Note that the ionization energies for the inert gases are higher than for the other elements. This is because the outermost level for each of the inert gases is a complete subshell and thus more energy is required to remove an electron than if this were not the case. As a counterexample, it is quite easy to remove the single outer shell electron from sodium.

TABLE 5.3 EXCITATION AND IONIZATION ENERGIES OF LIGHTING ELEMENTS

Element	First Excitation Energy (eV)	Ionization Energy (eV)
Helium	20.91	24.46
Neon	16.58	21.47
Sodium	2.10	5.12
Argon	11.56	15.68
Scandium	2.70	6.70
Krypton	9.98	13.93
Indium	2.75	5.76
Xenon	8.39	12.08
Mercury	4.66	10.39
Thallium	3.27	6.07

5.6 ENERGY LEVELS AND SPECTRA OF SODIUM, MERCURY, AND NEON

In this section, the energy level diagrams and emitted spectral lines of some of the elements commonly used in gaseous discharge light sources are examined. Sodium's atomic number is 11, indicating an inner core of two electrons in the $n = 1$ shell and eight electrons in the $n = 2$ shell, with one valence electron in the $n = 3$ shell. As previously discussed, the valence electron is the one most likely to be involved in excitation or ionization of the sodium atom. For example, it requires 5.12 eV to remove this electron from sodium, whereas to remove the second electron requires 47 eV, and to remove a third requires 71 eV.

In the normal state of sodium, the valence electron has the four quantum numbers, $n = 1$, $l = 0$, $m = 0$, $s = -\frac{1}{2}$. With this as the zero energy base level, the energy level diagram in Fig. 5.4 for sodium is built. There are several things to note about this diagram:

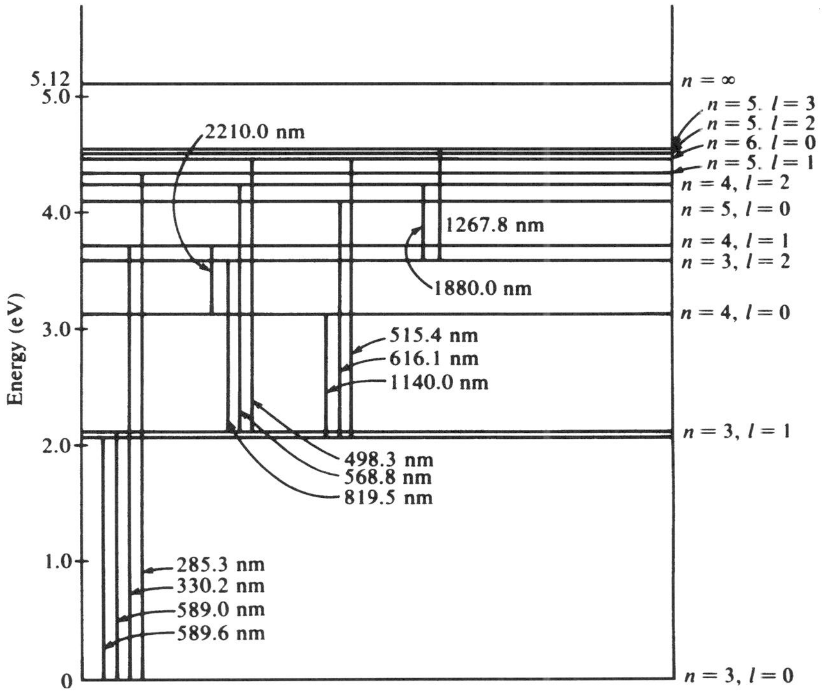

Figure 5.4 Energy level diagram of sodium.

1. The n = 3, $l = 0$ energy level is labeled 0 eV. It is common practice to assign 0 eV to the level in which the valence electron resides when the atom is in its normal state (if there is more than one valence electron we assign 0 eV to the one with highest energy). Then all excited levels are a plus number of electronvolts above the normal state.
2. Only four n = 3 subshells are shown, whereas it is known that there are 18. Also only two n = 3, $l = 1$ subshells are shown, whereas it is known that there are 6. Actually all of the subshells do exist, but the earth's magnetic field is so weak that the energy differences between the m = –1, 0, and +1 levels are very small, and they appear to merge into one level.
3. The n = 4, $l = 0$ level is below the n = 3, l = 2 level. This is a characteristic of the alkali metals (sodium, potassium, rubidium, cesium), which always have a single valence electron located in a $l = 0$ subshell immediately following a completed $l = 1$ subshell. What happens in many-electron atoms is that the main shells overlap. Actually the electrons are still in their lowest possible energy levels when the atom is in its normal state. Thus, for sodium, the n = 4, $l = 0$ level is in reality a lower energy level than the n = 3, $l = 2$ level.
4. In determining which transitions between energy levels can take place in an atom like sodium, certain "selection rules" apply regarding permitted changes

in angular momentum, the most important of which is that, in any transition, the value of l changes by either +1 or −1. The permitted transitions for sodium are shown in Fig. 5.4. The major transitions are the ones from n = 3, l = 1 to n = 3, l = 0, which produce the two predominant yellow lines in the sodium spectrum at 589.0 and 589.6 nm. No spectral lines occur between n = 4, l = 0 and n = 3, l = 0, and between n = 3, l = 2 and n = 3, l = 0, in agreement with the selection rule previously mentioned.

Mercury is discussed next, whose core is a 2 + 8 + 18 + 32 + 18 electron configuration. Outside of this are two valence electrons with opposite spins occupying the n = 6, l = 0 level. The energy diagram for mercury is shown in Fig. 5.5. The following points are noted:

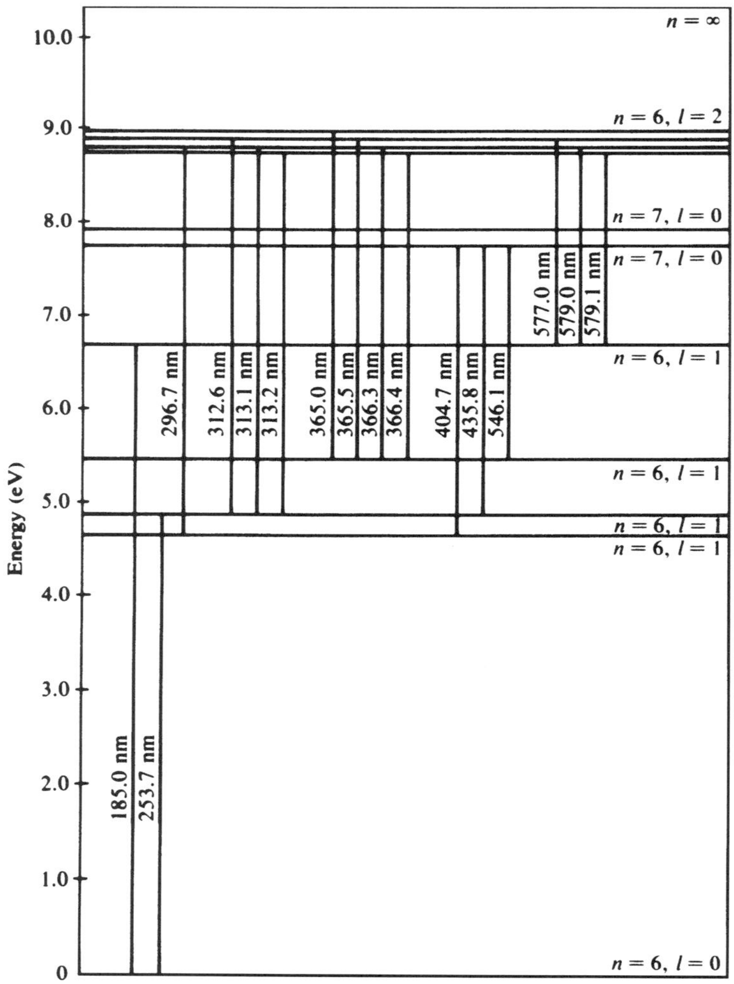

Figure 5.5 Energy level diagram of mercury.

1. The n = 6, $l = 0$ normal state is assigned the value 0 eV.
2. No transitions downward occur from the 5.43- and 4.66-eV levels. These are two metastable states and, as was previously noted, if an electron arrives in one or the other of these it will sit there until boosted to a higher level by an incoming electron or by other means.
3. The lowest excitation level that is not a metastable state is 4.88 eV. Transitions from this level to the normal state yield 253.7-nm radiation. This spectral line, well into the ultraviolet, is the predominant line in a low-pressure mercury spectrum. Germicidal lamps use this line directly for air sanitation in hospitals, nurseries, and the like, and to retard spoilage of food products. Fluorescent lamps depend on 253.7 nm photons to excite the phosphor coating on the inside of the bulb wall. Energy transitions downward within the phosphor then produce the visible energy that the lamp emits. Three levels close together, with two of them metastable, is a major reason that a fluorescent lamp is so efficient in generating 253.7-nm radiation.
4. Transitions from the 6.7-eV level to the ground state produce a line at 185 nm. Radiation at this wavelength produces ozone, which is a form of oxygen having three atoms in each molecule. So-called "ozone lamps" produce this spectral line and ozone is created from the oxygen in the air around the lamp.
5. Transitions from the n = 6, $l = 2$ levels to the two lowest levels at 4.66 and 4.88 eV produce a series of (erythemal) wavelengths in the 300-nm range to which human skin is sensitive. Sun lamps are designed particularly to produce these lines.
6. Transitions from n = 6, $l = 2$ to the 5.43-eV metastable state produce a group of lines in the 365-nm region. This near UV radiation is called "black light." When it impinges on fluorescent paints and dyes, they appear to "glow in the dark." Applications include fluorescent markers, fluorescent signs, observation of defects in machined parts, and detection of contaminants in foodstuffs.
7. The six remaining lines shown in Fig. 5.5 are the visible lines in the mercury spectrum. These are present in a fluorescent lamp but contribute minimally to the light output compared with the phosphor. However, in high-pressure mercury lamps, used in the industry in floodlighting of foliage and in some street lighting, these lines provide nearly all of the visible output. High pressure favors the production of these lines. This is so because high pressure means many mercury atoms are present, and this, in turn, maximizes the opportunity for multiple excitations in which an electron in one excited state is lifted to a higher state, from which it can radiate.

The inert gases, helium, neon, argon, krypton, and xenon, are used extensively in lighting. Argon and krypton are used as starting aids in fluorescent lamps, as is neon in low-pressure and some high-pressure sodium lamps. Xenon is used in flash tubes (for high-speed photography) and in high-pressure sodium lamps. Krypton and argon are used as filling gases in incandescent lamps. Neon is used in small glow lamps. All of these gases are used in gaseous discharge sign tubing (so-called "neon signs"). As an example of the inert gases, the energy level diagram of neon is shown in Fig. 5.6.

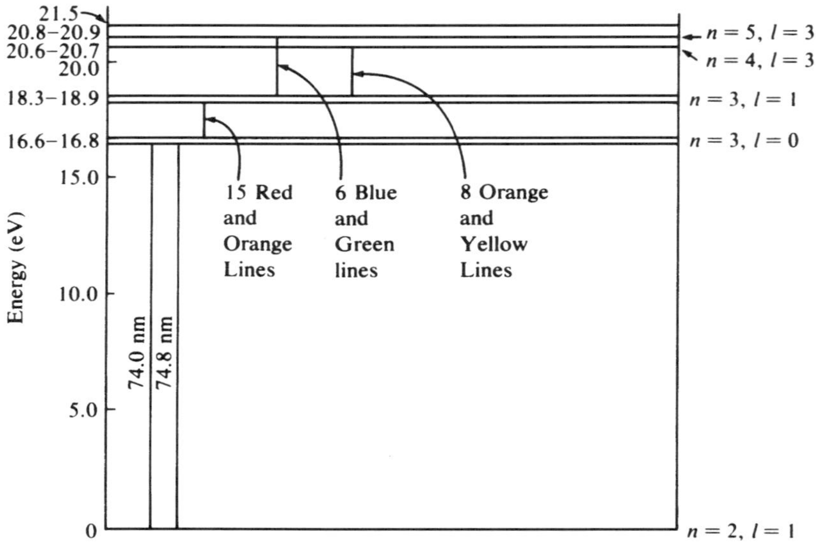

Figure 5.6 Neon.

Most of the orange-red light that is characteristic of neon comes from transitions from the n = 3, $l = 1$ level (18.3 to 18.9 eV) to the n = 3, $l = 0$ level (16.6 to 16.8 eV). The first excitation potential, as in the case of mercury, is a metastable state. The energy of this state (16.5 eV) is greater than the ionizing energy of either sodium (5.12 eV) or mercury (10.39 eV). Thus, when neon is mixed with sodium or mercury in a gaseous discharge tube, the metastable neon atoms can ionize sodium or mercury atoms by collision. Argon has the same attribute. Such gas mixtures are called Penning mixtures.

A Penning mixture of two gases occurs when the major gas (neon or argon) has a metastable state as its lowest excited state, and the energy of this state is greater than the ionization energy of the minor gas (sodium or mercury). A Penning mixture converts a high probability of excitation of the major gas into a high probability of ionization of the minor gas.

EXAMPLE 5.5

A neon atom in its first excited state collides with a mercury atom in its rest state and the mercury atom emits a 253.7 nm photon. The neon atom then collides with a second mercury atom in its rest state, ionizing the latter. How much energy in eV is left over?

Solution. From Table 5.3, the neon atom has an energy of 16.58 eV. From Eq. 5.7, a 253.7 nm photon requires an energy loss of 4.88 eV. Thus, the neon atom departs from the first mercury atom with

$$16.58 - 4.88 = 11.70 \text{ eV}$$

Its collision with the second mercury atom relieves it of an additional 10.39 eV, leaving

$$11.70 - 10.39 = 1.31 \text{ eV}$$

which is divided between the neon atom and the electron departing from the second mercury atom.

5.7 TYPES OF COLLISIONS IN GASES

Collisions in gases, where one particle, such as an electron, collides with another particle, such as a gas atom, are of two general types, namely, elastic and inelastic. In an elastic collision, both momentum and kinetic energy are conserved. Thus

$$m_1 \upsilon_{1i} + m_2 \upsilon_{2i} = m_1 \upsilon_{1f} + m_2 \upsilon_{2f} \tag{5.14}$$

$$m_1 \upsilon_{1i}^2 + m_2 \upsilon_{2i}^2 = m_1 \upsilon_{1f}^2 + m_2 \upsilon_{2f}^2 \tag{5.15}$$

where the subscript i denotes prior to the collision and the subscript f after the collision. An elastic collision does not result in either excitation or ionization of the gas atom.

In an inelastic collision only momentum is conserved and Eq. (5.15) does not hold. Rather some of the kinetic energy of the incoming particle (electron) is transferred to the gas atom, causing the latter to become excited or ionized.

Several of the types of collisions that can occur in a container filled with gas atoms, gas ions, free electrons, and photons will be examined.

Electron Impact

Gaseous discharge light sources (fluorescent and high-intensity discharge lamps) are generally tubular in shape with an electrode at each end. Through heating of the electrodes and application of a voltage between them, electrons are emitted and accelerated from one end of the tube to the other. When these electrons collide with gas atoms in the tube, several things may occur.

1. A relatively slow moving electron will likely have an elastic collision with a gas atom. In modeling this event, assume that the gas atom is standing still and that its mass is very much greater than that of the electron. An analogy would be a rapidly moving marble colliding with a stationary bowling ball.

   To reduce the problem to a single dimension, let the collision be head-on, rather than grazing. Then Eqs. (5.14) and (5.15) yield

   $$\upsilon_{1f} = -\upsilon_{1i}, \qquad \upsilon_{2f} = 2\frac{m_1}{m_2}\upsilon_{1i} \tag{5.16}$$

   where υ_{1i} is the velocity of the electron prior to the collision and υ_{1f} and υ_{2f} are the velocities of the electron and gas atom, respectively, after the collision. Equations (5.16) indicate that, in an elastic collision of these two particles, the electron rebounds with its kinetic energy essentially unchanged and the gas atom acquires only a very small amount of kinetic energy from the electron.

For example, in an elastic collision of an electron with a mercury atom, only about 0.0005% of the electron's energy is acquired by the mercury atom. As the electron continues on its way, it will tend to acquire as much kinetic energy between collisions from the applied electric field as it loses during a collision. Since it loses very little energy during a collision, its average kinetic energy will be relatively high.

2. An electron moving faster than in (1) may engage in an inelastic collision with a stationary gas atom in its ground state causing the latter to become excited. The incoming electron must possess a kinetic energy greater than the atom's first excitation potential for this to happen, and the energy balance relation is

$$KE_i = KE_f + W_{exc} \tag{5.17}$$

where KE_i = kinetic energy of the electron just prior to collision
KE_f = kinetic energy of the electron just after collision
W_{exc} = energy of one of the atom's excitation levels

Not all collisions where the incoming electron has kinetic energy greater than the first excitation potential of the atom result in the atom becoming excited. The probability of excitation is highest when the electron energy is just slightly greater than one of the excitation energies of the atom. Even then only about 3 out of 1000 such collisions are inelastic, resulting in excitation.

3. An electron moving more rapidly than in (2) may possess sufficient energy to ionize a gas atom. In this case the energy balance relation is

$$KE_i = KE_f + W_{ion} + KE_0 \tag{5.18}$$

where KE_i and KE_f are as before, W_{ion} is the ionization potential of the atom, and KE_0 is the kinetic energy of the electron freed from the atom. Once again, not all such collisions result in ionization. The probability of ionization is very small if the electron possesses just the ionization energy. It is a maximum if the electron's energy is $2W_{ion}$.

4. Quite often, in an electron-atom collision, the atom is already in an excited state. This is particularly likely if the atom has metastable states because the average lifetime in a metastable state is about a million times that in a normal excited state.

If an electron collides with an excited gas atom, two things may occur. The gas atom may become excited to a higher level from which it can radiate a photon and return to its rest state. This process is important in high-pressure mercury lamps. The visible lines in a mercury spectrum result from transitions from the higher energy levels. These higher energy levels are more easily reached by the two-step collision process just described, than by a single collision.

A second possibility is that the gas atom will become ionized. Such a result is important in low-pressure mercury arcs to create sufficient positive ions to sustain the discharge.

Positive Ion or Excited Atom Impact

Whenever an ionizing collision occurs, a positive ion is created, which is then accelerated by the electric field in a direction opposite to that of the flow of electrons. As it moves, the positive ion gains energy and ultimately collides with a gas atom. Such a collision is likely to be elastic if the gas atom and positive ion are the same element. The reason for this is that the positive ion and the gas atom have essentially the same mass. Thus, when a moving positive ion collides with a stationary gas atom, the former loses approximately half its energy to the latter. Just as with electrons, the positive ion tends to gain as much energy between collisions as it loses during a collision. Thus, its average kinetic energy is much lower than that of a free electron and sufficient for elastic collisions only. The result is that there is very little excitation or ionization of gas atoms by positive ions of the same element.

Within a mixture of gases, the situation is somewhat different. Excited atoms of one element can collide with unexcited atoms of another element, causing the latter to become ionized. This is possible only when the excitation energy of the first atom is greater than the ionization potential of the second atom. The possibility is enhanced if the excited atom is in a metastable state. This is the Penning mixture situation, discussed in Sect. 5.6.

Photon Impact

Unlike electrons and positive ions, photons are not influenced by electric fields. However, like these particles, photons can have excitation and ionization collisions with gas atoms.

1. Excitation by photon impact is similar to excitation by electron impact except for one important difference. The photon must possess exactly the difference in energy between two excitation levels in the gas atom for excitation to occur. Thus a neon photon, for example, cannot excite a mercury atom, but a sodium photon can excite a sodium atom. Suppose a sodium atom near the center of a gaseous discharge tube is excited by electron impact and emits a 589-nm photon. This photon impacts a second sodium atom and immediately ceases to exist, but the second atom emits a 589-nm photon. This absorption and emission process can repeat itself many times before a photon is emitted near the tube wall and escapes. The phenomenon is known as "imprisonment of radiation." The spectrum of a high pressure sodium lamp is called an "absorption spectrum" because the two dominant lines at 589 nm are very dim due to this imprisonment process.
2. In ionization by photon impact, the photon need not possess exactly the ionization energy but it must possess at least this amount. If it possesses more, the excess energy will appear as kinetic energy of the freed electron.

Thermal Agitation

The average kinetic energy of an assemblage of gas atoms is given by

$$W = \frac{3}{2}kT \text{ joules} = \frac{T}{7730} \text{ electronvolts} \tag{5.19}$$

where k is the Boltzmann constant and T is the temperature in kelvins. The most probable energy of an individual gas atom is one-third this amount. Thus at room temperature, gas atoms have an average kinetic energy of $\frac{300}{7730} \approx 0.04$ eV and a most probable energy of 0.013 eV, far less than the energies required for excitation and ionization.

However, statistically, there is a small percentage of electrons present with energies sufficient to excite mercury atoms. This requires an electron with a thermal energy of 4.88 ev, equivalent to T = 38,000 K. Thus, except in very high temperature, high-pressure gaseous discharges, thermal agitation does not play a significant part in the excitation and ionization process, and the assumption that the gas atoms are stationary is a reasonable one.

5.8 MODEL OF A GASEOUS DISCHARGE LAMP

An idealized voltage vs length graph for a gaseous discharge lamp is shown in Fig. 5.7(a). The lamp is filled with hydrogen, has an applied DC voltage of 120 V*, and is 48 in long.

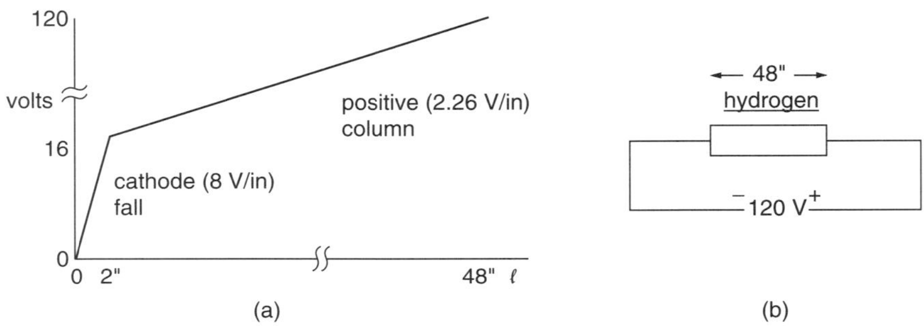

Figure 5.7 Excitation-ionization model for a gas discharge lamp.

Two distinct voltage regions are shown. These are the cathode fall, a narrow region of steep slope near the negative electrode and the positive column, a wide region of shallow slope occupying the remainder of the tube. Electrons emitted from the heated cathode with significant initial velocities are accelerated rapidly by the cathode fall voltage to values of energy that permit excitation and ionization. This process will be illustrated through an example.

EXAMPLE 5.6

In Fig. 5.7, assume an electron leaves the cathode with a thermal velocity of $\upsilon = 5 \times 10^5$ m/s and is accelerated by the cathode fall voltage. When it has a velocity of 2.1×10^6 m/s, it collides with a hydrogen atom.

a) Find the energy level to which the hydrogen electron is raised.
b) Find the wavelength of the emitted photon.
c) Find the distance from the cathode at which the collision occurs.
d) Find the distance from the cathode at which the probability of ionizing a hydrogen atom is highest.
e) Find the maximum number of excitations possible by one electron having the given initial velocity.

* For a fluorescent or high intensity discharge light source, the applied voltage is ac and thus the cathode (negative electrode) flip-flops from one end of the tube to the other many times per second.

Solution.

a) $KE = 0.5 \times 9.11 \times 10^{-31} \times 2.1^2 \times 10^{12} = 2.009 \times 10^{-18}$ J

$$\frac{2.009 \times 10^{-18}}{1.60 \times 10^{-19}} = 12.55 \text{ eV}$$

b) From Fig. 5.1, the hydrogen electron will most likely be elevated to the 12.03 eV level. Then from Eq. 5.7,

$$\lambda = \frac{1240}{12.03 - 0} = 103 \text{ nm}$$

The colliding electron departs with 12.55 – 12.03 = 0.52 eV

c) The electron has an initial kinetic energy of

$$KE = \frac{0.5 \times 9.11 \times 10^{-31} \times 5^2 \times 10^{10}}{1.60 \times 10^{-19}} = 0.71 \text{ eV}$$

Thus it gains 12.55 – 0.71 = 11.84 eV prior to the collision. From the definition of an electron volt, for an electron to gain 11.84 eV, it must pass through 11.84 V. The slope of the cathode fall region is 8 V/in. Thus the collision occurs $\frac{11.84}{8} = 1.48$ in from the cathode

d) The highest probability of ionization occurs when the colliding electron has twice the ionization energy of 13.53 eV (Fig. 5.1); thus, 27.06 eV is needed and

$$0.71 + 16 + 2.26\,(x - 2) = 27.06$$

where x is distance from the cathode, the first term is the initial eV, the second term is the eV to pass through the cathode fall, and the third term is the eV in the positive column. Solving gives x = 6.58 in.

e) An electron traversing the length of the tube without a collision would arrive at the anode with 120.71 eV. The first excitation level of hydrogen is 10.15 eV. The maximum number of excitations possible is

$$\frac{120.71}{10.15} = 11.9 \rightarrow 11 \text{ excitations}$$

EXAMPLE 5.7

Four electrons, each with an initial KE of 0.5 eV, leave the cathode of a fluorescent lamp, whose voltage vs distance curve is shown in Fig. 5.8. In each of the following cases, describe what is most likely to happen.

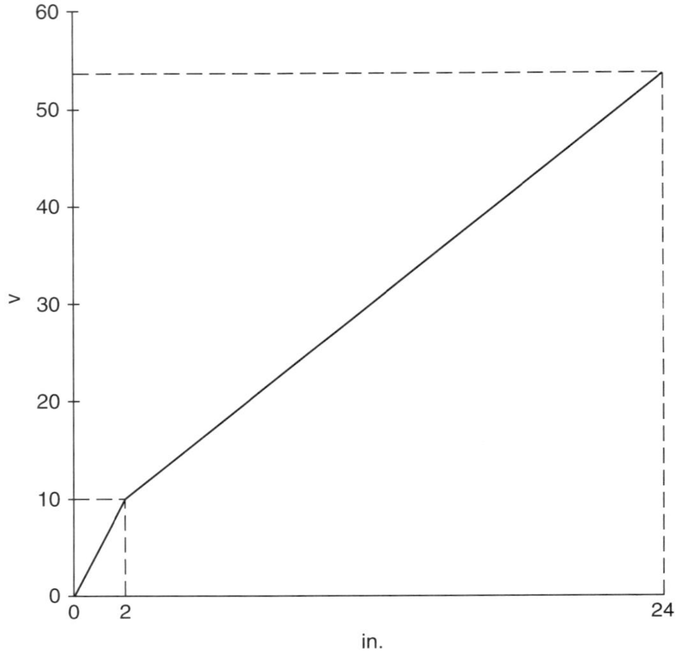

Figure 5.8

a) The first electron collides with a mercury atom 1.30 in from the cathode.
b) The second electron collides with a mercury atom 7.25 in from the cathode.
c) The third electron collides with an argon atom 2.75 in from the cathode.
d) The fourth electron collides with a mercury atom in its lowest metastable 0.75 in from the cathode.

Solution.

a) The slope of the cathode fall is 5 V/in. At 1.30 in, the electron has $5 \times 1.3 + 0.5 = 7.0$ eV. Excitation is most likely to occur when the electron energy is slightly greater than that required. Thus the atom will be excited to 6.70 eV (Fig. 5.5) and, from Eq. 5.7, a photon will be emitted with

$$\lambda = \frac{1240}{6.70} = 185\,\text{nm}$$

The departing electron has 0.3 eV.

b) The positive column slope is $\frac{44}{22} = 2\text{V/in}$. At 7.25 in, the electron has $10 + 2 \times 5.5 + 0.5 = 21$ eV. Ionization is most likely to occur when the electron energy is twice the ionization energy ($10.39 \times 2 = 20.78$ eV). Thus the atom will be ionized, leaving $21 - 10.39 = 10.61$ eV to be divided between the two electrons.

c) At 2.75 in, the electron has $10 + 2 \times 0.75 + 0.5 = 12$ eV. Argon's first excitation level is 11.6 eV and is metastable. The atom will be raised to this level, but no radiation will occur. The departing electron has 0.4 eV.

d) At 0.75 in, the electron has $5 \times 0.75 + 0.5 = 4.25$ eV. The mercury atom is at the 4.66 eV level (Fig. 5.5). It is lifted to the first level (8.84 eV) below 8.91 eV (4.66 + 4.25). Several possibilities for radiation exist. Using Eq. 5.7,

$$\frac{1240}{8.84-4.66} = 297 \text{ nm}, \qquad \frac{1240}{8.84-4.88} = 313 \text{ nm},$$

$$\frac{1240}{8.84-5.46} = 367 \text{ nm}, \qquad \frac{1240}{8.84-6.70} = 579 \text{ nm}.$$

5.9 NUMBERS OF COLLISIONS IN GASES

In Sect. 5.7 the types of collisions that can occur between particles in a gaseous medium were considered. In this section, the topics are how many collisions can occur, what the distance between collisions is, and how pressure, temperature, and tube dimensions affect these quantities.

The kinetic theory of gases describes a gas molecule as a small sphere in constant motion due to thermal agitation and continually making elastic collisions with its neighbors and the walls of its confining chamber. An electron fired into a gas at high speed is pictured as a much smaller sphere, colliding with and rebounding from the gas atoms and following a rather random path through the gas medium.

For an ideal gas,

$$p = nkT \tag{5.20}$$

where p = pressure in newtons per square meter (Pa).
n = concentration of gas molecules per cubic meter
k = Boltzmann constant in joules per degree kelvin
T = temperature in kelvins

This equation states that at a given temperature and pressure, all gases must contain the same concentration of gas molecules.

With this picture of an ideal gas established, the average number of collisions an electron makes while traversing a certain path length within the gas and the average distance an electron travels between such collisions can be calculated. This latter quantity is known as the mean free path (mfp) of the electron. To obtain these quantities, the model of the electron-gas medium shown in Fig. 5.9, is employed.

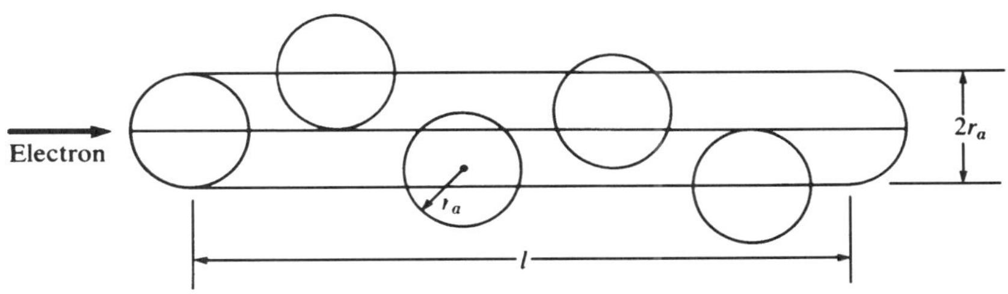

Figure 5.9 Model of electron-gas interaction.

Gas molecules have an average radius of $r_a = 10^{-10}$ m. The diameter of an electron is $d_e = 10^{-15}$ m. A cylinder of diameter $2r_a$ and arbitrary length l is placed within the gas and an electron is aimed into the cylinder along its center line. It is assumed that the gas atoms are stationary with respect to the moving electron, that the electron can be considered as a point with respect to the gas atoms, and that the actual path the electron travels may be replaced by a straight path of equal length. With these assumptions, the electron in Fig. 5.9 will collide with every gas molecule whose center lies within the cylinder. This leads to

$$n_c = \pi r_a^2 l n \tag{5.21}$$

where n_c is the number of collisions the electron makes in the distance l meters and n is the number of gas molecules per cubic meter. Inserting n from Eq. (5.20) into Eq. (5.21) yields

$$n_c = \frac{\pi r_a^2 p l}{kT} \tag{5.22}$$

Equation (5.22) shows that the number of collisions in a distance l meters varies directly with pressure and inversely with temperature. It also shows that, with temperature held constant, the number of collisions is a function of the pl product only.

To find the mean free path, l is divided by n_c in Eq. (5.22) yielding

$$\text{mfp} = \frac{kT}{\pi r_a^2 p} \tag{5.23}$$

The mfp varies directly with temperature and inversely with pressure. Values of n_c and mfp for electrons for some of the lighting gases at a pressure of 1 mm of mercury and a temperature of 273 K are shown in Table 5.4.

TABLE 5.4

Gas	mfp (cm)	n_c (collisions/cm)
Mercury	0.027	37
Argon	0.043	23
Neon	0.065	15
Krypton	0.036	28
Xenon	0.031	32

EXAMPLE 5.7

What must the radius of a mercury atom be for it to have the mean free path given in Table 5.4?

Solution. The temperature is 273 K and the pressure is $(1.013 \times 10^5)/760 = 133$ Pa. Inserting values into Eq. (5.23) yields

$$r_a = \sqrt{\frac{(1.38 \times 10^{-23})273}{\pi(2.7 \times 10^{-4})(133)}} = 1.8 \times 10^{-10} \text{ m}$$

5.10 TYPES OF GASEOUS DISCHARGES

Three major types of gaseous discharges can be identified. These are, in order of increasing current, Townsend, glow, and arc. A most convenient way to examine these three types is through a composite volt-ampere characteristic such as shown in Fig. 5.10. The gas tube contains two unheated parallel-plane electrodes. It is connected in series with a voltage source and limiting resistor, both variable. The circuit equation is

$$I = \frac{1}{R} V_L + \frac{V}{R} \tag{5.24}$$

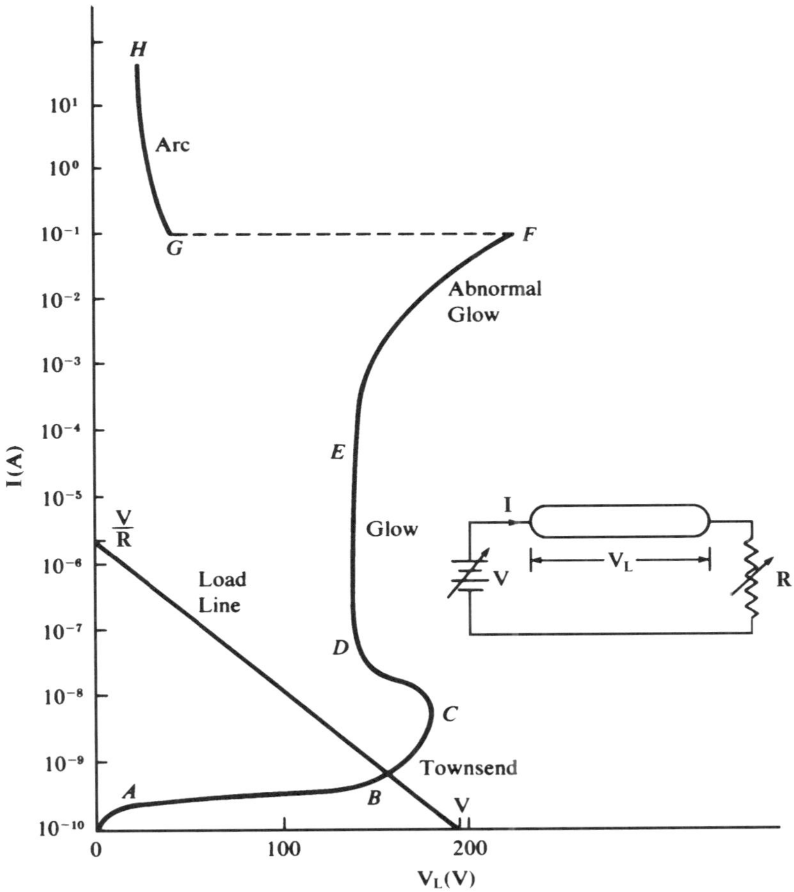

Figure 5.10

and the resulting "load line" is drawn on the volt-ampere characteristic from this equation. It is desired to slowly increase the supply voltage V and observe and explain the changes that take place in I and V_L, and within the tube.

Townsend Discharge

All gases contain ions and electrons because of external radiation present in the environment. Usually recombination occurs quickly, but in the case of the inert gases, it does not (this is one of the reasons for their use in gaseous discharge tubes). As the supply voltage is increased from zero in Fig. 5.10, the current I increases as the few electrons and positive ions in the gas tube are drawn to the electrodes. At A, these charges are being drawn away as fast as they are being created, and there is negligible increase in current as the tube voltage is further increased from A to B.

After B, a new phenomenon begins. The voltage V_L is now large enough so that the free electrons in the gas acquire sufficient kinetic energy to ionize additional gas atoms, creating positive ions and additional free electrons. The process is a geometric progression from B to C and results in a large increase in current, known as an electron avalanche. At point C, conditions are right for the discharge to become self-maintaining. For this to occur, another new phenomenon, secondary emission, must begin. The positive ions created by the ionization process are accelerated toward the cathode by the electric field. As the field strength increases (as V increases), some of these ions will have sufficient kinetic energy to dislodge electrons from the cathode upon impact. The condition for the discharge to become self-maintaining is that the electric field must be large enough so that, for each primary electron, enough positive ions are created so that one of these has sufficient energy to dislodge one secondary electron from the cathode. Then the primary electron is no longer needed and the discharge maintains itself.

The current at the anode in the presence of secondary emission can be shown to be

$$I_a = \frac{I_c e^{\alpha l}}{1 - \beta(e^{\alpha l} - 1)} \tag{5.25}$$

where I_c is the original electron current at the cathode (prior to secondary emission) l is the length of the tube, and α and β are called Townsend coefficients and are experimentally determined. The condition for self maintenance is that the denominator in Eq. (5.25) vanish, requiring

$$\beta(e^{\alpha l} - 1) = 1 \tag{5.26}$$

The voltage V_L at which a discharge becomes self-maintaining is several times larger than the ionizing potential of the gas. It has several names: breakdown, ignition, sparking, striking, starting. It will be called breakdown voltage and assigned the label V_B. V_B is a function of pressure, temperature, electrode spacing, and the type of gas. To relate these quantities, return to Eq. (5.22) and note that, for a specified gas and temperature, the number of collisions of electrons with atoms is a function of the product of the pressure and the electrode spacing. The breakdown voltage V_B is directly related to the number of collisions. This leads to Paschen's law, which states that, for parallel-plane electrodes and a particular gas at a specified temperature, V_B is a function of the pl product only. The general relationship is shown in Fig. 5.11.

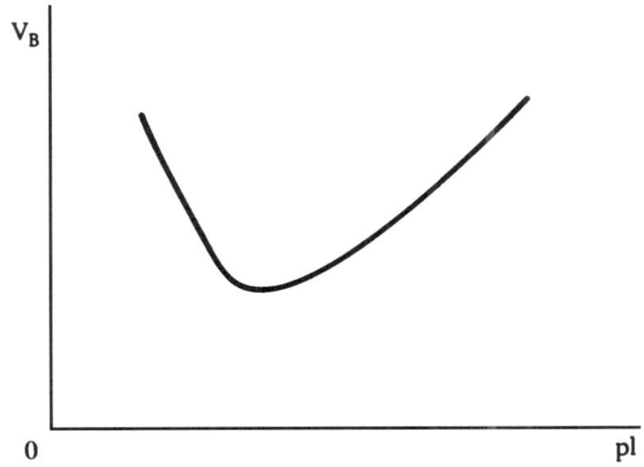

Figure 5.11 Illustration of Paschen's law.

At large values of pl, either the pressure is so high that the mfp is too short or the electrode spacing is so great that the electric field is too small. In either case, insufficient ionization occurs. If the pl product is too small, either the pressure is so low that there are insufficient gas atoms present or the electrode spacing is so small that an insufficient number of collisions is possible. In either case, there is once again insufficient ionization. In between these two extremes, there is a value of pl that requires the minimum V_B for breakdown. The minimum breakdown voltage for air is about 350 V, occurring at pl = 0.006, where p is in millimeters of mercury and l is in meters.

The discharge present from B to C in Fig. 5.9 is called a Townsend discharge*. Townsend discharge currents are measured in nanoamperes or microamperes, with voltages normally ranging from 50 to 150 volts. The Townsend discharge is important in gas-filled phototubes and Geiger tubes, but is not otherwise present directly in lighting-related devices.

Glow Discharge

Point C in Fig. 5.10 is not a stable operating point, and operation will move quickly from C to D and then to the intersection of the load line with the portion of the volt-ampere characteristic between D and E. During this movement, the current increases and the voltage across the tube decreases and then becomes constant. The discharge is now defined as a glow discharge and is self-maintained. The voltage across the tube is 200 to 300 V, of the order of the breakdown voltage of the gas, the pressure is a few millimeters of mercury, and the current is in the milliampere range. The voltage is essentially independent of current, with the latter being limited by the external circuit.

To gain further insight into a glow discharge, it is useful to consider the potential, electric field, and charge distributions present. The governing relation is Poisson's equation

$$\frac{d^2V}{dx^2} = -\frac{\rho}{\varepsilon} \tag{5.27}$$

* Named for J.S.E. Townsend.

where V is the voltage across the tube, ε is the permittivity, and ρ is the charge concentration. By including electric field intensity, Eq. (5.27) can be split into two first-order differential equations.

$$\mathscr{E} = -\frac{dV}{dx}, \qquad \frac{\rho}{\varepsilon} = \frac{d\mathscr{E}}{dx} \tag{5.28}$$

Graphs of V, $\mathscr{E}$, and ρ versus x appear in Fig. 5.12, along with a diagram showing the four major regions within a glow discharge.

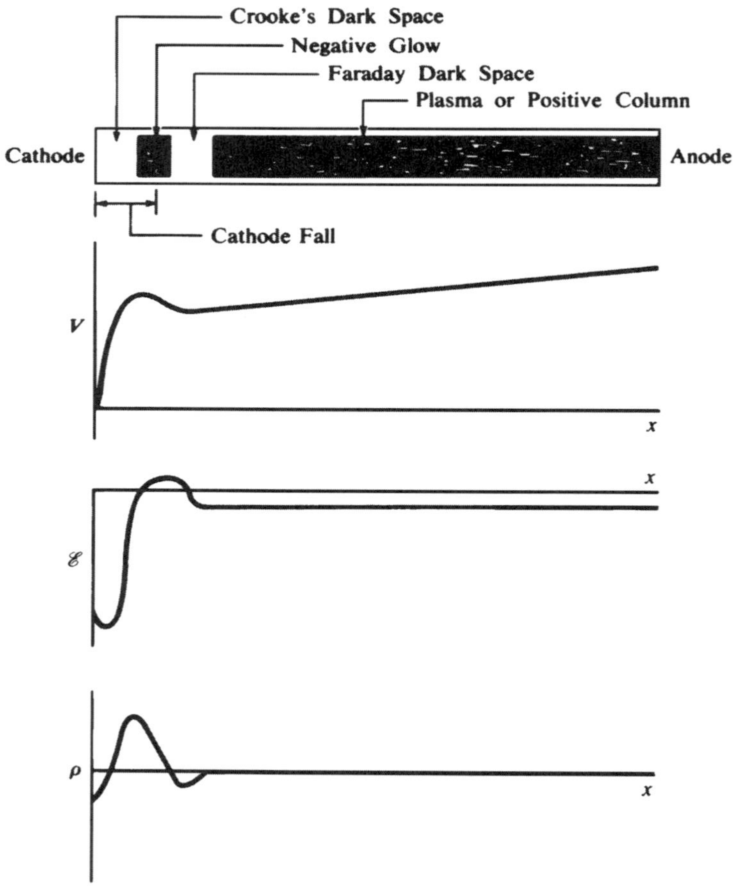

Figure 5.12 Potential, electric field, and charge distribution in a glow discharge.

Note, as was explained in connection with Fig. 5.7, that the voltage rises steeply in the cathode fall and much more gradually in the positive column.

There are large numbers of positive ions and free electrons in a glow discharge, and they produce a current given by

$$I = I_p + I_e = A(\rho_p v_p + \rho_e v_e) \tag{5.29}$$

where I_p is the positive ion current and I_e is the electron current, and these currents are expressed in terms of charge concentration, particle velocity, and cross-sectional area of the discharge. Both types of particles are accelerated by the electric field present and achieve kinetic energies given by

$$KE_p = q_e V = \frac{1}{2} m_p \upsilon_p^2, \qquad KE_e = q_e V = \frac{1}{2} m_e \upsilon_e^2 \tag{5.30}$$

Solving Eqs. (5.30) for the velocities and inserting into Eqs. (5.29) yields

$$\frac{I_e}{I_p} = \frac{\rho_e}{\rho_p}\sqrt{\frac{m_p}{m_e}} \tag{5.31}$$

Eq. (5.31) facilitates an explanation of the voltage gradients in the cathode fall and positive column. The positive column of a discharge contains slightly more positive ions than electrons, but is essentially a region of charge neutrality, as Fig. 5.12 shows. From Eq. (5.31), the current in the positive column will be predominantly electron current because $m_p >> m_e$. Thus 1 A of positive ion current can neutralize the effect of perhaps 100 A of electron current because the positive ions are traveling relatively slowly and remain in the positive column for a much longer time than do the electrons. This results in a large positive charge in the space near the cathode.

The positive column is a "plasma", a mixture of positive ions and electrons that has two unique properties. First, throughout the plasma the density of electrons in a small volume nearly equals the density of positive ions, so that the net charge density is essentially zero (see Fig. 5.12). If this were not the case, a local excess density of, say, positive ions, would cause mutual repulsion until a balance was reestablished. The second unique characteristic is the continuity of current at the plasma boundary. No net current can flow from the plasma to the glass wall of the tube. Since electrons are less massive than positive ions, they move faster toward the wall. The wall acquires a negative charge and repels electrons out of a narrow "sheath" around the plasma, giving the sheath a positive charge. An equality results between electron current to the wall and positive ion current from the wall. The thin sheath effectively isolates the plasma from the glass wall.

The situation in the cathode fall region is somewhat different. Because of the large positive space charge and resulting high electric field, most of the voltage drop across the tube appears here. For example, the cathode fall voltage for argon is of the order of 130 V. The current at the cathode is predominantly positive ion current and thus the ratio in Eq. (5.31) is much less than unity.

The cathode fall voltage is essentially independent of the gas pressure within the tube. What happens is that, as pressure is increased, the distance from the cathode covered by the cathode fall decreases so that the product of pressure and distance remains constant at a value to yield the minimum value of breakdown voltage, according to Paschen's law and Fig. 5.11.

In Fig. 5.12, Crooke's dark space is indicated as a region which is not luminous. The reason for this is that the predominant activity in this region is ionization. The electrons produced by secondary emission are in such a strong field that they acquire sufficient energy for ionization in a very short distance. At the anode end of Crooke's dark space, the electrons, after many ionizing collisions, are moving more slowly. They no longer can ionize gas atoms, but they can excite them. The result is a rather intense luminous region called the negative glow. Near the anode end of the negative glow, the electrons are traveling more slowly yet and do not have enough energy even for excitation. Thus a region of essentially elastic collisions, called the Faraday dark space, results. Because of their low velocity the electrons in this region congregate, producing a slightly negative space charge and a dip in potential distribution. At the anode end of this space the electrons are accelerated once again, this time by the small constant positive column field, and they acquire sufficient energy to excite gas atoms. Thus, the entire length of the positive column is luminous.

Glow discharges have several applications in lighting, particularly in cold-cathode (neon) sign tubing, fluorescent lamp starters, automobile taillights, and decorative displays. The first two will be discussed in Chap. 6.

Arc Dishcarge

Departure from the point D in Fig. 5.10 is accomplished by increasing the applied voltage again so that the intersection of the load line and the volt-ampere characteristic approaches E. If the cathode is observed as this is going on, it is noted that the glow covers an increasing portion of the cathode surface until, at E, all the surface is covered. What this means is that as one moves from D to E the current and the cross-sectional area of the cathode that it covers increase at the same rate, keeping the current density, and thus the voltage across the tube, constant.

Proceeding from E to F, the discharge becomes an abnormal glow. The current increases further, but now so does the current density and tube voltage. The glow is no longer uniform across the cathode, but rather begins to concentrate at a spot, which increases in temperature. At point F this cathode spot becomes hot enough (about 3300 K for tungsten and 3600 K for carbon) so that it begins to emit thermonically. As with point C, point F is not a stable operating point, and a rapid transition occurs to the intersection of the load line with the arc discharge portion of the volt-ampere characteristic. This is a negative resistance characteristic (voltage decreases as current increases) and thus the current must be limited by an external element, such as the resistor in Fig. 5.10 or the ballast in fluorescent and high-intensity discharge lamps.

In the arc discharge region, the current is high (amperes), and the voltage across the cathode fall region is low, of the order of the ionization potential of the gas (10 to 20 V). The latter condition occurs because it is no longer necessary to have a large cathode fall voltage to accelerate positive ions to the cathode so that a large number of secondary electrons are emitted. Rather, all that is needed is to have sufficient bombardment to keep the hot spot emitting thermionically. This requires much less voltage drop and ionization in the cathode fall region. As a consequence, the ratio of the positive column voltage drop to the cathode fall voltage increases substantially.

In all of the discharges discussed thus far, most of the electrons and positive ions are provided by the discharge itself. In an arc discharge there are three basic mechanisms for securing an abundance of charge carriers.

1. The cathode is heated to a temperature at which it can thermionically emit by the action of the discharge itself. This is the mechanism just described.

The carbon arc lamp, used for streetlighting in the late 1800s and early 1900s, is an example of this type of arc discharge. So also is the instant-start fluorescent lamp. In this lamp, the cathodes emit a few electrons per second thermionically at room temperature. These are accelerated by a high-applied voltage, thus creating positive ions which bombard the cathodes, rapidly raising their temperature. The result is that the discharge starts as a thermionic arc, bypassing the glow discharge stage.

2. The cathode is externally heated. This is the procedure used in contemporary preheat and rapid-start fluorescent lamps. The cathodes in these devices are coated with an emission material that makes them abundant suppliers of electrons when heated. In this case, the discharge is not truly self-maintained. It depends on the presence of a heating current for starting, but will continue if that current is removed or reduced after starting.
3. Electrons are extracted from a relatively cold cathode by application of a high field. This case differs from case (1) in that the mechanism for electron release is not bombardment by positive ions. Rather, the electrons are removed directly from the surface of the cathode by the force of the electric field. An example of this sort of arc is the cold-cathode fluorescent lamp, which is little used today.

5.11 GASEOUS DISCHARGE LAMP BASICS

In this section, the basics of operation of fluorescent and high intensity discharge lamps are discussed. These lamps have three important attributes:

- They are relatively efficient converters of electrical energy to light energy (up to 30%).
- They have long lives (greater than 10,000 hours).
- Their light outputs remain high over time (greater than 60% of initial lumen output at the end of life).

The basic guidelines for the operation of a fluorescent lamp emerged in the 1920s when it was discovered almost simultaneously in several locations that a mixture of mercury vapor at low pressure and an inert gas at higher pressure was approximately 60% efficient in converting electrical input power into a single spectral line at 253.7 nm. This, along with the development of efficiently-emitting and long-lived electrodes and appropriate fluorescent materials (phosphors), paved the way for the introduction of the first fluorescent lamp in the late 1930s. There has been little change in the basic components of the lamp to the present day.

The basic low-pressure gas discharge process begins when electrons are emitted from the electrodes at the ends of the tube, either by thermionic emission or by high-field emission. The former mechanism is used in all preheat fluorescent lamps (the ones introduced in 1938), where the external circuitry is arranged so that current flows through each electrode, heating them to incandescence prior to application of voltage across the lamp. The latter mechanism is used in instant-start fluorescent lamps, which first appeared in 1944. Voltage is applied across the lamp immediately and is sufficiently high to overcome the work function at the surface of the electrode and remove electrons directly, creating positive-ion bombardment that causes a portion of the electrode to heat instantly and emit thermionically. A combination of these two mechanisms is used in rapid-start fluorescent lamps, which were introduced in 1952 and dominate the market today. In these lamps, medium voltage and circulating filament current are applied simultaneously and the lamps light within 1 s.

As the free electrons emitted from the cathodes encounter argon atoms, their kinetic energy excites and ionizes them. The first excitation level of argon is metastable and forms a Penning mixture with mercury. Metastable argon at 11.56 eV can easily ionize mercury (10.39 eV). At low temperatures most of the mercury is in the form of small liquid droplets on the inside of the tube, but the heat from the argon discharge quickly vaporizes the mercury, which then is ionized. At room temperatures, there is enough vaporized mercury so that the ionization of mercury begins immediately. In both cases mercury is the predominant ionized constituent of the arc.

Of course it is not the ionized mercury atoms that produce the desired radiation; it is the excited ones. It requires 10.39 eV to ionize mercury and only 4.88 eV to reach the excited state of interest, so it is to be expected that considerable excitation will occur. The 4.88-eV level in mercury is not metastable, so the excited atoms revert immediately to their ground states and radiate 253.7-nm photons. These photons impinge on the phosphor on the inside of the tube and are converted into visible photons.

To explain further the underlying mechanisms in a fluorescent lamp, it is necessary to ask and answer several questions:

(a) Why Is It So Desirable to Use Mercury in Fluorescent Lamps?

The answer to this question is threefold. First, for high luminous, efficacy, it is necessary to have a high quantum ratio, defined as

$$\mathrm{QR} = \frac{\text{average energy of visible photons}}{\text{energy of UV photon}} \tag{5.32}$$

From Eq. (5.6), Eq. (5.32) can be rewritten as

$$\mathrm{QR} = \frac{\lambda\,\text{of UV photon}}{\text{average}\,\lambda\,\text{of visible photons}} \tag{5.33}$$

For mercury, the wavelength of the ultraviolet photon is 253.7 nm and the average wavelength of the visible photons, assuming the phosphor radiation is distributed reasonably uniformly throughout the visible spectrum, is 555 nm. This gives a relatively high quantum ratio of 0.46.*

Second, a single ultraviolet line with a high probability of reaching its excitation level is desired. The first three excitation levels of mercury are 4.66 eV, 4.88 eV, and 5.43 eV. The first and last of these are metastable and, thus, do not produce radiation directly. Also 4.88 eV is less than $\frac{1}{2} \times 10.39$ eV, which indicates a high probability of excitation. Thus, there is a situation in mercury where there are three easily-reached excitation levels, but only one is capable of radiation. The result is that 60% of the input watts appear as 253.7-nm radiation.

Third, it is essential to have a material with a vapor pressure at room temperature such that the lamp temperature for vaporization will not be excessive. By vapor pressure is

---

* The quantum ratio cannot exceed 1.0, i.e. the wavelength radiated cannot exceed the wavelength absorbed. This relationship is known as Stoke's Law.

meant the pressure exerted by a vapor when it is in equilibrium with its liquid, that is, when molecules are returned to the liquid at the same rate as they are leaving it. For mercury, the vapor pressure at room temperature is 1.8 μm.

Each fluorescent lamp has a slight excess of liquid mercury to insure an adequate supply during the life of the lamp. The mercury vapor pressure within the lamp is determined by the temperature of the coolest spot on the bulb, and the optimum situation occurs when this spot is about 40°C (104°F). At this temperature the mercury vapor pressure is about 8 μm.

(b) Why Is an Inert Gas Necessary for Operation of a Fluorescent Lamp?

Again, the answer is threefold. First, since the mercury vapor pressure at room temperature is relatively low, it is desirable to have an auxiliary agent to help establish the arc. Inert gases ionize at lower tube voltages than do other gases because the ratio of the first excitation potential and the ionization potential in an inert gas is relatively high, increasing the likelihood of ionization. For example, this ratio for argon is 11.56/15.68 = 0.74.

Second, it is necessary to have a means of ionizing vaporized mercury. As has been mentioned, the ability of some inert gases to form Penning mixtures with mercury greatly enhances the efficiency of this process.

Third, it is necessary to contain the mercury arc. Without the pressure of the inert gas, the diffusion of positive mercury ions and electrons out of the arc stream to the bulb wall, and the resultant recombination of these particles, would be excessive and the mercury arc would be inefficient.

(c) Why Should the Inert Gas Be Argon?

The answer to this question is in two parts. First, it is desired that the ratio of masses of the electron and the gas atom to be as small as possible so that electrons lose very little energy in elastic collisions with gas atoms. The inert gases, in order of increasing mass, are helium, neon, argon, krypton, and xenon. Based on the mass ratio criterion, xenon would be best and helium worst.

Second, it is necessary that the inert gas form a Penning mixture with mercury. It turns out that helium, neon, and argon do, but krypton and xenon do not. Also the metastable energy of argon (11.56 eV) is much nearer to the ionization energy of mercury than is that of helium (19.80 eV) or neon (16.62 eV). Thus, the first choice for the inert gas is argon.

(d) What Conditions Create Optimum Operation?

- High efficiency is obtained if mercury vapor at low current density is used to produce a single ultraviolet spectral line at 253.7 nm.
- Phosphors are needed whose peak spectral sensitivity is at 253.7 nm.
- To provide optimum arc temperature and pressure for generation of 253.7 nm, a bulb temperature of 100 to 110°F, a mercury pressure of 6 to 10 μm, and an inert gas pressure of 2 to 3 mm are required.
- A tubular lamp of appreciable length-to-diameter ratio is needed so that the fixed wattage losses at each end, in the cathode and anode fall regions, are a small percentage of the total watts input.
- Oxide-coated hot cathodes are required to provide an abundance of free electrons.

High pressure mercury lamps were developed at about the same time as the fluorescent lamp in the 1930s. The heart of this lamp is a quartz arc tube containing mercury operating at high temperature and pressure. The question might be asked: "Why not just remove the phosphor coating from the inside of a fluorescent lamp to create a mercury lamp?" The answer lies in the relative efficiency of generation of lines in the mercury spectrum. At low mercury pressure, such as exists in fluorescent lamps, conditions are optimum for a large amount (60%) of the input watts to be converted into a single line at 253.7 nm. As Fig. 5.5 shows, this transition requires the least amount of input energy from a colliding electron.

As pressure is increased, the likelihood of multiple collisions is enhanced, particularly those in which an outer-shell electron in either the 4.66 ev or 5.43 eV metastable levels is raised to a higher energy state. For example, an outer-shell electron at 4.66 eV could receive a 3.07-eV boost from a colliding electron, which would place it at 7.73 eV. From there, it might fall to 5.43 eV and emit a 546.1-nm photon; it might return to 4.66 eV, emitting a photon at 404.7 nm; or it might fall to 4.88 eV, in which case it would emit a 435.8-nm photon. In the last instance, it would then likely fall to the ground state and emit a 253.7-nm photon as well. It is also possible for an outer-shell electron at the 4.66- or 5.43-eV level to move to the 8.84 or 8.85-eV levels and then return to the 6.70-eV level, emitting a photon in the 577 to 579-nm range. Thus, for optimum generation of visible lines in mercury, the pressure should be high enough (1 to 2 atmospheres) so that many outer-shell electrons reside in metastable states and there are lots of free electrons taking part in multiple collisions that raise these outer-shell electrons to higher levels.

5.12 THE LASER

The first maser (*m*icrowave *a*mplification by *s*timulated *e*mission of *r*adiation) was completed in 1954 by C.H. Townes, J.P. Gordon, and H.J. Zeiger in an effort to generate electromagnetic waves at frequencies higher than was possible with electronic tubes. In 1958, Townes and A.L. Schawlaw published a paper showing the feasibility of an optical maser, later called a laser (with the "l" standing for "light").

The first successful laser was built by T.H. Maiman in 1960. Maiman's laser used a ruby crystal. Ruby is composed largely of aluminum oxide, with a few of the aluminum atoms replaced by chromium atoms, about 4% in Maiman's crystal. The ruby is machined into a rod 0.5 cm in diameter and 4 cm long. Its ends are polished to be perfectly flat and are silvered providing a fully reflecting mirror at one end and a partially reflecting mirror at the other end. The rod is placed inside a helically-coiled electronic flash tube, which is connected to a large capacitor-charged power supply.

When the power supply is activated, the capacitors charge and the voltage builds up. Then a firing button is pushed and the flash tube strikes, providing an intense flash of luminous flux for a very short period of time. This is called "pumping," and the pulse of light raises most of the chromium atoms to one of the excitation levels shown in the left portion of Fig. 5.13. The atoms immediately fall to a lower energy level, shown in the center portion of Fig. 5.13. This level is a metastable state. While they are in this state, they are likely to be impacted by chromium photons, which "stimulate" them to emit photons of the same wavelength and revert to the ground state, as shown in the right portion of Fig. 5.13. In the case of chromium, the photon wavelength is 694.3 nm.

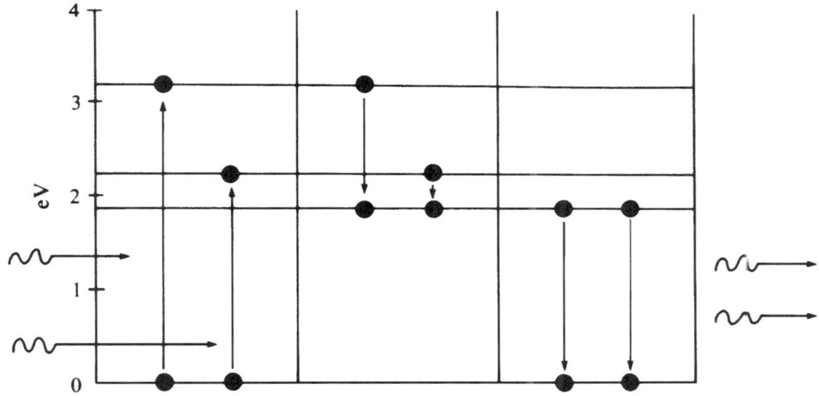

Figure 5.13 Ruby laser.

It is important to contrast what occurs in Fig. 5.13 with the phenomena of photon excitation and imprisonment of radiation that occur in high-intensity discharge lamps. Consider, for example, a sodium lamp. If a sodium photon at 589 nm collides with a sodium atom in its *ground* state, the atom is excited to the 2.1-eV level and the photon ceases to exist. When the atom reverts to its ground state, a new 589-nm photon is generated. In Fig. 5.13, the chromium atom is in an *excited* state when the chromium photon collides with it. The result is two chromium photons; the original photon does not cease to exist. This is how light amplification occurs.

From the foregoing, it is concluded that for the laser action to occur, a sizable number of atoms must already be in an excited state, rather than in a ground state. This situation is known as *population inversion*, and it is important that this inversion be maintained continually to make stimulation of radiation more likely than absorption of radiation.

Laser light is spatially and temporally coherent. Light from conventional sources is not. What is meant by coherence? In a conventional light source, the atoms radiate independently of one another, producing light waves that are both spatially incoherent (moving in all directions) and temporally incoherent (consisting of many wavelengths). A pinhole can be used to obtain nearly spatially coherent light from a spatially incoherent source, but most of the power is lost. A narrow band optical filter can be used to obtain nearly temporally coherent light, but again at a large sacrifice in power output.

When a chromium photon in the ruby laser encounters a chromium atom in the excited metastable state, not only is another photon of exactly the same frequency emitted, but the two photons are in phase. Some of these photons will travel to the side of the rod and escape, but many will proceed along the rod to its ends, where they are reflected back by the mirrored surfaces. The rod is an integral number of chromium wavelengths long and thus acts as a cavity resonator, building up the intensity of the laser beam. Ultimately an intense, collimated directional monochromatic beam emerges through the partially silvered mirror at one end of the rod.

In the ruby laser, beam intensities of 10^4 W/cm^2 have been produced, lasting for about 0.5 ms. If the beam is focused by an output lens, concentrations of 10 W/cm^2

can be achieved. The width of the red spectral line at 694 nm is about 0.002 nm, corresponding to a bandwidth of about

$$\Delta f = c\frac{\Delta\lambda}{\lambda^2} = 1200\ \text{MHz} \tag{5.34}$$

This is comparable to the bandwidth of the sharpest spectral line from a low-pressure gaseous discharge light source.

The divergence of the laser beam is less than 0.5° at high power outputs and 0.05° at low power outputs, the latter corresponding to 5 ft/mi. If the beam is sent through a telescope in reverse, almost negligible divergence can be achieved. A ruby laser beam sent to the moon in this manner illuminated a spot less than 2 mi in diameter.

All of the original lasers were operated in the pulsed mode, providing short bursts of energy. It is possible to build a solid continuous beam laser, but it is more feasible to use a gas laser. A widely used laser of this latter type is the helium-neon laser.

Consider Fig. 5.14, in which helium atoms are rings and neon atoms are solid dots. At the start, both are in the ground state. Through radio frequency excitation (not light), a glow discharge is created and helium atoms are raised to an excited state, as shown in the left portion of Fig. 5.14. The excited helium atoms collide with neon atoms in the ground state in the center portion of Fig. 5.14, exciting each of the latter to one of four distinct energy levels. The helium atoms revert to the ground state.

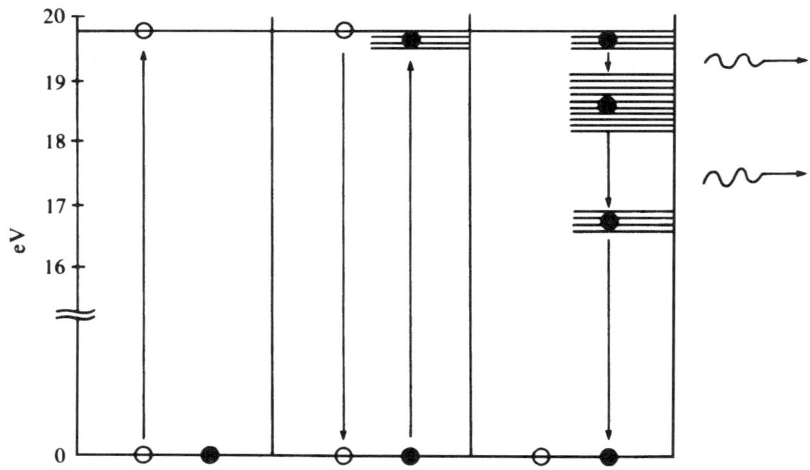

Figure 5.14 Helium-neon laser.

In the right portion of Fig. 5.14, the excited neon atoms are stimulated by neon photons, producing new photons. The neon atoms fall to one of 10 distinct energy levels, and then to the ground state in steps. The lower photon in Fig. 5.14 does not contribute to the laser beam. The upper photons reflect back and forth between end mirrors, as in the ruby laser, before emerging at one end.

The first helium-neon laser was constructed by A. Javan, W.R. Bennett, and D.R. Herriott in 1961. It operated at the very low power of 50 W and produced a beam with a wavelength of 1000 nm in the infrared and bandwidth of 1 MHz, much less than that of the ruby laser or of the best low-pressure gaseous discharge lamp. Because of its long life, low power and low cost, it has become the most widely used laser.

Most lasers, including the two already discussed, emit about the same amount of luminous energy as does a flashlight. The difference lies in the space and time coherence. The laser's output is highly directional and its output power per nanometer is very high.

There is one continuous laser which does provide high total power output. This is the carbon dioxide laser, invented by C.K.N. Patel. Carbon dioxide lasers produce infrared wavelengths and can generate up to 9 kW of continuous power. At the other extreme are semiconductor diode lasers the size of the head of a pin, composed of gallium arsenide. A summary of types of lasers and their properties is given in Table 5.5.

TABLE 5.5

Laser Type	Typical Operating Wavelengths (μm)	Principal Characteristics
Helium-neon	0.6328, 1.15, 3.39	Highly monochromatic, rugged, simple, long-life, inexpensive, low power
Helium-cadmium	0.3250, 0.4416	Highly monochromatic, visible and UV
Carbon dioxide	10.6	High power, high efficiency, IR
Argon	0.3511, 0.3638, 0.4579, 0.4765, 0.4880, 0.4965, 0.5017, 0.5145	CW, high power, highly monochromatic, inefficient
Krypton	0.3507, 0.3564, 0.4680, 0.4762, 0.5208, 0.5682, 0.6471, 0.6741	Similar to argon except for lines in the red portion of the spectrum
Nitrogen	0.3371	UV, inefficient, short pulses
Ruby	0.6943	High peak power, moderate spatial coherence
Glass	1.06	Low cost, excellent optical quality, high peak power, low average power, wide spectral line
Ga As	0.85–0.91	Very small, inexpensive, efficient, poor coherence, low power, temperature sensitive

The uses of lasers are many and are increasing. A brief summary of some of the important ones follows:

Optical Ranging

The spatial coherence of the laser makes it ideally suited for optical ranging and scanning applications. In 1962, laser echoes were received from the moon. In 1969, following the Apollo lunar landing, a laser was used to illuminate a two-mile-wide spot on the moon. This experiment provided an estimate of the moon's distance from the earth within an accuracy of 15 cm.

Searching lasers are used to scan the sky, in much the same way as radar does. At 500 mi, radar can determine distance within 100 ft. A ruby laser narrows this to 25 ft and, in addition, requires a smaller antenna and is less sensitive to interference by enemy signals.

Industrial Applications

The laser is used widely in industry. Computer-controlled CO_2 lasers spot-weld sheet metal parts on automotive production lines at the rate of 50 ft/min. Laser "microwelders" are used to spot-weld today's microelectronic circuits.

Lasers are used in heat-treating metals, raising temperatures over 2000°F in a few milliseconds. They can aid in the critical inspection of many products, particularly those with hard-to-see crevices, and can be used to cut extremely hard and brittle materials, such as tungsten.

In the furniture and garment industries, lasers are used for the nondestructive cutting of wood and fabric. In the aircraft industry, they are used to cut epoxy sheets.

Lasers can drill holes in diamonds, which then become dies through which fine wire can be drawn. A single drilling takes only minutes with a pulsed ruby laser, as compared to 2 days for the previous process which involved driving steel pins coated with olive oil.

Surveying and Measurement

Again, because of coherence, lasers are excellent surveying and aligning tools. They are widely used in laying pipe, boring tunnels, and creating plumb lines. Tunnels have been bored with a deviation of less than an inch in one and a half miles. In this application, photocells serve as a target for the laser, which is mounted on the boring machine. If the machine drifts off course, a signal is sent to the operator indicating what correction to make.

Laser interferometers have been built which can detect changes in length of 10^{-11} in. Gas lasers are used in seismometers to detect earth tremors. This is done by allowing one of the laser mirrors to move slightly during a tremor, thus altering the wavelength of the emitted laser beam.

Medical Applications

It is perhaps in the field of medicine that the laser has found its widest and most beneficial use. There are a myriad of medical laser applications. One of the most significant is the laser photocoagulator, a device used by eye surgeons to repair torn retinas. A weak laser beam, focused through the eye's lens, forms a set of tiny scars in the retinal tissues surrounding the tear, thus keeping it from expanding. The entire process takes only about 20 min and the patient feels essentially no pain.

In cancer treatment, the laser is being used to burn away skin tumors and blemishes. A CO_2 laser scalpel has shown success in selectively destroying inoperable brain tumors and in treating burn victims, where the surgeon must remove large amounts of tissue. Its major virtue is that it cauterizes as it cuts and thus reduces bleeding significantly.

In conjunction with fiber optics, lasers are being used to treat internal problems. In the case of ulcer treatment, two fiber optic cables are inserted into the stomach. Through one the doctor views the ulcer and through the second transmits a laser beam to close it. This, of course, avoids the major incision of conventional surgery.

Optical Communication

The visible spectrum extends from roughly 400 to 700 nm, representing a frequency spread of 321×10^6 MHz. A VHF television channel has a bandwidth of 6 MHz. Thus it is theoretically possible to insert about 50 million TV channels into the visible spectrum. This indicates the vast possibilities for communication at optical frequencies.

An example of optical communications is a fiber optic telephone system, driven by an aluminum gallium arsenide laser operating at 45 MHz. Such a system can handle 50,000 telephone conversions simultaneously and the signals travel up to 4 mi before amplification is required.

Holography

Holography is a photographic technique in which the patterns of the wavefronts of the reflected light waves from an object are recorded, rather than recording the image of the object itself. The result is startling three-dimensional imagery with depth and perspective. The holographic process was first conceived, quite accidentally, by Dennis Gabor, a Hungarian physicist, in 1948. However, it was not until the introduction of the laser that the process became practical.

The holographic recording process is shown schematically in Fig. 5.15. A laser beam is directed at a half-silvered mirror called a beam splitter, where it becomes two beams. One beam lights the object to be photographed, with the reflected light from the object proceeding to the holographic plate, usually a high-resolution photographic plate. The other beam, termed the reference beam, is reflected by a mirror directly to the holographic plate. This beam plays the key role of providing a phase as well as an amplitude, reference. When the two beams meet at the holographic plate, the information that is recorded is in the form of interference patterns. Specifically, a pattern of concentric alternating light and dark rings is formed on the plate for each point on the object that is illuminated. If the ring is light, the two beams are reinforcing; if it is dark, they are destructively interfering.

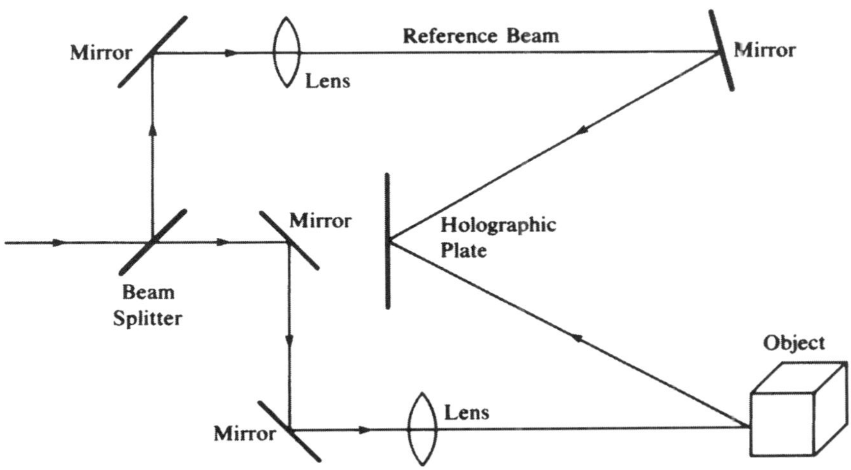

Figure 5.15 Holographic recording.

The exposed holographic plate is developed as in normal photography. If it is then illuminated by conventional light, nothing is seen and the plate appears as would a uniformly underexposed film. To reproduce the image, it must be illuminated by the exact laser reference beam that was used to make the hologram. Only then does the full three-dimensional fidelity appear.

Holograms are used in photomicrography to view and analyze extremely small objects. In photographic art, holographic techniques are used to recreate rare museum objects in three dimensions for public view. The objects themselves are stored safely away.

5.13 ELECTROLUMINESCENCE AND LIGHT EMITTING DIODES

Electroluminescence is the direct conversion of electrical energy into light by a phosphor without the intermediate step of excitation in a gas discharge. The phenomenon of electroluminescence was discovered in the 1930s, when it was observed that crystals of silicon carbide mounted between two metal probes emitted light when a DC voltage was applied. Current electroluminescent lamps consist of two flat plastic electrodes, one transparent and one opaque, with a phosphor of zinc sulfide powder in between, forming a "dielectric sandwich" to which an ac voltage is applied. Electrons in the electroluminescent phosphor are excited and emit light upon return to their ground state.

The efficacy of electroluminescent panels is very low, precluding their use in general lighting. The colors emitted are blue, green, yellow, and white. Luminances range from 2 to 10 cd/ft^2. The panels do not burn out. Rather their luminances fade away, from 40 to 70% of initial luminance at 2000 hours. Major uses of electroluminescent lamps include backlighting of liquid crystal displays, automobile dashboard lighting, lighting of aircraft instrument panels, and emergency and safety lighting.

A light emitting diode (LED)* is a forward-biased pn junction semiconductor that emits light through a process called "injection luminescence", in which holes and electrons are injected into the junction area, where they combine and radiate. In Fig. 5.16, a heavily-doped p-n junction is strongly forward-biased by a voltage V. Electrons are abundant on the n side

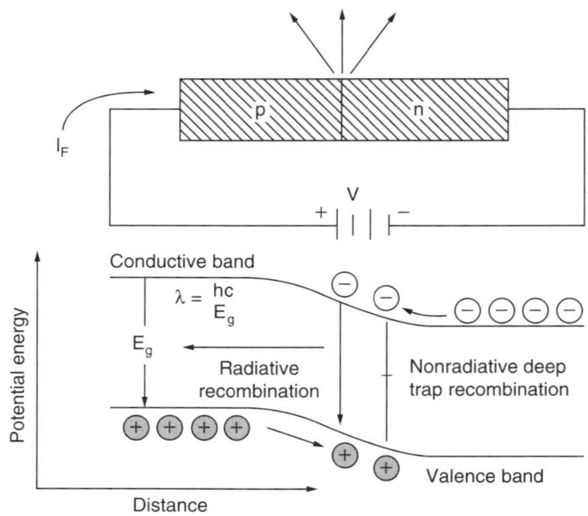

Figure 5.16 Operation of an LED p-n junction. ***Source: IESNA Lighting Handbook*****, 9th edition, 2000. Courtesy of the Illuminating Engineering Society of North America.**

* The first LED was red. It was invented by Nick Holonyak of GE in 1962.

and holes on the p side. Through the process of minority carrier injection, these excess charge carriers migrate to the junction, where many of them recombine and emit photons.*

The energy gap (E_g) between the conduction band and the valence band in a semiconductor is called the forbidden zone.** Its width, in electron volts, determines the wavelength of the emitted photons, which is given by solving Eq. (5.6) to obtain

$$\lambda = \frac{hc}{W} = \frac{hc}{E_g} = \frac{1240}{E_g}\ \text{nm} \tag{5.34}$$

LED chips are made from elements in Columns III and V of the Periodic Table.

It is the relative proportions of the elements in an LED that determines the energy gap width and thus the wavelength. For example, in a GaAsP LED, when the Ga/As ratio is 1.5, the gap width is 1.88 eV, giving a peak wavelength of 660 nm (red). When the ratio is 0.54, the gap width is 1.96 eV and the peak wavelength is 632 nm (orange). Finally, a ratio of 0.18 gives a gap width of 2.10 eV and a peak wavelength of 590 nm (yellow).

The two most common LEDs for producing light are aluminum indium gallium phosphide (AlInGaP), with an E_g of 1.95–2.10 eV, and indium gallium nitride (InGaN), with an E_g of 2.35–2.65 eV. From Eq. (5.34), these yield $\lambda = 590 - 630$ nm, and $\lambda = 470 - 530$ nm, respectively, the former in the red-orange-yellow region of the spectrum and the latter in the blue-green region.

An LED lamp consists of the semiconductor chip, a synthetic casing and a lens, as shown in Fig. 5.17. The chip is typically 0.25 nm square, one of 40,000 cut from a 2-in square wafer. It is mounted in a reflector cup on a cathode post. A gold wire connects

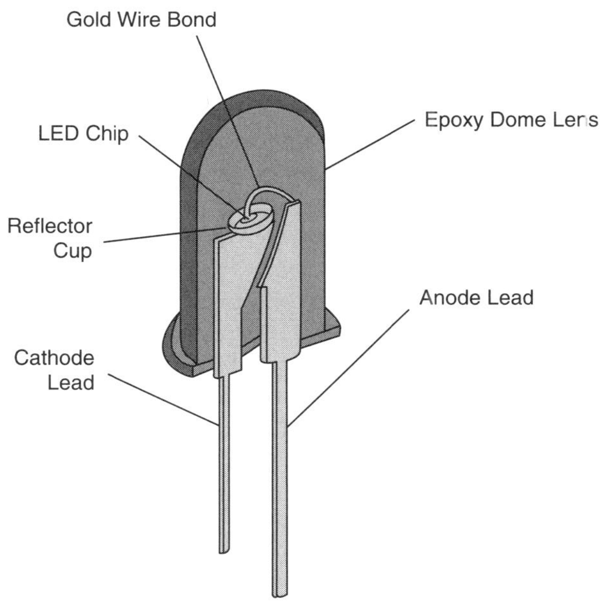

Figure 5.17 Typical LED lamp. ***Source*****: "Leds, Revolution or Evolution",** ***Lighting*****, November, 2000. Courtesy of Kerwil Publications.**

---

* If the voltage is increased further, a population inversion can occur in the junction region, as described in Sect. 5.12, creating a diode laser.

** Conductors have energy gaps <0.1 eV; insulators >3 eV.

from the anode post to the top of the chip. The device is encapsulated in a cylindrical epoxy casing 5.5 mm in diameter (T $1^3/_4$), with a clear, nondiffusing dome at the top, normally having a circular radiation pattern. Arrays are created by mounting several LEDs together on flat plates.

It was the development of the blue LED (InGaN) in 1993 by Shayi Nakamura in Japan that provided the opportunity to obtain white light from LEDs. There are two ways this is done. The beam from a blue LED can be combined with those from a green and red LED, much as one would combine the beams from spot lights with colored gels in a theater setting. This requires three separate lamps, whose intensities and aiming directions must be adjustable.

A second way is to add a phosphor coating to the inside of the bulb of a blue LED. Such an LED has been developed by Siemens and the Fraunhofer Institute. Blue (short wavelength) photons are absorbed by the phosphor and reemitted as yellow (longer wavelength) photons. The result is bluish-white light with a CCT of greater than 5000 K.

Luminous efficacies of LEDs are currently 30 lm/W for AlInGaP, 20 lm/W for green InGaN, 7 lm/W for blue InGaN and 10 lm/W for white. These have been rising by a factor of 10 per decade for the last three decades. LEDs typically have a forward voltage drop of 2–4 V, a current of 20 mA and a power of 40–80 mW. Peak intensities are given in millicandelas and range from 100 to 1100 mcd. Beam widths, between half-maximum intensity values, vary from 5 to 35° for T $1^3/_4$ lamps. LEDs are temperature sensitive, with the temperature of the pn junction the critical parameter. As junction temperature increases, light output decreases and the peak wavelength shifts to longer values.

LEDs offer several unique advantages. They are very shock-resistant (they have been fitted into a hockey puck to allow television viewers to follow the action from ice level). They have long life. As with electroluminescent panels, they don't fail but their light output decreases over time. Current estimates indicate 70% of initial light output at 50,000 hours. They are essentially point sources and thus can be controlled optically with relative ease (smooth beams as narrow as 6° are possible). Since they do not emit ultraviolet wavelengths, they can be used where degradation of materials is of concern, such as in museums. LEDs can be easily dimmed by controlling the forward current.

LED applications are varied and growing. Several examples are presented below.

Alphanumeric Displays

These are manufactured in 7-segment displays of LED rectangular bars, four vertically and three horizontally to form a figure "eight". They are also available in larger arrays.

Traffic Signals

T $1^3/_4$ LED lamps are used in arrays in traffic signals. AlInGaP lamps provide red and yellow signals, whereas InGaN lamps are used for green signals and white LEDs light the human in walking signals. Traffic signal lamps are narrow beam and high luminance. A 20-W LED array replaces a 135-W incandescent lamp, resulting in a large energy saving.

Automobile Taillights

Currently 30% of US rear high-mounted brake lights are LEDs (70% in Europe). These use high-intensity Al InGaP amber and red LED lamps.

Indicator Lamps
LED lamps are being widely used as replacements for incandescent indicator lamps. Multiple-color indicators can be provided by several LEDs of different peak wavelengths in a single package.

Display Screens
Large display screens for sporting events, concerts and outdoor (such as Times Square) utilize high intensity LEDs in large arrays.

Entertainment
The attractiveness of LEDs in the entertainment industry is that, unlike incandescent and fluorescent lamps, they are narrow-band sources and thus combinations of highly-saturated colors can be produced easily and be computer-program controlled. This accounts for their wide use in theaters, restaurants, amusement parks and theme parks.

REFERENCES

Bitter, F.: *Currents, Fields and Particles*, John Wiley, New York, 1957.
Birdsong, T.: "The Sun Rises in the West," *LD+A*, New York, July, 2001.
Coaton, J., and A. Marsden: *Lamps and Lighting*, Arnold, London, 1997.
Dow, W.G.: *Fundamentals of Engineering Electronics*, John Wiley, New York, 1952.
Ducharme, A.: "A Thousand Points of Light, at Least," *LD+A*, New York, July, 2001.
Elenbaas, W.: *Light Sources*, Macmillan Press Ltd., London, 1972.
Glasstone, S.: *Source Book on the Space Sciences*, D. Van Nostrand, New York, 1965.
Gray, T.: *Applied Electronics*, John Wiley and Sons, New York, 1954.
Gray, H.B.: *Chemical Bonds*, Benjamin, Menlo Park, Calif., 1973.
Halliday, D., and R. Resnick: *Physics-Third Edition*, John Wiley and Sons, New York, 1978.
Hughes Aircraft Co.: "What You Always Wanted to Know About Lasers," *Vectors*, Winter 1973.
Martin, T.L.: *Physical Basis for Electrical Engineering*, Prentice-Hall, Englewood Cliffs, N.J., 1957.
Meloy, T.: "The Laser's Bright Magic," *National Geographic*, December 1966
Pattison, K.: "Digital Lighting Sets the Stage," *LD+A*, New York, July, 2001.
Roush, M.: "LEDs, Revolution or Evolution, part 1," *Lighting*, Mississauga, Ontario, October–November, 2000.
Saleh, B., and M. Teich: *Fundamentals of Photonics*, John Wiley and Sons, New York, 1991.
Schawlow, A.L.: "Lasers and Light," *Readings from Scientific American*, Freeman, San Francisco, 1969.
Ward, B.: "Laser Technology," *Sky Magazine*, August 1979.
Waymouth, J.F.: *Electric Discharge Lamps*, MIT Press, Cambridge, Mass., 1978.
Whitney, K., K. Gailey, and R. Davis: *General Chemistry*, Saunders College Publishing, Philadelphia, 1988.

Chapter 6

Lamps

6.1 INTRODUCTION

There are nearly 6000 different lamps being manufactured today, most of which can be placed in the following six categories: incandescent, tungsten halogen, fluorescent, mercury, metal halide, and sodium. In this chapter, each of these categories will be considered in detail, tracing the history of development of the various lamp types, explaining their engineering design, and delineating their electrical and illumination parameters.

The major characteristics to be considered when choosing a lamp are its luminous efficacy, life, lumen depreciation, and color. Other characteristics that may also need to be considered are lamp temperature, burning position, controls (ballasting, dimming), and warm-up time. Luminous efficacy is the measure of the lamp's ability to convert input electric power, in watts, into output luminous power, in lumens, and is measured in lumens per watt (lm/W). If, in order to operate the lamp properly, auxiliary equipment is required, the watts dissipated in that equipment must be charged to the lighting system. This is always true in the case of fluorescent, mercury, metal halide, and sodium lamps, which require a ballast (inductor, capacitor, auto transformer, or electronic) for starting and operating the lamp. If a fluorescent lamp, for example, emits 3000 lm and consumes 40 W, and its ballast consumes 10 W, the lamp-ballast efficacy would be listed as 60 lm/W, not 75 lm/W.

The life of a lamp is the number of hours for approximately 50% of a large group of lamps of the same kind to fail. Failure generally implies that the lamp will no longer light, although in a few cases it means that the light output has dropped to a specific percentage value. It would be ideal if all lamps of the same kind burned out at the same time, but this is not the case. In a group of 100 lamps all rated at 1000 h, it is likely that 10 of them will have failed by 700 h, and 5 of them may last as long as 1400 h. However, at 1000 h, it is expected that about 50 of them will have failed. The graph showing the rate of failure of a large group of lamps is called a "mortality curve". It would be nice if such a curve resembled a vertical wall with all lamps burning for a certain number of hours and then failing—but it doesn't.

Lumen depreciation during life is also a characteristic of all lamps. This process of "aging" is caused by many factors and can result in a lamp emitting 70% or less of its initial lumens when it reaches rated life. It is important in lighting design to include lumen depreciation. If a lighting task requires 500 lux and the designer installs just enough lighting to provide that amount initially, the lighting level may fall to 350 lux as the lamps approach rated life. Often a compromise is sought which provides the design illumination at, say, 70% of life, which means that the installation will provide more than 500 lux initially and less than 500 lux at 100% life.

Finally, there is the matter of color, or more properly, color rendering and color temperature. Each of the lamp types does not provide the same nominal "white." Their differences in spectral distribution can produce two effects within a lighted space. First, some of the colors of objects within that space can appear unnatural or faded—reds can appear brown, violets nearly black, etc. This is measured in terms of color rendering index (CRI) and will be developed in Chap. 7. Second, the entire space may "feel" warm or cool. For example, a mercury lamp, because it is lacking in reds and oranges, can make a space seem cool, whereas an incandescent lamp, with deficiencies in the blues and violets, can make a space feel warm. Although personal preference, intended use of a space, and climate often govern the choice of warm versus cool, as a general guideline, a room that has a northern exposure should have "warmer" lamps—richer in reds and yellows—whereas a room with a southern exposure should be lit with "cooler" sources—richer in blues and greens. The "warmness" or "coolness" of a light source was quantified in Chap. 4 in terms of its correlated color temperature.

6.2 INCANDESCENT LAMP DEVELOPMENT

The incandescent lamp preceded all others in development and, aside from the carbon arc lamp and the Cooper Hewitt mercury pool lamp, was the only type available until fluorescent and mercury lamps appeared in the 1930s. Edison's group at his Menlo Park, N.J. laboratory began working on the development of an incandescent lamp in the fall of 1877. Edison was a relatively late starter in this work, as there were at least six other individuals in the 1870s working along similar lines—Joseph Swan and St. George Lane-Fox in England and Moses Farmer, Hiram Maxim, William Sawyer, and Albon Mann in the United States. Edison won the race in developing a commercially practical lamp for two main reasons:

1. He realized the need for an electrical system, not just a lamp. He rejected the series-circuit distribution system commonly used in arc lighting, reasoning that each lamp must be capable of operating independently of the others. This led him, in conjunction with Farmer, to develop a new type of dynamo, a shunt machine providing a constant voltage output of 110 V at close to 90% efficiency. The series dynamos used in arc lighting provided constant current and were roughly 50% efficient.
2. All incandescent lamps invented prior to 1879 had used thick low-resistance radiating elements. Edison, displaying considerably greater understanding of Ohm's law than his contemporaries, reasoned that a high-resistance filamentary element was necessary. This was a significant contribution to lamp design—the use of a thin, threadlike filament of high resistance that could be heated to incandescence.

With these decisions made, Edison turned to a consideration of what filament material to use. Before deciding that he needed a high-resistance element, Edison had rejected carbon and begun working with platinum. However, platinum was very expensive, and the temperature at which it became incandescent was very close to its melting temperature. Thus, in 1879, he returned to carbon, finally settling on carbonized cotton sewing thread as the filament material.

As is so often the case, one technological breakthrough must occur before another can transpire. Early lamp developers were plagued by being unable to produce a high vacuum inside the bulb. With some gases remaining, their filaments expired in a matter of minutes. The development of the vacuum pump by Herman Sprengel, a German chemist working in England, solved this problem in the late 1860s. Edison used Sprengel's pump to evacuate his lamps. He noticed that not only was it necessary to evaluate the lamp when the filament was cold, but it was also necessary to continue the evacuating process as the filament was raised to its operating temperature. This is because residual gases are released from both filament and glass bulb as temperature rises.

Edison's first practical carbonized sewing thread filament lamp was lit on October 19, 1879. It had an efficacy of 1.4 lm/W, had a color temperature of 2000 K and burned for 40 h, after which Edison raised the applied voltage slightly and it expired. Public announcement of Edison's success occupied the entire front page of the *New York Herald* on December 21, 1879, as a means of announcing a public "open house" at Menlo Park on New Year's Eve. Several thousand people visited the park that evening to see 60 of the new lamps lighting the buildings and pathways.

Since Edison's first lamp, there have been many improvements in lamp design leading to the incandescent lamps of today. A few of the most important of these are now presented.

Filament Materials Although carbon has a very high melting point (3870 K), it was necessary to operate carbon filaments much below this temperature because of rapid evaporation and resulting short life. A number of alternative materials were tried, but ultimately tungsten was chosen.

Tungsten is obtained from wolframite ore, an oxide of tungsten, manganese, and iron ($FeMnW0_4$), and Scheetite ore, an oxide of tungsten and calcium ($CaW0_4$). These compounds are reduced to oxides and then to tungsten powder by heating in a hydrogen atmosphere. In 1904, two chemistry assistants in a Vienna high school, A. Just and F. Hanaman, devised a procedure for converting tungsten powder into filaments. They made a paste of the powder, squirted it through diamond dies, then heated it in a hydrogen-nitrogen atmosphere to remove carbon and leave loops of pure tungsten. These early tungsten filaments (1907) provided 7.8 lm/W and had lives of 800 h, a vast improvement over Edison's first lamp. Today the tungsten oxides are doped with solutions of aluminum oxide, potassium oxide, and silicon dioxide to increase the strength of the filament.

Tungsten offers three major advantages over other filament materials. First is its high melting point (3650 K)—it can be operated at a higher temperature than any other metal, thus giving relatively high luminous efficacy (most incandescent filaments operate between 2800 and 3000 K). Second, its vapor pressure is the lowest of any metal at elevated temperatures, assuring minimum evaporation and, thus, longer life and better lumen maintenance (less bulb-wall blackening). Third, tungsten's emissivity is higher in the visible spectrum than in the infrared, again optimizing luminous efficacy.

Ductile Tungsten The problem with tungsten was that the filaments were very fragile and costly. The lamps were packed in cotton wadding for shipment but, even so, there was much filament breakage.

During the period from 1906 to 1910, Dr. William Coolidge, a General Electric research scientist noted also for his development of the x-ray tube, developed a process for converting crystalline tungsten into fibrous tungsten, with the latter being very ductile and having five times the tensile strength of steel. The process involves converting tungsten powder into ingots. These are sintered and swaged to produce tungsten rods. The rods are then drawn through a series of diamond dies, each slightly smaller in diameter than its predecessor, to create tungsten wire finer than a human hair. The new filament had an efficacy of 10 lm/W and, after 1911, replaced all previously-used filament materials.

Gas Filling Even the new tungsten filaments evaporated more rapidly than was desired, and a black tungsten deposit appeared on the inside of the bulb, which reduced light output. At first researchers felt the problem could be solved by creating an even better vacuum inside the bulb, but improvement was minimal.

Dr. Irving Langmuir, a General Electric chemist, began working on this problem in 1909. He first discovered that the presence of even minute amounts of water vapor (10 parts per million) inside the lamp greatly increased the amount of tungsten carried from the filament to the bulb wall. Next he examined the effect of adding gases, such as nitrogen, and found that these reduced evaporation significantly. The nitrogen formed a blanket around the filament, retarding evaporation and reducing the ability of the water vapor to pick up those tungsten particles that were evaporated. However, the nitrogen also took heat away from the filament, reducing its temperature. Langmuir found that the loss in light output was greater than the gain from reduced blackening.

Langmuir continued his work and noted that the increase in heat loss was proportionately less for larger filaments. This led him to coil the filament to make it more compact and resulted in lamps with much improved efficacy for the same life*. Today double-, and even triple-, coiled filaments are used, and all incandescent lamps of 40 watts and above are gas filled. Vacuum lamps are designated as Class B; gas-filled lamps as Class C.

Inside Frosting Early incandescent lamps used clear glass bulbs and produced disturbing glare. Several methods of treating the outside of the bulb were tried, but reduction in light output was too severe. An attempt was made to acid-etch the inside of the bulb but this produced many small jagged corners, which weakened the bulb and caused early breakage.

In 1925 Marvin Pipkin of the General Electric Lamp Development laboratory decided to re-etch the inside of a previously etched bulb with a dilute solution of acid. The effect was to round off the sharp corners and fill in the crevices, producing a strong bulb that diffused the light to such an extent that the glowing filament could barely be seen. Such lamps are called *inside frosted*. In recent years, the inside frosting has been replaced by a thin coating of silica powder, applied electrostatically. This spreads the filament image, thus giving partial diffusion of the light. In 1952, Pipkin devised a new treatment for the inside of the bulb, a thick fume coating of white silica powder. The "soft white" lamp provides nearly perfect diffusion, with negligible reduction in lumen output.

The element neodymium was discovered by Baron Carl Auer von Welsbachin in 1885. In the form of neodymium oxide, it has recently been added to the glass bulbs of some incandescent lamps, giving a light that is more bluish-white and cooler-appearing than that of a conventional lamp, as the result of filtering out the yellow wavelengths.

---

* In a coiled filament, heat loss is determined by the surface area of the coil, not the surface area of the filament, as was noted in Sect. 4.8.

The claim is that the neodymium lamp makes objects appear clearer and sharper, thus enhancing visibility.

Reflectorized Lamps The idea of one-piece, built-in reflectorized lamps occurred in the mid-1930s. Prior to that time, automobile head lamps, for example, consisted of a small lamp bulb mounted in an auxiliary reflector with a removable lens cover. Then Daniel Wright, a General Electric researcher, inserted two terminals into a common custard cup, aluminized the inner surface, mounted a filament, and sealed a glass cover to the cup. The result was the first sealed-beam lamp and has led to the PAR lamp introduced in Sect. 2.10, to the one-piece auto headlight, which appeared first on 1939 cars, and to the multifaceted reflector (MR) lamp, a tungsten-halogen capsule lamp that has become dominant in retail track and down lighting. PAR and MR lamps will be discussed in Sect. 6.7

Tungsten Halogen Lamps In 1958, E.G. Fridrich and E.H. Wiley discovered that introducing a halogen gas, originally iodine, into an incandescent lamp could increase the lumens per watt and improve the lumen maintenance significantly. The reason for this lies in the halogen regenerative cycle.

In an ordinary incandescent lamp, evaporated tungsten particles are transported to the inside of the bulb by gas convection currents. The bulb wall is relatively cool, and the particles adhere to it. The introduction of iodine causes the tungsten atoms to combine with the iodine vapor and, instead of tungsten depositing on the bulb wall, it recycles back to the filament. These lamps will be discussed in Sect. 6.6.

6.3 CONSTRUCTION OF AN INCANDESCENT LAMP

An incandescent lamp is shown diagrammatically in Fig. 6.1. Its parts will be considered in some detail.

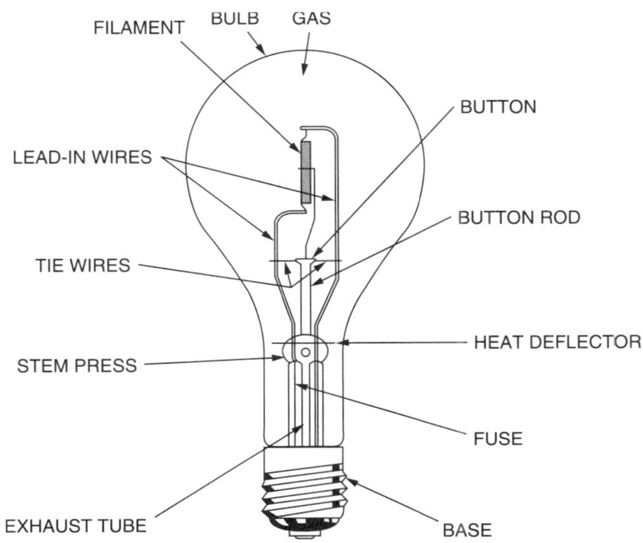

Figure 6.1 Parts of an incandescent lamp. ***Source*: *Incandesent Lamps*** **(IN003), 1996. Courtesy of Osram Sylvania.**

Bulbs

Bulbs are manufactured in two, quite different, ways. General service lamps have blown bulbs and are generally of soft, or lime, glass, composed primarily of silicon dioxide, with additives of soda or lime, and having a maximum temperature rating of about 650 K. They are mass produced by a ribbon machine, first introduced by Corning Glass Works in 1927. This unique machine takes molten glass from a furnace, presses it into a ribbon, and automatically blows it into molds as the ribbon moves along a conveyor belt. Current ribbon machines can produce up to 50,000 bulbs per hour.

The second manufacturing method is a pressed glass technique, used in the manufacture of sealed-beam lamps. These bulbs are made of hard, heat-resistant borosilicate (silicon dioxide and boric acid) glasses that can withstand temperatures up to 750–800 K. The glass is heated to soften it, cut into chunks, and placed between the two halves of a metal die. When the halves are closed, the glass is pressed into the shape of the die cavity. The process allows the accurate forming of reflectors and lenses, which are later fused together to complete the lamp.

A third glass is fused quartz, used in tungsten halogen capsules and in high-intensity discharge arc tubes. As with other glasses, quartz is basically silicon dioxide, but in its vitreous, rather than crystalline form. Its melting point is approximately 2000 K, and it can be used at working temperatures of 1300 K. Unlike other glasses, it transmits well in the ultraviolet, beginning at about 200 nm, and extending through the near infrared to 4000 nm. Also it has high resistance to thermal shock.

In addition, aluminosilicate glass is used in some single-ended tungsten halogen lamps, and lead-alkali silicate glasses are used for the internal glass parts in incandescent lamps, the latter being replaced in recent years by lead-free glasses containing barium oxide.

Lamp bulbs come in a variety of shapes and sizes. The lamp manufacturers use a letter-number code to distinguish between bulb types. The letter denotes the bulb shape and the number gives its diameter in eighths of an inch. As examples, a PS30 lamp has a pear-shaped bulb, 3.75 in in diameter; a T12 bulb is tubular with a diameter of 1.5 in; an R40 bulb is a reflector lamp envelope with a diameter of 5 in. The common household incandescent lamp, available in 25-, 40-, 60-, 75-, 100-, 150-, and 200-W sizes, has a tear-drop-shaped bulb, given the letter A for "arbitrary" shape.

Colored bulbs are manufactured in a variety of ways. Coated bulbs are made by either spraying or applying a fused enamel to either the inside or outside of the bulb. Dichroic lamps employ thin interference film coatings on the inside of the bulb that produces color by selective spectral transmission.

Some bulbs, designed for rough service, have rubberized silicone coatings.

Fill Gas

In his early work, Langmuir used pure nitrogen as the filling gas for incandescent lamps, primarily to prevent arcing between adjacent filament coils or lead-in wires. Later he found that the heat loss from the filament could be further reduced by the use of an inert gas such as argon, but pure argon produced a glow around the filament, which increased bulb blackening significantly.

All incandescent lamps undergo several flush/exhaust cycles to remove impurities before the fill gas is added to those of 40 W and above. Ninety-eight percent of today's lamps use a mixture of roughly 95% argon and 5% nitrogen. Occasionally krypton, with lower heat conductivity (reducing filament evaporation) but considerably greater cost, is used instead of argon in high-wattage lamps, and gives an increase of up to 10% in efficacy. The gases are introduced into the lamp at about 80% of atmospheric pressure.

It is mind-boggling to note that one drop of water, entering with the fill gas and distributed uniformly among 50,000 lamps, will produce severe bulb blackening in all of them. This illustrates the extremely close control required in incandescent lamp manufacture.

Filaments

Filaments are labeled S (straight wire), C (coiled), and CC (coiled-coil) A number after the letter indicates the filament configuration, i.e., a CC-8 filament is coiled-coil of configuration 8. Most incandescent lamp filaments today are either singly- or doubly-coiled. For example, the filament for a 120-V, 60-W lamp has a diameter of 0.0045 cm and an uncoiled length of 53 cm. After its first coiling (1130 turns), the filament length is about 8 cm. The second coiling reduces the length to about 2 cm.

In filament manufacture, a machine wraps tungsten wire in a single coil around a molybdenum mandrel. The coil and mandrel are then placed in a furnace fired with hydrogen to anneal the tungsten. Often the coil is then coiled again around a removable steel mandrel. Next it is cut to size, sent through the furnace again, and placed in an acid bath to remove the original molybdenum mandrel. If filament diameter at any point is 1% less than specified, filament life may be reduced by as much as 25%. This again indicates the critical need for close manufacturing controls.

For many years, filaments were mounted perpendicularly to the lamp axis, with numerous supports to prevent sagging. Contemporary filaments are mounted axially. With either base-up or base-down burning, this decreases bulb blackening, thus improving lumen maintenance and providing more lumens per watt.

Supports and Lead-In Wires

The assembly that holds the filament consists of a flare tube, an exhaust tube, and two lead-in wires. The exhaust tube and lead-in wires are inserted into the flare tube, and the unflared end is heated to soften the glass and then clamped to form a flat section called a "stem press", which seals the wires and exhaust tube in the flare tube. Just before the stem press solidifies, a small hole is blown to reopen the exhaust tube.

The lead-in wires are made of copper from the base to the seal, Dumet wire at the seal, (a nickel-iron alloy with a copper sleeve that has the same temperature coefficient of expansion as the glass), and nickel or copper from the seal to the filament. A fuse wire is included in one of the lead-in wires, which activates if the filament fails and arcs across the broken coil section. Following sealing, the lead-in wires are bent to accommodate the filament, which is mechanically clamped in place.

There are a variety of filament shapes, requiring varying numbers of support wires. In some lamps a glass rod, called a "button rod", projects outward from the stem press and holds these additional supports, which are usually of molybdenum. Molybdenum is particularly well-suited because of its high melting point (2900 K), and it is also used for lead-in wires in projection lamps.

Neck and Base

A heat-deflecting disc is placed just above the neck and base of the bulb in some higher-wattage lamps to reduce the circulation of hot gases into the stem press and base regions.

The exhaust tube, which projects beyond the base during manufacture, terminates in the stem press. Exhausting was done originally through the top of the lamp, leaving a rather dangerous glass tip. In contemporary lamps, the exhausting and adding of gases is done through the bottom of the lamp, after which the exhaust tube is sealed and cut off inside the base.

The base itself is of brass or aluminum. It may be cemented in place or mechanically anchored. For cemented bases, base temperature must not exceed 500 K. One lead-in wire is attached (soldered or welded) to the side of the base, the other to a contact in the center of the base, which is insulated from the rest of the base by a glass disc. There are four standard screw-type bases and several bayonet and prefocus bases, the latter to position filaments accurately.

The automatic ribbon machine has been mentioned in connection with bulb production. The rest of the incandescent lamp manufacturing process is also highly-automated and consists of three complex machines. The first is a mount machine, which prepares the stem press unit consisting of flare tube, button rod, exhaust tube, filament, supports, and lead-in wires. Stem press units are transferred by conveyor to a sealing and exhaust machine, where each one is inserted into a bulb and sealed. Next the lamp is exhausted, the fill gases are added, and then the exhaust tube is sealed. Each lamp is then sent to a third machine, which connects the lead-in wires to the base, attaches the base, and lights the completed lamp.

6.4 LIGHT, LIFE, AND LOSSES

The goals of any lamp design are that the lamp produce light most economically for the service intended and that lamps having the same ratings be uniform from one lamp to another with respect to lumen output, power input, and life.

In the design of an incandescent radiator, there is always a tradeoff between light and life. At one extreme, a photoflash lamp emits a million peak lumens, but lasts only 50 milliseconds. At the other extreme, a stove-top heating unit glows a dull red, giving negligible lumens, but a life measured in years. Between these extremes are the myriad of incandescent lamps in today's marketplace.

The maximum luminous efficacy of a blackbody radiator is 95 lm/W and occurs at an operating temperature of 6800 K. Tungsten melts at 3650 K and is a selective radiator. Thus, the maximum efficacy of an uncoiled tungsten filament, if it had zero life, would be 52 lm/W. To gain life, the filament temperature must be reduced well below tungsten's melting point. For example, a standard 100-W incandescent lamp operates at a filament temperature of 2850 K, giving it a maximum efficacy of about 20.5 lm/W if all of the input power is radiated beyond the bulb; but an incandescent lamp has losses in the form of heat conducted away from the filament by the lead-in and support wires (end loss), heat carried away from the filament by the gas (gas loss), and direct radiation absorbed by the bulb and base (bulb and base loss). In the 100-W lamp, these losses amount to 18% of the power input, leaving a realized initial efficacy of 16.9 lm/W.

The energy distributions of several 120-V, inside-frosted, general service incandescent lamps are shown in Table 6.1, and the operating data for these same lamps are shown in Table 6.2. All lamps in these tables have single-coiled filaments except the 60-, 75-, 100-, and 500-W sizes, whose filaments are double-coiled. The first three lamps are vacuum; the last seven are gas filled.

TABLE 6.1 GENERAL SERVICE INCANDESCENT LAMP ENERGY DISTRIBUTIONS

	% Radiation Beyond Bulb			% Energy Losses			
Watts	Light	Infrared	Total	End	Gas	Bulb and Base	Total
6	6.0	87.0	93.0	1.5	—	5.5	7.0
10	7.1	86.4	93.5	1.5	—	5.0	6.5
25	8.7	85.3	94.0	1.5	—	4.5	6.0
40	7.4	63.9	71.3	1.6	20.0	7.1	28.7
60	8.4	73.3	81.2	1.2	13.5	4.5	19.2
75	8.4	72.8	81.2	1.2	12.8	4.8	18.8
100	10.0	72.0	82.0	1.3	11.5	5.2	18.0
200	10.2	67.2	77.4	1.7	13.7	7.2	22.6
300	11.1	68.7	79.8	1.8	11.6	6.8	20.2
500	12.0	70.3	82.3	1.8	8.8	7.1	17.7

TABLE 6.2 GENERAL SERVICE INCANDESCENT LAMP CHARACTERISTICS

Watts	Filament Temp. (K)	Color Temp. (K)	Initial Lumens	Initial lm/W	Life (h)	Uncoiled Filament Length (cm)	Filament Diameter (μm)
6	2400	2370	40	6.7	1500	37.1	11.4
10	2420	2450	70	7.0	1500	43.2	16.3
25	2580	2550	210	8.4	2500	56.4	30.5
40	2740	2770	460	11.5	1500	38.1	33.0
60	2770	2800	890	14.8	1000	53.3	45.7
75	2840	2860	1210	16.1	750	54.9	53.3
100	2850	2870	1750	17.5	750	57.9	63.5
200	2900	2930	3900	19.5	750	63.5	96.5
300	2940	2940	5850	19.5	750	76.2	129.5
500	2950	2960	10,000	20.0	1000	87.4	180.3

An analysis of the losses in Table 6.1, as a function of power rating, is instructive. Consider end loss first. This is the percent of the input power conducted away from the filament by the various wires connected to it. End loss increases as wattage increases, except for a decrease for the three double-coiled filaments. This should be expected. The larger the wattage, the longer the filament and the greater the number of supports required for a given filament configuration. Double-coiling reduces the number of supports needed, thus reducing end loss.

The percentage of input energy conducted away from the filament by the gas decreases as the wattage increases, for a given filament configuration. This is because the filament, regardless of its size, is surrounded by a sheath of hot gas of essentially constant thickness. Larger-wattage lamps have thicker filaments, and thus their percentage heat loss to the gas is less. Also, double-coiled filaments have less percent gas loss than do single-coiled filaments, because of less surface area exposed to the gas sheath.

Bulb and base losses result from direct filament radiation absorbed by the bulb and base. A portion of this energy is reradiated at longer wavelengths, and the remainder is conducted and convected to the surrounding air. Percent bulb and base loss increases slightly with wattage because higher-wattage lamps operate at higher temperatures and thus emit more radiation, some of which is absorbed by the bulb and base.

Initial lumens and lumens per watt are given in Table 6.2. The light output of an incandescent lamp decreases over the life of the lamp because of tungsten evaporation, which decreases the diameter of the filament, lowering filament temperature, and, more importantly, causing the bulb to blacken. The decreases are approximately a straight line, with those lamps having double-coiled filaments showing the least depreciation. The effect of bulb blackening depends on bulb orientation. In base-up operation, the tungsten is deposited mainly in the neck, and the light loss is less than for other orientations. Typically lamp lumens at 100% life are 82 to 84% of initial lamp lumens.

The power consumed also decreases as the lamp ages because the filament evaporation narrows the filament, thus increasing its resistance. The wattage decrease is less than the lumen decrease, and thus the luminous efficacy decreases as the lamp ages.

The correlated color temperature (CCT) of an incandescent lamp is related to its luminous efficacy, over quite a wide range of wattages. This approximate relationship is illustrated in Table 6.3 for gas-filled lamps. Such data are useful in photographic and theatre applications because, from the lumens per watt figure, the designer can use Table 6.3 to estimate the CCT of a particular incandescent lamp.

TABLE 6.3 CCT VERSUS EFFICACY FOR INCANDESCENT LAMPS

CCT (K)	Approximate Lumens per watt
2160	4
2430	8
2650	12
2820	16
2990	20
3120	24
3260	28

The resistance of tungsten is a strong function of temperature. The "cold" resistance is only 6 to 8% of the "hot" resistance, resulting in an inrush current of greater than 10 times the rated current, but lasting for only a fraction of a second.

Typical catalog information for a 100-watt incandescent lamp and the meaning of each entry are shown below:

100A/W, 100, A19, Med., 120, C, CC-8, 750, 1710, $3\frac{1}{8}$, $4\frac{7}{16}$

100A/W	ordering abbreviation (W indicates soft white)
100	lamp wattage in watts
A19	tear-shaped bulb, 19/8 in. maximum diameter
Med.	medium screw base
120	lamp voltage in volts

C	class C (gas filled)
CC-8	coiled-coil filament of configuration 8 (axially mounted)
750	average rated life in hours
1750	initial lumens
$3^1/_8$	light center length in in. (distance from the center of the filament to the bottom of the base)
$4^7/_{16}$	maximum overall length in in. (distance from the top of the bulb to the bottom of the base)

Interest in conserving energy and prolonging lamp life has led to the development of long-life and energy-saving substitutions for standard incandescent lamps. A representative sample of these is given in Table 6.4, with the standard lamps included for comparison. The energy-saving lamps operate at 10 to15% lower wattages than their standard counterparts, and yield 82 to 88% of the lumens, resulting in a 2 to 4% decrease in efficacy. The long-life lamps show a 10 to 13% decrease in lumens and efficacy, whereas the long-life energy savers give a 23 to 27% decrease in lumen output and a 13 to15% decrease in efficacy. The penalty for using energy-saving lamps or long-life lamps, or both, is reduced lumens per watt. If it is determined that the illuminance level for a particular task is somewhat high, then an energy-saver lamp may be the answer, or, if it is determined that replacing a lamp is difficult, a long-life lamp may be appropriate. Otherwise, the standard lamps are the choice based on efficacy*.

TABLE 6.4 ENERGY-SAVING AND LONG-LIFE LAMPS

	40-watt				60-watt			
	STD	ES	LL	LLES	STD	ES	LL	LLES
watts	40	34	40	34	60	52	66	52
lumens	460	375	400	340	890	750	800	650
life (hrs)	1500	1500	3000	2500	1000	1000	2000	500
lm/W	11.5	11.0	10.0	10.0	14.8	14.4	13.3	12.5
	75-watt				**100-watt**			
	STD	ES	LL	LLES	STD	ES	LL	LLES
watts	75	67	75	67	100	90	100	90
lumens	1210	1065	1080	915	1750	1550	1580	1350
life (hrs)	750	750	1500	2500	750	750	1500	2500
lm/W	16.1	15.9	14.4	13.7	17.5	17.2	15.8	15.0

Note: STD = standard, ES = energy-saving,
LL = long-life, LLES = long-life, energy-saving

6.5 VOLTAGE EFFECTS

From the time of the first incandescent lamp, it was known that lamp performance, particularly lamp life, was seriously affected by changes in applied voltage. In 1932, Barbow and Meyer at the National Bureau of Standards used curve-fitting techniques to quantify this effect for variations in lumen output (Φ), and power input (P) with

* It should be understood that all incandescent lamps fail the "efficacy test" when compared with tungsten-halogen or gaseous-discharge sources.

applied voltage (V) in the following equations:

$$\log\frac{\Phi}{\Phi_0} = k_1 \log\frac{V}{V_0} + k_2 \log\left(\frac{V}{V_0}\right)^2$$
$$\log\frac{P}{P_0} = k_3 \log\frac{V}{V_0} + k_4 \log\left(\frac{V}{V_0}\right)^2 \qquad (6.1)$$

where the subscript 0 refers to rated values. Barbow and Meyer determined values for the constants experimentally, for certain types of both vacuum and gas-filled lamps. They also showed that for applied voltages within 77 to 130% of rated value, the squared terms in Eq. (6.1) could be neglected if errors in Φ and P of up to 5% could be tolerated.

Later, similar equations were developed relating life and voltage and CCT and voltage. The result is a set of four equations from which variations in all incandescent lamp parameters with voltage can be approximately determined:

$$\frac{\Phi}{\Phi_0} = \left(\frac{V}{V_0}\right)^k \qquad (6.2)$$

$$\frac{P}{P_0} = \left(\frac{V}{V_0}\right)^n \qquad (6.3)$$

$$\frac{L}{L_0} = \left(\frac{V_0}{V}\right)^d \qquad (6.4)$$

$$\frac{CCT}{(CCT)_0} = \left(\frac{V}{V_0}\right)^m \qquad (6.5)$$

The exponents in these equations are functions of lamp type, wattage, and voltage. Values are given in Table 6.5 and should be considered as averages over a typical operating range.

To illustrate how other relationships may be derived from these basic equations, suppose it is desired to know how lamp current varies with voltage. It can be written

$$\frac{I}{I_0} = \frac{P/V}{P_0/V_0} = \frac{P}{P_0}\left(\frac{V_0}{V}\right) = \left(\frac{V}{V_0}\right)^n\left(\frac{V_0}{V}\right) = \left(\frac{V}{V_0}\right)^{n-1} \qquad (6.6)$$

Thus the constants to use in expressing the current ratio are 0.54 for gas filled lamps and 0.58 for vacuum lamps.

From Table 6.5, it can be seen that life is the most sensitive parameter to voltage variation. Five percent overvoltage yields roughly 50% life. This represents 126 V for

TABLE 6.5 INCANDESCENT LAMPS—VOLTAGE VARIATIONS

Constant	Gas-Filled	Vacuum
k	3.38	3.51
n	1.54	1.58
d	13.1	13.5
m	0.42	0.42

% Rated Volts	% Rated Lumens	% Rated watts	% Rated Life	% CCT
80	47	71	1860	91
85	58	78	840	93
90	70	85	398	96
92	75	88	298	97
95	84	92	196	98
98	93	97	130	99
100	100	100	100	100
102	107	103	77	101
105	118	108	53	102
108	130	113	36	103
110	138	116	29	104
115	160	124	15	106
120	185	132	9.2	108
125	213	141	5.4	110

lamps rated at 120 V, and it is not at all uncommon for utility voltages to be this high to some residential customers.

EXAMPLE 6.1

Baseball fields used to be lit with incandescent lamps rated at 120 V, 1000 W, and 1000 h life. Assume 50 night games, each requiring 4 h of lamp burning, are played each season. At what voltage should the lamps be operated so that they will be at 70% of their rated life at the end of the season (which would mean about 90% survivors)? What percentage of the lamps required when 120 V is used would the owner use at this new voltage, while still maintaining the same initial illumination level? What percentage of the energy consumption at 120 V will be saved over the season?

Solution. The required lamp life is 200/0.7 = 286 h. From Table 6.5 this requires 110% of rated voltage, or 132 V. The lumen output at this voltage is 38% more than at 120 V. Thus only 100/1.38 = 72.5% as many lamps are required for the same initial illumination level. The wattage consumption per lamp will be 16% higher than at 120 V, but, with only 72.5% of the lamps, the net effect is a reduction in total initial power to 1.16(0.725) (100) = 84%. The average percentage of lamps surviving during the season is 95%. Thus the percentage energy saved over the season is 100[1 – 0.84(0.95)] = 20%.

EXAMPLE 6.2

It is quite common to find incandescent lamps rated at 130 volts used on 120-volt circuits. What are the results of doing this for a 100-watt gas-filled lamp?

Solution. In this case, $V/V_0 = 120/130 = 0.923$. From Eqs. 6.2–6.5,

$$\frac{\phi}{\phi_0} = 0.923^{3.38} = 0.76$$

$$\frac{P}{P_0} = 0.923^{1.54} = 0.88$$

$$\frac{L}{L_0} = \left(\frac{1}{0.923}\right)^{13.1} = 2.85$$

$$\frac{CCT}{(CCT)_0} = 0.923^{0.42} = 0.967$$

From Table 6.1, the lamp, at its rated voltage, provides 100 W, 1750 lm, 17.5 lm/W, and 750 hours life. These become 88 W, 1330 lm, 15.1 lm/W and 2140 hours. The result is a long-life lamp, consuming less energy, and providing less illumination at a lower efficacy, and it will appear warmer.

6.6 TUNGSTEN-HALOGEN LAMPS

Tungsten-halogen (TH) lamps were introduced in Sect. 6.2. They are governed by the halogen regenerative cycle, which is shown in Fig. 6.2.

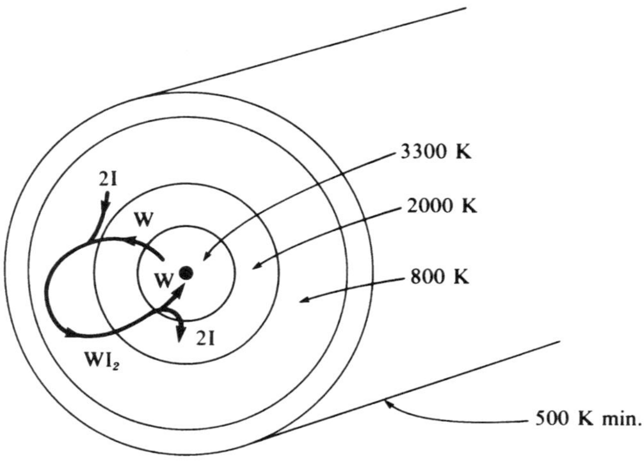

Figure 6.2 Halogen cycle.

When iodine (or bromine) is introduced inside the bulb, and the temperature near the filament is maintained at about 2000 K, the evaporated tungsten atoms combine with the iodine vapor to form tungsten iodide. The tungsten iodide is carried by the convection currents to the bulb wall, but if the bulb temperature is kept between 500 and 1500 K, it does not adhere to the bulb wall, but rather circulates back toward the filament. Very near the filament, where the temperature exceeds 2800 K, the tungsten iodide is reduced to tungsten and iodine vapor. The former redeposits on the filament, and the latter is free to acquire another tungsten atom and repeat the cycle. The chemical equations for these relations are:

$$W + 2I \xrightarrow{2000\ K} WI_2, \qquad WI_2 \xrightarrow{2800\ K} W + 2I \tag{6.7}$$

Whereas an incandescent lamp can be expected to provide only about 82 to 84% of its initial lumens at the end of life, the figure for a TH lamp is above 95%.

Because of the high bulb-wall temperatures required, it is necessary to use fused quartz for TH lamps. Quartz has a softening point of 1900 K, far above the temperature requirements of the halogen cycle. The stronger bulb and smaller volume mean that the lamp can be operated up to several atmospheres of internal gas pressure. For a given filament temperature, this higher pressure reduces the rate of filament evaporation, thus lengthening life; or, the lamp can be operated at a higher filament temperature for the same life, with resulting higher luminous efficacy. Parenthetically it should be noted that if each tungsten particle could be guided back to the spot on the filament from which it originated, the life would be infinite. Such is, of course, not the case, and thus life can be increased, but remains finite. Actually, the returning tungsten tends to deposit on the cooler portions of the filament, allowing the regions of higher temperature (hot spots) to continue to evaporate more rapidly.

Iodine was the halogen gas used in early TH lamps; bromine is used predominantly today. Bromine is colorless, whereas iodine can impart a purplish tint. Also bromine permits about 50 K lower bulb-wall temperature. In addition to the halogen gas, TH lamps are filled with a mixture of xenon and krypton. Correlated color temperatures of TH lamps range from 2800 K to 3400 K, and their lives are 2000 h or more. TH lamps may be dimmed over their full range of lumen output, but should be operated at full voltage for occasional brief periods to "clean up" any tungsten particles that have been deposited on the bulb wall.

TH lamps can have a number of shapes, but are most often tubular with the filament oriented axially. They are available in both double-ended and single-ended types, with the latter having a small screw base. The two types are illustrated in Fig. 6.3 and a summary of characteristics is given in Table 6.6.

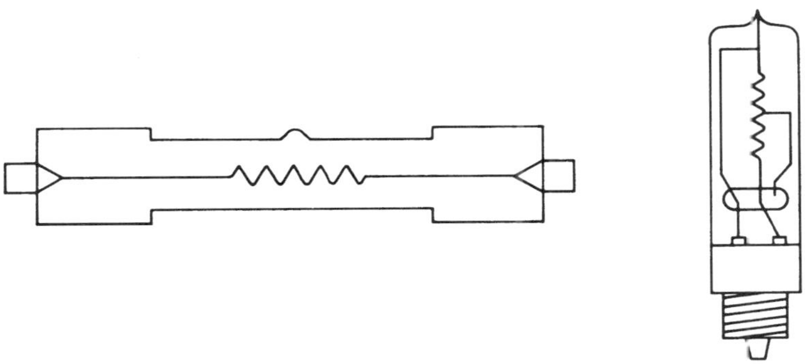

Figure 6.3 Double- and single-ended tungsten-halogen lamps.

TABLE 6.6 TUNGSTEN-HALOGEN LAMPS

Watts	Bulb	Light-Center Length* (in)	Life	Lumens	Lumens per watt
		Double-Ended			
100	T3	1	2000	1,600	16.0
200	T3	$2\frac{3}{8}$	2000	3,600	18.0
300	T4	$\frac{5}{8}$	2500	5,000	16.7
500	T3	2	2000	8,750	17.5
1000	T6	$1\frac{1}{8}$	2000	22,000	22.0
		Single-Ended			
100	T4	$1\frac{3}{8}$	1000	1,800	18.0
150	T4	$1\frac{1}{2}$	2000	2,800	18.7
250	T4	$1\frac{5}{8}$	2000	5,000	20.0
500	T4	2	2000	10,450	20.9
		A-Line			
42	MB19	$3\frac{1}{8}$	3500	570	13.6
52	MB19	$3\frac{1}{8}$	3500	770	14.8
72	MB19	$3\frac{1}{8}$	3500	1,500	16.0

* Lighted length for double-ended lamps.

Many TH lamps are a "lamp within a lamp", consisting of a tungsten filament mounted inside a small fused-quartz gas-filled capsule, which in turn, is mounted inside an outer glass and/or reflector envelope—A-line, PAR or MR. The A-line lamps, with a heavy "mid-break" bulb, are energy-efficient replacements for 60-W, 75-W and 100-W standard incandescent lamps, providing much longer life than their standard counterparts, but lower lumen output. Their properties are listed in Table 6.6. PAR and MR TH lamps will be discussed in the following section.

As shown in Table 6.1, a large percentage of the radiation beyond the bulb in an incandescent lamp is at infrared wavelengths (750–2000 nm). In halogen IR lamps, some of this energy is redirected to the filament, reducing the amount of external power that is required for a given filament temperature. A halogen IR lamp has a multilayer dichroic coating, consisting of alternating layers of tantalum pentaoxide and silicon dioxide applied to the outside of the double-ended bulged tubular quartz bulb. This coating absorbs ultraviolet wavelengths, passes the visible wavelengths, and reflects the infrared wavelengths to the cylindrical coiled filament.

Halogen IR lamps provide truly significant increases in lamp efficacy, as much as 35% above TH lamps and 70% above standard incandescent lamps, as the comparison in Table 6.7 shows. The increase achieved depends a great deal on the fraction of the optical bulb surface area that returns infrared radiation to the filament and the ability of those optics to overlay the reflected image of the filament on the real filament. Halogen IR lamps offer all of the other advantages of TH lamps—high lumen maintenance, long life, high color temperature—and are available in single-ended, double-ended, PAR, and MR bulbs.

TABLE 6.7 INCANDESCENT, TH AND HALOGEN IR COMPARISON

	Watts	Lumens	Life (hrs)	lm/W
Incandescent	85	930	2000	10.9
TH	75	1030	2500	13.8
Halogen IR	60	1110	3000	18.5

TH lamps have a variety of attributes. They are compact sources providing pleasing color temperature, excellent lumen maintenance and reasonable life Their filaments are mechanically stable and can be positioned with high precision. Their small envelopes permit optical components to be located close to the filament, permitting precise optical control. Because of higher filament temperatures, high filament luminances are obtained, making them ideal for use in optical systems such as spotlights, film projectors, and scientific instruments. They are also well-suited for sports lighting outdoors and for studio, theater, television, and retail lighting indoors.

6.7 REFLECTORIZED LAMPS

Reflectorized lamps, both one-piece blown bulbs and two-piece pressed bulbs, were introduced in Sect. 2.10. The former are available in R and BR* (paraboloid) and ER (ellipsoid) reflector shapes, with the filament at the focal point of the parabola and at one of the focal points of the ellipse. Thus, the rays from the R-lamp reflector are roughly parallel, whereas those from the ER-lamp reflector converge toward a second focal point, about two inches in front of the bulb face. When used in down lights, this focusing allows the light to exit from the luminaire through a small opening and then spread out. The amount of light trapped within the luminaire is minimized; thus, more light can reach the work surface than from a larger-wattage size R lamp.

In both R and ER lamps, the reflector portion of the inner-bulb surface is coated with vaporized aluminum, and the flat face is lightly frosted if a spot distribution of light is desired and more heavily frosted for a flood distribution. These lamps are generally made of soft-lime glass and intended for use indoors.

Pressed-bulb lamps are of two types: PAR and multifaceted reflector (MR). PAR lamps are made of hard, heat-resistant glass, and thus are suitable for outdoor as well as indoor use. These lamps are of two-piece construction: one piece for the reflector and the other for the lens cover. Prior to fusing the lens to the reflector, the latter is coated with a thin layer of vaporized aluminum, and the filament is mounted accurately at the parabola's focal point, allowing for more precise beam control than in R lamps. The lens also assists in providing better beam control, permitting a wide range of beam angles from very narrow spots (8°) to very wide floods (60°). Beam angles for reflector lamps are classified by the American National Standards Institute (ANSI). Angles less than 13° are rounded to the nearest whole number; those 13° or greater to the nearest 5°.

The evolution of the PAR 38 lamp is interesting. The original PAR 38 lamp consumed 150 watts, emitted 1740 lumens, gave 2000 hours life, and provided 11.6 lm/W. It was replaced by a 120-watt energy-saver, which emitted 1370 lumens, also had 2000 hours of life, and gave 11.4 lm/W. Next came a 90-watt halogen-capsule PAR 38, providing

* This reflector lamp has a bulge in the bulb between the neck and the reflector.

1270 lumens, 2000 hours life and 14.1 lm/W. Finally, a halogen IR PAR lamp was introduced, consuming 60 watts, emitting 1150 lumens, lasting 3000 hours, and giving 19.2 lm/W. This is a 90-watt reduction from the original PAR 38 and a 65% increase in luminous efficacy.

Early PAR lamps (Sect. 2.10) had only spot and flood distributions. These lamps had uniformly-frosted lenses for spots and stippled lenses for floods. Their beams were not as uniform as R lamp beams, and there was considerable stray light beyond the beam boundaries. Later generations of PAR lamps have corrected these deficiencies. Sprial lenticules in the reflector have smoothed the beam. Lenses have been redesigned to minimize stray light. Axially-mounted filaments permit tighter beam control. The tungsten-halogen capsule permits close control of the filament's environment and allows higher filament temperatures and whiter light. The PAR 38 size has been augmented in recent years by several smaller tungsten-halogen lamps—PAR 14, 16, 20, and 30, which are widely-used, especially in retail applications. Examples are shown in Fig. 6.4

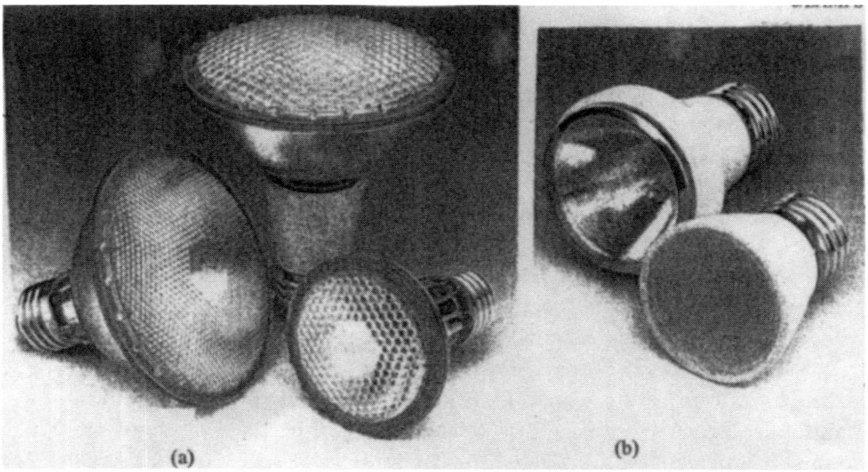

Figure 6.4 Examples of PAR lamps. a) PAR 30 and PAR 20, b) Ceramic PAR 16. Courtesy of Osram Sylvania.

Non-tungsten-halogen reflectorized lamps are the "gas guzzlers" of the incandescent lamp field. Because they were often located in hard-to-service places and were more costly than general service lamps, their lives were set at 2000 h, in contrast to the 750- and 1000-h life ratings of standard incandescent lamps. However, greater life usually means less light and reduced lumens per watt. Thus, in these times of energy conservation, the more efficient tungsten halogen and halogen IR reflectorized lamps are very welcome additions. Characteristics of selected R, BR, ER, and PAR lamps are shown in Table 6.8.

Multifaceted (MR) lamps are low-voltage (12 V), pressed-glass, tungsten-halogen reflector lamps. They are available in two sizes (MR 16 and MR 11) and operate from a 10 to 1 step-down transformer. They are available in wattages from 20- to 75-W and with beam categories from very narrow spot (7°) to very wide flood (55°). Each lamp is given a three-letter label. For example, a BAB lamp is a 20-W wide flood; an EYR lamp is a 42-W narrow spot. The letters have no innate significance.

TABLE 6.8 120-V REFLECTOR LAMPS

Watts	Bulb*	Type	CBCP (cd)	Beam** Angle (degrees)	Total Lumens	Lumens per watt	Life (h)
50	R20	FL	550	45	420	8.4	2000
75	R20	FL	650	45	690	9.2	2000
120	BR40	FL	900	120	1320	11.0	2000
120	BR40	SP	4,700	50	1320	11.0	2000
300	R40	FL	14,000	120	3030	10.1	2000
300	R40	SP	2,900	40	3030	10.1	2000
50	ER30	—	1,200	30	450	9.0	2000
75	ER30	—	1,500	30	750	10.0	2000
120	ER40	—	2,350	30	1370	11.4	2000
50	PAR20	NSP	5,000	10	530	10.6	2500
50	PAR20	NFL	1,250	30	530	10.6	2500
60	PAR16	NSP	5,000	10	650	10.8	2000
60	PAR16	NFL	1,300	30	650	10.8	2000
75	PAR30	NSP	14,000	9	1100	14.7	2500
75	PAR30	NFL	3,200	30	1100	14.7	2500
75	PAR30	FL	2,000	40	1100	14.7	2500
75	PAR38	SP	17,500	10	1040	13.9	2500
75	PAR38	FL	3,300	30	1040	13.9	2500
90	PAR38	SP	12,000	12	1300	14.4	2500
90	PAR38	FL	3,750	30	1300	14.4	2500
120	PAR38	SP	25,000	10	1900	15.8	3000
120	PAR38	FL	5,000	30	1900	15.8	3000
150	PAR38	SP	11,500	15	1660	11.0	2000
150	PAR38	FL	4,000	30	1660	11.0	2000
200	PAR46	NSP	31,000	8 × 11	2250	11.2	2000
200	PAR46	FL	12,500	11 × 25	2250	11.2	2000
300	PAR56	NSP	60,000	9 × 12	3720	12.4	2000
300	PAR56	FL	27,000	11 × 20	3720	12.4	2000
500	PAR64	NSP	100,000	7 × 12	6500	13.0	2000
500	PAR64	FL	49,000	9 × 25	6500	13.0	2000
60	PAR38	SP(HIR)	20,000	10	1110	18 5	3000
60	PAR38	FL(HIR)	3,600	30	1110	18 5	3000
100	PAR38	SP(HIR)	29,000	10	2070	20.7	3000
100	PAR38	FL(HIR)	6,300	25	2070	20.7	3000

* PAR lamps from 50 to 120 W contain tungsten halogen capsules.

** When two angles are listed, the beam is elliptical with the vertical angle given first.

MR lamps were originally developed for 35-mm slide projection systems operating on 120 V. For 12 V, the halogen capsule was redesigned for longer life, increased lumen output, better color, increased efficacy, and sharp cutoff of UV, reducing UVB and UVC radiation by 95%. With operation on low voltage, it was possible to use a thicker, shorter, axially-mounted filament* and incorporate more precise optics into the lamp. The all-glass reflector includes facets for better beam smoothing, reduced filament imaging and elimination of stray light. The lamps must be covered, either by a glass or screen on the lamp, or by a luminaire.

There are three types of MR 16 lamps, as illustrated in Fig. 6.5. The standard lamp, and first to appear, has a semi-hard dichroic reflector glass that acts like a cool beam mirror, reflecting visible wavelengths out the front while transmitting infrared wavelengths (and some visible) out the back. It has no cover glass. The reflector coating does degrade with time, causing marginal color consistency throughout life. The second type replaces the dichroic coating with an opaque coating of aluminum. This eliminates the cool beam feature, so that temperatures nearer the front of the bulb are higher than with the standard lamp. However, there is no back light nor heat to damage the transformer connected to the back of the lamp, and color consistency throughout life is excellent. These lamps have a cover lens. The third type returns to a dichroic coating, but this time it is of hard, titanium oxide. The cool beam feature reappears and the color consistency is excellent, with no degrading of the coating during the life of the lamp. There is no cover lens.

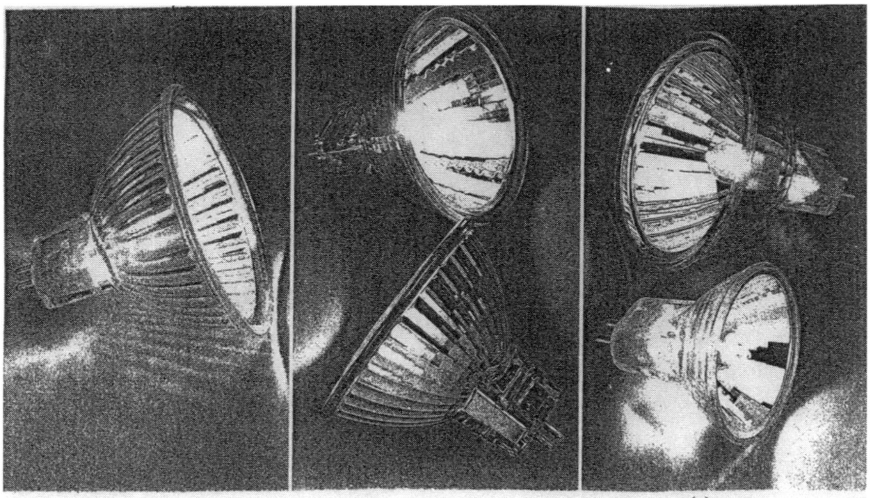

(a) (b) (c)

Figure 6.5 The three types of MR lamps. a) Standard Dichoric b) Aluminum-Coated c) Titanium Oxide-Coated. Courtesy of Osram Sylvania.

Characteristics of a representative group of MR 16 lamps are shown in Table 6.9. Each lamp has a rated life of 4000 h. All lamps in Table 6.9 have the titanium oxide dichroic coating, but the same wattages and beam angles are available in the other two types.

* For a 100-watt lamp, the filament diameter is 70 μm on 120 V and 270 μm on 12 V. Axial mounting increases center beam candlepower (CBCP) by 15%.

TABLE 6.9 CHARACTERISTICS OF 12-V MR 16 LAMPS

Watts	Description	CBCP (cd)	Beam Angle (degrees)
20	NSP(ESX)	6,000	8
20	FL(BAB)	700	40
35	NSP(FRB)	11,000	8
35	FL(FMW)	1,400	40
50	NSP(EXT)	11,000	12
50	NFL(EXZ)	3,200	25
50	FL(EXN)	2,000	40
65	NSP(FPA)	14,000	10
65	NFL(FPC)	4,000	25
65	F(FBB)	2,100	40

Source: SYLVANIA MR16 "TRU-AIM Application Guide".

Recent developments in MR lamps include the introduction of an MR 8 lamp and 24-V MR 16 lamps. In the former, optical performance is traded-off for size (the diameter of a reflector must be five times the diameter of a filament for good optical control). In the latter, more lamps can be connected to a given circuit but lamp optics are compromised.

6.8 FLUORESCENT LAMP DEVELOPMENT

As noted in Sect. 1.2, the basic principle of transforming ultraviolet radiation into visible radiation was discovered in 1852 by Sir George Stokes. From then until the 1920s, much work was done in the development of low- and high-pressure electric discharges in mercury and sodium, but very little work was done on fluorescence. Also the electric discharge devices that were developed were relatively inefficient in transforming the electrical input to the lamp into visible radiation. The electrodes emitted electrons inefficiently into the arc stream, many of the collisions of electrons with gas atoms were elastic, and the excitations that did result were not concentrated in one special line.

Then, as discussed in Sect 5.11, it was discovered in the 1920s that a mixture of mercury vapor at low pressure and an inert gas at higher pressure was approximately 60% efficient in converting electrical input power into a single spectral line at 253.7 nm. In 1934, Dr. W.L. Enfield, manager of General Electric's Lamp Development Laboratory in Cleveland, Ohio, received a report from England from Dr. A.H. Compton outlining the progress being made by the British in developing lamps containing mercury and using tubes coated with fluorescent materials. Enfield immediately created a research team composed of G.E. Inman, R.N. Thayer, A.E. Lemmers, and W.A. Roberts and charged them with the task of producing a commercially- practical fluorescent lamp.

By the summer of 1935, Inman's team had produced a prototype green fluorescent lamp with an efficacy of 60 lm/W. From their work emerged a set of guidelines for future fluorescent lamp development:

A. The basic discharge process is always in three steps:
   1. Free electrons are derived from the heated electrodes and accelerated by the applied field.

2. The kinetic energy of the free electrons is converted into excitation energy of the gas atoms.
3. The excitation energy of the gas atoms is converted into radiation.

B. High efficiency is obtained if mercury vapor at low current density is used to produce a single ultraviolet spectral line at 253.7 nm.

C. Phosphors are needed whose peak absorption is at 253.7 nm, and these must be applied to the inside of the tube because 253.7 nm does not pass through ordinary glass.

D. To provide optimum arc temperature and pressure for generation of 253.7 nm, a bulb temperature of 105 to 115°F (40–50°C) is necessary to provide a mercury pressure of 6 to 10 μm. An inert gas pressure of 2 to 3 mm is also required.

E. A tubular lamp of appreciable length-to-diameter ratio is needed so that the fixed wattage losses at each end, in the cathode and anode fall regions, are a small percentage of the total watts input.

F. Oxide-coated hot cathodes are required to provide an abundance of free electrons.

Late in 1935, a 2-ft-long green fluorescent lamp was demonstrated at the annual IESNA Technical Conference as "a laboratory development of great promise." It was $2^1/_2$ years later, in April 1938, that the fluorescent lamp was introduced commercially in white and six colors, obtained by using various mixtures of phosphor powders. The new lamps were displayed publicly in 1939 at the New York World's Fair and the Golden Gate Exposition in San Francisco.

The first lamps were 15-, 20-, and 30-W in 18-, 25-, and 36-in lengths, respectively. Soon after, the 40-W T12, 4-ft-long lamp, was introduced and took the lead rapidly for general lighting in offices, schools, industry, and commerce. As better phosphor combinations were found, the early yellowish-white lamp gave way to a 3500 K lamp, which became today's standard warm white, and a 6500 K daylight lamp, designed to simulate an average north skylight on an overcast day. Later a standard cool white, 4500 K lamp was introduced, and still later two deluxe lamps (cool and warm white) with improved color rendering, but lower efficacy. From early skepticism that the lamp was too long and bulky, sales increased until, in 1952, fluorescent lamps passed incandescent lamps as the major source of general lighting in the United States, and in the late 1960s fluorescent lamps provided two-thirds of the lumen-hours for commercial and industrial lighting in the U.S.

Since the introduction of the 4-ft long, $1^1/_2$-in diameter, 40-W lamp in 1940, there has been very little change in the basic arc discharge portion of the lamp. Argon is still used (occasionally an argon-krypton mixture), and the 253.7 nm line in mercury is still the phosphor energizer. However, in recent years there have been dramatic improvements in phosphors, electrode design, auxiliary circuits, and miniaturization. As a result, the 40-W T12 halophosphate phosphor lamp and its magnetic ballast are being phased out and replaced by 34-W T12 energy-saving lamps with energy-efficient magnetic ballasts, and, more recently, by T8 and smaller diameter lamps with rare earth phosphors and electronic ballasts, that offer improved color rendering, longer lives, and higher efficacies. In addition, a new family of fluorescent lamps, called compact fluorescent (CFL), has emerged to challenge the incandescent lamp in a variety of applications. These developments will be discussed in greater detail throughout the remainder of this chapter.

6.9 FLUORESCENT LAMP CONSTRUCTION AND TYPES

There are two major classes of fluorescent lamps: linear double-ended and compact single-ended. The first is described in this section; the second in Sect. 6.12.

A cutaway view of a linear fluorescent lamp is shown in Fig. 6.6a. It consists of a lime glass tube, which serves as an airtight enclosure for the drop of mercury, the argon filling gas, the phosphor coating, the electrodes, and their mount assemblies. The lamp also includes two bases, which have either two contacts (bipin) or one contact (single pin), depending on the starting method.

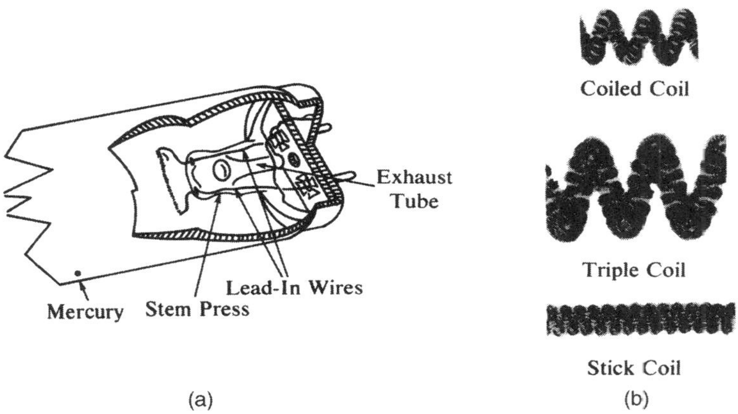

Figure 6.6 Fluorescent lamp construction. *Source*: "Fluorescent Lamps," (Bulletin TP-111R). Courtesy of General Electric Co.

The electrode mount assembly is very similar to the stem press unit in an incandescent lamp. The lead-in wires are copper outside, Dumet wire through the glass seal, and nickel inside, again as in the incandescent lamp. The filament functions as both cathode and anode, because the applied voltage is alternating. This results in electron bombardment of the filament and some wattage loss at the lamp ends.

Typical electrodes, shown in Fig. 6.6b, are of three types. The coiled-coil filament is similar to that used in an incandescent lamp. The other two consist of two tungsten wires instead of one, the second being a thin overwind over the thicker single or double coil.

The electrodes are dipped in a waffle-batter consistency mixture of barium, strontium, and calcium carbonates and baked during manufacture to become oxides, providing a low-work function coating that is highly emissive.

The manufacturing process consists of several steps. First, the glass tubes are washed and dried, and the phosphor is applied. The latter is usually a water suspension containing binders that is flushed into the tube and then dried with heated air to drive out the binders and enable the phosphor to adhere to the wall. In the newer double-coat lamps, a halophosphate phosphor is applied first, followed by a rare-earth triphosphor. Phosphor thickness is carefully controlled. If it is too thick, light is lost by absorption; if it is too thin, some 253.7-nm radiation passes through. In both cases, phosphor efficiency decreases.

Next mercury is added, either directly or in a capsule. Because mercury can be a hazardous material, lamp manufacturers have reduced significantly the mercury content of fluorescent lamps in recent years, from a high of 50 mg in a 4-ft lamp to a current level of less than 10 mg.* This has been possible in part due to improved phosphors that have reduced absorption at the bulb wall of mercury migrating from the arc stream, thus improving lumen maintenance during lamp life.

The electrode stem-press units are now mounted and sealed, except for an exhaust hole. Air is pumped from the lamp, and the inert gas is inserted, following which the exhaust tube is sealed. End caps are applied and, usually, the lamp is silicone coated, which aids in lamp starting in adverse humidity environments, by causing water vapor to bubble on the bulb surface, much as wax causes water to bubble on the hood of a car.

Linear fluorescent lamps are available in a variety of lengths, diameters, wattages, and correlated color temperatures. Lengths range from 4 in to 8 ft; diameters from T2 to T17 ($^1/_4$ to $2^1/_8$ in), and wattages from 4-215 W. Six CCTs are available—2700 K, 3000 K, 3500 K, 4100 K, 5000 K, and 6500 K—and three levels of color rendering.

Typical catalog information for a T12 and a T8 linear fluorescent lamp is given below:

F40T12/735, RS, Med. BiPin, 430, 20,000, 3300
F32T8/830, RS, Med. BiPin, 265, 20,000, 2950

F40T12/735 and F32T8/830 are ordering abbreviations. Lamp wattages are 40 and 32, and T12 and T8 are bulb diameters in eighths of an inch. The numbers 735 and 830 are codes for correlated color temperature (CCT) and color rendering index (CRI) of triphosphor lamps. The last two numbers give CCT, in this case 3500 K and 3000 K. The first numbers give CRI, 7 indicating a value between 70 and 79 (good color rendering) and 8 indicating a value between 80 and 89 (excellent color rendering). RS and Med. Bipin indicate that these lamps are rapid start** and have two pins in each endcap. The T12 lamp draws 430 ma, has a life of 20,000 hours, and an initial light output of 3300 lumens. For the T8 lamp, the figures are 265 ma, 20,000 hours, and 2950 initial lumens.

Other symbols in common use for describing linear fluorescent lamps are

RE	Rare-earth phosphor
ES	Energy-saving lamp
HO	High-output lamp
VHO	Very high-output lamp
CW	Cool-white phosphor
WW	Warm-white phosphor
FB	U lamp
IS	Instant-starting lamp

A summary of the electrical and lighting characteristics of several common linear fluorescent lamps is given in Table 6.10. Three important parameters in that table are life, lumens and lumens per watt.

---

* To maintain perspective, it should be noted that the largest man-made source of mercury in the atmosphere is from fossel-fuel combustion, not vapors from mercury in landfills.

** Starting methods are discussed in Sect. 6.15.

TABLE 6.10 REPRESENTATIVE FLUORESCENT LAMP DATA

Type	Lamp Label	Watts	Length (in)	Life in h	Initial Lumens	Lamp Current (mA)	Initial Lamp Lumens per watt	Lumens at 40% Life
Preheat	F14T8CW	14	15	7,500	685	420	49	580
	F14T12CW	14	15	9,000	650	390	46	550
	F15T8CW	15	18	7,500	825	300	55	725
	F15T12CW	15	18	9,000	760	330	51	685
	F20T12CW	20	24	9,000	1,200	380	60	1,150
Rapid start and preheat T12	F40T12CWX	40	48	20,000	2,200	425	55	1,670
	F40T12RE30	40	48	20,000	3,250	425	81	2,950
	F40T12RE30ES	34	48	20,000	2,750	425	81	2,475
Rapid start T8	F17TBRE30	17	24	20,000	1,400	265	82	1,290
	F25T8RE30	25	36	20,000	2,225	265	89	2,050
	F32T8RE30	32	48	20,000	2,950	265	92	2,715
	F40T8RE30	40	60	20,000	3,775	265	94	3,470
Instant start	F48T12CW	40	48	9,000	2,875	425	72	2,650
	F48T12RE30	40	48	9,000	3,050	425	76	2,870
	F72T12CW	55	72	12,000	4,500	425	82	4,140
	F72T12RE30	55	72	12,000	4,800	425	87	4,510
	F96T12CW	60	96	12,000	5,500	425	92	5,060
	F96T12RE30	75	96	12,000	6,500	530	87	6,110
	F96T12RE30ES	60	96	12,000	5,800	440	97	5,000
High output Rapid start	F48T12RE30HO	60	48	12,000	4,250	800	71	3,830
	F72T12RE30HO	85	72	12,000	6,650	800	78	5,990
	F96T12RE30HO	110	96	12,000	9,350	800	85	8,420
Very high output Rapid start	F48T12CW1500	110	48	10,000	6,200	1,500	56	4,030
	F72T12CW1500	165	72	10,000	9,000	1,500	55	6,300
	F96T12CW1500	215	96	10,000	13,500	1,500	63	10,125
	F96T12ES1500	185	96	9,000	13,500	1,500	68	9,380
T5	F14T5RE30	14	22	16,000	1,350	170	96	1,280
	F21T5RE30	21	33	16,000	2,100	170	100	1,995
	F28T5RE30	28	45	16,000	2,900	170	104	2,755
	F35T5RE30	35	57	16,000	3,650	170	104	3,470

Note 1: All RE lamps are also available in 3500 K, 4100 K and 5000 K.
Note 2: All ES and Rapid-start T8 lamps have a krypton-argon fill gas in the ratio of 3 parts krypton to 1 part argon. This lowers lamp voltage and electrode losses, but increases starting voltage.
Note 3: A4-ft ES lamp consumes 15% less watts but provides 15% less lumens. Thus there is no net efficacy gain.
Note 4: Lumen values for T5 lamps are based on operation at 35°C.

Life

The life rating of a fluorescent lamp is based on 3 hours of burning for each start. If a different burning cycle is chosen, life will change because each start has a deleterious effect on the electrodes, and electrode life is the primary determiner of lamp life. Life ratings for several burning cycles are shown in Table 6.11.

TABLE 6.11 LIFE VS. HOURS PER START

Hours/Start	% Life
1	63
2	87
3	100
8	123
20	148
∞	160

Lumens

The initial lumen rating is based on the lumens after 100 hours of burning. The depreciation during this first 100 hours may be as much as 10%. From then on, lumen depreciation is gradual, but steady, the main causes being (1) deterioration of the phosphor caused by mercury migration, (2) 185-nm radiation, degrading phosphor output, and (3) deposited electrode material. Tri-phosphor lamps show less lumen depreciation than halophosphate lamps.

Lamp manufacturers give two lumen maintenance factors, a mean lumen factor at 40% of life, and a lamp lumen depreciation factor at 70% life. These are, respectively, 0.84 and 0.80 for halophosphate lamps and 0.92 and 0.90 for triphosphor lamps.

Lumens Per Watt

Lumen per watt efficacy figures are based on initial lumens and lamp watts. For a true picture of efficacy, one must know the wattage loss in the ballast and then calculate system lumens per watt. For example, if a 20-W fluorescent lamp emits 1250 initial lumens and its ballast consumes 5W, its lamp efficacy is 63 lm/W, but its system efficacy is 50 lm/W.

The three methods of starting the fluorescent lamps shown in Table 6.10 are preheat, instant start, and rapid start. In preheat starting, current is passed through the electrodes before lamp starting voltage is applied. Starting occurs in 2–4 s. With instant starting, a high starting voltage is applied across the lamp without preheating, starting the lamp in under 100 ms. Rapid starting combines preheating with the instant application of starting voltage, resulting in starting in 0.5 to 1.5 s.

The high-output and very high-output fluorescent lamps in Table 6.10 have lower efficacies than their standard counterparts. A comparison is given in Table 6.12 for 4-ft T12 lamps. Note that a measure of the increased loading of the lamps is given, namely the lumens per linear foot of lamp length. A second measure, arc watts per square inch of bulb wall surface area, will be developed in Sect. 6.11.

TABLE 6.12 COMPARISON OF STANDARD, HO AND VHO LAMPS

Lamp	ma	Watts	Initial Lumens	Initial lm/W	lm/ft
STD	425	40	3250	81	812
HO	800	60	4250	71	1060
VHO	1500	110	6200	56	1550

The T5 miniature bipin lamps in Table 6.10 are a recent addition to the linear fluorescent lamp group. These are the first lamps to be made in metric lengths in the United States and Canada, and are about 2 in shorter than comparable existing lamps. This makes the new lamps compatible with metric ceiling grid dimensions. The lamps are designed to operate at 35°C, rather than at 25°C (T8 and T12 lamps). At 25°C, their lumen output is reduced by 10%. The lamps have rare-earth phosphors and operate from electronic ballasts. They have much higher luminances than T8 and T12 lamps and thus should not be used in luminaires where the lamps are exposed to view. However, their small diameter will permit excellent optical control. All four lamps in Table 6.10 are also available in high output, providing from 50 to 100% more lumens than their standard counterparts.

Although they are not linear, U lamps and circular lamps are formed from linear tubes and are included in this class. U lamps are made by bending 4-ft tubes in half so that they fit in 2-ft by 2-ft luminaires. They provide essentially the same lumen output as a comparable 4-ft linear lamp. Spacings of $3^5/_8$ in and 6 in between legs are available. The 40-W T12 U lamp with 6-in spacing delivers 3050 initial lumens, 2780 mean lumens, and has 14,000 hours of life. For the 32-W T8 lamp, the values are 2700 initial lumens, 2565 mean lumens, and 20,000 hours life.

T9 circular fluorescent lamps are formed by bending a linear tube into a circle, with a single, four-pin connection at the junction. They are rated at 12,000 hours and are available in four diameters of $6^1/_2$, $8^1/_4$, 12 and 16 in, with wattages of 20, 22, 32, and 40 W, and initial light outputs of 800, 1100, 1950, and 2700 lm. They have halophosphate phosphors and operate in the rapid start mode.

6.10 ENERGY DISTRIBUTION AND EFFICIENCY

The energy distribution of a typical linear fluorescent lamp is shown in Fig. 6.7. It is first noted that 3% of the input energy is converted directly into visible radiation. This represents transitions from higher levels on mercury's energy level diagram that result in the emission of the four visible mercury lines at 404.7 nm, 435.8 nm, 546.1 nm, and 577–579 nm. These lines poke through the phosphor and would provide a bluish-white light in the absence of the phosphor.

As mentioned earlier, 60% of the input energy goes directly into producing 253.7 nm radiation. Another 3% appears as 185 nm radiation. The remaining 34% of the input energy is dissipated as heat, in the positive column due to collisions of ions and atoms, at the bulb wall due to recombination of electrons and mercury ions, and at the electrodes due to bombardment.

Forty percent of the ultraviolet radiation is converted to visible energy by the phosphor, following Stoke's law as noted earlier. The remaining 60% is dissipated as heat within the phosphor.

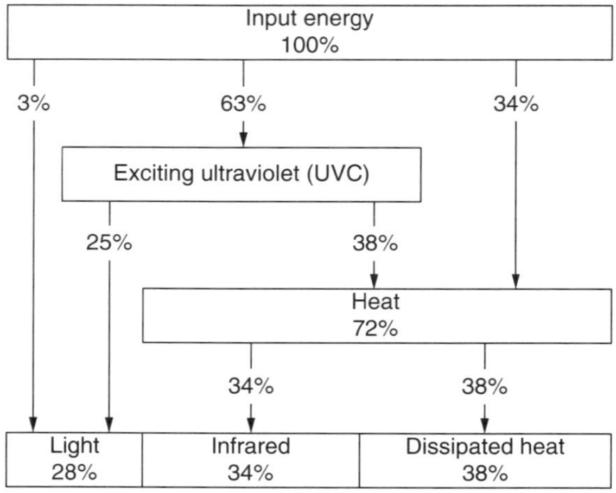

Figure 6.7 Typical fluorescent lamp energy distribution.

The distribution of the input energy from the point of view of realized efficacy can also be explored, much as was done for the incandescent lamp. If all of the input energy were converted into a single spectral line at 555 nm, the luminous efficacy would be 683 lm/W. But phosphors emit over a range of wavelengths, including some infrared energy. This reduces the maximum luminous efficacy by 49% to 348 lm/W. The quantum ratio is 46%, reducing the efficacy further to 160 lm/W. Next, only 60% of the input energy becomes 253.7-nm radiation, causing a further reduction to 96 lm/W maximum. Last, losses due to phosphor absorbtion, bulb absorbtion, end loss in luminance, and nonutilization of 253.7 nm result in a typical realized lamp efficacy of 80 lm/W.

6.11 FACTORS AFFECTING FLUORESCENT LAMP PERFORMANCE

The process of light production in a fluorescent lamp has two major steps, as illustrated in Fig. 6.8. The second step, the conversion of 253.7 nm radiation into visible radiation, is largely determined by Stoke's Law and the efficiency of the phosphor. The first step depends on five major factors in the design of the rest of the lamp and all of these are interrelated. They are:

1. Lamp loading
2. Argon pressure
3. Bulb-wall temperature (and associated mercury vapor pressure)
4. Tube length
5. Tube diameter

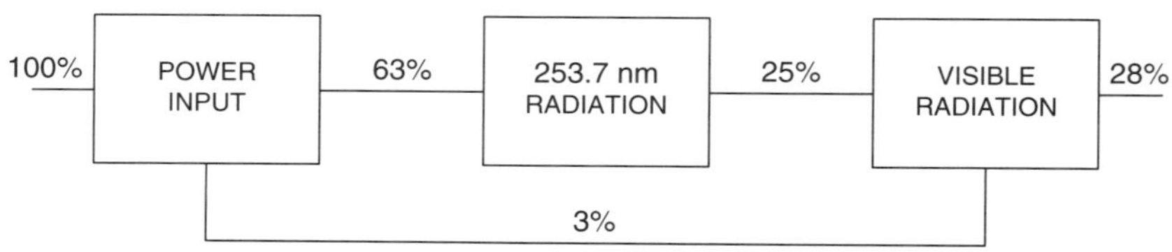

Figure 6.8 Fluorescent lamp, major steps in producing light.

The loading of a fluorescent lamp may be expressed as the arc watts per square inch of bulb-wall surface area according to the following equation:

$$\text{Loading} = \frac{P - 12I}{\pi d \ell} \tag{6.8}$$

where P is the input power, 12 is the average RMS electrode fall voltage (both ends added), I is the lamp current, and d and ℓ are the lamp's diameter and length. For the 40-W T12 lamp, Eq. (6.8) yields

$$\text{Loading} = \frac{40 - 12(0.425)}{226} = 0.15\ \text{W/in}^2$$

For a 32-W T8 lamp, the results from Eq. 6.8 are

$$\text{Loading} = \frac{32 - 12 \times 0.265}{151}$$

$$= 0.19\ \text{W/in}^2$$

Increasing the current for a given length and diameter of bulb tends to decrease efficacy. This is borne out by comparing the efficacies of the various 4-ft T12 lamps of different current ratings in Table 6.12. The loadings for these lamps from Eq 6.8, are 0.15 W/in^2, 0.22 W/in^2, and 0.43 W/in^2, respectively. Thus, as loading increases, efficacy decreases.

EXAMPLE 6.3

The question might be raised legitimately as to why higher wattage fluorescent lamps are not available. What length and diameter would be required for a 1000-W fluorescent lamp of reasonable efficacy?

Solution. Assume a loading of 0.18 W/in^2 is sought. Then the bulb surface area would be

$$A_s \cong \frac{1000}{0.18} = 5600\ \text{in}^2$$

neglecting the relatively-small cathode fall loss. If d were 6 in, ℓ would have to be 297 in $\simeq$ 25 ft, a totally unreasonable size for a lamp.

One may explain the fact that there is an optimum loading in the following way. As current increases, the probability of multiple collisions of electrons and mercury atoms increases, decreasing the probability of generation of the 253.7-nm line. Also, an increase in current tends to increase mercury vapor pressure, which increases the imprisonment of 253.7-nm radiation. On the other hand, as current decreases, mercury vapor pressure decreases, and this reduces the number of collisions of electrons and mercury atoms, reducing the number of 253.7-nm photons.

The second factor of importance is inert gas pressure. If it is too high, the mean free path is reduced and the number of elastic collisions increases. If it is too low, there is excessive migration of mercury ions and electrons out of the arc stream.

Fluorescent lamps are extremely sensitive to temperature changes. It was pointed out in Sect. 5.11 that optimum performance results when the coolest spot on the bulb is maintained at about 104°F. At this temperature, the mercury vapor pressure will be 1.07 Pa* (about 8 μm of mercury). The argon fill gas will be at a higher pressure, 100 to 400 Pa (0.75 to 3 mm). The situation when bulb temperature varies is shown in Fig. 6.9. As the temperature of the coolest part of the bulb decreases below optimum, mercury vapor pressure decreases. The result is a 1% decrease in efficacy for each 1.5°F decrease in bulb-wall temperature. As temperature increases above optimum, more mercury vaporizes, the mean free path decreases, and the proportion of elastic collisions increases. Also, there is more imprisonment of radiation. The net effect is less severe, representing a 1% decrease in efficacy for each 2.5°F increase in bulb-wall temperature.

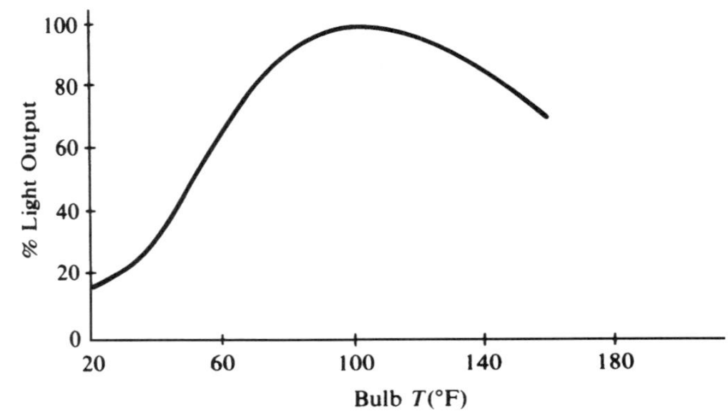

Figure 6.9 Bulb-wall temperature effects.

If mercury is inserted into the lamp in the form of an amalgam (an alloy of mercury and other metals), lumen output can be maintained better over a range of operating temperatures. The amalgam controls, and thus stabilizes, the amount of mercury vapor released into the arc stream. Amalgams are used extensively in compact fluorescent lamps, which tend to run hotter than linear lamps.

Lamp voltage is a linear function of positive column length for a given lamp loading. Thus, luminous efficacy increases as tube length increases because the constant wattage loss represented by the electrode fall potential becomes a smaller fraction of the total watts input. Also, the end loss in light output becomes a smaller portion of the total as tube length increases. For a T12 lamp operating at 425 mA, the theoretical maximum efficacy is about 75 lm per lamp watt when the length is infinite. This reduces to 65 lm per lamp watt when the length is 48 in (1200 mm).

Last, luminous efficacy is a function of tube diameter. As diameter is increased, luminous efficacy increases, then decreases. With increasing diameter, there is more imprisonment of 253.7-nm radiation. This means that fewer 253.7-nm photons reach the phosphor. This effect decreases luminous efficacy, but there is a counter effect. As tube diameter increases, the rate of loss of electrons and ions to the bulb wall decreases. This means that less new ionization is required to sustain the arc, and more electron energy

* Atmospheric pressure is 760 mm of mercury or 101,325 Pascals. Thus one Pascal corresponds to 7.5 μm of mercury pressure.

can be utilized in achieving excitation to the 4.88-eV level. The result is an increase in luminous efficacy. Reducing diameter from T12 to T8 improved efficacy, but also increased lamp voltage.

6.12 COMPACT FLUORESCENT LAMPS

Compact fluorescent lamps are small-diameter, single-ended, multiple-tube lamps, often containing "bridges" or "loops" to connect the arc stream from one portion of the lamp to another. Research to develop this lamp began in the 1970s, and it was introduced into the U.S. market in the early 1980s. There are two basic types: (1) four-pin T5 twin-tube (biaxial) lamps providing high lumen output in short-lengths and (2) a large family of curved, multiple-tube T4 and T5 lamps in a variety of shapes whose volumes and lengths are sufficiently small so that they may replace incandescent lamps.

The development of rare earth phosphors opened the way for the development of fluorescent lamps of reduced length and diameter having negligible reduction in lumen output. The new phosphors, unlike halophosphate phosphors, are very stable under high levels of ultraviolet radiation and provide the opportunity for high efficacy and excellent lumen maintenance.

The goal in creating a compact fluorescent lamp was to reduce lamp length without reducing luminous flux. This implies that the watts per unit length of the positive column must increase, which can be accomplished by increasing lamp current or by raising the positive column electric field (V/m). If lamp current is raised by reducing lamp length but keeping all other parameters constant, overall efficacy decreases because of increased electrode and ballast losses. If the electric field is raised, by whatever means, with lamp current constant, discharge efficacy, and thus overall efficacy, also decrease. These results led to the question: are there an optimum diameter and length to achieve high lumen output and efficacy?

The results of many experiments seeking an answer to this question showed:

1. A diameter of 10 mm (0.4 in) appears to be optimum for $\Phi < 1000$ lm, and 15 mm (0.6 in) for $1000 < \Phi < 2000$ lm.
2. Lamp voltage should be no more than 50% of line voltage.

To achieve these results and meet desired design criteria that efficacy be 3 to 4 times greater than that of an incandescent lamp and that overall lamp length be less than 20 cm (8 in), it became obvious that some form of bent-tube lamp was required. This resulted in two lamps being introduced in the early 1980s by Philips Lighting Co. The first lamp consisted of two parallel (biaxial) tubes connected by a small bridging tube. The bridging tube was near to, but not at, the opposite end of the lamp from the base. The small region above the bridging tube (the corners of the bends) was used as a point of mercury vapor pressure control, by holding (cold spot) bulb temperature there to 47°C (117°F). A mercury amalgam containing bismuth and indium was used to stabilize vapor pressure. The lamp had a glow switch starter integrated within the base, but a separate ballast.

In the second lamp, starter and ballast were integrated. The ballast was magnetic, and relatively long and heavy. Thus to achieve overall length less than 20 cm (8 in), it was necessary to bend the tube a second time, resulting in two U-shaped tubes side by side, and all parts were housed in an encapsulated screw base container (the bridge was in the base). Once again a mercury amalgam was used, to both reduce and stabilize mercury vapor pressure. It was supplied from an exhaust tube in the base. The lamp had an outer domed diffusing bulb cover.

From these beginnings, the number and variety of compact fluorescent lamps have expanded dramatically. T5 twin-tube (biaxial) high-lumen output lamps are made by all major lamp manufacturers. They operate on either magnetic or electronic ballasts, in preheat or rapid-start circuits. These lamps are designed to provide the same lumens as lamps twice as long, but lamp loading is high (1.3 to 1.8 W/in^2). Thus, they have high luminance and should be used only in well-shielded luminaires. A biaxial lamp is shown in Fig. 6.10 and characteristics of the family are given in Table 6.13.

Figure 6.10 Biaxial fluorescent lamp.

TABLE 6.13 BIAXIAL HIGH-LUMEN FLUORESCENT LAMPS

Watts	Length (in)	Initial Lumens	Mean Lumens	Life (h)	lm/W
18	10.5	1250	1130	20,000	69
27	12.8	1800	1620	12,000	67
39	16.5	2850	2510	12,000	73
40	22.5	3150	2850	20,000	79
50	22.5	4000	3400	14,000	80

Compact fluorescent lamps designed to replace incandescent lamps come in a variety of shapes and sizes. Wattages range from 5 to 55 W and initial light outputs from 250 to 4800 lm. The lamps are available in four CCTs, 2700 K, 3000 K, 3500 K, and 4100 K, and have 8000 to 10,000 hours life.

The lamps come in two types.

1. Integral
   Lamp, starting mechanism, and ballast are contained in one housing with a screw base. The ballast can be magnetic or electronic. If it is magnetic, power factor correction can be achieved with a capacitor. This will not work for electronic.
2. Separate
   The lamp and its starting mechanism are self-contained and the ballast is separate. If a starter is present, the base is four-pin, otherwise it is 2-pin. The ballast can either be incorporated into the luminaire or be separate with a screw base. Cutaway views of each type are shown in Fig. 6.11. In Fig. 6.11(a), an integrated lamp and electronic ballast unit with screw base are presented; in Fig. 6.11(b),

a 2-pin lamp containing a glow switch starter and a radio frequency suppression capacitor are shown. The unit plugs into a separate ballast.

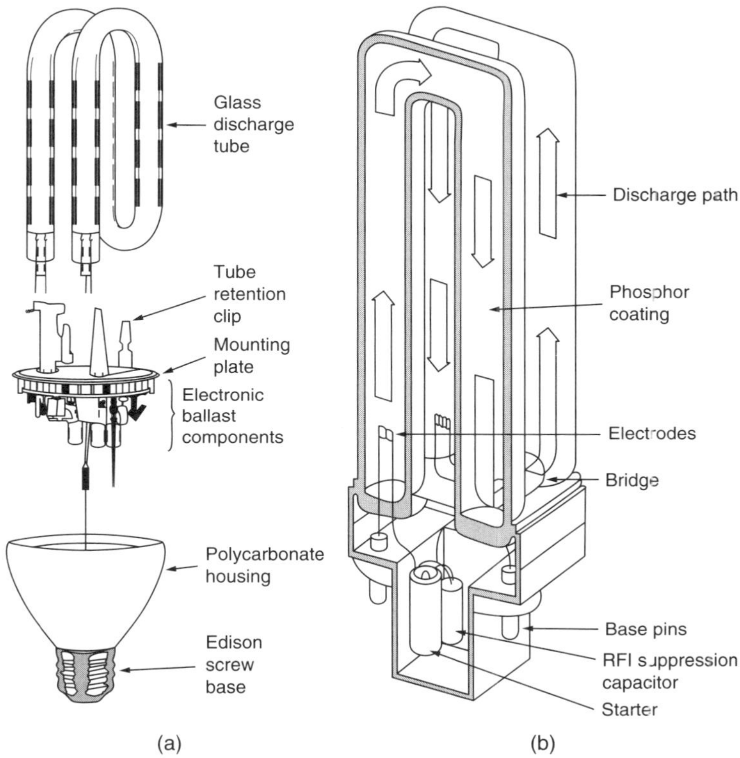

Figure 6.11 CFL Cutaway Views. ***Source*: *IESNA Lighting Handbook*, Ninth Edition, 2000. Courtesy of the Illuminating Engineering Society of North America.**

Characteristics of several types of CFLs appear in Table 6.14. For comparison purposes, the initial lumen outputs of A-line soft white incandescent lamps are also included. In general, one watt of compact fluorescent yields the same lumens as four watts of incandescent, i.e. a 26-W double twin tube CFL emits the same lumens as a 100-W incandescent. In all categories but the 2D lamp, either 2- or 4-pin lamps are available. The 2-pin versions operate on separate magnetic ballasts, contain an integral starter, and are not instant-on. The 4-pin varieties have built-in electronic ballasts, can be dimmed, and are instant on. The 2D lamps, named for their shape, are used with an external ballast (screw base) that provides instant on. These are very high output lamps, matching the lumens of 60- through 150-W incandescent lamps.

CFLs can be operated base up or base down. In the former case, mercury collects at the end opposite the base; in the latter case, near the electrodes and ballast. The greater temperature there can raise the mercury vapor pressure, reducing lumen output. CFLs are tested and rated base up.

Electrodeless lamps are a recent development by all three major lamp manufacturers. Because they have no electrodes, they offer long lives and excellent maintenance of lumen output and efficacy. They come in two basic types: a compact lamp smaller

TABLE 6.14 CHARACTERISTICS OF SELECTED CFL LAMPS

Watts	Length (in)	Life (h)	Initial Lumens	Mean Lumens	Pins	Bulb
5	4.2	10,000	250	210	2 or 4	Twin
7	5.3	10,000	400	330	2 or 4	Twin
9	6.6	10,000	600	500	2 or 4	Twin
13	7.3	10,000	825	710	2 or 4	Twin
13	5.3	10,000	900	765	2 or 4	Double Twin
18	6.1	10,000	1150	980	2 or 4	Double Twin
26	6.7	10,000	1710	1460	2 or 4	Double Twin
13	4.2	10,000	900	765	2 or 4	Triple Twin
18	4.8	10,000	1200	1020	2 or 4	Triple Twin
26	5.2	10,000	1800	1530	2 or 4	Triple Twin
32	5.8	10,000	2200	1870	2 or 4	Triple Twin
16	5.5	8,000	1050	880	4	2D
21	5.5	10,000	1350	1135	4	2D
28	8.1	10,000	2050	1720	4	2D
38	8.1	10,000	2850	2395	4	2D

Incandescent Lamps

Watts	25	40	60	75	100	150	200
Initial Lumens	180	445	870	1180	1710	2780	3800

than an R30 and completely self contained with a screw base; and an O-shaped lamp similar in size to a U lamp and requiring an auxiliary power supply.

The compact lamp is surrounded by an induction coil that is energized by a 2.5 MHz oscillator in the base. The induction coil excites and ionizes mercury vapor as in an ordinary fluorescent lamp. A rare-earth phosphor coating is on the inside of the reflector-shaped bulb. The lamp consumes 23 W, provides 1100 initial lumens, and has 15,000 h life.

The O-shaped lamp operates at 250 KHz . Two ferrite magnetic-core toroids surround the tube—one at the top of the O and the other at the bottom—and these are connected to an external electronic power supply. The electric magnetic field produced is inductively coupled from the toroids to the arc stream, and excites and ionizes mercury. Lamps are available, with triphosphors, in 107- and 157-W, giving 100,000 h life and providing 8,000 and 12,000 lumens, respectively.

6.13 PHOSPHORS

As noted earlier, the inside of a fluorescent lamp tube is coated with a mixture of phosphors, each of which has its peak absorption sensitivity near 253.7 nm. The earliest fluorescent lamps, developed by Inman's group at General Electric, contained zinc silicate phosphors activated with manganese and produced a green light. Soon they discovered that replacing a portion of the zinc with beryllium shifted the emission from green towards yellow. Concurrently, lamps were developed using magnesium tungstate phosphors that emitted blue light. A blending of the two phosphors enabled Inman's group to create the first white fluorescent lamp.

Beryllium is toxic, and in some fluorescent lamp manufacturing plants workers who got beryllium powder in cuts found that they wouldn't heal. In the 1940s, McKeag and Ranby developed a calcium halophosphate phosphor that was more efficient than zinc beryllium silicate (peak emission at 590 nm) and nontoxic as well. This phosphor, in combination with magnesium tungstate (peak emission at 480 nm), became the industry-wide standard for white fluorescent lamps for the next 40 years.

A phosphor is basically a semiconductor material to which activators have been added. The band theory model for a typical phosphor is shown in Fig. 6.12. It contains three zones, namely a valance band, a forbidden zone, and a conduction band.

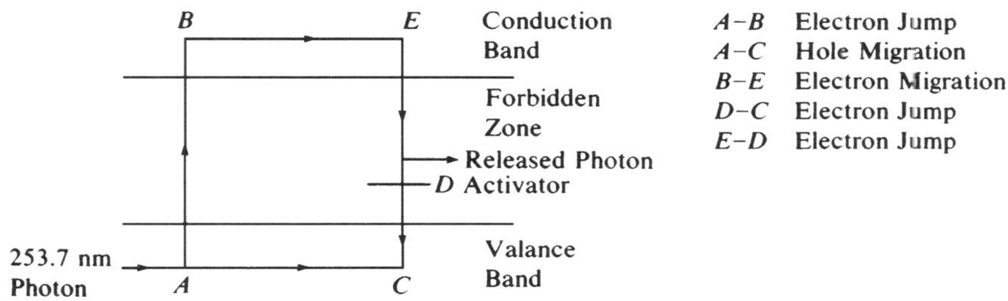

Figure 6.12 Phosphor-activator model for zinc sulfide.

In a pure semiconductor, the valance electrons are tied to their respective atoms, and there are very few free electrons and ions present. In its model, this is portrayed by having no permitted energy levels in the forbidden zone between the valance band and the conduction band. To free an electron from an atom, enough energy (measured in electonvolts) must be applied to cause an electron to leap directly from the valance band to the conduction band. In the case of a phosphor, activators are added that create permitted energy levels in the forbidden zone, as indicated in Fig. 6.12. For calcium halophosphate, the activators are antimony and manganese.

The process of converting ultraviolet energy to visible energy in the phosphor is explained diagrammatically in Fig. 6.12 for the relatively simple two-element zinc sulfide phosphor with a silver activator. A 253.7-nm photon impinges on the phosphor causing an electron to jump from divalent sulfur to divalent zinc, thus creating a univalent zinc ion (with an excess electron) in the conduction band and a univalent sulfur ion (deficient one electron) in the valance band. The excess electron in the conduction band migrates through the crystal lattice from one zinc ion to another. At the same time, the deficiency, or positive hole, in the valance band migrates from one sulfur ion to another. When the positive hole arrives near an activator, it captures an electron from it. This traps the positive hole in the activator. Then the migrating electron in the valance band falls into the positive hole and a photon is released.

The diagram in Fig. 6.12 justifies why the wavelength conversion is always from ultraviolet to visible, rather than vice-versa (Stoke's Law), i.e., the photon released (when an electron falls to the activator level) always has less energy than the incident photon.

In the 1980s, rare earth phosphors appeared, and today's phosphors are usually double-layered, with a coating of rare earth phosphor on top of an undercoat of calcium halophosphate. Work by several people in the early 1970s showed theoretically that three narrow-band spectra centered at 450 nm (short), 545 nm (medium), and 610 nm (long) could provide light that was perceived as white and that rendered most colors well. In the mid 1970s, it was discovered that there were indeed rare earth phosphors that would produce these bands if supplied with suitable activators, and triphosphor lamps were born.

The long wavelength phosphor is an yttrium oxide activated by europium, producing a strong spectral line at 611 nm. The medium wavelength phosphor is a compound of aluminum oxide, cerium and magnesium activated by terbium, with a spectrum centered at 545 nm. Finally, the short wavelength phosphor is also a compound of aluminum oxide, but with barium and magnesium. Europium again serves as the activator, and the center of the spectral band is near 450 nm.

Triphosphors offer significant advantages over halophosphors, providing improved color rendering and a wide range of CCTs. They are a major reason that T8 and smaller-diameter fluorescent lamps became feasible. Triphosphors are able to withstand high loading (arc watts per in^2) and not degrade significantly over time, thus maintaining lumen output and efficacy. These features are abetted by the application of a thin transparent coating of tin oxide and alumina to the inside of the bulb prior to applying the phosphor. This reduces mercury migration to the bulb wall and also decreases sodium migration from the glass to the arc stream. Triphosphors are much more expensive than halophosphors, but the opportunity to reduce lamp diameter meant less bulb wall surface area to cover, which made T8 triphosphor lamps economically viable.

6.14 BALLASTS

Arc discharges are negative resistance phenomena, which means that if lamp current increases, lamp voltage decreases. Thus it is essential to place some form of current-limiting device in series with the lamp. The simplest control element is a series resistor, but this leads to significant power loss. Consequently, resistive ballasting is used only when lamps are operated on dc; otherwise electromagnetic ballasting and, more recently, electronic ballasting are employed.

A ballast has four functions in addition to its fundamental one of limiting current in the negative resistance arc:

1. Provide sufficient open-circuit secondary voltage to initiate the arc
2. Regulate the lamp current against line voltage changes
3. Relight the lamp on each half cycle of applied voltage
4. Provide cathode heating.

A voltage several times the lamp operating voltage may be required to start the lamp. The peak of the applied voltage wave governs starting because, at the frequencies of the voltage supply, the time required for significant ionization to occur is small compared with the time of half a cycle, that is, the ionization growth rate is much greater than the applied voltage growth rate.

The ballast also regulates against line voltage changes. A general guideline is that the lamp operating voltage should be about one-half the required starting voltage if good voltage regulation is to be achieved. This will result in a 1% change in line voltage changing lumen output by about 1%.

The relighting of the lamp each $\frac{1}{120}$ second requires that the open circuit secondary voltage of the ballast be near its maximum when the lamp current passes through zero. This 90° phase shift is indeed what the ballast provides, so that reignition is not a problem.

For lamps that require preheating of the cathodes during starting, separate windings are provided on the ballast secondary to provide the necessary circulating currents.

A ballast is expected to perform these five functions with minimum power loss, very long life, consistent reliability, and minimal disturbance of the lamp and the electrical supply. Current electromagnetic and electronic ballasts perform these actions well.

There are four types of ballasts for fluorescent lamps:

a) Standard electromagnetic

This was the ballast used exclusively prior to 1980. It consisted of a core of steel laminations surrounded by two coils of aluminum wire, and possibly a capacitor for power factor correction. The capacitors contained toxic polychlorinated biphenyls (PCBs), an average of one ounce per capacitor. These were outlawed in 1976 and do not appear in ballasts after 1979.

b) Energy-saving electromagnetic

This consists of a higher grade core of steel laminations, annealed and treated to reduce hysteresis and eddy current losses, surrounded by two coils of copper wire. The completed core and coil unit is vacuum impregnated with an asphalt wax substance and mounted in a steel case. Also included on the frame are a capacitor and an automatic-reclosing thermal protector. (Thermally protected ballasts are labeled “Class P”). The entire unit is encased in a silica compound that conducts heat away from the coils and dampens vibrations, which can cause 60-Hz hum. Ballasts are sound rated from A to C, with A suitable for quiet locations, such as libraries, and C applicable to noisy environments, such as factories.

The innards of an energy-saving magnetic ballast are shown in Fig. 6.13. The upgrading of materials and manufacturing controls have made this ballast much more energy conserving than its predecessor. For example, two 40-watt rapid start lamps on a standard electromagnetic ballast consume 96 watts, whereas with an energy-saving ballast the total is 86 watts. Current electromagnetic ballasts have an estimated life of 12 years.

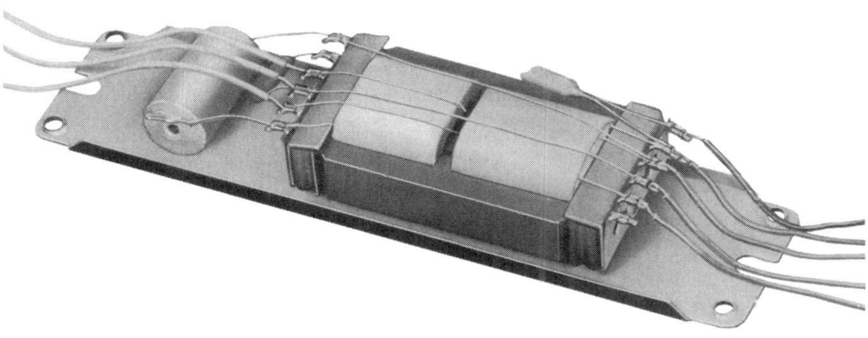

Figure 6.13 Electromagnetic ballast. Courtesy of Advance Transformer Co.

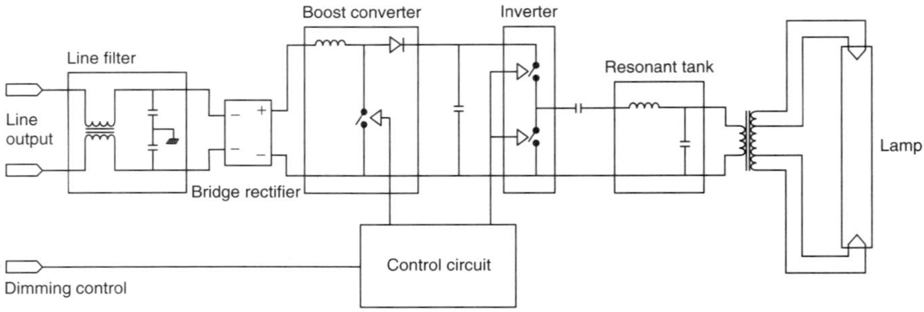

Figure 6.14 Electronic ballast circuit diagram. *Source*: *IESNA Lighting Handbook*, Ninth Edition, 2000. Courtesy of the Illuminating Engineering Society of North America.

c) Discrete electronic

This consists of separate electronic components that replace the core and coil in starting and controlling a fluorescent lamp. A circuit diagram is shown in Fig. 6.14 and has four basic parts.

- a line filter consisting of a capacitor, varistor (voltage-controlled resistor not shown), and common-mode inductor to block transient voltages in each direction.
- a bridge rectifier/boost converter that rectifies and filters the 60-Hz supply voltage and provides for power factor correction. The boost converter consists of a power transistor and nested resistors and capacitors.
- an inverter consisting of two MOSFET power transistors that chops the dc at a high-frequency rate (20–40 KHz)
- a resonant tank circuit and output transformer that provide open-circuit secondary voltage and cathode preheating current to start the lamp and control lamp current.

Electronic ballasts are more efficient than magnetic ballasts in converting input power into usable lamp power. Operation at high frequencies decreases electrode-related losses (providing increases in system efficacy of 15%) and reduces effects of flicker. Electronic ballasts run 30°C (86°F) cooler than electromagnetic ballasts and weigh significantly less. Estimates of the life of an electronic ballast are as high as twice that of an electromagnetic ballast.

d) Integrated electronic

In this ballast, a single silicon chip controls overall ballast operation. This integrated circuit controls cathode preheating, lamp starting, lamp power, lamp removal, and protection against voltage transients. It is the "brain" of the ballast and gives commands directly to the inverter and the power factor corrector.

New developments in electronic ballasts that are either available now or imminent include programmed starting scenarios to improve lumen maintenance and lengthen life, end of life sensing and lamp turn off mechanisms, on-off sensor controls and circuitry, and multiple-lamp ballasting arrangements.

Comparison data for several common T8 and T12 lamps on standard magnetic, energy saving magnetic, and electronic ballasts are given in Table 6.15.

TABLE 6.15 FLUORESCENT BALLAST COMPARISON DATA

Lamp Type	Lamp Label	No. of Lamps	Ballast Type	BF	Lamp Lumens	System Lumens	Input watts	System lm/W
Preheat	F1512CW	1	STD	0.95	750	712	20	36
	F20T12CW	1	STD	0.95	1175	1116	25	45
Rapid start	F40T12CW	2	STD	0.95	3050	5795	96	60
	F34T12CW	2	STD	0.88	2700	4752	82	58
	F40T12CW	2	EM	0.95	3050	5705	86	67
	F34T12CW	2	EM	0.88	2700	4752	72	66
	F40T12CW	2	EL	0.88	3050	5368	72	75
	F34T12CW	2	EL	0.88	2700	4752	62	77
	F032T8RE	2	EM	0.95	2800	5320	71	75
	F032T8RE	2	EL	0.88	2800	4928	62	79
	F96T12CW/HO/ES	2	STD	0.92	8000	14,720	210	70
	F96T12CW/HO/ES	2	EL	0.88	8800	15,488	210	74
Instant Start	F96T12CW	2	STD	0.93	6100	11,346	160	71
	F96T12CW/ES	2	STD	0.88	5400	9504	125	76
	F96T12CW	2	EL	0.88	6100	10,736	137	80
	F96T12CW/ES	2	EL	0.88	5400	9504	110	86
	F096T8RE	2	EL	0.88	5700	10,032	110	91

CW = cool white
RE = rare earth phosphor
RS = rapid start
IS = instant start
H0 = high output
ES = energy saving
STD = standard magnetic
EM = energy saving magnetic
EL = electronic
BF = ballast factor

Five metrics can be applied to quantify ballast performance. The first is ballast factor (BF). Lumen ratings published by manufacturers are based on operation of lamps on "reference ballasts" under closely-controlled conditions. When lamps are operated on commercial ballasts their lumen outputs are generally less than the rated values. The ballast factor is defined as:

$$BF = \frac{\phi_{\text{commercial ballast}}}{\phi_{\text{reference ballast}}} \tag{6.9}$$

Ballast factors are classed as "low" (0.75–0.85), medium (0.85–1.00), and high (1.00–1.20). Ballast factors for energy-efficient magnetic ballasts range from 0.90 to 0.95. For a ballast to receive the label of the Certified Ballast Manufacturers (CBM), it must have a ballast factor of at least 0.925 for standard 30- and 40-W rapid start lamps. However, this number can be misleading because a magnetic ballast is not allowed to thermally stabilize before readings are taken and, thus, light output may be too high. Correcting for this would yield a ballast factor of 0.875, rather than 0.925.

The reader should note in Table 6.15 that the system lumens do not equal the lamp lumens in single-lamp circuits and are not twice the lamp lumens in two-lamp circuits. This is because of the ballast factor. For example, in row 3, the value $5795 = 2 \times 0.95 \times 3050$.

A second ballast metric arises in conjunction with the National Appliance Energy Conservation Act of 1988. This metric is ballast efficacy factor (BEF), defined as

$$BEF = \frac{BF\,(\text{in}\,\%)}{\text{System Power Input}} \tag{6.10}$$

The 1988 law imposed minimum BEFs on certain lamp-ballast combinations, the most significant of which was BEF ≥ 1.06 for two 40-W T12 rapid start lamps. On a standard electromagnetic ballast (BF = 0.95), the input to this lamp-ballast system is 96 W, giving a BEF from Eq. (6.10) of 0.99. This combination fails. However, if the same pair of lamps operates on an energy-saving magnetic ballast, the power input is 86 W, the BEF is 1.10, and the combination passes, but only until 1992, when the National Energy Policy Act (EPACT) was passed, limiting lamp choice. This act eliminated the standard 40-W T12 lamp, as well as certain 8-ft standard and high output fluorescent lamps and several incandescent reflector lamps.

Two additional metrics, which evaluate distortion of the input voltage and current waves to a fluorescent lamp system, are crest factor (CF) and total harmonic distortion (THD). Each of these is a measure of the deviation from a pure sine wave of the voltage and current waves. Voltages in North America have a frequency of 60 Hz. Harmonics present are interger multiples of 60; the third harmonic, for example, is 180 Hz. This particular harmonic, which is often present, is especially troublesome in three-phase circuits, where the fundamental frequencies in each phase are separated by 120°, and thus the third harmonics by 360°, which means that the latter are directly additive in the neutral wire, often causing excessive current there.

Crest factor is defined as the ratio of the peak value to the root mean square (RMS) value of the voltage or current. For voltage

$$CF = \frac{V_{peak}}{V_{RMS}} \tag{6.11}$$

For a sinusoidal waveform the RMS value is 0.707 of the peak value, given CF = 1.4. Fluorescent lamp wave forms often have the shape shown in Fig. 6.15, indicating (through Fourier analysis) a strong third harmonic content.

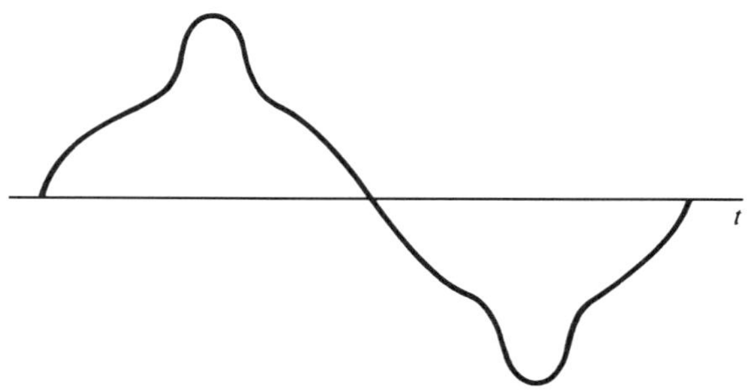

Figure 6.15 Typical ballast voltage waveform.

High current crest factor reduces lamp life. Lamp life values published by lamp manufacturers are invalid if current crest factor exceeds 1.7.

A second measure of distortion is total harmonic distortion (THD), defined for voltage in percent as

$$THD = \frac{\sqrt{\sum_{2}^{n} V_n^2}}{V_1} \tag{6.12}$$

where V_1 is the RMS value of the fundamental, and V_n is the RMS value of the nth harmonic. Early electronic ballasts had high THDs (35 to 40%), which caused excessive line current harmonics. For this reason, an international standard was developed limiting a fluorescent lamp ballast to ≤20% THD.

An additional metric is power factor. Power factor is defined as

$$PF = \frac{\text{watts}}{\text{RMS volts} \times \text{RMS amps}} = \cos\theta \Big|_{I}^{V} \tag{6.13}$$

where the first definition (watts divided by volt amperes) is general, but the second definition is valid only when the voltage and current waves are sinusoidal, or near so. For example, the square and triangular waves in Fig. 6.16 are in phase ($\theta = 0$ and $\cos\theta = 1$), but the power factor is 0.9 (the volt amperes exceed the watts).

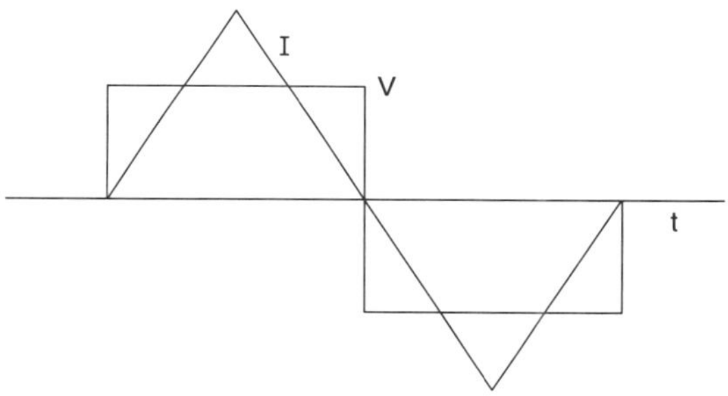

Figure 6.16 llustrating power factor.

Electromagnetic ballasts are inductive and, in series with fluorescent lamps, give power factors as low as 0.5. This means that the power company needs to provide twice the current for the same watts as would be the case for a power factor of 1.0. In electromagnetic ballast circuits, a capacitor is added in parallel with the series lamp-ballast to provide high power factor, defined as greater than 0.9. Power factor correction will be illustrated with a numerical example in the next section. Capacitor correction cannot be used with electronic ballasts. Rather, power factors of 0.9 are achieved through internal circuitry.

6.15 FLUORESCENT LAMP CIRCUITS AND STARTING METHODS

The original fluorescent lamp circuit consisted of a lamp, an inductor, and a starter button. One of the first improvements was to replace the starter button by an automatic starting device, as shown in Fig. 6.17. The starter is a small neon glow lamp, containing a fixed contact, a bimetallic strip, and a small capacitor. When the line switch is closed, the full line voltage appears across the neon glow tube, ionizing the gas and establishing a glow discharge, which then heats the bimetallic strip, causing it to bend and connect to the fixed contact. This creates a circuit for current to flow and preheats the electrodes. The current causes a voltage drop across the inductor, reducing the voltage across the glow tube, which causes the latter to deionize. The strip cools and breaks away from the fixed contact, producing a large L di/dt voltage surge across the inductor, which also appears across the lamp, causing

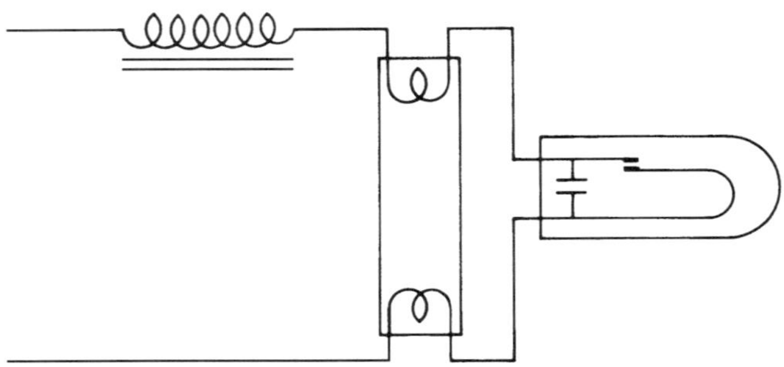

Figure 6.17 Glow switch starter.

the arc to strike. During lamp operation there is insufficient voltage across the glow tube to reestablish the glow discharge. The purpose of the capacitor is to short-out any high-frequency signals generated by the glow switch, which could produce radio interference.

The voltage and current waveforms in a fluorescent lamp circuit are not sinusoidal, due to nonlinearities in the ballast and lamp. However, for the simple series circuit of Fig. 6.17, a conventional ac circuit analysis based on sinusoidal waveforms gives a reasonable approximation. This will be illustrated by an example.

EXAMPLE 6.4

A fluorescent lamp and ballast are connected in series across a 200 V rms, 60 Hz supply and the following rms measurements are obtained: $V_{lamp} = 90$ V, $I = 500$ mA, $P_{lamp} = 40$ W, $P_{ballast} = 10$ W. Find the resistance and inductance of both the lamp and ballast.

Solution. A phasor diagram of voltages is shown in Fig. 6.18. The subscripts R and I refer to real and imaginary parts. The subscripts S, L, and B refer to supply, lamp, and ballast.

The calculations are as follows:

Lamp	Supply	Ballast
$\cos\theta = \frac{40}{90(0.5)} = 0.889$	$\cos\phi = \frac{50}{200(0.5)} = 0.500$	$V_{BR} = 100.0 - 80.0 = 20.0$ V
$\theta = 27.3°$, $\sin\theta = 0.458$	$\phi = 60°$, $\sin\phi = 0.866$	$V_{BI} = 173.2 - 41.2 = 132.0$ V
$V_{LR} = 90(0.889) = 80.0$ V	$V_{SR} = 200(0.500) = 100.0$ V	$R_{ballast} = \frac{20.0}{0.5} = 40\ \Omega$
$V_{LI} = 90(0.458) = 41.2$ V	$V_{SI} = 200(0.866) = 173.2$ V	$X_{ballast} = \frac{132.0}{0.5} = 264\ \Omega$
$R_{lamp} = \frac{80}{0.5} = 160\ \Omega$		$L_{ballast} = \frac{264}{377} = 0.70$ h
$X_{lamp} = \frac{41.2}{0.5} = 82.4\ \Omega$		
$L_{lamp} = \frac{82.4}{377} = 0.22$ h		

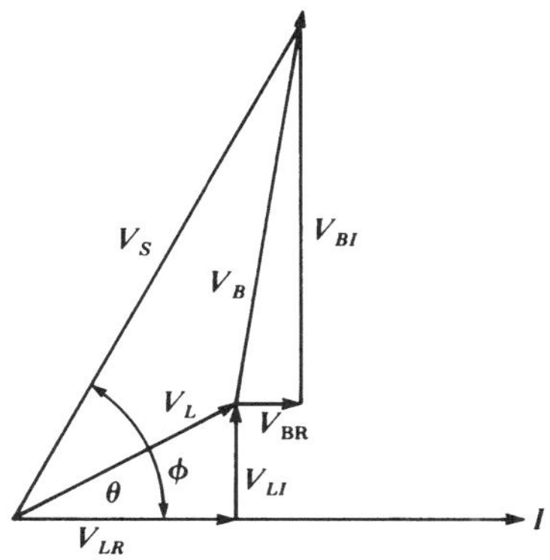

Figure 6.18

The procedure is to use Eq. (6.13) to obtain θ and ϕ and, from these, to calculate the in-phase and 90° out-of-phase voltage components for the lamp and the supply. Then the voltage components for the ballast are obtained by subtraction, and the resistances and inductive reactances are obtained by Ohm's Law. Finally, inductance is obtained from

$$X_L = 2\pi fL = \omega L \tag{6.14}$$

where $\omega = 2\pi f = 2\pi \times 60 = 377$ r/s.

EXAMPLE 6.5

In Sect. 6.14, power factor correction through the use of a capacitor was discussed. To illustrate this, what value capacitor, placed directly across the line in Fig. 6.17, will cause the overall circuit in Example 6.4 to operate at unity power factor?

Solution. A phasor diagram of the currents involved is shown in Fig. 6.19. The calculations are

$$I_c = 0.5\sin\phi = 0.5(0.866) = 0.433\,\text{A}$$

$$X_c = \frac{200}{0.433} = 462\ \Omega$$

$$C = \frac{1}{377(462)} = 5.7\,\mu\text{F}$$

What the capacitor has done is cancel the component of the supply current at −90° (0.433A) that was contributing nothing to the lamp watts. Thus, the supply current is reduced from 0.5 A to 0.5 cos 60 = 0.25 A, a considerable saving to the power company. In the last step, the value of capacitance is obtained from

$$X_c = \frac{1}{2\pi fc} = \frac{1}{\omega C} \tag{6.15}$$

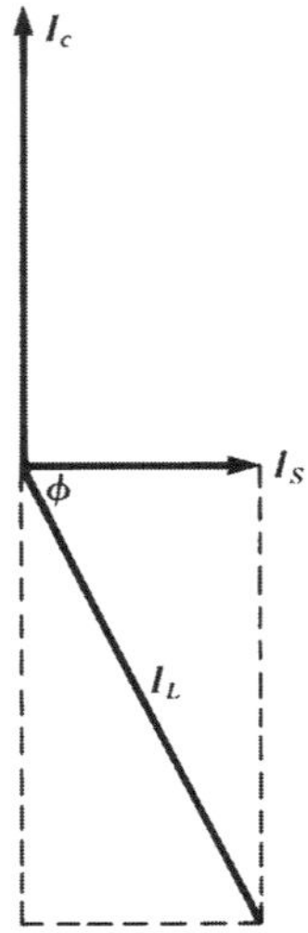

Figure 6.19

An alternative solution for correcting power factor is shown in Fig. 6.20. The autotransformer steps up the 120-V supply voltage. This is necessary for all fluorescent lamps of greater than 20 W in order to provide sufficient voltage to start the lamp. Lamp A operates with its current lagging the secondary voltage, whereas lamp B's current leads. The circuit elements are chosen so that a resulting power factor of 90% or more is achieved. It should be noted that the lead-circuit portion of Fig. 6.20 contains an inductor and capacitor in series, rather than simply a capacitor. At 60 Hz, pure capacitive ballasting produces a peaked lamp current, which is deleterious to the electrodes, resulting in shortened lamp life. This problem is not present at higher frequencies because the period of the applied voltage is now comparable to the time for ionization growth. The result is that there is insufficient time during each cycle for the lamp current to increase excessively. Thus, capacitor ballasts, with their advantages of small size, weight, and cost, are quite acceptable at higher frequencies. Also note that the lead circuit starter has a compensator winding in series to produce sufficient preheat current.

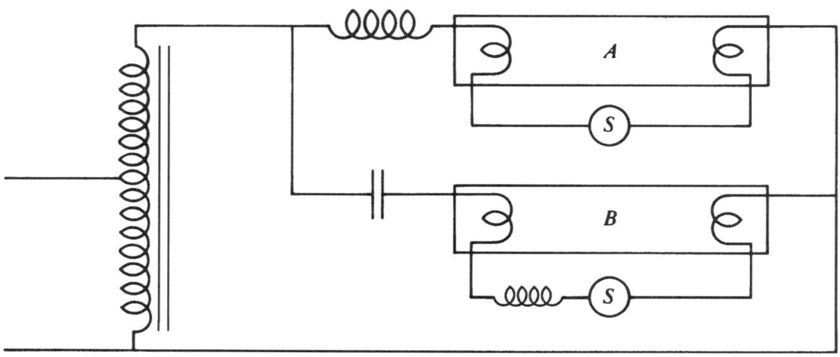

Figure 6.20 Two-lamp, lag-lead circuit.

Dating from its introduction in 1938, the preheat fluorescent lamp was criticized for taking too long to start. It was not until 1944 that instant-start fluorescent lamps appeared. These new lamps required no starters, had single-pin-bases that were stronger than the bipin bases of preheat lamps, were available in longer lengths of up to 8 ft, and could be operated over a range of currents, offering a 2:1 range of lumen output in a given lamp size. The major disadvantage was that, for a given lamp size, they required a much larger ballast. For example, a two-lamp ballast for 40 W T12 preheat lamps weighed 4 lbs and consumed 11 W. Its instant-start counterpart weighed $9^3/_4$ lb and consumed 20 W.

There were three major problems to overcome in designing an instant-start lamp, namely reduced life, early end blackening, and nonreliable starting. The first two are governed largely by electrode design. Preheat filaments are double coiled (Fig. 6.6). If sufficient voltage is applied to start them when cold, severe sputtering of emission material results. This blackens the lamp ends and reduces life to about one-third its normal value.

Attention was focused on designing an electrode that would deliver reasonable life and minimal end blackening despite some sputtering during each start. The result was the triple-coiled filament (also shown in Fig. 6.6). It consists of the normal double-coiled filament with a loose overwind of fine tungsten wire. The overwind reduces sputtering time by causing the filament to reach its emitting temperature more rapidly and it acts as a basket to hold a larger quantity of emission material.

The problem of reliable starting was a more baffling one. An experiment was performed in which the voltage across a 4-ft long T12 lamp without cathode preheating was gradually increased until the lamp started. The result was a wide range of starting voltages from a minimum of 250 V to a maximum of 750 V (a preheat lamp of this size requires 200 V to start). Even with the same lamp, the starting voltages on two different days could differ widely.

It was found that humidity was responsible for this large variation in starting voltages. As shown in Fig. 6.21, starting voltage increases as relative humidity increases, which corresponds to a reduction in the resistance of the bulb-wall surface. The reason for this effect

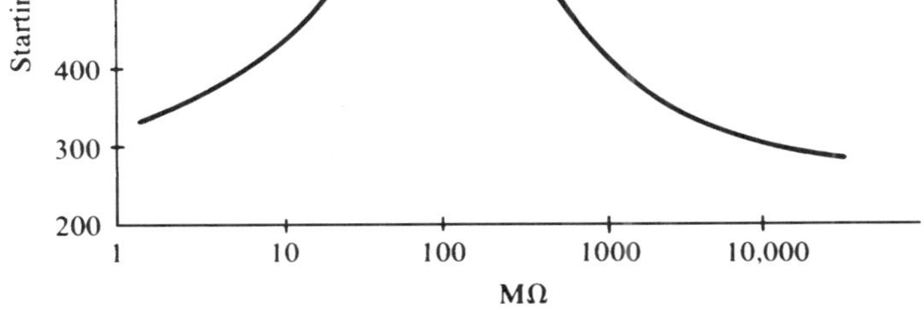

Figure 6.21 Starting voltage variation with humidity and bulb resistance.

is that the starting of discharges by combined field and thermionic emission requires a high electric field near each electrode. When the bulb resistance is very high, the metal fixture provides a low-resistance path between electrodes and thus concentrates most of the applied voltage near the lamp ends. When the bulb resistance is very low, the bulb surface performs this function of concentrating the voltage. At intermediate values of bulb resistance, which occur when relative humidity is 65 to 100%, neither the fixture nor the bulb concentrate the voltage sufficiently, and starting is difficult.

Several remedies for this situation were tried, but the most successful was a silicone coating of the exterior bulb wall, producing an effect much like that of car wax, in that any moisture present tended to form tiny droplets, rather than distributing itself uniformly over the surface. This caused the bulb resistance to remain at a high value independent of humidity, thus allowing the metal fixture to concentrate the voltage at the electrodes. The result was that reliable starting of the 4-ft T12 lamp could be achieved with 430 V, with starting time of typically 50 ms.

A single instant-start lamp can be operated in a circuit similar to that in Fig. 6.17, except that the starter is omitted and the filament leads at each end of the lamp are shorted together. Two instant-start lamps can be operated in the lag-lead circuit of Fig. 6.20, which places the lamps in parallel. However, it was found that ballast size, weight, and voltampere product are reduced if the lamps are operated in the series arrangement shown in Fig. 6.22. In this circuit, winding A is the primary and winding B the secondary. Winding C is a high-voltage low-current auxiliary winding. When the line switch is closed, the high voltage of winding C is impressed directly across lamp 1 and

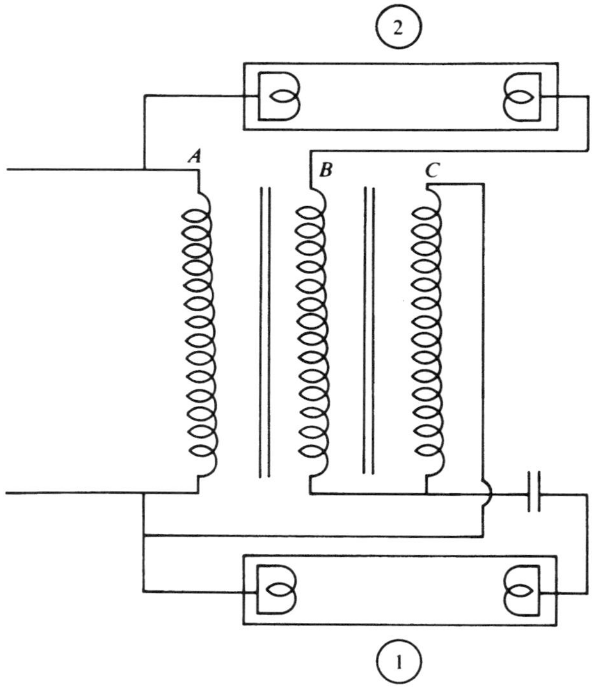

Figure 6.22 Sequence-start circuit.

that lamp starts. The voltage across lamp 2 is the phasor sum of the voltages of the three windings. The capacitor introduces a phase shift that causes the voltage of winding C to lag that of windings A and B by about 120°. The resultant voltage across lamp 2 is thus sufficient to start it a few milliseconds after lamp 1.

Although the advantages of instant-start lamps over preheat lamps were considerable, the disadvantages of large, inefficient ballasts and shorter lives caused lamp researchers to seek a new lamp that provided the best features of both preheating and instant starting. The result was the introduction of the rapid-start lamp in 1952. Rapid-start lamps require no starters. They start in less than a second. Their ballasts are smaller and more efficient than those of instant-start lamps. Their lives are long, in excess of 20,000 h at three or more burning hours per start.

In the development of the rapid-start lamp, it was decided first to investigate the effects on lamp performance when electrode heating and transformer secondary voltage were applied to a lamp simultaneously. In Fig. 6.23, the starting voltage for 40-W T12 fluorescent lamps is plotted against electrode heating current for several conditions. Curve A is for isolated lamps with no fixture present to concentrate the applied voltage at the electrodes. Curve B is for a lamp with a conductive strip running lengthwise of the lamp. Curve C is for a lamp with a silicone coating, using a metal fixture grounded to one end of the lamp.

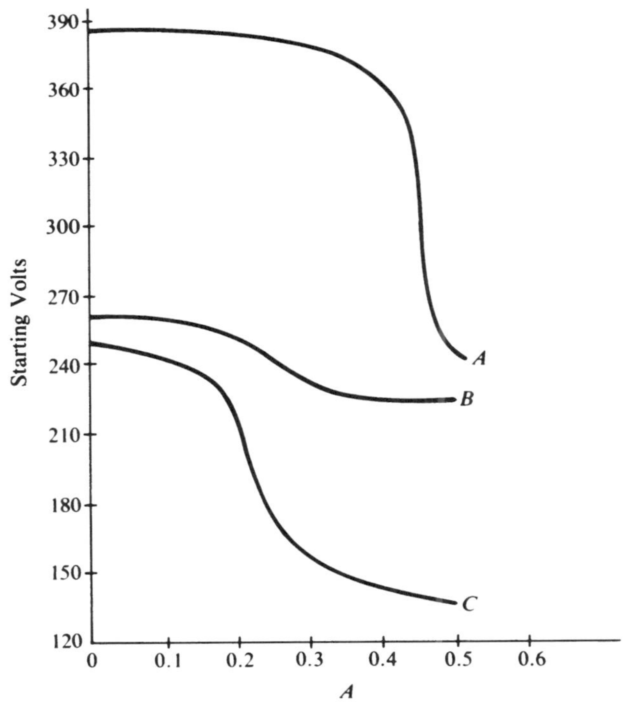

Figure 6.23 Starting volts versus heating current for 40-W T12 lamps.

Based on these curves, it was determined that a ballast circuit that could provide about 380 mA of heating current simultaneously with the application of applied voltage would start a lamp in about $^1/_2$ s. This presumes that the lamp is silicone coated on the outside and that it is housed in a metal fixture. In addition, it was decided to use shorter electrodes, having a voltage drop of 2 to 4 volts, to reduce the time required for heating and to decrease the wattage loss at the electrodes to about a watt at each end.

The basic rapid-start circuits are shown in Fig. 6.24. The one-lamp circuit is the same as that in Fig. 6.17 for the preheat lamp except for the replacement of the starter by the small auxiliary windings on the transformer that provide about 3.6 volts each. The two-lamp circuit is a sequence-start circuit similar to that for instant-start lamps in Fig. 6.22. The capacitor across the upper lamp aids in starting the lower lamp first. After starting, the voltage across the lower lamp is relatively low, placing a large voltage across the upper lamp, which then starts. The three auxiliary windings on the transformer provide the necessary electrode preheating during starting.

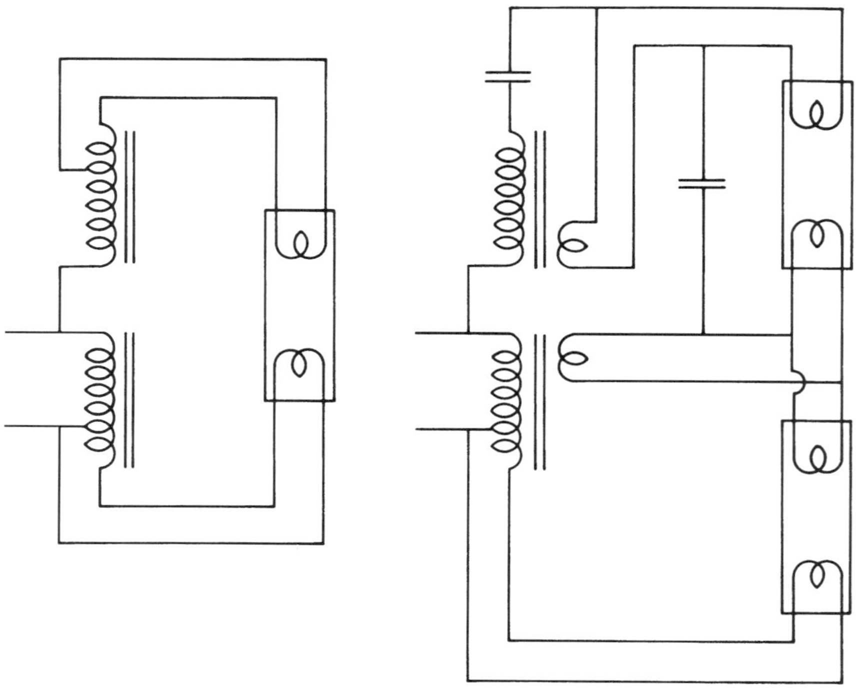

Figure 6.24 Rapid-start circuits.

For reliable starting, all rapid-start installations require that each lamp be mounted in a fixture that has a grounded metal plate extending the length of the lamps. Optimally, the strip should be 1 in wide and from $^1/_2$ to 1 in from the lamp. Even so, there is no guarantee each time that the cathodes will have reached their proper temperature prior to starting. If the lamp starts early, emission material can be sputtered off, shortening lamp life.

Normally the cathodes in Fig. 6.24 are heated continually during lamp operation by the three auxiliary windings, but some energy-saving ballasts have cutouts, which disconnect the windings after the cathodes are hot. This reduces the wattage of an energy-saving four-foot T12 lamp from 34 W to 32 W.

All three methods of starting fluorescent lamps are also available with electronic ballasts. Preheat electronic ballasts have internal timing circuitry that delays application of lamp voltage until cathode preheating has taken place. Rapid start designs are similar except that the preheating current continues after the lamp starts. There are electronic ballasts that start rapid start lamps instantly, but at a reduction in lamp life.

The newest innovation in starting is programmed start, used in conjunction with electronic ballasts. This technique can extend lamp life by as much as 50% through a starting sequence that is much less damaging to the cathodes.

In normal rapid starting, the voltage across the lamp creates a glow current during the preheating interval. This glow current causes emission material to be lost from the cathodes. In programmed starting, the voltage across the lamp is kept at a low level to reduce this glow current during cathode heating and to enable the cathodes to reach 700°C (1300°F) before lamp starting voltage is applied.

After lamp voltage is applied, there is a transition phase to full lamp current. In conventional rapid starting, transition time is 80 to100 ms; in programmed starting it is about 30 ms. This again reduces the amount of emission material lost.

Initial life testing has shown that conventional rapid starting with a 15-minute on, 5-minute off cycle provides about 16,000 starts before 50% of the lamps tested failed. With programmed starting, the total was 40,000 starts. Thus, programmed starting appears particularly well-suited to applications where lights are turned on and off frequently.

6.16 COLD CATHODE (NEON) LIGHTING

The five inert gases (helium, neon, argon, krypton, and xenon) were isolated from the atmosphere in the 1890s by William Ramsay, Baron Rayleigh, and M.W. Travers. When used as fill gases in early Geissler-Crookes discharge tubes, each gave its own distinctive color of light. Georges Claude developed an economical method of producing neon in 1902, formed the Claude Neon Co., and displayed neon lamps at the Paris Exposition in 1910. Neon sign tubing was introduced commercially in the U.S. shortly thereafter.

The distinction between "cold-cathode" lighting and "neon" lighting is blurry. Both are glow discharges, implying low currents (≈100 mA); both use neon (red) and argon (blue) as the fill gases; all other colors are from phosphors or colored glass tubing. The main difference appears to be the current drawn—30–60 ma implies "neon" whereas >120 ma denotes "cold cathode". The term "neon" will be used in the remainder of this section.

Neon sign tubing is usually created uniquely to a customer's specification. Lamps deliver from 100 to 500 lumens per foot and consume from 6.5 to 12.7 watts per foot, giving an efficacy of about 30 lm/W (electrode power consumption is only 10% of total power consumption). Lives are 10 years and systems are essentially maintenance-free and may be dimmed down to 10% of rated light output. Typical tube lengths are 1200 to1500 mm (48–60 in). These modular lengths are wired in series to create long lines for display and advertising. The series circuits are powered by high-voltage transformers, delivering 2,000–15,000 V (the limit per transformer is 1000 V in residential applications). Tube diameter, fill gas used, and overall system length determine the magnitude of the secondary voltage required.

Table 6.16 displays the maximum number of feet of tubing filled with neon that can be operated from three transformer secondary voltages, as a function of tube diameter. Note that as tube diameter decreases, the number of feet that can be operated by a given supply voltage decreases, i.e. the volts required per foot increases.

TABLE 6.16 OPERATING DATA FOR NEON

	Tube Diameter (mm)			Transformer Ratings		
	10	15	25	mA	V	W
max tube length* (ft)	33–38	60–69	102–118	0	15,000	235
	13–15	24–28	51–59	30	7,500	15
	6	10	17	30	3,000	50

* Deduct one foot for each pair of electrodes.

6.17 HIGH-INTENSITY DISCHARGE LAMP DEVELOPMENTS

As stated in Sect. 1.2, Peter Cooper Hewitt developed the first practical mercury lamp in 1901, a 385-W, 4-ft-long tubular lamp giving 12.4 lm/W. The first commercial use of this source was in the composing room of the *New York Evening Post* in 1903.

Moore tubes were introduced in 1904. These were 7 to 9-ft long tubes filled first with nitrogen, later with carbon dioxide, and still later with air freed of oxygen. Those filled with carbon dioxide, although inefficient, produced a white light whose correlated color temperature matched daylight, and they were widely used in color matching and grading applications.

Although some developmental work on high-intensity discharge lamps was carried on in Europe in the early 1900s, it was not until the 1930s that such lamps began to have an impact commercially. In 1932, Dr. Giles Holst in Holland developed a low-voltage, low-pressure sodium lamp, utilizing a special glass that could withstand the highly-alkaline effects of vaporized sodium. The lamp is still used extensively in Europe in street lighting installations. It was first introduced in the United States in 1933, but has received limited acceptance over the years, largely because of its low intensity, difficulty of optical control due to the size of the arc, and, particularly, its monochromatic spectrum (the double yellow line at 589.0 and 589.6 nm, which was discussed in Sect. 5.6).

Interest in mercury vapor sources quickened during the 1930s. Lamps at nearly atmospheric pressure (called high pressure) and rated at 400-W were introduced in Europe in the early 1930s and in the United States by the middle of the decade. These lamps consisted of an arc tube enclosed within an outer glass bulb and provided a lumen output about equal to that of a 750-W incandescent lamp. A 250-W mercury lamp was introduced in the late 1930s; 1000 and 3000-W mercury lamps appeared in the early 1940s, to satisfy a need for high-output, efficient lighting for industry during the war years.

Mercury arc discharges had two inherent problems. The first involved the bowing of the arc when the lamps were burned horizontally. This caused the arc to touch the glass tube, often melting it. All mercury lamps for street lighting were equipped with magnets to pull the arc away from the arc tube. This was the case until the early 1950s, when synthetic quartz was introduced. Contemporary mercury lamps have smaller quartz arc tubes, which can withstand the high temperatures generated by the discharge, and they do not require magnets.

The second problem with mercury lamps was their color. In Sect. 5.6, it was pointed out that a mercury discharge provides four visible lines at 404.7, 435.8, 546.1, and 577 to 579 nm. This produces a bluish-green white light with no reds or oranges and makes human complexions appear as though rigor mortis has set in. Mercury vapor lamps were reluctantly accepted for street lighting and a few other applications, but could not be used in any application where color rendition was important.

A first solution to this problem was simply to use a 400-W mercury lamp side-by-side with a 750-W incandescent lamp in industrial settings. The excess of reds and oranges in the incandescent lamp counterbalanced their absence in the mercury lamp. Then, in the 1950s, color-corrected mercury lamps were introduced. These consisted of a conventional mercury lamp with a phosphor added to the inside of the outer glass bulb, which converted ultraviolet mercury radiation into visible radiation in the red portion of the spectrum. These phosphors improved the overall color properties of the lamp somewhat, but not enough to make the lamps acceptable for a wide range of interior applications. Also, the phosphors increased the source size significantly, that is, the source was no longer the inner arc tube, but rather the outer glass bulb. This made optical control difficult. Today the most significant use of (clear) mercury lamps is in landscape lighting.

In the early 1960s work proceeded in a different direction to try to improve the color of mercury lamps, as well as their luminous efficacy. Dr. Gilbert Reiling found that the addition of certain metals, in the form of their halide salts (usually iodides), would produce sufficient concentration of metal atoms in the vapor to provide additional spectral lines in the discharge and enhance the overall spectral characteristics of the lamp. At the same time, the halides were able to recombine with the metals leaving the arc stream, thus preventing these metals from coming in contact with the wall of the arc tube, which would seriously degrade the latter. The result was the metal-halide lamp, introduced in the mid-1960s.

Metal-halide lamps are now available in a variety of sizes ranging from 10 to 1500 W. The original metallic additives were indium, thallium, and sodium, but these have been replaced by sodium and scandium in contemporary lamps. The lamps have excellent color characteristics when compared with other HID sources. Also they are relatively easy to control optically because of their small source size, and they have lamp efficacies of over 80 lm/W. Several have lives of 20,000 h. They suffer in that lumen depreciation is significant.

The most recent major milestone in lamp development was the introduction of the high-pressure sodium lamp, also in the mid-1960s. The lamp was developed by Dr. Kurt Schmidt and William Louden and was introduced commercially in 1965 in the 400-W size, with ratings of 6000 h and 42,000 initial lumens. The key to the introduction of the high-pressure sodium lamp was the development of a special ceramic material for the arc tube, a translucent polycrystalline alumina (PCA) that permits 92% light transmission and yet is free of minute pores that would allow the active sodium to pass through. The material is also chemically resistant to sodium and can withstand a central arc temperature of 1500 K.

High-pressure sodium lamps are available in wattages ranging from 35 to 1000 W with lamp efficacies of over 100 lm/W. Their spectra are deficient in reds, greens, and blues and they have a correlated color temperature of 2100 K, quite a bit lower than for incandescent lamps. Their lives are long, ranging from 12,000 to 20,000 h.

6.18 MERCURY LAMPS

A cutaway view of a mercury lamp is shown in Fig. 6.25. As mentioned earlier, the lamp consists of an inner quartz arc tube and an outer borosilicate glass envelope. The quartz (fused silica) tube must withstand an arc temperature up to 1200 K, whereas the outer tube can operate at a maximum of 700 K.

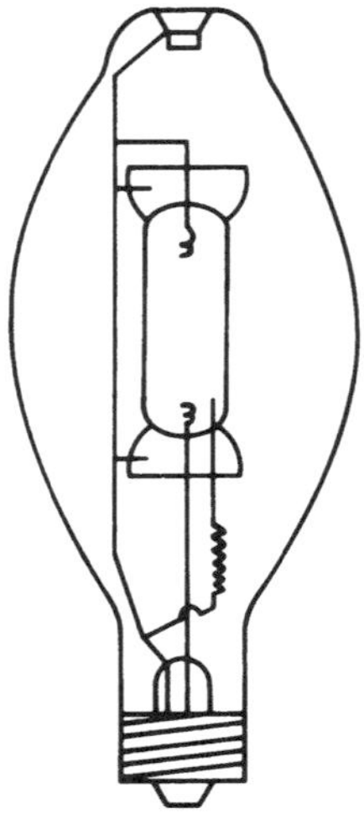

Figure 6.25 Mercury lamp construction.

Nitrogen is used as a filling gas between the inner and outer bulbs to thermally control the arc tube, to protect metal parts from oxidation and to absorb some of the UV radiation. The arc tube contains mercury and argon gas, the latter serving the same function as in a fluorescent lamp.

Within the arc tube are two main electrodes and a starting electrode. Each main electrode consists of a tungsten rod upon which a double layer of coiled tungsten wire is wound. The electrode is dipped into a mixture of thorium, calcium, and barium carbonates and then heated to convert these compounds to oxides. The electrodes are connected through the quartz tube by molybdenum foil leads.

When voltage is applied to the lamp, it appears between the starting electrode and the adjacent main electrode, as well as between the two main electrodes. This creates a local argon arc, with current limited by the starting resistor. The heat generated by the small argon discharge vaporizes some of the mercury present, creates ionization within the arc tube, and soon the main arc strikes. The resistance of the main arc circuit is much less than that of the starting arc circuit, and so the starting arc ceases. Following initiation of the main arc, mercury vaporization continues and, after 5 to 7 minutes, all of the mercury is vaporized, the pressure reaches 200–400 kPa (2–4 atmospheres), and the lamp reaches steady state.

Characteristics of typical clear-bulb mercury lamps are shown in Table 6.17. These five lamps are also available with vanadate phosphor coatings to provide improved color performance. The initial lumens of all HID lamps, as in the case of fluorescent lamps, are taken at 100 h, because the "clean-up" of impurities reduces light output significantly during the first 100 h of operation. The mean lumens are taken at 50% of life at 10 h per start, or about 12,000 h.

The mercury lamp life rating of 24,000 h is not the 50% survivor point. Rather, it is the point at which the lumen output has dropped so low that the lamp is no longer economically viable. At 24,000 h, the number of survivors is about 67%. Mercury lamps fade away, rather than die.

TABLE 6.17 HID LAMP DATA

Watts	Bulb	Clear Lamp Initial Lumens	Initial Lumens per Lamp watt	Life (h)	Mean Lumens
		Mercury			
100	E23½	4,000	40	24,000+	3,000
175	BT28	7,700	44	24,000+	7,150
250	BT28	11,600	46	24,000+	10,800
400	BT37	20,500	51	24,000+	18,700
1000	BT56	55,000	55	24,000+	44,000
		Standard Metal Halide (vertical)			
175	BT28	14,400	82	10,000	12,000
250	BT28	22,000	88	10,000	17,000
400	BT37	36,000	90	20,000	25,000
1000	BT56	110,000	110	15,000	88,000
1500	BT56	158,000	105	3,000	140,000
		Pulse Start Metal Halide			
50	ED17	3,900	78	5,000	2,200
70	ED17	5,500	79	12,000	3,500
100	ED17	9,000	90	15,000	6,200
150	BD17	12,500	83	15,000	8,600
175	ED23½	17,200	98	15,000	13,400
250	ED28	23,000	92	15,000	16,100
320	ED28	31,000	97	15,000	18,000
400	ED37	44,000	110	20,000	33,900
		Ceramic Metal Halide			
39	T6	3,300	85	9,000	2,600
70	ED17	6,200	89	10,000	4,960
100	ED17	9,300	93	10,000	7,500
150	T6	13,500	90	9,000	10,800
		Standard High-Pressure Sodium			
35	E17	2,250	64	16,000+	2,050
50	E17	4,000	80	24,000+	3,600
70	E17	6,300	90	24,000+	5,350
100	E17	9,500	95	24,000+	8,000
150	E17	15,800	105	24,000+	13,400
200	E18	22,000	110	24,000+	19,800
250	ET18	29,000	116	24,000+	26,100
400	ET18	50,000	125	24,000+	45,000
750	BT37	105,000	140	24,000+	94,500
1000	E25	130,000	130	24,000+	120,700
		Retrofit High-Pressure Sodium Lamps			
150	BT23	11,800	79	24,000	10,600
215	BT28	20,000	93	16,000	17,000
360	BT37	36,500	101	16,000	32,800

There are mercury lamps for use in open luminaires that are designed to self-extinguish when the outer bulb is broken. This is to protect against exposure to short-wavelength UV, which passes through the inner quartz arc tube.

6.19 METAL-HALIDE LAMPS

There had been interest since the 1930s in using metals other than mercury in arc discharge light sources to improve color. However, all efforts to use them in their elemental form had failed because of their low vapor pressures (less than 0.1 mm) at normal arc tube temperatures, which prevented the efficient production of visible radiation.

The discovery in 1960 by Dr. Reiling that such metals could be introduced into the arc stream via their iodide salts, which have vapor pressures of several millimeters at arc tube temperatures of 1000 K and do not react with quartz, made metal-halide lamps feasible. When the lamp is cold, the metal iodides reside on the bulb wall. As arc temperature rises, they vaporize, diffuse from the wall into the arc stream, and dissociate, yielding free metal and iodine atoms. The former are then excited and ionized in the same fashion as are the mercury atoms. They diffuse and convect outward toward the cooler bulb wall, recombine, and the process repeats.

In a multiple metal-halide lamp, the iodides do not all vaporize at the same time. In the lamps containing indium, thallium, and sodium iodides, the indium vaporizes first and forms a blue sheath around the mercury arc. Next, the thallium vaporizes and forms a green sheath around the indium. Last, the sodium vaporizes and forms a yellow sheath around the thallium. Such a lamp is very sensitive to changes in lamp wattage, largely because of sodium, the final iodide to vaporize. If the lamp watts are lower than rated, too little sodium will be vaporized, and the lamp will be deficient in yellow and red. Too high a lamp wattage causes a color shift toward pink.

This lamp has been replaced throughout the industry by the sodium-scandium metal-halide lamp. In this lamp, the scandium vaporizes first, followed by the sodium. The lamp is much less sensitive to lamp wattage changes than its three-iodide counterpart and thus gives much more stable color performance. By varying the proportions of sodium and scadium, CCTs of 2500–5000 K can be obtained. More recently, a combination of rare-earth iodides of dysprosium, holmium, and thulium has been introduced, with a CCT of about 5400 K.

The basic requirements for a metal additive are

1. The iodide of the metal must be stable at the bulb-wall operating temperature.
2. The iodide vapor pressure must be relatively high.
3. The metal atom must have excitation levels lower than the average of those of mercury (about 7.8 eV).
4. The metal must have an energy level configuration that encourages a high percentage of visible radiation.

Thallium is an example of a metal which satisfies these constraints. It has a strong line at 535 nm requiring only 3.3 eV for its generation, and the vapor pressure of its iodide salts is about 10 mm when the coolest spot on the arc tube is at about 800 K. Scandium, sodium, and indium also fulfill these requirements.

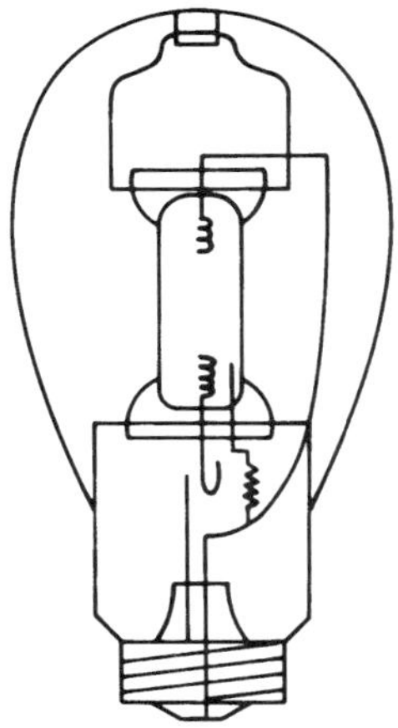

Figure 6.26 Metal-halide lamp.

A cutaway view of a conventional metal-halide lamp with pinched end seals is shown in Fig. 6.26. The lamp is very similar in construction to a mercury lamp with the following exceptions:

1. The arc tube ends are treated with a reflective white coating to redirect energy back into the tube so as to maintain a more uniform temperature over the entire arc tube length and increase vaporization of the halide additives.
2. A bimetal switch is included to short the starter electrode to the main electrode after starting. This prevents the build up of small electric fields that can destroy the seal between the lead-in wires and the glass.
3. The construction is frameless and the outer electric wire is of non-magnetic molybdenum. The reason for this is to prevent diffusion of sodium through the arc tube wall by electrolysis. This results from photoemission from the return lead wire and any frame present. The emission places a negative charge on the outside of the arc tube, attracting positive sodium ions.

The starting of a metal-halide lamp is identical with that of a mercury lamp up until the point where the additives begin to vaporize. Thus, the arc is initiated with the aid of a starter electrode, first in argon, then in mercury, and later in mercury plus the additives. The entire starting process to full light output takes from 3 to 5 minutes.

Some arc tubes are oval in shape (formed-body arc tubes) and have small, non-flat pinched ends. This shape of tube conforms better to the shape of the arc, allowing the halides to vaporize more rapidly and reducing warm-up time by as much as two-thirds. In these lamps, starting is accomplished, not through a starting electrode, but with a high voltage pulse of 3000 to 5000 V from an internal ignitor in the ballast. Such lamps are known as pulse-start metal halide.

Pulse-start technology emerged in 1984 as the mechanism for starting low wattage (<150 W) metal halide lamps, largely because there was no room for a starter electrode in the smaller lamps. The major aim of this technology was to reduce tungsten evaporation by increasing arc tube pressure.

Pulse-start lamps are now available in larger wattage sizes, but with the normal pinched-seal ends. The narrower ends used in low-wattage lamps cool the ends of the arc tube, thus lowering lamp efficacy. Pulse-start lamps have longer lives, better lumen maintenance, better color rendering, reduced color shift, better lamp-to-lamp color consistency, faster warm up, and hotter restrike capability than their predecessors.

Metal halide lamps are sensitive to burning position. Lamps with straight arc tubes are rated (life and lumens) based on operating vertically and give their best performance in that orientation. Operation horizontally causes the arc to bow upward, away from the halides, causing the vapor pressure of the latter to drop and lowering light output. Also, the greater proximity of the arc to the upper bulb wall raises temperature and decreases life by as much as 25%. Manufacturers have developed lamps with bowed arc tubes and lamps with electrodes located below the center line of the arc tube to circumvent this problem. A special base and socket are required to insure correct arc tube orientation.

A recent development is the substitution of a PCA ceramic arc tube for the traditional quartz tube. The major benefit of this arc tube is its ability to maintain CCT within ±200 K during the life of the lamp. Variation in standard metal halide lamps can be as much as ±600 K. In addition, the lamp provides excellent color rendering and stability, achieved by reduced loss of sodium to the arc tube wall. Ceramic arc tube metal halide lamps have no starting electrode and operate in the pulse start mode. Initial lamp efficiencies are 85–90 lm/W, due to operating at higher temperatures than is possible with quartz. These lamps are very popular in PAR configurations.

Most metal halide lamps require enclosed luminaires, to protect against the rupture of the arc tube, which occurs rarely near the end of life if the lamp has been burned continuously. Lamps designed for open luminaires have a protective shield (shroud) around the arc tube, to contain the hot quartz in case of tube failure.

Data for selected metal halide lamps are given in Table 6.17 in three categories: standard, pulse start, and ceramic. Note that for a given wattage size, because of the halide additives, the lumen output and luminous efficacy of a metal halide lamp are significantly greater than those of a mercury lamp. However, lamp life is shorter. The addition of the metallic iodides necessitates a higher voltage for reignition each half cycle than is required in mercury lamps. This increases electrode deterioration and has an adverse effect on lamp life. Also, metal halide lamps show greater lumen depreciation than the other HID types. This is caused by a shift in the chemical balance of the halide additives as the lamps age. Initial lumen ratings for metal halide lamps are given after 100 hours of burning. Mean lumens are at 40% of rated life.

Metal halide lamps are much like fluorescent lamps in that the life is dependent on the number of burning hours per start. The life ratings given in Table 6.17 are for 10 h per start. Life is reduced by 25% at 5 h per start and increased by 35% for continuous burning. Of the standard lamps, the 400-W metal halide lamp has the longest life (20,000 h); the 1500-W lamp the shortest life (3000 h). In the latter lamp, which is used in sports lighting, life is sacrificed in order to get greater lumen output per lamp, thus reducing the number of poles and fixtures required.

The standard metal halide lamps in Table 6.17 are designated "universal burn", meaning that they can be mounted in any orientation. The values given are for within 15° of vertical burning. For horizontal burning, lumen values are 15% less, and life values are 25% less. Lamps specifically designed for horizontal burning have lumen and life ratings comparable to those in the table. The operating position for the pulse start lamps in Table 6.17 is vertical, with the base up; for the ceramic lamps, any position.

6.20 HIGH-PRESSURE SODIUM LAMPS

It was pointed out earlier that the development of polycrystalline alumina (PCA) in 1959 paved the way for the introduction of the high-pressure sodium (HPS) lamp. The first commercial lamp appeared in 1965 and was rated at 400 W, 42,000 initial lumens, and 6000 h. Improvements since then have permitted these ratings to be raised to 50,000 initial lumens and 24,000 h at 10 h per start, giving a lamp that provides 2.4 times the lumen output of its mercury counterpart and, yet, has the same rated life.

A diagram of a high-pressure sodium (HPS) lamp appears in Fig. 6.27. The inner PCA arc tube is filled with xenon gas, for starting, and a sodium-mercury amalgam. Back wound and coated tungsten electrodes are mounted in each end. Originally, these were sealed in place with niobium end caps, but now a ceramic plug seal and niobium tube are used. The arc tube is inserted into a heat-resistant borosilicate outer bulb and supported by a sliding joint that permits the entire structure to expand and contract without distorting. The space between the tube and bulb is evacuated.

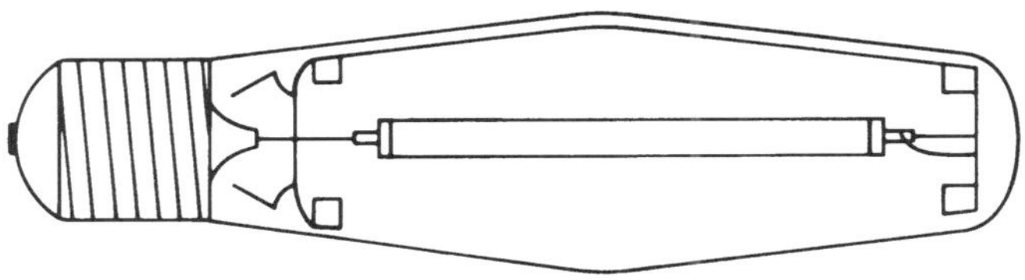

Figure 6.27 High-pressure sodium lamp.

There is no starting electrode in a high-pressure sodium lamp because the smaller diameter ($^3/_8$ in) arc tube prohibits it. Thus, the ballast must provide a high starting voltage. This is commonly done by superimposing a low-energy, high-voltage pulse on the ballast open-circuit voltage waveform. A typical pulse has a peak voltage of 2500 V and lasts for 1 μs. This pulse ionizes the xenon gas sufficiently for the open-circuit ballast voltage to then initiate and maintain the xenon arc. Initially, the arc is bluish-white, representing the effect of excited xenon and mercury. As temperature rises, sodium becomes excited and a low-pressure monochromatic yellow sodium spectrum results. The spectrum broadens as temperature rises further and becomes an absorption spectrum in that the sodium resonance line at 589 nm is self-absorbed. During this period, sodium pressure increases from 2 kPa (0.02 atm) in the monochromatic discharge to over 27 kPa (0.3 atm) in the final steady-state, broad-spectrum condition.

A high-pressure sodium lamp, like a metal halide lamp, is an "excess" discharge source. There is more amalgam in the reservoir formed inside the arc tube in back of one of the electrodes than is normally vaporized during lamp operation. During the first 100 h of burning, lamp voltage may rise or fall considerably as greater or lesser amounts of amalgam enter the arc stream. During the remainder of the lamp's life, the lamp voltage gradually increases, because of arc tube end blackening and loss of sodium through the arc tube wall, until a point is reached at which the ballast voltage is no longer sufficient to support the arc. This is the end of the useful life of the lamp and is manifested by the lamp "cycling," that is, repeatedly starting and extinguishing.

Reduced-mercury HPS lamps are available that do not cycle at the end of life, but have the same initial and mean lumen outputs as their standard counterparts. Limiting the amount of mercury in the sodium-mercury amalgam causes these lamps to change to bluish-white and then extinguish at the end of life. These lamps have higher xenon gas filled pressure and thus require an auxiliary starting coil wrapped around the arc tube and connected to one electrode. Initial ionization takes place between the starting aid and the opposite electrode.

There is a family of HPS lamps that contain two arc tubes in parallel in a single outer jacket. Following a power interruption, the second lamp ignites promptly when power is restored. Normally, with a single arc tube, warm-up time is 3 to 4 minutes and restrike time is at least one minute.

As pressure is raised in a high-pressure sodium lamp, the spectrum continues to broaden around the 589 mm "centerline". There is a family of high-pressure sodium lamps, designed to replace incandescent lamps in accent and display lighting, that has (warm) CCTs of 2600 to 2700 K and excellent color rendering. The price that is paid is reduced life and efficacy (10,000 hours and 45 lm/W) compared with conventional high pressure sodium lamps, but the life and efficacy are still considerably greater than those of the incandescent lamps replaced.

Data for several common high-pressure sodium lamps appear in Table 6.17. These lamps can be burned in any position, have the highest luminous efficacies of the HID family when used with ballasts specifically designed for them, and give excellent lumen maintenance (90% for 200 watts and above for the standard lamps). Also shown in the table is a group of high-pressure sodium lamps that can directly replace mercury lamps of comparable wattage size. Each uses the same ballast and fixture as does the comparable mercury lamp and gives a higher lumen output than its mercury counterpart. These lamps operate on mercury ballasts designed for 175, 250, and 400 W mercury lamps, respectively.

6.21 OTHER DISCHARGE SOURCES

In this section, two additional discharge light sources will be discussed—the low-pressure sodium lamp and the high-pressure xenon short arc lamp.

A low-pressure sodium lamp emits a pair of yellow spectral lines around 589 nm. It is a large tubular source, difficult to control optically, and terrible on color rendition. A graphic example is the inability of EMT personnel to distinguish between blood and oil at the scene of an accident under low-pressure sodium lighting. Nevertheless, the source is popular for use near observatories, because astronomers can remove the pair of spectral lines with a narrow-band optical filter, and in those rare outdoor applications where color identification is unimportant.

Low-pressure sodium lamps are available in both linear and U-shaped arc tubes. The starting gas is neon, admixed with argon and, on start up, the light output is red, turning to monochromatic yellow over 7 to 15 minutes as the sodium is vaporized. Sodium vapor pressure is critical for achieving high efficacy. It is held at 0.7 Pa (5 μm), which corresponds to a bulb wall temperature of 260°C. To hold this temperature, the arc tube is encased in an evacuated outer glass jacket.

Low-pressure sodium lamps have high initial overall efficacies of 150 to 180 lm/W. There is no lumen depreciation during the life of the lamp (18,000 h), but lamp wattage increases during life so efficacy decreases. Also, because the lamps have long arc tubes, as compared with the short arc tubes of HID sources, optical control is more difficult, resulting in less "delivered" efficacy from the luminaire. Properties of low-pressure sodium lamps are given in Table 6.18.

TABLE 6.18 LOW PRESSURE SODIUM LAMPS

Watts	Bulb	Initial and Mean Lumens	Initial Lumens per Lamp watt	Life (h)
18	T17	1,800	100	14,000
35	T17	4,800	137	18,000
55	T17	8,000	145	18,000
90	T21	13,500	150	18,000
135	T21	22,500	167	18,000
180	T21	33,000	183	18,000

COLOR-ENHANCED XENON SHORT ARC LAMP

Lamp Voltage	Lamp Current	Open Circuit Voltage	Ignition Voltage	Arc Gap	Life
25 V	10 A	125 V	15 kV	2.5 mm	200 h

A xenon short arc lamp is one of several short arc lamps that have similar characteristics. The term "short arc" means that the arc is short (1 to 10 mm) compared with the overall envelope, and thus the lamp may truly be considered as a point source. This also means that it has very high luminance, of the order of 2×10^8 cd/m^2.

The arc is contained in a quartz bulb that is spherical or elliptical. The lamp is double-ended, with rigid support rods connecting to the tungsten electrodes. It operates at high internal pressure (several atmospheres) even when cold. It may or may not have an outer jacket.

Short arc lamps are designed for either ac or dc operation, but generally not for both. For ac operation, each electrode is the same; for dc operation the positive electrode is much larger, since greater heat is dissipated there. For ac operation, conventional magnetic ballasting is used. For dc operation, a rectifier power supply is needed, which also provides the 50-kV pulse required to bridge the arc gap, ionize the gas, and heat the electrode. The discharge begins as a low-current, high-voltage spark and transforms to a high-current, low voltage arc, reaching 80% of light output instantly and full light output in 5 to 10 minutes. No cooling is required before restrike.

A xenon short arc lamp provides a continuous spectrum throughout the visible. It has a CCT of 6000 K, similar to daylight. Additives of zinc, thallium, bismuth, and mercury iodide can be included to enhance the output at certain wavelengths. These lamps have an evacuated outer jacket. Characteristics of a typical lamp are given in Table 6.18.

6.22 HID BALLASTS

An HID ballast has two major functions: (1) to provide starting voltage to initiate the arc and (2) to provide and control the current to the lamp, which, in turn will control lamp watts and light output.

Ballasts for HID lamps can be placed in four categories, as shown by the circuit diagrams of Fig. 6.28. The simplest is the reactor ballast in Fig. 6.28a, which is essentially a wire coil on an air-gapped iron core placed in series with the lamp to limit current. This is a "lag" ballast, causing current to lag line voltage by about 60° (PF = 0.5). Usually a capacitor is placed across the line to raise the power factor to ≥0.9. The reactor ballast provides a current crest factor of about 1.5, below the limiting value of 1.8, so that life and lumen maintenance are satisfactory. Its major limitation is that the starting voltage it can provide cannot exceed line voltage.

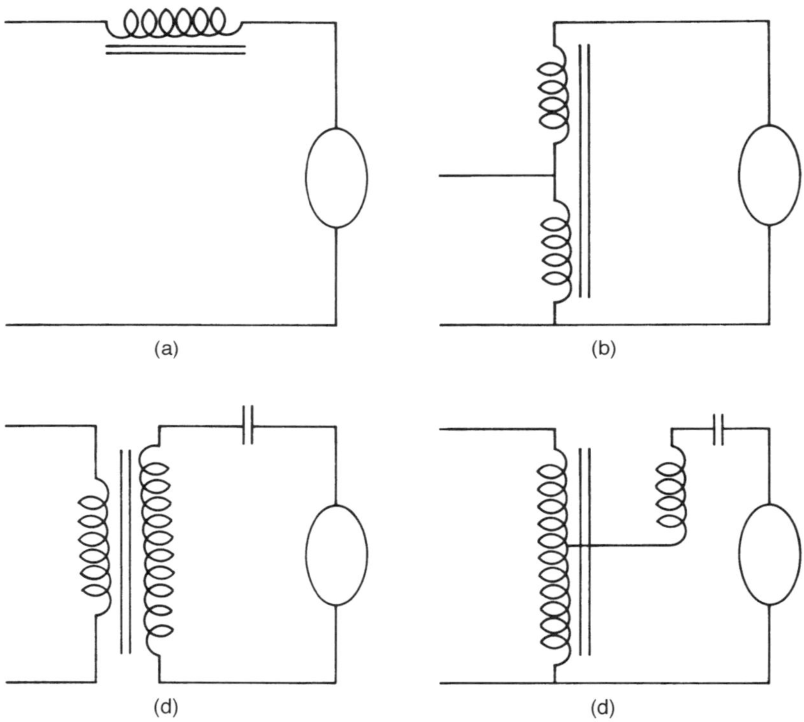

Figure 6.28 HID ballast circuits.

An important characteristic of a ballast is its regulation of lamp wattage, and thus lumen output, against line voltage changes. This relationship is shown in Fig. 6.29. For the reactor ballast, Fig. 6.29 shows that a line voltage change of ±5% changes lamp watts by ±12%. This is considered to be poor regulation.

The ballast shown in Fig. 6.28b is a combination autotransformer and high-reactance reactor. A leakage path for flux between the primary and secondary windings replaces the reactor of the previous ballast. This ballast has the same regulation characteristic as the reactor ballast, but overcomes the starting voltage limitation. However, it is larger, more costly, and has greater losses. As with the reactor ballast, a capacitor may be connected across the primary for power factor correction. This and the reactor ballast are the most commonly used for metal halide lamps below 175 W.

A constant wattage (CW) ballast is shown in Fig. 6.28c. It has isolated primary and secondary windings and achieves current limiting through a series capacitor that causes

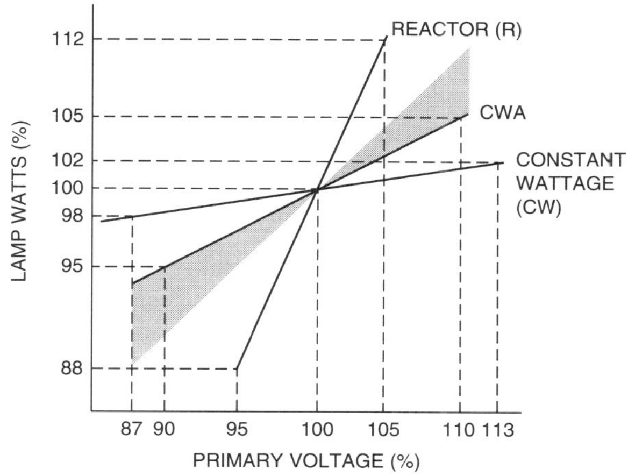

Figure 6.29 Regulation provided by HID ballasts. ***Source*****:** ***Pocket Guide to HID Lamp Ballasts*****, 1987. Courtesy of Advance Transformer Co.**

the current to lead the secondary voltage. This ballast provides excellent regulation (see Fig. 6.29). Line voltage changes of ±13% cause only a ±2% change in lamp watts. Power factor is >0.9. It suffers from a relatively high current crest factor of 1.65 to 2.0 and high cost.

Lastly, a constant wattage autotransformer (CWA) ballast is shown in Fig. 6.28d. This is also known as lead-peaked autotransformer and is the most commonly used circuit for metal halide lamps of 175 watts and higher. It has features of both the reactor and CW ballasts. It is smaller and less costly than a CW ballast, but does not provide isolation of the secondary from the primary. Its regulation (Fig. 6.29) is midway between the reactor and CW ballasts. Line voltage changes of ±10% cause ±5% changes in lamp watts for mercury and ±10% for metal halide.

High-pressure sodium lamps require ballasts that are different from those for other HID sources. Crest factors should not exceed 1.8 for proper lamp operation. Required starting voltages are much higher and an auxiliary (ignitor) circuit is necessary to provide the high-voltage spike to initiate the ionization of xenon. Also, lamp wattage must be closely controlled so as to control the amount of amalgam vaporized.

A high-pressure sodium lamp ballast has an operating locus, rather than an operating point, as shown in Fig. 6.30. This is because the lamp wattage and voltage change as the lamp ages. As the lamp is burned, the arc tube darkens because of evaporated electrode material, and the outer bulb darkens because of sodium leakage. This darkening causes arc tube temperature to increase, vaporizing more amalgam and raising lamp voltage by about 1 V for every 1000 h of burning. The ballast must compensate for this change to maintain reasonably constant lamp wattage over life.

In Fig. 6.30, curve A illustrates a proper design. Lamp wattage and lumen output start below rated values and increase as the lamp ages, finally decreasing at the end of life. With curve B, light output is decreasing for a significant portion of lamp life. The trapezoid boundary in Fig. 6.30 indicates the limits specified by the lamp manufacturer at 100 h, in this case between 84 and 140 lamp volts and between 300 and 475 lamp watts. The ballast designer must provide a ballast that causes the lamp to operate within these limits throughout life and to give peak lumens at approximately mid-life.

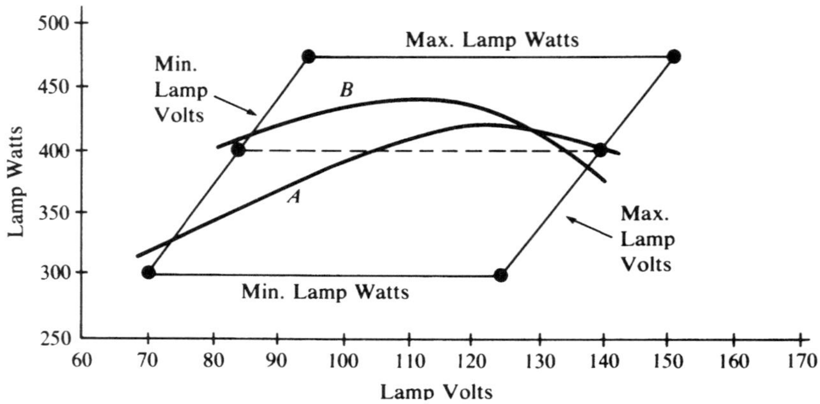

Figure 6.30 400-W HPS ballast locus.

Electromagnetic ballasts for HID lamps are rapidly being replaced by hybrid electronic/electromagnetic systems. Many features currently present in fluorescent lamp electronic ballasts are being included, such as automatic monitoring of input voltage and multiple-lamp operation on a single ballast. It should be noted, though, that the efficacy of HID lamps does not increase at high frequency, so electronic ballasts for HID lamps operate at a few thousand Hz.

A short summary of HID ballast wattage losses for electromagnetic ballasts is given in Table 6.19.

TABLE 6.19 HID BALLAST LOSSES

Lamp watts	Ballast watts	Total watts
	Mercury	
100	20	120
175	30	205
250	40	290
400	50	450
1000	75	1075
	Metal Halide	
175	35	210
250	50	300
400	60	460
1000	100	1100
1500	125	1625
	High-Pressure Sodium	
100	35	135
150	50	200
250	60	310
400	75	475
1000	100	1100

REFERENCES

Advanced Transformer Co.: *The ABC's of Electronic Ballasts*, Rosemont, IL, 1989.
Bennorth, G.: "Programmed Start: A New Way to Start a Fluorescent Lamp," *Lighting*, Kerwil, Mississauga, ON, October, 1999.
Briere, C.: "Electronic Ballasts 101," *Lighting*, Kerwil, Mississauga, ON, July, 2001.
Certified Ballast Manufacturers: *Fluorescent Lamp Ballast Primer*, New York, 1998.
Coaton, J.R., and A.M. Marsden: *Lamps and Lighting*, Fourth Edition, Arnold Press, London, 1997.
Gamble, M.: "New Technology Metal Halide Lighting," *Lighting*, Kerwil, Mississauga, ON, July, 2000.
Gamble, M.: "The Pulse of Pulse Start," *Lighting*, Kerwil, Mississauga, ON, July, 2001.
Inman, G.E.: "Fluorescent Lamps: Past, Present, Future," *G.E. Review*, July, 1954.
Lemmers, A.E., and W.W. Brooks: "New Fluorescent Lamp and Ballast Design for Rapid Starting," IES Technical Conference, 1952.
Oetting, R.L.: "Electric Lighting in the First Century of Engineering," *AIEE Transactions*, November, 1952.
OSRAM SYLVANIA: *Incandescent Lamps, Engineering Bulletin IN003*, Danvers, MA, 1996.
OSRAM SYLVANIA: *Incandescent Lamp Manufacture, Engineering Bulletin IN002*, Danvers, MA 1996.
OSRAM SYLVANIA: *Lamp and Ballast Catalog*, Danvers, MA, 2002.
OSRAM SYLVANIA: *High Intensity Discharge Metal arc Lamps, Engineering Bulletin HID-010*, Danvers, MA, 1998.
OSRAM SYLVANIA: *Lumalux/Unalux High Pressure Sodium Lamps, Engineering Bulletin, HID-009*, Danvers, MA, 1998.
Perszyk, P.: "Neon: The First 100 Years," *LD+A*, IESNA, New York, July, 2001.
Philips Lighting: *Guide to Incandescent Lamps*, 1986.
Rea, M.S., editor: *IESNA Lighting Handbook*, Ninth Edition, IESNA, New York, 2000.
Roberts, J., editor: *Advanced Lighting Guidelines*, 2001 Edition, New Buildings Institute, 2001.
Stambaugh, L.: "Cold-cathode and Neon Lighting," *Luminews*, North Texas Section of IESNA, September, 1994.
Thayer, R.N., and A.C. Barr: "Instant Start Slimline Lamps," *G.E. Review*, March, 1952.
Waymouth, J.F., W.C. Gungle, J.M. Harris, and F. Kodry: "A New Metal Halide Arc Lamp," IES Technical Conference, 1964.
Waymouth, J.F.: *Electric Discharge Lamps*, MIT Press, Cambridge, MA, 1971.
Waymouth, J.F., and R.E. Levin: *Designers Handbook*, GTE Sylvania, Danvers, MA, 1980.

Chapter 7

Color

7.1 COLOR THEORIES AND CONCEPTS

Much of the foundation for the current understanding of color stems from Sir Isaac Newton and is contained in his treatise "Opticks" (Wintringham, 1951). Newton asserted that the sensation of color arose because different wavelengths produced different visual effects. He concluded that the color appearance of objects was due to selective wavelength reflection or transmission of incident light. He also pointed out that it was impossible to distinguish the color components in a color mixture and that colors could appear the same, and yet have different spectral compositions.

Newton was able to show that mixtures of adjacent spectral colors produced intermediate spectral colors, but that a mixture of red and blue, which are far apart spectrally, produced purple, which is a nonspectral color. He further showed that several spectral colors could be added to produce white. He actually assumed seven primaries (red, orange, yellow, green, blue, indigo, and violet) and used these to calculate the colors of the other spectral wavelengths.

In 1792, C. E. Wunsch was able to show that four of Newton's primaries could be obtained from the other three. This was the beginning of tri-color colorimetry. Then, in 1801, Sir Thomas Young (1773–1829), English physician and physicist, published a paper "On the Theory of Light and Colours." This was the first serious attempt to explain color vision. Young hypothesized that the retina contained three types of "resonators", which responded selectively to different wavelengths. Initially, he described these resonators as red, yellow, and blue, essentially the pigment primaries. Later he changed to red, green, and violet, nearly the common contemporary light primaries, and concluded that all spectral colors could be produced by summations of these lights.

Young, in Lectures on Natural Philosophy in 1807, stated: "From three simple sensations, with their combinations, we obtain several primitive distinctions of colours, but the different proportions in which they may be combined afford a variety of tints beyond all calculations. The three simple sensations being red, green, and violet, the three binary combinations are yellow, consisting of red and green; crimson, of red and violet; and blue, of green and violet; and the seventh in order is white light, composed by all the three united."

Young's qualitative statement of the three-component color theory did not gain wide acceptance until H.L.F. von Helmholtz (1821–1894), German physicist, along with Maxwell, developed color-mixing equations. Helmholtz was one of the most brilliant scientists of the nineteenth century. He invented the opthalmoscope and clarified the mechanisms of sight, in addition to developing a theory of color vision based on Young's work. In 1856, he published *Physiological Optics*, a complete study of the physics and physiology of human vision.

The Young-Helmholtz theory states that any color can be synthesized by a unique mixture of three lights of different colors. It assumes that the eye contains three kinds of receptors that react to wavelengths selectively and in proportion to the light intensity at any given wavelength. Each of the three receptors is presumed to have its own set of nerves, and the optic nerve and brain produce the integrating effect to give the sensation of the particular color being viewed. The theory describes the transitions between blue and green (cyan) and red and blue (magenta), but cannot explain the transition between green and red (yellow). The color combination bluish-green and reddish-blue have meaning, but not greenish-red. Yellow is a distinct color, and one cannot see its red and green components. Also, the theory does not adequately explain perceived shifts in color when luminance is changed, nor does it explain color blindness well. However, quantitative work by G. Wald (1964) of Harvard, among others, has lent considerable credence to the Young-Helmholtz theory.

A second theory of color vision that is widely used today is the Hering Opponent Colors theory, advanced by the German psychologist, E. Hering in 1878. The Hering theory has a response rather than a stimulus basis. It assumes six basic colors—red, yellow, green, blue, black, and white—which act in three pairs of processes, rather than separately. There is a blue-yellow opponent pairing such that a color cannot appear bluish and yellowish simultaneously. In other words, blue cancels yellow and vice versa. The same is true of the red-green pair. For the white-black pair, a color may move toward the black or white from the gray, but not toward both at once. White does not cancel black, but rather blends with it to produce gray. The theory fails to explain two types of dichromatism—partial color blindness in which distinctions can be made between light and dark and either yellow and blue or green and red, but not both; also, it cannot explain the difference between brightness and lightness of a gray object. The Hering theory was quantified in the mid-1950s by L.M. Hurvich and D. Jameson through considering an excitation mechanism in the receptors and a response mechanism in the visual nerve center beyond the retina.

Refinements of these two color theories have been advanced in recent years. However, the present state of affairs is that both seem to be necessary for an adequate description of the color vision process, even though it is known that there are three types of cone receptors in the eye with peak sensitivities in the violet-blue, green, and yellow-red regions of the spectrum, as shown in Fig. 7.1. These are called the short (S), middle (M), and long (L) photoreceptors, and each contains its own characteristic photo pigment.

An attempt to blend the two color theories is shown in the model of Fig. 7.2. The inputs from the L and M cones are sent through a differencing channel to create the chromatic R-G opponent axis. These inputs are also sent through a summing channel to form the achromatic luminance signal. The achromatic signal in addition forms the Y signal and along with the S-cone input, is sent through a second differencing channel to create the chromatic Y-B opponent axis.

It is important for the lighting professional to distinguish between color, object color, and perceived object color. The IESNA defines color as "the characteristic of light by which a human observer may distinguish between two structure-free patches of light of the same size and shape." This definition implies the description of light in terms of the spectral distribution of radiant power leaving the object in question. The IESNA goes on to define object color as "the color of the light reflected or transmitted by an object when illuminated by a standard CIE light source." This definition assumes that

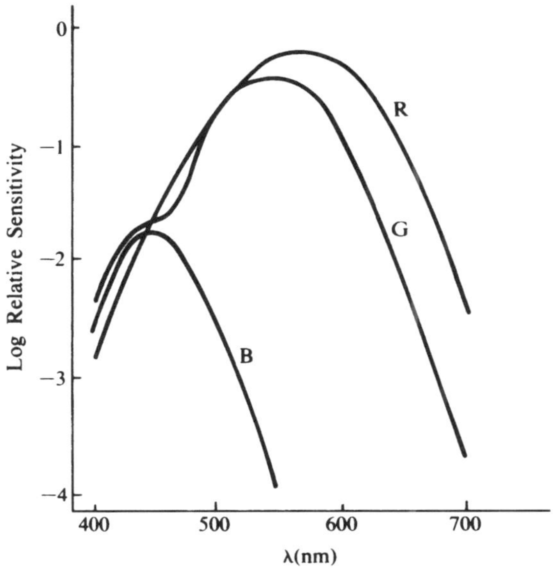

Figure 7.1 Spectral sensitivity curves of the cone receptors. ***Source*****:** ***IESNA Lighting Handbook, Reference Volume*****, 1981. Courtesy of the Illuminating Engineering Society of North America.**

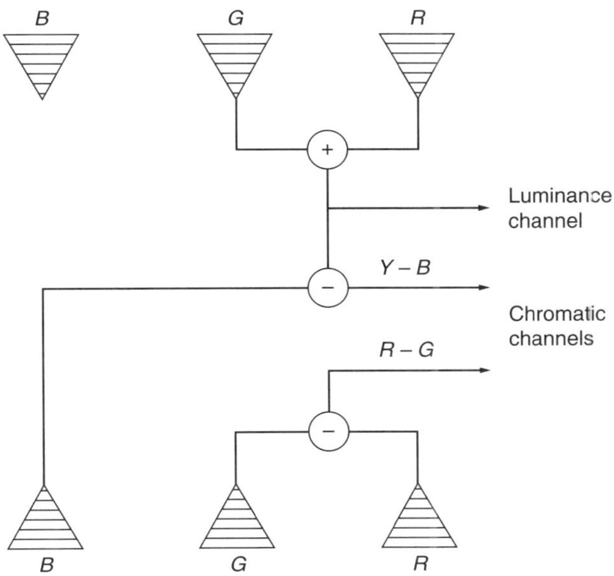

Figure 7.2 Model of the two color theories.

the radiation will be evaluated in terms of the standard CIE observer and under controlled conditions of viewing and adaptation. Thus object color does not define how an object will appear in a lighted space, but rather categorizes its appearance when lighted by certain standard sources. Last, the IESNA defines perceived object color as "the color perceived to belong to an object, resulting from characteristics of the object, the incident light, the surround, the viewing direction and observer adaptation." This is an attribute that is perceived almost instantly by an observer and is influenced by past memory. It can be altered by the environment, the color last seen, and by mood.

The word "color" is often used to mean all three of the above entities. It is perceived object color with which the lighting designer is most concerned, because it is this attribute of color that is a major factor in determining how occupants feel about a lighted space and perceive objects in that space. Although color is categorized in succeeding sections through the CIE system, such a mathematical representation will not completely describe the overall effect that a color produces on a viewer.

7.2 COLORIMETRY

Colorimetry is defined as the science of measuring color and designating it systematically. One of the fundamental aspects of colorimetry is color mixing and matching.

Any color may be matched by proper proportions of three component colors spaced far apart in the spectrum, subject only to the constraints that no one of the components can be matched by a combination of the other two and that, in some cases, it may be necessary to add one of the components to the color being matched in order to achieve a match (equivalent to adding two of the components and subtracting the third). This is usually true when matching spectral colors.

The three component colors are called *primaries*, and are illustrated in Fig. 7.3. In Fig. 7.3a, the additive system for lights is shown. The most common primaries of light are red, blue, and green. Two colors that, when mixed in the proper proportion, are achromatic (neutral) are called complementary colors. Addition of red and blue produces magenta, which is the complement of green. Blue and green form cyan, the complement of red. Red and green produce yellow, the complement of blue. The proper combination of all three primaries produces white.

The subtractive primaries for pigments are the complements of the additive primaries for lights, as shown in Fig 7.3b. Here the combination of the three primaries yields black. Subtractive color mixing is used for overlays in color printing, and with inks, dyes and paints.

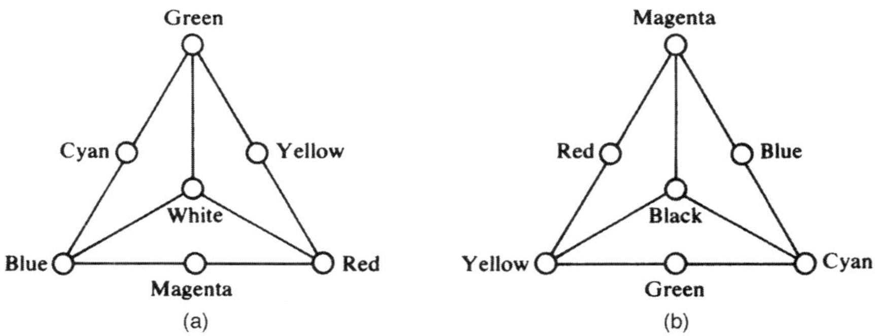

Figure 7.3 Additive and subtractive color mixtures.

The two most common color-measuring instruments in use today are the spectrophotometer, previously described in Chap. 4, and the colorimeter (see Fig. 7.5). The latter is a device for visually matching an unknown color with a combination of three (or more) colors of light, by varying the intensities of each. The unknown color is displayed in one-half of the visual field; the matching colors are superimposed in the other half. Occasionally one of the matching colors must be superimposed on the test field to obtain the match. Unlike the spectrophotometer, the colorimeter does not give the radiant power of a source in each wavelength band, but rather measures the overall radiant power for each primary component.

There are eight fundamental rules of colorimetry, which were formulated by Professor H. Grassman in "On the Theory of Compound Colors" in 1854. These may be stated as follows:

1. A color match is independent of luminance over a wide range.
2. The luminance of a mixture of colors is the sum of the individual luminances.
3. Any color can be matched by a mixture of no more than three colored lights (primaries), provided no two yield the third.
4. The components of a color mixture cannot be resolved by the eye.
5. A color match can be stated in equation form as

$$C = R(R) + G(G) + B(B) \tag{7.1}$$

   This means that R units of (R), G units of (G), and B units of (B), when summed, yield a match with color C. The coefficients R, G, and B are called tristimulus values.
6. Color matches obey the law of addition. Thus, if C_1 matches C_2 and C_3 matches C_4, then $C_1 + C_3$ matches $C_2 + C_4$. This rule is known as Grassman's law. However, the law of addition does not apply to brightness matches. For example, if a green light and a white light are adjusted to be equally bright, and then a red light is adjusted to be equally bright as the same white light, the yellow light resulting from the combination of the red and green lights is much less bright than the combination of two white lights. The reason for this derives from achromatic photometry in that heterochromatic brightness matches and matches using flicker photometry yield different results. Measurements based on flicker photometry (luminance) are additive, but brightness measurements are not.
7. Color matches obey the law of subtraction.
8. Color matches obey the law of transition. Thus if C_1 matches C_2 and C_2 matches C_3, then C_1 matches C_3.

7.3 COLOR MEASURING SYSTEMS

There are two major color systems used in North America, both based on three variables or descriptors. One is the Munsell color appearance system; the other is the CIE method of color specification. The former will be discussed in this section; the latter in Sect. 7.5.

In the Munsell system, a color is described in terms of hue, value, and chroma. These three variables are displayed in a cylindrical coordinate (r, θ, z) format in a color solid, as shown in Fig. 7.4. There are 10 named hues (green, blue-green, blue ...), positioned uniformly (in the θ direction) around a horizontal hue circle, with each hue subdivided into 10 finer divisions (a total of 100 hues). The value scale (along the z axis) consists of 10 intervals from black (0) at the bottom to white (10) at the top. This scale is intended to represent equal steps in lightness, where lightness is defined as the estimate

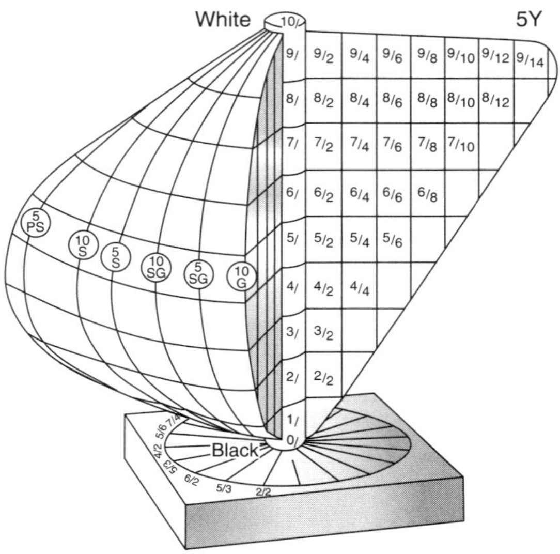

Figure 7.4 Munsell color solid. ***Source*****:** ***IESNA Lighting Handbook*****, Ninth Edition, 2000. Courtesy of the Illuminating Engineering Society of North America.**

of the brightness of an area relative to the brightness of a similarly-illuminated white area.

The relationship between Munsell value and luminous reflectance is shown in Table 7.1. These reflectance values have been calculated from empirically developed equations. However, as a rough approximation, Munsell value may be calculated from

$$V = \sqrt{\rho} \tag{7.2}$$

which clearly displays the non-linear relationship between value (perceived) and reflectance (measured). For example, a Munsell color with value 6 is nearer to white than to black, yet has a reflectance of only 30%.

TABLE 7.1 MUNSELL VALUE AND LUMINOUS REFLECTANCE

Munsell Value (V)	Luminous Reflectance (ρ in %)	$V = \sqrt{\rho}$
10.0	100.0	10.0
9.0	78.7	8.9
8.0	59.1	7.7
7.0	43.1	6.6
6.0	30.0	5.5
5.0	19.8	4.4
4.0	12.0	3.5
3.0	6.6	2.6
2.0	3.1	1.8
1.0	1.2	1.1
0.0	0	0.0

The third scale, in the r direction in the cylindrical coordinate system, is chroma. As one moves outward from the neutral gray z axis at constant value along a radial hue line, the color becomes increasingly saturated. It is as though one had a bucket containing a small amount of gray paint and began adding, for example, red paint. Upon stirring, the mixture would change from gray, to reddish-gray, to grayish-red, to red, becoming increasingly pure. The chroma scale has as many as 20 color chips from neutral gray to full saturation.

The 100 vertical hue profiles are not identical. In particular, maximum chroma does not occur at the same value for each hue. For example, the most saturated yellow occurs at a high value (nearer white) whereas a pure purple has a low value (nearer black).

In the Munsell system, a color chip is labeled with a letter/number code. For example, a chip labeled 5Y3/2 is in the middle of the yellow hue range, has a value of 3 (low reflectance), and has a chroma of 2 (unsaturated).

The Munsell scales are intended to represent equal visual increments, but this is true only under daylight viewing and with neutral surroundings. Also, the hue steps at high chroma are larger than those at low chroma. An accepted approximation, when the chroma is 5, is that 1 value step = 2 chroma steps = 3 hue steps for visual equality.

There are several additional color systems in use domestically and abroad. The Ostwald system is the color notation system used by the rest of the world outside North America. In this system, there are 24 identical equilateral triangles of uniformly-spaced hues, each standing on end with its apex to one side. Each triangle contains 36 color chips, 28 hue chips, and 8 common chips from white on the top to black on the bottom. Each hue has a number, and value and saturation are indicated by two letters, with the first denoting the percentage of white (a the most and p the least) and the second the percentage of black (p the most and a the least). For example, pale blue is denoted by 15 ca. (15 denotes blue).

The Ostwald system is based on ideal reflecting objects that have a constant spectral reflectance between two complementary visible wavelengths and a much lower constant spectral reflectance elsewhere in the visible spectrum. The lesser reflectance is called white content; the difference between unity and the greater reflectance is called black content; the difference between the two reflectances is called full-color content. The sum of these three numbers is unity, so only two are specified, usually black content and white content, and to these are added hue.

The DIN (Deutsche Industrie Norm) color appearance system is the official German standard and is widely used in Europe. Its variables are hue, saturation, and darkness, and it is based on the CIE color specification system.

The Pantone color matching system is widely used by the graphic arts, textile, fashion and interior design professions. It consists of 1225 color patches forming a vertical color cylinder. The axis of the cylinder has 10 levels of lightness, numbered from 10 to 19, with black at the bottom and white at the top. Colors radiate outward from the axis in 64 sectors, from 01 for yellow around to 64 for greenish yellow. Saturation also has 64 steps, from 00 for neutral gray to 64 for maximum color purity. Saturation increases with distance from the axis. Colors are designated with a three-number system. For example, the color 14-06-47 has a lightness of 14, on a hue spoke of 06 with a saturation of 47. All colors are given a name and this one is "celery". Pantone presents its colors on 175 $1\frac{1}{2}" \times 8"$ rectangles, seven colors to a rectangle, assembled in a fan, with a common pivot axis at the end of each rectangle.

The Inter-Society Color Council and the National Bureau of Standards (Kelly and Judd, 1976) have developed a table of 267 standardized names for describing paint, based on the Munsell system. The table utilizes 28 basic hue names, with up to 11 adverbs and adjectives for each hue to describe lightness and saturation (value and chroma). Each of the names defines an area in color space and the boundaries of each area are defined using Munsell notation.

None of the color systems can match all colors. For example, colors formed by desaturating a color by combining it with its complement are not accommodated. Recent extensions of the Ostwald and Munsell systems include this type of desaturation.

The problem of color matching is a critical one for lighting designers and engineers. Two objects may appear to be of the same color under one source, for example, incandescent, but appear to be different in color under a second source, such as daylight. Metamers are defined as lights of the same color, but of different spectral energy distributions. Two objects whose spectral reflectances differ, but which appear to have the same color under one or more sources, are called metameric objects. Two lamps which differ in spectral power distribution, but appear to have the same color, are called metameric sources.

There are several metrics that are used to describe the ability of one source or object to match the color of another source or object. Among these are correlated color temperature (CCT) and color rendering index (R or CRI). Correlated color temperature was introduced in Chap. 4. It and color rendering index will be discussed later in this chapter, after the mathematics of the CIE system for describing color has been developed.

7.4 EARLY MATHEMATICAL FORMULATIONS OF COLOR*

In this section, the mathematical formulation of a tristimulus color system that led to the current CIE color specification system is developed.

In the 1920s, W.D. Wright, J. Guild, and D. Judd conducted color-matching experiments that led to the "chromaticity characteristic of the standard observer" adopted by the CIE in 1931. The colorimeter used in these experiments is diagrammed in Fig. 7.5. The instrument has a divided visual field of 2 to 3 degrees. The spectral color to be matched is displayed in the left-half field; the mixture of the three primaries in the right-half field. Three calibrated attenuators are included to adjust the intensity of each primary.

The procedure used is similar to that employed by Gibson and Tyndall in obtaining the eye's spectral luminous efficiency curve. Monochromatic radiation of variable wavelength λ and constant power density impinges on the left-half field. Three monochromatic radiations of fixed wavelength and variable (through the attenuators) power density are superimposed on the right-half field. The primaries chosen were a red line at 700 nm, obtained by using a narrow filter with an incandescent source; a blue line at 435.8 nm and a green line at 546.1 nm, obtained from a mercury spectrum. The sources were defined as "standard" by the CIE in 1931.

The radiant fluxes of the CIE primaries obtained from these measurements that color match known amounts of radiant fluxes of spectral colors are shown in Fig. 7.6. These data are based on 17 observers and represent the average results of many observations

---

* For the reader more interested in the application of the CIE system of color specification, rather than in its mathematical development, this section may be omitted.

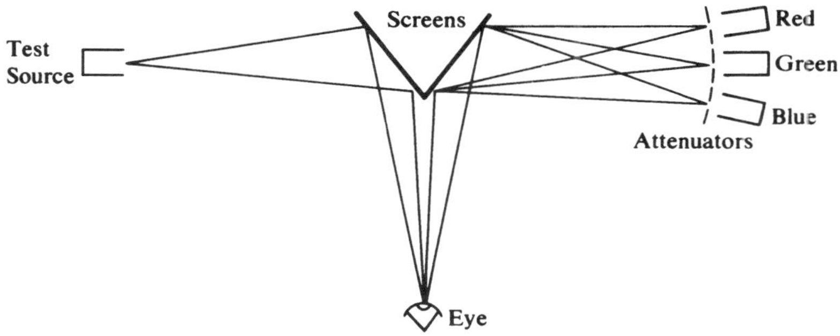

Figure 7.5 Colorimeter for basic color experiment.

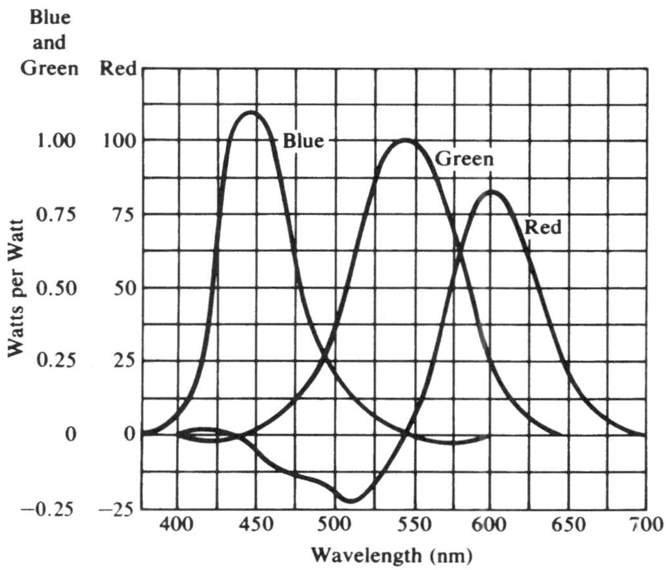

Figure 7.6 Wattages of CIE primaries to match spectral colors. ***Source*****: G.M. Glasford,** ***Fundamentals of Television Engineering*****, 1955. Courtesy of McGraw-Hill.**

by each one. For each spectral color, each observer adjusted the attenuators in Fig. 7.5 until the two halves of the field appeared alike in both luminance and color. The watts of the spectral color and of the attenuator outputs of each primary were then measured. For many of the spectral colors, it was necessary to transfer one of the primaries to the left-half field to obtain a match, corresponding to a negative amount of that primary.

As an example of the use of the curves in Fig. 7.6, if it is desired to match 1 W of pure spectral color at 500 nm, it requires approximately −18 W of the red primary, 0.38 W of the green primary, and 0.20 W of the blue primary. In Fig. 7.5, this particular match is achieved by adding 18 W of red to the spectral color in the left-half field.

It might appear that the data in Fig. 7.6 give a sound basis for the specification of color. Such is not the case for several reasons. First of all, the curves in Fig. 7.6 are difficult to use with any degree of accuracy. The watts of red far exceed the watts of green and blue, and negative amounts of red are required at several wavelengths. The first difficulty may be overcome by replotting the curves in Fig. 7.6 on a lumens per watt basis. This would entail multiplying each curve by 680* times its V(λ) value, 0.0041 for red, 0.9842 for green, and 0.0177 for blue. However, it would now be found that the luminous flux of the green primary far exceeds those of the red and blue primaries. For example at 500 nm, the lumens are approximately −50, 260 and 2, for red, green and blue, respectively. Thus, for either plot, if the same number of significant figures are to be used for each primary, different numbers of decimal places will be required, and this will make summations inaccurate.

An additional deficiency of Fig. 7.6 is that it is a plot of absolute, rather than relative, measurements and is quite sensitive to observer error. There is no reference base with which the observer is required to become calibrated at the outset to ensure that variations between observers are minimized and that sequential measurements by the same observer are consistent. A final problem with Fig. 7.6 is that luminance and color are locked together in the matching process. Observers tend to disagree more on luminance matches than they do on color matches.

As an alternative to Fig. 7.6, two additional sets of color specification curves have been developed, and these form the basis of modern colorimetry. The first is a set of tristimulus value curves. The tristimulus values are proportional to the number of units of the spectral primaries that will match a given quantity of spectral radiance (the radiometric equivalent of luminance) when the primaries are mixed.

A second basic way to specify color is to use chromaticity coordinates and luminance. Again three parameters are required: two normalized chromaticity values to describe the color of the object and a separate measurement or calculation to obtain its luminance. Each of these specification systems will be described in some detail.

Both types of color specification are based on calibrating the curves in Fig. 7.6 in terms of a "reference white." Equal energy white was chosen, having, by definition, the same radiance in watts per square meter per steradian per nanometer at each wavelength within the visible spectrum. Calculations were made to determine the luminance required of each primary that, when summed, would produce a match with unit radiance per nanometer of equal energy white.

The luminance of equal energy white with unit radiance per nanometer is

$$L_w = 680(1)\int_{380}^{770} V(\lambda)\,d\lambda \tag{7.3}$$

The area under the V(λ) curve was obtained in Sect. 4.6 and has the value of 106.858.** Thus

$$L_w = 680(1)(106.858) = 72{,}660 \text{ cd/m}^2 \tag{7.4}$$

* At the time of this work in the 1920s, the luminous efficacy constant was 680, rather than the current 683. Also, the long wave length unit of the visible spectrum was chosen as 770 nm, rather than 780 nm.

** In Sect. 4.6, unit radiance per 10 nanometers was used.

This luminance must be equated to the luminances provided by the three primaries. Thus

$$L_w = 680\left[V_r(\lambda)\int_{380}^{770} S_r(\lambda)d\lambda + V_g(\lambda)\int_{380}^{770} S_g(\lambda)d\lambda + V_b(\lambda)\int_{380}^{770} S_b(\lambda)d\lambda\right] \quad (7.5)$$

where the Ss inside the integrals are the spectral radiances* displayed in Fig 7.6. What Eq. (7.5) states is that, for each primary, to obtain the luminance provided by that primary, the watts per watt per nanometer throughout the spectrum must be summed and multiplied by 680 times the $V(\lambda)$ of that primary.

When the three integrals in Eq. (7.5) are evaluated, the ratios of the luminances of the primaries are:

$$L_r : L_g : L_b = 1:4.5907:0.0601 \quad (7.6)$$

Thus, from Eq. (7.5),

$$72{,}660 = L_r + 4.5907L_r + 0.0601L_r \quad (7.7)$$

giving the sizes of the luminances as

$$L_r = 12{,}860 \text{ cd/m}^2, \quad L_g = 59{,}030 \text{ cd/m}^2, \quad L_b = 722.8 \text{ cd/m}^2 \quad (7.8)$$

Now the curves in Fig. 7.6 can be scaled. The wattage peaks of these curves are at

$$\begin{aligned} \text{Red} &= 85.4 \text{ W/W} \\ \text{Green} &= 1.00 \text{ W/W} \\ \text{Blue} &= 1.09 \text{ W/W} \end{aligned} \quad (7.9)$$

To visually evaluate these curves requires first scaling each by the appropriate $V(\lambda)$ factor to obtain

$$\begin{aligned} \text{Red:} &\quad 0.0041(85.4) = 0.350 \\ \text{Green:} &\quad 0.9842(1.00) = 0.984 \\ \text{Blue:} &\quad 0.0177(1.09) = 0.0193 \end{aligned} \quad (7.10)$$

---

* The symbol S will be used to represent radiant flux, irradiance, radiant emittance, or radiance. This is to avoid confusion between radiometric and photometric quantities, which have the same letter designations.

and then scaling each curve by the luminance ratios in Eq. (7.6)

$$\text{Red:} \quad \frac{0.350}{1} = 0.350$$

$$\text{Green:} \quad \frac{0.984}{4.5907} = 0.214 \qquad (7.11)$$

$$\text{Blue:} \quad \frac{0.0193}{0.0601} = 0.321$$

The quantities in Eqs. (7.11) are the peaks of the tristimulus value curves. These curves, with the labels $\bar{r}(\lambda)$, $\bar{g}(\lambda)$ and $\bar{b}(\lambda)$, are plotted in Fig. 7.7.

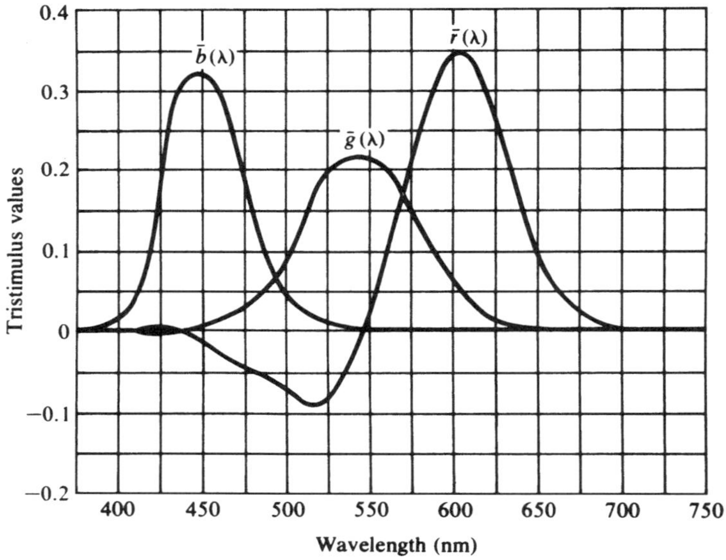

Figure 7.7 Tristimulus values of spectral stimuli of equal radiance. *Source:* G.M. Glasford, *Fundamentals of Television Engineering*, 1955. Courtesy of McGraw-Hill.

Each of the tristimulus curves is simply a scaled version of its counterpart in Fig. 7.6. The scaling was carried out so that equal energy white would require the same number of units of each primary for a match. This required that the areas under the tristimulus curves be equal, which is guaranteed by the luminance scaling that was done using Eq. (7.6). Thus

$$\int_{380}^{770} \bar{r}(\lambda)\,d\lambda = \int_{380}^{770} \bar{g}(\lambda)\,d\lambda = \int_{380}^{770} \bar{b}(\lambda)\,d\lambda \qquad (7.12)$$

For any color of radiance $S(\lambda)$, it can be written

$$R = K\int_{380}^{770} \bar{r}(\lambda)S(\lambda)\,d\lambda$$

$$G = K\int_{380}^{770} \bar{g}(\lambda)S(\lambda)\,d\lambda \qquad (7.13)$$

$$B = K\int_{380}^{770} \bar{b}(\lambda)S(\lambda)\,d\lambda$$

where R, G, and B are the numbers of units of the primaries required for a match. The areas in Eqs. (7.12) can be calculated as 18.91 each. If unit radiance of equal energy white is to be matched by one unit of each primary, the constant K must be assigned the value $1/18.91 = 0.05288$. Then, for equal energy white of 1 W/(sr · m² · nm), R, G, and B in Eqs. (7.13) are each unity.

EXAMPLE 7.1

Find R, G, and B for unit radiance of a spectral color at 500 nm. Find the luminance of this color.

Solution. From Fig. 7.7, the approximate tristimulus values are;

$$\bar{r}(\lambda) = -0.075, \quad \bar{g}(\lambda) = 0.085, \quad \bar{b}(\lambda) = 0.050$$

Then, from Eqs. (7.13),

$$R = 0.05288(-0.075) = -0.00397$$

$$G = 0.05288(0.085) = 0.00449$$

$$B = 0.05288(0.050) = 0.00264$$

The luminance of the color is obtained using Eqs. (7.8).

$$L = 12{,}860(-0.00397) + 59{,}030(0.00449) + 722.8(0.00264)$$

$$= 51.1 + 265.0 + 1.9$$

$$= 216 \text{ cd/m}^2$$

As a check, this luminance can also be found from $680\,V(500) = 680(0.323) = 220$ cd/m².

The calculation of luminance in Example 7.1 can be generalized for any spectral color as

$$680\,V(\lambda) = L_r R + L_g G + L_b B$$

$$= K[L_r\,\bar{r}(\lambda) + L_g\,\bar{g}(\lambda) + L_b\bar{b}(\lambda)] \qquad (7.14)$$

The second basic system for specifying color is through chromaticity coordinates. For each pure spectral color, R, G, and B are calculated from Eqs. (7.13). The following ratios are then formed

$$r = \frac{R}{R+G+B}$$

$$g = \frac{G}{R+G+B} \tag{7.15}$$

$$b = \frac{B}{R+G+B}$$

These are the chromaticity coordinates, and values of these at 5-nm intervals were standardized by the CIE in 1931 as the definition of the chromaticity characteristics of the standard observer. They are plotted in Fig. 7.8. In this system, reference white has coordinates ($^1/_3$, $^1/_3$, $^1/_3$). The coordinates of the primaries are (1, 0, 0), (0, 1, 0), and (0, 0, 1). Since from Eqs. (7.15)

$$r + g + b = 1 \tag{7.16}$$

there are only two independent chromaticity coordinates, and luminance information in Eqs. (7.15) has been completely lost. It may be regained from a separate calculation using Eq. (7.14).

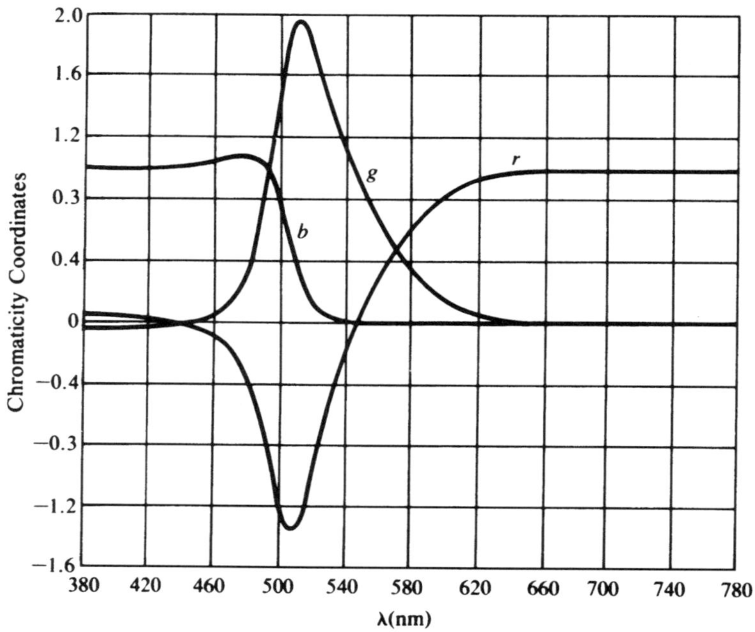

Figure 7.8 Chromaticity coordinates of spectral colors.

EXAMPLE 7.2

Find the chromaticity coordinates of the radiation in Example 7.1.

Solution. From Example 7.1 and Eqs. (7.15),

$$R + G + B = 0.00316$$

$$r = \frac{-0.00397}{0.00316} = -1.26$$

$$g = \frac{0.00449}{0.00316} = 1.42$$

$$b = \frac{0.00264}{0.00316} = 0.84$$

These values could also be read directly from Fig. 7.8.

There are two major deficiencies in the color system just presented. First, it contains both positive and negative tristimulus values. This makes summations awkward and subject to error. Second, the luminance of a color mixture depends on all three tristimulus values, again complicating the calculations.

As was stated at the outset, there is nothing unique about the set of primaries that was chosen, except that, because they are physical primaries, actual experiments can be carried out using them. Other than this, a set of nonphysical primaries would have done just as well.

The system ultimately adopted by the CIE in 1931 uses nonphysical primaries, denoted by X, Y, and Z. These are derived from the R, G, B physical primaries through a linear transformation, under the following conditions:

1. The primaries are chosen so that all real colors have positive tristimulus values.
2. Calculations of luminance involve only one of the primaries. This primary (Y) is chosen equal to $V(\lambda)$, making luminance calculations identical to those for a black and white system.
3. The scales of the new primaries are chosen so that one unit of each primary yields equal-energy white.
4. Two of the primaries (X and Y) are chosen such that the line connecting them is tangent to the spectral locus at 770 nm. This makes the value of the third primary (Z) negligible over a considerable range at the red end of the spectrum, simplifying calculations.

The transformation from one set of primaries to the other is straight-forward, but messy mathematically (see Wintringham, 1951). The final result is

$$\begin{aligned} X &= 2.7690R + 1.7518G + 1.1300B \\ Y &= 1.0000R + 4.5907G + 0.0601B \\ Z &= 0.0000R + 0.0565G + 5.5943B \end{aligned} \tag{7.17}$$

Note that since Y contains all of the luminance information, its coefficients are those originally derived in Eq. (7.6).

These equations are true for any color. Thus, they apply to the pure spectral colors. If the R, G, and B values are inserted for the spectral colors, obtained as in Example 7.1, into Eqs. (7.17), the X, Y, and Z values of the spectral colors will be obtained. These are the spectral tristimulus values for the new system of primaries and are labeled $\bar{x}$, $\bar{y}$, and $\bar{z}$. The chromaticity coordinates for the new system are labeled x and y and are obtained from the tristimulus values using the form of Eqs. (7.15).

7.5 THE CIE COLOR SPECIFICATION SYSTEM

In a black and white system, recall from Sect. 4.6 and Eq. (4.20) that luminous flux is given by

$$\phi = 683 \int_{380}^{780} P_\lambda V(\lambda) d\lambda \tag{7.18}$$

where $V(\lambda)$ is the eye's photopic sensitivity curve for black and white vision.

In a color system, three equations are required in the form

$$\left.\begin{aligned} X &= k \int_{380}^{780} S_\lambda \bar{x}(\lambda) d\lambda \\ Y &= k \int_{380}^{780} S_\lambda \bar{y}(\lambda) d\lambda \\ Z &= k \int_{380}^{780} S_\lambda \bar{z}(\lambda) d\lambda \end{aligned}\right\} \tag{7.19}$$

where S_λ is spectral irradiance, radiant flux, radiance or radiant exitance of the source or object; X, Y, and Z are the tristimulus values of the spectral distribution curve represented by S_λ; and $\bar{x}(\lambda)$, $\bar{y}(\lambda)$ and $\bar{z}(\lambda)$ are the tristimulus values of the pure spectral colors (derived in the previous section).

It is perhaps more intuitively useful to think of $\bar{x}(\lambda)$, $\bar{y}(\lambda)$ and $\bar{z}(\lambda)$ as the three eye sensitivity functions for color vision, similar to the one eye sensitivity function ($V(\lambda)$) for black and white vision. In fact $\bar{y}(\lambda)$ was chosen equal to $V(\lambda)$ so that it contains all of the luminance information in the color signal.

Chromaticity coordinates can be formed, which are dimensionless and do not contain luminance information. They simply give the proportions of X, Y, and Z present and are obtained from

$$\left.\begin{aligned} &D \equiv X + Y + Z \\ &x = \frac{X}{D}, \quad y = \frac{Y}{D}, \quad z = \frac{Z}{D} \\ &x + y + z = 1 \end{aligned}\right\} \tag{7.20}$$

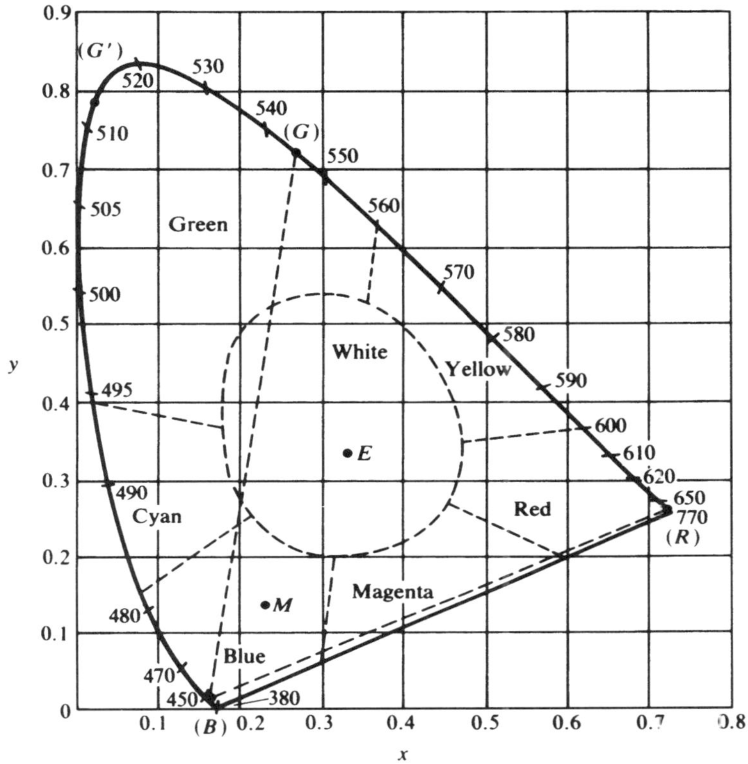

Figure 7.9 CIE chromaticity diagram.

The spectral tristimulus values and chromaticity coordinates of the pure spectral colors for the standard 1931 CIE observer (2° field)* are presented in Table 7.2, and the chromaticity coordinates are plotted in Fig 7.9. All real colors are included within this horseshoe locus and the line connecting its ends (the purple boundary). The physical primaries of Sect. 7.4 are located on the spectral locus. Their coordinates are interpolated from Table 7.2.

		x	y
(R)	700 nm	0.7347	0.2653
(G)	546.1 nm	0.2757	0.7147
(B)	435.8 nm	0.1661	0.0094

If these three points are connected by straight lines to form a triangle, as shown in Fig. 7.9, then all colors within the triangle can be matched by positive amounts of these three physical primaries. Colors outside the triangle require a negative amount of one primary, as was found in Sect. 7.4. In particular, note that many of the more saturated greens and cyans require negative amounts of the red primary.

* Values for the standard 1964 CIE observer (10° field) are given in the IESNA Lighting Handbook, Ninth Edition, 2000.

TABLE 7.2 CIE TRISTIMULUS VALUES AND CHROMATICITY COORDINATES FOR 2° FIELD

	Spectral Tristimulus Values on Equal-Energy Basis			Chromaticity Coordinates of Pure Spectral Colors	
λ (μm)	$\bar{x}$	$\bar{y}$	$\bar{z}$	x	y
0.38	0.0014	0.0000	0.0065	0.1741	0.0050
0.39	0.0042	0.0001	0.0201	0.1738	0.0049
0.40	0.0143	0.0004	0.0679	0.1733	0.0048
0.41	0.0435	0.0012	0.2074	0.1726	0.0048
0.42	0.1344	0.0040	0.6456	0.1714	0.0051
0.43	0.2839	0.0116	1.3856	0.1689	0.0069
0.44	0.3483	0.0230	1.7471	0.1644	0.0109
0.45	0.3362	0.0380	1.7721	0.1566	0.0177
0.46	0.2908	0.0600	1.6692	0.1440	0.0297
0.47	0.1954	0.0910	1.2876	0.1241	0.0578
0.48	0.0956	0.1390	0.8130	0.0913	0.1327
0.49	0.0320	0.2080	0.4652	0.0454	0.2950
0.50	0.0049	0.3230	0.2720	0.0082	0.5384
0.51	0.0093	0.5030	0.1582	0.0139	0.7502
0.52	0.0633	0.7100	0.0782	0.0743	0.8338
0.53	0.1655	0.8620	0.0422	0.1547	0.8059
0.54	0.2904	0.9540	0.0203	0.2296	0.7543
0.55	0.4334	0.9950	0.0087	0.3016	0.6923
0.56	0.5945	0.9950	0.0039	0.3731	0.6245
0.57	0.7621	0.9520	0.0021	0.4441	0.5547
0.58	0.9163	0.8700	0.0017	0.5125	0.4866
0.59	1.0263	0.7570	0.0011	0.5752	0.4242
0.60	1.0622	0.6310	0.0008	0.6270	0.3725
0.61	1.0026	0.5030	0.0003	0.6658	0.3340
0.62	0.8544	0.3810	0.0002	0.6915	0.3083
0.63	0.6424	0.2650	0.0000	0.7079	0.2920
0.64	0.4479	0.1750	0.0000	0.7190	0.2809
0.65	0.2835	0.1070	0.0000	0.7260	0.2740
0.66	0.1649	0.0610	0.0000	0.7300	0.2700
0.67	0.0874	0.0320	0.0000	0.7320	0.2680
0.68	0.0468	0.0170	0.0000	0.7334	0.2666
0.69	0.0227	0.0082	0.0000	0.7344	0.2656
0.70	0.0114	0.0041	0.0000	0.7347	0.2653
0.71	0.0058	0.0021	0.0000	0.7347	0.2653
0.72	0.0029	0.0010	0.0000	0.7347	0.2653
0.73	0.0014	0.0005	0.0000	0.7347	0.2653
0.74	0.0007	0.0003	0.0000	0.7347	0.2653
0.75	0.0003	0.0001	0.0000	0.7347	0.2653

In the early development of color television, it was thought that moving the green primary to a new location, such as (G′) in Fig. 7.9, might produce a better color picture. This change would include more of the saturated blues and greens, but would exclude more of the yellows and oranges. However, experience showed that the latter are more important for pleasing color reproduction than the former, particularly with regard to skin tones. Thus, the CIE primaries were a better choice.

Equal-energy white (E) is shown in Fig. 7.9, with coordinates x = y = 0.3333. This point is close to the center of the region labeled "white," where the unsaturated colors lie. Actually the color and white regions are not sharply defined, but rather blend, one into another. Also "perceived white" is a function of the current adaptation of the eyes and thus the "white point" can lie anywhere in the central region of the diagram. Nevertheless, it is possible to show roughly the regions of white, the three primary colors of light, and their three complements. This has been done in Fig. 7.9, with dashed lines radiating outward from point E.

The nonphysical primaries used to create Fig. 7.9 do not appear on the diagram. Their coordinates are at (1, 0), (0, 1), and (0, 0). As expected, these are located outside the region of real colors defined by the horseshoe locus.

When the chromaticity coordinates are formed in Eq. (7.20), the scaling parameter k in Eqs. (7.19) cancels out. Thus, this parameter may be chosen at will without altering x and y. One choice is to let k = 683 lm/W, which causes Y in Eq. (7.19) to yield lumens. A second choice is to choose k so that Y = 100 for a source. Then the Y of that source reflected from an object is the reflectance of that object in percent.

EXAMPLE 7.3

Verify that the chromaticity coordinates of the physical green primary of Sect. 7.4 are x = 0.2757 and y = 0.7147.

Solution. From Table 7.2 and Eqs. (7.20), after interpolation,

$$x = \frac{0.3776}{0.3776 + 0.9790 + 0.0132} = 0.2757$$

$$y = \frac{0.9790}{0.3776 + 0.9790 + 0.0132} = 0.7147$$

EXAMPLE 7.4

Find the chromaticity coordinates of a mercury lamp radiating 1.0, 0.8, 0.1 and 0.2 mw/cm^2, respectively, in its four visible lines.

Solution. The calculations will be done in tabular form, using Eqs. (7.19) and Table 7.2.

λ (nm)	**M_i(mW/cm^2)**	$\bar{x}_i$	$\bar{y}_i$	$\bar{z}_i$	$\bar{x}_iM_i$	$\bar{y}_iM_i$	$\bar{z}_iM_i$
404.7	1.0	0.0280	0.0008	0.1335	0.0280	0.0008	0.1335
435.8	0.8	0.3213	0.0182	1.5953	0.2570	0.0146	1.2762
546.1	0.1	0.3776	0.9790	0.0132	0.0378	0.0979	0.0013
578.0	0.2	0.8855	0.8864	0.0018	0.1771	0.1773	0.0004
					0.4999	0.2906	1.4114

Thus X = 0.4999, Y = 0.2906, and Z = 1.4114. Then, from Eqs. (7.20)

$$D = 2.2019 \quad x = 0.227 \quad y = 0.132 \quad z = 0.641$$

The point is labeled M (for mercury lamp) on Fig. 7.9, and lies in the blue region of the chromaticity diagram.

EXAMPLE 7.5

A spectral source has 1 W at 490 nm. What is the wavelength of a spectral source that can combine with this source to produce equal energy white? How many watts of the second source are required?

Solution. The resultant color of a mixture of two colors lies on a line connecting the two. Drawing a line on Fig. 7.9 from 490 nm through equal energy white intersects the horseshoe at about 600 nm. Letting P be the watts at 600 nm gives

λ (nm)	P_i(w)	$\bar{x}_i$	$\bar{y}_i$	$\bar{z}_i$
490	1	0.0380	0.2080	0.4652
600	P	1.0622	0.6310	0.0008

From Eqs. 7.20, and noting that the chromaticity coordinates of the mixture must be x = 0.3333, y = 0.3333,

$$x = \frac{1.0622P + 0.0380}{1.6950P + 0.7052} = 0.3333, \quad y = \frac{0.6310P + 0.2080}{1.6950 + 0.7052} = 0.3333$$

Solving either equation yields P = 0.41 W of 600 nm radiation.

7.6 COLOR SPECIFICATIONS OF SOURCES

Consider first a blackbody radiator. For each temperature, Eqs. (7.19) can be used, with S(λ) replaced by the M_λ of a blackbody from Eq. (4.8), and X, Y, and Z can be calculated. Then the chromaticity coordinates of the blackbody for each temperature may be obtained from Eqs. (7.20).

The chromaticies from such calculations are plotted on the CIE diagram in Fig. 7.10 as a function of temperature. The resulting curve is called the Planckian locus. It can be seen that near 1000 K, the blackbody is red. It is yellow near 3000 K. It must be pointed out, as was done earlier, that the perceived color of a blackbody at each of these temperatures may be different from the color indicted by its chromaticity coordinates because of viewer adaptation.

The chromaticity of a source lying on the Planckian locus is specified in terms of color temperature, whereas the chromaticity of a source lying near to the Planckian locus is specified in terms of correlated color temperature (CCT). Recall that in Chap. 4 CCT was defined as the absolute temperature at which a blackbody must operate if its output is to be the closest possible color match to the output of the light source under consideration. Stated another way, CCT is the absolute temperature of a blackbody whose chromaticity most nearly approximates that of the light source.

CCTs of sources whose chromaticity coordinates lie near the Planckian locus can be determined approximately by plotting lines of constant CCT on the chromaticity

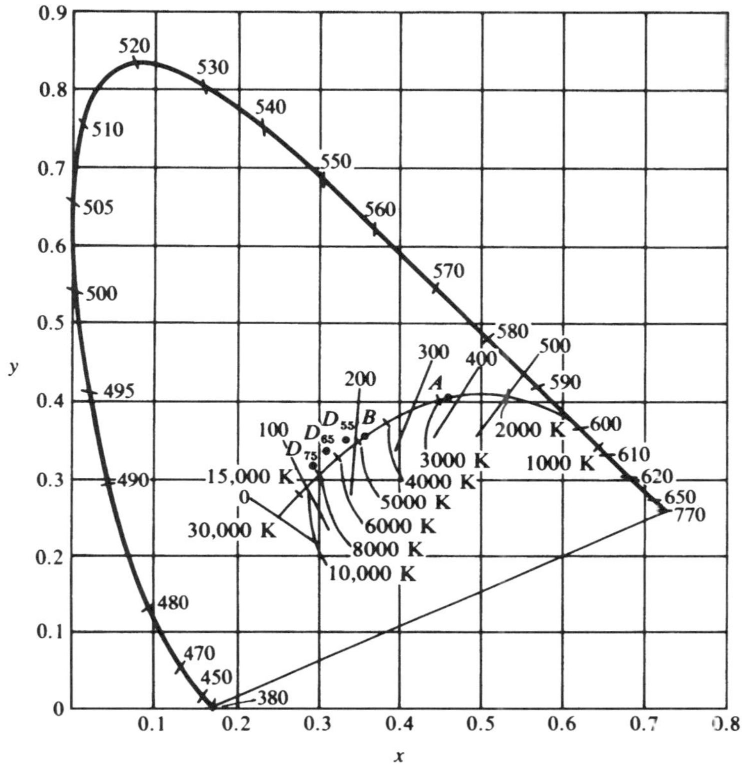

Figure 7.10 Planckian source locus.

diagram near the locus. Equal color differences on the Planckian locus are nearly in direct proportion to reciprocal color temperature, rather than color temperature itself. The customary unit is reciprocal megakelvins obtained by dividing 10^6 by the color temperature. Lines of constant CCT are shown in 100-$(MK)^{-1}$ intervals in Fig. 7.10.

In addition to specifying the standard observer and the chromaticity coordinate system, the CIE has also defined two standard sources and four standard illuminants,* designated A, C, D_{50}, D_{55}, D_{65}, and D_{75}. Source A is a tungsten filament incandescent lamp operating at 2856 K. It matches very closely a blackbody operating at that temperature. Source C is obtained by combining source A with liquid filters and has a CCT of 6774 K, representing a combination of direct sunlight and clear skylight.

The four standard illuminants represent daylight distributions with CCTs of 5000 K, 5500 K, 6500 K, and 7500 K. These illuminants are nonphysical sources specified in terms of their spectral power distributions. They were reconstituted from 622 spectral irradiance distributions of daylight measured by D.B. Judd, D.L. MacAdam, and G. Wysecki at three widely separated locations (Jerome, 1969). They found that all of the data taken could be represented closely by a chromaticity curve defined by

$$y = 2.870x - 3.000x^2 - 0.275 \tag{7.21}$$

* The former is a specified physical emitter of light; the latter is a specified spectral power distribution.

The current CIE recommendation suggests that sources A and C and illuminant D_{65} be used in evaluating the spectral reflectance or transmittance characteristics of colored objects. Relative spectral power distributions of these three CIE standard sources and illuminants are given in Table 7.3. They have been normalized so that each has the value 1000 at $\lambda = 560$ nm. The chromaticity coordinates for these emitters are also given in Table 7.3. The chromaticity coordinates and CCTs for several common lamp

TABLE 7.3 RELATIVE SPECTRAL POWER DISTRIBUTIONS OF CIE STANDARD EMITTERS

λ (nm)	A	C	D_{65}
380	98	313	500
390	121	450	546
400	147	601	828
410	177	765	915
420	210	932	934
430	247	1067	867
440	287	1154	1049
450	331	1178	1170
460	378	1169	1178
470	429	1176	1149
480	482	1177	1159
490	539	1146	1088
500	599	1065	1094
510	661	972	1078
520	725	920	1048
530	791	931	1077
540	859	970	1044
550	929	999	1040
560	1000	1000	1000
570	1072	972	963
580	1144	929	958
590	1217	885	887
600	1290	852	900
610	1363	840	896
620	1436	837	877
630	1508	836	833
640	1580	834	837
650	1650	838	800
660	1720	835	802
670	1788	820	823
680	1854	798	783
690	1919	762	697
700	1983	725	716
710	2044	688	743
720	2104	649	616
730	2161	612	699
740	2217	584	751
750	2270	562	636
760	2321	552	464
770	2370	553	668

CHROMATICITY COORDINATES OF CIE STANDARD EMITTERS

	A	C	D_{65}
x	0.4476	0.3101	0.3129
y	0.4074	0.3163	0.3290

types are given in Table 7.4, and the spectral power distribution curves for common fluorescent and HID lamps are shown in Fig. 7.11.

TABLE 7.4 CIE CHROMATICITY COORDINATES AND CCT OF COMMON SOURCES

Lamp	x	y	CCT (K)
100 W Incandescent	0.460	0.410	2700
Tungsten halogen	0.424	0.399	3190
Cool white fluorescent	0.373	0.385	4250
Warm white fluorescent	0.436	0.406	3020
Triphosphor fluorescent	0.463	0.415	2700
Triphosphor fluorescent	0.440	0.403	3000
Triphosphor fluorescent	0.413	0.393	3500
Triphosphor fluorescent	0.376	0.387	4100
Triphosphor fluorescent	0.346	0.359	5000
Daylight fluorescent	0.316	0.345	6250
Clear mercury	0.326	0.390	5710
Color-improved mercury	0.373	0.415	4430
Clear metal halide	0.396	0.390	3720
Color-improved metal halide	0.388	0.379	3800
High-pressure sodium	0.519	0.418	2100
Low-pressure sodium	0.569	0.421	1740
Xenon	0.324	0.324	5920

EXAMPLE 7.6

Verify the numbers given in Table 7.3 for the chromaticity coordinates of source C.

Solution. The integrals in Eqs. (7.19) must be evaluated when $S(\lambda)$ is the spectral power distribution of source C. This has been done for all of the standard emitters, and the results are given in Table 7.5, where values have been normalized so that

$$\int_0^\infty S(\lambda)\bar{y}(\lambda)\,d\lambda = 100 \tag{7.22}$$

From that table, and using Eqs. (7.20),

$$D = 98.041 + 100.000 + 118.103 = 316.144$$

$$x = \frac{98.041}{316.144} = 0.310$$

$$y = \frac{100.00}{316.144} = 0.316$$

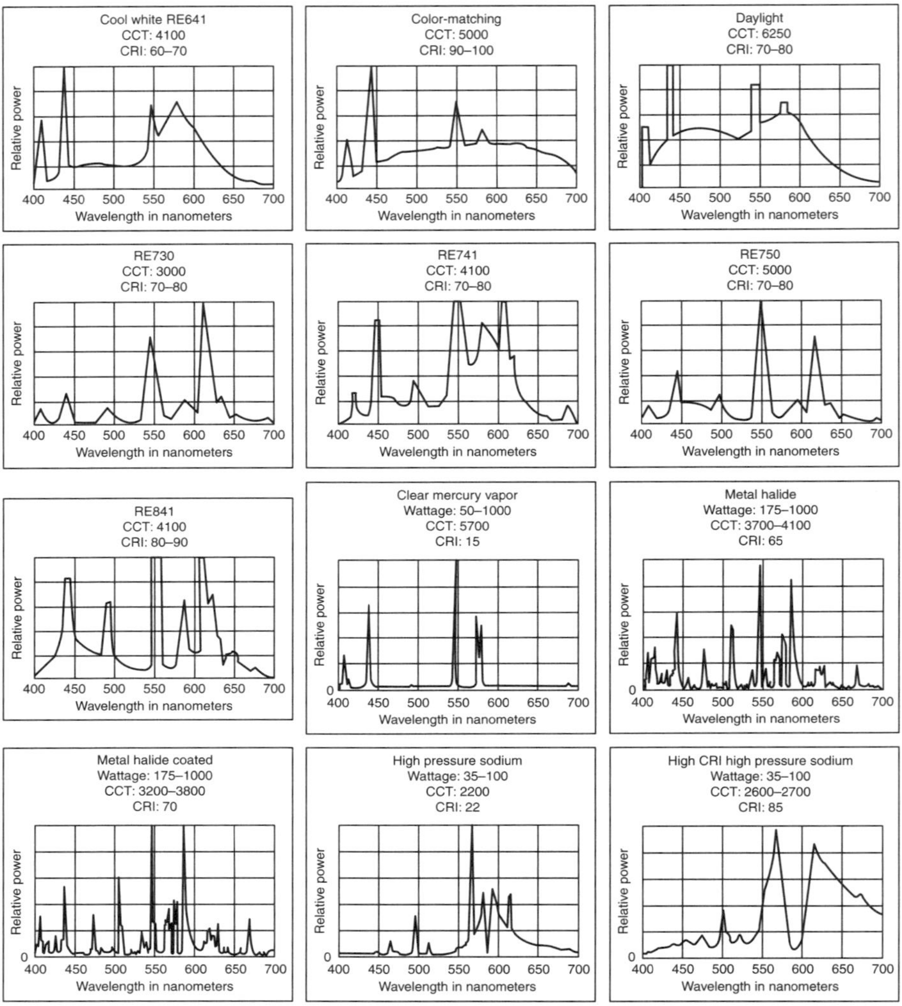

Figure 7.11 Fluorescent and HID SPD curves. ***Source*: *IESNA Lighting Handbook*, 9th edition, 2000. Courtesy of the Illuminating Engineering Society of North America.**

7.7 COLOR SPECIFICATIONS OF OBJECTS

Consider now that radiant flux from one of the standard emitters or from a common light source impinges on a colored object and is selectively reflected and/or transmitted. The chromaticity coordinates of the reflected or transmitted light are to be calculated.

For this situation, Eqs. (7.19) are modified so that $S(\lambda)$ is either $\rho(\lambda)\ E(\lambda)$ or $\tau(\lambda)\ E(\lambda)$, where $E(\lambda)$ is the spectral irradiance of the surface and $\rho(\lambda)$ and $\tau(\lambda)$ are the spectral reflectance and transmittance. The procedure will be illustrated by an example.

TABLE 7.5 PRODUCTS OF TRISTIMULUS VALUES AND SOURCE RADIANCES

Wave-length	A			C			D_{65}		
(nm)	$\bar{x}(\lambda)S(\lambda)$	$\bar{y}(\lambda)\bar{S}(\lambda)$	$\bar{z}(\lambda)S(\lambda)$	$\bar{x}(\lambda)S(\lambda)$	$\bar{y}(\lambda)S(\lambda)$	$\bar{z}(\lambda)S(\lambda)$	$\bar{x}(\lambda)S(\lambda)$	$\bar{y}(\lambda)S(\lambda)$	$\bar{z}(\lambda)S(\lambda)$
380	0.001	0.000	0.006	0.004	0.000	0.020	0.007	0.000	0.031
390	0.005	0.000	0.023	0.019	0.000	0.089	0.022	0.001	0.104
400	0.019	0.001	0.093	0.085	0.002	0.404	0.112	0.003	0.532
410	0.071	0.002	0.340	0.329	0.009	1.570	0.377	0.010	1.796
420	0.262	0.008	1.256	1.238	0.037	5.949	1.189	0.035	5.711
430	0.649	0.027	3.167	2.997	0.122	14.628	2.330	0.095	11.370
440	0.926	0.061	4.647	3.975	0.262	19.938	3.458	0.228	17.343
450	1.031	0.117	5.435	3.915	0.443	20.638	3.724	0.421	19.627
460	1.019	0.210	5.851	3.362	0.694	19.299	3.243	0.669	18.614
470	0.776	0.362	5.116	2.272	1.058	14.972	2.124	0.989	13.998
480	0.428	0.622	3.636	1.112	1.618	9.461	1.048	1.524	8.915
490	0.160	1.039	2.324	0.363	2.358	5.274	0.330	2.142	4.791
500	0.027	1.792	1.509	0.052	3.401	2.864	0.051	3.343	2.815
510	0.057	3.080	0.969	0.089	4.833	1.520	0.095	5.132	1.614
520	0.425	4.771	0.525	0.576	6.462	0.712	0.628	7.041	0.775
530	1.214	6.322	0.309	1.523	7.934	0.338	1.687	8.785	0.430
540	2.313	7.600	0.162	2.785	9.149	0.195	2.869	9.425	0.201
550	3.732	8.568	0.075	4.282	9.832	0.086	4.267	9.796	0.086
560	5.510	9.222	0.036	5.880	9.841	0.039	5.625	9.415	0.037
570	7.571	9.457	0.021	7.322	9.147	0.020	6.947	8.678	0.019
580	9.719	9.228	0.018	8.417	7.992	0.016	8.304	7.885	0.015
590	11.579	8.540	0.012	8.984	6.627	0.010	8.612	6.352	0.009
600	12.704	7.547	0.010	8.949	5.316	0.007	9.046	5.347	0.007
610	12.669	6.356	0.004	8.325	4.176	0.002	8.499	4.264	0.003
620	11.373	5.071	0.003	7.070	3.153	0.002	7.089	3.101	0.002
630	8.980	3.704	0.000	5.309	2.190	0.000	5.062	2.088	0.000
640	6.558	2.562	0.000	3.693	1.443	0.000	3.547	1.386	0.000
650	4.336	1.637	0.000	2.349	0.886	0.000	2.147	0.810	0.000
660	2.628	0.972	0.000	1.361	0.504	0.000	1.252	0.463	0.000
670	1.448	0.530	0.000	0.708	0.259	0.000	0.680	0.249	0.000
680	0.804	0.292	0.000	0.369	0.134	0.000	0.347	0.126	0.000
690	0.404	0.146	0.000	0.171	0.062	0.000	0.150	0.054	0.000
700	0.209	0.075	0.000	0.082	0.029	0.000	0.077	0.028	0.000
710	0.110	0.040	0.000	0.039	0.014	0.000	0.041	0.015	0.000
720	0.057	0.019	0.000	0.019	0.006	0.000	0.017	0.006	0.000
730	0.028	0.010	0.000	0.008	0.003	0.000	0.009	0.003	0.000
740	0.011	0.006	0.000	0.004	0.002	0.000	0.005	0.002	0.000
750	0.006	0.002	0.000	0.002	0.001	0.000	0.002	0.001	0.000
760	0.004	0.002	0.000	0.001	0.001	0.000	0.001	0.000	0.000
770	0.002	0.000	0.000	0.001	0.000	0.000	0.001	0.000	0.000
Total	109.828	100.00	35.547	98.041	100.00	118.103	95.018	100.000	108.845

EXAMPLE 7.7

The spectral reflectance characteristic of a sample of green paint is known. The sample is to be lighted by illuminant A. Determine the chromaticity coordinates of the light reflected from the sample.

Solution. From Eqs. (7.19), Table 7.5, and the given spectral reflectance, the following tabulation can be prepared.

λ (nm)	$\rho(\lambda)$	$\bar{x}(\lambda)S(\lambda)$	ΔX	$\bar{y}(\lambda)S(\lambda)$	ΔY	$\bar{z}(\lambda)S(\lambda)$	ΔZ
400	0.200	0.019	0.004	0.001	—	0.093	0.019
410	0.150	0.071	0.011	0.002	—	0.340	0.051
420	0.125	0.262	0.033	0.008	0.001	1.256	0.157
430	0.100	0.649	0.065	0.027	0.003	3.167	0.317
440	0.085	0.926	0.079	0.061	0.005	4.647	0.395
450	0.075	1.031	0.077	0.117	0.009	5.435	0.408
460	0.080	1.019	0.082	0.210	0.017	5.851	0.468
470	0.130	0.776	0.101	0.362	0.047	5.116	0.665
480	0.330	0.428	0.141	0.622	0.205	3.636	1.200
490	0.465	0.160	0.074	1.039	0.483	2.324	1.081
500	0.550	0.027	0.015	1.792	0.986	1.509	0.830
510	0.590	0.057	0.034	3.080	1.817	0.969	0.572
520	0.600	0.425	0.255	4.771	2.863	0.525	0.315
530	0.580	1.214	0.704	6.322	3.667	0.309	0.179
540	0.530	2.313	1.226	7.600	4.028	0.162	0.086
550	0.470	3.732	1.754	8.568	4.027	0.075	0.035
560	0.410	5.510	2.259	9.222	3.781	0.036	0.015
570	0.340	7.571	2.574	9.457	3.215	0.021	0.007
580	0.280	9.719	2.721	9.228	2.584	0.018	0.005
590	0.230	11.579	2.683	8.540	1.964	0.012	0.003
600	0.180	12.704	2.287	7.547	1.358	0.010	0.002
610	0.140	12.669	1.774	6.356	0.890	0.004	0.001
620	0.110	11.373	1.251	5.071	0.558	0.003	—
630	0.080	8.980	0.718	3.704	0.296	—	—
640	0.070	6.558	0.459	2.562	0.179	—	—
650	0.055	4.336	0.238	1.637	0.090	—	—
660	0.053	2.268	0.120	0.972	0.052	—	—
670	0.050	1.448	0.072	0.530	0.027	—	—
680	0.040	0.804	0.032	0.292	0.012	—	—
			21.843		33.164		6.811

Then, from Eqs. (7.20)

$$D = 61.818, \quad x = 0.353, \quad y = 0.536$$

The coordinates are located at point S on Fig. 7.12. The coordinates of the source are shown at point A. It is observed that the sample's chromaticity falls in the greenish-yellow region of the diagram when illuminated by source A.

The coordinates on the chromaticity diagram of the radiation incident on and reflected (transmitted) from an object can be used to determine dominant wavelength and excitation purity. Dominant wavelength (λ_d) is defined as the wavelength of radiant

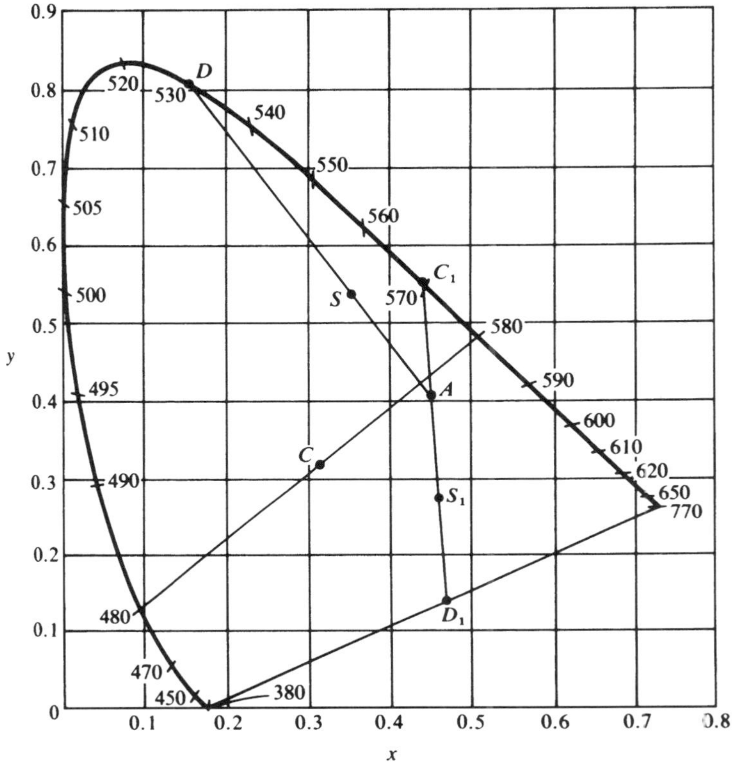

Figure 7.12 Object loci.

energy of a single frequency that, when combined in suitable proportion with the radiant energy of a reference standard, matches the color of the light reflected (transmitted) from an object. Excitation purity is defined as the ratio on the CIE chromaticity diagram of the distance between the reference point and the light point to the distance in the same direction between the reference point and the spectrum locus (including the straight line closing the horseshoe).

On the chromaticity diagram, a line drawn from the chromaticity coordinates of the source through the chromaticity coordinates of the sample will intersect the spectral locus at the dominant wavelength. For the case in Example 7.7, $\lambda_d = 530$ nm, as shown in Fig. 7.12. The spectral locus represents 100% purity; the source point 0% purity. From Fig. 7.12,

$$\% \text{ excitation purity} = 100\frac{\text{AS}}{\text{AD}} \tag{7.33}$$

giving 32% for the green sample of Example 7.7.

Assume a sample S_1 when illuminated by source A gives chromaticity coordinates as shown in Fig. 7.12. What are its dominant wavelength and excitation purity? The intersection with the straight line closing the spectral locus is at point D_1, a nonspectral color. In this case, a new quantity is required, the complementary wavelength (λ_c), defined as the wavelength of radiant energy of a single frequency that, when combined in suitable proportion with the light reflected (transmitted), matches the color of the reference standard. For S_1, the complementary wavelength is 570 nm in Fig. 7.12. The % excitation purity is 100 (AS_1/AD_1), in this case 51%.

7.8 COLOR MIXING

According to Grassman's law, stated in Sect. 7.2, color mixtures obey the law of addition. Thus, if the tristimulus values of two colors are known, they may be added to obtain the tristimulus values of the resultant.

In Example 7.4, the overall tristimulus values are calculated as the sums of the tristimulus values of the four spectral lines. Then, from these sums, the chromaticity coordinates are obtained. This is one way to do color mixing calculations, but it requires that the spectral power distribution curve of each source to be combined is known. Then the tristimulus values of each source are obtained and summed to get the overall tristimulus values.

$$X = \Sigma X_i, \quad Y = \Sigma Y_i, \quad Z = \Sigma Z_i \tag{7.34}$$

The chromaticity coordinates are calculated from

$$x = \frac{\Sigma X_i}{\Sigma(X_i + Y_i + Z_i)}, \quad y = \frac{\Sigma Y_i}{\Sigma(X_i + Y_i + Z_i)} \tag{7.35}$$

Rather than the spectral power distribution curves, usually the chromaticity coordinates of a color and some measure of its intensity (luminance L, luminous flux ϕ, luminous exitance M, or illuminance E) are provided. This measure is directly proportional to Y, because $\bar{y}(\lambda)$ and $V(\lambda)$ are identical. Thus x_i , y_i and Y_i are given for each color i being mixed. Knowing these quantities, the overall x, y, and Y can be obtained.

$$x_i = \frac{X_i}{X_i + Y_i + Z_i} = \frac{X_i}{D_i}, \quad y_i = \frac{Y_i}{X_i + Y_i + Z_i} = \frac{Y_i}{D_i} \tag{7.36}$$

The second Eq. (7.36) can be solved for D_i and substituted into the first Eq. (7.36).

$$D_i = \frac{Y_i}{y_i}, \quad X_i = x_i D_i = x_i \frac{Y_i}{y_i} \tag{7.37}$$

Then, from Eq. (7.34),

$$x = \frac{\Sigma x_i \dfrac{Y_i}{y_i}}{\Sigma \dfrac{Y_i}{y_i}}, \quad y = \frac{\Sigma Y_i}{\Sigma \dfrac{Y_i}{y_i}} \tag{7.38}$$

Sometimes a source is specified in radiometric, rather than photometric, terms. If a source consumes P_i watts and its luminous efficacy is η_i, the product $P_i\eta_i$ can be used for this source in place of Y_i in Eqs. (7.38).

EXAMPLE 7.8

Two colors are specified as follows:

x	y	M (lm/ft²)
0.25	0.45	10
0.30	0.20	15

Find the chromaticity coordinates and luminance exitance of the mixture of the two colors.

Solution. From Eqs. (7.38)

$$x = \frac{0.25(10/0.45) + 0.30(15/0.20)}{(10/0.45) + (15/0.20)} = 0.29$$

$$y = \frac{10+15}{(10/0.45) + (15/0.20)} = 0.26$$

$$M = 10 + 15 = 25 \text{ lm/ft}^2$$

EXAMPLE 7.9

Determine the ratio of radiant fluxes from two spectral (single-wavelength) sources that will produce a match with source C, if one of the spectral radiances has a wavelength of 480 nm.

Solution. The second wavelength must be found. The chromaticity coordinates are

	x	y
Source C	0.3101	0.3163
480 nm	0.0913	0.1327

These two points are located on Fig. 7.12, and a line is drawn between them and extended to the locus. The intersection is at $\lambda \simeq 580$ nm, with coordinates $x = 0.5125$ and $y = 0.4866$.

For a spectral line, $y_i = \bar{y}_i P_i$, where P_i is the radiated watts of source i. Then, from Eqs. (7.38),

$$x = \frac{(x_1\bar{y}_1P_1/y_1) + (x_2\bar{y}_2P_2/y_2)}{(\bar{y}_1P_1/y_1) + (\bar{y}_2P_2/y_2)}, \quad y = \frac{\bar{y}_1P_1 + \bar{y}_2P_2}{(\bar{y}_1P_1/y_1) + (y_2P_2/y_2)} \qquad (7.39)$$

Everything is known in these two equations except P_1 and P_2. Thus, either equation may be solved to yield $P_1/P_2 = 1.7$.

EXAMPLE 7.10

A conference room is lighted by a combination of daylighting, equivalent to CIE D65, and cool white fluorescent lighting. At night the average illuminance in the room is 50 fc; during the day, with the electric lighting on, it is 75 fc. What are the chromaticity coordinates of the combination?

Solution. Using Tables 7.5 and 7.4,

	x	y	E
D65	0.3129	0.3290	25
CW	0.373	0.385	50

Using Eqs. (7.35),

$$x = \frac{0.3129\dfrac{25}{0.3290} + 0.373\dfrac{50}{0.385}}{\dfrac{25}{0.3290} + \dfrac{50}{0.385}} = 0.351$$

$$y = \frac{25 + 50}{\dfrac{25}{0.3290} + \dfrac{50}{0.385}} = 0.364$$

7.9 COLOR RENDERING

In Sect. 7.3, it was stated that various metrics are used to describe the ability of one source or object to match the color of another source or object. One of these, correlated color temperature, was discussed earlier and is a means of specifying chromaticities of sources lying near the Planckian locus. Another metric, to be discussed now, is color rendering index (CRI or R_a). The CRI of a light source is a measure of the shift in chromaticity of an object in color space when it is lighted by the source as compared to being lighted by a reference source of comparable color temperature.

The resultant perceived color shift seen by an observer when an object is lighted first by one source and then by another is a combination of the actual colormetric shift, as measured by changes in luminance and chromaticity, and an adaptive color shift, caused by changes in chromatic adaptation. CRI permits the calculation of the former and a partial accounting for the latter, about which a great deal is still unknown. CRI provides a rough, but incomplete, measure of the color-rendering properties of light sources.

CRI is based on a test-color method that was proposed to the CIE by the IES Color Committee in 1961. In 1965, CIE Publication No. 13, "Method of Measuring and Specifying Color Rendering Properties of Light Sources," based on the U.S. proposal, was approved. With minor improvements (1971), which partially take into account chromatic adaptation of the observer, this is the accepted contemporary method for rating the ability of a lamp to give a true representation of the color of an object.

The determination of CRI is carried out on a uniform uv chromaticity diagram, rather than on the xy chromaticity diagram. It was found that the xy diagram has a non-linear chromaticity scale. The perceived color difference between two chromaticity points on the diagram separated by a given distance varies with the location of the two points, i.e., equal perceived color differences about a point have an elliptical, rather than a circular locus.

For example, a change in chromaticity in the yellow-red portion of the CIE diagram results in a much greater perceived color change than the same change in chromaticity in the blue-green region of the diagram.

The uv diagram has more uniform perceptible color differences than the xy diagram. It is called "the CIE 1960 UCS Diagram" (uniform color space) and is based on a transformation described in 1937 by D.L. MacAdam. The diagram is derived from the xy diagram through the coordinate transformations:

$$u = \frac{4x}{-2x+12y+3} = \frac{4X}{X+15Y+3Z}$$
$$v = \frac{6y}{-2x+12y+3} = \frac{6Y}{X+15Y+3Z} \tag{7.40}$$

and is displayed in Fig. 7.13.

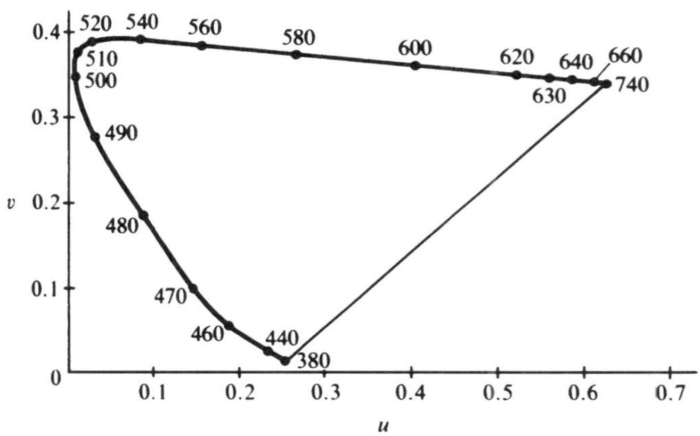

Figure 7.13 CIE 1960 UCS diagram.

The uv diagram, like the xy diagram, is two-dimensional and thus contains no luminance information. In 1964, the CIE extended this diagram to three dimensions with its "CIE 1964 UCS (uniform chromaticity scale) Diagram." The coordinates of this latter diagram are derived from

$$W^* = 25Y^{1/3} - 17 \tag{7.41a}$$

$$U^* = 13W^*(u - u_n) \tag{7.41b}$$

$$V^* = 13W^*(v - v_n) \tag{7.41c}$$

where u_n and v_n are the uv coordinates of the light source and are placed at the origin of the U^*, V^*, W^* system.

This new system, developed by G. Wyszecki, converts Y by Eq. (7.41a) into what is called a lightness index. W^* approximates the value function in the Munsell color system in the range $1 \leq Y \leq 100$. U^* and V^* are called chromaticness indexes, and the system is designed so that a lightness difference (ΔW^*) of 1 corresponds to a chromaticness difference $(\Delta U^{*2} + \Delta V^{*2})^{1/2}$ of 13. Color differences are calculated from

$$\Delta E = (\Delta W^{*2} + \Delta U^{*2} + \Delta V^{*2})^{1/2} \tag{7.42}$$

based on one unit of ΔE corresponding to one National Bureau of Standards unit of color difference, about one tenth of a Munsell value step.

The procedure for determining the CRI of a light source involves several steps. First, reference light sources are selected. To minimize effects due to chromatic adaptation, it was decided that the reference source should have nearly the same chromaticity coordinates as the source being evaluated. Thus the reference source for any light source whose CCT is less than 5000 K is a Planckian radiator of that color temperature. For a source whose CCT exceeds 5000 K, the reference source is the nearest in CCT of a series of mathematically derived daylight spectral power distributions based on data obtained in England, Canada, and New York.

Next, the spectral power distribution curve for the light source being evaluated is determined and its CCT calculated. This determines the reference source to be used and its spectral power distribution.

The third step is to select color samples. The CIE has chosen eight Munsell chip color samples, distributed uniformly around the Munsell hue circle, with colors red (R), yellow (Y), greenish-yellow (GY), green (G), bluish-green (BG), purplish-blue (PB), purple (P), and reddish-purple (RP). Each of these samples has a Munsell value of 6 and a medium chroma (4 to 8) so that there is some reflectance from each sample at each wavelength. Each of the samples has a specified and known spectral reflectance.

In addition, to the eight standard samples, there are six special samples suggested by the CIE. Four of these have high chroma (8 to 13) so that the effect of the test source on saturated colors can be estimated. The other two are colors that represent a Caucasian complexion and green foliage. The CRI calculation procedure includes only the eight standard samples. A CRI for each sample is calculated. Then a general CRI is found as the average of the eight samples. This general index is the single number most often used.

The fourth step is to calculate the tristimulus values and chromaticity coordinates for the eight colored samples when lighted by the reference source and by the source being evaluated. The procedure is identical to that used in Example 7.7. These sets of coordinates are then transformed to uv space using Eqs. (7.40). Then a perturbation is made of the uv coordinates of the color samples when lighted by the source being evaluated to partially account for chromatic adaptation by correcting for any difference in the chromaticity coordinates of the source being evaluated and the reference source.

The uv coordinates for the samples under the reference source and the altered uv coordinates for the samples under the source being evaluated are transformed into $U^*V^*W^*$ coordinates by Eqs. (7.41). Then the resultant color shift of each test color is determined from Eq. (7.42).

The final step involves deciding on a CRI scale and calibrating the results obtained from Eq. (7.42). It was decided to have an index in which a better performance was represented by a larger number [in Eq. (7.42) a larger number would represent a poorer performance]. Also it was decided to place the CRI of the reference source at 100 and to make the CRIs of all other sources be less than 100. Finally, it was agreed that a standard fluorescent white lamp at 3000 K would have a CRI of 50. With these decisions made, the CRI for each color sample can be written as

$$R_i = 100 - 4.6\Delta E_i \tag{7.43}$$

and the general CRI for all samples as

$$R_a = \frac{1}{8}\sum_{1}^{8} R_i \tag{7.44}$$

The 4.6 in Eq. (7.43) arises from the calibration to a CRI of 50 for the 3000 K white fluorescent lamp. Thus

$$50 = \frac{1}{8}\sum_{1}^{8}(100 - k\Delta E_i) \tag{7.45}$$

yields k = 4.6 when ΔE_i are the vectors for the 3000 K lamp.

The vector shifts in position of the sample colors when illuminated by a test source of $R_a = 63$ are shown in the U*V* diagram in Fig. 7.14. CRIs for common lamp types are given in Table 7.6.

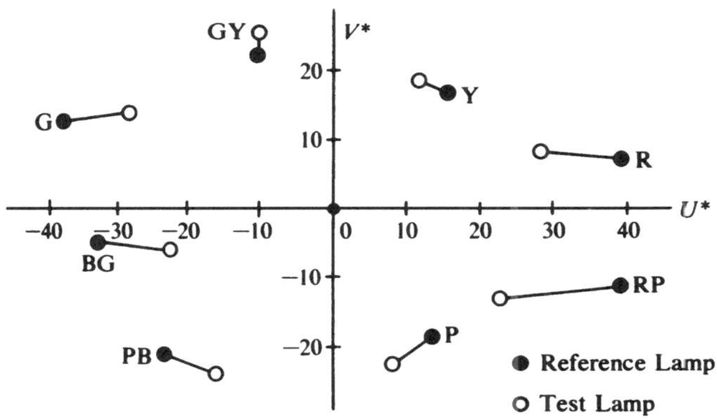

Figure 7.14 Graphic representation of CRI. ***Source*****: C.W. Jerome. "Basic Colorimetry for the Lighting Industry," Salem, Mass., 1969. Courtesy of GTE Lighting Products.**

The color rendering index is not a panacea for insuring color quality of lighting. It has several deficiencies. First, since each light source has a specific CCT reference, two light sources of differing color temperatures cannot be compared directly. Second, since CRI is an average of eight responses, two lamps with the same CCT and CRI may differ drastically in their ability to render one or several colors. For example, a lamp

TABLE 7.6 COLOR RENDERING DATA FOR COMMON LAMPS

Lamp	CCT (K)	CRI
100W A19	2700	95
Tungsten halogen	3190	100
Cool white fluorescent	4230	64
Warm white fluorescent	2940	51
Deluxe cool white fluorescent	4100	89
Deluxe warm white fluorescent	2940	73
Daylight fluorescent	6430	76
Triphosphor fluorescent	3000–5000	65–90
26W CFL	2700	85
Clear mercury	5710	15
Clear metal halide	3720	60
Coated metal halide	3800	70
High-pressure sodium	2100	21

giving six indexes of 80 and two of 40 has the same CRI as a lamp with eight indexes of 70. Third, some of the individual indices are negative for some lamps, for example, low-pressure sodium and clear mercury. Fourth, even if an R_i is the same for two sources, this does not mean that the sample will appear the same under each source, because the R_is do not include the directions of color shifts. Finally, CRI is based on Planckian radiators and sources of reconstituted daylight. There is nothing inherent in these sources that says they are perfect in a color-rendering sense. In fact, they may render certain colors quite differently. For example, two sources at 3000 K and 5500 K, having the same CRI, may render colors differently because their reference sources do.

7.10 LAB, LUV AND LINKING CIE AND MUNSELL

There are two additional color coordinate systems, which were adopted in 1976 by the CIE to replace the U*, V*, W* system. These are known as CIELUV and CIELAB. They arose because color differences calculated by the U*, V*, W* system, while working well for Munsell color differences, did not correlate well with the smaller color differences encountered often in industrial colorant processes. Although the new systems supercede the U*, V*, W* system, the latter is still used for the calculation of CRI. The reason for introducing two new systems rather than one was that the CIE Colorimetry Committee did not feel that a single color-difference system fit all of the experimental data and industrial applications.

The coordinates of CIELUV are:

$$\left.\begin{aligned} L^* &= 116\left(\frac{Y}{Y_n}\right)^{1/3} - 16 \quad \text{for } \frac{Y}{Y_n} > 0.008856 \\ L^* &= 903.29\frac{Y}{Y_n} \quad \text{for } \frac{Y}{Y_n} \le 0.008856 \\ u^* &= 13L^*(u' - u'_n), \quad v^* = 13L^*(v' - v_n) \end{aligned}\right\} \tag{7.46}$$

where

$$u' = \frac{4X}{X + 15Y + 3Z}$$

$$v' = \frac{9Y}{X + 15Y + 3Z}$$

and u'_n, v'_n, and Y_n are values of u′, v′ and Y for a standard source or illuminant, with $Y_n = 100$.

The changes from U^*, V^*, W^* to L^*, u^*, v^* are minor, namely a slight change in W^* to get L^* and setting $v' = 1.5v$.

The coordinates for CIELAB are:

$$\left.\begin{aligned} L^* &= 116\left(\frac{Y}{Y_n}\right)^{1/3} - 16 \\ a^* &= 500\left[\left(\frac{X}{X_n}\right)^{1/3} - \left(\frac{Y}{Y_n}\right)^{1/3}\right] \\ b^* &= 200\left[\left(\frac{Y}{Y_n}\right)^{1/3} - \left(\frac{Z}{Z_n}\right)^{1/3}\right] \end{aligned}\right\} \frac{X}{X_n}, \frac{Y}{Y_n}, \frac{Z}{Z_n} > 0.008856 \tag{7.47}$$

$$\left.\begin{aligned} L^* &= 903.29\left(\frac{Y}{Y_n}\right) \\ a^* &= 3893.5\left(\frac{X}{X_n} - \frac{Y}{Y_n}\right) \\ b^* &= 1557.4\left(\frac{Y}{Y_n} - \frac{Z}{Z_n}\right) \end{aligned}\right\} \frac{X}{X_n}, \frac{Y}{Y_n}, \frac{Z}{Z_n} \le 0.008856 \tag{7.48}$$

As before, "n" implies a standard source or illuminant, with $Y_n = 100$.

Both the L^*, a^*, b^* and L^*, u^*, v^* systems are opponent color systems. This is illustrated in Fig. 7.15 for the former. In that figure, the Hering opponent-color perpendicular axes (yellow-blue and red-green) are shown at 90° with the lightness axis. From that figure,

$$\left.\begin{aligned} h^* &= \tan^{-1}\frac{b^*}{a^*} \\ c^* &= \sqrt{a^* + b^*} \end{aligned}\right\} \tag{7.49}$$

as expressions for the hue (h^*) and saturation (c^*) of a color in the a^* b^* system.

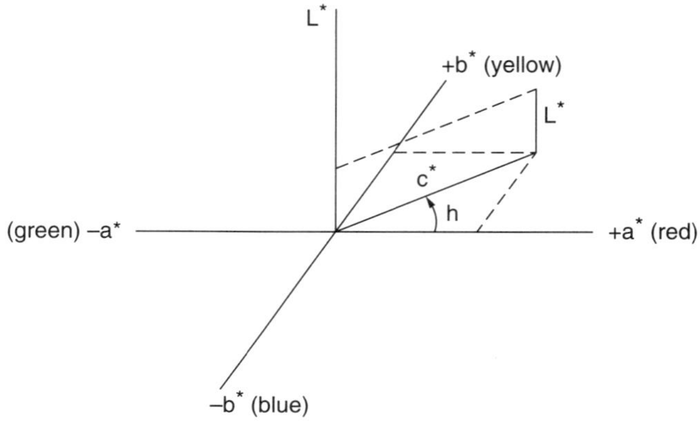

Figure 7.15 Opponent colors representation of L*, a*, b*.

The interrelationship of the x, y, Y, the L*, a*, b* and the Munsell systems will be illustrated through an example.

EXAMPLE 7.11

Assume a red object has the following values when lighted by illuminant D65:

$$Y = 15.00, \quad x = 0.440, \quad y = 0.280$$

It is required to convert these to L*, a*, b* coordinates:

Solution.

Step 1 Find X and Z (Y = 15). From Eqs. (7.37),

$$D = \frac{Y}{y} = \frac{15.0}{0.280} = 53.57$$

$$X = xD = 0.440 \times 53.57 = 23.57$$

$$Z = zD = (1 - 0.440 - 0.280)53.57 = 15.00$$

Step 2 Look up values for D65. From Table 7.3, x = 0.3129, y = 0.3290 for D65. Then, from Eqs. (7.37),

$$D_n = \frac{100}{0.3290} = 303.95, \quad (Y_n = 100)$$

$$X_n = 0.3129 \times 303.95 = 95.11$$

$$Z_n = 0.3583 \times 303.95 = 108.91$$

Step 3 Find the tristinulus ratios.

$$\frac{X}{X_n} = \frac{23.57}{95.11} = 0.2478$$

$$\frac{Y}{Y_n} = \frac{15.00}{100.00} = 0.1500$$

$$\frac{Z}{Z_n} = \frac{15.00}{108.91} = 0.1377$$

Step 4 Find L*, a*, b*.
Since all of the ratios in Step 3 exceed 0.00856, Eqs. (7.47) apply.

$$L^* = 116(0.1500)^{1/3} - 16 = 45.64$$

$$a^* = 500[(0.2478)^{1/3} - (0.1500)^{1/3}] = 48.39$$

$$b^* = 200[(0.1500)^{1/3} - (0.1377)^{1/3}] = 2.99$$

Step 5 Find hue and saturation. From Eqs. (7.49),

$$h^* = \tan^{-1}\frac{2.99}{48.39} = 3.54°$$

$$c^* = \sqrt{48.39^2 + 2.99^2} = 48.48$$

A diagram relating L*, a*, b* to Munsell is shown in Fig. 7.16. To form this diagram, first create the a*, –a* and b*, –b* axes, as in Fig. 7.15. Scale each axis in steps of 10 from –50 to +50, and draw six circles of radii 10, 20, 30, 40, 50, and 60. These circles indicate the Munsell saturation (chroma) of the colors. Next draw radial lines at angles such that the 10 major Munsell hues and the 4 subsets of each hue are represented in terms of a* and b*.

It remains to include L*, which is directly related to Munsell value. For example, an L* of 50 is a Munsell value of 5. The three outer circles in Fig. 7.16 represent three ranges of L*. The inner most is for L* values from 60–90, the middle for L* values of 40–60 and the outer most for L* values of 20–40. The short slanted lines are included because, for a given a* and b*, Munsell hue is somewhat a function of lightness.

Locate the color in the example on the diagram in Fig 7.16. First mark the point a* = 48.39 and b* = 2.99. Next note that L* = 45.64 (a Munsell value of approximately 5) placing it on the middle circle (L* = 40 to 60). Now draw a line from the center of the diagram, through the marked point, to the middle value circle, and mark a second point. In this case, the drawn line coincides almost exactly with a Munsell hue of 7.5 RP, a purplish-red.

To obtain Munsell chroma, refer to Table 7.7 which relates c*, and Munsell hue and chroma.

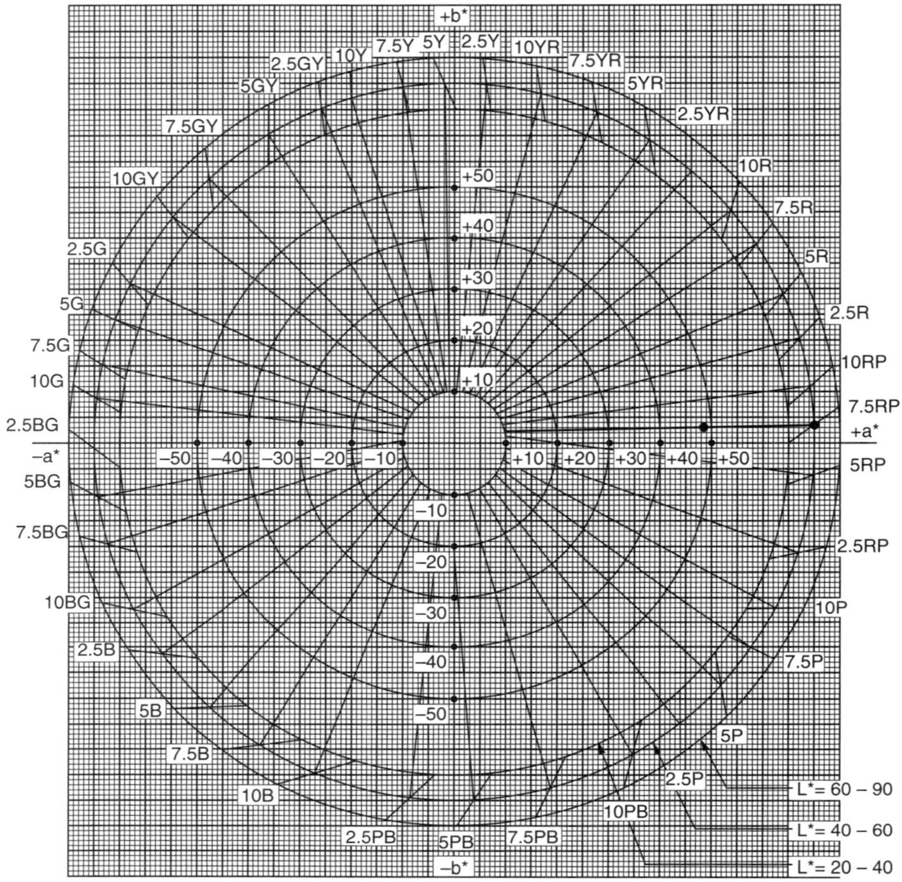

Figure 7.16 Relating L*, a*, b* and Munsell.

TABLE 7.7 VALUES OF c*

Munsell Chroma	Munsell Hue									
	R	YR	Y	GY	G	BG	B	PB	P	RP
1	5	7	8	7	6	5	4	4	4	4
2	9	12	15	13	11	10	9	9	9	9
3	15	17	22	21	16	15	13	13	13	13
4	19	24	29	27	22	20	17	17	18	18
6	29	36	44	41	33	28	26	26	27	26
8	39	48	58	55	44	38	34	34	36	35
10	49	60	72	69	54	49	42	43	45	44
12	59	72	86	82	—	—	—	52	53	51
14	69	85	98	—	—	—	—	66	62	60
16	84	98	109	—	—	—	—	75	—	—

For the example, c* = 48.48. Referring to the RP column in Table 7.7, the nearest Munsell chroma value is 12. Thus the color can be represented in Munsell notation as 7.5 RP5/12.

7.11 S/P RATIO

In Sect. 1.7, the two types of photoreceptors in the eye were introduced, and their properties were summarized in Table 1.3. Then in Sect. 3.2 it was noted that there are three overlapping levels of vision involving these receptors, namely photopic, mesopic, and scotopic. There has been significant interest in recent years by vision researchers and light-source manufacturers to learn more about the mesopic level of vision and about the role played by the rods and cones at all levels of vision.

Whether vision is peripheral, central, or both is governed by (a) the distribution of rods and cones throughout the retina and (b) their relative sensitivity to light. The distribution of photoreceptors in the retina is shown in Fig. 7.17. Rods are not present in the fovea (the 2° central field). They rise to a maximum concentration of 16×10^4 per mm^2 at 20° from the center of the retina and decline as the visual angle is increased further. Cones are concentrated in the fovea, but are also present outside, at a level of 0.5×10^4 per mm^2, about 5% of the average rod concentration.

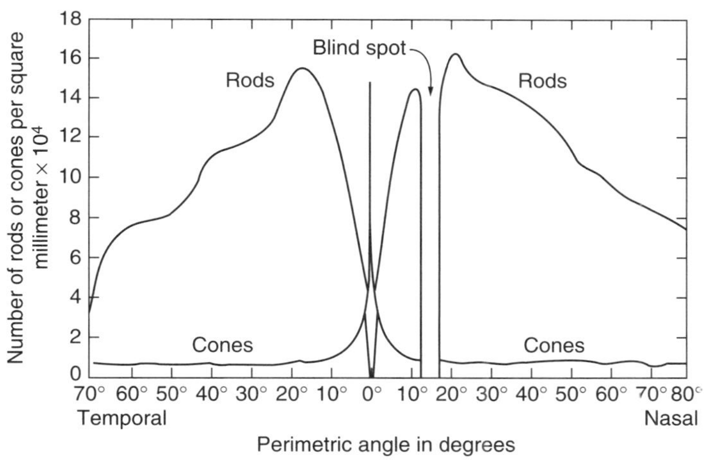

Figure 7.17 Distribution of rods and cones. ***Source*: *IESNA Lighting Handbook*, 9th edition, 2000. Courtesy of the Illuminating Engineering Society of North America.**

At luminance levels below 0.01 cd/m^2*, cones are not energized. Thus there is no central (foveal) vision and no perception of color. In the periphery, rods are energized, but because several hundred rods feed signals to a single ganglion cell, resolution of detail is poor, even though sensitivity to light level is high.

As luminance is increased to mesopic levels, between 0.01 and 3 cd/m^2, cones become energized and three things happen. First, central (foveal) vision commences

* As a rough rule of thumb, a luminance of 0.01 cd/m^2 corresponds to a luminous exitance of 0.003 lm/ft^2, which, for a floor or sidewalk reflectance of 30%, gives an illuminance of 0.01 fc, i.e. L in cd/m^2 and E in fc may be assumed numerically equal in such cases.

and increases until it dominates peripheral vision. Second, color discrimination commences and strengthens. Third, the peak of the spectral sensitivity curve of the eye shifts from 507 nm for scotopic vision to 555 nm for photopic vision, resulting in greater relative luminance of warmer colors (reds), as compared with cooler colors (blues). It should be noted that the spectral luminous efficiency curve for the eye for photopic vision was obtained from central 2° viewing (cones only), at levels of luminance above 3 cd/m^2, whereas the scotopic curve was measured for off-axis viewing about 3 mm from the fovea at detection levels (rods only).

The interest in mesopic vision and the performance of rods has quickened in two areas—typical levels of indoor lighting and low levels of nighttime exterior lighting. At interior levels, S. Berman and D. Jewett have proposed that, even though cone vision dominates, it is the rods that establish pupil diameter. This, in turn, influences visual acuity and the depth of field that can be focused, and, because the spectral luminous efficiency curve for scotopic vision is shifted toward the blue end of the spectrum from that for photopic vision, sources with higher color temperature (more bluish-white) cause smaller pupil size for a given luminance. The metric for this relationship is called S/P ratio (scotopic over photopic). This is the ratio of the photopic lumens a source produces to its scotopic lumens, where

$$\left.\begin{aligned} \phi_p &= 683\int_{380}^{780} P_\lambda\, V(\lambda)\, d\lambda \\ \phi_s &= 1700\int_{380}^{780} P_\lambda\, V'(\lambda)\, d\lambda \end{aligned}\right\} \tag{7.50}$$

The first Eq. (7.50) is identical with Eq. (4.20) and uses the data in Table 3.2 to give photopic lumens, and the second Eq. (7.50) uses the data from Table 3.3 to obtain scotopic lumens. The 683 constant for photopic vision is the lumens per watt obtained at 555 nm, and the 1700 constant for scotopic vision is the lumens per watt obtained at 507 nm.*

From Eqs (7.50), the following relationship is obtained by definition:

$$\frac{S}{P} = \frac{\phi_s}{\phi_p} \tag{7.51}$$

Values of S/P for typical sources are given in Table 7.8. The inference is that the higher the S/P ratio, the greater the visual acuity and depth of field. Also, although no one has been able to quantify it, spaces illuminated by cool sources (high CCT) appear to have better visual clarity than those illuminated by warmer sources (low CCT) and the sources appear brighter, for the same illuminance level.

From their research, Berman and Jewett have developed an empirical relationship to compare lamps in their ability to produce clear vision in various interior environments. To this end, the term "visually effective lumens" is defined as

$$VE\phi = \phi_p (S/P)^k \tag{7.52}$$

The parameter k depends on the task environment. For example, in a space dominated by paper tasks, where the lighting provides both ambient and task illumination, scotopic content is of some importance, and research has shown empirically that $k = 0.78$

* The peak of the scotopic curve is 2.489 times larger than the peak of the photopic curve. Thus $683 \times 2.489 = 1700$.

TABLE 7.8 S/P RATIOS

Source	S/P
Low pressure Sodium	0.23
High Pressure Sodium	0.62
White High Pressure Sodium	1.14
Clear Mercury	0.80
Metal halide	1.49–2.10
Incandescent	1.41
Halogen	1.50
Warm White Fluorescent	1.00
Cool White Fluorescent	1.46
Daylight Fluorescent	2.22
4100K Triphosphor Fluorescent	1.54–1.62
5000K Triphosphor Fluorescent	1.96
6500K Triphosphor Fluorescent	2.14

is optimum to provide clear vision. On the other hand, in a computer space, where glare must be minimized but the surround should not be so dimly-lit that pupils dilate, scotopic content is dominant, leading to an empirical value of $k = 1.0$.

EXAMPLE 7.11

Consider the first of the two spaces just discussed, and compare the lighting provided by two 32-watt T8 fluorescent lamps:

a) 3500 K, 85 CRI, 2950 lumens, S/P = 1.40
b) 5000 K, 85 CRI, 2950 lumens, S/P = 1.96

Next consider the second of the two spaces and compare the lighting provided by the same two lamps.

Solution. For the first space, $k = 0.78$ and Eq. (7.52) yields

a) $VE\phi = 2950(1.40)^{0.78} = 3835$ lm
b) $VE\phi = 2950(1.96)^{0.78} = 4985$ lm

Thus lamp (b), while providing the same photopic lumens as lamp (a), produces 30% more visually-effective lumens.

For the second space, $k = 1.0$, giving

a) $VE\phi = 2950 \times 1.40 = 4130$ lm
b) $VE\phi = 2950 \times 1.96 = 5780$ lm,

a 40% increase, again in favor of lamp (b).

The implications for energy saving from using higher CCT sources are significant. If the research results can be replicated elsewhere, they could lead to reductions in illuminance levels in interior spaces without impairing seeing.

The foregoing was presented with the view toward indoor lighting at photopic levels. Do the same conclusions apply at night to outdoor lighting? For straight-ahead viewing, lighting levels at night are generally low-level photopic, cone vision is active and the conclusions reached for indoor lighting apply. For the peripheral field, mesopic levels usually pertain, with both rods and cones participating. Thus a mix of the photopic and scotopic spectral luminous efficiency curves applies, and use of conventional photopic illuminance as a measure of seeing is questionable. Cooler sources would seem to be preferred outdoors when peripheral viewing is important. It has been suggested (Advanced Lighting Guidelines) that a lumen multiplier of 1.2 to 1.4 be applied to cool arc discharge sources (metal halide and high CCT fluorescent) to represent more properly their effectiveness in producing nighttime visibility.

REFERENCES

Berman, S.: "The Coming Revolution in Lighting Practice," *Energy User News*, October, 2000.

Berman, S.M., D.L. Jewett, B.R. Benson and T.M. Law: "Despite Different Wall Colors, Vertical Scotopic Illuminance Predicts Pupil Size," *Journal of IESNA*, New York, Summer, 1997.

Berman, S.M.: "Energy Efficiency Consequences of Scotopic Sensitivity," *Journal of IESNA*, New York, Winter, 1992.

Committee on Colorimetry: *The Science of Color*, Optical Society of America, Washington, D.C., 1953.

General Electric Co.: *Specifying Light and Color*, GE Lighting, Nela Park, Cleveland, 1978, 1994.

Gerritsen, F.: *Color*, Van Nostrand Reinhold, New York, 1975.

Glasford, G.M.: *Fundamentals of Television Engineering*, McGraw-Hill, New York, 1955.

GTE Sylvania: "The Color Rendering Index," GTE Products Corp., Danvers, Mass., 1972.

Hazeltine Corporation: *Principles of Color Television*, Hazeltine Corp., Little Neck, Long Island, New York, 1952.

Jerome, C.W.: "Basic Colorimetry for the Lighting Industry," GTE Lighting Products, Salem, Mass. 1969.

Kaiser, P.K.: "Photometry of Brightness: Problems, Physiology and Models," *Journal of Light and the Visual Environment*, 1984.

Kelly, K.L., and D.B. Judd: *Color-Universal Language and Dictionary of Names*, National Bureau of Standards, Washington, D.C., 1976.

Michaels, E.L.: "Basic Colorimetry for Color Television," *Electrical Engineering*, November 1956.

Minolta: *Precise Color Communication*, Minolta, Osaka, Japan.

Munsell: *Munsell Color*, Macbeth, Baltimore, MD, 1990.

Pantone: *Wherever There's Color*, Pantone Inc., Moonachie, NJ, 1990.

Rea, M.S. editor: *IESNA Lighting Handbook*, Ninth Edition, IESNA, New York, 2000.

Roberts, J., editor: *Advance Lighting Guidelines*, New Buildings Institute, White Salmon, WA, 2001.

Wald, G.: "The Receptors of Human Color Vision," *Science*, vol. 145, September 1964.

Waymouth, J.F., and R.E. Levin: *Designers Handbook—Light Source Applications*, GTE Products Corp., Danvers, Mass., 1980.

Wintringham, W.T.: "Color Television and Colorimetry," *Proceedings of IRE*, vol. 39, no. 10, October 1951.

Wright, W.D.: *The 1931 and 1964 CIE Systems of Colorimetry: Their Significance to Colorimetrists Today*, AATCC International Conference, Charlotte, NC, 1987.

Wyszecki, G., and W.S. Stiles: *Color Science*, Wiley, New York, 1967.

Chapter 8

Luminaires and Controls for Interior Lighting

8.1 INTRODUCTION

A luminaire is a complete lighting unit, consisting of lamp(s), ballast(s), optical components, support hardware, and an enclosure and wiring. Its function is to position the lamp so as to control and distribute the light, placing the light where it is wanted and keeping it from where it is not wanted. It must do this in an efficient manner.

Luminaires may be integrated with room architecture, recessed in a ceiling, suspended from a ceiling, surface mounted on a ceiling or wall, or mounted in room furniture. They can distribute light upward, downward, up and down, and laterally, the latter to provide wall lighting.

In today's luminaire design, efficiency, in both an electrical and an optical sense, is of prime importance. High-efficacy lamps, efficient (electronic) ballasts, and common-sense electrical controls are required to minimize energy consumption. Quality reflective materials, well-designed refractors, and judicious shielding mechanisms are necessary to insure that the light is placed properly and does not produce glare.

In the sections to follow, the components and properties of luminaires will be explored in detail.

8.2 INTERIOR LIGHTING: PAST AND PRESENT

The earliest electric lighting in school rooms, offices, and industry was simply a bare-bulb in a socket, the latter dangled from a ceiling, with a wire festooned across the ceiling to whichever wall possessed an electrical outlet. This was replaced in the 1920s by prismatic or translucent globes, suspended from the ceiling and housing incandescent lamps. Such units provided lumens both directly and indirectly (through reflections from room surfaces) to the work plane. Glass-enclosing globes tended to have high luminances and produced considerable glare in workers' eyes, but less than from bare bulbs.

Totally indirect incandescent lighting appeared in the 1930s. Pan-shaped or concentric-ring luminaires, each with a half-silvered lamp mounted base up in a hole in the center of the unit, redirected the lamp lumens to the ceiling. The ceiling, in essence, became the light source. These indirect units produced high-quality glare-free lighting, but were inherently very inefficient, both from their use of incandescent lamps and from the fact that no lumens traveled directly to the work plane. Also, because many rather large lamps were required in a given space to provide adequate work-plane illuminance, considerable heat (infrared) was generated, which often caused the space to be uncomfortable thermally.

The advent of the fluorescent lamp in the late 1930s initiated a renaissance in interior lighting after the Second World War. These lamps had much lower luminances than their incandescent counterparts and thus it was no longer necessary to send all of the lamp lumens to the ceiling for redirection downward. Instead, with suitable reflectors, baffles, and lenses most of the lumens could be sent directly downward, with minimal glare. In addition, the fluorescent lamp had about four times the efficacy of the incandescent lamp. As a result, 70 fc of fluorescent lighting could be provided more efficiently than 30 fc of incandescent lighting. Fluorescent lighting gradually replaced incandescent lighting until, by the 1960s, almost all general interior lighting in offices and schools was fluorescent. Illuminance levels were high, electricity was inexpensive, and power consumption of 3 to 4 W/ft^2 was the norm. The most popular lighting fixture was the four-foot long recessed troffer, containing up to four 40-W T12 fluorescent lamps.

The advent of metal halide and high-pressure sodium lamps in the 1960s, coupled with the energy crisis in the early 1970s, caused several changes in interior lighting. These lamps, although concentrated and of high luminance like incandescent, had efficacies seven or more times as great. Thus it became economically feasible once again to use indirect lighting in interior spaces. Also, suspended direct-indirect two-lamp fluorescent luminaires appeared. Concurrently, in an effort to reduce energy consumption, illuminance levels were lowered, and interest in so-called task-ambient lighting systems began. Such systems, instead of providing a reasonably uniform illuminance over the entire work-plane area, sought rather to provide a base level of illuminance throughout a space, supplemented by additional illuminance at the task locations.

Among the many trends in interior luminaires and lighting from the 1980s to the present are the following:

1. Reduction in watts per square foot, from 3–4 to 1–2. This has been possible because of higher-efficacy fluorescent lamps, electronic ballasts, reduced illuminance levels, more efficient luminaires, and improved electrical controls.
2. With the advent of the computer in the workplace, a shift from horizontal to vertical tasks, with the latter self-illuminated. The importance of reducing reflections in computer screens calls for the use of indirect or louvered-direct luminaires.
3. Utilization of daylight, with sensors to extinguish the electric light when not needed.
4. Improved luminaire optical design, including reflectors, louvers, baffles, lenses, fritted glass, optical coatings, and perforated shields.
5. Integration of lighting into the architecture through cornices, valances, wall brackets, soffits, canopies, and coves, and into office furniture (kiosks).
6. Utilization of lamps with high CRIs and a range of CCTs from 3000 to 5000 K.
7. Sparing use of incandescent, hardly at all in offices, schools, and industry and only for displays in retail. PAR, R, and A lamps in downlights and wall sconces have been replaced by compact fluorescent lamps.
8. Wall lighting, which makes a space feel more spacious and less confining. It is accomplished through asymmetric down lighting, indirect lighting, and architectural elements.
9. An (unwelcome) use of bare-lamp luminaires with high output fluorescent lamps in high-ceiling retail establishments, such as discount houses and supermarkets. This lighting suggests to customers high-volume, low-price, and quick in-and-out shopping.
10. Large, open-plan office areas with partitioned spaces having a myriad of louvered fluorescent luminaires uniformly arranged and spaced. It is difficult to

custom design lighting for such spaces because the workstations often face in several directions and are not always fixed in location.

11. Use of T5 and smaller diameter fluorescent lamps, which have high luminances, and thus should never be viewed directly. Suitable fixtures are indirect and perforated direct-indirect.
12. Limited use of high pressure sodium lighting. Its use in schools during the early days of the energy crisis has not continued and its use in interiors is limited now to gymnasiums and in high-bay industrial settings.

8.3 LAMPS FOR INTERIOR LIGHTING

Lamps were discussed in detail in Chap. 6. The purpose here is to present briefly those attributes of lamps that the lighting designer should consider in choosing a lamp for an interior lighting application. Among these are:

1. What is its luminous efficacy? The lamp, in conjunction with the lighting system, must provide the required illuminances economically.
2. What is the life of the lamp? How difficult is it to replace burned-out lamps? Is group replacement of lamps the better choice economically?
3. What is the lumen maintenance of the lamp? Is it important that the level of illuminance never drop below a prescribed minimum?'
4. Is color a factor? Although all the lamps produce "white" light, their CCTs and CRIs differ. How important is it for the colors of the seeing task and its surroundings to be faithfully reproduced?
5. What auxiliary equipment is required? As has been seen, all gas discharge light sources require ballasts, whereas incandescent sources do not. The ballast used can affect lamp output, life, and starting reliability, as well as system efficiency and occupant comfort.
6. What other miscellaneous factors are present in the particular environment? Is temperature a problem? Must the area be free from stroboscopic effects? Will electromagnetic interference disturb the activities going on in the space? Is dimming included? Are fumes present that could produce corrosion or an explosive atmosphere?

Table 8.1 gives a comparison of the first four of these factors for the common lamp types used in interior lighting. Several comments are included on each listing. As noted in Sect. 6.1, lamp efficacy does not include ballast losses and thus, except for incandescent and halogen lamps, system efficacy will be lower than lamp efficacy. Also the efficacies given are based on initial lumens and watts. Except for halogen lamps, efficacy decreases as a lamp ages, because of lumen depreciation.

Lamp life normally implies the point at which 50% of a large group of lamps of the same kind fail. But it can also be taken as the end of "useful" life, the point at which the lamp efficacy and/or lumen output of a lamp have dropped so far that the lamp is no longer viable economically and/or from an illuminance standpoint. Also lamp life says nothing about the rate at which lamps fail, i.e., a lamp's mortality curve.

All lamps suffer lumen depreciation. It is particularly severe for standard incandescent and metal halide lamps, but is very slight for tungsten halogen lamps because

TABLE 8.1 COMPARISON OF LAMP TYPES

Lamp Type	Lamp Efficacy in lm/W	Life (h)	% Lumen Depreciation	CCT (K)	CRI
Standard Incandescent	5–18	750–2000	12–18	2650–2950	100
Tungsten Halogen	15–25	2000–6000	<5	2750–3050	100
High-wattage CFL	50–80	10000–20000	14–17	3000–5000	82
Low-wattage CFL	20–55	10000	14–17	2700–5000	82
Linear Fluorescent	65–95	7500–20000	5–10	2700–5000	60–90
Metal Halide	45–100	7500–20000	22–36	2900–4400	65–85
High Pressure Sodium	45–120	16000–24000	9–12	1900–2100	22

of the halogen cycle. Lumen depreciation is based on initial lumens, which are at zero hours for incandescent sources and at 100 hours for gaseous discharge sources, and on lumens at a certain percentage of rated lamp life: 50% for incandescent, tungsten halogen, and high pressure sodium lamps, and 40% for fluorescent and metal halide lamps.

Correlated color temperatures are very low for high pressure sodium lamps, manifested by their yellowish-white appearance. The CCTs of standard incandescent lamps increase as lamp wattage increases. The 40-watt lamp operates at 2650 K, while the 500-watt lamp operates at 2950 K. Tungsten halogen lamps are usually operated a few hundred degrees hotter than incandescent lamps, providing slightly higher CCTs. Low wattage CFLs generally have CCTs of 2700 K, 3500 K, 4100 K, or 5000 K. High wattage CFLs are essentially the same, except that the lowest CCT is 3000 K. Low-wattage metal halide lamps have CCTs of 2900–3000 K; higher wattage lamps have CCTs of 3600–4400 K.

Because incandescent and tungsten halogen lamps essentially lie on the black-body locus on the CIE diagram, their CRIs are 100. CFLs, regardless of wattage and CCT, have CRIs of 82. On the other hand, linear fluorescent lamps are available in several CRIs. Triphosphor lamps have CRIs in the 70s, 80s, and 90s, whereas halophosphate lamps tend to be lower, in the 50s to 70s. However all manufacturers have a 5000 K non-triphosphor lamp with a CRI of 90 for exacting color work. The CRIs of metal halide lamps are 65, 70, or 75, but there is a high color rendering metal halide lamp with a CRI of 85. High pressure sodium lamps (CRI of 22) render colors very poorly.

8.4 LUMINAIRE CHARACTERISTICS

What is it that the designer needs to know about a luminaire in order to decide intelligently which luminaire to choose for a given interior lighting application?

Luminaire Flux Distribution

Recall that in Sect. 2.9 the use of a goniophotometer was explained in obtaining the intensity distribution curve of a lamp or luminaire. It was also shown how Eqs. (2.43–2.45) can be used to calculate the overall lumens emitted and the lumens emitted in certain angular zones.

Generally, in luminaire selection, the factors of interest are the percentage of the total lumen output that is directed upward toward the ceiling, the percentage that is directed downward toward the work plane, and the directivity of the downward component. The latter factor is generally quantified in terms of the percentage of the luminaire lumens in the 0 to 40° zone.

The CIE has devised a six-category classification system for interior luminaires, which is presented in Table 8.2. This system classifies luminaires according to the percent of the luminaire output emitted above and below the horizontal. The shape of the upward and downward distributions is not considered, except in the cases of general diffuse and direct-indirect, where the latter classification implies very little luminous flux at angles near the horizontal. With this one exception, the intensity distribution curves of luminaires within a given category may take many forms, as long as the overall upward and downward lumen percentages fall within the category's bounds.

TABLE 8.2 LUMINAIRE CLASSIFICATION

Category	% Upward Lumens	# Downward Lumens
Direct	0–10	100–90
Semidirect	10–40	90–60
General diffuse	40–60	60–40
Direct-indirect	40–60	60–40
Semi-indirect	60–90	40–10
Indirect	90–100	10–0

Examples of representative luminaires and their intensity distributions are shown in Figs. 8.1–8.4. The distributions in Fig. 8.1 provide all of the luminaire lumen output directly downward. In Fig. 8.1a, a diffuse downward cosine distribution is shown. The luminaire consists of a white recessed half-cylinder above a smaller perforated half-cylinder. A cosine distribution can also be obtained from a recessed incandescent luminaire with a translucent glass lens.

The baffled and louvered linear fluorescent luminaires in Figs. 8.1b and c, the HID metal-reflector luminaire in Fig. 8.1d, and the compact fluorescent downlight in Fig. 8.1e all have similar batwing distributions, with less intensity directly downward compared with that at 30°–45°. The luminaire in Fig. 8.1b is 2 ft by 4 ft; that in Fig. 8.1c 1 ft × 4 ft. The HID luminaire in Fig. 8.1d is open, with a spun-aluminum reflector. The compact fluorescent downlight has a spun specular reflector.

The luminaire in Fig. 8.1f is a recessed troffer with a prismatic lens; that in Fig. 8.1g is a pendant downlight containing either an incandescent or compact fluorescent lamp. Each has a distribution that is narrower than a cosine and is non-batwing. The distribution in Fig. 8.1h is narrower still and slightly batwinged. It is generated by the narrow parabolic louvered fluorescent troffer shown, with good shielding for VDT applications, or by a deep recessed downlight.

The last two luminaires, in Figs. 8.1i and j, generate an asymmetric distribution for wall washing. The luminaire in Fig. 8.1i is a downlight with a specular reflector on one side to direct lumens to the wall area. The fluorescent luminaire in Fig. 8.1j has a linear asymmetric specular reflector to accomplish the same purpose.

The luminaires in Fig. 8.2 have both upward and downward intensity components. A bare fluorescent strip, such as shown in Fig. 8.2a, directs 10% to 40% of its light upward, and its large downward distribution is nearly a cosine. The fluorescent wraparound luminaire in Fig. 8.2b has a prismatic lens that creates a downward batwing

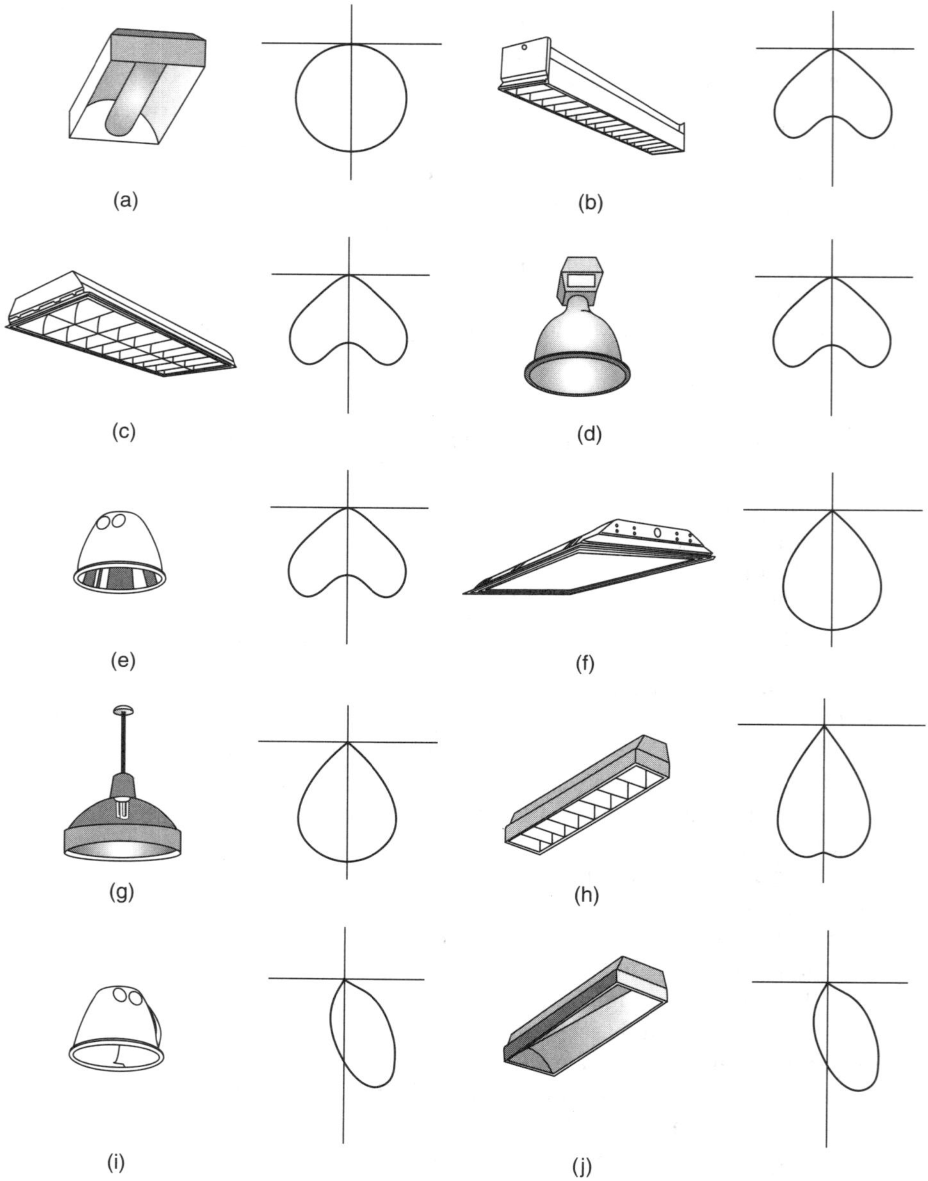

Figure 8.1 Direct luminaires. ***Source*: *IESNA Lighting Handbook*, 9th edition, 2000 and *Advanced Lighting Guidelines*, 2001 edition. Courtesy of the Illuminating Engineering Society of North America and New Buildings Institute.**

distribution. It also has a small upward batwing component. The pendant translucent globe in Fig. 8.2c generates a nearly uniform intensity distribution. The luminaire in Fig. 8.2d has the inverse distribution of that in Fig. 8.2b, with the large batwing portion upward. Its bottom is louvered, and its top is open.

This luminaire and the luminaire in Fig. 8.2e are suspended direct-indirect units, the difference being that the luminaire in Fig. 8.2e has a largely enclosed top and thus

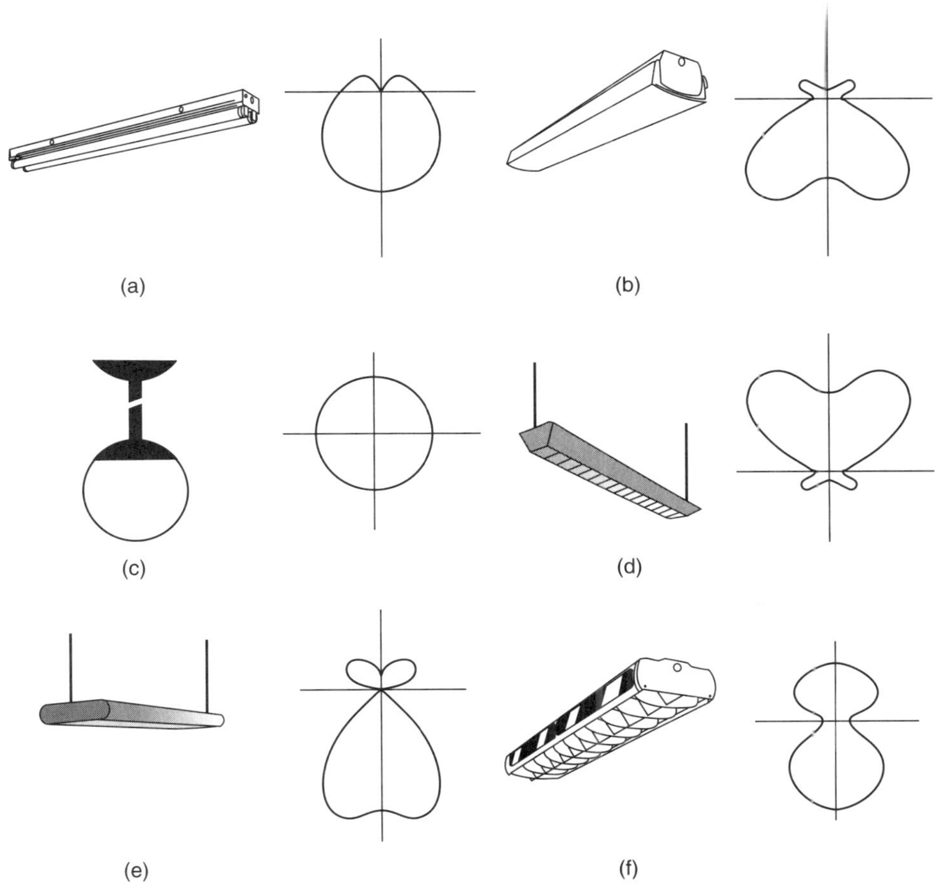

Figure 8.2 General diffuse and direct-indirect luminaires. ***Source: IESNA Lighting Handbook*, 9th edition, 2000 and *Advanced Lighting Guidelines*, 2001 edition. Courtesy of the Illuminating Engineering Society of North America and New Buildings Institute.**

directs most of its batwing distribution downward. The luminaire in Fig. 8.2f is a louvered commercial unit, with translucent plastic sides. It is a direct-indirect luminaire with a small intensity component at 90°.

The luminaires in Fig. 8.3 are completely indirect. The unit in Fig. 8.3a is a suspended linear fluorescent luminaire and can produce either a cosine or wide batwing intensity distribution. The luminaire in Fig. 8.3b is a decorative indirect pendant unit, housing compact or circular fluorescent lamps. Its distribution is an upward batwing. A wall-mounted up light containing T5 or T8 fluorescent lamps is shown in Fig. 8.3c. It provides an asymmetric upward intensity distribution for wall washing.

Six architectural lighting elements are shown in Fig. 8.4, each designed to hold one or more fluorescent lamps. The cornice produces a downward wall-washing intensity distribution, similar to that provided by the luminaires in Figs. 8.2i and j. A broader form of this distribution is produced by a soffit, which is wider than a cornice and has a bottom diffuser, and by a canopy, with the latter providing also a small upward component.

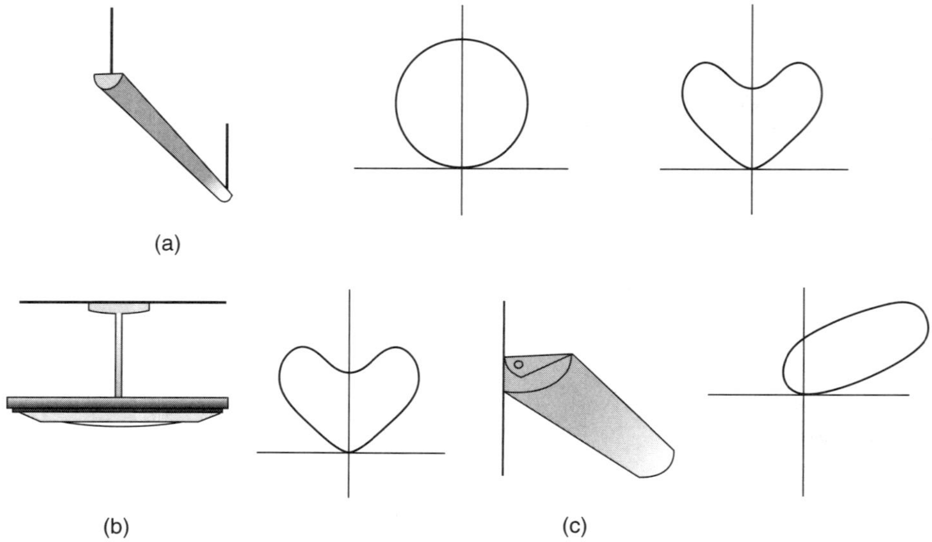

Figure 8.3 Indirect luminaires. ***Source*: *IESNA Lighting Handbook*, 9th edition, 2000 and *Advanced Lighting Guidelines*, 2001 edition. Courtesy of the Illuminating Engineering Society of North America and New Buildings Institute.**

A cove provides upward light, to the wall and ceiling, similar to that from the wall bracket in Fig. 8.3c. Finally, the valance and the wall bracket provide an asymmetric up-down distribution for wall or drapery lighting.

Luminaire Efficiency

This is the ratio of the lumens emitted by the luminaire to those emitted by the lamps. The lower this number, the more luminous flux is being trapped and absorbed by the luminaire. Generally efficiencies of less than 65% should cause one to take a second look before selecting that luminaire. Lenses, baffles, louvers, shields, reflectors, and refractors all tend to reduce luminaire efficiency. Whereas a bare fluorescent strip will likely have an efficiency of 95%, the efficiency of a prismatic lens wraparound luminaire will lie between 70% and 75% and that of a deep-cell louver fluorescent troffer between 55% and 65%. It is important to understand the distinction between luminaire efficiency and the percentages in Table 8.2. The percent upward and downward lumens exiting the luminaire always sum to 100, regardless of the luminaire efficiency.

The National Electric Manufacturers Association (NEMA) has developed a metric to aid in determining the energy efficiency of fluorescent luminaires.* It is called Luminaire Efficacy Rating (LER) and is defined by

$$\mathrm{LER} = \frac{\eta \times \phi_L \times \mathrm{BF}}{P_{IN}} \ \mathrm{lm/W} \tag{8.1}$$

* This metric also includes HID industrial luminaires and non-residential downlights.

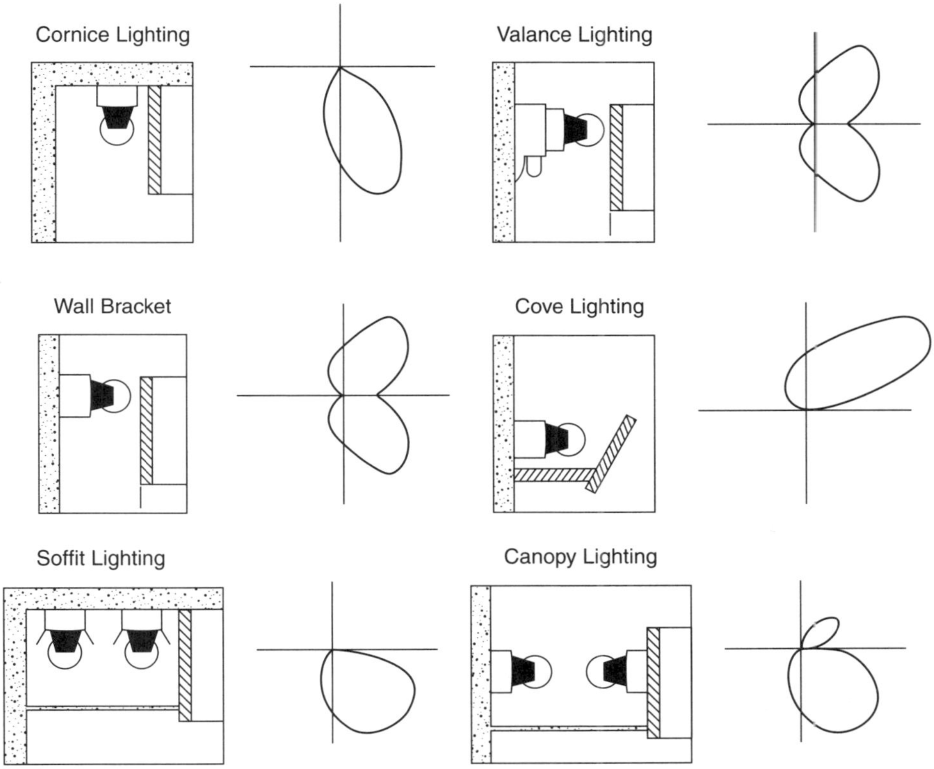

Figure 8.4 Architectural lighting elements. *Source*: *IESNA Lighting Handbook*, 9th edition, 2000 and *Advanced Lighting Guidelines*, 2001 edition. Courtesy of the Illuminating Engineering Society of North America and New Buildings Institute.

where η is luminaire efficiency (decimal)
ϕ_L is total initial rated lamp lumens
BF is ballast factor
P_{IN} is luminaire input watts.

The numerator of this expression is the total rated lumens leaving the luminaire initially. Thus LER is the ratio of the total rated initial lumen output of the luminaire to the total power input in watts.

Once LER is determined, an economic metric, yearly lighting cost may be calculated. This is the annual cost of energy per 1000 lumens from the given luminaire. It is based on 3000 hours of burning and an energy rate of 8 cents per kilowatt-hour. The formula for the calculation is

$$C = \frac{240}{LER} \tag{8.2}$$

where C is the annual lighting energy cost in dollars. The numerator of Eq. (8.2) is obtained from

$$\frac{0.08 \times 3000 \times 1000}{1000} = 240$$

In using Eqs. (8.1) and (8.2), it is important that similar luminaires are compared. To this end, NEMA has identified five categories of fluorescent luminaires and denoted each with a two-letter identification label. These are shown in Table 8.3.

TABLE 8.3 LUMINAIRE CATEGORIES

Type	Designation	Benchmark LER
2 × 4 Lensed troffers (2, 3 and 4 lamp)	FL	41–47
2 × 4 Parabolic troffers (2, 3 and 4-lamp)	FP	47–53 (2-lamp) 41–47 (3-lamp) 37–46 (4-lamp)
4-ft Wraparounds (2 and 4-lamp)	FW	46–48
4-ft Strips (1 and 2-lamp)	FS	50–56
4-ft Industrials (2-lamp)	FI	53

These five categories account for 80% of the sales of commercial and industrial fluorescent luminaires. LER values should be compared only within a given category. For example, it is appropriate to compare the LER values of two wraparounds, but not the LER values of a wraparound and a lensed troffer.

For each category, NEMA has established benchmark LERs for acceptable energy efficiency. These are to be considered minimum required LERs. Within a category, the benchmark LER may vary based on the number of lamps in the luminaire and, in the case of louvered units, the shielding angle.* For example, in all categories (except FI) the luminaire with the fewest lamps has the highest minimum LER. In addition, in the FP category, as the shielding angle decreases, the minimum LER increases. The reason for these effects is that more light is trapped in a luminaire as the number of lamps increases and as the shielding angle increases.

The lighting community has expressed concern that LER might be applied indiscriminately without sufficient concern for the quality of the lighting in a space. The only quality metric currently available is visual comfort probability (VCP)** and it has several deficiencies. A joint committee of the Illuminating Engineering Society of North America (IESNA) and the International Association of Lighting Designers (IALD) is working to develop a new quality metric to accompany LER.

A final point to be made is that LER is for comparing ceiling-mounted fluorescent luminaires. As a case in point, it would make no sense to compare the LERs of two luminaires whose optics, and therefore usage, are entirely different, for example an MR16 tracklight and a CFL wall sconce.

* This is the greatest angle between the horizontal and the line of sight at which the lamps are not visible through the louvers.

** To be covered in Sect. 8.8.

EXAMPLE 8.1

Consider two 2-lamp FP luminaires, one containing 32-watt T8 lamps and the other 34-watt T12 lamps. Each luminaire uses an energy-saving magnetic ballast. The data are

$$32\text{ W} \qquad \eta = 80\%, \quad BF = 0.94, \quad \phi = 2850\text{ lm}, \quad P_{in} = 70\text{ W}$$

$$34\text{ W} \qquad \eta = 78\%, \quad BF = 0.87, \quad \phi = 2650\text{ lm}, \quad P_{in} = 70\text{ W}$$

Find LER and C for each luminaire.

Solution. From Eq. (8.1)

$$\underline{32\text{ W}} \qquad LER = \frac{0.80 \times 5700 \times 0.94}{70} = 61.2\text{ lm/W}$$

$$\underline{34\text{ W}} \qquad LER = \frac{0.78 \times 5300 \times 0.87}{70} = 51.4\text{ lm/W}$$

Then from Eq. (8.2)

$$\underline{32\text{ W}} \qquad C = \frac{240}{61.2} = \$3.92/\text{yr}$$

$$\underline{34\text{ W}} \qquad C = \frac{240}{51.4} = \$4.67/\text{yr}$$

It is seen that the 32-W system is about 20% more energy-efficient than the 34-W system.

Luminaire Luminance

Bare lamps are usually too bright and produce too much glare. Thus one of the functions of a luminaire is to filter and shield the bare lamp from the viewer so as to provide a visually comfortable seeing environment.

Luminaire manufacturers are encouraged to include maximum and average luminances in the data provided to the customer. Maximum luminance is a measured quantity and is the luminance of the brightest square inch of a luminaire at a particular viewing angle. Generally several values of maximum luminance are given within the 45 to 90° zone, where direct glare is likely to be bothersome, for both crosswise (perpendicular to the luminaire) and lengthwise (along the luminaire) viewing. The direct and indirect glare zones are shown in Fig. 8.5.

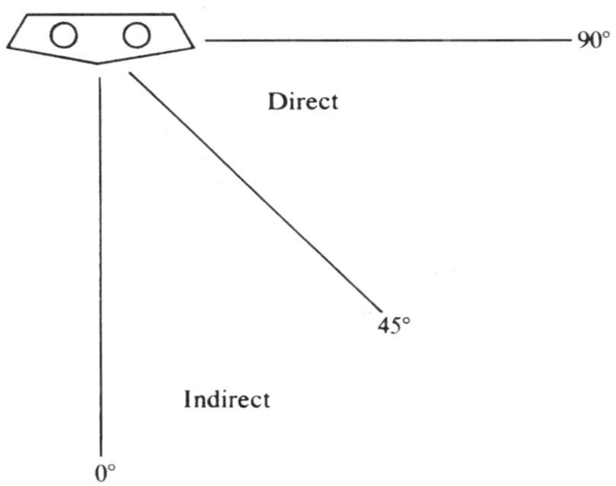

Figure 8.5 Glare zones.

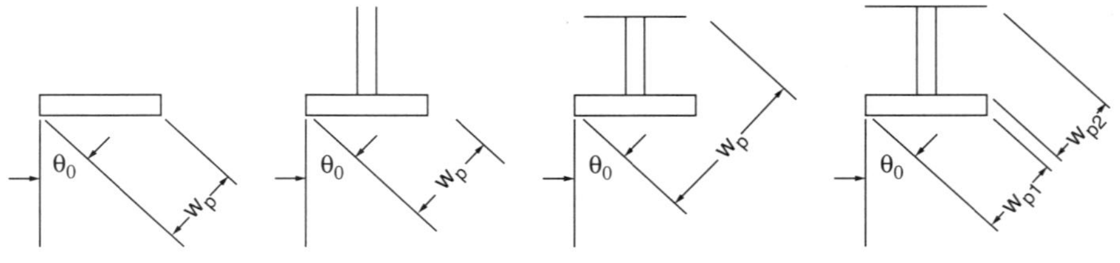

Figure 8.6 Projected areas of basic luminaires.

Average luminance in a particular viewing direction is a calculated quantity. It is obtained by dividing the intensity of the luminaire in the given direction by the projected area of the lighted portion of the luminaire in that direction.

Four types of luminaires that can cause glare are shown in Fig. 8.6. Luminaire (a) is surface mounted and has dark side panels. Luminaire (b) is suspended, but has no upward component. It does have luminous side panels. Luminaire (c) is also suspended, but it has an upward component and thus its effective lighted width is greater than that of the luminaire itself. Luminaire (d) is like luminaire (c) except that it has dark side panels. Thus, its effective lighted width is split into two parts, a luminaire portion and a ceiling portion.

Assume each luminaire in Fig. 8.6 has a length l and a projected lighted width w_p in the direction of an observer who is viewing the luminaires crosswise. If the average intensity of the luminaire lighted area is I in the direction of the observer, then the average luminance presented in that direction is

$$L_{ave} = \frac{I}{w_p \times l} \equiv \frac{I}{A_{p0}} \tag{8.3}$$

where A_{p0} is the crosswise projected area (in Fig. 8.6a, $A_{p0} = A \cos\theta_0$, where A is the lighted area of the luminaire).

A more general case is shown in Fig. 8.7. The viewer is at point O. Parallel rows of luminaires, one of which is indicated by the line from S_0 to S, are located on the ceiling. Consider a single luminaire centered at point S_0 and having a projected area A_{p0} in the direction of O. This is the same situation as in Fig. 8.6 and yields the luminance given by Eq. (8.3).

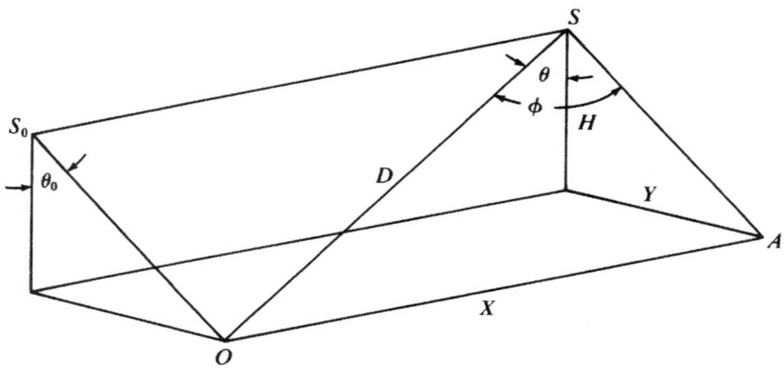

Figure 8.7 General case of luminaire viewing.

Now let the luminaire slide down the row from S_0 to S. The width of the projected area does not change, but its length does. The new projected area in the direction of O is

$$A_p = A_{p0} \cos \phi \tag{8.4}$$

It is possible to express ϕ in terms of the polar angles at S_0 and S.

$$\cos \phi = \frac{\sqrt{Y^2 + H^2}}{D}, \quad \cos \theta_0 = \frac{H}{\sqrt{Y^2 + H^2}}, \quad \cos \theta = \frac{H}{D} \tag{8.5}$$

which yields

$$\cos \phi = \frac{\cos \theta}{\cos \theta_0} \tag{8.6}$$

Assuming the luminaire at S has an average intensity I in the direction of O, the average luminance of that luminaire as seen at O is

$$L_{ave} = \frac{I}{A_p} = \frac{I \cos \theta_0}{A_{p0} \cos \theta} \tag{8.7}$$

It is usually considered that direct glare will not be a problem in a lighting installation if the ratio of the maximum to average luminaire luminance for both crosswise and lengthwise viewing does not exceed 5:1 within the 45 to 90° glare zone, and if the maximum luminaire luminances are less than the values in Table 8.4.*

TABLE 8.4

Angle from Nadir	Max L cd/m^2 (cd/in^2)
45	7710 (5.0)
55	5500 (3.5)
65	3860 (2.5)
75	2570 (1.7)
85	1695 (1.1)

EXAMPLE 8.2

Assume a 2-ft × 4-ft lensed troffer has the following intensity values when viewed crosswise:

Angle above Nadir (degrees)	Intensity (cd)
45	1048
55	874
65	685
75	508
85	244

* There is an additional requirement relating to visual comfort that is presented in Sect. 8.8.

Does this luminaire meet the criteria for direct glare not to be a problem?

Solution. Consider $\theta_0 = 45°$ first. From Eq. (8.3), $w_p \times l = A \cos \theta_0$, giving

$$L_{ave} = \frac{1048}{8 \times 144 \cos 45} = 1.28 \text{ cd/in}^2 = 1990 \text{ cd/m}^2$$

Similar calculations yield 2050 cd/m^2 at 55°, 2180 cd/m^2 at 65°, 2640 cd/m^2 at 75°, and 3760 cd/m^2 at 85°. The luminaire meets the criteria at all but 85°, indicating that those toward the rear of a room might find a luminaire in the front of the room disturbing.

Glare in areas where computers are viewed needs to be addressed. In such spaces, nonuniformity of ceiling luminance can produce disturbing brightness in computer screens. The IESNA recommends that the L_{max}/L_{min} ratio on the ceiling not exceed 4 to 1 in VDT intensive spaces and 8 to 1 in VDT and paper spaces. In addition, the maximum ceiling luminance should not exceed 850 cd/m^2 (0.55 cd/in^2), or 10 times the average screen luminance, whichever is smaller. If maximum ceiling luminance is less than 425 cd/m^2 (0.27 cd/in^2), there should be no problem with reflections in computer screens, and thus uniformity will not be an issue. With regard to ceiling-mounted direct-lighting luminaires in computer spaces, the average luminaire luminances should not exceed 1/5 to 1/10 of the values in Table 8.4.

8.5 LUMINAIRE OPTICS AND LIGHT CONTROL

It has been stated that bare-lamp lighting installations are undesirable. They are sources of glare and do not distribute luminous flux effectively. There are many methods of light control used in contemporary luminaires. It is the purpose here to identify them and describe their attributes. The technical (ray-tracing) aspects of optical control will be discussed in Chap. 10.

Diffusers

These are clear prismatic lenses, translucent sheets, or combinations thereof. They are used on the bottom and sides of a luminaire to redirect and scatter the light, to obscure the lamps inside and to control luminances outside, particularly within the 45 to 90° glare zone.

Diffusers are made of glass or plastic, with the latter more prevalent. Glass is durable and does not yellow throughout life; however, it is heavier, more costly, and more susceptible to breakage than plastic. Plastics for diffusers are of three types: polycarbonate, polystyrene, and acrylic. Polycarbonate lenses have very high impact strength, but can yellow and become brittle over time from exposure to ultraviolet radiation. Polystyrene and acrylic are very similar. Both have high transmittance (92% at normal incidence). Acrylic is more costly than polystyrene, but has somewhat better ultraviolet resistance.

Clear prismatic lenses use the optical principle of refraction to produce desired downward intensity distributions. Plastic sheets of 1/8 or 5/32 inch thickness are embossed with 3/16-inch* square-base conical prisms (25 per square inch) and

* 3/16 inch appears to be optimal. Smaller and larger prisms do not obscure lamps as well.

mounted in the luminaire with the embossed side facing down. The most common form, denoted by the lighting industry as K12, consists of recessed (female) cones (tips projecting upward) oriented at 45° with the luminaire axis.* Sheet thickness for 2 ft × 4 ft is 1/8 inch and for 3 ft × 3 ft, 3/16 inch.

A second popular pattern is K19, which is 5/32-inch thick, with the cone tips projecting downward (male) and oriented at 90° with the luminaire axis. It provides somewhat better control of luminances at high angles from the normal (55 to 75°). Both of these patterns have efficiencies of 71%. Luminance values for 2 × 4 troffers using these lenses and containing four T12 3200-lumen fluorescent lamps are presented in Table 8.5.

TABLE 8.5 PRISMATIC LENS LUMINANCES

Degrees from Nadir	K12 L cd/m² (cd/in²)		K19 L cd/m² (cd/in²)	
	Parallel	Crosswise	Parallel	Crosswise
45	1381 (0.89)	1570 (1.01)	1363 (0.88)	1552 (1.00)
55	1112 (0.72)	1086 (0.70)	640 (0.41)	578 (0.37)
65	651 (0.42)	611 (0.39)	315 (0.20)	296 (0.19)
75	375 (0.24)	357 (0.23)	379 (0.24)	256 (0.17)
85	347 (0.22)	296 (0.19)	504 (0.33)	444 (0.29)

The inside of a troffer is generally painted white and has a diffuse reflectance characteristic. Thus it is only the lamps that require the optical control. The increasing use of T8 and smaller diameter lamps has made it more difficult to obscure the lamps and control their luminances, because these sources have higher luminances and are more concentrated than their T12 counterparts. One new development addressing this problem is the reverse apex lens, called A12. It turns out that male and female conical prisms image a lamp on opposite sides of the cone. The reverse apex lens combines these two prism types, thus increasing the ability of the lens to hide the lamps without altering the 3/16-inch base dimension.

There are also white translucent diffusers. They spread light uniformly in all downward directions (cosine intensity distribution), rather than refracting it selectively to particular zones. Since these diffusers generally have negligible absorption, most of the luminous flux that is not transmitted initially is reflected back into the luminaire and returns to the diffuser, increasing transmission. Overall transmission varies with pigment density and sheet thickness. Transmittances of 1/8-inch thick sheets are typically 45 to 70%, and will drop 20% if thickness is doubled, for the same pigment density. Newer, non-pigment diffusers have sand-blasted finishes or encapsulated bubbles to spread the light, with efficiencies approaching 90%.

Baffles and Louvers

A baffle is a single, usually V-shaped, shielding element, often placed parallel to and between lamps in a two-lamp luminaire, as shown in Fig. 8.8a. A louver is a group of baffles in an egg-crate arrangement, as diagrammed in Fig. 8.8b. The louvers may be either straight or parabolic as shown in Fig. 8.8c. The latter are normally made of specular

* This orientation produces a slightly rectangular intensity distribution, providing more uniform horizontal illumination than 90° orientation.

reflecting material to provide control of the directivity of the downward lumens and insure low luminances in the field of view.

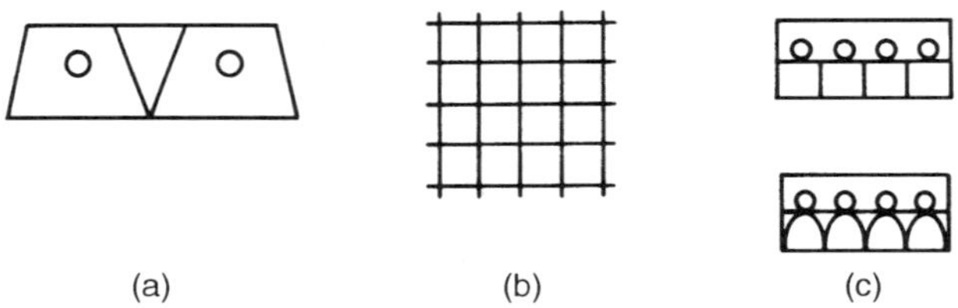

Figure 8.8 Baffles and louvers.

The significant property of a specular parabolic reflector is that all light rays emanating from its focal point reflect from the parabola in a parallel beam.* This principle is used in the design of a parabolic louver. Consider the Fig. 8.9a, where two dashed lines have been drawn at 45° from two points, A and B, bounding the opening between two parabolic louvers. It is desired that all of the flux emerging from the opening AB from the luminaire above be contained within the dashed lines. Place a fluorescent lamp at point B, and make point B the focal point of the portion of a parabola containing point A, as shown in Fig. 8.9b. Now all rays emanating from point B, after reflection, will emerge from the reflector at 45°. The rays from any lamp inside the luminaire, such as at point C, will leave the reflector at an angle less than 45°. This is because a mirrored surface obeys the law that the angle of reflection equals the angle of incidence (angle with the normal to the reflector), and the angle of incidence for the ray from C is greater than that of the ray from B.

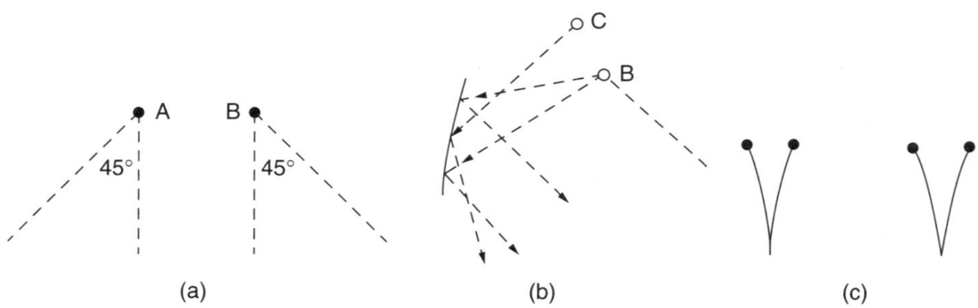

Figure 8.9 Parabolic louver development.

It remains to use exactly the same reasoning to complete the other half of the parabolic opening. The result is shown in Fig. 8.9c, where the portions of the parabolas for adjacent louvers are also shown. It is easy to see in Fig. 8.9c why this device is called a parabolic wedge louver.**

* The equation of a parabola will be developed in Chap. 10.

** These louvers can also be made with a small parabolic wedge on top of the larger wedge. Luminaire efficiency is improved somewhat.

The various cutoff and shielding angles associated with louvers are shown in Fig. 8.10. There are four parameters to define:

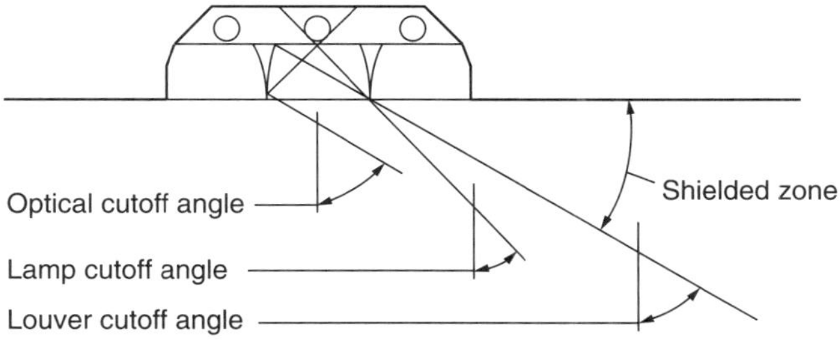

Figure 8.10 Louver cutoff and shielding angles. *Source*: *IESNA Lighting Handbook*, 9th edition, 2000. Courtesy of the Illuminating Engineering Society of North America.

Lamp cutoff angle
: The angle measured from nadir at which the lamp is just not seen.

Optical cutoff angle
: The angle measured from nadir at which the reflection of the lamp in the parabolic louver is just not seen.

Shielding angle
: The angle measured from the horizontal at which the lamp is just visible.

Louver cutoff angle
: The angle measured from nadir to the line passing through the tip of one louver and the base of the adjacent louver.

Louvers come in a variety of sizes, shapes, materials, and finishes. In areas where low ceiling brightness is required, such as with visual display units, small hexagonal or square cell louvers are often used. The cells are typically $^1/_2$ in deep and $^1/_2$ to $^3/_4$ in wide. Each vertical cell wall is parabolic, and the shielding angle is typically 45°. The louver finish can be specular silver, gold, or bronze, translucent white, or opaque black. A disadvantage of these units is that the resulting luminaire efficiency is less than 45%.

At the other extreme are deep, open cell parabolic louvers. These are generally made of specular or semispecular aluminum, with a silver, gold, copper, or bronze finish. Cells are typically 5 to 7 in square by 3 to 4 in deep and provide 30 to 35° shielding. Luminaire efficiencies are higher than for small cell louvers, of the order of 55 to 70%. Lamps are generally mounted directly above the cell edges.

Reflectors

Reflectors are formed from coated metal or plastic sheets and are placed above the lamps to redirect the upward component of luminous flux. Their surface finishes are specular, semi-specular, spread, or diffuse. Basic specular contours are circular, parabolic, elliptical, or combinations thereof. General specular contours may be tailor-made to produce a particular overall luminous intensity distribution and are frequently designed graphically, rather than mathematically. Both types will be discussed in detail in Chap. 10.

Specular and semispecular reflectors are generally used with incandescent and HID sources. They are normally made of aluminum, which is formed and then etched, polished, brushed, plated, or anodized to produce the reflecting surface. The anodizing procedure is an electrochemical process that deposits a uniform, permanent, and highly-reflecting surface of aluminum oxide on the aluminum, with a reflectance of 75 to 95%. Downlight housings are often of spun aluminum and may have concentric grooves for tight beam control.

With spread reflectors, the reflecting surface is corrugated, etched, or hammered, spreading an incident ray into a cone of reflected rays. Thus, these reflectors are effective in smoothing beam patterns. Reflectance is 50 to 75%.

Diffuse reflectors, commonly used in fluorescent luminaires, serve simply to gather and redirect as much of the upward luminous flux as possible downward to the lens, diffuser, or louver. Because of the diffusion, the contour of the reflector is relatively unimportant, and the reflected intensity distribution curve will be approximately a cosine distribution, except that at angles near the horizontal it will decrease more rapidly than does the cosine of the angle because the reflecting surface is further from the light sources than for angles near the vertical. Matte white paint and porcelain enamel are typical finishes, with the latter showing specularity at high angles of incidence. Reflectance is 65 to 90%.*

The use of reflectors in luminaires requires several cautions. First, the redirecting of reflected power back through the lamp may cause the lamp to operate at higher than normal temperature with the possible consequences of reduced lumen output and lamp life. Also, concentration of reflected power on a lens or diffuser can cause sag, thermal stress, and failure. Finally, in specular reflectors, the location of the lamp with respect to the reflector is critical if the desired intensity distribution pattern is to be achieved. If the sockets are slightly out of position or the lamp base is slightly askew, the center of the light source can be displaced from the reflector focal point.

A problem with coated luminaire materials that can be particularly prevalent with triphosphor lamps is iridescence. Iridescence occurs when incident white light from the lamp reflects from the coating and also from the underlying base. This can cause some of the wave-lengths in the two reflected waves to interfere and annul each other.** One then sees the remaining colors, rather than white light, reflected. Recently, polyester noniridescent coatings have been introduced for application to aluminum sheet material.

As a contribution to the efforts to conserve lighting energy, specular reflector inserts were introduced in the 1980s. These attempted to redirect lumens that would otherwise be trapped and absorbed within the luminaire. Contemporary luminaires are more efficiently designed than their predescessors, thus removing the need for corrective reflectors. However, many older lensed troffers are still in service, and their efficiencies can be increased from 70% to 80% by inserting a specular reflector carefully

* There are many new high-reflectance specular and diffuse coating materials entering the marketplace with reflectances, in several cases, above 90%.

** Interference with thin films is discussed in Chap. 10.

designed for that luminaire type. Particularly important is to rephotometer the luminaire to insure that the intensity distribution has not been altered significantly.

Other Considerations

There are several additional factors that the lighting designer should consider when choosing a luminaire.

Appearance The lighting of interior spaces should be both functional and aesthetically pleasing. These goals need not be incompatible, and the designer should make every effort to use luminaires that are attractive and that complement the architecture of the space being lighted.

Maintenance Luminaires should be easy to service, both for lamp replacement and for periodic cleaning. In choosing a luminaire, the designer should also be concerned about degradation of luminaire materials over time. In adverse environments, luminaires with dissimilar metals may cause problems because of the possibility of electrolytic action at connection points. Susceptibility of the luminaire to dust and dirt accumulation should also be considered.

Ventilation The ability of the luminaire to properly dissipate the build-up of heat should be considered. In some situations, it may be desirable to integrate the luminaire with the air-handling equipment of the building.

Acoustics Arc discharge light sources require ballasts, which, if electromagnetic, generate noise. The luminaire can transmit and, in some cases, amplify this noise. Luminaires should be designed so as not to resonate with the vibration frequencies of the ballast. Occasionally, when noise levels are critical, ballasts should be located remotely from their lamps and luminaires.

Mounting Luminaires may be suspended, surface-mounted, recessed (flush-mounted), connected in rows, or individually mounted. The choice affects both the functional and aesthetic aspects of the lighting. Luminaires that are not connected end-to-end in rows can be difficult to align. Spacing-to-mounting height guidelines for luminaires will be presented in Chap. 9.

Radiation Electromagnetic radiation from arc discharge lamps can cause interference with nearby electronic equipment. The interference is of two types: direct radiation from the luminaire and feedback through the electric supply. The former may be minimized by the use of an all-metal enclosed luminaire; the latter can be controlled by electrical line filters located at each luminaire.

Hazardous Locations Luminaires for hazardous locations should have the UL label for the particular environment in question, indicating that they comply with certain safety requirements stated in the National Electric Code. If corrosive atmospheres are present, the designer should give careful consideration to the luminaire materials used.

8.6 PHOTOMETRIC COORDINATE SYSTEMS

There are three categories of photometric coordinate systems, known as Types A, B, and C. These systems use variations of the standard spherical coordinate system shown in Fig. 8.11a. In that system, the three orthogonal variables are r, θ, and ϕ. The system is built from the conventional x, y, z Cartesian axes. r is the distance from

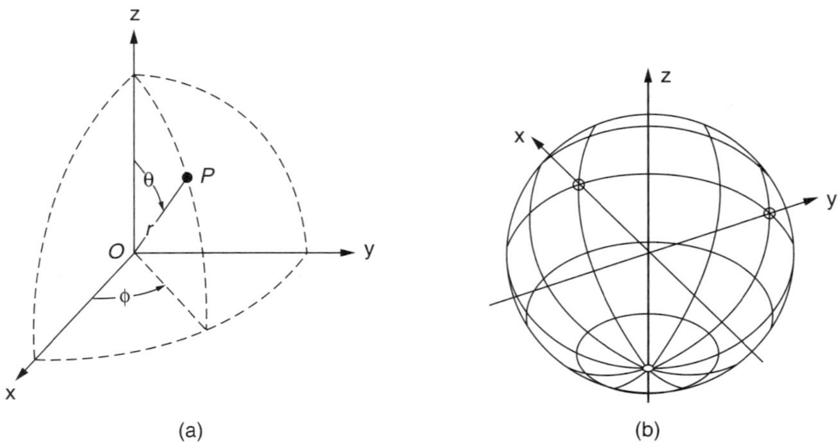

Figure 8.11 Standard spherical coordinate system.

the origin 0 to any point P. θ is the (polar) angle between the vertical polar (z) axis and the line drawn from 0 to P. The surface θ = constant is a vertical cone centered at 0. ϕ is the (azimuth) angle between the x axis and the projection of the line OP on the z = 0 plane. The surface ϕ = constant is a vertical half plane containing the z axis. The analogy with the earth and its polar axis is shown in Fig. 8.11b.

The luminaire orientations for the three photometric coordinate systems are shown in Fig. 8.12 and the various angles are illustrated in Fig. 8.13. In Type A photometry, the polar axis is vertical and the luminaire is aimed along the +x axis. Angles measured in vertical planes (analogous to θ) are called Y angles and vary from −90° (nadir) to +90° (zenith). Horizontal angles from the x axis to the vertical planes (analogous to ϕ) are called X angles and vary from −180 to +180°. The initial aiming of the luminaire is at an X angle of 0°. In performing the actual Type A photometry, the photo detector is fixed and the luminaire is rotated, first about the z axis to a desired X angle. Then it is rotated vertically from −90° to +90°, and intensity readings are taken at (usually) 10° increments. Type A photometry is used for optical systems and automotive lighting, and occasionally for parking lot luminaires.

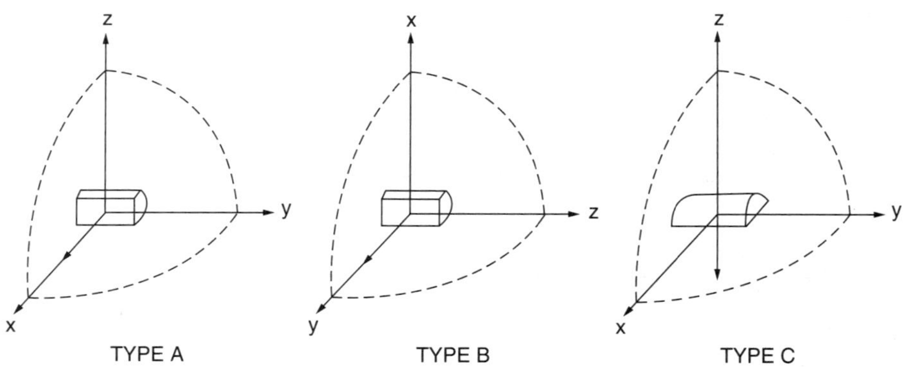

Figure 8.12 Type A, B, and C coordinate systems.

Floodlights are usually photometered using Type B photometry. As shown in Fig. 8.12, Type B photometry looks very much like Type A photometry, except that the z (polar) axis of the coordinate system is now horizontal. The luminaire is aimed along the +y axis. Horizontal (H) angles are measured in y z plane and range from −90° to +90°. Vertical (V) angles, measured in vertical planes containing the x axis, range from −180° to +180°, with −90° at nadir and +90° at zenith. As in Type A photometry, the photo detector is fixed. In obtaining photometric data, the luminaire is first rotated to a selected V angle. Then intensities are read at H angles from −90° to +90°, again usually in 10° steps. Type B photometry will be described in greater detail in connection with the discussion of floodlights in Chap. 11.

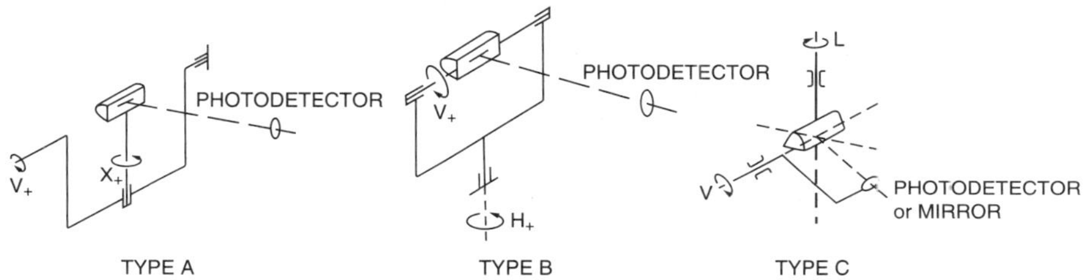

Figure 8.13 Goniophotometers for Types A, B, and C. *Source*: *Goniometer Types and Photometric Coordinates* (LM-75), 2001. Courtesy of the Illuminating Engineering Society of North America.

In Type C photometry, the polar axis is again vertical. The photometry is very similar to Type A, except in this case the photo detector moves. The luminaire faces downward (nadir) and can be rotated, but not tilted. Vertical (V) angles, analogous to θ in Fig. 8.11, vary from 0° at nadir to 180° at zenith. Lateral (L) angles, analogous to ϕ in Fig. 8.11, vary from 0° to 360°. To perform a photometric test the luminaire is rotated to the desired L angle and the photo detector is then moved in degree increments (usually 5° or 10°) from V = 0° to 180°. Type C photometry is always used for measuring lamps that are position-sensitive to tilting. It is the most widely-used type of photometry, particularly for roadway luminaires and for luminaires used in interior lighting. It was Type C photometry that was employed in our initial discussion of the measurements of intensity and luminous flux in Sect. 2.9.

8.7 PHOTOMETRIC REPORTS

A photometric report is a manufacturer's specification sheet that provides quantitative information about a luminaire's characteristics. It includes some or all of the following:

- Luminous intensity distribution, determined from laboratory measurements. Data are generally provided from Type C photometry at polar angle increments of 5°, from zero to 180°, in vertical planes that lie at 0°, 45°, and 90° with respect to the luminaire axis. Sometimes intensities at polar angle increments of 1° and in 22.5° and 67.5° vertical planes are required. Luminaire intensity data are presented either tabularly or in polar-coordinate distribution curves, or both ways.

- Coefficients of utilization (CU). Each of these is the ratio of the lumens reaching the work plane* in a given rectangular space to the total lamp lumens. CU is the product of luminaire efficiency, discussed in Sect. 8.4, and room efficiency, which is the ratio of the lumens reaching the work plane to those exiting the luminaire. Usually a table of CUs as functions of room dimensions and room surface reflectances is presented. CU will be used in Chap. 9 in calculating average work plane illuminance.
- Average luminaire luminances in the 45 to 90° glare zone, as described in Sect. 8.4. Recall from that section that maximum luminance is a measured quantity, whereas average luminance is calculated from Eq. (8.7). Average luminance is meaningful only if luminaire luminances are reasonably uniform.
- Zonal lumens from the luminaire in concentric vertical-axis cones with their apexes at the center of the luminaire. Generally zonal lumens are given for 0–30°, 0–60°, 0–90°, and 90–180° cones. The sum of the last two gives the total lumens emanating from the luminaire. Dividing this number by total lamp lumens gives luminaire efficiency.
- Visual comfort probability (VCP), which is a measure of the likelihood that individuals in a given space will not experience discomfort glare. The elements contributing to discomfort glare and a formula for calculating VCP will be presented in Sect. 8.8.
- Spacing criterion (SC), which is a metric that serves as a guide to luminaire spacing. SC is the ratio of the maximum spacing between luminaires to the luminaire mounting height above the work plane that is likely to result in reasonable uniformity of work plane illuminance. The method of determining SC will be presented in Sect. 9.11.
- Wall exitance coefficients, which are similar to CUs and permit one to calculate the average luminous exitance of a wall surface. Then, knowing the average wall reflectance, one can calculate the average illuminance on the walls, and the exitances of tack boards, chalk boards, and the like. The procedure is covered in Sect. 9.13.

Three examples of manufacturer's luminaire data are given in Figs. 8.14–8.16. The "a" portion of each figure gives a description of the luminaire and ordering information, including options. The "b" portion is the photometric report. The lumen and candela data in this report are based on specified test lamp types being operated on a prescribed reference ballast that provides full light output.

Actual operation can differ from test operation in two ways. First, the conditions during testing usually do not match those in the real world, necessitating a ballast factor, as given in Eq. (6.9). Second, the lamps being used may not be the same type as those in the photometric report, requiring a lumen adjustment. The total correction factor, called a lumen ratio, is given by

$$\mathrm{LR} = \frac{\text{Rated lamp lumens} \times \mathrm{BF}}{\text{Test lamp lumens}} \tag{8.8}$$

* An imaginary horizontal plane, usually 30 inches above the floor.

FEATURES

OPTICAL SYSTEM

- Twin matte white polyester powder paint finished reflectors provide uniform light distribution. Optional diffuse aluminum stepped reflector available.
- All diffusers control direct light distribution and glare by shielding lamps from direct view.
- All shieldings snap into place by pivoting on light trap for easy lamp access.
- Injection molded light traps prevent light leaks between shielding and endplates.

SHIELDING OPTIONS

- Metal Diffuser staggered Round holes (MDR) 52% open perforated metal with .075" diameter holes backed with white acrylic diffuser.
- Straight Blade Louver (SBL) sides of perforated metal with staggered round holes and solid blade louvered center. Sides and louver backed with white acrylic diffuser.
- Metal Diffuser aligned Mini slots (MDM) 46% open perforated metal backed with white acrylic diffuser.
- Acrylic Diffuser Prismatic lens (ADP) extruded acrylic lens backed with white acrylic diffuser.
- Metal Diffuser staggered Linear slots (MDL) 45% open perforated metal backed with white acrylic diffuser.

ELECTRICAL SYSTEM

- Class P, Thermally protected, resetting, HPF, Non- PCB, UL Listed, CSA-certified electromagnetic ballast is standard. Energy saving and electronic ballast are sound rated A. Standard combinations are CBM approved and conform to UL 935.

HOUSING

- Housing is powder painted cold rolled steel. All edges hemmed or rounded.
- Trims available for standard 1" tee bar, mini-tee bar, screw slot grids.
- Drywall ceiling adapters available.

LISTING

- UL listed and labeled. Listed and labeled to comply with Canadian and Mexican Standards (see options).

Specifications subject to change without notice.

Catalog Number	Type

Direct/Indirect General Lighting System

AV 2' × 2'

Compact Fluorescent
2 lamp
T8, T5 or T5HO
2 or 3 lamp

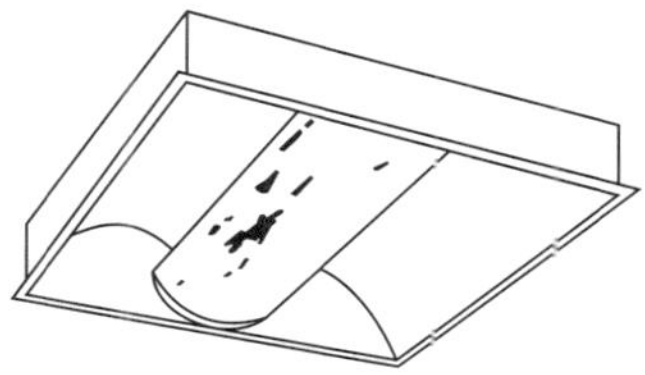

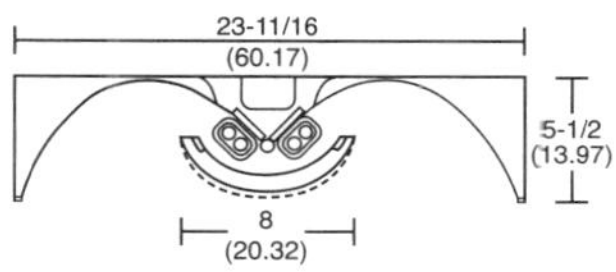

Avante
Recessed Direct Indirect Lighting

ORDERING INFORMATION

Example: **2AVG 2 CF40 MDR 120 GEB**

2AV

Series
2AV 2' wide symmetric distribution

Trim Type
G Grid trim
ST Screw slot

Number of Lamps
2, 3[1]

Lamp Type
17 17W T8 (24")
14T5 14W T5 (22")
CF40 40W TT5 (24")
CF50 50W TT5 (24")
CF55 55W TT5 (24")
24HOT5 24W T5HO (22")

Diffuser
MDR Metal diffuser, round holes.
SBL Straight blade louver, round holes.
MDM Metal diffuser, mini slots.
ADP Acrylic diffuser, linear prismatic lens.
MDL Metal diffuser, staggered linear slots.

Voltage
120, 277, 347
Others available

Options
GEB Electronic ballast, <20% THD.
GEB10IS Electronic ballast, ≤10% THD, Instant Start.
GEB10RS Electronic ballast, ≤10% THD, Rapid Start.
ADEZ Advance Mark X two-wire dimming ballast. (CF40 only)
EL Emergency battery pack (nominal 300 lumens, see Life Safety section).
GLR Internal fast-blow fuse.
GMF Internal slow-blow fuse.
LP Lamped. Specify lamp type and color.
PWS1836 6" prewire, 3/8" dia.,13-gauge, 3 wires.
RIF Radio interference filter.
HTC T-bar safety clips (snap-on)
LATC T-bar safety clips (screw-on)
CSA Listed and labeled to comply with Canadian Standards.
NDM Listed and labeled to comply with Mexican Standards.

Reflector Option
ASR Aluminum stepped reflector.

Accessories
Order as separate catalog number.
DGA22[2] Flanged grid to drywall adapter, unit installation.

Notes:
1 3 lamp available only with 14W T5, 17W T8 and 24W T5 HO.
2 Use G trim plus DGA accessory for fixture trim flange and fixture support in plaster or plasterboard ceilings.

Figure 8.14(a) AV 2′ × 2′ compact fluorescent luminaire description. Courtesy of Lithonia Lighting.

The luminaire in Fig. 8.14 is a recessed direct unit, whose generic type is shown in Fig. 8.1a, with a diffuse distribution. It houses two or three T5 or T8 lamps. The features are described in Fig. 8.14a, and ordering information is given. Note the many options available, in particular that the diffusing half cylinder can be perforated, louvered,

CATALOG NUMBER: 2AV G 2 CF40 MDR

CANDELA PLOT

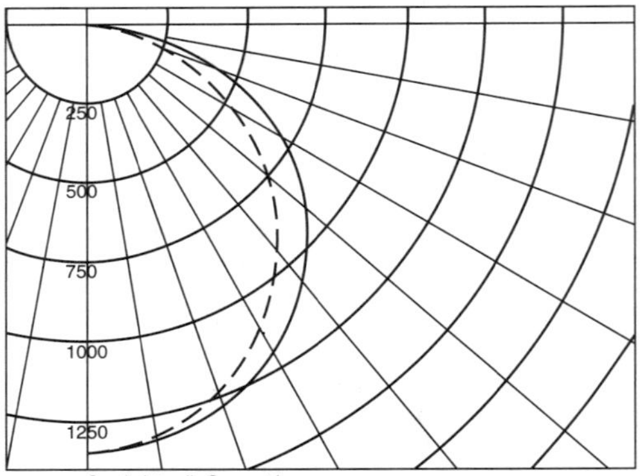

Quadralaterally Symmetric
Dashed: 0 Degrees Solid: 90 Degrees

SUMMARY DATA

EFFICIENCY (Total):	62.4%
EFFICIENCY (Downlight):	62.4%
EFFICIENCY (Uplight):	0.0%
CIE CLASSIFICATION:	DIRECT
SPACING CRITERION (0-Deg.):	1.23
SPACING CRITERION (90-DEG.):	1.32
LUMENS/LAMP:	3150
NO. OF LAMPS:	2
LUMINOUS OPENING:	RECTANGULAR
Width:	1.90(Feet)
Length:	1.90
Height:	0.00
INPUT WATTS:	85
RP-1—93 VDT CONFORMANCE:	NON-CONFORMING

ZONAL LUMEN SUMMARY

Zone	Lumens	%Lamp	%Luminaire
0—30	1051.5	16.7	26.8
0—40	1730.7	27.5	44.1
0—60	3092.6	49.1	78.7
60—90	835.7	13.3	21.3
0—90	3928.3	62.4	100.0
90—180	0.0	0.0	0.0
0—180	3928.3	62.4	100.0

AVERAGE LUMINANCE

(Candelas/Square Meter).

Angle	0	22.5	45	67.5	90
0	3998	3998	3998	3998	3998
45	3572	3681	3909	4116	4179
55	3270	3441	3810	4107	4221
65	2787	3118	3655	4050	4141
75	2223	2604	3064	3260	3352
85	1334	1539	1916	2087	2087

CANDELA TABLE

Angle	0	22.5	45	67.5	90
0	1341	1341	1341	1341	1341
2.5	1340	1342	1338	1344	1349
5	1335	1335	1338	1341	1346
7.5	1327	1327	1332	1338	1341
10	1315	1316	1319	1329	1335
12.5	1300	1303	1310	1320	1324
15	1281	1286	1297	1305	1314
17.5	1261	1266	1277	1291	1299
20	1235	1243	1258	1273	1280
22.5	1210	1218	1237	1256	1264
25	1182	1190	1213	1234	1241
27.5	1148	1159	1183	1210	1221
30	1111	1127	1154	1182	1196
32.5	1072	1088	1121	1152	1168
35	1033	1056	1086	1124	1138
37.5	988	1008	1048	1090	1104
40	943	965	1009	1053	1070
42.5	895	918	969	1014	1032
45	847	873	927	976	991
47.5	798	823	882	933	952
50	742	771	834	885	905
52.5	689	722	786	840	860
55	629	662	733	790	812
57.5	572	612	683	741	765
60	510	555	628	688	710
62.5	454	498	574	635	653
65	395	442	518	574	587
67.5	341	385	463	512	518
70	289	333	405	435	427
72.5	240	281	337	343	353
75	193	226	266	283	291
77.5	150	172	199	222	227
80	109	124	148	166	169
82.5	70	78	97	111	111
85	39	45	56	61	61
87.5	18	18	21	24	22
90	0	0	0	0	0

COEFFICIENT OF UTILIZATION TABLE

Effective Floor cavity Reflectance = 20%

Pcc...	80				70				50			30			10			0
Pw...	70	50	30	10	70	50	30	10	50	30	10	50	30	10	50	30	10	0
RCR																		
0	.74	.74	.74	.74	.73	.73	.73	.73	.69	.69	.69	.66	.66	.66	.64	.64	.64	.62
1	.68	.65	.62	.60	.66	.63	.61	.59	.61	.59	.57	.58	.57	.55	.56	.55	.53	.52
2	.62	.56	.52	.49	.60	.55	.51	.48	.53	.50	.47	.51	.48	.46	.49	.47	.45	.43
3	.56	.49	.44	.40	.55	.48	.44	.40	.47	.42	.39	.45	.41	.38	.43	.40	.38	.36
4	.51	.44	.38	.34	.50	.43	.38	.34	.41	.37	.33	.40	.36	.33	.38	.35	.32	.31
5	.47	.39	.33	.29	.46	.38	.33	.29	.37	.32	.29	.36	.32	.28	.35	.31	.28	.27
6	.43	.35	.29	.25	.42	.34	.29	.25	.33	.29	.25	.32	.28	.25	.31	.28	.25	.23
7	.40	.32	.26	.22	.39	.31	.26	.22	.30	.26	.22	.29	.25	.22	.29	.25	.22	.21
8	.37	.29	.24	.20	.36	.29	.23	.20	.28	.23	.20	.27	.23	.20	.26	.22	.19	.18
9	.35	.27	.21	.18	.34	.26	.21	.18	.25	.21	.18	.25	.21	.18	.24	.20	.18	.16
10	.33	.24	.20	.16	.32	.24	.19	.16	.24	.19	.16	.23	.19	.16	.22	.19	.16	.15

Figure 8.14(b) AV 2′ × 2′ photometric report. Courtesy of Lithonia Lighting.

Catalog Number	Type

Optimax® Parabolic Light Control System

PMO 2' × 4'

2 or 3 Lamps

27 Cells

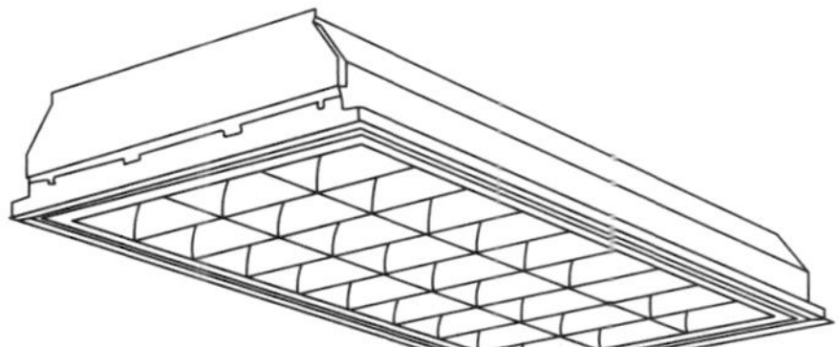

FEATURES

- Full family of light controlling parabolic luminaires designed to control screen glare in VDT open office environments.
- Meets IES RP-1 minimum luminance criteria for office lighting systems in VDT applications.
- Choice of diffuse or specular louvers utilize the latest developments in louver finishing for minimized louver iridescence. Ideal for use with triphosphor lamps.
- Black reveal provides floating louver appearance, conceals optional air-supply slots.
- Overlapping flange and modular ceiling trims factory-installed with standard swing-gate hangers or field convertible with optional trim and hanger kits.
- Optional heat-removal dampers and air-pattern control blades allow airflow control.
- T-hinges die-formed for maximum strength. Latches spring-loaded, concealed in reveal.
- Guaranteed for one year against mechanical defects in manufacture.

SPECIFICATIONS

BALLAST — Thermally- protected, resetting, Class P, HPF, non-PCB, UL listed, CSA certified ballast is standard. Energy-saving and electronic ballasts are sound rated A. Standard combinations are CBM approved and conform to UL 935.

WIRING & ELECTRICAL — Fixture conforms to UL 1570 and is suitable for damp locations, AWM, TFN or THHN wire used throughout, rated for required temperatures.

MATERIALS — Housing formed from cold-rolled steel. Louvers formed from anodized aluminum. No asbestos is used in this product.

FINISH — Five-stage iron-phosphate pretreatment ensures superior paint adhesion and rust resistance. Painted parts finished with high-gloss, baked white enamel.

LISTING —UL listed and labeled. Listed and labeled to comply with Canadian and Mexican Standards (see options).

Specifications subject to change without notice.

ENERGY

- Luminaire Efficacy Rating (LER) and Annual Energy Cost/1000 Lumens:

  2-lamp LER.FP (LS louver) = 49. Annual energy cost = $4.90.
  3-lamp LER.FP (LS louver) = 59. Annual energy cost = $4.07.Based on 32W T8 lamps, 2850 lumen lamp and electronic ballast with ballast factor =.88 and input watts = 62 (2-lamp) and 86 (3-lamp).

Calculated in accordance with NEMA standard LE.5.

ORDERING INFORMATION

Example: **2PMO G B 3 32 27LD 120 GEB**

2PMO ___ ___ ___ ___ ___ ___ ___

Series

2PMO Optimax Light Control System,2' Wide

Trim type

G Grid
F Overlapping flanged
MT Modular fit-in
ST Screw slot

Air function

A Air supply/return (slots in side trim)
H Heat removal (through lamp cavity, dampers available)
D Combination A & H
B Static (no air function, matching appearance)

Number of lamps

2, 3
Not included.

Lamp type

32 32W T8 (48")
40 40W T12(48")

Number of cells

27 3 rows of 9

Louver finish

LD Low iridescent diffuse silver
LS Low iridescent specular silver

Voltage

120, 277, 347
MVOLT[1]
Others available.

Options

1/3 One 3-lamp ballast
GEB Electronic ballasts, ≤20% THD
GEB10IS Electronic ballast, ≤10% THD Instant Start
GEB10RS Electronic ballast, ≤10% THD Rapid Start
EL Emergency battery pack (nominal 300 lumens)
PWS1836 6' prewire, 3/8" dia., 18-gauge, 3 wires
GLR Internal fast-blow fuse
GMF Internal slow-blow fuse
LST Tandam wired fixture pairs (shared ballasts)
LP__ Lamped, specify lamp type and color
CRE Flanged trim for continuous row mounting (end)
CRM Flanged trim for continuous row mounting (middle)
HRD Heat-removal dampers
APB Air-pattern control blades (A and D models only)
JP Palletized and stretch-wrapped without individual cartons, G and MT trim only
CSA Listed and labeled to comply with Canadian Standards
NOM Listed and labeled to comply with Mexican Standards

NOTES:
1 MVOLT available with GEB10IS only.

Figure 8.15(a) PMO 2′ × 4′ parabolic louver luminaire description. Courtesy of Lithonia Lighting.

slotted, or be a prismatic lens. The objective, whatever diffuser is used, is to have the luminances of the diffuser and the two background half cylinders be nearly the same.

The photometric report is given in Fig. 8.14b, for a two-lamp 40-watt T5 luminaire with a metal diffuser having staggered round holes. From the summary data,

CATALOG NUMBER: 2PMO 3 32 27LS

CANDELA PLOT

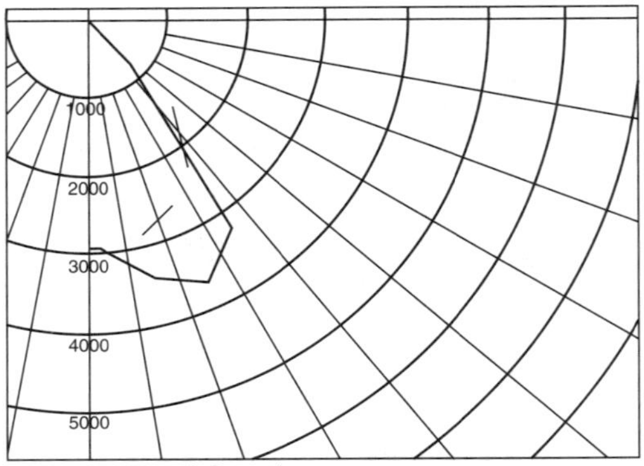

Quadralaterally Symmetric
Dashed: 0 Degrees Solid: 90 Degrees

SUMMARY DATA

EFFICIENCY (Total):	67.3%
EFFICIENCY (Downlight):	67.3%
EFFICIENCY (Uplight):	0.0%
CIE CLASSIFICATION:	DIRECT
SPACING CRITERION (0-Deg.):	1.24
SPACING CRITERION (90-Deg.):	1.51
LUMENS/LAMP:	2900
NO. OF LAMPS:	3
LUMINOUS OPENING:	RECTANGULAR
Width:	1.76 (Feet)
Length:	3.78
Height:	0.00
INPUT WATTS:	105
RP-1-93 VDT CONFORMANCE:	YES-INTENSIVE USE

ZONAL LUMEN SUMMARY

Zone	Lumens	% Lamp	% Luminaire
0—30	2668.5	30.7	45.6
0—40	4545.8	52.3	77.6
0—60	5841.9	67.1	99.7
60—90	15.5	0.2	0.3
0—90	5857.4	67.3	100.0
90—180	0.0	0.0	0.0
0—180	5857.4	67.3	100.0

CANDELA TABLE

Angle	0	22.5	45	67.5	90
0	2949	2949	2949	2949	2949
5	2941	2962	2976	2976	2966
15	2811	2912	3148	3330	3379
25	2604	2895	3330	3597	3688
35	2274	2762	3136	3300	3237
45	1545	1892	1894	906	786
55	191	266	230	108	46
65	10	10	10	9	9
75	5	6	4	5	4
85	1	1	1	1	1
90	0	0	0	0	0
95	0	0	0	0	0
105	0	0	0	0	0
115	0	0	0	0	0
125	0	0	0	0	0
135	0	0	0	0	0
145	0	0	0	0	0
155	0	0	0	0	0
165	0	0	0	0	0
175	0	0	0	0	0
180	0	0	0	0	0

AVERAGE LUMINANCE
(Candelas/Square Meter)

Angle	0	22.5	45	67.5	90
0	4771	4771	4771	4771	4771
45	3535	4329	4334	2073	1798
55	539	750	649	305	130
65	38	38	38	34	34
75	31	38	25	31	25
85	19	19	19	19	19

COEFFICIENT OF UTILIZATION TABLE

Effective Floor cavity Reflectance = 20%

Pcc...	80				70				50			30			10			0
Pw...	70	50	30	10	70	50	30	10	50	30	10	50	30	10	50	30	10	0
PCR																		
0	.80	.80	.80	.80	.78	.78	.78	.78	.75	.75	.75	.72	.72	.72	.69	.69	.69	.67
1	.76	.74	.72	.70	.74	.72	.70	.69	.69	.68	.67	.67	.66	.65	.65	.64	.63	.62
2	.71	.67	.64	.62	.70	.66	.63	.61	.64	.62	.60	.62	.60	.58	.60	.58	.57	.56
3	.67	.62	.58	.55	.65	.61	.57	.54	.59	.56	.53	.57	.55	.52	.56	.53	.52	.50
4	.62	.56	.52	.49	.61	.56	.51	.48	.54	.51	.48	.53	.50	.47	.51	.49	.47	.45
5	.58	.52	.47	.44	.57	.51	.47	.43	.50	.46	.43	.49	.45	.43	.47	.45	.42	.41
6	.55	.47	.43	.39	.53	.47	.42	.39	.46	.42	.39	.45	.41	.39	.44	.41	.38	.37
7	.51	.44	.39	.36	.50	.43	.39	.35	.42	.38	.35	.41	.38	.35	.41	.37	.35	.34
8	.48	.40	.36	.32	.47	.40	.35	.32	.39	.35	.32	.38	.35	.32	.38	.34	.32	.31
9	.45	.37	.33	.29	.44	.37	.32	.29	.36	.32	.29	.36	.32	.29	.35	.32	.29	.28
10	.42	.35	.30	.27	.42	.34	.30	.27	.34	.30	.27	.33	.29	.27	.33	.29	.27	.26

Figure 8.15(b) PMO 2′ × 4′ photometric report. Courtesy of Lithonia Lighting.

its CIE classification is direct (Table 8.2), and its efficiency is 62.4%. This number appears in two places: in the summary data and also in the zonal lumen summary.

Two spacing criteria are given, for 0° (parallel to the half cylinders) and for 90° (perpendicular to the half cylinders). The 90° value is slightly greater than the 0° value. The reason for this appears in the polar candela plot, where the 90° candela curve has a slightly greater bulge than the 0° candela curve. Thus the same uniformity can be achieved crosswise (90°) with slightly greater spacing than lengthwise (0°).

The zonal lumen summary gives the luminaire lumens in the various cones described earlier in this section. The numbers in the "% Lamp" column are obtained by dividing the numbers in the "Lumens" column by 6300 total lamp lumens. The numbers

FEATURES

- High performance, specification static luminaires provide general illumination for recessed applications.
- Designed for optimal performance with T8 lamps, low profile electronic ballasts and T8 sockets.
- Innovative fixture design yields superior lamp obscuration and improved visual comfort.
- A12 lens features reverse apex technology.
- Embossed housing ribs provide fixture rigidity and redirect light out from behind each lamp.
- Fixture sides feature smooth hemmed edges for safe handling. End flanges formed inward also have smooth edges.
- Standard steel door frame features unique cross-section providing a premium extruded appearance. Double-locking tabs and corner keys yield precision flush mitered corners and superior structural integrity.
- Steel door allows easy lens replacement without frame disassembly (for lenses upto .156" thick).
- Powder-painted, steel latches provide easy, secure door closure.
- Door frame latches and hinges from either side.
- Housing and doorframe interface creates a superior mechanical light seal—no foam gasketing required.
- Easily-positioned integral T-bar clips secure fixture to T-bar system.
- Socket features unique rotor system providing contact protection.
- Aluminum doorframes available—flush or regressed.
- Overlapping flange and modular ceiling trims available factory- installed including swing-gate hangers.
- Lens options, modular wiring, prewire and battery packs available.
- Guaranteed for one year against mechanical defects in manufacture.

Catalog Number	Type

Specification Premium T8 Troffer

SP8 2' × 4'

2, 3 or 4 Lamps

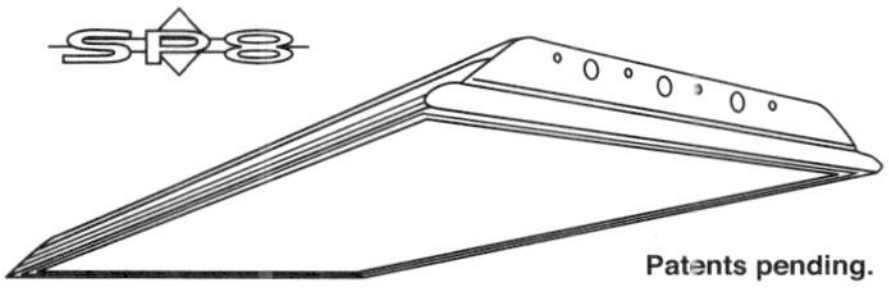

Patents pending.

FINISH — Five-stage iron-phosphate pretreatment ensures superior paint adhesion and rust resistance. Painted housing and ballast cover finished with 90 percent reflective matte white enamel.

LISTING — UL Listed and labeled. Listed and labeled to comply with Canadian and Mexican Standards (see options).

Specifications subject to change without notice.

SPECIFICATIONS

BALLAST — Thermally-protected, resetting, Class P, HPF, non-PCB, UL listed, CSA certified ballast is standard. Standard combinations conform to UL 935.

WIRING & ELECTRICAL — Fixture conforms to UL 1570 and is suitable for damp locations. AWM, TFN or THIN wire used throughout,rated for required temperatures.

MATERIALS — Housing formed from cold-rolled steel. Acrylic shielding material 100% UV stabilized. No asbestos used in this product.

ENERGY

- Luminaire Efficacy Rating (LER) and Annual Energy Cost/1000 Lumens:
  2-lamp LER.FL = 77. Annual energy cost = $3.12. Based on 32W T8 lamps, 2850 lumens, and electronic ballast. Ballast factor = .90, input watts = 57.
  3-lamp LER.FL = 74. Annual energy cost = $3.24. Based on 32W T8 lamps, 2850 lumens, and electronic ballast. Ballast factor = .87, input watts = 85.
  4-lamp LER.FL = 77. Annual energy cost = $3.12. Based on 32W T8 lamps, 2850 lumens, and electronic ballast. Ballast factor = .88 input watts = 107.

Calculated in accordance with NEMA standard LE.5.

ORDERING INFORMATION

Example: **2SP8 G 3 32 FW A12 120 1/3 GEB**

2SP8

Series
2SP8 2" wide

Trim type
G Grid
F Overlapping flanged
MT Modular fit-in

Number of lamps
2, 3, 4"
Not included

Lamp type
32 32W T8 (48")

Door frame
(blank) Flush steel, white
FN Flush aluminum, natural
FM Flush aluminum, matte black
FW Flush aluminum, white
RN Regressed aluminum, natural
RM Regressed aluminum, matte black
RW Regressed aluminum, white

Diffuser type
A12 #12 pattern acrylic, reverse apex
A12125 #12 pattern acrylic, 0.125" thick
A19 #19 pattern acrylic, 0.156" thick
A15 #15 pattern acrylic, 0.2" thick
PC1S 1/2" × 1/2" × 1/2" plastic cube louver, silver
PC2S 1-1/2" × 1-1/2" × 1" plastic cube louver, silver w/ flange
PC3S 3/4" × 3/4" × 1/2" plastic cube louver, silver

Voltage
120, 277, 347, MVOLT[1]
Others available.

Options
1/4 One 4-lamp ballast
1/3 One 3-lamp ballast
GEB Electronic ballasts, ≤20% THD
GEB10IS Electronic ballast, ≤10% THD, Instant Start
GEB10RS Electronic ballast, ≤10% THD, Rapid Start
EL Emergency battery pack (nominal 300 lumens; see Life Safety Section)
GLR Internal fast-blow fuse
GMF Internal slow-blow fuse
LST Tandem-wired fixture pairs (shared ballasts)
PWS1836 6' prewire, 3/8" dia, 18-gauge, 3 wires
LP__ Lamped, specify lamp type and color
CRE Flanged trim for continuous row mounting (end)
CRM Flanged trim for continuous row mounting (middle)
JP Palletized and stretch-wrapped without individual cartons
CSA Listed and labeled to comply with Canadian Standards
NOM Listed and labeled to comply with Mexican Standards

NOTES:
1 MVOLT available with GEB10IS only.

Figure 8.16(a) SP8 2′ × 4′ lensed troffer luminaire description. Courtesy of Lithonia Lighting.

in the "% Luminaire" column are obtained by dividing the numbers in the "Lumens" column by 3928.3 total luminaire lumens.

Values of average luminaire luminance are obtained using the technique described in Sect. 8.4. The left-hand column gives values of polar angle; the five other columns

CATALOG NUMBER; 2SP8 3 32 A12 1/3 ADDE

CANDELA PLOT

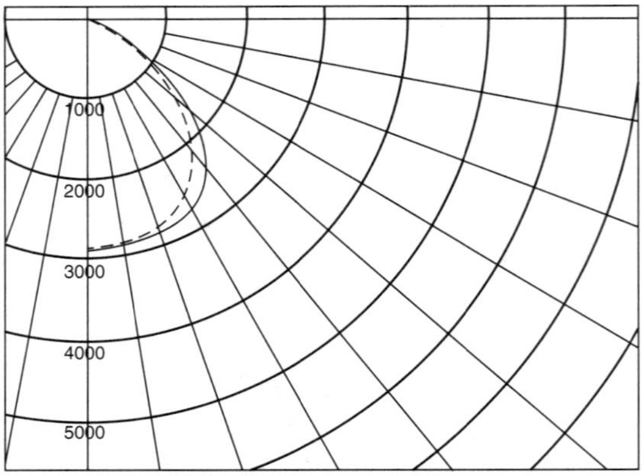

Quadralaterally Symmetric
Dashed: 0 Degrees Solid: 90 Degrees

SUMMARY DATA

EFFICIENCY (Total):	83.6%
EFFICIENCY (Downlight):	83.6%
EFFICIENCY (Uplight):	0.0%
CIE CLASSIFICATION:	DIRECT
SPACING CRITERION (0-Deg.):	1.25
SPACING CRITERION (90-Deg.):	1.37
LUMENS/LAMP:	2850
NO. OF LAMPS:	3
LUMINOUS OPENING:	RECTANGULAR
Width:	1.77 (Feet)
Length:	3.76
Height:	0.00
INPUT WATTS:	85
RP-1–93 VDT CONFORMANCE:	NON-CONFORMING

ZONAL LUMEN SUMMARY

Zone	Lumens	% Lamp	% Luminaire
0–30	2318.5	27.1	32.4
0–40	3833.2	44.8	53.6
0–60	6164.1	72.1	86.2
60–90	983.3	11.5	13.8
0–90	7147.4	83.6	100.0
90–180	0.0	0.0	0.0
0–180	7147.4	83.6	100.0

AVERAGE LUMINANCE
(Candelas/Square Meter)

Angle	0	22.5	45	67.5	90
0	4695	4695	4695	4695	4695
45	3776	3936	4234	4380	4435
55	2524	2783	3040	3107	2916
65	1986	2124	2051	2235	2124
75	2193	1837	1675	1850	2268
85	1893	1763	1522	1763	1930

CANDELA TABLE

Angle	0	22.5	45	67.5	90
0	2903	2903	2903	2903	2903
2.5	2874	2889	2890	2921	2928
5	2867	2879	2880	2912	2924
7.5	2851	2867	2870	2903	2912
10	2838	2850	2856	2893	2903
12.5	2812	2828	2839	2879	2894
15	2777	2798	2817	2864	2880
17.5	2742	2766	2792	2849	2867
20	2700	2726	2764	2829	2857
22.5	2653	2682	2730	2808	2831
25	2599	2633	2692	2775	2803
27.5	2534	2571	2642	2732	2766
30	2461	2506	2592	2688	2720
32.5	2371	2425	2527	2625	2655
35	2272	2340	2453	2549	2571
37.5	2139	2222	2349	2433	2449
40	1998	2089	2218	2290	2305
42.5	1828	1914	2041	2111	2127
45	1651	1721	1851	1915	1939
47.5	1461	1521	1642	1709	1732
50	1250	1317	1434	1492	1491
52.5	1066	1146	1255	1297	1262
55	895	987	1078	1102	1034
57.5	763	852	928	939	860
60	655	735	782	792	722
62.5	580	640	655	682	630
65	519	555	536	584	555
67.5	471	478	440	501	498
70	430	411	364	424	450
72.5	391	349	306	354	404
75	351	294	268	296	363
77.5	301	249	226	243	308
80	237	199	179	195	247
82.5	160	148	127	147	175
85	102	95	82	95	104
87.5	50	43	39	41	49
90	0	0	0	0	0

COEFFICIENT OF UTILIZATION TABLE

Effective Floor cavity Reflectance = 20%

Pcc...	80				70				50			30			10			0
Pw...	70	50	30	10	70	50	30	10	50	30	10	50	30	10	50	30	10	0
RCR																		
0	1.00	1.00	1.00	1.00	.97	.97	.97	.97	.93	.93	.93	.89	.89	.89	.85	.85	.85	.84
1	.92	.88	.85	.82	.90	.86	.83	.81	.83	.80	.78	.80	.78	.76	.77	.75	.74	.72
2	.84	.78	.73	.69	.82	.77	.72	.68	.74	.70	.66	.71	.68	.65	.69	.66	.63	.62
3	.78	.70	.63	.58	.76	.68	.62	.58	.66	.61	.57	.64	.59	.56	.61	.58	.55	.53
4	.72	.62	.55	.50	.70	.61	.55	.50	.59	.54	.49	.57	.52	.49	.55	.51	.48	.46
5	.66	.56	.49	.44	.64	.55	.49	.44	.53	.48	.43	.52	.47	.43	.50	.46	.42	.41
6	.61	.51	.44	.39	.60	.50	.43	.38	.49	.43	.38	.47	.42	.38	.46	.41	.38	.36
7	.57	.46	.39	.34	.56	.46	.39	.34	.44	.38	.34	.43	.38	.34	.42	.37	.34	.32
8	.53	.42	.36	.31	.52	.42	.35	.31	.41	.35	.31	.40	.34	.30	.39	.34	.30	.29
9	.50	.39	.32	.28	.49	.38	.32	.28	.38	.32	.28	.37	.31	.28	.36	.31	.27	.26
10	.47	.36	.30	.25	.46	.36	.29	.25	.35	.29	.25	.34	.29	.25	.33	.28	.25	.24

Figure 8.16(b) SP8 2′ × 4′ photometric report. Courtesy of Lithonia Lighting.

are headed by azimuth angles. For example at $\theta = 55°$ and $\phi = 0°$, I = 629 cd from the candela table. The luminous opening, from the summary data, is $1.9 \times 1.9 = 3.61$ ft^2, and its projected areas is 3.61 cos 55 = 2.07 ft^2. Then Eq. (8.3) yields $L = \frac{629}{2.07} =$ 304 cd/ft^2 = 3270 cd/m^2, the value in the average luminance table.

In the candela table, values are given for every 2.5° in polar angle and 22.5° in azimuth angle. No values need be given for azimuth angles greater than 90° because the luminaire possesses quadrilateral symmetry, i.e., the intensities in all four quadrants are the same.

The final item in the photometric report is a coefficient of utilization table. The values in this table are for 20% floor reflectance. In the Lumen Method, to be covered in Chap. 9, a table of multipliers is presented for other floor reflectances. ρ_{cc} is the average reflectance of the ceiling cavity, the cavity above the luminaire plane. If luminaires are recessed or surface mounted, ρ_{cc} is simply the average ceiling reflectance. ρ_w is average wall reflectance and RCR is room cavity ratio. Calculation of ρ_{cc} and RCR will be covered in Chap. 9, and the CU table will be explained in detail there.

The luminaire data sheets in Figs. 8.15 and 8.16 are very similar to those in Fig. 8.14. The luminaire in Fig. 8.15 is 27-celled parabolic-louver unit for either two or three 32-W T8 or 40-W T12 fluorescent lamps. It is designed particularly to control glare in VDT screens. For this luminaire, LER ratings and annual energy costs are given, as calculated using Eqs. (8.1) and (8.2). The luminaire category is FP and, from Table 8.3, the benchmark LER is 47–53 for a 2-lamp unit and 41–47 for a 3-lamp unit. The luminaire meets the standard for 2 lamps (49) and exceeds it for 3 lamps (59).

The photometric report for this luminaire is based on three 32-W T8 lamps. This luminaire also has no upward lumens and its efficiency of 67.3% is greater than that of the previous luminaire. Note that nearly all of the lumens lie in the 0 to 60° cone. Its candela curve has a slight bat wing, but not as much as the generic luminaire in Fig. 8.1c.

The final luminaire in Fig. 8.16 is a recessed lensed troffer with 2, 3, or 4 32-W T8 fluorescent lamps. It easily exceeds the LER minimum in Table 8.3 for FL lensed units. There are several lens options—reverse apex and K12, K15, and K19, as well as small plastic cube (not parabolic) louvers. The photometric report is done for a 3-lamp unit, with a K12 reverse apex lens. As with the other luminaires, this is a direct unit, but it has a much higher efficiency of 83.6%. Its prototype is in Fig. 8.1f, with an intensity distribution that is narrower than a cosine distribution but still quite diffuse.

The lighting professional should have only cautious confidence in data such as provided in Figs. 8.14–8.16. The photometric testing of luminaires is, at best, an inexact science. The actual luminaire, even though it has been tested by an independent laboratory, will probably not provide the values stated on the catalog sheet, for a variety of reasons. First, often only one sample of a luminaire is sent to a testing laboratory for photometering. The unit to be tested is chosen by the manufacturer, not by the testing laboratory; thus usually a "good" sample is chosen. Second, there are bound to be statistical variations as a function of the quality control exercised by the manufacturer. Third, there is no guarantee that subsequent production runs of the same unit will yield the same photometric data. Small changes in materials, fabrication method, finishes, and assembly techniques all have an effect on photometric parameters. Fourth, the luminaire and lamps may be installed improperly. For example, a slight misorientation of a lamp with respect to a reflector can alter a luminaire intensity distribution.

Finally, there is no guarantee that the parameters of the lamps and ballasts in the luminaire tested will be replicated in the commercial product; nor is there any guarantee that the luminaire will be installed in the same type of environment as it was tested.

8.8 VISUAL COMFORT

In Sect. 3.10, glare was defined as excess luminances in the field of view that produce discomfort, loss of visibility, or both. Disability glare was discussed in some detail, but discussion of discomfort glare was delayed until now.

Discomfort glare, by definition, produces annoyance, sometimes leading to pain, but does not impair visibility or visual performance. Discomfort glare can result from both direct glare, from luminaires, windows, and skylights, and from reflected glare, arising from polished or glossy room surfaces. The viewing zones for direct and reflected glare were shown in Fig. 8.5. Reflected glare can be particularly disturbing from a ceiling luminaire mounted directly in front of an observer at an angle of 25° from the vertical, which is the usual viewing angle for printed material on a desktop. The ratio of the illuminance on the printed task from the offending luminaire to the total illuminance on the task should not exceed 0.3 if discomfort is to be avoided. If the ratio exceeds 0.7, discomfort is likely to result. Generally, luminaire luminances should not exceed those of surrounding surfaces by more than 100 to 1 if discomfort glare is to be minimized.

Investigations into the matter of discomfort glare have been going on for the past 40 years. Throughout this time the objective has been to develop a universally applicable, easy-to-use procedure for evaluating quantitatively the discomfort produced by glare sources (luminaires) in the field of view. The problem is a difficult one. What is desired is a reasonably simple empirical formula that yields numerical values of glare sensation that can be related to estimates of glare effect by a variety of observers under a variety of seeing conditions. Such a formula has been derived and a procedure has been developed to permit computation of the percentage of observers who will be visually comfortable in a given seeing environment. The metric is called visual comfort probability (VCP), the probability that a normal observer does not feel discomfort when viewing an array of ceiling luminaires under specified conditions. It is the purpose of this section to describe this formula and procedure, and its limitations.

Discomfort glare results from direct glare from luminaires that are too bright, inadequately shielded, or too great in area. It is also caused by reflected glare from specular surfaces lighted by sources having concentrated beams. The procedure described herein considers the former, but not the latter. There are five major factors influencing the amount of discomfort glare and resulting visual discomfort. These are field luminance (F) in candelas per square meter; position index (P) of each glare source; visual size (ω) in steradians of each glare source; luminance (L) in candelas per square meter of each glare source; and the number of glare sources (n).

Visual comfort can be evaluated for any location within a space and for any direction of viewing from that location. It has been found that the worst viewing condition is typically near the center of a wall with the viewer looking horizontally across the room. In order to have a common basis for the comparison of lighting systems, it has been decided to base visual comfort evaluations on a standard viewing position 4 ft out from the center of a wall and 4 ft above the floor with the viewer looking forward horizontally. This evaluation is normally made both lengthwise and crosswise in the room, to include effects of room shape and luminaire orientation.

Field Luminance

This is the background luminance against which the glare sources are viewed. It is the solid angle weighted average of all luminances in the field of view, given by:

$$F = \frac{L_w \omega_w + L_f \omega_f + L_c \omega_c + \sum L_s \omega_s}{5} \tag{8.8}$$

where w, f, c, and s denote walls, floor, ceiling, and source, respectively, and the 5 in the denominator is based on the assumption that the total field of view is 5 steradians (a cone half angle of about 78°). From the standard viewing position, the average luminance of the opposite wall is sometimes used in place of F in Eq. (8.8) as an approximation of field luminance, particularly in small rooms with high ceilings.

Logically, a higher field luminance should mean a lesser glare effect, because the background (adaptation) luminance is higher and the luminance of the glare source is closer in value to the luminance of its background.

Position Index

This is a measure of the location of the glare source with reference to the horizontal line of sight of the observer. Position index versus location is graphed in Fig. 8.17, Note from the graph that P increases as H or Y increases, but decreases as X increases. This means that a glare source making a large angle at the eye with the horizontal line of sight has a larger position index than one making a smaller angle. Such a source should produce less glare effect, and thus it is anticipated that the glare effect is inversely proportional to position index.

The data in Fig. 8.17 may be represented by the equation:

$$\ln P = \left(\frac{317}{9} - \frac{287}{900}\tau - \frac{11}{9}\varepsilon^{-2\tau/9}\right)10^{-3}\sigma + \left(21 + \frac{4}{15}\tau - \frac{2}{675}\tau^2\right)10^{-5}\sigma^2 \tag{8.9}$$

where ε is the Naperian log base and

$$\tau = \tan^{-1}\frac{Y}{H}, \quad \sigma = \tan^{-1}\frac{\sqrt{H^2 + Y^2}}{X} \tag{8.10}$$

The angles τ and σ in Eqs. (8.10) are to be entered in degrees in Eq. (8.9).

Visual Size

The visual size of a glare source is the solid angle in steradians that it subtends at the eye. From the diagram in Fig. 8.17, this angle is given by

$$\omega = \frac{A_p}{H^2 + X^2 + Y^2} \tag{8.11}$$

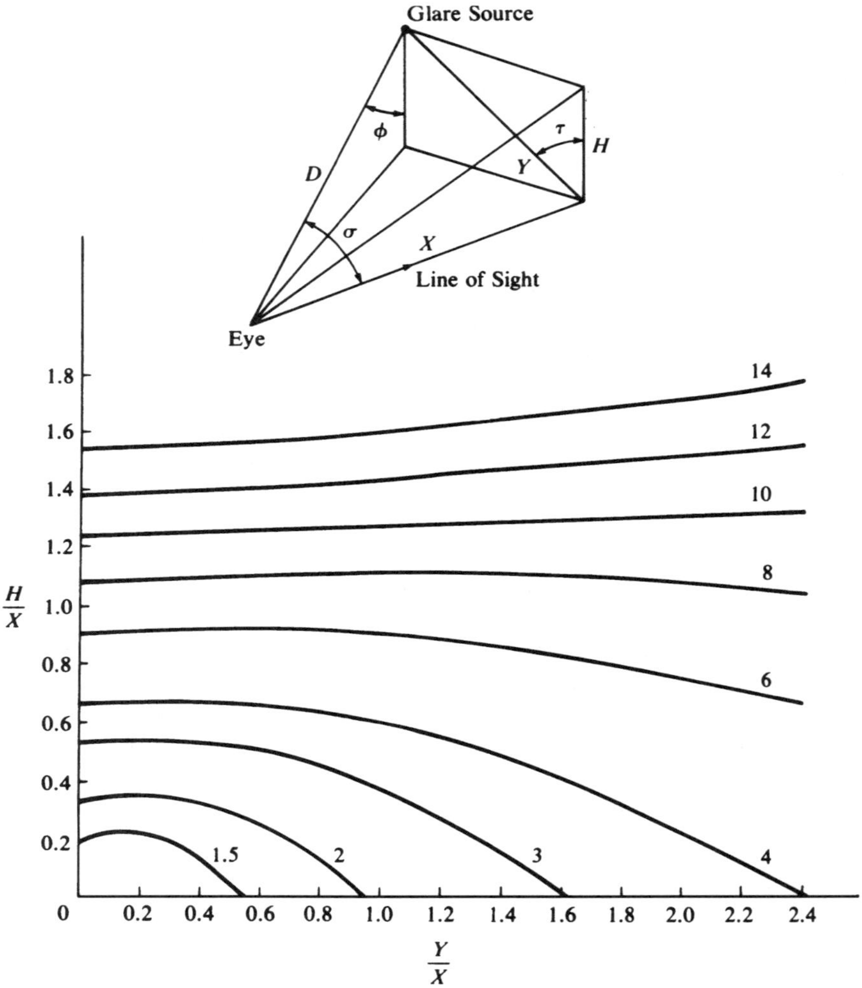

Figure 8.17 Position index P for discomfort glare evaluation.

where A_p is the projected area of the glare source in the direction of the observer, given by

$$A_p = A\cos\phi = A\frac{H}{\sqrt{H^2 + X^2 + Y^2}} \tag{8.12}$$

with A the actual glare source area (sources assumed flat and horizontal) and ϕ the angle between the vertical and the line between glare source and observer.

Data taken from many observers indicate that the glare effect is related to ω in a complicated way. An empirical formula has been developed to express this relationship. The result is that the glare effect is directly proportional to a quantity Q, given by

$$Q = 20.4\omega + 1.52\omega^{0.2} - 0.075 \tag{8.13}$$

A plot of Q versus ω is shown in Fig. 8.18.

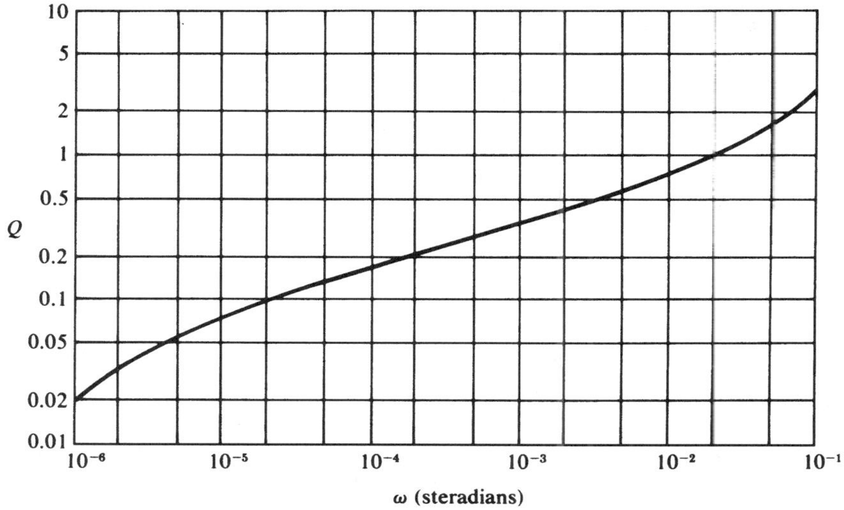

Figure 8.18

Table 8.6 is included to relate a solid angle to its half-angle at the tip of a cone. For example, a solid angle of 0.1 steradian means an angular spread subtended at the observer's eye of $2 \times 10.2 = 20.4°$.

TABLE 8.6

Solid Angle (steradians)	Cone Half-Angle (degrees)
5.0	78.2
3.0	58.5
2.0	47.0
1.0	32.7
0.5	23.0
0.2	14.5
0.1	10.2
0.05	7.2
0.02	4.6
0.01	3.2
0.005	2.3
0.002	1.4
0.001	1.0
0.0001	0.3

Source Luminance

This is the average luminance L_s of the glare source in the direction of the observer and is usually computed from the candlepower distribution data of the luminaire. If I is the luminaire intensity in candelas in the direction of the observer and A_p is the projected area of the luminaire in that direction, the luminance is given by I/A_p. It is assumed that the glare effect will be directly proportional to L_s.

Number of Sources

Certainly the glare effect will increase as the number of glare sources increases, but because glare sources differ in size, luminance, and position, the relationship should not be expected to be a linear one.

The fundamental glare formula was developed by S.K. Guth (1963) and gives the glare index of sensation M of a single source as

$$M = \frac{0.5L_sQ}{PF^{0.44}} \tag{8.14}$$

where L_s and F are in candelas per square meter. It is noted in this formula that the glare sensation varies directly with source luminance (L_s) and source size (Q) and inversely with position index (P) and field luminance (F), the latter to the 0.44 power.

In determining the glare effect of a source, it has been determined that the most meaningful measure is that value of source luminance that produces a sensation at the borderline between comfort and discomfort for the average observer. Such a luminance is called a BCD luminance. The relationship between BCD luminance and source solid angular size for sources on the line of sight and with constant field luminance has been determined experimentally and is shown in Fig. 8.19. This rather unusual looking curve can be described as follows: For small sources (less than 10^{-2} steradian), the curve is reasonably linear on a log-log plot. For larger sources, source area becomes increasingly more important and the curve slopes downward. In other words, as source area becomes quite large, an observer can tolerate considerably less luminance. The curve then reaches a minimum and turns upward, showing that the source is now becoming a significant part of the visual field and is less identified as a spot of glare by an observer. Eventually the curve levels when the entire field is filled by the source.

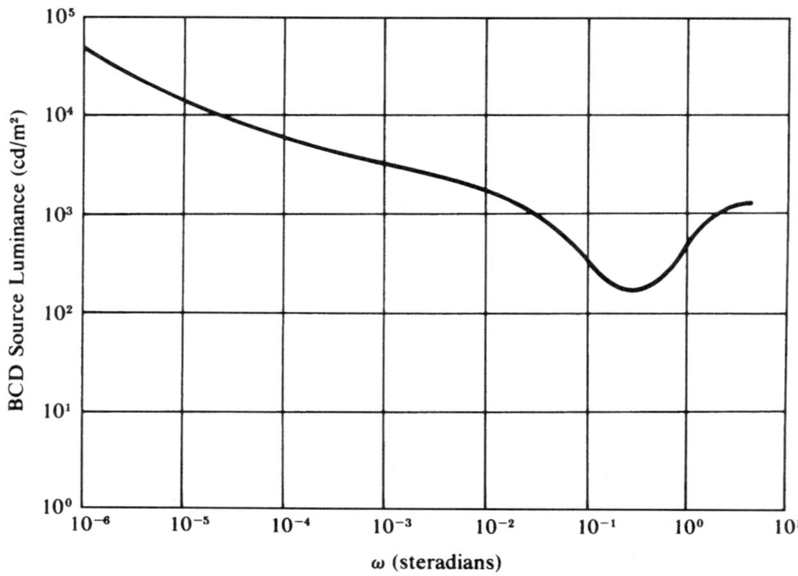

Figure 8.19 BCD luminance versus source solid angle.

The basic curve can be expanded into families of parallel curves by including field luminance and position index as additional variables. Either increasing field luminance or displacing the source from the line of sight gives higher values of BCD luminance for sources of 10^{-1} steradian or less. For example, increasing field luminance by 5:1 or moving the source 20° above the line of sight doubles the BCD luminance. For sources larger than 10^{-1} steradian, the curves are no longer parallel and actually come together when the source fills the field. The region up to 10^{-1} sr (10°) includes most common luminaires and it is this region for which the glare sensation formula and the equation for Q apply. Large area glare sources remain a subject for further research inquiry.

It remains to explore multiple sources. In developing a procedure for these, two approaches are possible:

1. Derive a new formula for multiple sources.
2. Develop a summation procedure using the single source formula.

It is the latter course that has been pursued.

It is tempting to simply sum the glare sensations of the several glare sources in a given installation to get the overall glare sensation, but this procedure does not yield correct results, largely because of the nonlinear relationship between glare sensation and source size ω. Rather it is necessary to develop experimental data in which observers are asked to establish their comfort or discomfort for groups of luminaires of varying sizes, luminances, and positions, much as was done originally for a single glare source. This was done by Guth, using 210 observers and 48 lighting conditions. Furthermore, Guth developed a variable exponential method of computing the overall glare sensation that gives good agreement with experimental results. First the individual glare sensations are summed to obtain

$$M_t = M_1 + M_2 + \cdots + M_n \tag{8.15}$$

Then an overall discomfort glare rating is defined by

$$DGR = (M_t)^a \tag{8.16}$$

in which the exponent a is a function of the number of glare sources present in the area being evaluated. This relationship has been determined experimentally to be

$$a = n^{-0.0914} \tag{8.17}$$

Values of a as a function of n are given in Table 8.7. Note that a decreases rapidly at first as n increases but then levels off.

Glare sensations for a single source and discomfort glare ratings for many sources are numbers to which it is hard to give physical interpretation. Higher values of M and DGR indicate greater discomfort, but how much greater is difficult to say.

Guth devised a method of expressing discomfort glare in terms of the percentage of a group of observers that would likely find a given M or DGR at, or more comfortable than, the borderline between comfort and discomfort. The experimental relationship

TABLE 8.7

n	a
1	1
2	0.939
3	0.904
4	0.881
6	0.849
8	0.827
10	0.810
20	0.760
50	0.699
100	0.656
200	0.616

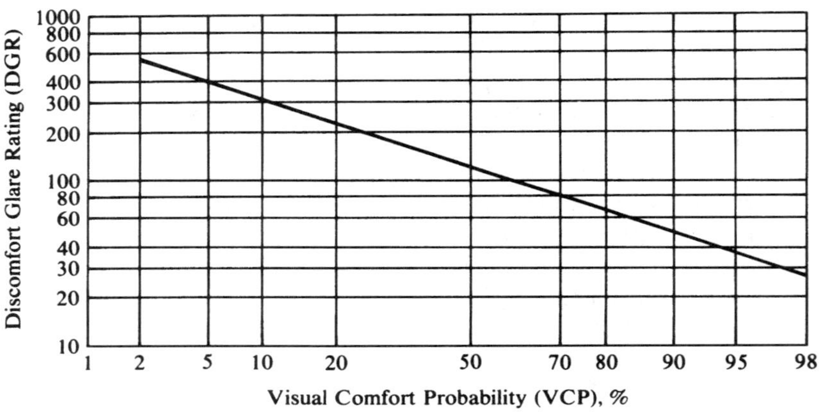

Figure 8.20 Conversion of DGR to VCP.

between discomfort glare rating (DGR) and visual comfort probability (VCP) is shown in Fig. 8.20, which is based on 107 observers in 35 different experimental conditions. It may be expressed analytically by (Rea, 2000)

$$\mathrm{VCP} = \frac{100}{\sqrt{2\pi}} \int_{-\infty}^{K} \varepsilon^{-k^2/2}\, dk \tag{8.18}$$

where k is simply a dummy variable of integration and

$$K = 6.374 - 1.3227 \ln \mathrm{DGR} \tag{8.19}$$

The graph in Fig. 8.20 shows that 50% of the observers are likely to find a given installation at or more comfortable than BCD when the DGR is 120. For 90% of the observers to be at or more comfortable than BCD, the DGR must be 45, about 37% of the value for 50% VCP. A summary of glare sensation descriptions and DGR and VCP values is given in Table 8.8.

TABLE 8.8

Glare Description	DGR	VCP
Unnoticeable	35	95
Perceptible	50	87
Acceptable	65	75
Distracting	90	64
BCD	120	50
Barely uncomfortable	160	34
Perceptibly uncomfortable	220	20
Uncomfortable	300	11
Intolerable	400	5

The IESNA has established an alternative glare criterion. Their statement is that direct glare will not pose a visual comfort problem in a lighting installation if the following three conditions are met:*

1. The VCP is 70 or greater in offices, and 80 or greater for direct luminaires in VDT spaces.
2. The ratio of maximum (brightest square inch) to average luminaire luminance does not exceed 5:1 at angles of 45°, 55°, 65°, 75°, and 85° from the vertical for both lengthwise and crosswise viewing.
3. Maximum luminaire luminances lengthwise and crosswise do not exceed the values in Table 8.4.

EXAMPLE 8.3

Assume a glare source is located 30° above the line of sight and has a luminance of 2000 cd/m^2 in the direction of an observer in the rear of a room. Further assume that projected source size is 10^{-2} steradian and that the field luminance is 100 cd/m^2. Compute the glare sensation M and the visual comfort probability VCP.

Solution. First, from Eq. (8.9) with $\tau = 0$ and $\sigma = 30°$, a position index of 3.35 is obtained. Next, from Eq. (8.13), Q = 0.734 for $\omega = 0.01$ steradian. Using Eq. (8.14),

$$M = \frac{0.5(2000)(0.734)}{3.35(100^{0.44})} = 29$$

and entering Fig. 8.20 shows that about 97% of the occupants of the area will be comfortable.

To see the relative effect of the various factors, first halve the field luminance to 50 cd/m^2. M becomes 39 and VCP 94. Thus a 50% decrease in F produces only a 34% increase in M and a 3% decrease in VCP. The mathematical reason for this relatively small influence of F is that it is raised to the 0.44 power in the glare sensation formula.

* The VCP requirement is newly stated. The other two conditions were stated in Sect. 8.4.

If the glare source is located only 10° above the line of sight instead of 30°, P becomes 1.40. Then M becomes 69 and VCP about 75. Thus a 3:1 decrease in angle of elevation results in a 2.4:1 increase in M and a 23% decrease in VCP.

Next, if the projected source size increases to 0.05 steradian, Q becomes 1.78, M becomes 70, and VCP 75. Lastly, if source luminance is increased 2.4 times, M is 69 and VCP is again 75.

Thus a 3:1 decrease in elevation angle, a 5:1 increase in source size and a 2.4:1 increase in source luminance all produce a 2:1 increase in M and a 16% decrease in VCP. For field luminance to produce this same effect, F would have to decrease from 100 to about 20 cd/m^2, a 5:1 decrease.

EXAMPLE 8.4

Assume four identical sources are located on the ceiling of a room in the following positions:

Source	X(ft)	Y(ft)	H(ft)
1	8	4	6
2	8	–4	6
3	14	4	6
4	14	–4	6

Furthermore assume each source is a square panel 1.5 ft on a side with a luminance in the direction of the observer of 1700 cd/m^2 and that the field luminance is 170 cd/m^2. Compute the glare sensation (M), disability glare rating (DGR), and the visual comfort probability (VCP).

Solution. First the position indexes are obtained. The computations are tabulated as follows:

Source	τ(°)	σ(°)	P
1	33.7	42.0	4.5
2	33.7	42.0	4.5
3	33.7	27.3	2.4
4	33.7	27.3	2.4

Next values are obtained for Q. Again, in tabulated form

Source	D(ft)	A_p(ft^2)	ω	Q
1	10.8	1.25	0.011	0.77
2	10.8	1.25	0.011	0.77
3	15.8	0.85	0.0034	0.48
4	15.8	0.85	0.0034	0.48

Now the glare sensation is calculated from Eq. (8.14) as

$$M = \frac{0.5(1700Q)}{P(170^{0.44})} = \frac{88.7Q}{P}$$

This yields M = 15.2 for the first two sources, and M = 17.7 for the last two. Summing gives $M_t = 65.8$.

Noting that $4^{-0.0914} = 0.881$, the discomfort glare rating is $DGR = 65.8^{0.881} = 40$ and, from Fig. 8.20, the visual comfort probability is about 93%.

In order to systematize VCP calculations and permit comparisons of luminaires under similar conditions in a space, the IESNA has developed a set of standard conditions for VCP determination. These are:

1. An initial average horizontal illuminance of 1000 lx (100 fc)
2. Reflectance values of $\rho_{cc} = 80\%$, $\rho_w = 50\%$, and $\rho_{fc} = 20\%$
3. Mounting heights of luminaires of 8.5, 10, 13, and 16 ft above the floor
4. Standardized room lengths and widths, to include square, wide and narrow rooms
5. A standard layout of uniformly-distributed luminaires
6. A limit to the field of view at the standard observer location of 53° upward from the forward viewing direction
7. An observation point 1.2 m (4 ft) in from the center of the rear wall and 1.2 m (4 ft) above the floor, with the observer looking forward horizontally.

Even with these standardizations, the VCP system should be used with caution. It has already been mentioned that its use should be limited to glare sources of less than 0.1 steradian. In addition, it should be noted that the equations developed to calculate VCP were from observations of simple lighting patterns in a test situation. The system was validated only for direct lensed fluorescent troffers. Thus VCP should not be applied to large area sources (windows) and small area sources (incandescent, CFL or HID), and should be used only with caution with non-uniform luminance sources (parabolic fluorescent). Finally, two lighting systems should differ by more than 5 in VCP if they are to be judged different in visual comfort.

The VCP system has been used in North America for many years, while the rest of the world has used other systems. Comparisons have shown that all these systems predict the average discomfort of a group of observers fairly well. In an effort to bring the various systems together, the CIE has proposed a new consensus system called the Unified Glare Rating (UGR) system.

The UGR and VCP systems are quite similar, but do have some differences. The UGR system is based on the following formula:

$$UGR = 8 \log_{10}\left[\frac{0.25}{L_F}\sum_{1}^{n}\frac{L_i^2\omega_i}{P_i^2}\right] \quad (8.20)$$

where L_F = luminance of the field of view (cd/m²), not including luminaire luminances
L_i = luminance of the ith luminaire in the direction of the observer
ω_i = solid angle in steradians of the ith luminaire subtended at the observer
P_i = position index of the ith luminaire

The major differences between VCP and UGR are:

1. A higher UGR represents more glare, whereas a higher VCP indicates less glare.
2. The UGR system has larger exponents on P and L than the VCP system (second power vs first power), giving greater emphasis to these quantities in the UGR system.
3. Source luminances are included in the field luminance in the VCP system, but not in the UGR system. Also the field luminance has a 0.44 exponent in the VCP system.
4. Solid angle is treated differently, through a quadratic equation in the VCP system and raised to a power in the UGR system.
5. The effects of multiple sources are summed linearly in the UGR system, whereas in the VCP system, their sum is raised to a non-linear exponent.

UGR values range generally from 10 to 25, VCP values from 90 to 40. This relationship is plotted in Fig. 8.21.

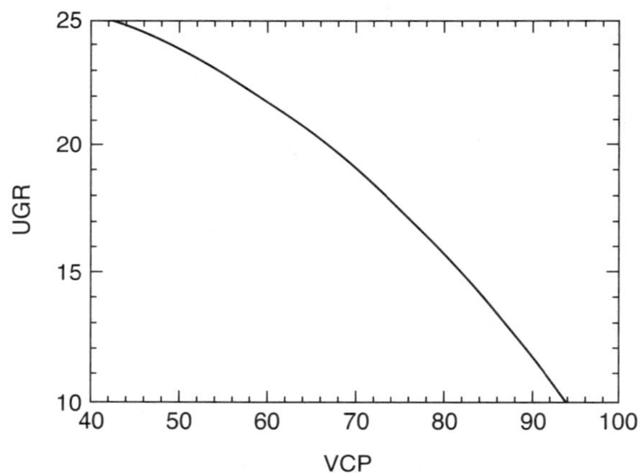

Figure 8.21 Relationship between UGR and VCP. *Source*: *IESNA Lighting Handbook*, 9th edition, 2000. Courtesy of the Illuminating Engineering Society of North America.

A study was done recently (Mistrick and Choi, 2000) to compare the VCP and UGR systems. First a computer analysis of a variety of fluorescent troffers was done in a standardized room. Results showed that as the glare level decreases (VCP increases and UGR decreases), the separation of VCP and UGR values increases. At a VCP of 50, the spread in UGR was 19 to 22; at a VCP of 90, it was 12 to 19.

A further study attempted to explain these differences by assessing the effect of two luminaires, a sharp cut-off parabolic troffer and a perfectly-diffusing lensed troffer, at four ceiling locations. It was found that in the VCP system, all luminaires in a space contribute significantly to the glare effect, whereas in the UGR system a few luminaires create most of the glare sensation. For a perfectly-diffusing luminaire, VCP values increase when the source is near the point of observation, whereas the UGR shows little change with source location throughout the room. For the parabolic source, VCP varies only slightly as the luminaire is moved about in the room, but the UGR system gives a lower rating when the source is close to the observer.

EXAMPLE 8.5

For the luminaire in Ex. 8.3, calculate the UGR.

Solution. From Ex. 8.3, the following information is tabulated:

$$L_s = 2000 \text{ cd/m}^2$$

$$\omega = 10^{-2} \text{ sr}$$

$$F = 100 \text{ cd/m}^2$$

$$P = 3.35$$

First F in Eq. (8.8) needs to be modified to obtain L_F, where L_F is the luminance of the field without source luminances.

$$F = \frac{L_F\omega_F + L_s\omega_s}{5} \tag{8.21}$$

Assuming $\omega_F = 5$ sr and using the values in Ex. 8.3,

$$L_F = 96 \text{ cd/m}^2$$

Then Eq. (8.20) yields

$$\text{UGR} = 8 \log_{10}\left[\frac{0.25}{96} \times \frac{(2\times10^3)^2 \times 10^{-2}}{3.35^2}\right]$$

$$= 8 \log_{10} 9.28 = 7.7$$

corresponding to a VCP of 97 in Ex. 8.3. These values are off the graph in Fig. 8.21. To be able to compare two values that are on the graph, let L_s increase to 4000 cd/m^2. This decreases L_F to 92 cd/m^2 and increases UGR to 12.7. From Eq. (8.14), M becomes 58 giving a VCP of about 84 from Fig. 8.20. The graph in Fig. 8.21 indicates a VCP of about 88.

In recent years, researchers concerned with lighting quality have begun to assess the discomfort resulting from overhead glare from sources above the 53° field used in computing VCP (Ngai and Boyce, 2000). The inference in using VCP has been that luminaires above the 53° viewing angle do not produce significant discomfort glare. Though this may have been true for diffusing-lensed troffers, it is not true for the current generation of parabolic-louvered luminaires, with their more concentrated intensity distributions, and for CFL downlighting units. Light from such luminaires is likely to reflect into the eyes from portions of the face, and the effect is particularly severe for people wearing glasses, with light reflecting into their eyes from the lenses and rims. Experiments are being conducted, involving both experts and lay people as subjects, in an effort to identify and quantify the ramifications of overhead glare, with the purpose ultimately of creating a quality metric to describe the effect.

8.9 LIGHTING CONTROLS

A major motivation for using electrical controls in lighting installations is to save energy and power, or perhaps more honestly, to save cost. A reduction in the number of kilowatt hours consumed by a lighting installation obviously results in a lower utility bill, but utilities also impose demand charges based on kilowatt requirements, so it also behooves a user to smooth electrical loads by removing peaks in power demand. Studies have shown that electrical controls can reduce lighting energy usage by up to 50% in existing buildings.

There are several additional benefits from the utilization of electrical controls. Worker productivity can be enhanced, not from maximum illumination in a space, but from optimal illumination. There is a variety of visual tasks in the work place and a range of vision differences in the workers performing them, particularly as a function of age. It is important for workers to be able to adjust their lighting levels to accommodate these needs. This will permit them to do their jobs better and will provide them satisfaction and comfort.

There are several other benefits of electrical lighting controls:

- Reduction in energy consumption reduces energy production, which reduces the emissions of air pollutants from power plants—carbon dioxide, sulfur dioxide, and nitrous oxides.
- Accommodation of daylight in a space is facilitated and made economical if lighting can be turned off or lowered selectively.
- Multi-function spaces, such as conference rooms, benefit from electrical control so that lighting appropriate to the function being carried on can be provided. This need is both a visibility and an aesthetic requirement.
- In retail spaces, lighting must create an appropriate atmosphere, attract shoppers to the merchandise and provide sufficient light for its appraisal. Lighting controls are a must to carry out these functions.
- Workers often move from one work station to another, or a worker's task at a given work station can vary throughout the day. This is especially true of computer work stations and requires lighting flexibility.
- Large area spaces must be reorganized as requirements change, for example, the need for new or reoriented work stations in an open plan office.
- The opportunity for efficient overall facility management practices is enhanced by having control of lighting and HVAC through standard building/energy management systems.
- In order to meet the many industry and governmental energy codes, it is essential that lighting controls be utilized. Many states, as well as the ASHRAE/IESNA 90.1 and the Canadian energy code, suggest or mandate lighting controls through building standards.

There are seven basic functions performed by electrical lighting controls. These are:

- Occupancy sensing
- Automatic scheduling

- Task tuning and luminance balancing
- Controlled response to daylight
- Lumen maintenance
- Load shedding for demand control
- Day-night compensation.

Each of these will be discussed in some detail, noting in each case that control can be manual or automatic, that it can be achieved by switching or dimming or both, that it can involve predictable or unpredictable scheduling of space, that it can be achieved by local manual or central microprocessor control, and that it can be accomplished through hard wiring or radio-controlled linkages.

Occupancy Sensing

Occupancy sensing is used in spaces that are occupied intermittently, to turn on lights when people are present and to turn them off when they are not. There are two principal types, passive infrared (PIR) and ultrasonic (US). PIR sensors respond to the motion of heat sources, i.e., the movement of people. Their peak sensitivity is at 1000 nm, which is also the peak of the heat emission spectrum of people. They utilize a pyroelectric detector, which converts IR energy into a voltage, mounted behind a Fresnel lens. The lens has an etched optical pattern that presents a field of wedge-shaped segments to the detector, as shown in Fig. 8.22. The motion of a person through the pattern triggers the sensor to turn the lights on. A timing control continues light for a period of time (3 sec to 30 min) after the space is vacated. Normal movements of people should sustain the lighting when the space is occupied. This can be a problem if activities in a space do not require motion.

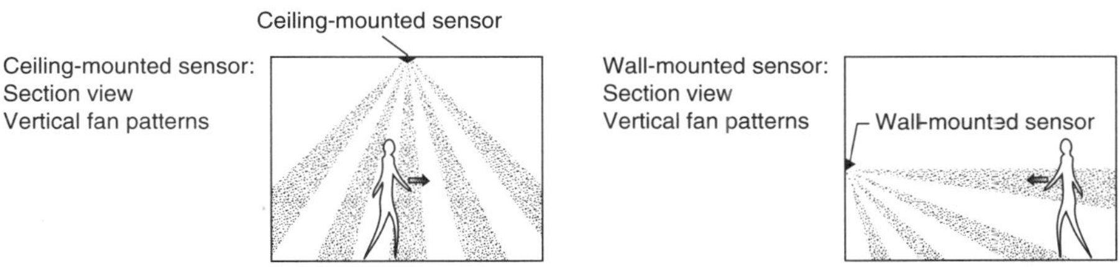

Figure 8.22 Motion detection with infrared sensing. *Source: IESNA Lighting Handbook*, 9th edition, 2000. Courtesy of the Illuminating Engineering Society of North America.

PIR sensors are strictly line of sight, i.e., they cannot "see" around corners. Thus they are most useful in small areas, such as one-person offices and conference rooms, and are not as useful in open-plan offices and partitioned spaces. As distance from the sensor increases, the gaps between the lens wedges increase (see Fig. 8.22). PIR sensors are sensitive generally to hand motion (small) up to 10 ft, arm motion (medium) up to 20 ft, and full-body movement (large) up to 30 ft and are most sensitive to motion perpendicular to the direction of detector viewing.

The second major type of occupancy sensor is ultrasonic. A US sensor contains a quartz crystal oscillator that generates frequencies of 32 kHz or higher. As shown in Fig. 8.23, the signal transmits into the space and reflects from objects, including people, in the space. The reflected wave has the same frequency as the incident wave if the object is stationary, but is higher or lower if the object is moving (Doppler shift). It is the changes in the signal's return time to the detector that create a voltage to control the lights. In Fig. 8.23, reflection from a stationary object (wall) to a wall-mounted sensor is shown in the left figure, and from a moving person in the center and right figures.

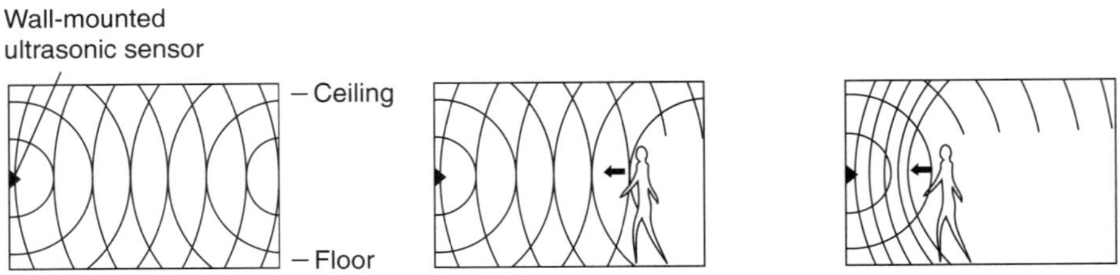

Figure 8.23 Motion detection with ultrasonic sensing. *Source*: *IESNA Lighting Handbook*, 9th edition, 2000. Courtesy of the Illuminating Engineering Society of North America.

US sensors have no blind spots; they can "see" around corners and behind barriers. Thus they are more effective in detecting occupancy in areas such as partitioned spaces. Hand motion can be detected up to 25 ft, arm motion up to 30 ft, and full-body motion up to 40 ft. However, US sensors are more likely to turn the lights on in the absence of occupants, for example from the movement of a curtain.

Both types of sensors can be mounted in several ways: on the ceiling, high and in the center of a wall, at the intersection of two walls, in a wall switch box. The floor area covered by a single sensor depends upon the beam pattern of the sensor and its signal strength and can vary from 150 ft^2 to as much as 2000 ft^2. Beam patterns can be elliptical, circular, fan-shaped or rectangular. Sensors combining both PIR and US technologies are now available.

Automatic Scheduling

Scheduling may be defined as predetermining when the lighting is to be on, off, or dimmed. Scheduling assumes a reasonably-predictable and repetitive occurrence of lighting requirements, such as for lunch periods and cleaning operations, where work and occupancy schedules are pre-defined. In such cases, it may be desirable to have centralized automatic control. This control strategy divides a building into zones, independently controlled and sized by building geometry and functions performed therein. The control is by a central microprocessor, which often handles motors and HVAC equipment, as well as lighting. With such a system all electrical loads are controlled from one central location. This makes it possible to achieve efficient building energy management. A lighting control system consists of a master lighting power controller, a programmed logic circuit, and one or more sensing devices. Such a system is diagrammed in Fig. 8.24.

For many worker activities in buildings, it is absolutely necessary to have local overriding capability of automatic central control if worker serenity is to be maintained. Manual local control is customarily accomplished by switching or dimming. Switching control can be accomplished in a variety of ways: by turning off a row of luminaires, by turning off selected luminaires in a uniform grid layout, by turning off luminaires

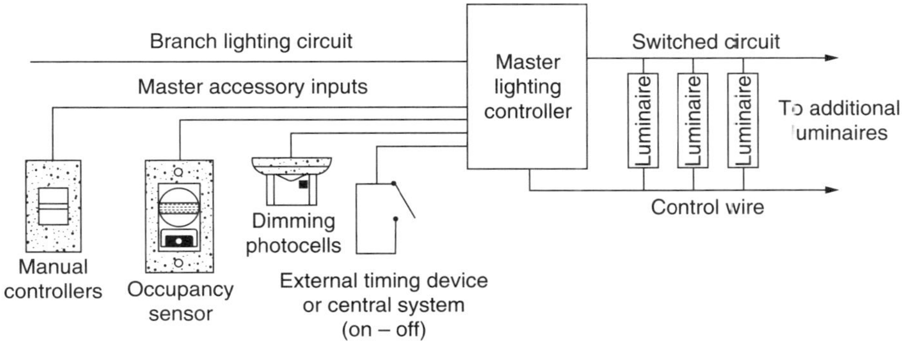

Figure 8.24 Lighting control system. ***Source*: *IESNA Lighting Handbook*, 9th edition, 2000. Courtesy of the Illuminating Engineering Society of North America.**

in a section of a room, or by turning off one or more lamps in a multi-lamp luminaire. This latter method, known as split-level wiring, is illustrated in Fig. 8.25a and b for four-lamp and three lamp luminaires. Note that this requires bi-level switching (two wall switches). It is also possible to tandem-wire two luminaires to reduce the number of ballasts required. This is shown in Fig. 8.25c. In Fig. 8.25a, three lighting levels are possible (0, 50, and 100%), while in Figs. 8.25b and c, four levels are provided (0, 33, 67, and 100%). In Fig. 8.25, the lighting settings actually shown are 50%, 33%, and 67%, for a, b, and c, respectively. An alternative to split-level and tandem-wiring is a multilevel ballast, designed to operate lamps at two or more light output levels. The most common use is with metal halide lamps in high bay applications, providing two light levels to compensate for daylight variations.

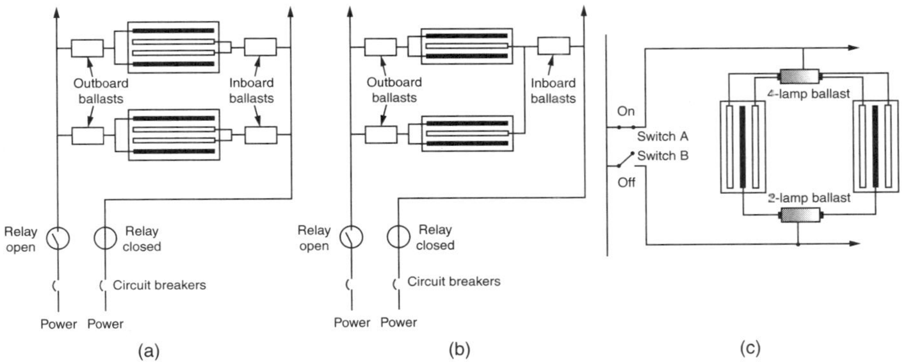

Figure 8.25 Split-level and tandem wiring. ***Source*: *IESNA Lighting Handbook*, 9th edition, 2000, and Advanced Lighting Guidelines, 2001 Edition. Courtesy of the Illuminating Engineering Society of North America and New Buildings Institute.**

Dimming systems reduce the power applied to lamps, and thus their light output. The relationship between the measured illuminance produced by this light output and the perceived illuminance is nonlinear. Figure 8.26 shows the relationship. Note that at a measured illuminance of 25% of full illuminance, the perceived brightness sensation

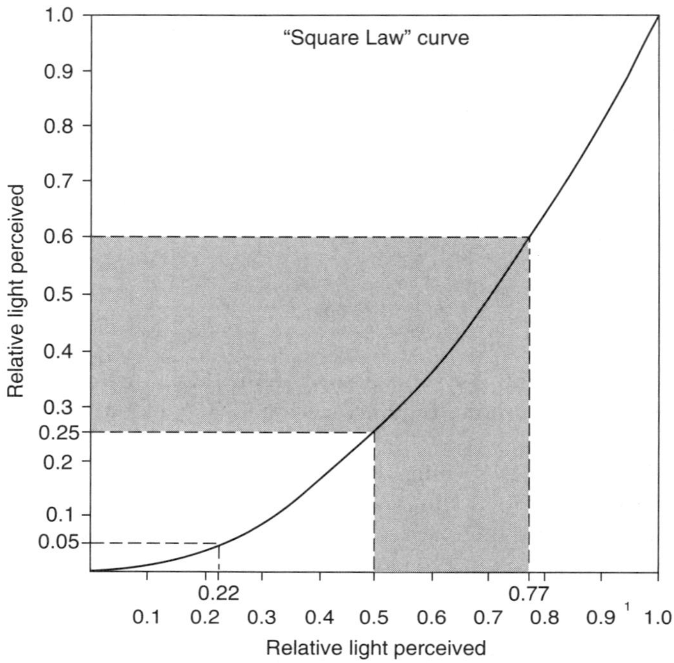

Figure 8.26 Relationship between measured illuminance and perceived brightness. ***Source*: *IESNA Lighting Handbook*, 9th edition, 2000. Courtesy of the Illuminating Engineering Society of North America.**

is 50% of what it is at full illuminance. At 5% of full illuminance, it is 22% of full perceived. What this shows is that dimming must be extensive to produce a significant decrease in perceived brightness.

Dimming of an incandescent lamp in earlier years was done by a variable resistor in series with the lamp. This scheme wasted power in the resistor. Next came the autotransformer, which was 80% efficient, but was bulky, heavy, and costly. Dimming today is accomplished by interrupting (chopping) the ac wave 120 times per second (as shown in Fig. 8.27), the wider the chop, the greater the dimming. This width is controlled by a knob or slider. Today solid-state dimmers called "triacs" are used, which are small, light, and low cost. Incandescent lamps can be dimmed from 100% to 0% of lumen output. Dimming affects lamp life of incandescent sources (see Table 6.4). It also changes a lamp's color temperature to a lower value (also Table 6.4). As noted in Sect. 6.6, tungsten halogen lamps may be dimmed from 100% to 0° of lumen output, but require operation at full voltage periodically to remove any tungsten particles that have deposited on the bulb wall. Life can be reduced if the lamp is dimmed for a long period of time.

Fluorescent lamps require special dimming ballasts, either magnetic or electronic. With magnetic ballasts, the dimming range is from 100 to 20% of light output. The ballasts become noisier as lamps are dimmed, and lamps may "drop out" at the low end of dimming. Dimming with electronic ballasts has a range of 100 to 1% of light output for linear fluorescent lamps and 100 to 5% for CFLs. Because operation is at

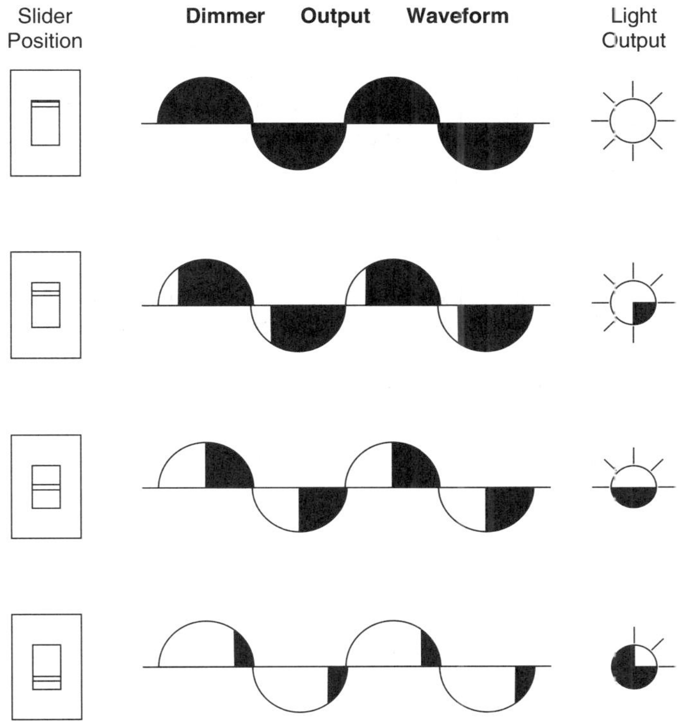

Figure 8.27 Dimmer output voltage waveform (courtesy of Lutron Electronics Co.).

high frequency (27 kHz), the light output is flicker-free and stable at low light levels. A block diagram of an electronic dimming ballast was shown in Fig. 6.14. Note that the control circuit is fed by the dimming control, which can be a sensor, a personal computer, or a central energy management system. Control is by one of three methods: phase control, pulse width modulation (PWM), and low-voltage DC. Phase control permits dimming to 1%, PWM to 10% and low-voltage DC to 20%.

HID lamps can be dimmed, but with greater difficulty than incandescent and fluorescent lamps. The lamps should not be dimmed until they reach full light output, and they respond much more slowly to dimming instructions, taking as long as 10 min. to readjust. Lamp efficacy and color hold reasonably well down to 50% of full light output for most lamps, but low wattage metal halide lamps begin to change color at 80% of full light output. High pressure sodium lamps tend to behave like low pressure sodium lamps below 50% of full light output, exhibiting the single yellow line at 590 nm.

Task Tuning and Luminance Balancing

In many spaces, particularly large open-plan offices, uniform illuminances are provided throughout. Task tuning permits adjustments in such a lighting system to provide greater illumination where the visual tasks occur and lesser illumination in areas such as aisles and reception areas. Such a lighting scheme is often labeled "task-ambient," an ambient lighting level throughout a space supplemented by task lighting. Task tuning can be accomplished by either switching or dimming. In the case of switching, individual luminaires may be switched on or off, or multilevel switching of a small group of luminaires may be used. Control should be local and manual, to allow the individual worker to optimize the illumination for the particular task.

Task tuning can also achieve luminance balancing in a space, the balancing of luminance levels so that extreme luminance ratios are avoided, between windows and their adjacent walls, between luminaires and surrounding ceiling space, and between visual tasks and their immediate backgrounds. Some of this balancing can be provided by window controls (blinds or louvers), by using luminaires that have significant uplight, and by avoiding dark colored work surfaces. But additional balancing can be achieved by altering selectively the illuminances produced by the electric lights, through switching and dimming.

The IESNA recommends the following luminance ratio guidelines in an office environment:

Between task and adjacent VDT screen	3 to 1 or 1 to 3
Between task and adjacent surround	3 to 1 or 1 to 3
Between task and remote surround	10 to 1 or 1 to 10

Small remote luminous areas that exceed these values are acceptable, for example on art work.

Controlled Response to Daylighting

The subject of daylighting will receive a full chapter of coverage (Chap. 12). The purpose here is to explore how electric lighting is controlled when daylight has significant presence in a space. Daylight enters a space from above (skylights)* or from the side (windows). Daylighting from windows lights room perimeters, penetrating only one-to-two window heights into the space, and varies significantly throughout the day due to the location of the sun. Daylighting from skylights can be distributed more uniformly throughout a space and is much less variable during the day.

The illuminance that daylight produces can be monitored by strategically-placed photosensors. A photosensor, by definition, consists of a photocell, optics to guide the light to the cell and circuitry to conduct the photocell output current to a luminaire ballast. The current flow creates a standardized 0–10 V control voltage. If the photosensor detects a low daylight illuminance, it lowers the control current, which raises the control voltage and the luminaire output. The opposite occurs when the daylighting increases. Usually the daylight illuminance above which the lamps are turned off is made greater than the daylight illuminance below which they are switched on. This "dead zone" prevents on-off cycling of the electric lighting near the threshold level. Photocells utilize four basic forms of photoelectric phenomena; photoemissive, photoconductive junction, photoconductive bulk, and photovoltaic. These were discussed

* 62% of the floor space in commercial buildings in the U.S. is in single-story or the top floor of multistory buildings.

in Sect. 2.13. Photocells to monitor illuminance in a daylit space use the photoconductive principle—their electrical conductivity increases (resistance decreases) when exposed to light. There are two types:

- Bulk devices use a cadmium sulfide crystal as the active element. These devices are inexpensive, non-linear, sensitive to variations in illumination, low in accuracy and subject to change over time.
- Junction devices use a silicon diode as the active element. These devices are more costly, give faster response, are less sensitive to variations in illumination (high in accuracy), and more stable over time.

Photocells should be color-corrected to the eye's spectral luminous efficiency curve for photopic vision. If not, errors will result in evaluating the daylighting, which will affect the timing of turning the electric lighting on and off. The field of view of the photosensor is also a factor in this timing. This field is a function of both the optics and the location of the photosensor. The optics consist of lens, prism or fiber, and determine the solid angle of the field of view. A cone with a half angle of 30° (0.84 sr) is considered optimum. If too wide, daylight from outside the space may be detected. If too narrow, unfair emphasis may be given to a bright or dark area.

The placement of the photosensor in the space should be such that changes in illuminance at the work stations relate directly to changes in illuminance at the photosensor. The most obvious location of the sensor is at the work station, where it sees exactly what the task sees. However, wiring is awkward and task materials can obscure the sensor. Alternatively, the photosensor may be mounted on the ceiling over the work station and close to the luminaires to be controlled. In other instances, the photosensor may "look" directly at a wall and monitor the total illuminance (daylighting and electric lighting) there, or the photosensor may view the sky directly and respond just to the amount of daylighting. In general, a photosensor should not view direct light from the luminaires it is to control and it should not be shadowed from the daylighting it is receiving. It has been suggested to use a 5 to 1 ratio of task illuminance to ceiling illuminance (for a ceiling mounted photosensor) as a measure of required sensor illuminance for a given task illuminance.

As with other functions of electrical control, daylight control can be accomplished by switching or dimming. A typical situation in a rectangular space is three rows of fluorescent luminaires parallel to a window wall. The rows should be individually switched, for on-off control, multi-level control, or dimming control, with the outer row most often off or at reduced output level.

Lumen Maintenance

The illumination provided by a lighting installation decreases with time. There are four major causes of this decrease: lamp lumen depreciation, luminaire dirt, room surface dirt, and unreplaced lamp burnouts. Together they can often cause the average illminance in a space to fall to 70% of its initial value over time. This light loss is usually compensated for in lighting design by providing more illumination than is required when the installation is first turned on and falling to the design illumination at approximately 70% of lamp life, at which time the lamps are all replaced at once. Some control systems will keep track of the number of hours the lamps have been burned.

An alternative to over designing is to install electrical control that compensates for lumen and dirt depreciation by gradually increasing power to the lamps throughout life, so that full lamp power is not applied until relamping is about to occur. This lumen maintenance control can be achieved with a dimming system and photosensor monitoring, in much the same manner as was described in the section on daylighting control.

Load Shedding for Demand Control

As mentioned at the start of this section, utilities charge not only for kilowatt hours of energy consumption but also for kilowatts of power required. Load shedding is a means of controlling these latter demand charges, by reducing selectively the lighting power in certain, less critical areas during periods of peak demand. This can be particularly effective in the summer, when air conditioning requirements are high but there is also an excess of "free" daylight. There is a double benefit from reducing the electric lighting load in the summer—the savings because of reduced electric lighting power and the savings because the cooling system does not have to remove the heat generated by the lighting.

The most common demand interval is 15 minutes. A demand meter records the energy consumption for each period during a day. At the end of each billing cycle, the utility assigns a demand change based on the maximum energy consumed in one 15-minute interval.

Demand control equipment features communication between a power monitoring system and the lighting control system that sets a maximum demand and sheds load during times when demand exceeds this maximum. Control is predetermined and is automatic. Some luminaires are dimmed while others are turned off, using the electrical control methods previously discussed.

Day-Night (Dusk-Dawn) Compensation

A final control strategy is day-night compensation, by which is meant that light levels are reduced in the evening based on the assumption that people prefer less illumination to perform non-critical seeing tasks at night than during the day. The control devices used are either photosensors or timers. The mechanisms employed are either switching or dimming, with the latter preferred so as to gradually reduce the electric lighting during the evening hours. Typical applications include shopping malls, public buildings, travel concourses, supermarkets, and discount houses. In addition to day-night compensation, timers are useful in interior spaces during the day in those cases where the occupancy is very predictable. Otherwise local overrides are essential.

Time controls (clocks or switches) are either electromechanical or electronic and are operated on a fixed, user-determined schedule. A basic unit is a 24-hour time switch that can be set to turn a lighting load on and off or operate split ballasts two or more times per day. Some timers have built-in compensation for changes in the number of daylight hours, eliminating the need for photosensor control.

At the other end of the spectrum from basic time clocks are microprocessor time controls that permit detailed day-by-day programming of luminaire light output. These electronic devices provide a greater flexibility than electromechanical timers. The most advanced varieties give early warning to occupants when luminaires are about to be turned off or dimmed and allow the occupants to override the pending action.

The procedures discussed above can also be applied to weekend and holiday time schedules.

The latest development in fluorescent lighting controls is Digital Addressable Lighting Interface (DALI). DALI permits independent control of luminaires that are on the same control circuit, with minimal wiring changes. It does this through addressing each individual ballast digitally by means of a control wire. Its basic components are simply a DALI ballast, a control switch, and a source of power. DALI is versatile, permitting local room control, as well as interconnection with overall building systems.

REFERENCES

CIBSE: *CIBSE Code for Interior Lighting*, Chartered Institution for Building Services, London, 1994.

CIE: *Discomfort Glare in Interior Lighting*, CIE, Vienna, Austria, 1995.

Coaton, J., and A. Marsden: *Lamps and Lighting*, Fourth Edition, Arnold, London, 1997.

Davis, R.: *Specifier Reports: Photosensors*, National Lighting Product Information Program, Lighting Research Center, Troy, NY, March, 1988.

English, C.: "Rating the Luminaires," *LD + A*, IESNA, June, 1997.

Gordon, G., and J. Nuckolls: *Interior Lighting for Designers*, Third Edition, John Wiley and Sons, New York, 1995.

IESNA Testing Procedures Committee: *Goniophotometer Types and Photometric Coordinates, LM-75*, IESNA, New York, 2001.

IESNA Testing Procedures Committee: *Approved Method for Photometric Testing of Indoor Fluorescent Luminaires, LM-41*, IESNA, New York, 1998.

IESNA Testing Procedures Committee: *IESNA Guide for Determination of Average Luminance for Indoor Luminaires, LM-37*, IESNA, New York, 1997.

Guth, S.: "Computing Visual Comfort Ratings for a Specific Interior Lighting Installation," *Illuminating Engineering*, October 1966.

Guth, S.: "A Method for the Evaluation of Discomfort Glare," *Illuminating Engineering*, May 1963.

Leadford, K.: A12 Reverse Apex Lens, Lithonia Lighting, Conyers, GA, 2001.

Levin, R.: "The Photometric Connection," Parts 1–4, *LD + A*, IESNA, September–December, 1982.

Levin, R.: "Position Index in VCP Calculations," *Journal of the IES*, January 1975.

Levin, R.: "An Evaluation of VCP Calculations," *Journal of the IES*, July, 1973.

Lighting Control Council: *NEMA Guide to Lighting Controls*, National Electrical Manufacturers Association, Washington, DC, 1992.

Lightolier: *Comprehensive Lighting Handbook*, Lightolier, Fall River, MA, 1999.

Lindsey, J.: *Applied Illumination Engineering*, Second Edition, The Fairmont Press, Lilburn, GA, 1998.

Lithonia Lighting: *The Basics of Efficient Lighting*, Lithonia Lighting, Conyers, GA, 2000.

Maniccia, D.: *Specifier Reports: Occupancy Sensors*, National Lighting Product Information Program, Lighting Research Center, Troy, NY, May 1997.

McGowan, T., and S. Guth: "Extending and Applying the IES Visual Comfort Rating Procedure," *Illuminating Engineering*, April 1969.

Mistrick, R., and A. Choi: "A Comparison of the Visual Comfort Probability and Unified Glare Rating Systems," *IESNA Journal*, Summer, 1999.

National Electrical Manufacturers Association: *Procedure for Determining Luminaire Efficacy Ratings for Fluorescent Luminaires*, NEMA Publication LE5, Washington, DC, 1995.

Ngai, P., and P. Boyce: "The Effect of Overhead Glare on Visual Discomfort," *IESNA Journal*, Summer, 2000.

OSRAM SYLVANIA: *Reading the Photometric Report*, OSRAM SYLVANIA, 1996.

Pitsor, K.: "What is LER?," National Lighting Collaborative, Rosslyn, VA, 1999.

Purdy, S.: "DALI is Here to Stay", *LD + A*, January, 2003.

Rea, M. editor: *IESNA Lighting Handbook*, Ninth Edition, IESNA, New York, 2000.

Roberts, J. editor: *Advanced Lighting Guidelines*, New Buildings Institute, White Salmon, WA, 2001.

RQQ Committee: "Report No. 2" and "Appendix to report No. 2," *Journal of the IES*, April and July 1973.

Steffy, G.: *Architectural Lighting Design*, Van Nostrand Reinhold, New York, 1990.

Chapter 9

Interior Lighting Design: Average Illuminance

9.1 INTRODUCTION

This chapter considers in detail one important facet of interior lighting design, the provision for an adequate average maintained illuminance on the work plane in an interior space to insure that all visual tasks may be carried out with high accuracy and reasonable speed. This is not the only facet that the lighting designer must consider, but it is a significant one, and it impacts the others. In spaces where visual work is done, sufficient illuminance, properly applied, is a key ingredient in a successful overall lighting design.

Average illuminance is essentially a quantitative metric. What other factors does the lighting designer need to consider? The additional factors are largely qualitative and address whether the lighting is such that workers can perform their visual tasks without physical or mental duress in a space that is pleasant and enjoyable. There are two major factors to consider:

1. Health, Safety, and Well-Being

   This factor addresses the degree of physical and mental comfort of the worker. Is there sufficient vertical illuminance to see people's faces and interpret facial expressions? Is emergency lighting provided to enhance workers' feelings of safety? Do lamps flicker and do ballasts hum, producing annoyance? Are discomfort or disability glare present, particularly with regard to computer screens which may cause workers to assume awkward neck and back positions? Are stair edges, corridor boundaries, and wall-to-floor intersections clearly discernible (of particular importance to older people)? Are vertical and oblique visual tasks properly lighted? Is the lighting appropriate to the task surface characteristics, for example unidirectional grazing lighting to accent the variations in a rough surface or multidirectional area lighting to provide shadow-and glare-free lighting on a smooth shiny surface? Are aisle ways and machines lighted adequately for worker safety?

2. Atmosphere

   The space and its lighting should have amenity, that is it should be pleasing aesthetically. People should want to be in the space and should find it comfortable. There are many factors affecting atmosphere. What is the appearance of the space, of its surface finishes, and its luminaires? Does the space accommodate daylighting properly—its variability, its intensity, its directivity (sunlight), and its potential for producing glare? How is color utilized and integrated throughout the space? Does the space seem warm, cool, or neutral? Are luminance ratios in the space within IESNA guidelines? Does the space have vitality and sparkle without being glary? Is the lighting controllable, to create different levels, patterns, and moods?

9.2 PSYCHOLOGICAL ASPECTS OF LIGHTING

In the 1970s, Dr. John Flynn did an extensive series of studies on the psychological aspects of lighting (Flynn, et al., 1979). Flynn identified seven subjective visual impressions:

- visual clarity
- spaciousness
- color tone
- public/private
- relaxed/tense
- pleasantness
- preference

He also identified four opposing lighting modes that affect these impressions:

- overhead/peripheral
- uniform/non-uniform
- bright/dim
- visually warm/visually cool

These became known as the peripheral factor, uniformity factor, level factor, and color factor.

Flynn created a seven-step bipolar (semantic differential) rating scale for each of the seven visual impressions. An example is shown in Fig. 9.1. For a variety of lighting installations, observers were asked to place a check mark on one of the seven lines for each visual impression. Subjects were told to place a mark on the left-most (1) or right-most(7) line if they felt very strongly about an impression and on the center (4) line if they were neutral.

Lighting Installation ______________________________

Clear	____	____	____	____	____	____	____	Hazy
Confined	____	____	____	____	____	____	____	Spacious
Warm	____	____	____	____	____	____	____	Cool
Public	____	____	____	____	____	____	____	Private
Relaxed	____	____	____	____	____	____	____	Tense
Pleasant	____	____	____	____	____	____	____	Unpleasant
Like	____	____	____	____	____	____	____	Dislike

Figure 9.1 Visual impressions questionnaire.

Visual impression profiles were prepared for the average of the results for a group of subjects for each of the seven visual impressions. An example is shown in Figure 9.2 for the clear-hazy impression. The vertical numbers refer to the six (in this case) lighting installations evaluated.

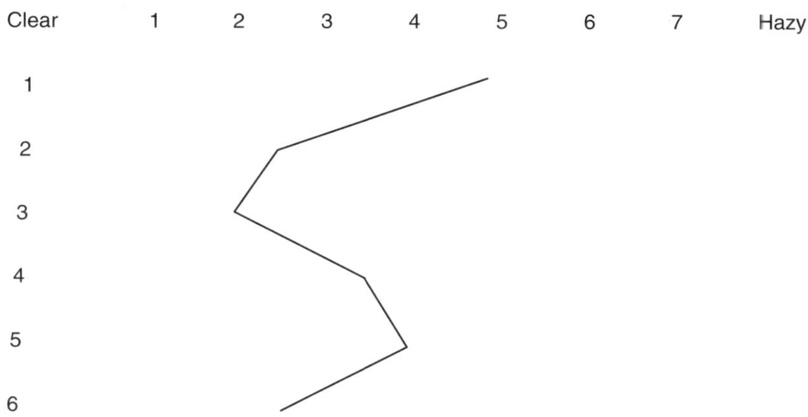

Figure 9.2 Visual impression profile—one characteristic.

Alternatively, a visual impression profile for a single lighting installation for all seven visual impressions can be presented. An example is shown in Figure 9.3.

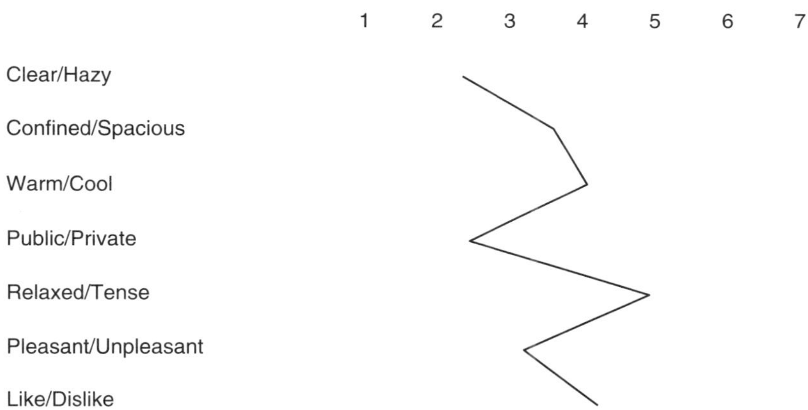

Figure 9.3 Visual impression profile—one lighting installation.

General conclusions from the Flynn work are that the following visual impressions are reinforced by the bulleted items:

Visual Clarity

- high luminance on the horizontal plane
- high luminance in the center of the space
- wall lighting
- cool sources

Spaciousness

- uniform lighting
- wall lighting

Relaxation

- nonuniform lighting
- wall lighting
- warm sources

Privacy

- nonuniform lighting
- low luminance in the vicinity of the occupant
- higher luminances in the zones surrounding the occupant
- wall lighting

Pleasantness

- nonuniform lighting
- wall lighting

After more than 20 elapsed years, this work continues to be valuable and to attract attention (Prof. Flynn died prematurely in 1980). It has been reinforced by the results from British experiments (Hawkes, et al., 1979) and continues to be utilized in lighting design. It is not only the conclusions from this work (previously stated) that are important, but also that it established a research methodology for examining the psychological aspects of lighting.

In the Flynn experiments, lamp color temperature (coolness or warmness) did not turn out to play a major role. Other experiments examining the role of CCT have postulated a direct psychological relation between illuminance and CCT, the so-called Kruitof effect, shown in Figure 9.4. This effect implies that low light levels will make a space feel cold and dim, whereas high light levels will cause a space to seem warm and colorful. In Figure 9.4, the intermediate white area is the implied region of acceptable

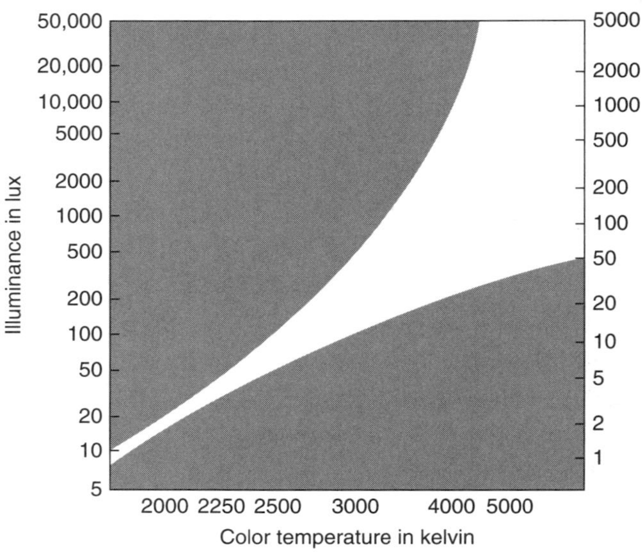

Figure 9.4 The Kruitof effect. ***Source*: *IESNA Lighting Handbook*, 9th edition, 2000. Courtesy of the Illuminating Engineering Society of North America.**

CCT and illuminance. This result has been replicated by some and not by others. The latter (Davis, et al., 1990) have found that after observers have spent sufficient time in a space to become color-adapted, illuminance, and not CCT, is the dominant parameter governing how people feel.

9.3 ILLUMINANCE SELECTION

Charles Steinmetz, in "Excerpts from 1916 Lectures on Illuminating Engineering," quotes two statements on school lighting and factory lighting. The first, attributable to F.A. Vaughn, states:

> "A committee of the society, having in charge the preparation of a Code of School Lighting, is practically ready to submit for general adoption a splendid compilation of scientific and common-sense suggestions, the general adoption and enforcement of which would be of inestimable benefit to mankind."

The recommendations that were put forth appear in Table 9.1.

TABLE 9.1 1916 SCHOOL LIGHTING RECOMMENDATIONS

Area	Minimum Footcandles	Desirable Footcandles
Storage, stairways	0.25	0.25–5.0
Classrooms, libraries	3.0	3.0–5.0
Serving, drafting	5.0	5.0–10.0
Blackboards	3.0	3.0–5.0
Auditoriums	2.0	2.0–4.0

The second quote, attributable to C.E. Clewell, is:

> "In all of these recommendations, an effort is apparent to make sure of sufficient illumination over every factory area so that operators may carry on their work without risk of accident and without loss of time."

These recommendations appear in Table 9.2.

These two tables, and the accompanying prose have been included, to point out that lighting recommendations, while one would like to believe that they are based solely on sound science, are usually based on consensus, which considers science but reflects the times—what lighting levels are people used to, what light sources are available, how

TABLE 9.2 1916 FACTORY LIGHTING RECOMMENDATIONS

Area	Footcandles
Bakery	2–3
Drafting room	7
Garments—light	5
Garments—dark	7
Offices	3
Paint shop	4–8
Woodworking	3–5

much do they cost, how much does energy cost, what visual work is there to do, what does society desire,..?

The evolution of a formalized IESNA illuminance selection procedure begins with the first IES lighting handbook in 1947, in which a table of "Levels of Illumination—Good Current Practice" is presented with no rationale, other than to state that casual tasks should have 10 fc, ordinary tasks 30 fc, and difficult tasks 50 fc. At this time, the British IES had a well-developed code for interior lighting based on laboratory data (Weston's work), with well thought out illuminance recommendations that included both size and contrast. This is shown in Figure 9.5.

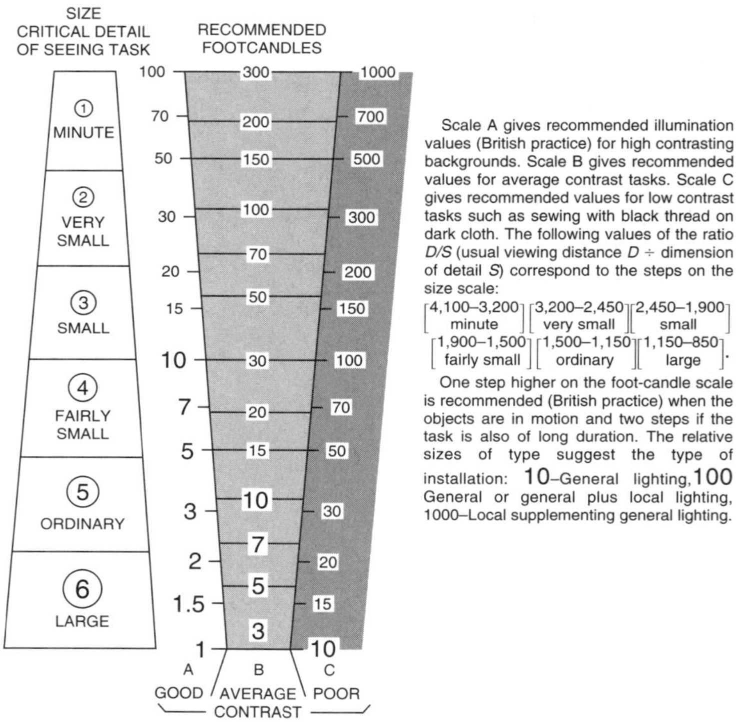

Figure 9.5 1947 British IES interior lighting code.

Note the detail of this presentation. Three levels of contrast are shown, appropriately darkened. There are six sizes, and their encircled numbers are sized accordingly. Assuming a viewing distance of 15 in., the D/S ratios given transform to sizes ranging from about 1 minute of arc to 3.4 minutes of arc. The footcandle levels are in logarithmic intervals, as the eye responds. The three sizes of footcandle numbers indicate the need for general lighting, general plus local lighting, and supplementary lighting.

The next change in North American illuminance recommendations occurs in the third IESNA handbook (1959). By then Blackwell's vision research (Sect. 3.14) was being utilized. An extensive table of illumination levels is given, with values considerably higher than in previous handbooks and essentially consensus-based. For example, office lighting recommendations rose from 5 fc to 20 fc for corridors and stairways and from 30 fc to 100 fc for regular office work. As a second example, store lighting levels went from 20 fc to 30 fc for circulation areas and from 50 fc to 100 fc for merchandising areas. The reasons for these increases are not completely clear, but it should be noted that the IES was beginning to move away from Weston's recommendations and toward Blackwell's findings. Also, energy was cheap and fluorescent lamps and luminaires were plentiful.

There is essentially no change in the table of recommended illuminances in the fourth handbook (1966), and it is again consensus based. However, it is stated that the IES had adopted the Blackwell method for prescribing task illuminances. In the fifth handbook (1972), the illuminance table included values for equivalent sphere illumination (ESI) for some tasks. For example, the illuminance numbers for offices didn't change, but now they were ESI numbers, rather than average illuminances. For the first time, there was an illuminance selection procedure, based on Blackwell's visual performance metric (VL8). Instructions to the designer were to "use the Lumen and Point calculation methods and consider the effects of veiling reflections". These instructions were, at best, confusing.

It is in the sixth handbook (1981) that a formalized illuminance selection procedure first appears. It organized visual tasks into nine categories (A through I) according to Table 9.3. Occasional visual tasks are A-C, normal tasks D-F, and critical tasks G-I. In each of the nine categories, three levels (max, mid, and min) are given. The Handbook states, "In recent years it became apparent, through on-going research and design experience, that it was time to move away from the single-value recommendation to a range approach—illuminance ranges accompanied by a weighting-factor guidance system reflecting lighting-performance trends found in research". The weighting system asks the user to consider occupant age, speed and accuracy requirements, and task background reflectance in determining which of the three ranges to assign.

The illuminance selection procedure included four steps: (1) define the visual task, (2) select the illuminance category, (3) determine the illuminance range, and (4) establish an illuminance target value. An illuminance table was included that assigned illuminance categories to a myriad of visual tasks. Also included was a table of equivalent contrasts to replace illuminance, if desired, in the selection of an illuminance category.

This procedure was offered as "interim" in 1981, but became the accepted procedure and remained so until the ninth handbook (2000). There were minor changes in the seventh (1987) and eighth (1993) handbooks, the most important being the elimination of the equivalent contrast table, which effectively removed the last vestiges of the Blackwell-based method introduced in 1959.

The current IESNA illuminance selection procedure (2000) differs significantly in two ways from the 1981 version:

- The recommended illuminance categories are reduced to seven and presented in three subsets, as shown in Table 9.4, with each subset described in some detail. There is no general table of illuminance values for specific applications, but rather an IESNA Lighting Design Guide that includes horizontal illuminance as one of 18 design considerations. The illuminances are consensus values.
- The three ranges within each illuminance category are abandoned, largely because it is felt by some that the parameters defining these ranges are not well-defined and readily quantifiable and thus should not be included. As a result, the issues of speed and accuracy, the effect of errors on health and safety, worker age, worker visual capacity, and the effect of prolonged intense viewing, for example, are not considered directly by the designer in illuminance selection.

TABLE 9.3 1981 IES ILLUMINANCE CATEGORIES

Type of Activity	Illuminance Category	Ranges of Illuminances: Lux	Ranges of Illuminances: Footcandles	Reference Work-Plane
Public spaces with dark surroundings	A	20-30-50	2-3-5	
Simple orientation for short temporary visits	B	50-75-100	5-7.5-10	General lighting throughout spaces
Working spaces where visual tasks are only occasionally performed	C	100-150-200	10-15-20	
Performance of visual tasks of high contrast or large size	D	200-300-500	20-30-50	
Performance of visual tasks of medium contrast or small size	E	500-750-1000	50-75-100	Illuminance on task
Performance of visual tasks of low contrast or very small size	F	1000-1500-2000	100-150-200	
Performance of visual tasks of low contrast and very small size over a prolonged period	G	2000-3000-5000	200-300-500	
Performance of very prolonged and exacting visual tasks	H	5000-7500-10000	500-750-1000	Illuminance on task, obtained by a combination of general and local (supplementary lighting)
Performance of very special visual tasks of extremely low contrast and small size	I	10000-15000-20000	1000-1500-2000	

TABLE 9.4 2000 IESNA ILLUMINANCE CATEGORIES

Orientation and simple visual tasks. Visual performance is largely unimportant. These tasks are found in public spaces where reading and visual inspection are only occasionally performed. Higher levels are recommended for tasks where visual performance is occasionally important.

A	Public spaces	30 lx (3 fc)
B	Simple orientation for short visits	50 lx (5 fc)
C	Working spaces where simple visual tasks are performed	100 lx (10 fc)

Common visual tasks. Visual performance is important. These tasks are found in commercial, industrial and residential applications. Recommended illuminance levels differ because of the characteristics of the visual task being illuminated. Higher levels are recommended for visual tasks with critical elements of low contrast or small size.

D	Performance of visual tasks of high contrast and large size	300 lx (30 fc)
E	Performance of visual tasks of high contrast and small size, or visual tasks of low contrast and large size	500 lx (50 fc)
F	Performance of visual tasks of low contrast and small size	1000 lx (100 fc)

Special visual tasks. Visual performance is of critical importance. These tasks are very specialized, including those with very small or very low contrast critical elements. Recommended illuminance levels should be achieved with supplementary task lighting. Higher recommended levels are often achieved by moving the light source closer to the task.

G	Performance of visual tasks near threshold	3000 to 10,000 lx (300 to 1000 fc)

The current procedure consists simply of defining the visual tasks and then selecting one of the seven illuminance categories to determine the appropriate horizontal illuminance. It is presumed, but is not stated, that these are minimum values of average illuminance. The designer is given some latitude in whether to accept the recommended illuminance. It is stated in the Handbook that "other criteria in the Design Guide may be more important than (task) illuminance." Thus the designer is invited to deviate up to ±1/3 from the design illuminance (two standard deviations), for reasons that are not well delineated.

The new procedure has received a mixed response. There are many who feel that the procedure should include an evaluation and inclusion of the factors listed earlier. One proposed scheme (Murdoch, 1998) includes two ranges, normal and difficult, within each illuminance category (this procedure was initially approved by the IESNA Board of Directors, and subsequently disapproved). The normal value would most often apply. The difficult value, with an illuminance twice that of the normal value, would apply only if a numerical assessment and summation of the extenuating factors yielded a number greater than a certain value.

As has been seen in Chap. 3, the determination of visual performance as a function of size, contrast, and luminance has many ramifications and many proposed solutions. Many feel (Boyce, 1996) that the consensus-based approach cannot be improved upon.

Others feel that the development and testing of visual performance models should continue. At the very least, it seems that all those factors that affect visual performance, particularly worker age and visual capacity, should be considered and included in lighting designs to the best of a designer's ability.

9.4 THE ILLUMINANCE DESIGN PROCESS

There are seven steps in the illuminance design process.

1) Define the visual tasks. Determine the types of tasks for which illuminance is to be prescribed. For each task, consider:

   - its relative importance
   - its location in the workplace
   - its duration of viewing
   - its orientation—horizontal, vertical, oblique
   - the intensity of viewing
   - its relationship to other tasks
   - its size
   - its contrast
   - its details and complexity
   - its requirement for speed
   - its requirement for accuracy
   - the ages of its viewers
   - the visual capacity of its viewers
   - the time for seeing
   - its repetitiveness
   - its colors
   - its background
   - its accessibility to daylight

2) Assess the characteristics of the space. Draw plan and elevation views. Show a floor plan with furniture, etc., locations. Describe for the space,

   - its shape
   - the reflectances of its surfaces
   - window and door locations
   - locations of items on walls—tackboards, logos, ...
   - room colors and finishes
   - ceiling form—pitched, trussed, grid dimensions, ...
   - work stations
   - compass orientation of room
   - susceptibility to dirt
   - luminaires
   - lamps
   - electrical control
   - window description and controls
   - skylights (if present)
   - partitions (if present)

3) Using the procedure outlined in the previous section, select appropriate illuminance levels, bearing in mind that if there are multiple tasks in the space, it is likely that more than one illuminance level needs to be chosen. From the requirements of the tasks just delineated, decide if the illuminance levels chosen should be increased by 1/3. Decreasing them by 1/3 is discouraged.
4) From the information in Chaps. 6 and 8 and from lamp and luminaire manufacturers' catalogs, select lamps and luminaires for general lighting, being mindful that luminaire choice, and ultimately layout, are influenced by the ceiling grid pattern. Also select tentatively lamps and luminaires for supplementary lighting, at task locations and for tackboards, logos, and the like. Consider the items in Sect. 9.1 when making these selections, for both general and supplementary lighting.
5) Estimate light losses over time. These were introduced in Chap. 8 and will be considered in detail in a later section in this chapter. The quantifiable ones are lamp burnouts, lamp lumen depreciation, luminaire dirt depreciation, and room surface dirt depreciation, and they often, collectively, reduce the illuminance level over time to 70% of its initial value.
6) Perform the lumen method calculation to obtain the average maintained illuminance in the space. The team "average" means with regard to the space and the term "maintained" means with regard to time. The basic lumen method will be developed in the next section. It gives the average illuminance on the workplane in a rectangular space with a horizontal ceiling (a rectangular parallelepiped). Extension of the method, permitting calculation of illuminances in non-rectangular rooms, rooms with non-horizontal ceilings and partitioned spaces, will be presented in later sections.
7) Execute a luminaire layout. This should be an iterative process. As each lighting layout is developed, it should be assessed with regard to the characteristics of the tasks and of the space. It is not unusual to begin with a simple uniform layout and then refine it to include consideration of the location of room furniture, doors, windows, file drawers, tackboards, and the like. All electrical controls (switches, dimmers, sensors) should be indicated. Luminaires, lamps, and ballasts should be listed in a "Schedule". This is especially important to do for lamps, so that replacement lamps will duplicate the lamp being replaced. A reflected ceiling plan should be drawn, which is the reflection of the ceiling that would be seen in a mirrored plane at the location of the workplane, 30 in. above the floor. The plan should include luminaires (and their dimensions), the ceiling grid layout, and any beams or supports present.
8) Address the issues raised in the two major factors in Sect. 9.1 to determine whether the designed installation provides quality lighting.

9.5 BASIC LUMEN METHOD

The procedure used to determine the average illuminances within a space is called the lumen method of interior lighting design. In Chap. 2, the illuminance at a point in a plane from a point source of light was calculated. Attention was restricted to light coming directly from the source to the point in question (direct component), and reflections from room surfaces (reflected component), which also result ultimately in light reaching the point, were not considered. The lumen method includes both direct and reflected components of luminous flux and deals with average illuminance on a horizontal plane, rather than with illuminance at a point.

The idea of a lumen method of interior lighting design had been proposed earlier, but it was not until 1920 that W. Harrison and E.A. Anderson introduced a standardized procedure that included both direct and reflected lumens. Their method involved separating the luminaire flux into three components (direct, indirect, and horizontal), which were then weighted by experimentally determined utilization factors that took into account room dimensions and room surface reflectances.

The Harrison-Anderson method served as the standard IES procedure until it was replaced in 1956 by the zonal factor interflectance method, an analytical procedure based on the calculation of luminous flux transfer between the surfaces in an interior space. This, in turn, was refined to the zonal cavity method, later renamed the lumen method, which is more versatile than its predecessor and has served as the standard IES procedure for determining average horizontal illuminance since 1964.

The analytical basis for the coefficients of utilization, cavity reflectances, and luminous exitance coefficients of the lumen method, based on flux transfer, will be developed in Chap. 13. The purpose of the next several sections is to become familiar with the use of the method in predicting the average maintained illuminance on the work plane in an interior space.

The lumen method is based on a three-surface model of a space, consisting of ceiling, walls, and floor, with the walls considered as a single surface. Each of the three surfaces is assigned a single diffuse reflectance, and it is assumed that each surface is uniformly lighted.

To obtain the average illuminance on the work plane, the total luminous flux reaching that plane needs to be determined. This flux is composed of two components: flux that comes directly from the luminaires to the work plane (direct component) and flux that is (repeatedly) reflected from room surfaces and then reaches the work plane (reflected component). In the lumen method, the fraction of the initial lamp lumens that ultimately reaches the work plane, both directly and from reflection, is called the coefficient of utilization (CU), and it is this number that is published in tables of luminaire data for various room dimensions and room surface reflectances (see Figs. 8.14–8.16). It should be realized that CU includes both the efficiency of the luminaire in releasing lamp lumens into the interior space and the efficiency of the room surfaces in redirecting these lumens to the work plane. Thus CU may be expressed as:

$$\mathrm{CU} = \frac{\phi_{\mathrm{wp}}}{\phi_{\mathrm{i}}} = \eta_{\mathrm{lum.}} \times \eta_{\mathrm{room}} \tag{9.1}$$

where ϕ_i is total initial lamp lumens, ϕ_{wp} is the initial lumens reaching the work plane, $\eta_{lum.}$ is luminaire efficiency, and η_{room} is room efficiency.

In the definition of CU, initial lamp lumens are used. It was noted previously that lumen output decreases as lamps age. Also luminaire and room surfaces become less clean with time, and some lamps may burn out and not be immediately replaced. The net effect of all these factors is lumped into what is known as a light loss factor (LLF). LLF is included in the calculations because it is usually of interest to provide an average illumination level maintained over time, not an average initial illumination level.

The four light losses just discussed are recoverable. There is an additional light loss factor that is not recoverable and is present from the moment lamps are initially lit.

This is ballast factor (BF), introduced in Sect. 6.14. As noted there, BF can vary considerably. A conservative estimate, which is used in lumen method calculations, is 0.93.

With CU and LLF defined an expression can be written for the average maintained work-plane illuminance as

$$E = \frac{\Phi_i (CU)(LLF)}{A} \tag{9.2}$$

where Φ_i is the total initial lamp lumens, and A is the work-plane area. The numerator of Eq. (9.2) gives the average maintained lumens reaching the work plane. The problem is now reduced to finding CU and LLF for a given interior space and luminaire.

It should be understood that LLF = BF when calculating initial average illuminance and that LLF contains, in addition, the product of the other four factors when average maintained illuminance is sought.

9.6 CAVITY RATIOS

The lumen method subdivides an interior space into three cavities as shown in Fig. 9.6a. The space beneath the work plane is termed the *floor cavity*. In the rare event that the work plane is the floor (for example, a shuffle board court), the floor cavity ceases to exist. The space above the luminaires is called the *ceiling cavity*. This space disappears if the luminaires are mounted on (surface-mounted)* or in (flush-mounted) the ceiling. The third space is named the *room cavity* and extends from the work plane to the luminaires.

Luminous flux distribution in a cavity is strongly related to the ratio of vertical surface area (vsa) to horizontal surface area (hsa)**. The former is $2h(\ell + w)$, the latter $2\ell w$, where ℓ, w, and h are the length, width, and height of the cavity. Then cavity ratio (CR) can be expressed as

$$CR = 5 \times \frac{vsa}{hsa} = \frac{5h(\ell + w)}{\ell \times w} \tag{9.3}$$

where h is h_c for the ceiling cavity, h_r for the room cavity, and h_f for the floor cavity.

The ratio of vertical surface area to horizontal surface area is generally less than 1. It was decided to use a multiplier of 5 in defining cavity ratio so as to obtain whole numbers greater than one for most rooms. Thus a 10 × 10 × 10 ft cavity has a cavity ratio of 10, whereas a 100 × 100 × 10 ft cavity has a cavity ratio of 1. As an example of the use of Eq. (9.3), a 30 × 20 × 8 ft schoolroom with flush-mounted luminaires has a room cavity ratio of 2.3, assuming a 30-in work plane, ($h_r = 8 - 2.5 = 5.5$ ft).

The smaller the value of h in Eq. (9.3), the smaller the cavity ratio; the smaller the values of ℓ and w, the larger the cavity ratio. A long, wide room with a low ceiling height would be expected to utilize lumens better than a smaller room with a high ceiling.

---

* Surface-mounted luminaires project slightly into the room, but this is ignored.

** Cavity ratio is also dependent on the length-to-width ratio, but less strongly. In determining CUs, a middle-range ℓ/w ratio of 1.6 has been assumed.

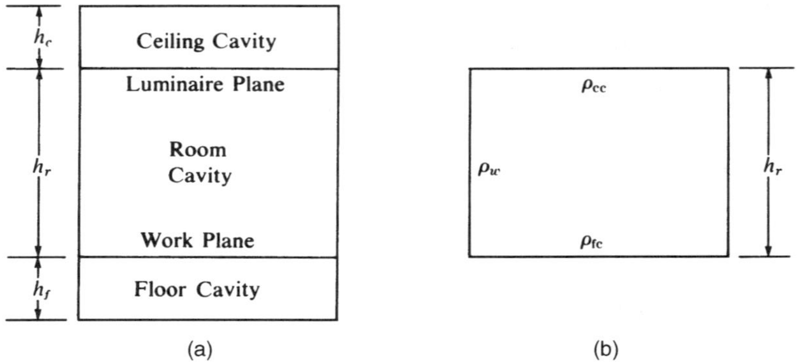

Figure 9.6 The three cavities used in the lumen method.

Thus as CR decreases, CU is expected to increase, assuming all other variables remain the same.

It is interesting to explore the meaning of a cavity ratio of 0. From Eq. (9.3), this implies that the walls are very widely separated such that luminous flux simply bounces back and forth between the ceiling of the cavity and its floor. Let

η_u = fraction of the lamp flux going upward
η_d = fraction of the lamp flux going downward

Then for a room with widely-separated walls,

$$\phi_{wp} = \phi_i \times \eta_u \rho_c (1 + \rho_c \rho_f + \rho_c^2 \rho_f^2 + \cdots) + \phi_i \times \eta_d (1 + \rho_c \rho_f + \rho_c^2 \rho_f^2 + \cdots) \quad (9.4)$$

where ρ_c an ρ_f are ceiling and floor reflectances, respectively. The series in Eq. (9.4) converge ($\rho_c \rho_f < 1$) and can be expressed in closed form as

$$(1 + \rho_c \rho_f + \rho_c^2 \rho_f^2 + \cdots) = \frac{1}{1 - \rho_c \rho_f} \quad (9.5)$$

Then the coefficient of utilization is

$$CU = \frac{\phi_{wp}}{\phi_i} = \frac{\eta_u \rho_c + \eta_d}{1 - \rho_c \rho_f} \quad (9.6)$$

For example, if a very wide room has a ceiling reflectance of 70% and a floor reflectance of 20%, and if 25% of the lamp flux is upward and 45% is downward (luminaire efficiency of 70%), the coefficient of utilization is

$$CU = \frac{0.25 \times 0.70 + 0.45}{1 - 0.70 \times 0.20} = 0.73$$

A square room has a CU, from Eq. (9.3) of $10h/\ell$. For such a room to be considered very wide, such that Eq. (9.6) applies, it is sufficient that $\ell \geq 20h$, giving a cavity ratio of ≤ 0.5.

The lumen method was developed for spaces that were rectangular parallelepipeds, six-surface spaces with all sides rectangular. It will be shown later that the method can be extended to spaces with nonrectangular floors and sloped ceilings. Thus, the lumen method is indeed versatile and robust.

9.7 CAVITY REFLECTANCE

The next step in the lumen method is to replace the original three-cavity space by its room cavity only. The reasoning here is that if effective reflectances for the two boundary planes between the room cavity and the ceiling and floor cavities can be found, there is no need to retain these latter cavities in the calculations. This is illustrated in Fig. 9.6(b) where ρ_{cc} and ρ_{fc} are defined as the effective reflectances of the ceiling cavity and floor cavity, respectively.

The effective reflectance of a cavity is the ratio of the total flux out of the cavity opening to the total flux into the cavity opening, as shown in Fig. 9.7 for a floor cavity. The resulting expression, involving multiple reflections and flux transfers within the cavity, is developed in Chap. 13. Calculated values of effective cavity reflectance for cavities with horizontal bases are given in Table 9.5. This table handles all flat-ceiling (and floor) rooms. For rooms with non-flat ceilings, an approximate expression for cavity reflectance will be presented later in the chapter.

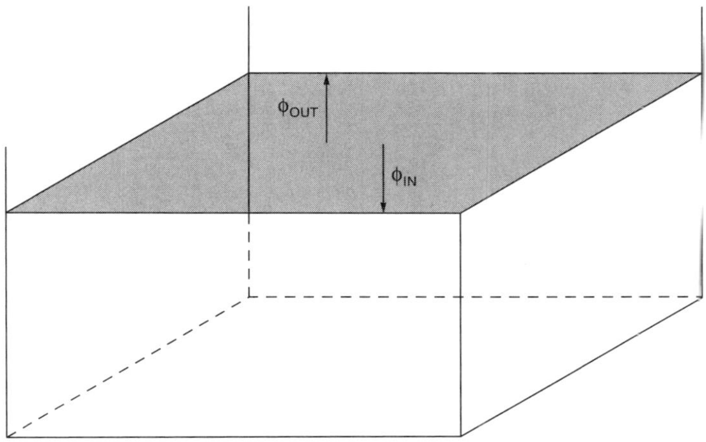

Figure 9.7 Defining cavity reflectance.

Reflectances of typical room finishes are given in Table 9.6. In most cases, diffuse (matte) or, at worst, semi-gloss finishes should be used on walls and ceilings. Floor tiles and woods should preferably be non-glossy.

EXAMPLE 9.1

A school room 40 × 20 × 12 ft has reflectances of 70%, 50%, and 30% (ceiling, walls, floor). Luminaires are suspended 2 ft below the ceiling. Work plane height is 30 in. Find the room cavity ratio and the ceiling and floor cavity reflectances.

TABLE 9.5 PERCENT CEILING OR FLOOR CAVITY REFLECTANCES (VALUES IN THIS TABLE ARE BASED ON A LENGTH-TO-WIDTH RATIO OF 1.6)

Percent Base Reflectance: 90

Cavity Ratio \ Percent Wall Reflectance	90	80	70	60	50	40	30	20	10	0
0.2	89	88	88	87	86	85	85	84	84	82
0.4	88	87	86	85	84	83	81	80	79	76
0.6	87	86	84	82	80	79	77	76	74	73
0.8	87	85	82	80	77	75	73	71	69	67
1.0	86	83	80	77	75	72	69	66	64	62
1.2	85	82	78	75	72	69	66	63	60	57
1.4	85	80	77	73	69	65	62	59	57	52
1.6	84	79	75	71	67	63	59	56	53	50
1.8	83	78	73	69	64	60	56	53	50	48
2.0	83	77	72	67	62	56	53	50	47	43
2.2	82	76	70	65	59	54	50	47	44	40
2.4	82	75	69	64	58	53	48	45	41	37
2.6	81	74	67	62	56	51	46	42	38	35
2.8	81	73	66	60	54	49	44	40	36	34
3.0	80	72	64	58	52	47	42	38	34	30
3.2	79	71	63	56	50	45	40	36	32	28
3.4	79	70	62	54	48	43	38	34	30	27
3.6	78	69	61	53	47	42	36	32	28	25
3.8	78	69	60	51	45	40	35	31	27	23
4.0	77	69	58	51	44	39	33	29	25	22
4.2	77	62	57	50	43	37	32	28	24	21
4.4	76	61	56	49	42	36	31	27	23	20
4.6	76	60	55	47	40	35	30	26	22	19
4.8	75	59	54	46	39	34	28	25	21	18
5.0	75	59	53	45	38	33	28	24	20	16
6.0	73	61	49	41	34	29	24	20	16	11
7.0	70	58	45	38	30	27	21	18	14	08
8.0	68	55	42	35	27	23	18	15	12	06
9.0	66	52	38	31	25	21	16	14	11	05
10.0	65	51	36	29	22	19	15	11	09	04

Percent Base Reflectance: 80

Cavity Ratio \ Percent Wall Reflectance	90	80	70	60	50	40	30	20	10	0
0.2	79	78	78	77	77	76	76	75	74	72
0.4	79	77	76	75	74	73	72	71	70	68
0.6	78	76	75	73	71	70	68	66	65	63
0.8	78	75	73	71	69	67	65	63	61	57
1.0	77	74	72	69	67	65	62	60	57	55
1.2	76	73	70	67	64	61	58	55	53	51
1.4	76	72	68	65	62	59	55	53	50	48
1.6	75	71	67	63	60	57	53	50	47	44
1.8	75	70	66	62	58	54	50	47	44	41
2.0	74	69	64	60	56	52	48	45	41	38
2.2	74	68	63	58	54	49	45	42	38	35
2.4	73	67	61	56	52	47	43	40	36	33
2.6	73	66	60	55	50	45	41	38	34	31
2.8	73	65	59	53	48	43	39	36	32	29
3.0	72	65	58	52	47	42	37	34	30	27
3.2	72	65	57	51	45	40	35	33	28	25
3.4	71	64	56	49	44	39	34	32	27	24
3.6	71	63	54	48	43	38	32	30	25	23
3.8	70	62	53	47	41	36	31	28	24	22
4.0	70	61	53	46	40	35	30	26	22	20
4.2	69	60	52	45	39	34	29	25	21	18
4.4	69	60	51	44	38	33	28	24	20	17
4.6	69	59	50	43	37	32	27	23	19	15
4.8	68	58	49	42	36	31	26	22	18	14
5.0	68	58	48	41	35	30	25	21	18	14
6.0	66	55	44	38	31	27	22	19	15	10
7.0	64	53	41	35	28	24	19	16	12	07
8.0	62	50	38	32	25	21	17	14	11	05
9.0	61	49	36	30	23	19	15	13	10	04
10.0	59	46	33	27	21	18	14	11	08	03

Percent Base Reflectance: 70

Cavity Ratio \ Percent Wall Reflectance	90	80	70	60	50	40	30	20	10	0
0.2	70	69	68	68	67	67	66	66	65	64
0.4	69	68	67	66	65	64	63	62	61	58
0.6	69	67	65	64	63	61	59	58	57	54
0.8	68	66	64	62	60	58	56	55	53	50
1.0	68	65	62	60	58	55	53	52	50	47
1.2	67	64	61	59	57	54	50	48	46	44
1.4	67	63	60	58	55	51	47	45	44	41
1.6	67	62	59	56	53	47	45	43	41	38
1.8	66	61	58	54	51	46	42	40	38	35
2.0	66	60	56	52	49	45	40	38	36	33
2.2	66	60	55	51	48	43	38	36	34	32
2.4	65	60	54	50	46	41	37	35	32	30
2.6	65	59	54	49	45	40	35	33	30	28
2.8	65	59	53	48	43	38	33	30	28	26
3.0	64	58	52	47	42	37	32	29	27	24
3.2	64	58	51	46	40	36	31	28	25	23
3.4	64	57	50	45	39	35	29	27	24	22
3.6	63	56	49	44	38	33	28	25	22	20
3.8	63	56	49	43	37	32	27	24	21	19
4.0	63	55	48	42	36	31	26	23	20	17
4.2	62	55	47	41	35	30	25	22	19	16
4.4	62	54	46	40	34	29	24	21	18	15
4.6	62	53	45	39	33	28	24	21	17	14
4.8	62	53	45	38	32	27	23	20	16	13
5.0	61	52	44	36	31	26	22	19	16	12
6.0	60	51	41	35	28	24	19	16	13	09
7.0	58	43	38	32	26	22	17	14	11	06
8.0	57	46	35	29	23	19	15	13	10	05
9.0	56	45	33	27	21	18	14	12	09	04
10.0	55	43	31	25	19	16	12	10	08	03

Percent Base Reflectance: 60

Cavity Ratio \ Percent Wall Reflectance	90	80	70	60	50	40	30	20	10	0
0.2	60	59	59	59	58	57	56	56	55	53
0.4	60	59	59	58	57	55	54	53	52	50
0.6	60	58	57	56	55	53	51	51	50	46
0.8	59	57	56	55	54	51	48	47	46	43
1.0	59	57	55	53	51	48	45	44	43	41
1.2	59	56	54	51	49	46	44	42	40	38
1.4	59	56	53	49	47	44	41	39	38	36
1.6	59	55	52	48	45	42	39	37	35	33
1.8	58	55	51	47	44	40	37	35	33	31
2.0	58	54	50	46	43	39	35	33	31	29
2.2	58	53	49	45	42	37	34	31	29	28
2.4	58	53	48	44	41	36	32	30	27	26
2.6	58	53	48	43	39	35	31	28	26	24
2.8	58	53	47	43	38	34	29	27	24	22
3.0	57	52	46	42	37	32	28	25	23	20
3.2	57	51	45	41	36	31	27	23	22	18
3.4	57	51	45	40	35	30	26	23	20	17
3.6	57	50	44	39	34	29	25	22	19	16
3.8	57	50	43	38	33	29	24	21	19	15
4.0	57	49	42	37	32	28	23	20	18	14
4.2	56	49	42	37	32	27	22	19	17	14
4.4	56	49	42	36	31	27	22	19	16	13
4.6	56	49	41	35	30	26	21	18	16	13
4.8	56	48	41	34	29	25	21	18	15	12
5.0	56	48	40	34	28	24	20	17	14	11
6.0	55	45	37	31	25	21	17	14	11	07
7.0	54	43	35	30	24	20	15	12	09	05
8.0	53	42	33	28	22	18	14	11	08	04
9.0	52	40	31	26	20	16	12	10	07	03
10.0	51	39	29	24	18	15	11	09	07	02

Percent Base Reflectance: 50

Cavity Ratio \ Percent Wall Reflectance	90	80	70	60	50	40	30	20	10	0
0.2	50	50	49	49	48	48	47	46	46	44
0.4	50	49	48	48	47	46	45	45	44	42
0.6	50	48	47	46	45	44	43	42	41	38
0.8	50	48	47	45	44	42	40	39	38	36
1.0	50	48	46	44	43	41	38	37	36	34
1.2	50	47	45	43	41	39	36	35	34	29
1.4	50	47	45	42	40	38	35	34	35	27
1.6	50	47	44	41	39	36	33	32	30	26
1.8	50	46	43	40	38	35	31	30	28	25
2.0	50	46	43	40	37	34	30	28	26	24
2.2	50	46	42	38	36	33	29	27	24	22
2.4	50	46	42	37	35	31	27	25	23	21
2.6	50	46	41	37	34	30	26	23	21	20
2.8	50	46	41	36	33	29	25	22	20	19
3.0	50	45	40	36	32	28	24	21	19	17
3.2	50	44	39	35	31	27	23	20	18	16
3.4	50	44	39	35	30	26	22	19	17	15
3.6	50	44	39	34	29	25	21	18	16	14
3.8	50	44	38	34	29	25	21	17	15	13
4.0	50	44	38	33	28	24	20	17	15	12
4.2	50	43	37	32	28	24	20	17	14	12
4.4	50	43	37	32	27	23	19	16	13	11
4.6	50	43	36	31	26	22	18	15	13	10
4.8	50	43	36	31	26	22	18	15	12	09
5.0	50	42	35	30	25	21	17	14	12	09
6.0	50	42	34	29	23	19	15	13	10	06
7.0	49	41	32	27	21	18	14	11	08	05
8.0	49	40	30	25	19	16	12	10	07	03
9.0	48	39	29	24	18	15	11	09	07	03
10.0	47	37	27	22	17	14	10	08	06	02

TABLE 9.5 *(continued)*

Percent Base Reflectance: 40

Cavity Ratio \ Percent Wall Reflectance	90	80	70	60	50	40	30	20	10	0
0.2	40	40	39	39	39	38	38	37	36	36
0.4	41	40	39	39	38	37	36	35	34	34
0.6	41	40	39	38	37	36	34	33	32	31
0.8	41	40	38	37	36	35	33	32	31	29
1.0	42	40	38	37	35	33	32	31	29	27
1.2	42	40	38	36	34	32	30	29	27	25
1.4	42	39	37	35	33	31	29	27	25	23
1.6	42	39	37	35	32	30	27	25	23	22
1.8	42	39	36	34	31	29	26	24	22	21
2.0	42	39	36	34	31	28	25	23	21	19
2.2	42	39	36	33	30	27	24	22	19	18
2.4	43	39	35	33	29	27	24	21	18	17
2.6	43	39	35	32	29	26	23	20	17	15
2.8	43	39	35	32	28	25	22	19	16	14
3.0	43	39	35	31	27	24	21	18	16	13
3.2	43	39	35	31	27	23	20	71	15	13
3.4	43	39	34	30	26	23	20	17	14	12
3.6	44	39	34	30	26	22	19	16	14	11
3.8	44	38	33	29	25	22	18	16	13	10
4.0	44	38	33	29	25	21	18	15	12	10
4.2	44	38	33	29	24	21	17	15	12	10
4.4	44	38	33	28	24	20	17	14	11	09
4.6	44	38	32	28	23	19	16	14	11	08
4.8	44	38	32	27	22	19	16	13	10	08
5.0	45	38	31	27	22	19	15	13	10	07
6.0	44	37	30	25	20	17	13	11	08	05
7.0	44	36	29	24	19	16	12	10	07	04
8.0	44	35	28	23	18	15	11	09	06	03
9.0	44	35	26	21	16	13	10	08	05	02
10.0	43	34	25	20	15	12	08	07	05	02

Percent Base Reflectance: 30

Cavity Ratio \ Percent Wall Reflectance	90	80	70	60	50	40	30	20	10	0
0.2	31	31	30	30	29	29	29	28	28	27
0.4	31	31	30	30	29	28	28	27	26	25
0.6	32	31	30	29	28	27	26	26	25	23
0.8	32	31	30	29	28	26	25	25	23	22
1.0	33	32	30	29	27	25	24	23	22	20
1.2	33	32	30	28	27	25	23	22	21	19
1.4	34	32	30	28	26	24	22	21	19	18
1.6	34	33	29	27	25	23	22	20	18	17
1.8	35	33	29	27	25	23	21	19	17	16
2.0	35	33	29	26	24	22	20	18	16	14
2.2	36	32	29	26	24	22	19	17	15	13
2.4	36	32	29	26	24	22	19	16	14	12
2.6	36	32	29	25	23	21	18	16	14	12
2.8	37	33	29	25	23	21	17	15	13	11
3.0	37	33	29	25	22	20	17	15	12	10
3.2	37	33	29	25	22	19	16	14	12	10
3.4	37	33	29	25	22	19	16	14	11	09
3.6	38	33	29	24	21	18	15	13	10	09
3.8	38	33	28	24	21	18	15	13	10	08
4.0	38	33	28	24	21	18	14	12	09	07
4.2	38	33	28	24	20	17	14	12	09	07
4.4	39	33	28	24	20	17	14	11	09	06
4.6	39	33	28	24	20	17	13	10	08	06
4.8	39	33	28	24	20	17	13	10	08	05
5.0	39	33	28	24	19	16	13	10	08	05
6.0	39	33	27	23	18	15	11	09	06	04
7.0	40	33	26	22	17	14	10	08	05	03
8.0	40	33	26	21	16	13	09	07	04	02
9.0	40	33	25	20	15	12	09	07	04	02
10.0	40	32	24	19	14	11	08	06	03	01

Percent Base Reflectance: 20

Cavity Ratio \ Percent Wall Reflectance	90	80	70	60	50	40	30	20	10	0
0.2	21	20	20	20	20	20	19	19	19	17
0.4	22	21	20	20	20	19	19	18	18	16
0.6	23	21	21	20	19	19	18	18	17	15
0.8	24	22	21	20	19	19	18	17	16	14
1.0	25	23	22	20	19	18	17	16	15	13
1.2	25	23	22	20	19	17	17	16	14	12
1.4	26	24	22	20	18	17	16	15	13	12
1.6	26	24	22	20	18	17	16	15	13	11
1.8	27	25	23	20	18	17	15	14	12	10
2.0	28	25	23	20	18	16	15	13	11	09
2.2	28	25	23	20	18	16	14	12	10	09
2.4	29	26	23	20	18	16	14	12	10	08
2.6	29	26	23	20	18	16	14	11	09	08
2.8	30	27	23	20	18	15	13	11	09	07
3.0	30	27	23	20	17	15	13	11	09	07
3.2	31	27	23	20	17	15	12	11	09	06
3.4	31	27	23	20	17	15	12	10	08	06
3.6	32	27	23	20	17	15	12	10	08	05
3.8	32	28	23	20	17	15	12	10	07	05
4.0	33	28	23	20	17	14	11	09	07	05
4.2	33	28	23	20	17	14	11	09	07	04
4.4	34	28	24	20	17	14	11	09	07	04
4.6	34	29	24	20	17	14	11	09	07	04
4.8	35	29	24	20	17	13	10	08	06	04
5.0	35	29	24	20	16	13	10	08	06	04
6.0	36	30	24	20	16	13	10	08	05	02
7.0	36	30	24	20	15	12	03	07	04	02
8.0	37	30	23	19	15	12	08	06	03	01
9.0	37	29	23	19	14	11	08	06	03	01
10.0	37	29	22	18	13	10	07	05	03	01

Percent Base Reflectance: 10

Cavity Ratio \ Percent Wall Reflectance	90	80	70	60	50	40	30	20	10	0
0.2	11	11	11	10	10	10	10	09	09	09
0.4	12	11	11	11	11	10	10	09	09	08
0.6	13	13	12	11	11	10	10	09	08	08
0.8	15	14	13	12	11	10	10	09	08	07
1.0	16	14	13	12	12	11	10	09	08	07
1.2	17	15	14	13	12	11	10	09	07	06
1.4	18	16	14	13	12	11	10	09	07	06
1.6	19	17	15	14	12	11	09	08	07	06
1.8	19	17	15	14	13	11	09	08	06	05
2.0	20	18	16	14	13	11	09	08	06	05
2.2	21	19	16	14	13	11	09	07	06	05
2.4	22	19	17	15	13	11	09	07	06	05
2.6	23	20	17	15	13	11	09	07	06	04
2.8	23	20	18	16	13	11	09	07	05	03
3.0	24	21	18	16	13	11	09	07	05	03
3.2	25	21	18	16	13	11	09	07	05	03
3.4	26	22	18	16	13	11	09	07	05	03
3.6	26	22	19	16	13	11	09	06	04	03
3.8	27	23	19	17	14	11	09	06	04	02
4.0	27	23	20	17	14	11	09	06	04	02
4.2	28	24	20	17	14	11	09	06	04	02
4.4	28	24	20	17	14	11	08	06	04	02
4.6	29	25	20	17	14	11	08	06	04	02
4.8	29	25	20	17	14	11	08	06	04	02
5.0	30	25	20	17	14	11	08	06	04	02
6.0	31	26	21	18	14	11	08	06	03	01
7.0	32	27	21	17	13	11	08	06	03	01
8.0	33	27	21	17	13	10	07	05	03	01
9.0	34	28	21	17	13	10	07	05	02	01
10.0	34	28	21	17	12	10	07	05	02	01

Percent Base Reflectance: 0

Cavity Ratio \ Percent Wall Reflectance	90	80	70	60	50	40	30	20	10	0
0.2	02	02	02	01	01	01	01	00	00	0
0.4	04	03	03	02	02	02	01	01	00	0
0.6	05	05	04	03	03	02	02	01	01	0
0.8	07	06	05	04	04	03	02	02	01	0
1.0	08	07	06	05	04	03	02	02	01	0
1.2	10	08	07	06	05	04	03	02	01	0
1.4	11	09	08	07	06	04	03	02	01	0
1.6	12	10	09	07	06	05	03	02	01	0
1.8	13	11	09	08	07	05	04	03	01	0
2.0	14	12	10	09	07	05	04	03	01	0
2.2	15	13	11	09	07	06	04	03	01	0
2.4	16	13	11	09	08	06	04	03	01	0
2.6	17	14	12	10	08	06	05	03	02	0
2.8	17	15	13	10	08	07	05	03	02	0
3.0	18	16	13	11	09	07	05	03	02	0
3.2	19	16	14	11	09	07	05	03	02	0
3.4	20	17	14	12	09	07	05	03	02	0
3.6	20	17	15	12	10	08	05	04	02	0
3.8	21	18	15	12	10	08	05	04	02	0
4.0	22	18	15	13	10	08	05	04	02	0
4.2	22	19	16	13	10	08	06	04	02	0
4.4	23	19	16	13	10	08	06	04	02	0
4.6	23	20	17	13	11	08	06	04	02	0
4.8	24	20	17	14	11	08	06	04	02	0
5.0	25	21	17	14	11	08	06	04	02	0
6.0	27	23	18	15	12	09	06	04	02	0
7.0	28	24	19	15	12	09	06	04	02	0
8.0	30	25	20	15	12	09	06	04	02	0
9.0	31	25	20	15	12	09	06	04	02	0
10.0	31	25	20	15	12	09	06	04	02	0

Source: *IESNA Lighting Handbook*, 9th Edition, 2000. Courtesy of the Illuminating Engineering Society of North America.

TABLE 9.6 REFLECTANCES OF ROOM SURFACE FINISHES

Material	% Reflectance
Clear glass or plastic	8–10
White porcelain enamel	60–80
White paint	75–90
Cream paint	70–80
Light oak	25–35
Dark oak	10–15
White plaster	90–92
Black structural glass	5–10
White structural glass	75–80
Fritted glass	30–50
Limestone	35–65
Red brick	13–15
Cement	20–30
Light gray paint	55–60
Dark gray paint	25–30
Sandstone	20–40
Buff paint	60–65
Acoustic tile	60–80

Solution. From Eq. (9.3), the room cavity ratio is

The ceiling cavity reflectance is obtained from Table 9.5. The base reflectance is 70% and the wall reflectance is 50%. From Eq. (9.3) the ceiling cavity ratio is

$$\mathrm{CCR} = \frac{5\times 2(40+20)}{40\times 20} = 0.75$$

Then Table 9.5 gives $\rho_{cc} = 0.61 = 61\%$ (using linear interpolation between CCRs of 0.60 and 0.80). The floor cavity reflectance is also obtained from Eq. (9.3) and Table 9.5.

$$\mathrm{FCR} = \frac{5\times 2.5(40+20)}{40\times 20} = 0.94$$

and

$$\rho_{fc} = 0.27 = 27\%$$

It is noted that the effect of the cavity in each case is to reduce the reflectance below that of the base reflectance.

9.8 COEFFICIENTS OF UTILIZATION

As described earlier, the coefficient of utilization (CU) is the fraction of the lamp lumens that ultimately reaches the work plane. Values of CU for a large number of luminaires used in interior spaces have been calculated using flux transfer theory (see Chap. 13). The results for 10 typical ones are given in Table 9.7.

TABLE 9.7 TYPICAL LUMINAIRES

Typical Luminaire	Typical Intensity Distribution	ρcc →	80			70			50			30			10			0
		ρw →	70	50	30	70	50	30	50	30	10	50	30	10	50	30	10	0
		RCR ↓																
1. 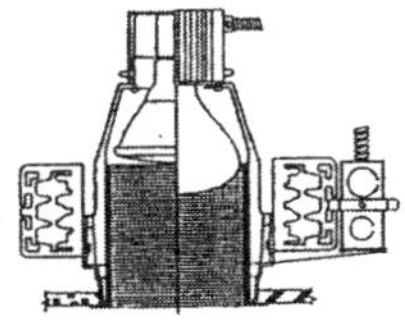 PAR-38 Flood with black baffle			EFF = 66.2%			% DN = 100			% UP = 0			Lamp = 150PAR38FL*, SC (along, across, 45°) = 0.6, 0.6, 0.6						
		0	0.79	0.79	0.79	0.77	0.77	0.77	0.74	0.74	0.74	0.70	0.70	0.70	0.68	0.68	0.68	0.66
		1	0.76	0.75	0.73	0.75	0.73	0.72	0.71	0.70	0.69	0.68	0.68	0.67	0.66	0.66	0.65	0.64
		2	0.74	0.71	0.69	0.72	0.70	0.68	0.68	0.67	0.65	0.66	0.65	0.64	0.64	0.63	0.63	0.62
		3	0.71	0.68	0.66	0.70	0.67	0.65	0.66	0.64	0.62	0.64	0.63	0.61	0.63	0.61	0.60	0.60
		4	0.69	0.65	0.63	0.68	0.65	0.62	0.63	0.61	0.60	0.62	0.60	0.59	0.61	0.60	0.58	0.58
		5	0.67	0.63	0.60	0.66	0.62	0.60	0.61	0.59	0.57	0.60	0.58	0.57	0.59	0.58	0.56	0.56
		6	0.65	0.61	0.58	0.64	0.60	0.58	0.59	0.57	0.55	0.59	0.57	0.55	0.58	0.56	0.55	0.54
		7	0.63	0.59	0.56	0.62	0.58	0.56	0.57	0.55	0.53	0.57	0.55	0.53	0.56	0.54	0.53	0.52
		8	0.61	0.57	0.54	0.60	0.56	0.54	0.56	0.53	0.52	0.55	0.53	0.52	0.55	0.53	0.51	0.51
		9	0.59	0.55	0.52	0.59	0.55	0.52	0.54	0.52	0.50	0.54	0.51	0.50	0.53	0.51	0.50	0.49
		10	0.58	0.53	0.51	0.57	0.53	0.50	0.53	0.50	0.49	0.52	0.50	0.48	0.52	0.50	0.48	0.48
2. 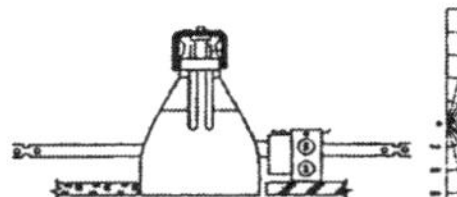 8″ Open reflector downlight (32W CFL)	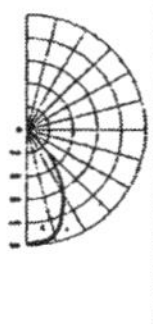		EFF = 64.9%			% DN = 100%			% UP = 0%			Lamp = CF6/32, SC (along, across, 45°) = 1.1, 1.1, 1.1						
		0	0.77	0.77	0.77	0.75	0.75	0.75	0.72	0.72	0.72	0.69	0.69	0.69	0.66	0.66	0.66	0.65
		1	0.73	0.72	0.70	0.72	0.70	0.69	0.67	0.66	0.65	0.65	0.64	0.63	0.63	0.62	0.61	0.60
		2	0.69	0.66	0.63	0.68	0.65	0.62	0.63	0.61	0.59	0.61	0.59	0.58	0.59	0.58	0.57	0.55
		3	0.66	0.61	0.58	0.64	0.60	0.57	0.59	0.56	0.54	0.57	0.55	0.53	0.55	0.54	0.52	0.51
		4	0.62	0.57	0.53	0.61	0.56	0.52	0.55	0.51	0.49	0.53	0.51	0.48	0.52	0.50	0.48	0.47
		5	0.58	0.53	0.49	0.57	0.52	0.48	0.51	0.48	0.45	0.50	0.47	0.45	0.49	0.46	0.44	0.43
		6	0.55	0.49	0.45	0.54	0.48	0.45	0.47	0.44	0.41	0.46	0.43	0.41	0.46	0.43	0.41	0.40
		7	0.52	0.46	0.41	0.51	0.45	0.41	0.44	0.41	0.38	0.44	0.40	0.38	0.43	0.40	0.38	0.37
		8	0.49	0.43	0.38	0.48	0.42	0.38	0.42	0.38	0.35	0.41	0.38	0.35	0.40	0.37	0.35	0.34
		9	0.47	0.40	0.36	0.46	0.40	0.36	0.39	0.35	0.33	0.38	0.35	0.33	0.38	0.35	0.33	0.32
		10	0.44	0.37	0.33	0.44	0.37	0.33	0.37	0.33	0.31	0.36	0.33	0.30	0.36	0.33	0.30	0.29
3. 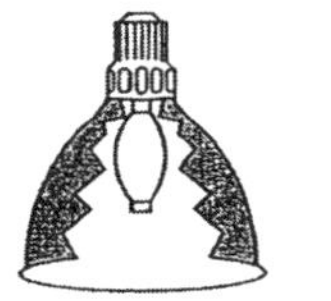 High bay, open metal reflector, medium	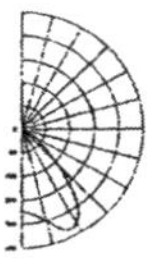		EFF = 83.9%			% DN = 95.2%			% UP = 4.8%			Lamp = M400/C/U, SC (along, across, 45°) = 1.6, 1.6, 1.4						
		0	0.99	0.99	0.99	0.96	0.96	0.96	0.91	0.91	0.91	0.86	0.86	0.86	0.82	0.82	0.82	0.80
		1	0.93	0.90	0.87	0.90	0.88	0.85	0.83	0.81	0.80	0.80	0.78	0.77	0.76	0.75	0.74	0.72
		2	0.86	0.81	0.77	0.84	0.79	0.75	0.76	0.73	0.70	0.73	0.70	0.68	0.70	0.68	0.66	0.64
		3	0.80	0.73	0.68	0.78	0.72	0.67	0.65	0.65	0.61	0.66	0.63	0.60	0.64	0.61	0.58	0.57
		4	0.75	0.67	0.61	0.73	0.65	0.60	0.63	0.58	0.54	0.61	0.57	0.53	0.58	0.55	0.52	0.51
		5	0.70	0.61	0.54	0.68	0.59	0.54	0.57	0.52	0.48	0.55	0.51	0.48	0.54	0.50	0.47	0.45
		6	0.65	0.55	0.49	0.63	0.54	0.48	0.53	0.47	0.43	0.51	0.46	0.43	0.49	0.45	0.42	0.41
		7	0.60	0.51	0.44	0.59	0.50	0.44	0.48	0.43	0.39	0.47	0.42	0.39	0.45	0.41	0.38	0.37
		8	0.56	0.47	0.40	0.55	0.46	0.40	0.44	0.39	0.35	0.43	0.38	0.35	0.42	0.38	0.34	0.33
		9	0.53	0.43	0.37	0.52	0.42	0.36	0.41	0.36	0.32	0.40	0.35	0.32	0.39	0.35	0.31	0.30
		10	0.50	0.40	0.34	0.48	0.39	0.33	0.38	0.33	0.29	0.37	0.32	0.29	0.36	0.32	0.29	0.27
4. 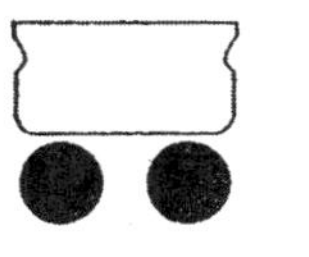 2-Lamp bare strip	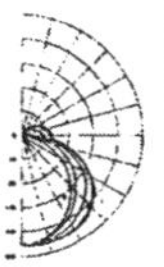		EFF = 89.3%			% DN = 86.4%			% UP = 13.6%			Lamp = (2) F40T12, SC (along, across, 45°) = 1.3, 1.5, 1.6						
		0	1.03	1.03	1.03	1.00	1.00	1.00	0.92	0.92	0.92	0.86	0.86	0.86	0.80	0.80	0.80	0.77
		1	0.93	0.88	0.83	0.89	0.84	0.80	0.78	0.75	0.72	0.73	0.70	0.68	0.67	0.65	0.63	0.61
		2	0.83	0.75	0.68	0.80	0.72	0.66	0.67	0.62	0.58	0.62	0.58	0.55	0.58	0.55	0.52	0.49
		3	0.75	0.65	0.57	0.72	0.63	0.56	0.58	0.52	0.47	0.54	0.49	0.45	0.50	0.46	0.43	0.40
		4	0.69	0.57	0.49	0.65	0.55	0.47	0.51	0.45	0.40	0.48	0.42	0.38	0.44	0.40	0.36	0.34
		5	0.63	0.51	0.42	0.60	0.49	0.41	0.46	0.39	0.34	0.43	0.37	0.32	0.40	0.35	0.31	0.29
		6	0.58	0.45	0.37	0.55	0.44	0.36	0.41	0.34	0.29	0.38	0.32	0.28	0.36	0.31	0.27	0.25
		7	0.53	0.41	0.33	0.51	0.40	0.32	0.37	0.30	0.26	0.35	0.29	0.25	0.33	0.27	0.24	0.22
		8	0.50	0.37	0.29	0.47	0.36	0.29	0.34	0.27	0.23	0.32	0.26	0.22	0.30	0.25	0.21	0.19
		9	0.46	0.34	0.26	0.44	0.33	0.26	0.31	0.25	0.20	0.29	0.24	0.19	0.27	0.22	0.19	0.17
		10	0.43	0.31	0.24	0.41	0.30	0.23	0.29	0.22	0.18	0.27	0.21	0.18	0.25	0.20	0.17	0.15
5. 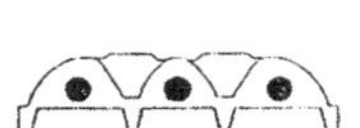 2 × 4, 3-Lamp parabolic troffer with 3″ semi-spec. louvers, 18 cells	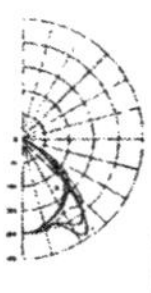		EFF = 72.7%			% DN = 100			% UP = 0			Lamp = (3) F32T8, SC (along, across, 45°) = 1.3, 1.6, 1.6						
		0	0.87	0.87	0.87	0.85	0.85	0.85	0.81	0.81	0.81	0.77	0.77	0.77	0.74	0.74	0.74	0.73
		1	0.81	0.78	0.76	0.79	0.77	0.74	0.74	0.72	0.70	0.71	0.69	0.68	0.68	0.67	0.66	0.65
		2	0.75	0.70	0.66	0.73	0.69	0.65	0.66	0.63	0.61	0.64	0.61	0.59	0.62	0.60	0.58	0.57
		3	0.69	0.63	0.58	0.68	0.62	0.57	0.60	0.56	0.52	0.58	0.54	0.52	0.56	0.53	0.51	0.49
		4	0.64	0.56	0.51	0.62	0.55	0.50	0.54	0.49	0.46	0.52	0.48	0.45	0.51	0.47	0.44	0.43
		5	0.59	0.51	0.45	0.58	0.50	0.44	0.48	0.44	0.40	0.47	0.43	0.40	0.46	0.42	0.39	0.38
		6	0.55	0.46	0.40	0.53	0.45	0.40	0.44	0.39	0.35	0.43	0.38	0.35	0.42	0.38	0.35	0.33
		7	0.51	0.42	0.36	0.50	0.41	0.36	0.40	0.35	0.31	0.39	0.35	0.31	0.38	0.34	0.31	0.30
		8	0.47	0.38	0.32	0.46	0.38	0.32	0.37	0.32	0.28	0.36	0.31	0.28	0.35	0.31	0.28	0.27
		9	0.44	0.35	0.29	0.43	0.35	0.29	0.34	0.29	0.25	0.33	0.29	0.25	0.32	0.28	0.25	0.24
		10	0.41	0.32	0.27	0.40	0.32	0.27	0.31	0.26	0.23	0.31	0.26	0.23	0.30	0.26	0.23	0.22
6. 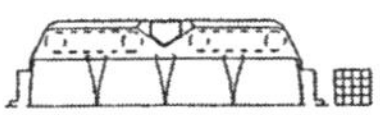 2 × 2, 2-Lamp (U) parabolic troffer with 3″ semi-spec. louver, 16 cells	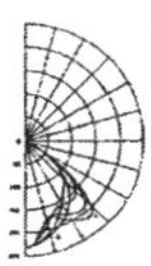		EFF = 50.8%			% DN = 100%			% UP = 0%			Lamp = (2) F31T8/U/6, SC (along, across, 45°) = 1.2, 1.5, 1.4						
		0	0.61	0.61	0.61	0.59	0.59	0.59	0.56	0.56	0.56	0.56	0.56	0.56	0.54	0.54	0.54	0.52
		1	0.57	0.55	0.53	0.55	0.54	0.52	0.52	0.50	0.49	0.52	0.50	0.49	0.50	0.49	0.48	0.47
		2	0.53	0.49	0.47	0.51	0.48	0.46	0.47	0.45	0.43	0.47	0.45	0.43	0.45	0.43	0.42	0.42
		3	0.49	0.44	0.41	0.48	0.43	0.40	0.42	0.39	0.37	0.42	0.39	0.37	0.41	0.38	0.37	0.38
		4	0.45	0.40	0.36	0.44	0.39	0.36	0.38	0.35	0.32	0.38	0.35	0.32	0.37	0.34	0.32	0.34
		5	0.42	0.36	0.32	0.41	0.35	0.32	0.34	0.31	0.29	0.34	0.31	0.29	0.33	0.31	0.28	0.30
		6	0.39	0.33	0.29	0.38	0.32	0.28	0.31	0.28	0.25	0.31	0.28	0.25	0.30	0.27	0.25	0.27
		7	0.36	0.30	0.26	0.35	0.29	0.25	0.29	0.25	0.23	0.29	0.25	0.23	0.28	0.25	0.22	0.24
		8	0.33	0.27	0.23	0.33	0.27	0.23	0.26	0.23	0.20	0.26	0.23	0.20	0.26	0.22	0.20	0.22
		9	0.31	0.25	0.21	0.31	0.25	0.21	0.24	0.21	0.18	0.24	0.21	0.18	0.24	0.20	0.18	0.20
		10	0.29	0.23	0.19	0.29	0.23	0.19	0.22	0.19	0.17	0.22	0.19	0.17	0.22	0.19	0.17	0.19

TABLE 9.7 (*continued*) TYPICAL LUMINAIRES

7. Typical Luminaire | Typical Intensity Distribution

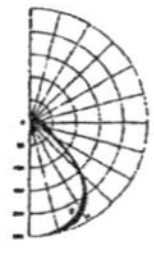

1 × 4, 2-Lamp lensed wrap-around, surface mounted

EFF = 68.9% % DN = 91.4% % UP = 8.6% Lamp = (2) F32T8 SC (along, across, 45°) = 1.3, 1.5, 1.5

ρcc →	80			70			50			30			10			0
ρw →	70	50	30	70	50	30	50	30	10	50	30	10	50	30	10	0
RCR ↓																
0	0.81	0.81	0.81	0.78	0.78	0.78	0.73	0.73	0.73	0.69	0.69	0.69	0.65	0.65	0.65	0.63
1	0.74	0.71	0.68	0.71	0.68	0.66	0.64	0.62	0.60	0.61	0.59	0.58	0.57	0.56	0.55	0.53
2	0.68	0.62	0.58	0.65	0.61	0.56	0.57	0.54	0.51	0.54	0.51	0.49	0.51	0.49	0.47	0.45
3	0.62	0.55	0.50	0.60	0.54	0.49	0.51	0.47	0.44	0.48	0.45	0.42	0.46	0.43	0.41	0.39
4	0.57	0.50	0.44	0.55	0.48	0.43	0.46	0.41	0.38	0.44	0.40	0.37	0.41	0.38	0.36	0.34
5	0.53	0.45	0.39	0.51	0.43	0.38	0.41	0.37	0.33	0.39	0.35	0.32	0.38	0.34	0.31	0.30
6	0.49	0.40	0.35	0.47	0.39	0.34	0.38	0.33	0.29	0.36	0.32	0.29	0.34	0.31	0.28	0.26
7	0.46	0.37	0.31	0.44	0.36	0.31	0.34	0.30	0.26	0.33	0.29	0.26	0.31	0.28	0.25	0.24
8	0.42	0.34	0.28	0.41	0.33	0.28	0.32	0.27	0.23	0.30	0.26	0.23	0.29	0.25	0.23	0.21
9	0.40	0.31	0.26	0.38	0.30	0.25	0.29	0.24	0.21	0.20	0.24	0.21	0.27	0.23	0.20	0.19
10	0.37	0.29	0.23	0.36	0.28	0.23	0.27	0.22	0.19	0.26	0.22	0.19	0.25	0.21	0.19	0.17

8.

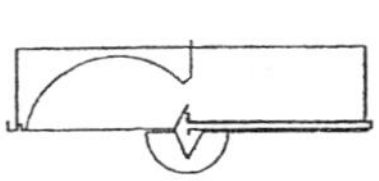

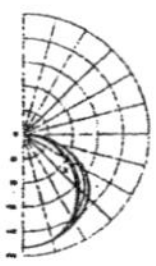

2 × 2, semi-recessed troffer

EFF = 54.2% % DN = 99.2% % UP = 0.5% Lamp = (2) FT36* SC (along, across, 45°) = 1.3, 1.5, 1.5

RCR																
0	0.65	0.65	0.65	0.63	0.63	0.63	0.60	0.60	0.60	0.58	0.58	0.58	0.55	0.55	0.55	0.54
1	0.58	0.55	0.53	0.57	0.54	0.52	0.52	0.50	0.48	0.49	0.48	0.46	0.47	0.46	0.45	0.44
2	0.53	0.48	0.44	0.51	0.47	0.43	0.45	0.41	0.39	0.43	0.40	0.38	0.41	0.39	0.37	0.36
3	0.48	0.42	0.37	0.46	0.41	0.36	0.39	0.35	0.32	0.37	0.34	0.31	0.36	0.33	0.31	0.30
4	0.43	0.37	0.32	0.42	0.36	0.31	0.34	0.30	0.27	0.33	0.29	0.27	0.32	0.29	0.26	0.25
5	0.40	0.33	0.27	0.39	0.32	0.27	0.31	0.26	0.23	0.30	0.26	0.23	0.28	0.25	0.23	0.21
6	0.37	0.29	0.24	0.36	0.29	0.24	0.28	0.23	0.20	0.27	0.23	0.20	0.26	0.22	0.20	0.19
7	0.34	0.26	0.21	0.33	0.26	0.21	0.25	0.21	0.18	0.24	0.20	0.18	0.23	0.20	0.17	0.16
8	0.32	0.24	0.19	0.31	0.24	0.19	0.23	0.19	0.16	0.22	0.18	0.16	0.21	0.18	0.15	0.14
9	0.29	0.22	0.17	0.29	0.22	0.17	0.21	0.17	0.14	0.20	0.17	0.14	0.20	0.16	0.14	0.13
10	0.28	0.20	0.16	0.27	0.20	0.16	0.19	0.15	0.13	0.19	0.15	0.13	0.18	0.15	0.13	0.12

9.

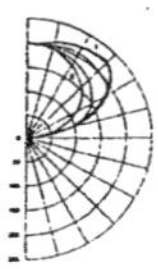

V-shaped, completely indirect

EFF = 88.3% % DN = 0 % UP = 100 Lamp = (2) F32T8 SC (along, across, 45°) = N/A

RCR																
0	0.84	0.84	0.84	0.72	0.72	0.72	0.49	0.49	0.49	0.28	0.28	0.28	0.09	0.09	0.09	0.00
1	0.76	0.73	0.70	0.65	0.62	0.60	0.43	0.41	0.40	0.25	0.24	0.23	0.08	0.08	0.07	0.00
2	0.70	0.64	0.59	0.59	0.54	0.50	0.37	0.35	0.33	0.22	0.20	0.19	0.07	0.07	0.06	0.00
3	0.63	0.56	0.50	0.54	0.48	0.43	0.33	0.30	0.27	0.19	0.17	0.16	0.06	0.06	0.05	0.00
4	0.58	0.49	0.43	0.49	0.42	0.37	0.29	0.26	0.23	0.17	0.15	0.14	0.05	0.05	0.05	0.00
5	0.53	0.43	0.37	0.45	0.37	0.32	0.26	0.22	0.20	0.15	0.13	0.12	0.05	0.04	0.04	0.00
6	0.48	0.39	0.32	0.41	0.33	0.28	0.23	0.19	0.17	0.13	0.11	0.10	0.04	0.04	0.03	0.00
7	0.45	0.35	0.28	0.38	0.30	0.24	0.21	0.17	0.15	0.12	0.10	0.09	0.04	0.03	0.03	0.00
8	0.41	0.31	0.25	0.35	0.27	0.22	0.19	0.15	0.13	0.11	0.09	0.08	0.03	0.03	0.03	0.00
9	0.38	0.28	0.22	0.32	0.24	0.19	0.17	0.13	0.11	0.10	0.08	0.07	0.03	0.03	0.02	0.00
10	0.35	0.26	0.20	0.30	0.22	0.17	0.15	0.12	0.10	0.09	0.07	0.06	0.03	0.02	0.02	0.00

10.

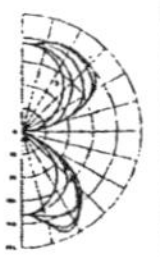

Direct/indirect, 2-lamp, thin profile, parabolic baffles, 60% up

EFF = 88.3% % DN = 40.3 % UP = 59.7 Lamp = (2) F32T8 SC (along, across, 45°) = N/A

RCR																
0	0.93	0.93	0.93	0.84	0.84	0.84	0.69	0.69	0.69	0.55	0.55	0.55	0.42	0.42	0.42	0.36
1	0.85	0.82	0.79	0.77	0.75	0.72	0.61	0.60	0.58	0.49	0.48	0.47	0.38	0.37	0.37	0.31
2	0.78	0.72	0.67	0.71	0.66	0.62	0.54	0.51	0.49	0.44	0.42	0.40	0.34	0.33	0.32	0.27
3	0.71	0.64	0.58	0.65	0.58	0.53	0.48	0.45	0.42	0.39	0.37	0.34	0.31	0.29	0.28	0.24
4	0.65	0.57	0.50	0.60	0.52	0.46	0.43	0.39	0.36	0.35	0.32	0.30	0.28	0.26	0.24	0.21
5	0.60	0.50	0.44	0.55	0.46	0.40	0.39	0.34	0.31	0.32	0.28	0.26	0.25	0.23	0.21	0.18
6	0.55	0.45	0.38	0.50	0.42	0.36	0.35	0.30	0.27	0.28	0.25	0.23	0.23	0.20	0.19	0.16
7	0.51	0.41	0.34	0.47	0.38	0.32	0.32	0.27	0.24	0.26	0.23	0.20	0.21	0.18	0.16	0.14
8	0.47	0.37	0.30	0.43	0.34	0.28	0.29	0.24	0.21	0.24	0.20	0.18	0.19	0.16	0.15	0.13
9	0.44	0.34	0.27	0.40	0.31	0.25	0.26	0.22	0.19	0.22	0.18	0.16	0.17	0.15	0.13	0.11
10	0.41	0.31	0.25	0.37	0.28	0.23	0.24	0.20	0.17	0.20	0.17	0.14	0.16	0.14	0.12	0.10

Source: IESNA Lighting Handbook, 9th Edition, 2000. Courtesy of the Illuminating Engineering Society of North America.

Examine the various parts of each luminaire's data in Table 9.7. On the far left, a cross-section of the luminaire is given, showing the lamp(s) and any optical control. The intensity distribution is given in polar form. When the luminaire does not have azimuthal symmetry, more than one intensity curve is given. Several items appear above the CU table. The luminaire efficiency, is given. The quantities % DN and % UP are the percentages of luminaire (not lamp) lumens that exit the luminaire downward and upward. To obtain these percentages in terms of lamp lumens, % DN and % UP must be multiplied by luminaire efficiency. This yields for Luminaire 4, for example, 77% of lamp lumens going down and 12% going up. The final item is SC, spacing criterion, which is discussed later in the chapter.

The CU table provides values for 10 room cavity ratios, five ceiling cavity reflectances and three wall reflectances. It also gives CUs for the case of zero ceiling cavity and wall reflectances. All of these are for a floor cavity reflectance of 20%. CU multipliers for other ρ_{fc} values are given in Table 9.8.

Consider one of the luminaire CU tables in Table 9.7, for example that for Luminaire 5. As RCR increases, with reflectances fixed, the CU would be expected to decrease

TABLE 9.8 FLOOR CAVITY CU MULTIPLIERS

% Effective Ceiling Cavity Reflectance, ρ_{cc}	80				70				50			30			10		
% Wall Reflectance, ρ_w	70	50	30	10	70	50	30	10	50	30	10	50	30	10	50	30	10
For 30% Effective Floor Cavity Reflectance (20% = 1.00)																	
Room Cavity Ratio																	
1	1.092	1.082	1.075	1.068	1.077	1.070	1.064	1.059	1.049	1.044	1.040	1.028	1.026	1.023	1.012	1.010	1.008
2	1.079	1.066	1.055	1.047	1.068	1.057	1.048	1.039	1.041	1.033	1.027	1.026	1.021	1.017	1.013	1.010	1.006
3	1.070	1.054	1.042	1.033	1.061	1.048	1.037	1.028	1.034	1.027	1.020	1.024	1.017	1.012	1.014	1.009	1.005
4	1.062	1.045	1.033	1.024	1.055	1.040	1.029	1.021	1.030	1.022	1.015	1.022	1.015	1.010	1.014	1.009	1.004
5	1.056	1.038	1.026	1.018	1.050	1.034	1.024	1.015	1.027	1.018	1.012	1.020	1.013	1.008	1.014	1.009	1.004
6	1.052	1.033	1.021	1.014	1.047	1.030	1.020	1.012	1.024	1.015	1.009	1.019	1.012	1.006	1.014	1.008	1.003
7	1.047	1.029	1.018	1.011	1.043	1.026	1.017	1.009	1.022	1.013	1.007	1.018	1.010	1.005	1.014	1.008	1.003
8	1.044	1.026	1.015	1.009	1.040	1.024	1.015	1.007	1.020	1.012	1.006	1.017	1.009	1.004	1.013	1.007	1.003
9	1.040	1.024	1.014	1.007	1.037	1.022	1.014	1.006	1.019	1.011	1.005	1.016	1.009	1.004	1.013	1.007	1.002
10	1.037	1.022	1.012	1.006	1.034	1.020	1.012	1.005	1.017	1.010	1.004	1.015	1.009	1.003	1.013	1.007	1.002
For 10% Effective Floor Cavity Reflectance (20% = 1.00)																	
Room Cavity Ratio																	
1	.923	.929	.935	.940	.933	.939	.943	.948	.956	.960	.963	.973	.976	.979	.989	.991	.993
2	.931	.942	.950	.958	.940	.949	.957	.963	.962	.968	.974	.976	.980	.985	.988	.991	.995
3	.939	.951	.961	.969	.945	.957	.966	.973	.967	.975	.981	.978	.983	.988	.988	.992	.996
4	.944	.958	.969	.978	.950	.963	.973	.980	.972	.980	.986	.980	.986	.991	.987	.992	.996
5	.949	.964	.976	.983	.954	.968	.978	.985	.975	.983	.989	.981	.988	.993	.987	.992	.997
6	.953	.969	.980	.986	.958	.972	.982	.989	.977	.985	.992	.982	.989	.995	.987	.993	.997
7	.957	.973	.983	.991	.961	.975	.985	.991	.979	.987	.994	.983	.990	.996	.987	.993	.998
8	.960	.976	.986	.993	.963	.977	.987	.993	.981	.988	.995	.984	.991	.997	.987	.994	.998
9	.963	.978	.987	.994	.965	.979	.989	.994	.983	.990	.996	.985	.992	.998	.988	.994	.999
10	.965	.980	.989	.995	.967	.981	.990	.995	.984	.991	.997	.986	.993	.998	.988	.994	.999
For 0% Effective Floor Cavity Reflectance (20% = 1.00)																	
Room Cavity Ratio																	
1	.859	.870	.879	.886	.873	.884	.893	.901	.916	.923	.929	.948	.954	.960	.979	.983	.987
2	.871	.887	.903	.919	.886	.902	.916	.928	.926	.938	.949	.954	.963	.971	.978	.983	.991
3	.882	.904	.915	.942	.898	.918	.934	.947	.936	.950	.964	.958	.969	.979	.976	.984	.993
4	.893	.919	.941	.958	.908	.930	.948	.961	.945	.961	.974	.961	.974	.984	.975	.985	.994
5	.903	.931	.953	.969	.914	.939	.958	.970	.951	.967	.980	.964	.977	.988	.975	.985	.995
6	.911	.940	.961	.976	.920	.945	.965	.977	.955	.972	.985	.966	.979	.991	.975	.986	.996
7	.917	.947	.967	.981	.924	.950	.970	.982	.959	.975	.988	.968	.981	.993	.975	.987	.997
8	.922	.953	.971	.985	.929	.955	.975	.986	.963	.978	.991	.970	.983	.995	.976	.988	.998
9	.928	.958	.975	.988	.933	.959	.980	.989	.966	.980	.993	.971	.985	.996	.976	.988	.998
10	.933	.962	.979	.991	.937	.963	.983	.992	.969	.982	.995	.973	.987	.997	.977	.989	.999

Source: IESNA Lighting Handbook, 9th Edition, 2000. Courtesy of the Illuminating Engineering Society of North America.

because the room is becoming taller and more of the luminaire flux is reflected from the walls, where there is some absorption. This trend is seen in Table 9.7. The reduction is greater for Luminaire 5 than it is for Luminaire 1, which has a narrow downward distribution. If the effect of ceiling cavity reflectance variations is then examined, with RCR and ρ_w fixed, it is noted that CU decreases as ρ_{cc} decreases. The effect is less pronounced in Luminaire 5, whose flux is entirely downward, than it is in Luminaire 9, a totally indirect luminaire. Finally, it is noted that CU decreases as ρ_w decreases, with RCR and ρ_{cc} fixed. Once again, the decrease is greater for Luminaire 5 than for Luminaire 1, because of the latter's narrower intensity distribution.

Is it possible to have a CU greater than 1? The answer is "yes", as a few of the entries in Table 9.7 show. How does this arise? Consider Luminaire 4 in a very wide room (RCR = 0). The luminaire is 89.3% efficient, with $89.3 \times 0.136 = 12.1\%$ of the lamp lumens going up and $89.3 \times 0.864 = 77.2\%$ going down. For an 80% ceiling and 20% floor and assuming no wall reflections, Eq. (9.6) yields $\text{CU} = \frac{0.121 \times 0.8 + 0.772}{1 - 0.80 \times 0.20} = 1.03$. What happens is that the initial lumens incident on the work plane reflect from the floor, and then the ceiling, and thus a portion of them are incident more than once on the work plane.

Consider next the utility of CUs for zero ceiling cavity and wall reflectances. If these reflectances are zero, all indirect luminaires will have $CU_o = 0$, and Eq. (9.2) will yield $E = 0$. Such is the case for Luminaire 9 in Table 9.7. For luminaires with a downward component, there will be no reflections from walls, and the E in Eq. (9.2) will be the result of luminous flux coming directly to the work plane from the luminaire. Thus

$$E_{DIR} = \frac{\phi_i (CU_o)(LLF)}{A} \tag{9.7}$$

Then, for a given set of wall and ceiling cavity reflectances,

$$E_{IND} = E - E_{DIR} \tag{9.8}$$

gives the component of illuminance due to interreflected light within the space.

Choosing Luminaire 10 as an example, with $RCR = 2$, $\rho_{cc} = 70\%$, and $\rho_w = 50\%$, $CU_o = 0.27$, and $CU = 0.66$. From Eqs. (9.2) and (9.7), $E_{DIR}/E = 0.27/0.66 = 0.41$. Thus, from Eq. (9.8), $E_{IND} = E - 0.41E = 0.59E$. The workplane receives 41% of its luminous flux directly and 59% indirectly, giving $E_{DIR}/E_{IND} = 1.4$. Keeping this ratio under 1.5 reduces the probability of shadows being disturbing.

In Sect. 9.6, a cavity ratio of zero was examined. All of the CUs in Table 9.7 for zero RCR may be obtained from Eq. (9.6). Consider, for example, Luminaire 4, with $\rho_{cc} = 70\%$ and $\rho_{fc} = 20\%$.

$$CU = \frac{(0.136 \times 0.70 + 0.864)0.893}{1 - 0.70 \times 0.20} = 1.00$$

where the 0.893 is included to convert luminaire percentages to lamp percentages. 1.00 is the value that appears in the RCR = 0 row and $\rho_{cc} = 70\%$ column for Luminaire 4. The value 0.77 for $RCR = \rho_w = \rho_{cc} = 0$, is simply the product of the luminaire efficiency and % DN. For Luminaire 4, it is $0.893 \times 0.864 = 0.77$.

The CU tables in Table 9.7 have several caveats:

- All room surfaces are perfectly diffusing (Lambertian).
- Incident luminous flux on a room surface (ceiling, walls, and floor) is uniformly distributed over that surface.
- Luminaires are uniformly distributed throughout the room.
- The room is empty. In particular, the floor cavity is empty.
- Room surfaces are spectrally neutral.
- CU values are derived from luminaire intensities at each polar angle that are averaged for azimuthal variations according to $I_{av} = (I_o + 2I_{45} + I_{90})/4$.
- CU values are based on a spacing to mounting height ratio of 0.4 for fluorescent luminaires and 0.7 for HID and incandescent luminaires.
- The CUs are linearly related to luminaire efficiency and can be applied to other luminaires with similarly-shaped intensity distribution curves, with a multiplier used to account for the efficiency change.

- Spacing criterion values are only general guides and are not given for semi-indirect, indirect, and bat wing luminaires.
- The CUs are for the number of lamps shown. Multipliers are available from luminaire manufacturers for increasing or decreasing this number. Typically a multiplier of 0.95 is used if a two-lamp luminaire is changed to four lamps.
- If reduced-wattage fluorescent lamps replace standard lamps, CU values are likely to rise by typically 5% because of lower lamp operating temperature.
- CU values are based on a ℓ/w ratio of 1.6 (as noted earlier).

9.9 INITIAL AVERAGE ILLUMINANCE

To use Table 9.7, one first calculates the cavity ratios for the given room dimensions from Eq. (9.3). Next, the effective reflectances of the ceiling and floor cavities (if these exist) are obtained from Table 9.5. Next a luminaire is chosen, Table 9.7 is entered with the particular room cavity ratio (RCR), ρ_{cc}, and ρ_w, and the CU is determined based on $\rho_{fc} = 20\%$ (often interpolation between adjacent rows and columns in Table 9.7 is required to obtain the CU). If ρ_{fc} is not 20%, a correction factor for the CU may be obtained from Table 9.8.

EXAMPLE 9.2

A 28 × 32 × 10 ft school room is to be lighted by Luminaire 5 in Table 9.7. Each luminaire contains three 32-watt T8 fluorescent lamps. Reflectances for the ceiling, walls, and floor are 80%, 50%, and 30%, respectively. An average initial illuminance of 50 fc is desired. How many luminaires are required?

Solution. From Eq. (9.3)

$$\text{RCR} = \frac{5(7.5)(60)}{28(32)} = 2.51$$

$$\text{FCR} = 2.51\left(\frac{2.5}{7.5}\right) = 0.84$$

$$\text{CCR} = 0$$

Since there is no ceiling cavity, Table 9.7 can be entered directly and a CU = 0.66 for $\rho_{fc} = 20\%$ (the 0.66 value is midway between 0.70 for RCR = 2 and 0.63 for RCR = 3).

It now must be checked whether the effective floor cavity reflectance was 20%. From Table 9.5, $\rho_{fc} = 0.28$. Then from Table 9.8, a multiplier of 1.06 is obtained for a 30% floor cavity. For 28%, 1.05 will be used. Thus, the final CU value is 1.05(0.66) = 0.69. (In many cases when $\rho_f = 30\%$, it is reasonable to assume $\rho_{fc} = 20\%$. because the floor cavity is not empty and absorbs flux. Thus, the multiplier would be 1.0 (i.e., no correction). Inserting CU = 0.69 into Eq. (9.2), with LLF = BF = 0.93, and solving for Φ_i gives

$$\Phi_i = \frac{50(896)}{0.69 \times 0.93} = 69{,}800\,\text{lm}$$

Each luminaire houses three lamps. From Table 6.10, each lamp emits 2950 initial lumens. Thus, the number of luminaires required is

$$n = \frac{69,800}{3 \times 2950} = 8$$

EXAMPLE 9.3

Repeat Example 9.2 for Luminaire 9 suspended 1.5 ft from the ceiling. All other parameters are unchanged.

Solution. Because the luminaire is suspended, some of the cavity ratios change. From Eq. (9.3),

$$\text{RCR} = \frac{5(6)(60)}{28(32)} = 2.0$$

$$\text{CCR} = 2.0\left(\frac{1.5}{6}\right) = 0.5$$

$$\text{FCR} = 0.84 \qquad \text{(as before)}$$

In this case $\rho_{cc} \neq \rho_c$. Table 9.5 gives $\rho_{cc} = 0.72$. Entering Table 9.7 gives a CU of 0.56, where a linear interpolation has been performed between CU values of 0.54 and 0.64 $(0.54 + 0.20 \times 0.10)$ for a 20% floor cavity reflectance. The value of ρ_{fc} is unchanged. ρ_{cc} and RCR have changed slightly, but not enough to alter the multiplier. Thus, CU = $0.56 \times 1.05 = 0.59$.

Insertion of CU = 0.96 into Eq. (9.2) yields

$$\Phi_i = \frac{50(896)}{0.59 \times 0.93} = 81,600\,\text{lm}$$

The lamp is the same, but only two are required. Thus the number of luminaires is

$$n = \frac{81,600}{5900} = 14\,\text{luminaires}$$

EXAMPLE 9.4

In the previous two examples, only a single interpolation in Table 9.7 was required. Consider now the case of $\rho_w = 40\%$, $\rho_{cc} = 73\%$, $\rho_{fc} = 20\%$, and RCR = 1.6. Find the CU for Luminaire 5.

Solution. When multiple interpolations are required, it is desirable to set up a small table showing the quantities involved:

$\rho_{cc} \rightarrow$	80		70	
$\rho_w \rightarrow$	50	30	50	30
RCR				
1	0.78	0.76	0.77	0.74
2	0.70	0.66	0.69	0.65

The easier interpolation is ρ_w so it is done first (a single interpolation), yielding

ρ_{cc}	80	73	70
RCR			
1	0.770		0.755
1.6		0.71	
2	0.680		0.670

Each of these numbers is just halfway between the coefficients for $\rho_w = 30\%$ and 50% in the first chart. An extra decimal place is retained temporarily. Now, from the second chart, where the intermediate RCR and ρ_{cc} are included, start with the smallest of the four entries and write

$$0.670 + 0.3 \times 0.01 + 0.4 \times 0.085 = 0.71$$

In this equation, it is noted that ρ_{cc} is 3/10 of the CU "distance" from $\rho_{cc} = 0.70$ to 0.80, and 4/10 of the CU "distance" from RCR = 2 to 1.

EXAMPLE 9.5

Assume the room of Ex. 9.2 has three walls with $\rho = 50\%$ and one (28-ft) wall with $\rho = 30\%$. Further, assume that there is a 3 × 8 ft tackboard mounted on one of the $\rho =$ 50% walls and that its reflectance is 40%. Find the weighted average wall reflectance.

Solution. In working this problem, it is not correct to simply average the three reflectances. Rather they must be weighted according to their areas. A general equation is:

$$\rho_{wa} = \frac{\rho_1 A_1 + \rho_2 A_2 + \cdots + \rho_n A_n}{A_1 + A_2 + \cdots + A_n} \tag{9.9}$$

For this problem of a 28 × 32 × 7.5 ft room cavity, it follows that

$$\rho_{wa} = \frac{0.50(92 \times 7.5 - 3 \times 8) + 0.30 \times 28 \times 7.5 + 0.40 \times 3 \times 8}{120 \times 7.5}$$

$$= 0.45$$

The 50% walls still dominate, but the tackboard and dark wall have reduced the over reflectance by 10%.

9.10 LIGHT LOSS FACTOR

It has been mentioned in several places that if the interest is in average maintained illuminance, rather than average initial illuminance, a light loss factor must be included in the calculation. Total light loss is divided conventionally into nonrecoverable and recoverable losses, and the light loss factor should include as many of each of these as are quantifiable. Nonrecoverable losses are those not generally subject to correction by lighting maintenance procedures. Up to now only one of these has been mentioned—ballast factor. The rest will be discussed later in this section.

There are four recoverable losses:

1. Lamp burnouts (LBO)
2. Lamp lumen depreciation (LLD)
3. Luminaire dirt depreciation (LDD)
4. Room surface dirt depreciation (RSDD)

Lamp Burnouts

It may not be feasible to replace a lamp every time one burns out. If burned-out lamps are not replaced immediately, the average illuminance level will decrease. A decision should be made as to how many burned-out lamps will be tolerated in a given installation before they are replaced. Then the LBO factor is the ratio of the number of lamps remaining lighted to the total number of lamps, when the maximum number of burnouts is reached.

Lamp Lumen Depreciation

This factor is the ratio of the lumens produced at a certain time in a lamp's life to the initial lumens. As discussed earlier and shown in Table 8.1, the lumen output of a lamp decreases as the lamp ages. The LLD factor is generally taken at the 40% life point for fluorescent and metal halide lamps and at 50% life for incandescent, tungsten halogen, and high pressure sodium lamps. The base lumens are at 0 h for incandescent sources and 100 h for discharge sources.

Recommended illuminance levels, as determined in Sect. 9.3, are to be considered as maintained values. Thus, they should be thought of as the minimum illuminance on the task at any time. The 70% lifepoint is the recommended point for group lamp replacement. If a group relamping program is adopted, LLD should be determined based on 70% life, rather than 40% or 50%.

Luminaire Dirt Depreciation

Dirt accumulation on a luminaire will reduce the reflectance of luminaire surfaces and the transmittance of luminaire lenses and diffusers. In the 1960s, two significant papers on light loss from luminaire dirt depreciation were published (Clark, 1963, 1966). The first paper was an extensive field study that resulted in graphs of LDD vs. time for various luminaire types and dirt conditions. The second paper did an analysis of the kinds of dirt that can adhere to a luminaire and defined five dirt conditions, from very clean to very dirty. As the result of this work, the IESNA developed a procedure (Rea, 2000) that classified luminaires into six categories based on their top and bottom enclosures, as shown in Table 9.9, and then presented six families of curves, one for each luminaire category, that displayed LDD vs. time for each of the five dirt conditions.

A decaying exponential equation with two adjustable variables has been developed to fit the plotted data:

$$\text{LDD} = \varepsilon^{-At^{B}} \tag{9.10}$$

where t is the time in years between cleanings and A and B are constants whose values are given in Table 9.10 for the six luminaire maintenance categories.

TABLE 9.9 LUMINAIRE MAINTENANCE CLASSIFICATIONS

Category	Top Enclosure	Bottom Enclosure
I	1. None	1. none
II	1. none 2. transparent, translucent or opaque with ≥15% uplight through apertures	1. none 2. louvers or baffles
III	1. transparent, translucent or opaque with <15% uplight through apertures	1. none 2. louvers or baffles
IV	1. transparent, translucent or opaque unapertured	1. none 2. louvers
V	1. transparent, translucent or opaque unapertured	1. transparent or translucent unapertured
VI	1. none 2. transparent, translucent or opaque unapertured	1. transparent, translucent or opaque unapertured

TABLE 9.10

Luminaire Maintenance Category	B	A				
		Very Clean	Clean	Medium	Dirty	Very Dirty
I	0.69	0.038	0.071	0.111	0.162	0.301
II	0.62	0.033	0.068	0.102	0.147	0.188
III	0.70	0.079	0.106	0.143	0.184	0.236
IV	0.72	0.070	0.131	0.216	0.314	0.452
V	0.53	0.078	0.128	0.190	0.249	0.321
VI	0.88	0.076	0.145	0.218	0.284	0.396

In the past it was customary for a manufacturer to indicate the Roman numerical maintenance category on each luminaire specification sheet, but unfortunately this practice is followed less often today. Thus, the user of Table 9.7 must refer to the generic information in Table 9.9 to determine the appropriate Roman numerical

There has been growing concern that the LDD factors resulting from Eq. (9.10) and based on research done in the 1960s are too low. Environmental conditions in buildings were poorer then (smoking was the norm), air conditioning was unusual, luminaires were not as well constructed, their finishes were not as durable, and measurements were not as precise.

Recently an extensive field study of selected fluorescent luminaires was conducted (Levin, et al., 2002). The study was carried out at 217 office, school, and retail sites over a three-year period. Four luminaire types were tested: 2 × 4 ft lensed, 2 × 4 ft louvered, 2 × 4 ft louvered air handling, and 2 × 2 ft louvered. All luminaires were direct fluorescent, surface-mounted, or recessed. They represented luminaire maintenance Category IV (louvered) and Category V (lensed). Collectively, these four luminaire types represent 90% of the recessed and surface mounted luminaires installed in the U.S.

At the conclusion of the field study, and after all data were analyzed and plotted, new A and B coefficients were calculated for Eq. (9.10). It was concluded that all of the data could be pooled, i.e., that there were not significant differences among the four types of luminaires. Thus, only one set of new A and B coefficients was presented.

$$A = 0.072, \quad B = 0.44 \tag{9.11}$$

EXAMPLE 9.6

To compare the new with the old, consider a clean room with a 3-year cleaning cycle. Calculate LDD for Categories IV and V and compare with the new value of LDD.

Solution. From Table 9.10,

Category IV	$A = 0.131,$	$B = 0.72$
Category V	$A = 0.128,$	$B = 0.53$

Insertion into Eq. (9.10), with $t = 3$, yields LDD = 0.75 for Category IV and 0.80 for Category V. Using the values in Eq. (9.11), the new LDD = 0.89. What this means is that, for this set of conditions, the initial illuminance can be reduced by 16% and 10%, when compared to Categories IV and V, respectively.

Room Surface Dirt Depreciation

Dirt on room surfaces also reduces the reflected component of luminous flux and thus the illuminance on the work plane. The resulting RSDD factor is a function of the dirtiness of the environment, the time between cleanings, the luminaire flux distribution, and the room proportions.

Studies have shown that room surface dirt depreciation is essentially the same function of room cleanliness and time as the luminaire dirt depreciation for maintenance Category V. Thus, the percent depreciation can be obtained by using the constants for Category V from Table 9.10 in the following modified form of Eq. (9.10).

$$\%\,\text{Dirt depreciation} = 100\left(1 - e^{-At^{0.53}}\right) \tag{9.12}$$

Next, one must determine the type of luminaire flux distribution from Table 8.2 and the manufacturer's intensity distribution data. For example, Luminaire 4 in Table 9.7 delivers 86.4% of the total luminaire output downward and 13.6% upward. From Table 8.2, the luminaire is classified as semidirect.

With this information in hand, one can enter Table 9.11 and obtain the RSDD. Table 9.11 relates percent dirt depreciation to luminaire flux distribution and room proportions.

EXAMPLE 9.7

A lighting installation in a large office area 160 × 50 × 10 ft consists of 80 three-lamp luminaires (Luminaire 5 in Table 9.7) containing F32T8 fluorescent lamps. Maintenance policy is to clean luminaires and room surfaces annually and to replace burned-out

TABLE 9.11 RSDD FACTORS

Percent Expected	Direct				Semidirect				Direct-Indirect				Semiindirect				Indirect			
Dirt Depreciation	10	20	30	40	10	20	30	40	10	20	30	40	10	20	30	40	10	20	30	40
Room cavity ratio																				
1	98	96	94	92	97	92	89	84	94	87	80	76	94	87	80	73	90	80	70	60
2	98	96	94	92	96	92	88	83	94	87	80	75	94	87	79	72	90	80	69	59
3	98	95	93	90	96	91	87	82	94	86	79	74	94	86	78	71	90	79	68	58
5	97	94	91	89	94	90	84	79	93	86	78	72	93	86	77	69	89	78	66	55
7	97	94	90	87	93	88	82	77	93	84	77	70	93	84	76	68	89	76	65	53
10	96	92	87	83	93	86	79	72	93	84	75	67	92	83	75	67	88	75	62	50

lamps when 12 failures have occurred. Find the light loss factor for this installation, due to recoverable factors.

Solution.

$$\text{LBO} = \frac{228}{240} = 0.95$$

From Table 6.10, the lamp emits 2950 initial lumens and 2715 lumens at 40% life. Thus

$$\text{LLD} = \frac{2715}{2950} = 0.92$$

No maintenance category is given in Table 9.7. From Table 9.9, Category IV seems most appropriate. Using the new A and B values and t = 1 in Eq. (9.10) gives

$$\text{LDD} = \varepsilon^{-0.072(1)^{0.44}} = 0.93$$

Luminaire 5 delivers all of its output lumens downward. From Table 8.2, it is classified as direct. The percent expected dirt depreciation is, from Eq. (9.12), assuming a clean room

$$\%\,\text{Dirt depreciation} = 100(1 - e^{-0.128(1)^{0.53}}) = 12\%$$

Assuming a 30-in work plane, the room cavity ratio is, from Eq. (9.3),

$$\text{RCR} = \frac{5(7.5)(210)}{160(50)} = 0.98$$

Then Table 9.11 yields RSDD = 0.98. Thus, the recoverable light loss factor is

$$\text{LLF}_\text{R} = 0.95 \times 0.92 \times 0.93 \times 0.98 = 0.80$$

As noted earlier, there are several nonrecoverable, as well as recoverable, losses. Ballast factor has already been discussed. Some of the other nonrecoverable causes of lowered light output are:

Luminaire Ambient Temperature (LAT) Fluorescent lamp and luminaire ratings are established in still air at 25°C. Furthermore, fluorescent luminaire photometry is relative in that the luminaire output is adjusted for a 25°C (77°F) ambient still air temperature. The cold-spot temperature on the lamp increases 0.6°C (1°F) for each °C rise above ambient. In an actual installation, the luminaire operating temperature is likely to differ from the test condition. Operation above or below the test value decreases lumen output by about 1.5% per degree C. For example, if the plenum temperature for a fluorescent luminaire is 30°C, the lamp output would decrease by about 7.5%, giving a temperature correction factor of 0.925. There is no significant variation with ambient temperature for incandescent and HID lamps.

Voltage at the Luminaire (VL) Deviations in supply voltage from rated value can affect the light output of all types of lamps. A 1% change in voltage causes a 3% change in lumen output for incandescent lamps and a 0.4% change for fluorescent lamps on magnetic ballasts.

Luminaire Surface Depreciation (LSD) Effects such as yellowing of plastic with age and pitting and discoloration of enamel over time can cause a reduction in light reflected within the luminaire and transmitted through the diffuser. No numbers are available for this factor, and present-day luminaires are much less susceptible to degradation than their ancestors.

Heat Extraction (HE) This is a measure of the lumen increase or decrease due to temperature changes resulting from air flowing through a luminaire, and pertains specifically to air-handling luminaires linked with HVAC systems.

9.11 SPACING OF LUMINAIRES

One additional piece of information that has not yet been considered is given in Table 9.7. This is the spacing criterion (SC) for each luminaire. SC is a ratio of the maximum spacing between luminaires to the mounting height of the luminaires above the work plane that is likely to give reasonable uniformity of horizontal illuminance throughout the space. SC is simply a guide to the lighting designer, and should not be interpreted as a specification of the actual spacing-to-mounting-height ratio to be used in a given installation. In many cases, it is desirable to space luminaires closer together than the SC criterion suggests.

The SC numbers in Table 9.7 were determined using the luminaire layouts in Fig. 9.8. In Fig. 9.8a, two identical luminaires with symmetrical intensity distributions, separated by a distance S_1, are shown. It is assumed that the illuminances on the work plane at point P directly under one luminaire and at point Q midway between the two luminaires are produced by just the direct components of luminous flux of just those luminaires. Thus, the direct components of other luminaires in the space are neglected and the reflected components from room surfaces produced by all luminaires are also neglected. The distance S_1 is set so that the illuminances at P, due to source 1, and at Q,

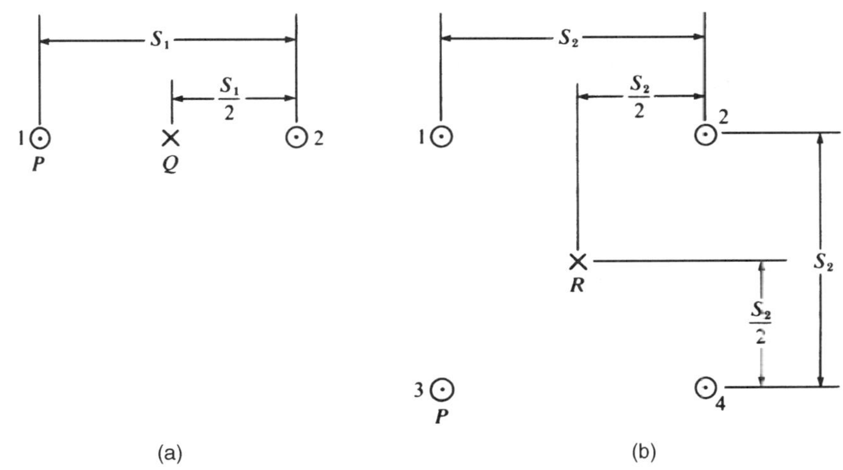

Figure 9.8 Layouts for determining SC.

due to both sources, are equal. The experiment is then repeated for the configuration in Fig. 9.8b with the same assumptions. The distance S_2 is set so that the illuminances at P, due to source 1, and at R, due to all four sources, are equal. The spacing criterion is then obtained from

$$SC = \frac{S}{MH} \tag{9.13}$$

where MH is the mounting height above the work plane and S is the smallest of the two values, S_1, and S_2.

If the luminaire has an asymmetric intensity distribution, as is often the case with fluorescent units, only the test in Fig. 9.8a is performed, but it is done twice—once with the luminaires perpendicular to the line between them and once with them parallel to this line. Two values of SC are then listed in Table 9.7, the first denoting the maximum spacing between rows (across) and the second giving the maximum spacing between luminaires in a row (along). In all cases, measurements are taken from the luminaire centers.

Certain luminaires, such as those with "batwing" intensity distributions, may not fit the spacing criterion. These units may produce reasonably uniform illuminance only within a range of spacings, rather than for spacings less than some maximum value. In such cases, the manufacturer's particular spacing guideline should be used. Furthermore, the spacing criterion is not applicable for large spacings ($SC > 1.5$). Here again, a specific manufacturer's spacing criterion is required.

9.12 AN ELEMENTARY LIGHTING DESIGN

All of the necessary tools to carry out an elementary lighting design are now known. Consider a 45 × 20 × 10 ft school room, with reflectances $\rho_c = 80\%$, $\rho_w = 50\%$, and $\rho_f = 30\%$, and with windows the length of one 45-ft north-facing wall. Assume steps 1 and 2 in Sect. 9.4 have been completed, with a decision to use Luminaire 7 in Table 9.7, housing two F32T8 fluorescent lamps with a CCT of 3500K and a CRI of 70. Assume an average of 10 fc of daylighting is present during daytime room occupancy. The task is to determine the number of luminaires required and to propose a lighting layout.

The first step is to determine the required illuminance level, using the guidelines in Sect. 9.3. The illuminance category from Table 9.3 is D or E, most likely the latter because school room visual tasks are often of "medium" contrast. Thus, the appropriate illuminance is 50 average maintained footcandles. This can be reduced in the daytime if the daylighting provides reasonable uniformity throughout the space. Since the room width is only 20 ft and the window wall is north facing, this is likely to be the case. But we should anticipate some nighttime use, so the design level will be set at 50 fc, with switching of individual rows of luminaires included.

Next, the cavity ratios are calculated from Eq. (9.3), assuming a 30-in work plane. There is no ceiling cavity. Thus

$$RCR = \frac{5(7.5)(65)}{45(20)} = 2.7$$

$$FCR = 2.7\left(\frac{2.5}{7.5}\right) = 0.9$$

Knowing RCR, the CU for a 20% floor reflectance can be obtained from Table 9.7, in this case with a single interpolation. The result is CU = 0.65. The floor cavity reflectance is 27% from Table 9.5. From Table 9.8, the CU multiplier is 1.05 for a 30% floor and 1.04 for a 22% floor, and the overall CU = 0.68.

Turning next to the light loss factor, 5% burnouts are allowed before replacement. Therefore, LBO = 0.95. LLD is 0.92, as in Ex. 9.7. Luminaire 7 falls in maintenance category V. Assuming a clean space cleaned annually, Eq. (9.10) yields LDD = 0.93 (A = 0.072). Luminaire 7 is a direct luminaire. Thus, the percent dirt depreciation will be the same as in Ex. 9.7 (12%), and entering Table 9.11 yields RSDD = 0.98. Thus, the recoverable LLF = 0.95 × 0.92 × 0.93 × 0.98 = 0.80. A ballast factor of 0.93 is assumed.

To determine the number of luminaires required, Eq. (9.2) is used.

$$\Phi_i = \frac{50(900)}{0.68(0.80)(0.93)} = 88{,}900\,\text{lm}$$

From Ex. 9.7, each lamp delivers 2950 initial lumens. Therefore,

$$N = \frac{88{,}900}{2 \times 2950} = 15 \text{ luminaires}$$

The spacing criteria for Luminaire 7 are 1.5 (11 ft) between rows and 1.3 (10 ft) between luminaires in the same row. Three possible layouts are shown in Fig. 9.9. These are all symmetrical layouts that do not consider the location of windows, work stations, and furniture within the space. Each has slightly more than 15 luminaires. These layouts, should be considered as only a first cut at a viable design.

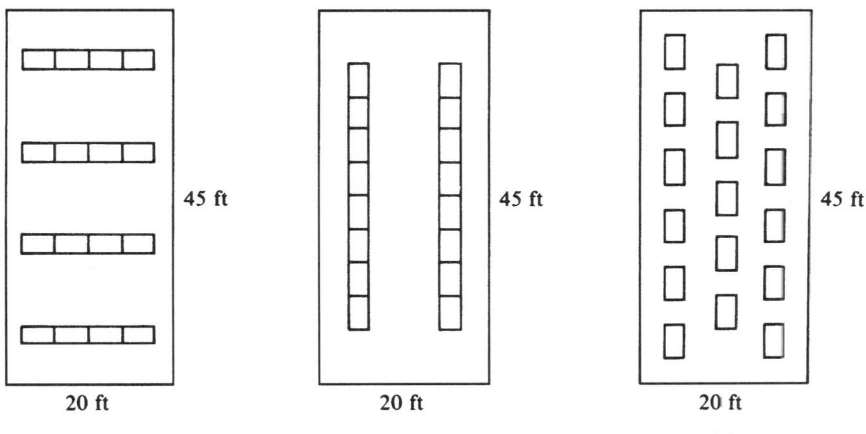

Figure 9.9 Possible school room lighting layouts.

The design that was just performed can be organized into a worksheet, as shown in Table 9.12. The user first enters the room dimensions and reflectances and the lamp and luminaire information; then maintenance information and the desired maintained illuminance level is recorded.

The remainder of the worksheet involves calculations. Cavity ratios and reflectances are determined and, from these, the CU values are found. Light loss information is calculated and entered. Finally, the required initial lumens and number of luminaires are

TABLE 9.12 ZONAL CAVITY METHOD WORKSHEET

Room dimensions (ft):	ℓ 45, *w* 20, *h* 10 WP height 2.5, WP area (A_{wp}) 900
Reflectances (%):	ρ_c 80, ρ_w 50, ρ_f 30
Lamp: F32T8	Initial lumens 2950 Lumens at 40% life 2715 Lamp watts (P_L) 32 Luminaire ballast watts (P_B) 7 (2 lamp ballast)
Luminaire: 7	Spacing/MH (SC) 1.5/1.3 Downward flux 91% Suspension distance (ft) 0 Number of lamps/luminaire (n) 2
Maintenance:	Room classification clean Luminaire maintenance category V Cleaning schedule 1/yr Interim lamp replacement none
Desired average maintained footcandles (E):	45 (50 total)
Calculations:	RCR 2.7, CCR 0, FCR 0.9 ρ_{cc} 80, ρ_{fc} 27, CU_{20} 0.65, CU 0.68 LBO 0.95, LLD 0.92, BF 0.93 LDD 0.93, RSDD 0.98, LLF_R 0.80
Number of luminaires:	N 15
Maximum spacing (ft):	Between units 11, Between rows 10
Initial lumens (16 luminaires)	ϕ_{ia} 94,400
Initial footcandles:	$E_i = \dfrac{\phi_{ia}(CU)(BF)}{A_{wp}}$ 66
Maintained footcandles:	$E = E_i(LLF)$ 53
Watts per square foot.	$W = \dfrac{(nP_L + P_B)N}{A_{wp}}$ 1.3

determined, the spacing criteria are entered and the initial and maintained illumination levels and watts per square foot of power consumption are calculated. The final illuminances are based on the total initial lumens (ϕ_{ia}) for the number of luminaires actually used. The watts per square foot factor is a "benchmark" of whether the designer has created an energy-efficient lighting design. The value of 1.3 is reasonably good, although some contemporary fluorescent lighting designs are being accomplished at around 1 W/ft^2.

9.13 WALL AND CEILING LUMINOUS EXITANCE COEFFICIENTS

It is often desired to calculate the average luminous exitance of the wall and ceiling surfaces within a space, and, from these, the illuminances on a tackboard or logo, for example. Equations similar in form to that used to find the average maintained horizontal illuminance in a space [Eq. (9.2)] have been developed for this purpose. These are:

$$M_w = \frac{\phi_i(WEC)(LLF)}{A} \tag{9.14}$$

$$M_c = \frac{\phi_i(CEC)(LLF)}{A} \tag{9.15}$$

where M_w, M_c = average maintained wall and ceiling cavity luminous exitances
Φ_i = total initial lamp lumens
WEC, CEC = wall and ceiling cavity luminous exitance coefficients
LLF = total light loss factor
A = work-plane area

If the luminaires are recessed or surface-mounted, M_c is the average luminous exitance of the ceiling plane; if they are suspended, of the luminaire plane. In either case, the luminous exitances of the luminaires are excluded from M_c.

If the wall surfaces do not all have the same reflectance factor, the luminous exitance of a particular wall can be found from

$$M_i = M_w \frac{\rho_i}{\rho_{wa}} \tag{9.16}$$

where the subscript i refers to the ith wall section and "wa" means weighted average wall reflectance. Since all room surfaces are assumed to be perfectly diffusing, the luminance of a wall or ceiling surface can be found by dividing the luminous exitances in Eqs. (9.14) and (9.15) by π. If the luminous exitances are in lumens per square foot, the luminances are in candelas per square foot.

Wall and ceiling luminous exitance coefficients, as well as coefficients of utilization, for two luminaires are given in Table 9.13.*

EXAMPLE 9.9

It is desired to find the initial average illuminances on the workplane, walls, and ceiling for the room in Ex. 9.2, when it is lighted by 24 of the fluorescent luminaires in Table 9.13, each containing one F32T8 fluorescent lamp (2950 initial lumens).

Solution. From Ex. 9.2, RCR = 2.51, $\rho_c = 80\%$, $\rho_w = 50\%$ and $\rho_f = 30\%$. Then from Table 9.13, CU = 0.54, WEC = 0.139 and CEC = 0.096 for a 20% floor. The floor correction factor from Ex. 9.2 is 1.05, giving CU = 0.57, WEC = 0.175 and CEC = 0.101. Next Eqs. (9.2), (9.14) and (9.15) yield, with BF = 0.93,

$$E = \frac{24 \times 2950 \times 0.57 \times 0.93}{28 \times 32} = 42\,\text{fc}$$

$$M_w = \frac{24 \times 2950 \times 0.175 \times 0.93}{28 \times 32} = 13\,\text{lm/ft}^2$$

$$M_c = \frac{24 \times 2950 \times 0.101 \times 0.93}{28 \times 32} = 7\,\text{lm/ft}^2$$

* Because calculations involving WEC and CEC often involve subtracting two numbers close together in size, values of these coefficients are given to three decimal places.

TABLE 9.13 EXITANCE COEFFICIENTS

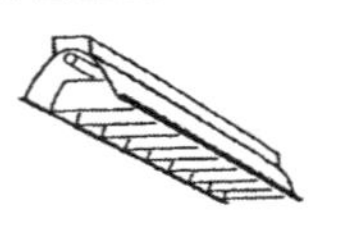

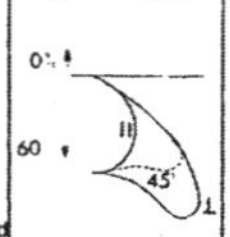

Luminaire	Typical Intensity Distribution and Per Cent Lamp Lumens		ρ_{CC} →	80			70			50			30			10			0
			ρ_W →	50	30	10	50	30	10	50	30	10	50	30	10	50	30	10	0
	Maint. Cat.	SC	RCR ↓	Coefficients of Utilization for 20 Per Cent Effective Floor Cavity Reflectance (ρ_{FC} = 20)															
Bilateral batwing distribution—louvered fluorescent unit	IV	N.A.	0	.71	.71	.71	.70	.70	.70	.66	.66	.66	.64	.64	.64	.61	.61	.61	.60
			1	.64	.62	.60	.63	.61	.60	.60	.59	.58	.58	.57	.56	.56	.55	.54	.53
			2	.57	.54	.51	.56	.53	.51	.54	.52	.50	.52	.50	.48	.51	.49	.47	.46
			3	.51	.47	.44	.50	.46	.43	.49	.45	.43	.47	.44	.42	.46	.43	.41	.40
			4	.46	.41	.38	.45	.41	.37	.44	.40	.37	.42	.39	.36	.41	.38	.36	.35
			5	.41	.36	.33	.40	.36	.32	.39	.35	.32	.38	.35	.32	.37	.34	.31	.30
			6	.37	.32	.28	.36	.32	.28	.35	.31	.28	.34	.31	.28	.34	.30	.28	.27
			7	.33	.29	.25	.33	.28	.25	.32	.28	.25	.31	.27	.25	.30	.27	.24	.23
			8	.30	.26	.22	.30	.25	.22	.29	.25	.22	.28	.25	.22	.28	.24	.22	.21
			9	.28	.23	.20	.27	.23	.20	.27	.23	.20	.26	.22	.20	.25	.22	.19	.18
			10	.25	.21	.18	.25	.21	.18	.25	.20	.18	.24	.20	.18	.23	.20	.18	.17

ρ_{CC} →	80			70			50			30			10		
ρ_W →	50	30	10	50	30	10	50	30	10	50	30	10	50	30	10
RCR ↓	Wall Exitance Coefficients for 20 Per Cent Effective Floor Cavity Reflectance (ρ_{FC} = 20)														
1	.144	.082	.026	.140	.080	.025	.132	.076	.024	.125	.072	.023	.118	.069	.022
2	.142	.078	.024	.139	.076	.024	.132	.073	.023	.126	.071	.022	.120	.068	.021
3	.137	.073	.022	.134	.072	.022	.128	.069	.021	.123	.067	.021	.118	.065	.020
4	.131	.068	.020	.128	.067	.020	.123	.065	.019	.118	.063	.019	.113	.062	.019
5	.124	.063	.018	.122	.062	.018	.117	.061	.018	.113	.059	.018	.109	.058	.017
6	.117	.059	.017	.115	.058	.017	.111	.057	.017	.107	.056	.016	.104	.055	.016
7	.111	.055	.016	.109	.054	.015	.105	.053	.015	.102	.052	.015	.099	.051	.015
8	.105	.051	.014	.103	.051	.014	.100	.050	.014	.097	.049	.014	.094	.048	.014
9	.100	.048	.013	.098	.048	.013	.095	.047	.013	.092	.046	.013	.089	.045	.013
10	.094	.045	.013	.093	.045	.013	.090	.044	.012	.088	.044	.012	.085	.043	.012

ρ_{CC} →	80			70			50			30			10		
ρ_W →	50	30	10	50	30	10	50	30	10	50	30	10	50	30	10
RCR ↓	Ceiling Cavity Exitance Coefficients for 20 Per Cent Floor Cavity Reflectance (ρ_{FC} = 20)														
	.114	.114	.114	.097	.097	.097	.066	.066	.066	.038	.038	.038	.012	.012	.012
1	.105	.094	.084	.090	.080	.072	.061	.055	.050	.035	.032	.029	.011	.010	.009
2	.099	.079	.063	.085	.068	.054	.058	.047	.038	.033	.027	.022	.011	.009	.007
3	.094	.068	.048	.080	.059	.042	.055	.041	.029	.032	.024	.017	.010	.008	.006
4	.089	.060	.038	.077	.052	.033	.053	.036	.023	.030	.021	.014	.010	.007	.004
5	.085	.054	.030	.073	.046	.026	.050	.032	.019	.029	.019	.011	.009	.006	.004
6	.082	.049	.025	.070	.042	.022	.048	.029	.015	.028	.017	.009	.009	.006	.003
7	.078	.044	.021	.067	.039	.018	.046	.027	.013	.027	.016	.008	.009	.005	.003
8	.075	.041	.018	.064	.036	.016	.044	.025	.011	.026	.015	.007	.008	.005	.002
9	.071	.038	.016	.062	.033	.014	.043	.023	.010	.025	.014	.006	.008	.005	.002
10	.068	.036	.014	.059	.031	.012	.041	.022	.009	.024	.013	.005	.008	.004	.002

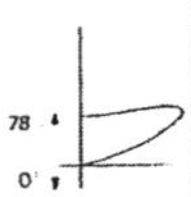

Luminaire	Typical Intensity Distribution and Per Cent Lamp Lumens		ρ_{CC} →	80			70			50			30			10			0
			ρ_W →	50	30	10	50	30	10	50	30	10	50	30	10	50	30	10	0
	Maint. Cat.	SC	RCR ↓	Coefficients of Utilization for 20 Per Cent Effective Floor Cavity Reflectance (ρ_{FC} = 20)															
Open top, indirect, reflector type unit with HID lamp (mult. by 0.9 for lens top)	VI	N.A.	0	.74	.74	.74	.63	.63	.63	.43	.43	.43	.25	.25	.25	.08	.08	.08	.00
			1	.64	.62	.59	.55	.53	.51	.38	.36	.35	.22	.21	.20	.07	.07	.07	.00
			2	.56	.52	.48	.48	.45	.42	.33	.31	.29	.19	.18	.17	.06	.06	.06	.00
			3	.49	.44	.40	.42	.38	.35	.29	.26	.24	.17	.15	.14	.05	.05	.05	.00
			4	.43	.38	.34	.37	.33	.29	.26	.23	.20	.15	.13	.12	.05	.04	.04	.00
			5	.38	.33	.28	.33	.28	.25	.23	.20	.17	.13	.12	.10	.04	.04	.03	.00
			6	.34	.28	.24	.29	.25	.21	.20	.17	.15	.12	.10	.09	.04	.03	.03	.00
			7	.31	.25	.21	.26	.22	.18	.18	.15	.13	.11	.09	.08	.03	.03	.03	.00
			8	.28	.22	.18	.24	.19	.16	.16	.13	.11	.10	.08	.07	.03	.03	.02	.00
			9	.25	.20	.16	.21	.17	.14	.15	.12	.10	.09	.07	.06	.03	.02	.02	.00
			10	.23	.17	.14	.20	.15	.12	.14	.11	.09	.08	.06	.05	.03	.02	.02	.00

ρ_{CC} →	80			70			50			30			10		
ρ_W →	50	30	10	50	30	10	50	30	10	50	30	10	50	30	10
RCR ↓	Wall Exitance Coefficients for 20 Per Cent Effective Floor Cavity Reflectance (ρ_{FC} = 20)														
1	.201	.114	.036	.172	.098	.031	.117	.067	.022	.068	.039	.013	.022	.013	.004
2	.184	.101	.031	.158	.087	.027	.108	.060	.019	.062	.035	.011	.020	.011	.004
3	.170	.091	.027	.146	.078	.024	.100	.054	.017	.058	.032	.010	.019	.010	.003
4	.158	.082	.024	.135	.071	.021	.093	.049	.015	.054	.029	.009	.017	.009	.003
5	.147	.075	.022	.126	.065	.019	.087	.045	.013	.050	.027	.008	.016	.009	.003
6	.137	.069	.020	.117	.059	.017	.081	.042	.012	.047	.024	.007	.015	.008	.002
7	.128	.063	.018	.110	.055	.016	.076	.038	.011	.044	.023	.007	.014	.007	.002
8	.120	.059	.016	.103	.051	.014	.071	.036	.010	.042	.021	.006	.013	.007	.002
9	.113	.054	.015	.097	.047	.013	.067	.033	.009	.039	.020	.006	.013	.006	.002
10	.106	.051	.014	.091	.044	.012	.063	.031	.009	.037	.018	.005	.012	.006	.002

ρ_{CC} →	80			70			50			30			10		
ρ_W →	50	30	10	50	30	10	50	30	10	50	30	10	50	30	10
RCR ↓	Ceiling Cavity Exitance Coefficients for 20 Per Cent Floor Cavity Reflectance (ρ_{FC} = 20)														
	.743	.743	.743	.635	.635	.635	.433	.433	.433	.249	.249	.249	.080	.080	.080
1	.737	.721	.707	.631	.619	.609	.431	.426	.421	.248	.247	.245	.080	.079	.079
2	.732	.706	.685	.627	.608	.592	.430	.421	.413	.248	.245	.242	.079	.079	.079
3	.727	.695	.670	.623	.600	.581	.428	.417	.408	.247	.243	.240	.079	.079	.079
4	.723	.687	.660	.620	.594	.573	.426	.414	.404	.247	.242	.239	.079	.079	.079
5	.718	.681	.653	.617	.589	.568	.425	.411	.401	.246	.242	.238	.079	.079	.078
6	.714	.676	.648	.614	.585	.564	.423	.410	.399	.246	.241	.237	.079	.079	.078
7	.710	.671	.644	.611	.582	.562	.422	.408	.398	.245	.240	.237	.079	.079	.078
8	.706	.668	.642	.608	.579	.559	.421	.407	.397	.245	.240	.236	.079	.079	.078
9	.703	.665	.639	.605	.577	.558	.419	.406	.396	.244	.240	.236	.079	.079	.078
10	.699	.662	.638	.603	.575	.556	.418	.405	.395	.244	.239	.236	.079	.079	.078

Source: IESNA Lighting Handbook, 8th Edition, 1993. Courtesy of the Illuminating Engineering Society of North America.

The respective illuminances on the walls and ceiling are:

$$E_w = \frac{13}{0.5} = 26\,\text{fc} \quad \text{and} \quad E_c = \frac{7}{0.8} = 9\,\text{fc}$$

Assume a tackboard of reflectance 35% hangs on one of the walls. Its luminous exitance is

$$M_{tb} = E_w \times \rho_{tb} = 26 \times 0.35 = 9\,\text{lm}/\text{ft}^2$$

9.14 NON-HORIZONTAL CEILINGS

In Sect. 9.7, Table 9.5 was presented for obtaining cavity reflectances for cavities with flat horizontal bases. In this section, a procedure for obtaining cavity reflectances for non-horizontal ceilings, such as sloped ceilings and ceilings with beams present, will be presented. An approximating expression for cavity reflectance, developed from flux transfer analysis, is

$$\rho_{eff} = \frac{1}{1+\left(\frac{A_s}{A_0}\right)\left(\frac{1-\rho}{\rho}\right)} \tag{9.17}$$

where A_s is the area of the cavity surfaces, A_0 is the area of the cavity opening, and ρ is the reflectance of the cavity surfaces. If the cavity is made up of surfaces having different reflectances, ρ must be the weighted-average reflectance of these surfaces.

For a rectangular cavity with walls of area A_w and reflectance ρ_w, and a base of area A_b and reflectance ρ_b, the weighted average reflectance is given by

$$\rho_{ba} = \frac{\rho_w A_w + \rho_b A_b}{A_w + A_b} \tag{9.18}$$

Insertion of Eq. (9.18) into Eq. (9.17) yields

$$\rho_{eff} = \frac{1}{\frac{[1+(A_w / A_b)]^2}{\rho_b + \rho_w (A_w / A_b)} - \frac{A_w}{A_b}} \tag{9.19}$$

For a ceiling cavity $A_b = A_c$ and $\rho_b = \rho_c$. For a floor cavity $A_b = A_f$ and $\rho_b = \rho_f$.

Eq. (9.17) gives exact values for ρ_{eff} when $\rho_w = \rho_c$ (or $\rho_w = \rho_f$). For the more general case when $\rho_w < \rho_c$ ($\rho_w > \rho_f$), the equation gives values which are slightly low (high), but these small errors do not affect seriously the ultimate values of horizontal work-plane illuminance that result.

Two examples of non-horizontal ceilings are a hemispherical dome and a cylindrical barrel vault. These are shown in Fig. 9.10.

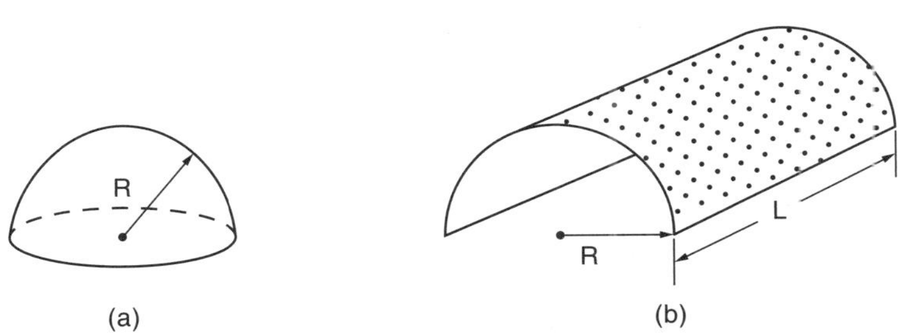

Figure 9.10 Examples of non-horizontal ceilings.

For the hemispherical dome of reflectance ρ_c and radius r, Eq. (9.17) yields

$$\rho_{eff} = \frac{1}{1+\frac{2\pi r^2}{\pi r^2}\left(\frac{1-\rho_c}{\rho_c}\right)} = \frac{\rho_c}{2-\rho_c} \tag{9.20}$$

and for the cylindrical barrel vault, again with reflectance ρ_c, radius r, and length P, and ignoring the ends, there is

$$\rho_{eff} = \frac{1}{1+\frac{\pi r\ell}{2r\ell}\left(\frac{1-\rho_c}{\rho_c}\right)} = \frac{1}{1+\frac{\pi}{2}\left(\frac{1-\rho_c}{\rho_c}\right)} \tag{9.21}$$

Note that neither of these effective reflectances is a function of the dimensions of the ceiling.

EXAMPLE 9.10

Calculate ρ_{eff} for a ceiling cavity 40 × 40 × 4 ft, with $\rho_c = 80\%$ and $\rho_w = 30\%$. Repeat for a floor cavity of the same size with $\rho_f = 20\%$ and $\rho_w = 50\%$.

Solution. First use Eq. (9.18) to obtain

$$\rho_{ca} = \frac{0.30\times 640+0.80\times 1600}{640+1600} = 0.657$$

$$\rho_{fa} = \frac{0.50\times 640+0.20\times 1600}{640+1600} = 0.286$$

Then Eq. (9.17) yields

$$\rho_{eff} = \frac{1}{1+\left(\frac{2240}{1600}\right)\left(\frac{1-0.657}{0.657}\right)} = 0.58\,(\text{ceiling})$$

$$\rho_{eff} = \frac{1}{1+\left(\frac{2240}{1600}\right)\left(\frac{1-0.286}{0.286}\right)} = 0.22\,(\text{floor})$$

EXAMPLE 9.11

A sawtooth skylight, such as is often found in industrial and warehouse settings, is shown in Fig. 9.11. The dimensions are a = 4 ft, b = 3 ft, w = 5 ft, ℓ = 20 ft. Surface a and the two ends have reflectances of 70%. Surface b is glass, with a reflectance of 10%. Find the effective cavity reflectance at the horizontal plane denoted by the dashed line.

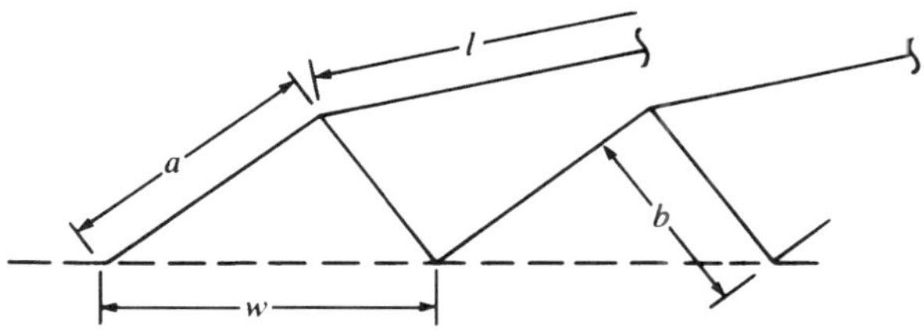

Figure 9.11

Solution. It is necessary first to obtain the average reflectance of the cavity surfaces. Using Eq. (9.9) gives

$$\rho_{wa} = \frac{80(0.7)+12(0.7)+60(0.1)}{80+12+60} = 0.463$$

Then, Eq. (9.17) yields

$$\rho_{cc} = \frac{1}{1+\frac{152}{100}[(1-0.463)/0.463]} = 0.36$$

EXAMPLE 9.12

A room with ceiling beams and suspended luminaires is shown in Fig. 9.12. The ceiling has a reflectance of 80%. The walls, and the sides and bottoms of the beams, have reflectances of 30%. Find ρ_{cc} at the luminaire plane.

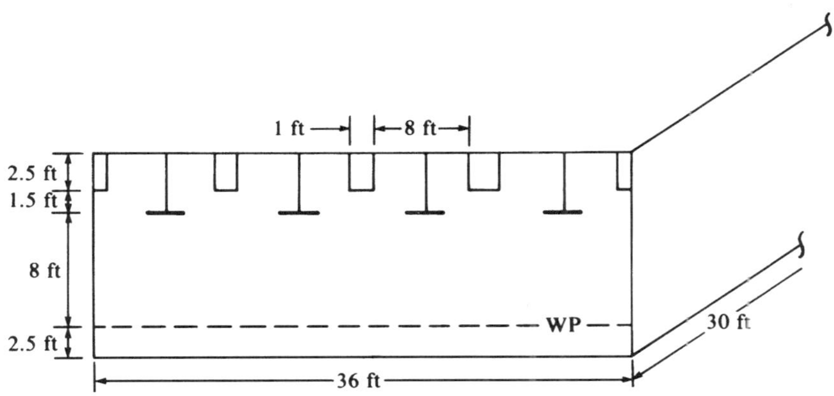

Figure 9.12

Solution. The problem can be worked in two ways, first using Table 9.5, and second using Eq. (9.17). The cavity ratio of the small cavities formed by the beams is

$$\text{CR} = \frac{5(2.5)(38)}{240} = 2.0$$

yielding $\rho_{eff} = 0.48$ from Table 9.5.

The average reflectance of the “ceiling” of the shallow cavity between the bottom of the beams and the luminaires is

$$\rho_{ca} = \frac{1(0.30) + 8(0.48)}{9} = 0.46$$

Its cavity ratio is

$$CR = \frac{5(1.5)(66)}{1080} = 0.46$$

From Table 9.5, after double interpolation,

$$\rho_{cc} = 0.41$$

at the luminaire plane.

Now solving the problem using Eq. (9.17), the weighted average reflectance of the small cavities is

$$\rho_{ca} = \frac{240(0.80) + 190(0.30)}{240 + 190} = 0.579$$

Then Eq. (9.17) yields

$$\rho_{eff} = \frac{1}{1 + \frac{430}{240}[(1 - 0.579)/0.579]} = 0.434$$

For the shallow cavity

$$\rho_{ca} = \frac{0.434(960) + 0.30(120) + 0.30(198)}{960 + 120 + 198} = 0.401$$

and again using Eq. (9.17) gives

$$\rho_{cc} = \frac{1}{1 + \frac{1278}{1080}[(1 - 0.401)/0.401])} = 0.36$$

As was noted earlier, the answer using Eq. (9.17) is somewhat conservative (12% low), largely because the equation weights the 30% walls a bit too heavily.

9.15 NON-RECTANGULAR FLOOR AREAS

To consider non-rectangular floor areas, the second form of Eq. (9.3) must be modified

$$CR = 5\frac{VSA}{HSA} = 5\frac{hP}{2A} = 2.5\frac{hP}{A} \tag{9.22}$$

where A is the work plane (horizontal) area, which is also the floor and ceiling area, P is the perimeter of work plane, and h is again cavity height.

EXAMPLE 9.13

Assuming a cavity height of 12 ft, find the cavity ratio for each of the spaces in Fig. 9.13.

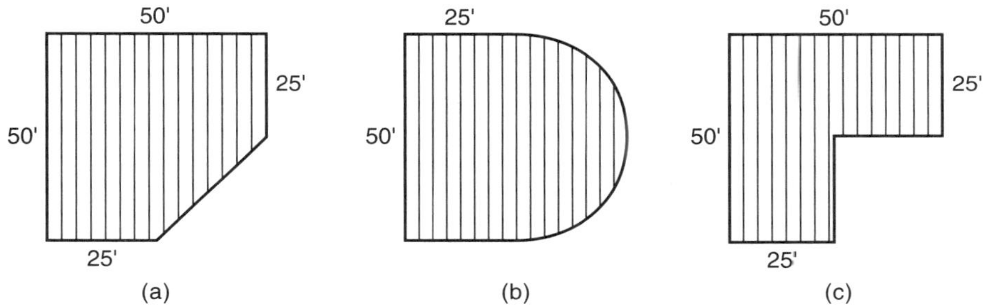

Figure 9.13 Non-rectangular floor areas.

Solution. Using Eq. (9.22), we obtain

$$\text{(a)} \qquad CR = 2.5\frac{12\times 185.4}{2187.5} = 2.54$$

$$\text{(b)} \qquad CR = 2.5\frac{12\times 178.5}{2231.7} = 2.40$$

$$\text{(c)} \qquad CR = 2.5\frac{12\times 200}{1875} = 3.20$$

The results in Ex. 9.13 illustrate a point. If the room in Fig. 9.11 were 50 ft square, its CR would be 2.40. Use of Eq. (9.22) gives quite accurate results for rooms such as shown in Fig. 9.13a and b, where the deviations from a rectangle (square in this case) are modest. It is not as accurate for L-shaped rooms, such as shown in Fig. 9.13c, especially if each wing of the L is pronounced so that the room begins to look more like two corridors intersecting. In going from a rectangular room to an L-shaped one with the same long walls, the perimeter remains the same, but the area decreases. Thus, the CR increases significantly, as the third result in Ex. 9.13 shows.

There is an alternative approach for L-shaped rooms. Consider Fig. 9.14 and proceed as follows:

- Create two rectangular spaces (dashed line).
- Consider one of the spaces (A) as the primary space.

- Find the effective cavity reflectance of space B at the dashed boundary.
  — with lights on in B
  — with lights off in B
- Find the weighted average wall reflectance for space A.
- Carry out the lumen method procedure for space A.
- Repeat the procedure with space B as the primary space.

The third step requires further explanation. Refer to Fig. 9.15. It shows the dashed boundary between spaces A and B. Assume that luminaires are spaced uniformly throughout spaces A and B. If the dashed boundary were a wall, it would reflect luminous flux from luminaire A to the work plane at point O, as shown. The wall reflectance would determine the percent reflected. In the absence of a wall, the luminous flux from luminaire B goes to point O, without reflection. Thus, if the lights are on in B, the dashed boundary appears as though it were a wall of $\rho = 100\%$.

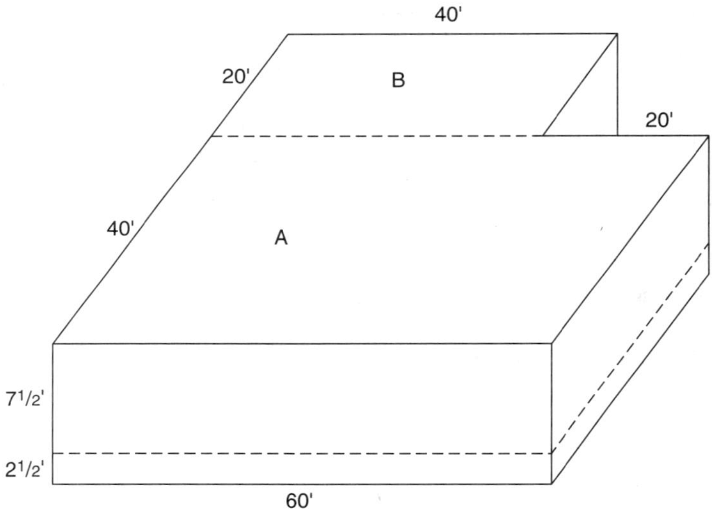

Figure 9.14 L-shaped room.

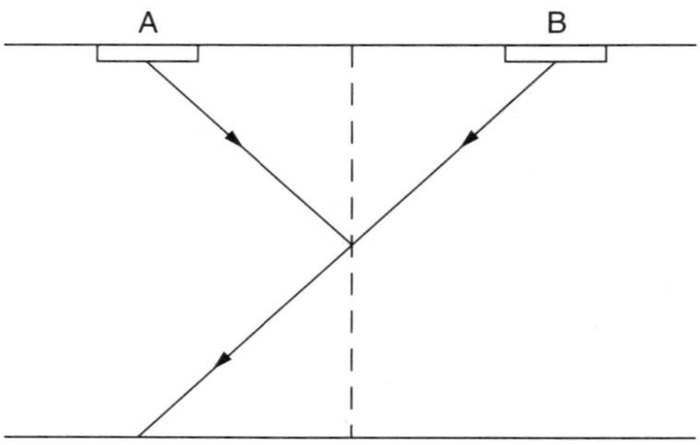

Figure 9.15 Condition at space boundary.

If the lights are off in B, it is the case of light entering a cavity and multiply reflecting. The effective cavity reflectance needs to be found from Table 9.5 or Eq. (9.17).

EXAMPLE 9.14

For the L-shaped room in Fig. 9.14, with $\rho_c = 70\%$, $\rho_w = 50\%$, $\rho_f = 20\%$, and with luminaires flush mounted, find the weighted average reflectance of the walls in space A with the lights on in space B and off in space B.

Solution. If the lights are on in space B, the reflectance of the dashed wall is 100%. Then, from Eq. (9.7), the weighted average wall reflectance in space A is

$$\rho_{wa} = \frac{(40+60+40+20)0.5+40\times 1.0}{200} = 0.60$$

(Since the height is constant, it is not included it in the calculation).

If the lights are off in space B, first find the weighted average "wall" reflectance of the space B cavity. Its base is the far 40-ft wall. Its "walls" are the ceiling, floor, and two walls of space B. Thus,

$$\rho_{wa}(\text{walls}) = \frac{80\times 0.2+40\times 0.5+80\times 0.7}{200} = 0.46$$

Then, the overall weighted average reflectance of space B is

$$\rho_{wa} = \frac{400\times 0.5+2000\times 0.46}{2400} = 0.47$$

and Eq. (9.17) yields

$$\rho_{\text{eff}} = \frac{1}{1+\dfrac{2400}{400}\left(\dfrac{1-0.47}{0.47}\right)} = 0.13$$

Alternatively, Table 9.5 can be used to find ρ_{eff}. First the cavity ratio of space B is needed. From Eq. (9.3),

$$\text{CR} = 5\frac{20(40+10)}{400} = 12.5$$

This exceeds the largest CR in Table 9.5. For a CR of 10, a base reflectance of 50% and a wall reflectance of 47%, Table 9.5 gives $\rho_{\text{eff}} = 0.16$; graphically extrapolating gives about 0.15, comparing quite well with the calculated value of 0.13.

Finally, the weighted average wall reflectance for space A, using the lesser ρ_{eff}, is

$$\rho_{wa} = 5\frac{(40+60+40+20)0.5+40\times 0.13}{200} = 0.43$$

As was mentioned earlier, the lumen method was developed for rectangular parallelpipeds. Thus, the developments in this and the previous section are approximations and should be treated as such. They do, however, yield quite accurate results for most situations.

9.16 PARTITIONED SPACES

There are many open-plan offices with partitioned cubicles. In this section, it is shown how the lumen method can be applied to such spaces to obtain the average maintained illuminance on the work plane.

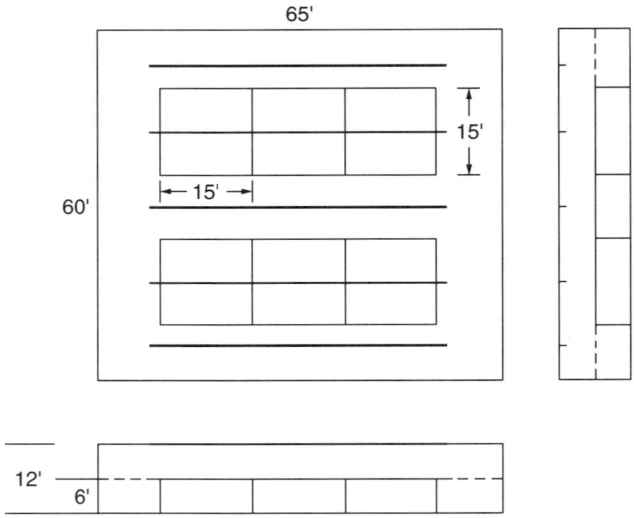

Figure 9.16 A partitioned space.

Consider the plan and elevation views of a partitioned space shown in Fig. 9.16. The room is 65 ft × 60 ft with a 12-ft ceiling height. There are six cubicles, each 15 ft × 15 ft and 6 ft high. Reflectances are $\rho_c = 80\%$, $\rho_w = 50\%$, $\rho_f = 20\%$, and $\rho_{cub} = 30\%$. Five rows of 12 surface-mounted luminaires per row provide the lighting. It is desired to find the average initial illuminance at a work plane height of 30 in inside each cubicle. The steps in the procedure are as follows:

a) Place a work plane at the top of the partitions and covering the entire room. Find the fraction of total lamp flux reaching this work plane, i.e., find the CU for the space above the partitions. Assume $\rho_{fc} = 0$ for this calculation (an approximation). Call the decimal CU obtained K_1.
b) Relocate the luminaires to the top of the partitions. Find the fraction of the total lamp flux entering the lower space (K_2) given by:

$$K_2 = \rho_{cc} \times \eta_u + \eta_d \tag{9.23}$$

where ρ_{cc} is the effective reflectance of the upper cavity, η_u is the fraction of the lamp flux going upward, and η_d is the fraction of the lamp flux going downward.

c) Form the ratio K_1/K_2, which indicates that the luminaires are K_1/K_2 as effective in providing luminous flux to the lower cavity when they are mounted on the ceiling, as opposed to being mounted at the top of the partitions.

d) Find the CU for one cubicle (CU_{cub}). The height of the cubicle room cavity is the cubicle height minus the work plane height. In this calculation, assume that the cubicle ρ_{fc} is 20% (an approximation). Since one wall is usually missing in a cubicle ($\rho = 0$), assume $\rho_{wa} = 0.75\rho_w$.

e) Find the average initial illuminance in a cubicle from

$$E = \frac{\phi_i \times \frac{A_{cub}}{A_{wp}} \times CU_{cub} \times BF \times \frac{K_1}{K_2}}{A_{cub}} \tag{9.24}$$

where ϕ_i is total initial lamp lumens, A_{cub} is the work plane area within a cubicle, and A_{wp} is the entire work plane area. The numerator of this expression gives the luminous flux on the work plane within a cubicle. Note that A_{cub} can be canceled out in Eq. (9.24).

EXAMPLE 9.15

Assume Luminaire 7 housing two F32T8 fluorescent lamps is used in the room in Fig. 9.16. Find the initial average illuminance on the work plane in a cubicle.

Solution. The total initial lamp lumens are $\phi_i = 2950 \times 2 \times 60 = 354{,}000$ lm

a) The room cavity ratio for the upper cavity is

$$RCR_u = \frac{5 \times 6(65+60)}{65 \times 60} = 1.0$$

From Table 9.7, CU = 0.71 for $\rho_c = 80\%$ and $\rho_w = 50\%$. From Table 9.8 for $\rho_{fc} = 0$, the CU multiplier is 0.87. Thus, $CU_o = K_1 = 0.87 \times 0.71 = 0.62$, implying that 62% of the lamp flux enters the space below the tops of the partitions.

b) The luminaires are now at the tops of the partitions. From Table 9.7, the values of η_u and η_d are:

$$\eta_u = 0.086 \times 0.689 = 0.059$$
$$\eta_d = 0.914 \times 0.689 = 0.630$$

The ceiling cavity ratio is 1.0 (the room cavity ratio of part a). From Table 9.5 for $\rho_c = 80\%$ and $\rho_w = 50\%$, $\rho_{cc} = 0.67$. Thus, from Eq. (9.23) $K_2 = 0.67 \times 0.059 + 0.630 = 0.71$.

c) The reduction factor is

$$\frac{K_1}{K_2} = \frac{0.62}{0.71} = 0.87$$

Thus the luminaires are 87% as effective on the ceiling as they are at the top of the partitions.

d) The room cavity ratio for a cubicle is given by

$$RCR_{CUB} = \frac{5 \times 3.5(15+15)}{15 \times 15} = 2.33$$

The cubicle ceiling cavity and wall reflectances are 0.67 and $0.30 \times 0.75 = 0.23$. From Table 9.7, with interpolation,

$$CU_{cub} = 0.52 \quad \text{for } \rho_{fc} = 20\%$$

e) The average initial illuminance is, from Eq. (9.24),

$$E = \frac{354{,}000 \times 0.52 \times 0.93 \times 0.87}{65 \times 60} = 38 \text{ fc}$$

From the plan and elevation views of the space in Fig. 9.16, it is possible to determine the portion of the ceiling that will be "seen" at a point on the work plane within a cubicle. This enables one to determine those luminaires, or portions thereof, that provide luminous flux directly to that point.

The situation is depicted in Fig. 9.17, where a task location is shown by a "dot" within a cubicle. In the elevation views, lines from the dot are projected upward

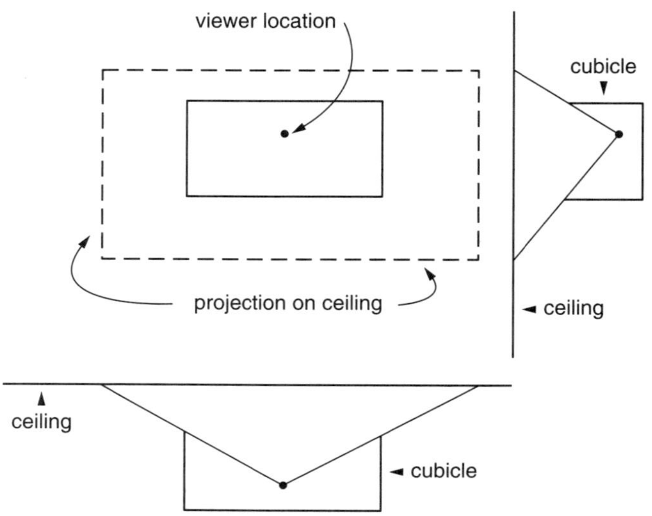

Figure 9.17 Direct illuminance in a cubicle.

through the tops of the partitions to the ceiling. These lines define the dashed rectangle in the plan view, which is the portion of the ceiling visible from the point.

9.17 TOTALLY INDIRECT LUMINAIRES

When totally indirect luminances are present in a space, there are two ways to carry out lumen method calculations. The first is the conventional method of replacing the ceiling cavity with an effective reflectance at the luminaire plane. The second is to assume that the ceiling, rather than the luminaire, is the source and thus that the room cavity extends to the ceiling. This is the preferred method when the ceiling cavity is relatively deep and/or the luminaire distribution is not wide, because in these cases, the luminaire flux is not uniformly distributed on the ceiling, as the lumen method requires.

EXAMPLE 9.16

Consider a room 20 ft × 20 ft × 10 ft, with a 2.5-ft work plane, and lighted by the indirect luminaire in Table 9.13. Reflectances are 80-30-20% (ceiling, walls, floor). Luminaires are suspended 2 ft. Find CU by the two methods and compare.

Solution. By the conventional method,

$$\text{RCR} = \frac{5 \times 5.5 \times 40}{400} = 2.75$$

$$\text{CCR} = 2.75 \times \frac{2}{5.5} = 1.00$$

$$\rho_{cc}\,(\text{from Table 9.5}) = 0.62$$

$$\text{CU}_{20}\,(\text{from Table 9.13}) = 0.34$$

By the second method

$$\text{RCR} = \frac{5 \times 7.5 \times 40}{400} = 3.75$$

$$\text{CCR} = 0$$

$$\rho_{cc} = \rho_c = 0.80$$

$$\text{CU}_{20}\,(\text{from Table 9.13}) = 0.39$$

$$\text{Ratio} = \frac{0.39}{0.34} = 1.15\ (15\%\ \text{increase})$$

This is a general result. The conventional method yields a ceiling cavity reflectance that is too low, because most upward luminous flux encounters the ceiling mostly, rather than the ceiling cavity, on its first reflection.

All totally indirect luminaires have CUs that are directly proportional to their efficiencies. This is because the lumen method assumes that the luminous flux leaving the ceiling cavity does so in a uniform diffuse manner. Thus, if there is a CU table and efficiency for one luminaire, the CUs for a second luminaire can be calculated if the latter's efficiency is known.

To illustrate this concept, revisit Eq. (9.6) for rooms with widely separated walls and rewrite it for a totally indirect luminaire as

$$CU = \frac{\eta_u \rho_c}{1 - \rho_f \rho_c} \tag{9.25}$$

With all luminaire flux upward, η_u is simply η, the luminaire efficiency. Thus

$$\frac{CU}{\eta} = \frac{\rho_c}{1 - \rho_f \rho_c} \tag{9.26}$$

This ratio is a function only of room reflectances.

There is one totally indirect luminaire in Table 9.7 and a second in Table 9.13. Forming the ratio in Eq. (9.26) for RCR = 0, $\rho\phi_c = 0.8$, and $\rho_f = 0.2$ gives

$$\frac{0.84}{0.883} = \frac{0.74}{0.78} = \frac{0.8}{1 - 0.16} = 0.95$$

which is indeed the same number for both luminaires.

REFERENCES

Boyce, P.: "Illuminance Selection Based on Visual Performance—and Other Fairy Stories," *JIES*, Summer, 1996.

Davis, R., and D. Ginthner: "Correlated Color Temperature, Illuminance Level and the Kruithof Curve," *JIES*, 1990.

Flynn, J., C. Hendrick, T. Spencer, and O. Martyniuk: "A Guide to Methodology Procedures for Measuring Subjective Impressions in Lighting," *IES Journal*, January, 1979.

Flynn, J.: "A Study of Subjective Responses to Low Energy and Nonuniform Lighting Systems," *Lighting Design & Application*, February, 1977.

Flynn, J.: "The Psychology of Light," 8-part series, Electrical Consultant, December 1972 to August, 1973.

Hawkes, R., D. Loe, and E. Rowlands: "A Note Towards the Understanding of Lighting Quality," *JIES*, 1979.

Jones, J., and B. Jones: "Using the Zonal-Cavity System in Lighting Calculations," *Illuminating Engineering*, Part 1, May 1964; Part 2, June 1964; Part 3, July 1964; Part 4, August 1964.

Kaufman, J.: *IESNA Lighting Handbook*, Fourth Edition, Illuminating Engineering Society, 1966.

Levere, R., R. Levin, and W. Primrose: "Spacing Criteria for Interior Luminaires—the Practices and Pitfalls," *IES Journal*, October 1973.

Levin, R.: "Revision of the S/MH Concept," *Lighting Design and Application*, August, 1977.
Levin, R.: "Cavities, Coefficients and Direct Ratios," *IES Journal*, April, 1982.
Levin, R.: "The Photometric Connection," Parts 1–4, *Lighting Design and Application*, September-December, 1982.
Levin, R.: "Fluorescent Light Loss Factors," *Lighting Design & Application*, November, 1985.
Levin, R., W. Brackett, J. Burke, and N. Frank: "Field Study of Luminaire Dirt Depreciation," *JIES*, Summer, 2002.
Mistrick, R., and D. DiLaura: "On the Error of the Zonal Cavity Calculations for Indirect Lighting," *JIES*, 1986.
Moon, P.: *The Scientific Basis for Illuminating Engineering*, McGraw-Hill, New York, 1936.
Murdoch, J.: "Step by Step," *Lighting Design & Application*, IESNA, New York, June, 1998.
Rea, M. editor: IESNA *Lighting Handbook*, Eighth Edition, Illuminating Engineering Society, 1993.
Rea, M. editor: *IESNA Lighting Handbook*, Ninth Edition, Illuminating Engineering Society, New York, 2000.
Steffy G.: *Architectural Lighting Design*, Van Nostrand Reinhold, New York, 1990.

Chapter 10

Optics and Control of Light

10.1 INTRODUCTION

In this chapter, the fundamentals of geometric and physical optics will be reviewed, and their application to the optical control of luminous flux distributions in interior and exterior spaces will be discussed.

Light may be controlled optically in a number of ways, the major of which are by reflection, diffusion, transmission, absorption, refraction, interference, diffraction, and polarization. The first five of these are based on geometric optics; the last three are phenomena from physical optics. Each of these control mechanisms will be discussed in some detail, including both their overall effect and their selective effect on certain wavelengths.

An introduction to geometric optics is based customarily on ray tracing from point sources, and this analysis will be developed. However, incandescent filaments and HID arc tubes are finite in size. It will be shown how basic optical analysis can be modified to deal with such sources.

Lastly, the use of fiber optics in lighting has expanded greatly in the past decade. The optical principles of fiber optics and light guides will be examined and some of their lighting applications will be described.

10.2 REFLECTION

Much of geometric optics is based on Huygens' principle, which states that as an electromagnetic wavefront advances, each point on it may be treated as a source of secondary wavefronts, which then advance in all directions in the same fashion as did the original wavefront. The new overall wavefront is then obtained by constructing the envelope of the secondary wavefronts. The process is illustrated in Fig. 10.1 to demonstrate a plane light wave reflecting from a plane polished surface. The effect of the reflecting surface is to reverse the direction of travel of the secondary wavefronts as they impinge upon it.

Reflection is a generic term that describes the process by which a portion of the electromagnetic radiation (luminous flux) incident on a surface leaves that surface without passing through it. The different types of reflection are shown in Fig. 10.2. Specular reflection is illustrated in Fig. 10.2a, and is, by definition, the situation in which the law of reflection for plane waves is obeyed. This law states that the incident and reflected rays and the normal to the surface lie in the same plane, and that the angle of reflection (θ_r) is equal to the angle of incidence (θ_i). The law may be derived from the construction in Fig. 10.1.

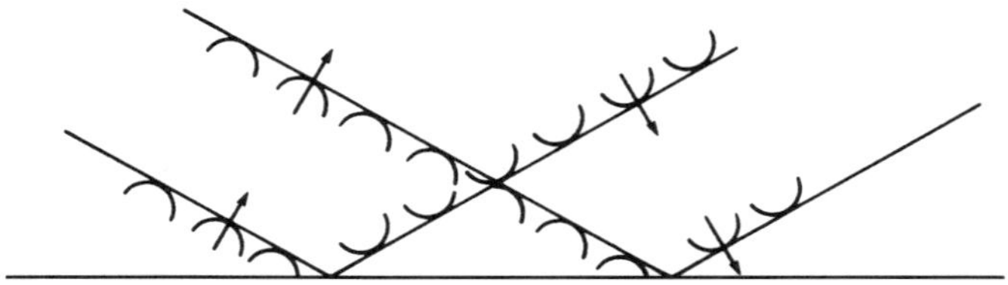

Figure 10.1 Plane wave reflection.

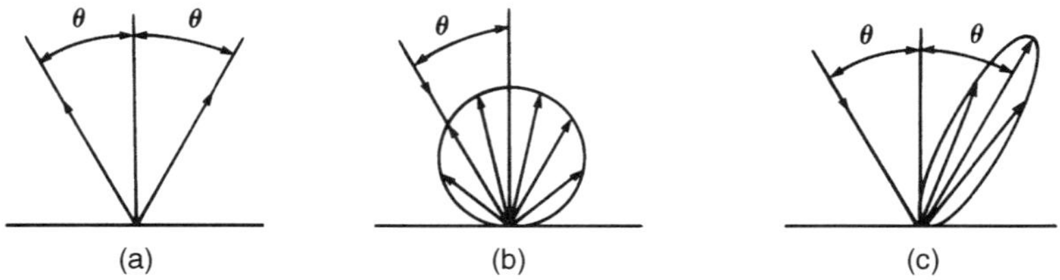

Figure 10.2 Types of reflection.

Examples of specular materials are silvered mirrors, polished metals, aluminum foil, and stainless steel. Reflectances of specular surfaces range from 90 to 95% for vaporized aluminum and silver, and 85–90% for aluminum foil, to 50% for stainless steel.

A clear glass surface reflects approximately 4% of the incident light for angles of incidence <30°, but reflectance increases markedly for larger angles, as shown in Fig. 10.3a. With a piece of clear plate glass, there are two interfaces, air-glass and glass-air.

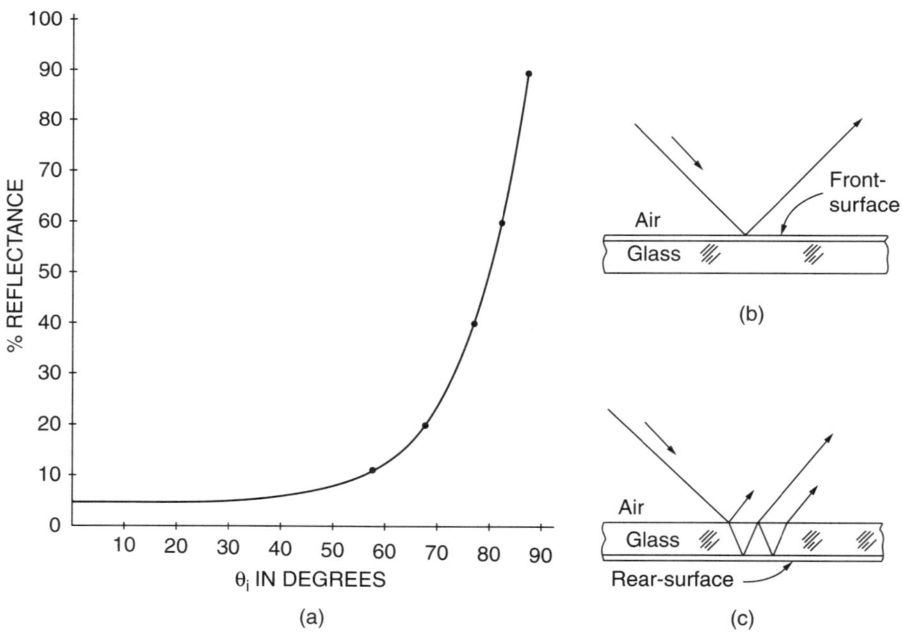

Figure 10.3 Reflection at an air-glass interface. ***Source*: *IESNA Lighting Handbook*, 9th edition, 2000. Courtesy of the Illuminating Engineering Society of North America.**

If a ray of light is incident normally on the front surface, 4% is reflected from that surface and another 4% from the rear surface. Thus, a piece of clear plate glass transmits approximately 92% of the incident light. Often one of the glass surfaces is silvered-coated. The situation for a front-surface mirror is shown in Fig. 10.3b; that for a rear-surface mirror in Fig. 10.3c. In the rear surface case, multiple internal reflections and refractions can occur.

At the other extreme from specular reflection is diffuse reflection, which is the characteristic of matte-finished surfaces and surfaces composed of small crystal particles. It is illustrated in Fig. 10.2b. A perfectly diffusing surface, as was noted in Chap. 1, is defined as one for which luminance is constant with angle of observation. Flat paints and white plaster closely approximate this condition, except at large angles of incidence; magnesium oxide and magnesium carbonate approximate this condition, as well.

Between perfectly diffuse and specular lie all other types of reflection, one of which is spread reflection, shown in Fig. 10.2c. Here there is no mirror image of the source, as in the specular case, but the angle of maximum reflected intensity does equal the angle of incidence. There is some diffusion, but insufficient to create constant luminance. Spread reflection occurs when polished aluminum or similar material is brushed, etched, hammered, pebbled, or corrugated. Etched aluminum has a reflectance of 60 to 80%; brushed aluminum 40 to 50%.

It is also possible to have combinations of the three types, called compound reflection. Thus, diffuse-specular, diffuse-spread, and specular-spread surfaces exist. These may provide almost any pattern of reflected light, and the pattern is often a function of angle of incidence. Porcelain enamel, semigloss and gloss paint, and glossy paper are examples of diffuse-specular materials, particularly at high angles of incidence.

10.3 MIRRORS AND RAY TRACING

Mirrors may be plane or curved, with the curvature of the latter in either one or two dimensions. The curved reflecting surfaces may be circular, parabolic, elliptical, hyperbolic, or some other contour to suit a particular application.

Inherent in any reflector design is the ability to do ray tracing. The concave spherical mirror in Fig. 10.4 will be used as a vehicle for illustrating how this is done. An object AX is located a distance p from the vertex O of the mirror, whose radius is r. Three rays are shown emanating from point A. The first is directed through the mirror's center at C, encounters the mirror at D, and reflects back along itself through C. The second is directed parallel to the axis of the mirror toward E, where it obeys the law of reflection and departs through point F. The third, directed toward the mirror's vertex at O, also obeys the law of reflection and proceeds along the line OB.

For small θ, the three reflected rays cross at point B, and thus the image of A will lie at B. Similarly, the image of X will lie at Y, and the object AX is imaged as BY, an inverted image smaller than the original object.

The point F needs to be investigated more fully. To obtain its distance f from O, it is observed from Fig. 10.4 that angle OCE = θ and angle OFE = 2θ.
Thus

$$EF = \frac{AX}{\sin 2\theta} = \frac{r\sin\theta}{\sin 2\theta} = \frac{r}{2\cos\theta} = CF \tag{10.1}$$

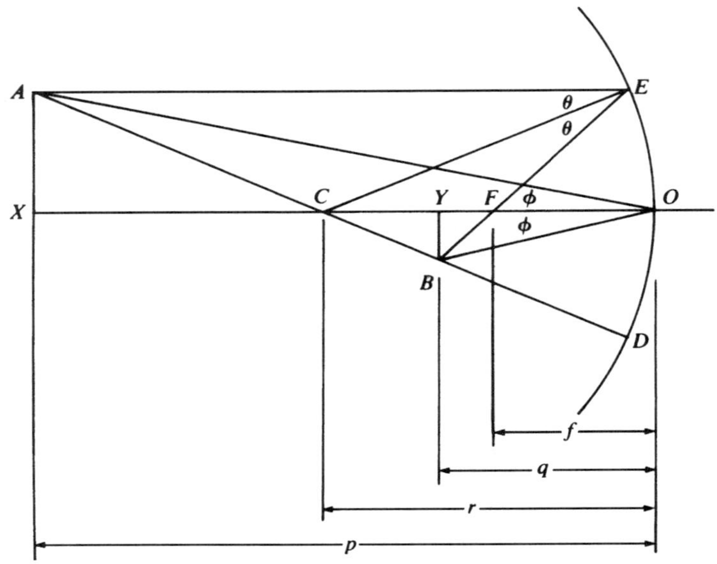

Figure 10.4 A concave spherical mirror.

and

$$f = CO - CF = r\left(1 - \frac{1}{2\cos\theta}\right) \tag{10.2}$$

For small θ, cos θ ≈ 1 and

$$f \simeq r\left(1 - \frac{1}{2}\right) = \frac{r}{2} \tag{10.3}$$

Thus, for a small-aperture mirror, all parallel rays, such as AE, reflect through point F, midway between C and O. By definition, F is the focal point of the mirror, and f is its focal length.

As the ray AE moves farther from the axis XO, the focal length from Eq. (10.2) decreases. For θ = 10°, f = 0.49r; for θ = 20°, f = 0.47r; and for θ = 30°, f = 0.42r. This property of a spherical mirror is called spherical aberration and can cause such a mirror to form a blurred image of an object.

For the small-aperture case, it is useful to relate p, q, and r. From Fig. 10.4

$$\begin{aligned} \frac{AX}{BY} &= \frac{XC}{CY} = \frac{p-r}{r-q} \\ \frac{AX}{BY} &= \frac{XO}{YO} = \frac{p}{q} \end{aligned} \tag{10.4}$$

Combining yields

$$\frac{1}{p} + \frac{1}{q} = \frac{2}{r} = \frac{1}{f} \tag{10.5}$$

Thus if p and r are known, q and f can be found. If $r = \infty$, $q = -p$, which is the equation for a plane mirror and states that the image is located behind the mirror at a distance equal to the object distance in front of the mirror.

In doing ray tracing, it is desirable to establish certain conventions. These are generally stated as follows:

1. Draw all diagrams with incident rays from left to right.
2. Assume object and image distances are positive if the object and image are to the left of the mirror.
3. Assume radius and focal length are positive if the center of curvature and focal point lie to the left of the mirror.
4. Assume images have positive size if they are above the horizontal axis.

This last stipulation, the geometry in Fig. 10.4, and the second Eq. (10.4) allows the magnification M of the mirror to be stipulated as

$$M = \frac{BY}{AX} = -\frac{q}{p} \tag{10.6}$$

It is instructive to plot magnification from Eq. (10.6) versus object distance for a concave spherical mirror. The result is shown in Fig. 10.5, where object distance is

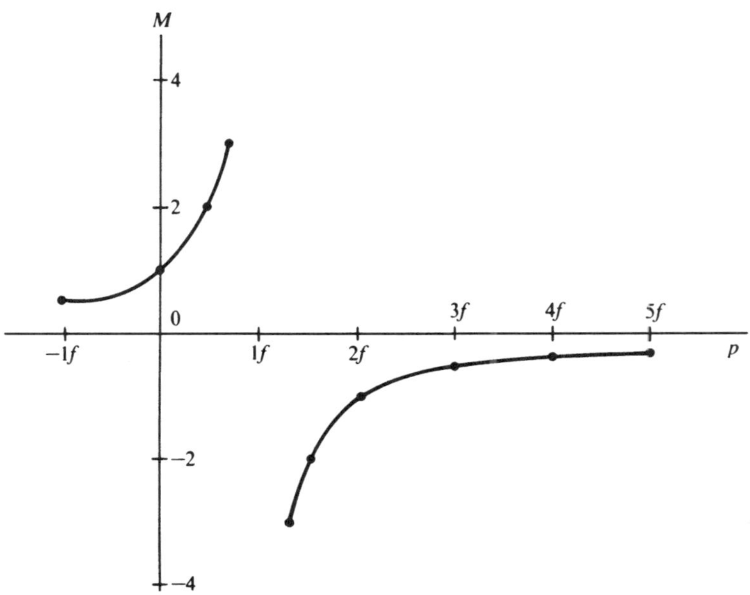

Figure 10.5 Magnification versus object distance for a concave spherical mirror.

expressed in focal lengths. If the object is an incandescent lamp filament, then the graph shows the relative size of the image of the filament as a function of filament distance from the vertex of the spherical reflecting surface.

EXAMPLE 10.1

In Fig. 10.4, let p = 30 cm and r = 20 cm. Find the image distance and magnification. Repeat for a convex mirror with the same radius and object distance. Find the focal length for each mirror.

Solution. For the concave mirror, from Eq. (10.5),

$$\frac{1}{30}+\frac{1}{q}=\frac{2}{20}$$

$$q=15 \text{ cm}$$

A positive value for q indicates that the image in Fig. 10.4 is to the left of the vertex. From Eq. (10.6) the magnification is –0.5 and from Eq. (10.3), the focal length is 10 cm.

For the convex mirror case, the center of curvature is to the right of the vertex. Thus, from Eq. (10.5)

$$\frac{1}{30}+\frac{1}{q}=\frac{2}{-20}$$

$$q=-7.5 \text{ cm}$$

$$f=\frac{-20}{2}=-10 \text{ cm}$$

$$M=-\frac{-7.5}{30}=0.25$$

In this case, the image is to the right of the mirror and, with a positive magnification, is erect and above the horizontal axis. The construction is shown in Fig. 10.6.

The image in Fig. 10.4 is called *real*, while that in Fig. 10.6 is called *virtual*. When the rays originating at a point converge after reflection, as in Fig. 10.4, a real image is produced. When they diverge, a virtual image results. A virtual image does not exist physically; it would not be formed on a screen placed at the image location. It is simply the apparent origin of light rays leaving the mirror.

EXAMPLE 10.2

Devise a concave mirror to form an image on a screen 3m away of a 1-cm long lamp filament 10 cm from the mirror. Find the radius and focal length of the mirror and the length of the filament on the screen.

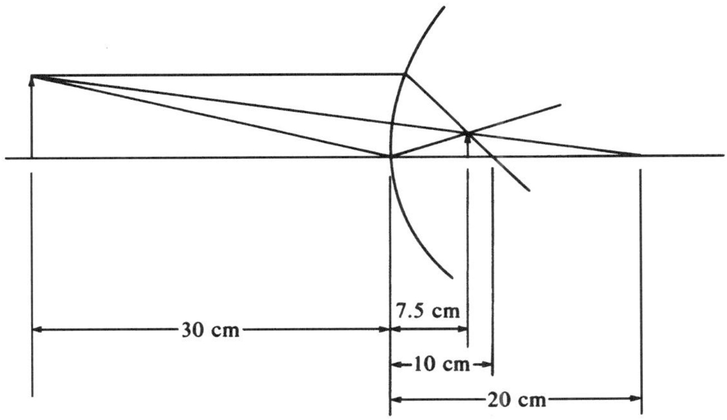

Figure 10.6 A convex spherical mirror.

Solution. From Eq. (10.5),

$$\frac{1}{10}+\frac{1}{300}=\frac{2}{r}=\frac{1}{f}$$

giving r = 19.4 cm and f = 9.7 cm. The magnification from Eq. (10.6) is

$$M=-\frac{300}{10}=-30$$

and thus the filament will appear 30 cm long, and inverted, on the screen.

10.4 CONIC SECTIONS

Reflector contours can be placed into two major categories, conic and general. Conic sections, often called basic contours, can be described easily mathematically and include the circle, ellipse, parabola, hyperbola, and combinations thereof. General contours are those designed through general mathematical analysis or through computerized ray-tracing techniques to provide specific reflected intensity distributions.

The basic contours are often used in reflector design, at least as a starting point, because they are relatively easy to understand and analyze and because they often yield ray patterns close to those that are ultimately desired.

All of the basic contours are formed as the intersection of a plane and a right circular cone, as shown in Fig. 10.7. In Fig. 10.7a, a plane parallel to the base of the cone intersects the cone in a circle. In Fig. 10.7b, a plane parallel to the slant face of the cone intersects the cone in a parabola. Any plane of intermediate orientation intersects the cone in an ellipse, as shown in Fig. 10.7c. If the plane in Fig. 10.7b becomes more upright than for a parabola, a hyperbola results, as shown in Fig. 10.7d. Each of these basic contours will now be considered in further detail.

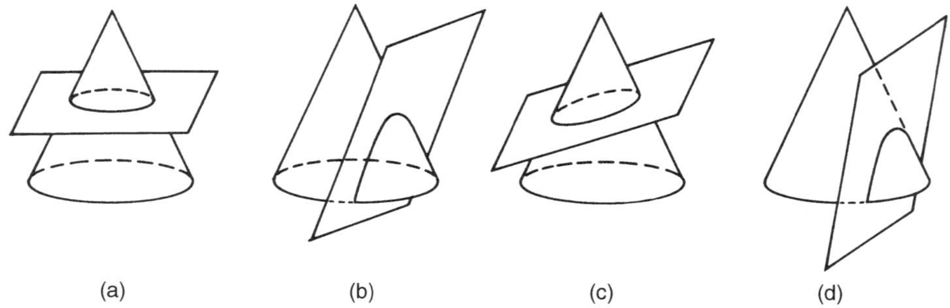

Figure 10.7 Conic sections.

A circle is the locus of a point at a constant distance from a fixed point. If the fixed point (the center of the circle) is placed at point F in an xy coordinate system, as shown in Fig. 10.8a, the locus equation is

$$(x - a)^2 + y^2 = a^2 \tag{10.7}$$

The point F is the focal point of the circle, and a is the focal length.

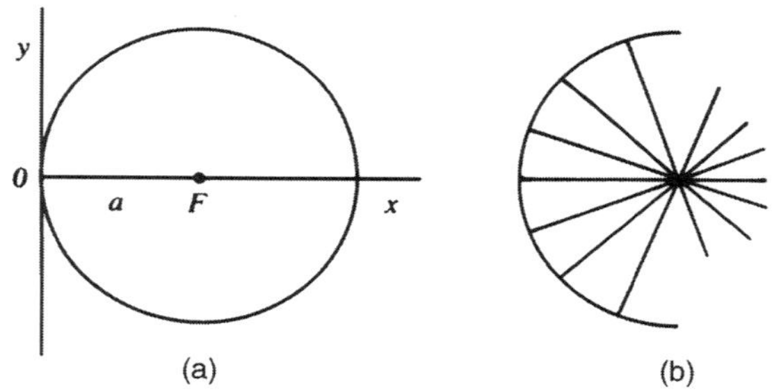

Figure 10.8 Ray patterns of a circular reflector.

Assume that a point source of light is placed at the center of a half-circular reflector. The rays incident on the reflector will be directed back through the source, as shown in Fig. 10.8b. If there were no losses, the intensity of the source in all directions to the right of the reflector would be doubled.

Spherical reflectors, which are in reality three-dimensional circular reflectors, are widely used in projection systems to increase the intensity of the light collected by the lens system. Cylindrical reflectors are used in tubular incandescent showcase lamps to redirect light from the linear filament to the items on display.

A lamp filament is not a point source. Thus, the ray pattern in Fig. 10.8b is idealized. Very often, as in projection lamp systems, the filament is planar, and lies in the position indicated in Fig. 10.9. As shown in that figure, reflected rays do not pass through the focal point and are scattered widely. In a projection lamp system, these rays emerging from the reflector are gathered and focused by the lens system. If the lamp filament becomes too large, rays from its ends will not be intercepted by the lens. Thus, there is a maximum size filament that a given lens system can accommodate, and increasing filament size beyond this point simply results in wasted light.

The optical gain of a spherical reflector system depends on the opacity of the source and how well it is centered. For open filament forms, the use of a spherical

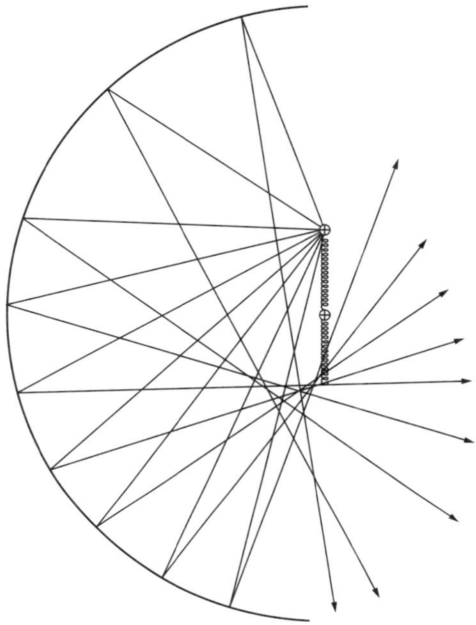

Figure 10.9 Reflections from a finite filament in a circular reflector.

reflector to enhance intensity is quite effective; in a compact form it is relatively ineffective. Also, the reflected rays can increase the temperature of the filament, increasing light output but decreasing life.

EXAMPLE 10.3

In Fig. 10.10, let the radius of a spherical mirror be 11 cm and the length of a filament be 6 cm. Find the point C at which the reflected ray BC, from incident ray AB, emerges from the filament.

Solution.

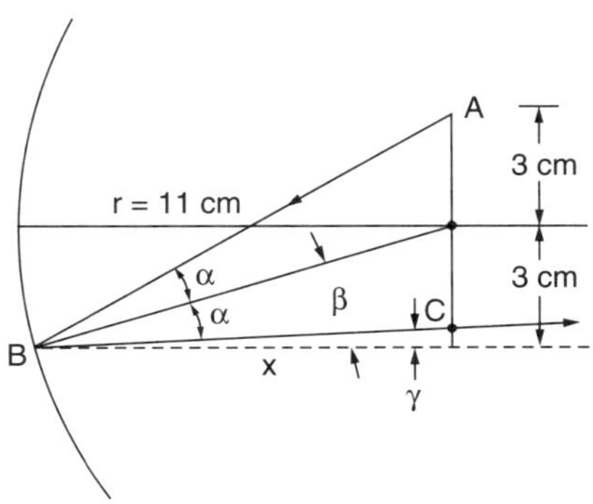

$$\beta = \sin^{-1}\tfrac{3}{11} = 15.827°$$

$$x = 11\cos\beta = 10.583 \text{ cm}$$

$$\alpha = \tan^{-1}\frac{6}{10.583} - \beta$$

$$= 29.551 - 15.827 = 13.724°$$

$$\gamma = \beta - \alpha = 2.103°$$

$$d = x\tan\gamma = 10.583 \times 0.0367$$

$$= 0.39 \text{ cm}$$

Figure 10.10

Thus, point C is located 2.61 cm from the center of the filament in Fig. 10.10.

The parabola is the next contour to consider. It is the locus of a point equally distant from a fixed point and a fixed straight line. The situation is shown in Fig. 10.11. The fixed point F is called the focal point, located a distance f, the focal length, from the origin. The fixed straight line NM is also located a distance f from the origin. This line is called the directrix.

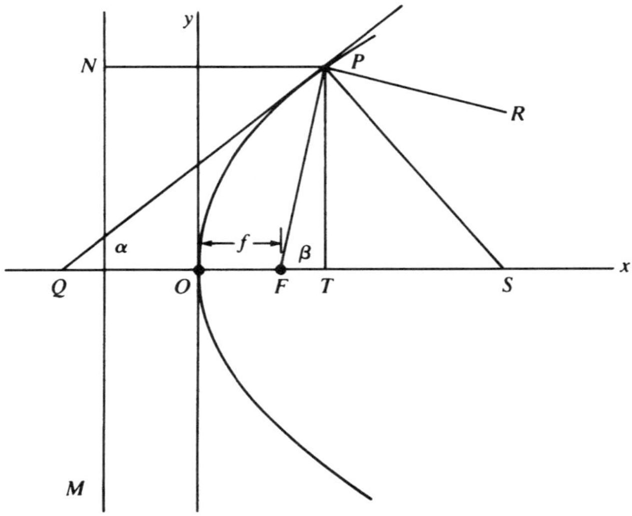

Figure 10.11 Parabolic contour.

From the definition of a parabola, the criterion in Fig. 10.11, is that FP = NP. Thus, the locus equation is

$$(x - f)^2 + y^2 = (x + f)^2 \tag{10.8}$$

which, upon simplification, yields

$$y^2 = 4fx \tag{10.9}$$

Assume a point source of light is located at the focus of the parabola in Fig. 10.11. It will be shown that any ray of light from the source will be reflected horizontally from the reflector. Consider the fictitious ray FPR. Draw a tangent PQ and a normal PS to the parabola and a perpendicular TP to the x axis, all at point P. For ray PR to be horizontal, it must be shown that angle TPR = 90°. Define angle PQO as α and angle TFP as β. From Eq. (10.9),

$$\tan\alpha = \frac{dy}{dx} = \frac{4f}{2y} = \sqrt{\frac{f}{x}} \tag{10.10}$$

From Fig. 10.11

$$\tan\beta = \frac{y}{x-f} = \frac{2\sqrt{fx}}{x-f} \tag{10.11}$$

From the law of tangents

$$\tan 2\alpha = \frac{2\tan\alpha}{1-\tan^2\alpha} = \frac{2\sqrt{fx}}{x-f} \tag{10.12}$$

Therefore
$$\beta = 2\alpha \tag{10.13}$$

and the tangent line QP bisects the angle NPF. From similar triangles, angle TPS = α and, from the law of reflection, angle SPR = angle FPS. Angle FPS = Angle FPT + Angle TPS = $90 - \beta + \alpha = 90 - \alpha$. Thus, the sum of angles TPS and SPR is 90°, completing the proof.

It is relatively easy to design a parabola to fit into a given rectangular space. Consider Fig. 10.12a, in which a rectangle of depth d and half-width w is shown. The parabola to be inserted is to pass through points A and B. At point B, Eq.(10.9) yields

$$w^2 = 4fd \tag{10.14}$$

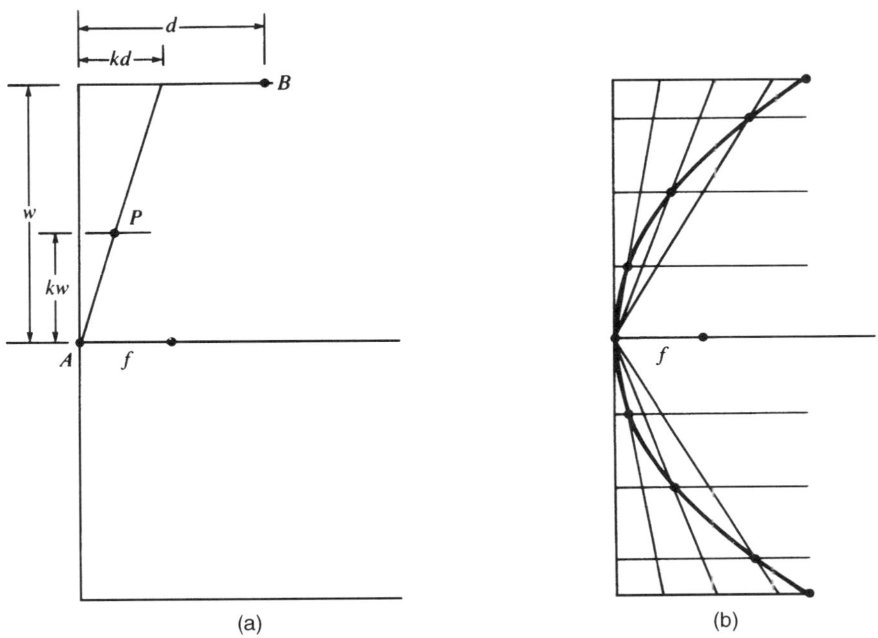

Figure 10.12 Designing a parabolic reflector.

Thus, the focal length of the desired parabola is

$$f = \frac{w^2}{4d} \tag{10.15}$$

and Eq. (10.9) can be rewritten as

$$y^2 = \frac{w^2}{d}x \tag{10.16}$$

For any other point P on the parabola for which $y = kw$ $(k < 1)$, Eq. (10.16) gives

$$\begin{gathered} k^2w^2 = \frac{w^2}{d}x \\ x = k^2d \end{gathered} \tag{10.17}$$

Thus at P

$$\frac{y}{x} = \frac{kw}{k^2d} = \frac{w}{kd} \tag{10.18}$$

The result in Eq. (10.18) suggests a way of constructing the parabola through similar triangles. At $y = kw$, a horizontal line is drawn. Next, a distance kd on d is marked off and a line is drawn to A. The intersection of the two lines is at P. In Fig. 10.12b, the entire parabolic construction is shown.

A parabola with a point source of light at its focal point will produce parallel reflected rays. This is the ideal case and can be closely approximated if a compact incandescent lamp filament is the light source. If the source, or part of it, is not at the focal point, the reflected rays are somewhat nonparallel. This is illustrated in Fig. 10.13a where the source is in front of the focal point, and in Fig. 10.13b where it is behind it. In Fig. 10.13a, the reflected rays collectively converge, but individually diverge. The opposite is true in Fig. 10.13b.

In Fig. 10.13c, the effects shown in Fig. 10.13a and b are combined for a small spherical source whose center is at the focal point, but whose surface is not. The solid angle of the cone of reflected rays is seen to decrease as the distance from the source to the point of reflection increases.

Parabolic contours are used widely in incandescent lighting (in PAR and R lamps, search lights, automobile headlamps, photoflood lamps, and infrared lamps). They are also used as reflectors in fluorescent luminaires, in the parabolic wedge louver, and in fluorescent cove lighting systems. With HID lamps, they are used to create high-density floodlights and spotlights.

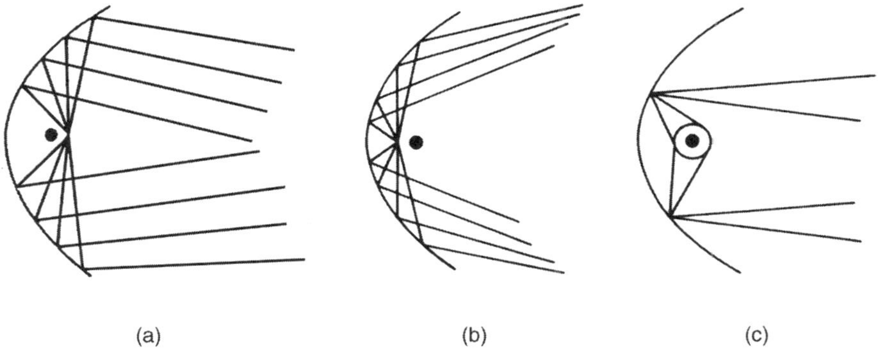

Figure 10.13 Parabola: Source in front of or behind focus.

Two interesting combinations of circular and parabolic reflectors are shown in Fig. 10.14. In each combination, the focal points of the circle and the parabola are the same. The circular reflector intensifies the rays from the source prior to their impinging on the parabolic reflector. The unit in Fig. 10.14b is particularly useful in lighting vertical surfaces. If the unit, in the form of a fluorescent cove, is mounted above eye level, the source will be shielded from viewers in the space. In the unit in Fig. 10.14a, the intensity at the periphery of the reflector will be greater than that in the center of the beam, because the rays of the former are non-divergent.

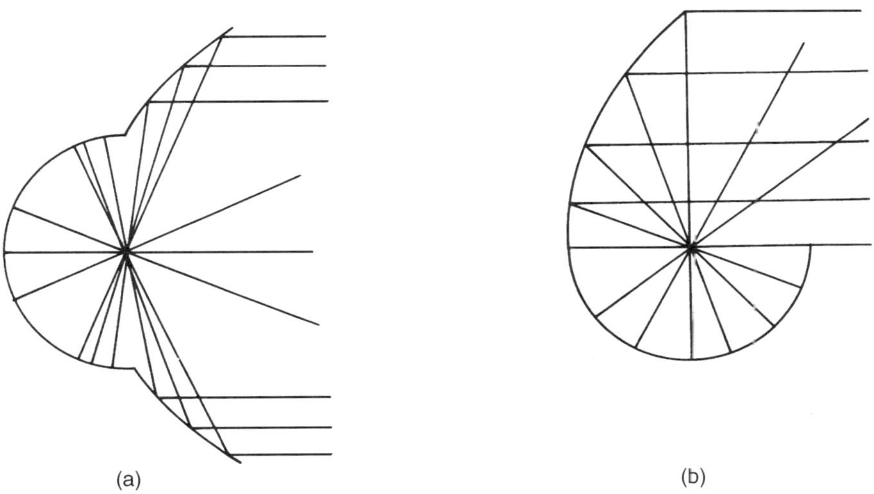

Figure 10.14 Circular-parabolic reflectors.

EXAMPLE 10.4

Design a parabolic reflector to fit exactly inside a cylinder of radius 10 in. and length 8 in. Calculate its focal length and the length of its latus rectum.

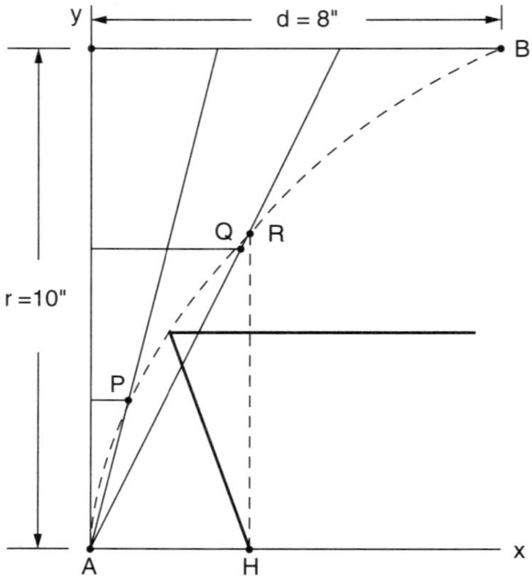

Figure 10.15

Solution. In Fig. 10.15, four points on the parabola are shown, for k = 0, 0.3, 0.6 and 1. The focal point is H and, from Eq. (10.15), the focal length is

$$f = \frac{r^2}{4d} = 3.1 \text{ in}$$

A typical ray is shown.

The latus rectum is, by definition, twice the length of the line HR, perpendicular to the x axis at the focal point. In Eq. (10.9), with x = f, y = ±2f. Thus, the latus rectum is 4 × 3.1 = 12.4 in. It should be noted that the slope of the parabola is 1 at point R, as may be seen by differentiating Eq. (10.9) and setting x = f.

Elliptical reflector contours will be considered next. An ellipse is the locus of a point, the sum of whose distances from two fixed points is constant. In Fig. 10.16, the two fixed points are labeled F and F′ and are termed the foci of the ellipse, with x coordinates f and f + 2c. If P is any point on the ellipse and 2a is defined as the constant sum of its distances from the foci, then

$$FP + F'P = 2a \tag{10.19a}$$

$$\sqrt{(x-f)^2 + y^2} + \sqrt{(x-f-2c)^2 + y^2} = 2a \tag{10.19b}$$

The quantity 2a can be given physical meaning. Let y = 0 in Eq. (10.19b). The result is

$$x = 0,\ 2a \tag{10.20}$$

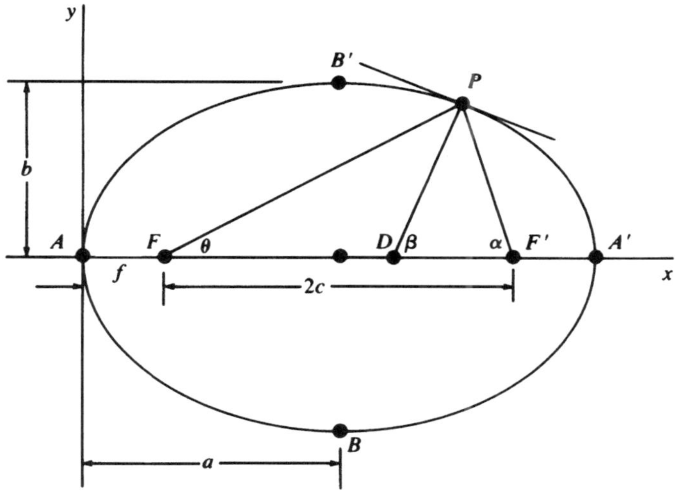

Figure 10.16 Elliptical contour.

Thus, the quantity 2a is the distance AA′ in Fig. 10.16 and

$$a = c + f \tag{10.21}$$

The distance AA′ is called the major axis of the ellipse.

The minor axis of the ellipse if BB′. To find its length, substitute x = c + f in Eq. (10.19b). This yields

$$y = \pm\sqrt{a^2 - c^2} \equiv \pm b \tag{10.22}$$

giving

$$BB' = 2\sqrt{a^2 - c^2} = 2b \tag{10.23}$$

Finally, using Eq. (10.21), and after considerable algebraic simplification, Eq. (10.19b) can be placed in the more useful form

$$\frac{(x-a)^2}{a^2} + \frac{y^2}{a^2 - c^2} = 1 \tag{10.24}$$

An elliptical reflector has the unique property that if a point source of light is placed at one focal point all reflected rays pass through the second focal point. To prove this property, it is necessary to show that the three angles indicated in Fig. 10.16 are related through the law of reflection by

$$\beta - \theta = 180 - \beta - \alpha \tag{10.25}$$

The angle β is defined from the line DP, which is drawn perpendicular to the ellipse at P. The slope of this line, which is tan β, is the negative reciprocal of the slope of the tangent to the ellipse at P. To obtain the latter, Eq. (10.24) is differentiated to obtain

$$\frac{2(x-a)dx}{a^2}+\frac{1}{a^2-c^2}(2y)dy=0 \tag{10.26}$$

$$\frac{dy}{dx}=\frac{(x-a)(c^2-a^2)}{a^2y} \tag{10.27}$$

Thus β is given by

$$\tan\beta=\frac{a^2y}{(x-a)(a^2-c^2)} \tag{10.28}$$

The other two angles in Fig. 10.16 are expressed by

$$\tan\theta=\frac{y}{x-a+c} \tag{10.29}$$

$$\tan\alpha=\frac{y}{a+c-x} \tag{10.30}$$

Then, noting that

$$\tan(180-\beta-\alpha)=-\tan(\beta+\alpha) \tag{10.31}$$

Eq. (10.25) can be rewritten as

$$\tan(\beta-\theta)=-\tan(\beta+\alpha) \tag{10.32}$$

From the law of tangents, each side of Eq. (10.32) can be expressed as

$$\tan(\beta-\theta)=\frac{\dfrac{a^2y}{(x-a)(a^2-c^2)}-\dfrac{y}{x-a+c}}{1+\dfrac{a^2y^2}{(x-a)(x-a+c)(a^2-c^2)}} \tag{10.33}$$

$$-\tan(\beta+a)=\frac{\dfrac{a^2y}{(x-a)(a^2-c^2)}-\dfrac{y}{x-a-c}}{1+\dfrac{a^2y^2}{(x-a)(x-a-c)(a^2-c^2)}} \tag{10.34}$$

After algebraic simplifications, Eqs. (10.33) and (10.34) can be shown to be equal, completing the proof.

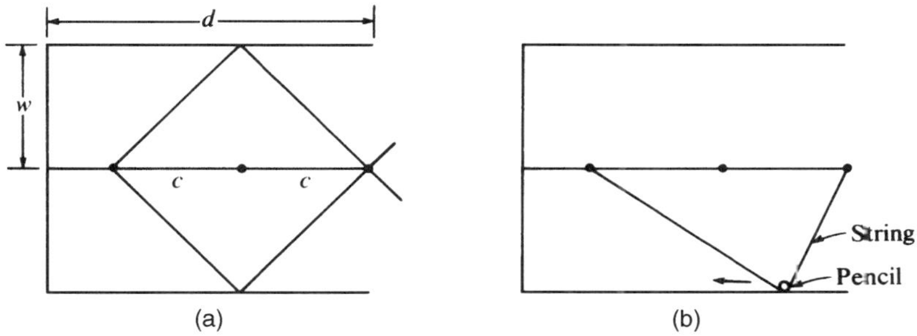

Figure 10.17 Designing an elliptical reflector.

In Fig. 10.17a, it is desired to place an elliptical reflector inside a rectangular enclosure of half-width w and depth d such that one focal point is in the plane of the enclosure opening. First, the location of the other focal point and the length of the major axis must be deteremined. It is noted that

$$\begin{aligned} d &= 2c + f \\ 2a &= d + f \\ w &= \sqrt{a^2 - c^2} \end{aligned} \tag{10.35}$$

where the last of these relations comes from Eq. (10.22). Solving Eqs. (10.35) for f and 2a yields

$$f = \frac{w^2}{d} \tag{10.36}$$

$$2a = d + \frac{w^2}{d} \tag{10.37}$$

To draw the elliptical contour, the points from Eq. (10.24) can be calculated. A simpler procedure is illustrated in Fig. 10.17b. Two pins are placed at the focal points, and a string of length 2a is anchored between them. The ellipse is obtained by placing a pencil against the string to keep it taut and then tracing the locus.

Suppose it is desired that the emerging divergent beam not extend beyond a certain number of degrees above and below the horizontal. What must be the distance x to the ends of the elliptical contour? The angle of interest is α in Fig. 10.16. y is solved for in terms of x in Eq. (10.30) and the result is inserted in Eq. (10.24).

$$y = (a + c - x)\tan\alpha = (d - x)\tan\alpha \tag{10.38}$$

$$\begin{aligned} &(x-a)^2 + \frac{a^2}{w^2}(d-x)^2\tan^2\alpha = a^2 \\ &\left(1 + \frac{a^2\tan^2\alpha}{w^2}\right)x^2 - 2a\left(1 + \frac{ad\tan^2\alpha}{w^2}\right)x + \frac{a^2d^2\tan^2\alpha}{w^2} = 0 \end{aligned} \tag{10.39}$$

All quantities in Eq. (10.39) are known except x.

EXAMPLE 10.5

Design an elliptical reflector to fit into a rectangular enclosure of depth 24 cm and half-width 12 cm, as in Fig. 10.17. It is desired that the beam angle α be 60°.

Solution. From Eqs. (10.36) and (10.37),

$$f = \frac{144}{24} = 6 \text{ cm}, \quad 2a = 24 + 6 = 30 \text{ cm}$$

Thus, in Fig. 10.17, the two focal points are at 6 cm and 24 cm and the string length for the locus tracing is 30 cm. The ends of the ellipse are obtained from Eq. (10.39).

$$5.69x^2 - 255x + 2700 = 0$$

$$x = 27.7 \text{ cm}, 17.1 \text{ cm}$$

The first answer is rejected. It comes about when x is greater than a + c in Eq. (10.30), giving a negative value for tan α. However, it is $\tan^2 \alpha$ that appears in Eq. (10.39). Thus, there are two values of x satisfying Eq. (10.39), but only one of these satisfies Eq. (10.38).

The design in Ex. 10.5 illustrates the two major attributes of an elliptical reflector, namely its ability to converge all reflected rays to a small region and its ability to provide a divergent beam of controllable angular size.

As with the other reflector contours, the ray patterns of an ellipse are often not the ideal shown in Fig. 10.17a because the source is usually not a point source. The resulting patterns when the source is ahead of or behind a focal point are shown in Fig. 10.18a and b, respectively. In Fig. 10.18c, the effect of finite source size is demonstrated. Here, a small spherical source is centered at the left focal point. The rays from each point on the reflector diverge somewhat upon reflection, and the rays from all points on the reflector converge in the vicinity of the focal point, but not at it.

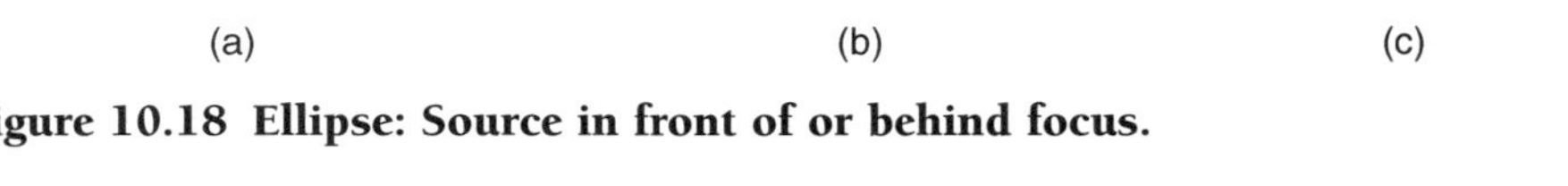

Figure 10.18 Ellipse: Source in front of or behind focus.

The elliptical reflector is widely used in light control. A few examples are shown in Fig. 10.19. Figure 10.19a shows the very common pinhole downlighting application, used extensively in architectural lighting. The light source is at one focal point and a small hole in the ceiling at the other. The source and reflector unit, which may be quite large, is nearly completely concealed from the viewer.

In Fig. 10.19b, a combination elliptical and circular reflector unit is shown. The lamp is at the focal point of the circle and at one focal point of the ellipse. Rays hitting

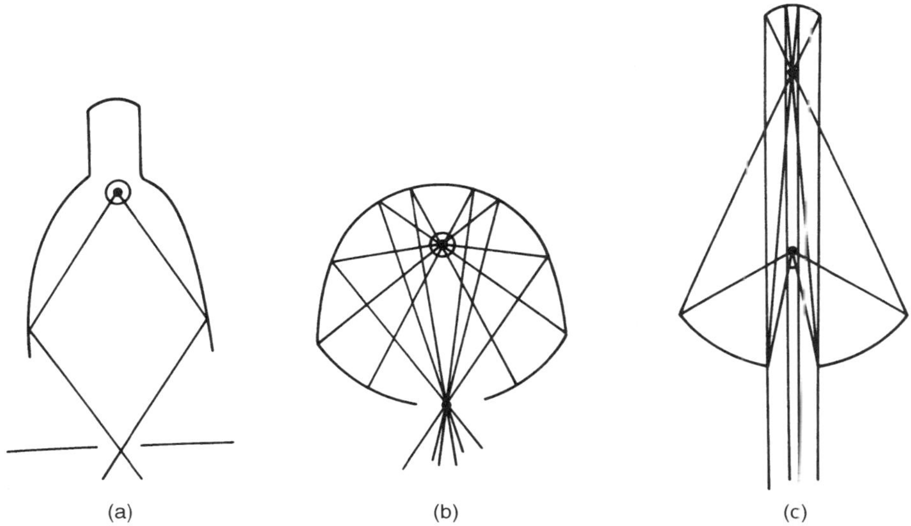

Figure 10.19 Applications of elliptical reflectors.

the circular reflector are reflected back through the source (assuming a noncompact filament), as in a projection lamp system. All reflected rays ultimately pass through the vicinity of the second focal point of the ellipse.

In Fig. 10.19c, a special surgical spotlight is shown, a combination elliptical and parabolic reflector unit. The source is at one focal point of the ellipse. Rays reflected from the ellipse pass through the second focal point, which is also the focal point of the parabola. An intense parallel-ray concentrated beam emerges from the parabola and passes through a hole in the ellipse and then to the operating table. A small shield blocks diverging direct rays from the source reaching the operating table.

The last basic reflector contour to be considered is the hyperbola, which can be used in place of the ellipse to produce a controlled diverging beam. A hyperbola is the locus of a point, the difference of whose distances from two fixed points is constant. In Fig. 10.20, the two fixed points (the two foci) are labeled F and F′, with x coordinates f + 2c and –f. For any point P on the right locus

$$FP - F'P = 2a \tag{10.40}$$

$$\sqrt{(x+f)^2+y^2} - \sqrt{(x-2c+f)^2+y^2} = 2a \tag{10.41}$$

where

$$c = f + a \tag{10.42}$$

For the left locus, the quantity 2a becomes –2a in Eq. (10.41).

Equation (10.41) can be algebraically simplified to yield

$$\frac{(x-a)^2}{a^2} - \frac{1}{c^2-a^2}y^2 = 1 \tag{10.43}$$

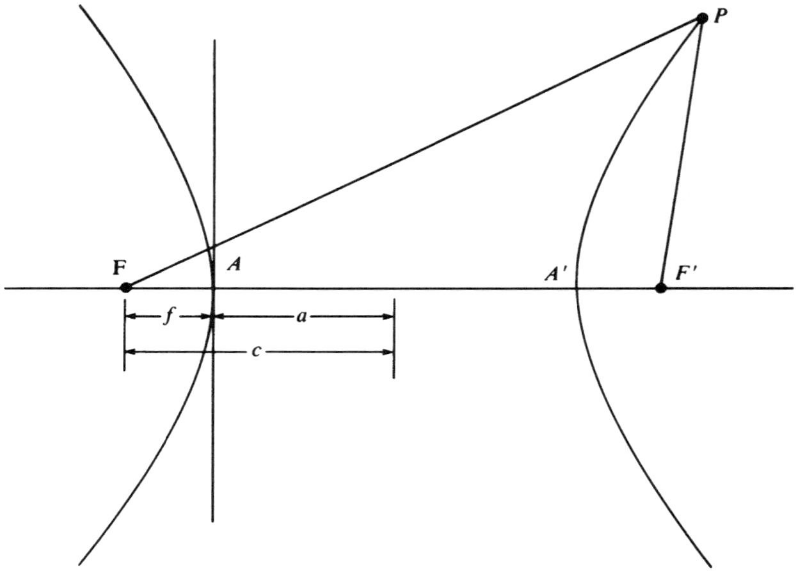

Figure 10.20 Hyperbolic contour.

A hyperbolic reflector can produce only a diverging beam. The rays from a point source of light at F′ reflect from the hyperbola as thought they had originated at F, as shown in Fig. 10.20. Thus, instead of passing through the second focal point, as in the ellipse, they appear to emanate from it.

It remains to be shown that the equations for the four basic reflector contours are related. Consider Eq. (10.24) for the ellipse. If the major and minor axes are equal, $a^2 = a^2 - c^2$, and, thus, $c = 0$ and $a = f$, from Eq. (10.21); the two focal points in Fig. 10.16 merge and Eq. (10.24) becomes Eq. (10.7), the equation of a circle.

The ellipse becomes a parabola if the focal point F′ in Fig. 10.16 moves to infinity. This causes both a and c to become infinite. From Eqs. (10.21) and (10.24),

$$\frac{x^2}{a} - 2x + \frac{a}{a^2 - c^2} y^2 = 0 \tag{10.44a}$$

$$\frac{x^2}{a} - 2x + \frac{a}{f(a + c)} y^2 = 0 \tag{10.44b}$$

As a, and thus c, approach infinity, Eq. (10.44b) becomes

$$-2x + \frac{1}{2f} y^2 = 0 \tag{10.45}$$

which is identical with Eq. (10.9) for the parabola.

Last, Eq. (10.43) for the hyperbola is identical in form to Eq. (10.24) for the ellipse, except that $c > a$ for the hyperbola and $c < a$ for the ellipse.

10.5 DESIGN OF NON-CONIC SECTION REFLECTORS

In situations where basic conic section reflectors are unsuitable, there are mathematical-graphical procedures for determining an appropriate reflector contour. One such procedure will be described here, adapted from the method of Jolley, Waldram, and Wilson (1930).

In a general reflector design, the given quantities are the intensity distribution curve of the lamp and the desired intensity distribution curve of the lamp-reflector combination. The design involves determining what reflector contour will produce the latter from the former. It is accomplished by treating the reflector as an assemblage of small flat reflecting elements, with each element oriented so as to reflect the flux incident on it into the appropriate zone. The procedure will be developed in a step-by-step manner with the aid of a numerical example for a reflector with a 60° cutoff angle.

Step 1 Tabulate the intensity distribution data for the bare lamp and the desired intensity distribution data for the lamp-reflector combination by zones. For the example to be worked, this information is presented in the first three columns of Table 10.1. The choice of angular zone size is dependent on the available lamp data. Here 10° zones have been used. The use of 5°, or $2^1/_2$°, zones can, of course, improve the accuracy of the design.

Step 2 For each zone, the bare lamp lumens and the required total lumens from the lamp-reflector combination are calculated, using the method of Sect. 2.9 and the zonal constants of Table 2.1. The results appear in columns 5 and 6 of Table 10.1.

For the procedure to work, it is necessary that the bare lamp lumens in each of the zones out to cutoff (60° in this case) be less than the required total lumens in those zones. This is the case in the example chosen. If it is not the case, there are two remedies possible: Choose a lamp of lesser intensity in each of the zones out to cutoff or scale the desired lamp-reflector intensity data by a factor $k > 1$, which will increase the lamp-reflector intensity values in each zone, but not alter the shape of the distribution curve.

Step 3 For each zone out to cutoff, subtract the bare lamp lumens from the total lumens to obtain the reflected lumens required in each zone (column 7 in Table 10.1). Then sum these to obtain the cumulative reflected lumens (column 8). For the example, the total reflected lumens required are 1681.3 lm.

Step 4 It is desired that the total bare lamp lumens from cutoff to 180°, when multiplied by the reflectance ρ of the reflector, equal the cumulative reflected lamp lumens from zero to cutoff.

Thus, it is necessary that the last number in column 10 equal the last number in column 8. In equation form, the requirement is

$$\phi_{\text{CO-180}}\,(\rho) = \phi_{\text{LR}} - \phi_{\text{0-CO}} \tag{10.46}$$

where $\phi_{\text{CO-180}}$ is the total lamp lumens from cutoff to zenith, ϕ_{LR} is the desired total lamp-reflector lumens, and $\phi_{\text{0-CO}}$ is the total lamp lumens from nadir to cutoff.

TABLE 10.1 GENERAL REFLECTOR DESIGN

(1) Zone (Degrees)	(2) Bare Lamp I_{mid} (cd)	(3) Lamp Reflector I_{mid} (cd)	(4) Zonal Constant (Table 2.1)	(5) Zonal Bare Lamp Lumens	(6) Zonal Total Lumens	(7) Zonal Reflected Lumens	(8) Cumulative Reflected Lumens	(9) Zonal Bare Lamp Lumens x ρ	(10) Cumulative Bare Lamp Lumens x ρ
0–10	90	1536	0.095	8.6	145.9	137.3	137.3		
10–20	114	1350	0.283	32.3	382.1	349.8	487.1		
20–30	152	1094	0.463	70.4	506.5	436.1	923.2		
30–40	188	856	0.628	118.1	537.6	419.5	1342.7		
40–50	216	570	0.774	167.2	441.2	274.0	1616.7		
50–60	228	300	0.897	204.5	269.1	64.6	1681.3		
60–70	228		0.993	601.1 226.4	2282.4	1681.3		192.4	1685.1
70–80	228		1.058	241.2				205.0	1492.7
80–90	228		1.091	248.7				211.4	1287.7
90–100	228		1.091	248.7				211.4	1076.3
100–110	228		1.058	241.2				205.0	864.9
110–120	228		0.993	226.4				192.4	659.9
120–130	228		0.897	204.5				173.8	467.5
130–140	165		0.774	127.7				108.5	293.7
140–150	182		0.628	114.3				97.2	185.2
150–160	143		0.463	66.2				56.3	88.0
160–170	117		0.283	33.1				28.1	31.7
170–180	44		0.095	4.2				3.6	3.6
				1982.6				1685.1	

Solving Eq. (10.46) for reflectance gives

$$\rho = \frac{\phi_{LR} - \phi_{0\text{-}CO}}{\phi_{CO\text{-}180}} \tag{10.47}$$

For the given example,

$$\rho = \frac{2282.4 - 601.1}{1982.6} = 0.848$$

Step 5 For each of the zones from cutoff to zenith, multiply the bare lamp zonal lumens by the reflectance ρ (column 9) and sum the results to obtain cumulative bare lamp lumens multiplied by reflectance (column 10). The summation is done from 180° to cutoff, rather than the reverse, because usually the design of the reflector begins with the 170–180° zone. A value of 0.85 was used in calculating the values in columns 9 and 10 in Table 10.1.

It is possible that Eq. (10.47) will give a required reflectance greater than unity. When this happens, it becomes necessary to reduce the desired intensity values in column 3. If each of these values is multiplied by a single constant $k < 1$, the final intensity distribution will be unaltered. Alternatively, if a slightly changed distribution and cutoff can be tolerated, the rim and neck of the reflector can be altered.

Step 6 A reflector pattern is chosen. There are four basic ray patterns for reflectors, as shown in Fig. 10.21. The contours in Fig. 10.21a and b reflect flux across the plane at the opening of the reflector; those in Fig. 10.21c and d do not. For the contours in Fig. 10.21a and c, flux reflected within the 170 to 180° zone is directed to the 0 to 10° zone; for the contours in Fig. 10.21b and d, to the 80 to 90° zone.

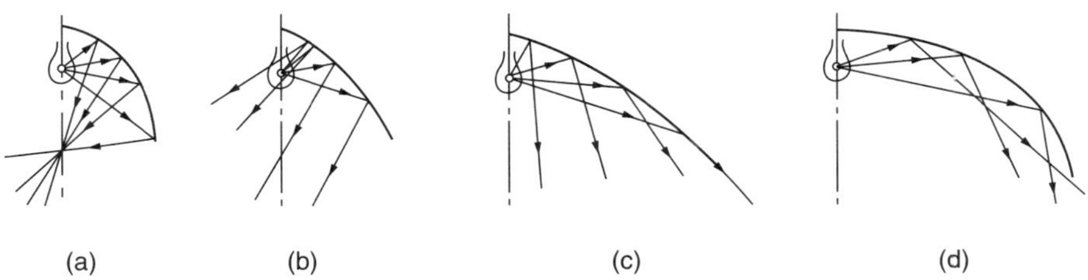

Figure 10.21 Four basic reflector patterns. *Source*: *IES Lighting Handbook, Reference Volume*, 1981 edition. Courtesy of the Illuminating Engineering Society of North America.

The patterns in Fig. 10.21c and d require larger diameter reflectors, as shown. The pattern in Fig. 10.21b has the disadvantage that a considerable amount of flux passes back through the bulb. Thus, the pattern in Fig. 10.21a is usually preferred, yielding a small reflector and minimum flux reflected through the lamp. This pattern will be used in the current example.

Step 7 Curves of required cumulative reflected lumens and of available cumulative bare lamp lumens times ρ are plotted from the data in columns 8 and 10 in Table 10.1. These curves are shown in Fig. 10.22. The left curve is plotted first, from the data in column 8 and with θ, the angle from nadir, increasing to the right along the x axis. The plot ends at θ = 60° when 1681.3 lm have been accumulated. Then the right

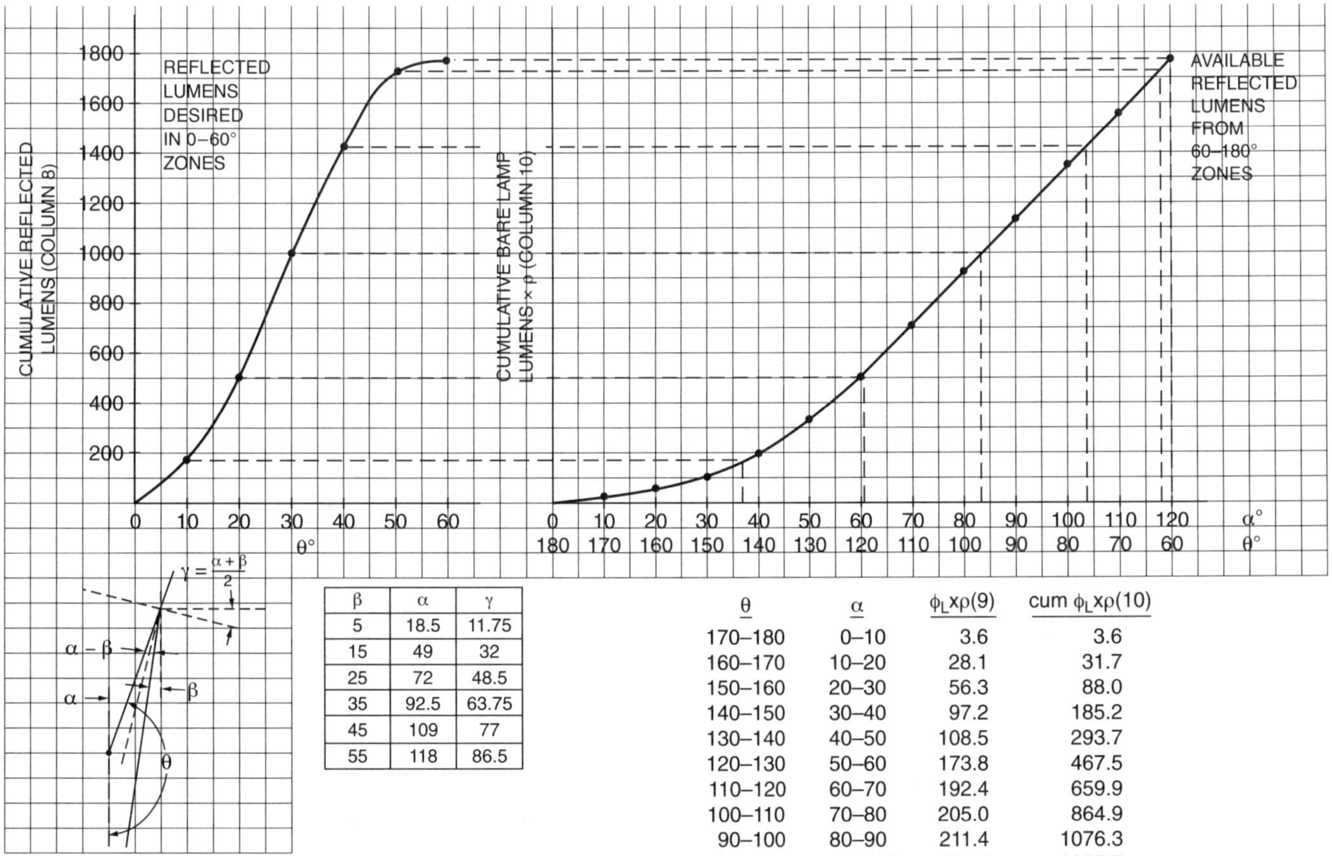

β	α	γ
5	18.5	11.75
15	49	32
25	72	48.5
35	92.5	63.75
45	109	77
55	118	86.5

θ	α	$\phi_L x \rho (9)$	cum $\phi_L x \rho (10)$
170–180	0–10	3.6	3.6
160–170	10–20	28.1	31.7
150–160	20–30	56.3	88.0
140–150	30–40	97.2	185.2
130–140	40–50	108.5	293.7
120–130	50–60	173.8	467.5
110–120	60–70	192.4	659.9
100–110	70–80	205.0	864.9
90–100	80–90	211.4	1076.3
80–90	90–100	211.4	1287.7
70–80	100–110	205.0	1492.7
60–70	110–120	192.4	1685.1

Figure 10.22 Graphical determination of reflector zones.

curve is plotted from the data in column 10, as a function of both θ and α, the angle from zenith, and extends from zenith to cutoff.

Next lines are drawn vertically on the left curve at 10°, 20°, 30°, 40°, and 50°. At the points of intersection with the curve, lines are drawn horizontally to the right curve, and then vertically to its x axis. To explain these lines, consider the leftmost one. It is drawn vertically at $\theta = 10°$ and intersects the curve at 137.3 lm, the required reflected lumens in the 0–10° zone. The horizontal line intersects the right curve at a zonal angle of 143°. This shows that a flat reflector element of 37° angular width should be inserted between 180 and 143° and oriented so as to direct the 137.3 lm reflected from it into the 0 to 10° zone. The orientation is determined from the diagram of angles in Fig. 10.22. For the 180-143° zone, the center angle is $x = 18.5°$ ($\theta = 161.5°$). Thus, the angle of tilt of the reflector segment must be such that a ray incident at this angle is reflected to the center of the 0-10° zone, at $\beta = 5°$. The general equation for the tilt angle γ is

$$\gamma = \frac{\alpha + \beta}{2} \tag{10.48}$$

which, for $\alpha = 18.5°$ and $\beta = 5°$, yields $\gamma = 11.75°$. Values of γ for the six angles are given in Fig. 10.22.

The entire construction is shown approximately in Fig. 10.23, where a smooth curve has been drawn through the linear segments. The six reflected rays cross the vertical axis close to each other, but not necessarily at a point.

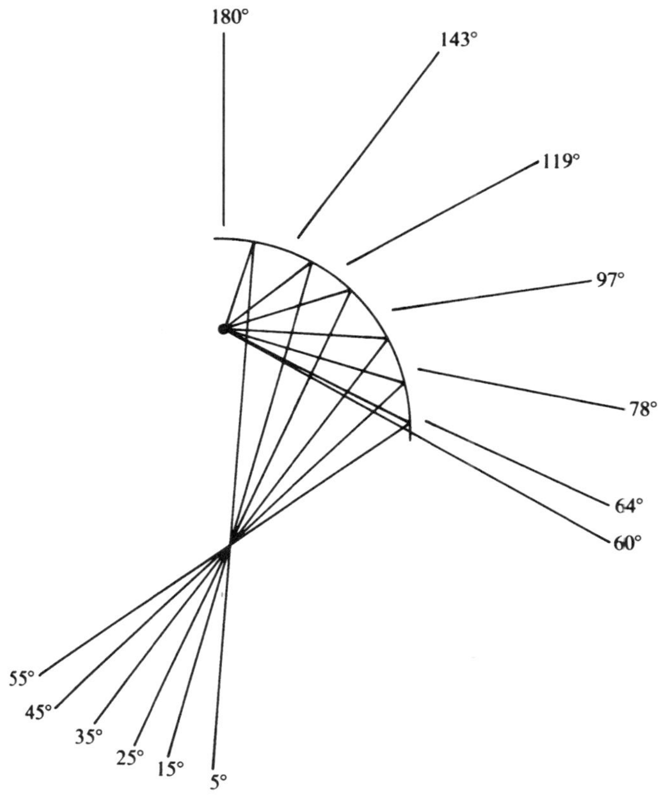

Figure 10.23 Development of reflector contour.

The procedure just presented should be considered as providing a first estimate of the desired reflector contour. It does not consider finite source size and, as was shown in connection with conic sections, this can lead to deviations in reflected ray paths. Also, as was discussed in Chap. 2, intensity, which is a metric from afar, may not be a reliable indicator in the near field of lamp performance.

10.6 FLASHING

Consider a specular parabolic cylindrical reflector housing a 48-in fluorescent lamp centered at its focal point. The lamp is a T12 and emits 3000 lm. The reflector is 16 in wide and $5^1/_3$ in deep with a reflectance of 85%. It is shown in Fig. 10.24.

Since the source is not a point source, the reflection from any point on the reflector will be a triangular wedge centered on the normal to the reflector, rather than a single plane parallel to the normal. The extremes of the beam are defined by the two maximum-angle rays from the tube wall shown in Fig. 10.24 for the case of a 20° beam angle α (10° above and below the normal). Other pairs of maximum-angle rays from the tube wall to other points on the reflector will create similar triangular-wedge beams, but of different beam angles. Beam angles decrease as the distance of point P from the center of the reflector increases.

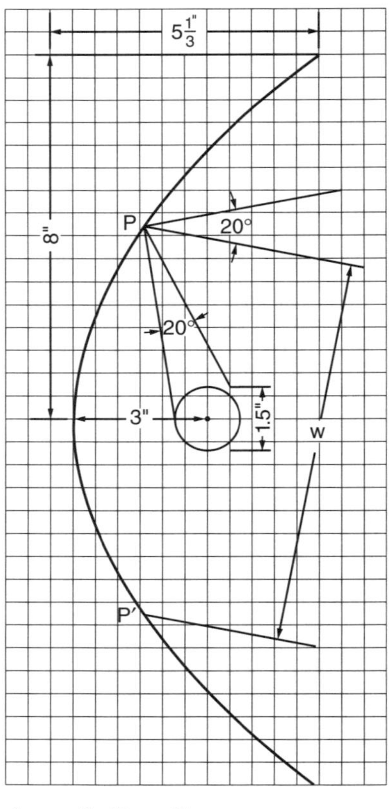

Figure 10.24 Cylindrical parabolic reflector.

Reflected rays from P′ will also create a 20° wedge beam. An observer standing far away from the reflector and looking upward at a viewing angle $\theta = \frac{\alpha}{2} = 10^\circ$ with the horizontal will see only the portion of the reflector from P to P′ lighted (a projected width of w). It is said that this portion of the reflector is "flashed" or "filled".

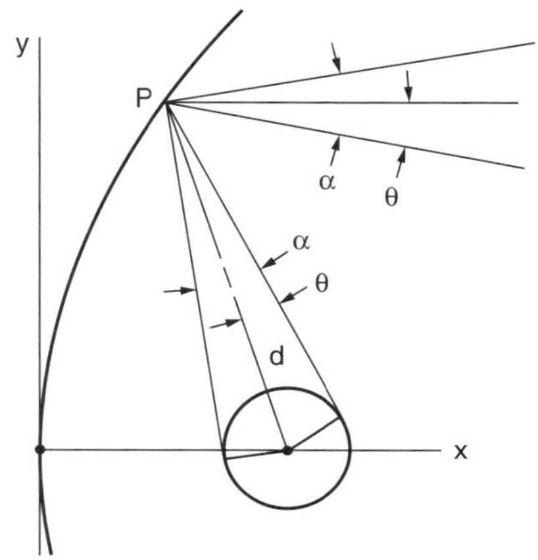

θ	d	x	y	w
14.48	3.000	0	0	0
14	3.100	0.100	1.10	2.13
13.5	3.213	0.213	1.60	3.11
12.5	3.465	0.465	2.36	4.61
11	3.931	0.931	3.34	6.56
10	4.319	1.319	3.98	7.84
9	4.794	1.794	4.64	9.17
7.5	5.746	2.746	5.74	11.38
6	7.175	4.175	7.08	14.08
5.16	8.339	5.339	8.00	15.94

Figure 10.25 Viewing angles and flashed widths.

To calculate viewing angles (θ) and flashed widths (w), a relationship is needed between θ and the coordinates of P. Refer to the diagram in Fig. 10.25. It is noted that

$$\sin\theta = \frac{0.75}{d}, \quad d = \frac{0.75}{\sin\theta} \tag{10.49}$$

and

$$y^2 + (3 - x)^2 = d^2 \tag{10.50}$$

From Eq. (10.9), $y^2 = 12x$ and substituting into Eq. (10.50) yields

$$\left.\begin{aligned} x &= d - 3 = \frac{0.75}{\sin\theta} - 3 \\ y &= \sqrt{12x} \end{aligned}\right\} \tag{10.51}$$

Eqs. (10.51) allow the preparation of the table in Fig. 10.25, where the projected width is

$$w = 2y\cos\theta \tag{10.52}$$

The two beam angle extremes are shown in Fig. 10.26. For α_1, θ, (which is $\frac{\alpha}{2}$) is found by noting that $d = 8^1/_3$" and then using Eq. (10.49) to obtain θ = 5.16°. The observer from afar, looking upward at this angle from the horizontal, would see the reflector fully flashed with a projected width of nearly 16 in. At all viewing angles less than 5.16°, the reflector would also appear fully flashed with its intensity following

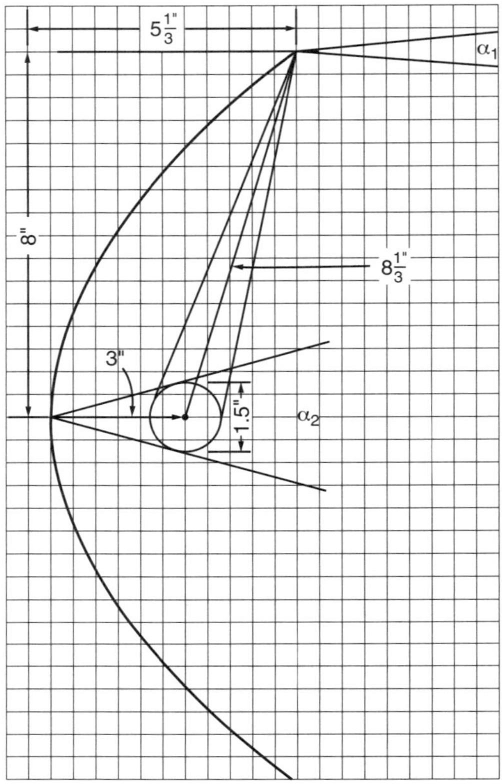

Figure 10.26 Beam angle extremes.

a cosine curve (constant luminance times the projected area of the entire reflector). Because cos 5.16 ≈ cos 0, the intensity will be essentially constant at its maximum value.

At the other extreme (α_2) is the pair of maximum-angle rays from the tube wall to the center of the reflector in Fig. (10.26). These form a beam wedge of 28.96°. If the observer views the reflector looking upward from the horizontal and an angle of 14.48° or greater, the entire reflector will appear unlighted, and only the light from the bare lamp will be seen.

EXAMPLE 10.6*

For the lamp-reflector unit in Fig. 10.24, determine and plot the flashed width and the luminous intensity distribution in a plane perpendicular to the luminaire axis at the center (lengthwise) of the luminaire for $0 \le \theta \le 90°$.

Solution. The flashed width plot is dealt with first. It appears in Fig. 10.27. A portion of this, for $\theta = 5.16°$ to 14.48°, is obtained directly from the table in Fig. 10.25. For $\theta \le 5.16°$, the reflector is fully flashed, and the projected width is $w = 16 \cos \theta$. For $\theta > 14.48°$, the reflector contributes nothing, and the only light is from the lamp. For $5.16° < \theta < 14.48°$ the projected flashed area decreases somewhat nonlinearly as θ increases.

* The author is indebted to Dr. Robert Levin for introducing him to this example of flashing.

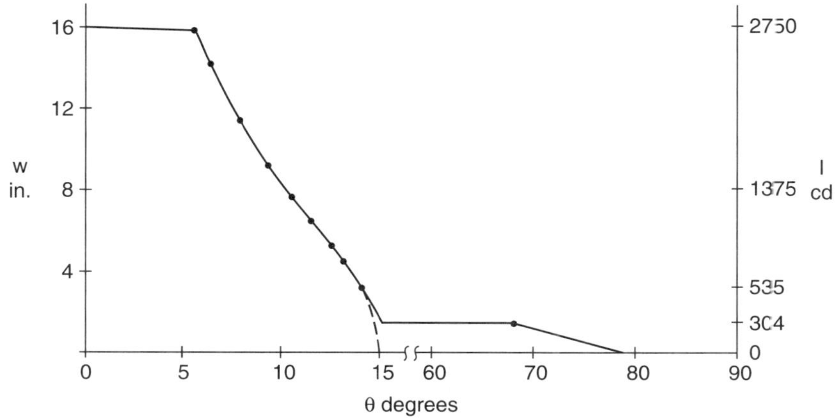

Figure 10.27 Intensity and flashed width vs viewing angle.

Actually the lamp projected area begins to overlap the reflector flashed projected area before $\theta = 14.48°$ is reached. This overlapping begins when one of a pair of maximum-angle rays from the tube wall reflects back on itself, as shown in Fig. 10.28a for ray PB.

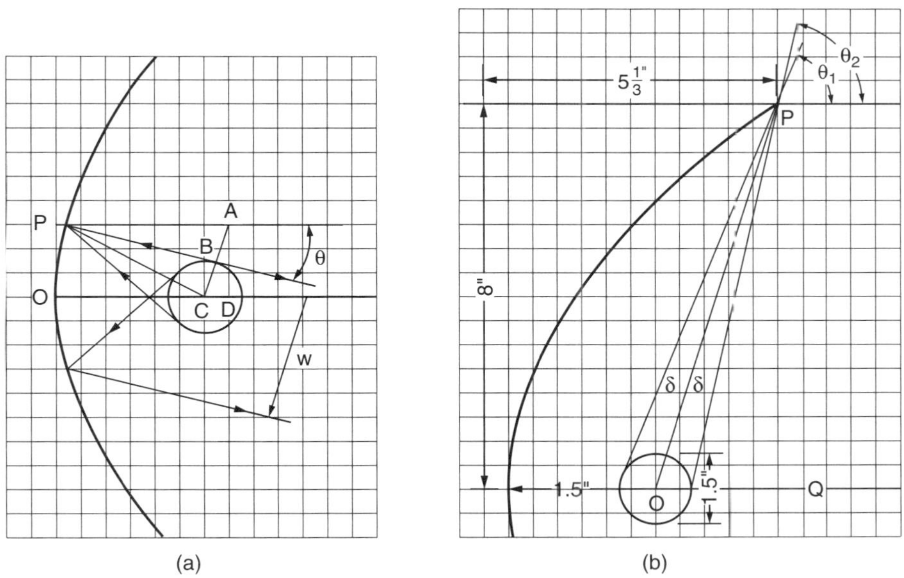

Figure 10.28 Angles at which lamp affects w.

BP is the bisector of angle CPA; thus CB = AB = 0.75. Noting from Eq. (10.50) that PC = x + 3 = PA, it follows that OD = 3 + 2x and thus CD = 2x. Then from triangle ACD,

$$(2x)^2 + y^2 = 1.5^2$$

$$4x^2 + 12x - 2.25 = 0$$

giving

$$x = 0.175 \quad \text{and} \quad 2x = 0.35$$

Thus θ = angle CAD $= \sin^{-1}\frac{0.35}{1.5} = 13.5°$ is the angle at which the lamp begins to obscure the flashed reflector. At that point the value of w is 3.11 in. from Fig. 10.25. The short dashed line in Fig. 10.27 is the flashed width if the lamp were transparent; the solid line with the lamp opaque.

Beyond $\theta = 14.48°$, the projected width is 1.5 in. until the reflector begins to obscure the lamp. As shown in Fig. 10.28b, this occurs at $\theta = \theta_1$. The lamp will be totally obscured at $\theta = \theta_2$. From Fig. 10.26, OP in Fig. 10.28b = $8^1/_3$". Angle POQ = $\tan^{-1}\frac{8}{2^1/_3} = 73.7°$. The angles labeled δ are given by $\sin^{-1}\frac{0.75}{8^1/_3} = 5.2°$. Thus $\theta_1 = 68.2°$ and $\theta_2 = 78.9°$.

Luminous intensity is now discussed. The lamp emits 3000 lm. Its exitance is given by

$$M = \frac{\phi}{\pi d\ell} = \frac{3000}{\pi \times 1.5 \times 48} = 13.26 \text{ lm/in}^2$$

Assuming a perfectly diffuse source,

$$L = \frac{M}{\pi} = 4.22 \text{ cd/in}^2$$

Except for losses, this luminance will remain invariant within the optical system. This is a general result known as Abbe's Law.

Considering the reflector next, its projected area is $A_p = 48w \text{ in}^2$. Then its luminous intensity (excluding lamp luminance) is

$$I_{refl} = L \times \rho \times A_p = 4.22 \times 0.85 \times 48w$$

$$= 172w \text{ cd}, \ 0 < \theta < 13.5°$$

where for $0 < \theta < 5.16°$, $w = 16 \cos\theta$, and for $5.16 < \theta \le 13.5°$, $w = 2y\cos\theta$, with values given in Fig. 10.25. Thus I varies directly with w from 172 × 16 = 2750 cd at $\theta = 0°$ to 172 × 3.11 = 535 cd at $\theta = 13.5°$.

For $\theta > 14.48°$, it is only lamp intensity that is seen, given by

$$I_{lamp} = L \times A_p = 4.22 \times 1.5 \times 48$$

$$= 304 \text{ cd}$$

The reflectance is not included, nor is $\cos\theta$, the latter because the lamp, being cylindrical, appears 1.5 in wide from all viewing angles.

10.7 REFRACTION

Refraction is the bending of light as it passes from one medium to another. As in the case of reflection, Huygens' principle may be applied to describe the situation. In Fig. 10.29a, a plane wavefront impinges at an angle ϕ on the boundary between two transparent media. Assuming the lower medium to be denser than the upper medium,

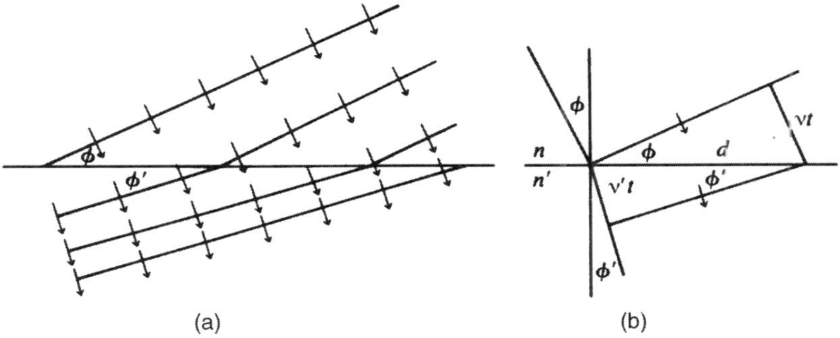

Figure 10.29 Refraction at a plane surface.

the wavefront will be bent toward the normal and enter the lower medium at an angle $\phi' < \phi$.

The result in Fig. 10.29a may be quantified. First, define the index of refraction of a medium as the ratio of the velocity of light in a vacuum to that in the medium. Thus

$$n = \frac{c}{v} \tag{10.53}$$

and is a number greater than one. Indexes of refraction are a function of the wavelength of the incident light. This phenomenon is called dispersion and will be discussed in connection with prisms later. Values of n at 589 nm (low-pressure sodium line) for common materials are shown in Table 10.2.

TABLE 10.2 INDEXES OF REFRACTION

Crown glass	1.48–1.61
Flint glass	1.53–1.96
Soda lime glass	1.51
Lead glass	1.53–1.67
Borosilicate glass	1.48
Aluminosilicate glass	1.54
Acrylic	1.49
Quartz	1.54
Diamond	2.42
Calcium carbonate	2.66
Calcium fluorite	1.43
Sodium chloride	1.54
Water	1.33
Carbon disulfide	1.63
Carbon dioxide	1.00045
Air	1.00029

With n defined, consider Fig. 10.29b. A wavefront advancing at a velocity v in a medium of refractive index n impinges on a second medium of refractive index $n' > n$ at an angle ϕ. The velocity in the second medium is $v' < v$. In the time it takes for the extreme right of the wavefront in the upper medium to advance a distance $s = vt$, the extreme left of the wavefront in the lower medium will advance a distance $s' = v't$.

From Fig. 10.29b,

$$\sin\phi = \frac{vt}{d}, \quad \sin\phi' = \frac{v't}{d} \tag{10.54}$$

Then, from Eqs. (10.53) and (10.54),

$$\frac{\sin\phi}{\sin\phi'} = \frac{v}{v'} = \frac{n'}{n} \tag{10.55}$$

The result in Eq. (10.55) is known as Snell's law.*

Velocity, frequency, and wavelength are related by Eq. (1.9). In passing from one medium to another, the frequency of the light waves does not change. Thus, as velocity changes, wavelength must change. In particular, in going from air to a denser medium, such as glass, velocity decreases, and thus so does wavelength, within the glass.

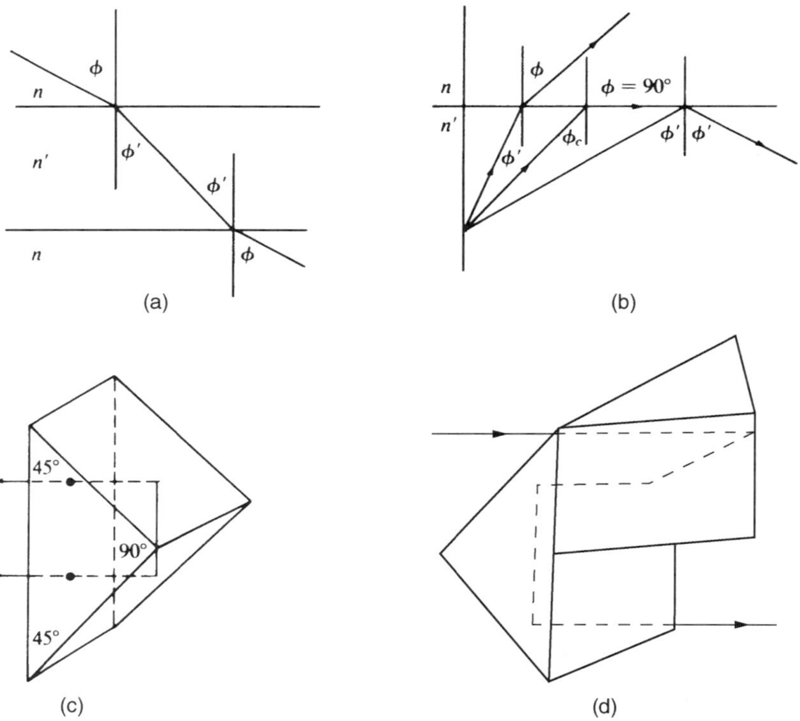

Figure 10.30 Applications of Snell's law.

Refraction by a flat glass plate ($n' > n$) with parallel faces is illustrated in Fig. (10.30a). Here the net effect is to displace the light ray rather than alter its direction. In Fig. 10.30b. (again $n' > n$), the possibility of total internal reflection is examined.

* Named for Willebrod Snell in 1621. Snell studied refraction, but died at the age of 35 with the results of his work unpublished. Rene Descartes published the law of refraction 11 years later.

At the critical angle ϕ_c, determined from Eq. (10.55) as

$$\sin\phi_c = \frac{n}{n'} \tag{10.56}$$

the refracted ray will graze the boundary. For all angles greater than ϕ_c, the ray will be totally reflected internally.

For an air-glass surface, with the latter having a typical index of refraction of 1.50, $\phi_c = 42°$. The fact that this angle is less than 45° makes it possible to construct totally reflecting prisms, as shown in Fig. 10.30c and d. These have the advantages over mirrored reflector surfaces of being 100% reflecting and of having no degradation over time. An extension of Fig. 10.30c is shown in Fig. 10.30d. This is called a Porro prism, two right angle prisms oriented at 90°. Each prism produces a 180° reversal; the combination produces an image inversion, as well as folding optical path length. Porro prisms are used in binoculars.

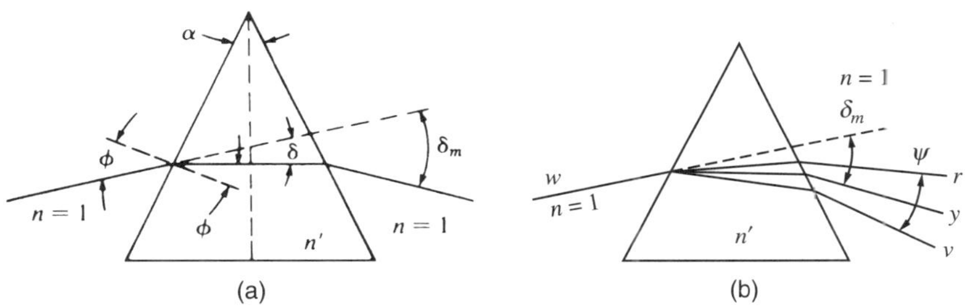

Figure 10.31 Deviation by a prism.

There are two properties of prisms that should be discussed, namely deviation and dispersion. Consider the isosceles prism in Fig. 10.31a. A ray of light is incident on the left face at such an angle ϕ that it passes through the prism horizontally and exits symmetrically from the right face. The apex angle α of the prism is shown and it is desired to calculate the deviation δ_m of the ray in passing through the prism. From the geometry and Eq. (10.55), and noting that half the deviation takes place at each surface,

$$\phi' = \frac{\alpha}{2}, \quad \delta = \frac{\delta_m}{2} \tag{10.57a}$$

$$\phi = \phi' + \delta = \frac{\alpha + \delta_m}{2} \tag{10.57b}$$

From Eq. (10.55),

$$\sin\phi = \sin\frac{\alpha + \delta_m}{2} = n'\sin\phi' = n'\sin\frac{\alpha}{2} \tag{10.58a}$$

$$n' = \frac{\sin[(\alpha + \delta_m)/2]}{\sin(\alpha/2)} \tag{10.58b}$$

Since it is possible to measure the angles α and δ_m very accurately, Eq. (10.58b) can be used to determine indexes of refraction with considerable precision.

The reason for the symbolgy δ_m requires explanation. The letter "m" denotes minimum. As the angle of incidence ϕ in Fig. 10.31a is decreased from a large value, the deviation first decreases and then increases. It is a minimum when the ray passes through the prism symmetrically, as is the case in Fig. 10.31a. It is this minimum deviation that is customarily measured for a prism.

A second parameter of interest with prisms is dispersion. As noted earlier, the index of refraction is a function of wavelength. Within the visible spectrum, $n_{violet} > n_{red}$ and thus the shorter visible wavelengths undergo greater deviation in passing through a prism than do the longer wavelengths. Because of dispersion, it is necessary to specify the wavelength at which prism deviation is measured. As noted earlier in connection with Table 10.2, 589 nm is customarily chosen.

Prism dispersion is illustrated in Fig. (10.31b). White light (w) impinges on the left prism face, is dispersed, and emerges from the right prism face with the spectral components separated. The measure of dispersion is usually taken as the angular spread between the red (r) and the violet (v) wavelengths.

For small prism apex angles, simple expressions for both dispersion and minimum deviation can be derived. From Eq. (10.58b), if α is small, the sine of the angle can be replaced by the angle itself, giving

$$\delta_m = (n'_y = -1)\alpha \tag{10.59a}$$

Then the dispersion angle can be written as

$$\psi = (n'_v - n'_r)\alpha \tag{10.59b}$$

where y, v, and r denote yellow, violet, and red, respectively, δ_m and ψ are indicated in Fig. 10.31b. Typical values for δ_m and ψ for prisms with 10° apex angles are 6.1° and 0.50° for flint glass, 5.4° and 0.16° for quartz, and 5.1° and 0.18° for crown glass.

EXAMPLE 10.7

A water tank is 100 cm deep and a coin has been dropped into it. How far below the surface does the coin appear to be?

Solution. The situation is diagrammed in Fig. 10.32. From Snell's Law (Eq. 10.55) and Table 10.2,

$$1.0 \sin\theta_1 = 1.33 \sin\theta_2$$

For small angles, the sine and tangent are approximately equal. Thus

$$1.0 \tan\theta_1 \approx 1.33 \tan\theta_2$$

$$\frac{x}{y} = 1.33\frac{x}{100},$$

$$y = 75 \text{ cm}$$

This result can be generalized to state that an object submerged in water is 33% deeper than it appears to be.

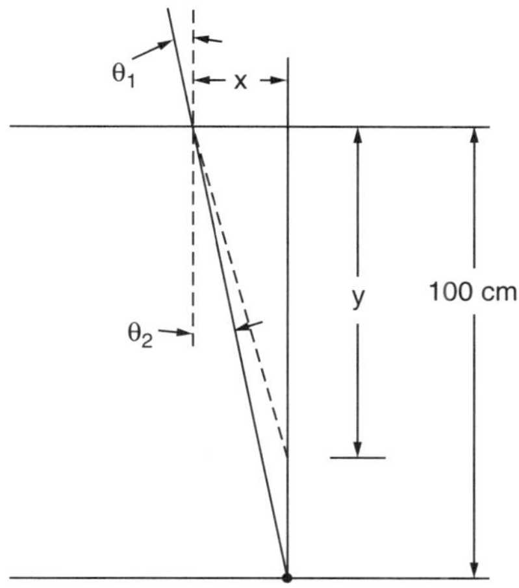

Figure 10.32

For an exact analysis of Example 10.7, let d be tank depth. Then

$$\frac{d}{y} = \frac{\tan\theta_1}{\tan\theta_2}, \quad 1.33 = \frac{\sin\theta_1}{\sin\theta_2}$$

$$\frac{d^2}{y^2} = \frac{\sin^2\theta_1/1-\sin^2\theta_1}{\sin^2\theta_2/1-\sin^2\theta_2} = 1.33^2\frac{1-\sin^2\theta_2}{1-\sin^2\theta_1} \tag{10.60}$$

$$\frac{d}{y} = 1.33\sqrt{\frac{1-\sin^2\theta_2}{1-\sin^2\theta_1}}$$

Thus d/y is 1.33 for $\theta_1 = \theta_2 = 0$. For $\theta_1 = 10°$, θ_2 is 7.5°, and d/y is 1.34. For $\theta_1 = 20°$, θ_2 is 14.9° and d/y is 1.37. In Example 10.7, for d = 100 cm, y = 75 cm, 74.6 cm, and 73.0 cm.

10.8 LENSES

Lenses involve refraction at two curved surfaces, generally air to glass and glass to air. As a preliminary to their discussion, it is necessary to consider refraction at a single spherical surface, as diagrammed in Fig. 10.33. An object AX is located a distance p from the vertex O of a spherical refractor of radius r. As in the spherical mirror in Fig. 10.4, three rays are shown emanating from point A. One is directed normally through the center of the spherical surface. The second is directed parallel to the refractor's axis and is bent upon entering the refractor. The third is directed toward the refractor's vertex and is also bent upon entering the refractor. All refractions obey Snell's law. For small ϕ, the three rays intersect at point B, producing the image of A. Thus the object AX is imaged as BY, an inverted image reduced in size for the dimensions in Fig. 10.33.

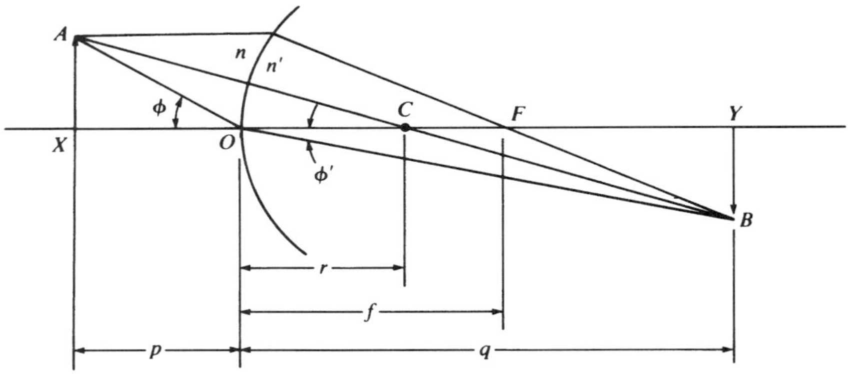

Figure 10.33 Refraction at a convex spherical surface.

To quantify the relations in Fig. 10.33, it is noted that

$$\frac{AX}{BY} = \frac{XC}{CY} = \frac{p+r}{q-r} \tag{10.61}$$

$$\frac{AX}{OX} = \tan\phi, \quad \frac{BY}{OY} = \tan\phi' \tag{10.62}$$

For small angles, the sine and tangent are approximately equal and Eqs. (10.62) can be combined with the aid of Snell's law to yield

$$\frac{AX}{BY} = \frac{OX \sin\phi}{OY \sin\phi'} = \frac{pn'}{qn} \tag{10.63}$$

Then Eqs. (10.61) and (10.63) give

$$\frac{n}{p} + \frac{n'}{q} = \frac{n'-n}{r} \tag{10.64}$$

and

$$M = -\frac{BY}{AX} = -\frac{qn}{pn'} \tag{10.65}$$

where M is "magnification".

Sign conventions for Fig. 10.33 and Eqs. (10.64) and (10.65) are a bit different than those for a spherical mirror. In the present case, the following rules apply:

1. Object distances are positive if the object is to the left of the refractor vertex.
2. Image distances and radii are positive if the image and the center of spherical surface are to the right of the refractor vertex.
3. Images are positive in size if they lie above the horizontal axis.

As in the mirror case, the focal point of a refractor is the location of the image when the object is located at infinity. From Eq. (10.64),

$$\frac{n'}{f} = \frac{n'-n}{r}, \quad f = \frac{r}{1-(n/n')} \tag{10.66}$$

The focal point and the focal length are labeled as F and f, respectively, in Fig. 10.33. Note that for $n' > n$, $f > r$.

EXAMPLE 10.8

For a spherical refractor with $n' = 2$ and $r = 3$ cm, find the image distance, focal length, and magnification for object distances of $p = 6$ cm, 1 cm, and −6 cm.

Solution. From Eqs. (10.64), (10.65), and (10.66), the following tabulation can be made, where all distances are in centimeters:

p	q	f	M
6	12	6	−1
1	−3	6	+3/2
−6	4	6	+1/3

In the first case, the image is inverted and to the right of the vertex. In the second case, the image is erect and magnified and to the left of the vertex. In the third case, the object is inside the refractor, placed there by a previous refraction. The image is erect, reduced in size, and to the right of the vertex.

A lens may be considered as the superposition of two spherical refractors. Thus problems involving lenses can be solved by two applications of Eq. (10.64), as follows:

$$\left.\begin{aligned} \frac{1}{p_1} + \frac{n'}{q_1} &= \frac{n'-1}{r_1} \\ \frac{n'}{p_2} + \frac{1}{q_2} &= \frac{1-n'}{r_2} \end{aligned}\right\} \tag{10.67}$$

If it is assumed that the lens has negligible thickness compared with the other dimensions involved, $p_2 = -q_1$ and Eqs. (10.67) may be added to obtain

$$\frac{1}{p_1} + \frac{1}{q_2} = (n'-1)\left(\frac{1}{r_1} - \frac{1}{r_2}\right) \tag{10.68}$$

There are two focal lengths associated with a lens. From Eq. (10.68), with either p_1 or q_2 infinite,

$$\frac{1}{f} = (n'-1)\left(\frac{1}{r_1} - \frac{1}{r_2}\right) \tag{10.69}$$

Then substituting Eq. (10.69) into Eq. (10.68) gives

$$\frac{1}{p}+\frac{1}{q}=\frac{1}{f} \tag{10.70}$$

where the intermediate subscripts used in the development have been dropped.

The ray-tracing construction for both convex and concave lenses is shown in Fig. 10.34. The location of the object for an image at infinity is called the first focal point (F1); the location of the image for an infinitely distant object is named the second focal point (F2). These points are labeled in Fig. 10.34.

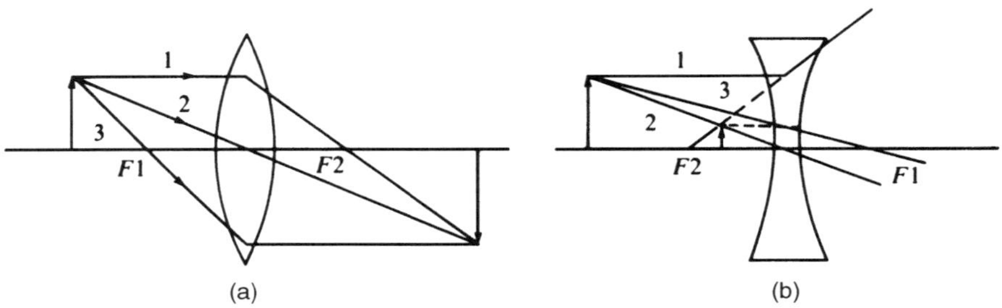

Figure 10.34 Ray tracing for lenses.

The three rays used in determining graphically the image location are also shown in Fig. 10.34. In either diagram, ray 1 from an object at infinity, exits the lens through the second focal point. Ray 2 passes undeviated through the vertex of the lens. Ray 3, from the image at infinity, emerges from the lens through the first focal point. The intersection of the three rays marks the location of the image.

In Fig. 10.34a, the image is real and inverted. In Fig. 10.34b, it is virtual and not inverted. The virtual image is the apparent location of the object when viewed from the right side of the lens.

EXAMPLE 10.9

An object lies 6 cm to the left of a convex lens with $r_1 = 3$ cm, $r_2 = 4$ cm, and $n' = 2$. Find the location of the image produced.

Solution. The problem will be worked in two ways, first by use of Eq. (10.68) for thin lenses and second by two applications of Eq. (10.64) for a refracting surface. Each solution is illustrated in Fig. 10.35.

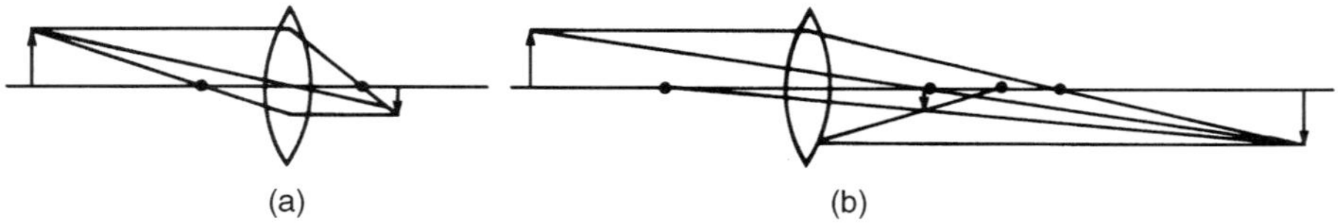

Figure 10.35 Graphical solution of example 10.9.

Application of Eq. (10.68) yields

$$\frac{1}{6}+\frac{1}{q_2}=(1)\left(\frac{1}{3}-\frac{1}{-4}\right), \quad q_2=\frac{12}{5}\text{ cm}$$

From Eq. (10.69)

$$\frac{1}{f}=(1)\left(\frac{1}{3}-\frac{1}{-4}\right), \quad f=\frac{12}{7}\text{ cm}$$

f could also be obtained from Eq. (10.70):

$$\frac{1}{6}+\frac{5}{12}=\frac{1}{f}, \quad f=\frac{12}{7}\text{ cm}$$

Using Eq. (10.64) gives

$$\frac{1}{6}+\frac{2}{q_1}=\frac{2-1}{3}, \quad q_1=12\text{ cm}$$

Also Eq. (10.66) yields

$$f_1=\frac{3}{1-\frac{1}{2}}=6\text{ cm}$$

A second application of Eq. (10.64), this time with an object distance of −12 cm, yields

$$\frac{2}{-12}+\frac{1}{q_2}=\frac{1-2}{-4}, \quad q_2=\frac{12}{5}\text{ cm}$$

From Eq. (10.66)

$$f_2=\frac{-4}{1-\frac{2}{1}}=4\text{ cm}$$

The diagram in Fig. 10.35a is the thin-lens construction and duplicates that in Fig. 10.34a. The diagram in Fig. 10.35b is based on considering the problem as two refracting surfaces in which the image from the first refraction serves as the object for the second refraction.

The magnification provided by a thin lens has yet to be considered. It is simply the product of the magnifications at each of its surfaces. From Eq. (10.65),

$$M=\frac{q_1 n}{p_1 n'}\left(\frac{q_2 n'}{p_2 n}\right)=-\frac{q_2}{p_1}\equiv-\frac{q}{p} \tag{10.71}$$

where use is made of the fact that $p_2=-q_1$ for a thin lens.

EXAMPLE 10.10

A filament and its reflector can be represented by a 9 mm diameter luminous disk having a luminance of 45 cd/mm^2. The unit is centered at the focal point of a plano convex lens of 100 mm diameter, 200 mm focal length, and 90% transmittance. Find the maximum intensity that this filament-reflector-lens unit can produce. At what point on the optic axis will the intensity be half this value (the reflector appearing to be half filled)?

Solution. The situation is diagrammed in Fig. 10.36. Letting A be the area of the fully flashed lens, the maximum intensity is

$$I = \tau LA = 0.9 \times 45 \times \pi \times 50^2$$
$$= 318 \times 10^3 \text{ cd}$$

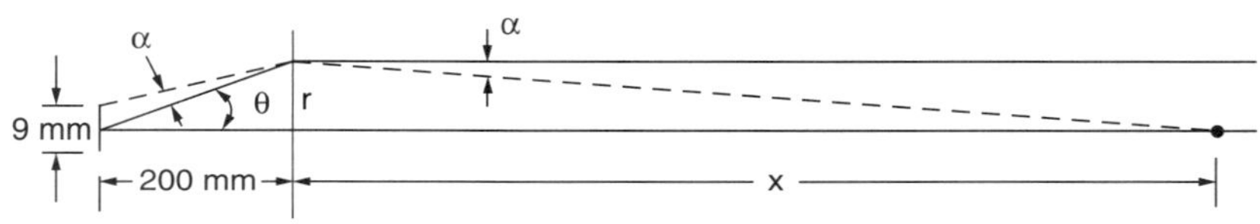

Figure 10.36

Half intensity implies the lens half filled or $\pi r^2 = 0.5 \times \pi \times 50^2$, giving r = 35.36 mm.

To find x, the point at which the intensity will be 50% of maximum, α must be calculated. First note that

$$\theta = \tan^{-1}\frac{35.36}{200} = 10.03°$$

Then α, in radians, is given by

$$\alpha = \frac{4.5 \cos 10.03}{200/\cos 10.03} = 0.0218 \text{ rad.}$$

Finally,

$$x \approx \frac{35.36}{0.0218} = 1620 \text{ mm} = 1.62 \text{ m}$$

10.9 LENS APPLICATIONS AND DESIGN

In this section, a few applications of lenses in the field of light and vision will be considered and a graphical procedure for the design of a refractive surface will be developed.

Perhaps the most common use of lenses is in eyeglasses. The lens system of a normal human eye forms an image on the retina of an object at infinity. Two common deviations from this are hyperopia (farsightedness) and myopia (nearsightedness). In the hyperopic eye, the image of an object at infinity lies behind the retina; in a myopic eye, in front of it.

The far point of an eye is the farthest point from the eye at which an object will be imaged on the retina. The near point is the nearest point at which this occurs. In the myopic eye, the far point is nearer than infinity, and the near point is closer to the eye than for a normal eye.

Both hyperopia and myopia can be corrected by eyeglasses, as can presbyopia, the increase in the near point of a normal eye with aging that is described in Sect. 3.10. These corrections will be illustrated through an example.

EXAMPLE 10.11

(a) The near point of an eye is 50 cm in front of the cornea. What lens should be used to move the near point to 20 cm in front of the cornea?

(b) If the far point of an eye is 2 m from the cornea, what lens should be used to move the far point to infinity?

Solution. Equation 10.70 is the appropriate equation to use for both parts of the problem.

(a) A convex (converging) lens is needed to make it appear to the eye that the object is at 50 cm. Thus p = 20 cm, q = −50 cm, and

$$\frac{1}{f} = \frac{1}{20} + \frac{1}{-50}, \quad f = 33 \text{ cm}$$

(b) A concave (diverging) lens is required that will make an object at infinity appear to the eye to be located at 2 m. Thus, p = ∞, q = −200 cm, and the calculation yields

$$\frac{1}{f} = \frac{1}{\infty} + \frac{1}{-200}, \quad f = -200 \text{ cm}$$

Perhaps the second most common use of lenses is in photography. Actually, pictures can be taken without lenses using a pinhole camera, but one is restricted to stationary objects because the pinhole admits very little light and thus exposures of seconds, or even minutes, are required. If the opening is enlarged, more light is admitted but the picture becomes out of focus. With a lens, the opening can be enlarged to admit more than 50,000 times the amount of light passed by a pinhole. Thus, the primary function of a camera lens is to focus the light rays so that the diaphragm opening can be large enough to avoid long exposures.

The fundamental parameters of a camera lens are its focal length and diameter. From Eqs. (10.70) and (10.71), for a given object size and distance, the longer the focal length of the lens, the larger the image. Thus, whether the lens is telephoto, normal, or wide-angle, the image size is determined by the focal length.

The diameter of a camera lens governs its light-gathering power and, along with focal length, establishes its maximum angular field of coverage. The actual field of coverage is set by the field stop and is generally stated in terms of the area of acceptable sharpness and definition, rather than the total area seen by the lens. A normal lens used in still photography covers a field of about 50°.

It is customary to provide a single number to describe a camera lens. This is the f-number (or f-stop) and is the ratio of focal length to diaphragm diameter.

$$f_n = \frac{f}{d} \tag{10.72}$$

The f-numbers commonly used form a series 1, 1.4, 2, 2.8, 4, 5.6, 8, 11, 16, and 22. Each succeeding f-number provides one-half of the light-gathering power of the one preceding it. For example, for a given lens, the ratio of diameters for $f_{5.6}$ and f_8 is 8:5.6. The ratio of areas is 64:32 = 2. Thus, an $f_{5.6}$ lens admits twice the luminous flux as an f_8 lens of the same focal length in a given setting.

A 35-mm slide camera generally has a focal length of 2 in. From Eq. (10.72), for $f_{1.4}$, its diaphragm opening is 1.4 in and, at the other extreme, for f_{22}, the diaphragm opening is 0.09 in. Thus the range of its light-gathering power is about 250:1. The image location for different object distances for this lens is shown in Table 10.3. Note that until the object gets quite close to the camera (<10 in), the image distance is very nearly constant. At these closer distances, the lens can be moved farther from the film or a lens of slightly different focal length can be used. For example, at p = 10 in, a focal length of 1.67 would be required to place the image 2 in from the lens, from Eq. (10.70).

TABLE 10.3

p(in)	q(in)
∞	2.0
200	2.02
100	2.04
50	2.08
25	2.17
10	2.50
5	3.33
4	4.00
3	6.00

A third application of lenses, as well as of prisms, is in conjunction with lamps and luminaires in lighting systems. One example is the Fresnel* lens, illustrated in Fig. 10.37a. This is essentially a plano-convex lens, in which portions of the convex surface have been eliminated to reduce weight and cost while still providing essentially the same optics. It can also be constructed in cylindrical form, as shown in Fig. 10.37b.

The use of a clear prismatic lens in a fluorescent luminaire is illustrated in Fig. 10.37c. The prisms redirect lamp lumens downward and upward and reduce luminances in and near the 45° glare zone. Clear prismatic lenses, in either glass or plastic, transmit 70 to 92% of the flux incident on them with very little diffusion. Several prism patterns are used, two common ones being hexagonal pyramids and circular cones, with densities of 16 to 64 prisms per square inch.

* Named for Augustin Fresnel (1788–1827), French engineer and mathematician, who made several contributions in geometric and physical optics.

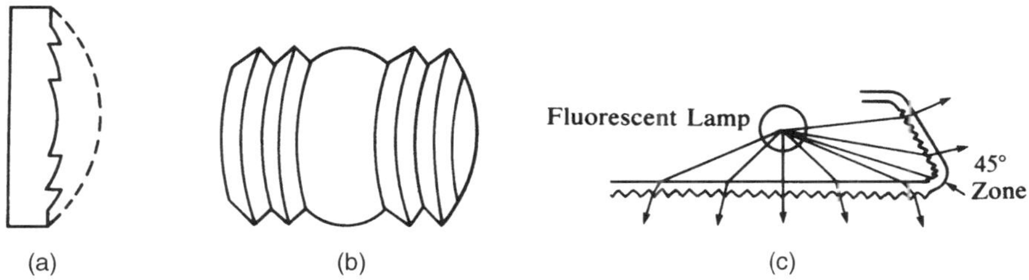

Figure 10.37 Lens and prism applications. ***Source of (c):*** **"Light Measurement and Control," (Bulletin TP-118), 1971. Courtesy of General Electric Co.**

To close this section, a graphical procedure will be presented that has been developed for the design of a refractive surface to redirect a ray from a light source (Committee on Light Control and Equipment Design of the IES, 1959, 1967, 1970). The procedure is illustrated in Fig. 10.38 and will be presented in steps.

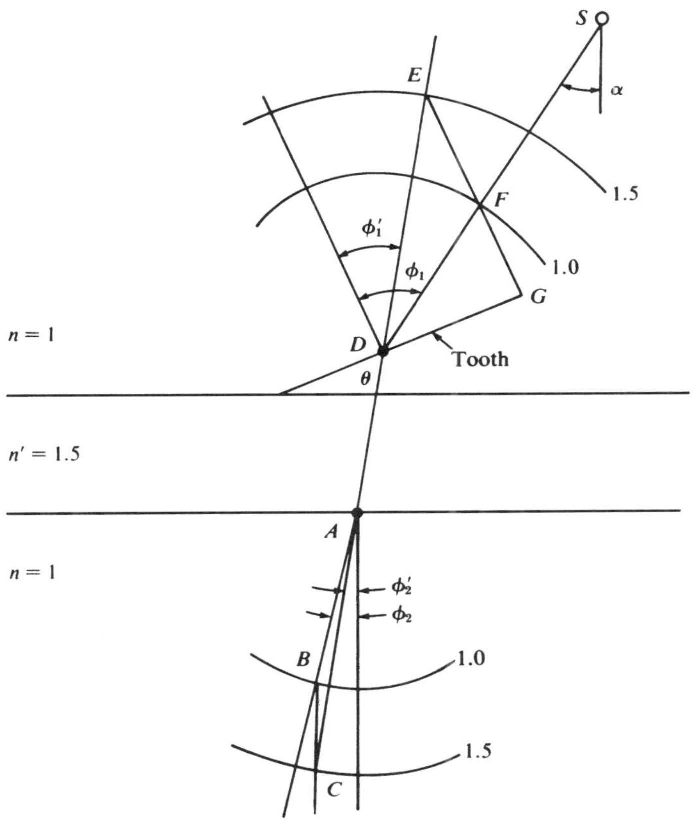

Figure 10.38 Design of a refractor surface.

Step 1 Establish the known quantities. In Fig. 10.38, the refractive plate has a flat bottom surface and its top surface is to be formed into a sawtooth pattern. One such "tooth," inclined at an angle θ (to be determined) is shown. The light source is located such that a ray from it to the tooth makes an angle α with the normal to the plate. It is desired that the refracted ray emerge from the bottom surface at an angle ϕ_2 with the normal. The index of refraction of the plate (n′) is known and the medium above and below the plate is air (n = 1). In Fig. 10.38 a value of 1.5 for n′ is used.

Step 2 The graphical construction begins at point A. With A as a center, arcs are struck with radii proportional to n and n′. The line AB is drawn at the given angle ϕ_2. From point B a perpendicular is dropped to point C. Then the line CAE is drawn, defining the angle ϕ_2', which is the angle of the refracted ray within the plate. This construction is justified by noting that

$$\frac{\sin\phi_2'}{\sin\phi_2} = \frac{AB}{AC} = \frac{n}{n'} \tag{10.73}$$

in accordance with Snell's law.

Step 3 Estimate the angle θ required for the tooth and locate the point D on line CAE. With D as a center, strike two additional arcs whose radii are again proportional to n and n′. Locate point F on line SD and draw line EFG, which is the required normal to the slant face of the tooth. If angle FGD is not 90°, repeat the step with a new estimate of θ.

Step 4 To design additional teeth repeat steps 1 through 3.

The design can also be done analytically, again by repeating estimates. ϕ_2 is known and thus ϕ_2' can be found from Eq. (10.73). Then using Snell's law and the known geometry,

$$\left.\begin{aligned} \phi_1' &= \phi_2' + \theta \\ \phi_1 &= \alpha + \theta \\ \sin\phi_1 &= \frac{n'}{n}\sin\phi_1' \end{aligned}\right\} \tag{10.74a}$$

and combining these equations yields

$$\sin(\alpha+\theta) = \frac{n'}{n}\sin(\phi_2'+\theta) \tag{10.74b}$$

All quantities except θ are known in Eq. (10.74b). θ may be displayed explicitly, using trig identities.

$$\tan\theta = \frac{\sin\alpha - \dfrac{n'}{n}\sin\phi_2'}{\dfrac{n'}{n}\cos\phi_2' - \cos\alpha} \tag{10.75}$$

EXAMPLE 10.12

It is desired to design a borosilicate glass refractor lens to convert diverging rays from an incandescent point source S into parallel rays by creating teeth at appropriate angles. Calculate the tooth angles required in Fig. 10.39 for source rays at $\alpha = 15°$, 30°, and 45° from the normal.

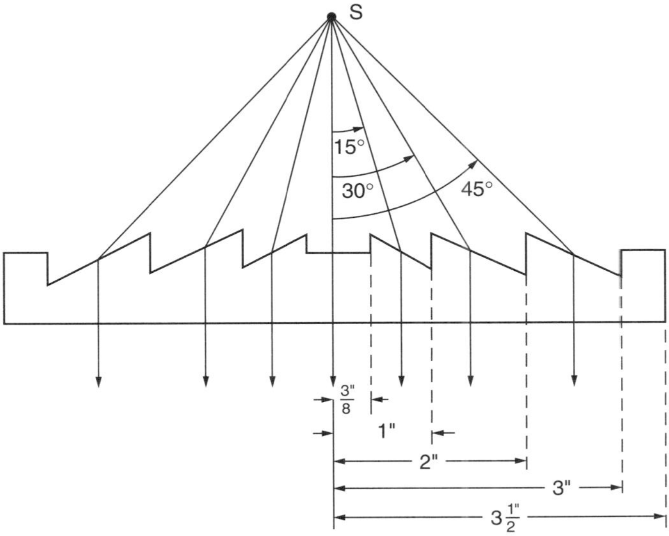

Figure 10.39

Solution. From Fig. 10.38, it is desired that $\phi_2 = 0$ for each tooth. Therefore $\phi_2' = 0$, and from Eq. (10.75),

$$\tan\theta = \frac{\sin\alpha}{1.48 - \cos\alpha}$$

This yields, $\theta = 26.7°$ for $\alpha = 15°$. Similarly, for $\alpha = 30$, $\theta = 39.2°$, and for $\alpha = 45°$, $\theta = 42.5°$.

10.10 INTERFERENCE

The first part of this chapter dealt with geometric optics; the remainder deals mostly with physical optics, treating the subjects of interference, diffraction, and polarization, and is followed by a concluding section on fiber optics.

What is the distinction between these two branches of optics? In geometric optics, rays of light reflected or refracted at definite angles at a boundary between two media are considered. The problems are geometrical and only the amplitude of the light is considered, not its phase.

In physical optics, the notion of rays is abandoned for the more fundamental idea that light is electromagnetic wave motion and that the superposition of more than one coherent wavefront depends on the time phase relationship between the waves, as well as their amplitudes. If, upon superposition, the amplitude of the resultant wave is greater than that of either of the original waves, the waves are said to constructively interfere. If it is less, destructive interference is said to occur. The extreme case of destructive interference is annulment, when the original waves are equal in amplitude and 180° out of time phase.

The concept of constructive interference of two light waves is illustrated in Fig. 10.40. Two coherent monochromatic point sources of light are placed a certain number of wavelengths apart (3λ) at S_1 and S_2. The requirement for the wavefronts from these sources to constructively interfere at point P is

$$PS_1 - PS_2 = \frac{m\lambda}{2} \tag{10.76}$$

where m is an even integer (for destructive interference, m is an odd integer). The loci in Fig. 10.40 are hyperbolic.

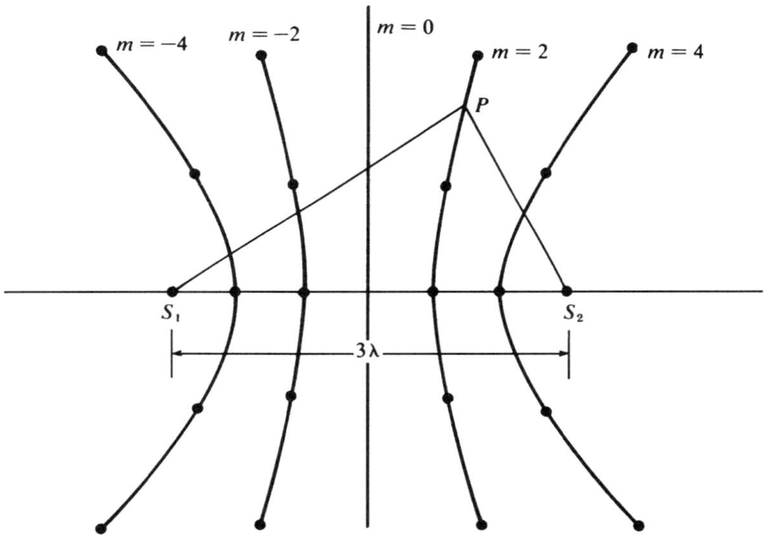

Figure 10.40 Interference loci.

As pointed out in Sect. 1.3, one of the earliest demonstrations of interference was Thomas Young's famous experiment in 1827. The experiment is diagrammed in Fig. 10.41. Light from the source S falls on the slit at the left. By Huygens' principle, secondary wavefronts emanate from this slit and travel equal distances to the slits at A and B, arriving in phase at these points. Thus, the three slits serve to provide coherent radiation at points A and B, even though the source S is not coherent.

Again by Huygens' principle, second sets of secondary wavefronts leave A and B and travel toward the screen. Alternating light and dark bands appear on the screen, as a function of the difference in the path lengths AP and BP. To derive an expression for the location of these bands, it is observed that, if R >> d, the arc BC centered at P is essentially a straight line. Then the triangles ACB and DOP are similar, and thus angle ABC = θ. It follows that, for constructive interference

$$AC = d \sin\theta = m\lambda, \quad \sin\theta = \frac{m\lambda}{d} \tag{10.77}$$

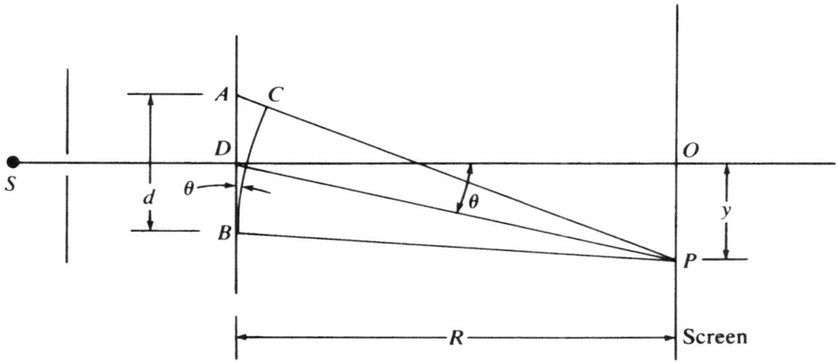

Figure 10.41 Young's interference experiment.

where m is an integer. A typical value for d is 10^{-2} cm. Generally 5 to 10 interference bands can be seen, so $m \leq 10$. For a λ at the middle of the visible spectrum

$$\sin\theta = \frac{10(550\times10^{-9})}{10^{-4}} = 0.055, \quad \theta \simeq 3°$$

Thus θ is very small and

$$y = R\tan\theta \simeq R\sin\theta \tag{10.78}$$

From Eqs. (10.77) and (10.78),

$$y = \frac{mR\lambda}{d}, \quad \lambda = \frac{yd}{mR} \tag{10.79}$$

The second Eq. (10.79) permits λ to be determined from simple length measurements.

EXAMPLE 10.13

Assume in Fig. 10.41 that the two slits are separated by 0.15 mm. The screen is 2 m from the slits and the angle θ of the fifth bright band is 1.2°. What is the wavelength of the (monochromatic) light source?

Solution. From Eq. (10.78),

$$y = 2000 \sin 1.2 = 41.9 \text{ mm}$$

Then, from Eq. (10.79)

$$\lambda = \frac{(41.9\times10^{-3})(0.15\times10^{-3})}{5(2)} = 0.630\times10^{-6}\text{ m}$$

$$= 630 \text{ nm}$$

The phenomenon of interference permits an explanation of thin-film dielectrics, semiconductors, and metals that are used as optical coatings to control spectral and overall reflectance and transmittance in lamps and lighting control equipment. Consider the thin film of thickness t in Fig. (10.42a). A light beam (a) is incident on the film's upper surface. A portion is transmitted (b), and the second portion is reflected (c). The transmitted portion impinges on the lower surface of the film, where again a portion is transmitted (d) and the second portion is reflected (e). The reflected portion, upon reaching the upper surface of the film, is party transmitted (f) and reflected (g), and so on.

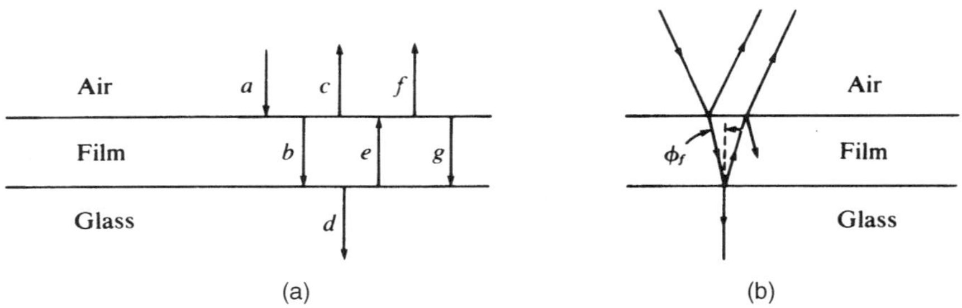

Figure 10.42 Thin-film interference.

For certain wavelengths, the thickness of the film may be such that the reflected waves (c and f) interfere (the transmitted waves can interfere also). From Eqs. (1.9) and (10.55), wavelengths and refractive indexes are related in general by

$$\lambda n = \lambda' n' \tag{10.80}$$

For the air-thin film interface in Fig. 10.42a,

$$\lambda = \lambda_f n_f, \quad \lambda_f = \frac{\lambda}{n_f} \tag{10.81}$$

The number of wavelengths in two film thicknesses is (assuming normal incidence)

$$\text{No. of } \lambda_f\text{s} = \frac{2tn_f}{\lambda} \tag{10.82}$$

If there were no other factors involved, the requirement for constructive interference would be that the quantity in Eq. (10.82) be an integer. However, it has been shown, analytically from Maxwell's equation and experimentally with the aid of a device known as Lloyd's mirror (Jenkins and White, 1937), that when a wavefront impinges on the surface of a medium of higher refractive index than that in which it is traveling, there is a 180° shift in phase between the incident and reflected waves. This corresponds to one-half wavelength, and thus the requirement for constructive interference in Fig. 10.42a is that the quantity in Eq. (10.82) be an odd number of half-wavelengths. For destructive interference, an even number of half-wavelengths (an integral number of wavelengths) is required.

An application of the principles just presented is nonreflecting glass, illustrated in Fig. 10.42b. The film is generally a dielectric magnesium fluoride coating, either single or multilayered, and is chosen so that its refractive index lies between that of air and the glass. Thus, there will be a 180° phase reversal at both the upper and lower film boundaries. If, from Eq. (10.82), the thickness of the film is

$$t = \frac{\lambda_f \cos\phi_f}{4} = \frac{\lambda \cos\phi_f}{4n_f} \tag{10.83}$$

the reflected waves will destructively interfere with each other. Since t can only satisfy the constraint in Eq. (10.83) at one wavelength, only partial annulment occurs. Generally t is designed for 550 nm. Also, Eq. (10.83) assumes that all waves strike the surface at the same angle, which is not usually the case. Finally, there is the matter that the amplitudes of the two reflected waves are not necessarily equal. To explore the latter issue, the reflectances at the air-film and film-glass surfaces must be calculated.

The minimum reflectance from a boundary between two media occurs when the angle of incidence is 0° (at highly grazing angles, the reflectance may approach 100%). It can be shown (Jenkins and White, 1937; Waymouth and Levin, 1980) that the minimum reflectance is given by

$$\rho_{min} = \left(\frac{n_1 - n_2}{n_1 + n_2}\right)^2 \tag{10.84}$$

For an air-glass surface, with an average refractive index of glass of 1.5, ρ_{min} is 4%.

For the present air-film-glass problem, it is desired that the reflectances at the two boundaries be equal. From Eq. (10.84), the requirement is

$$\frac{n_a - n_f}{n_a + n_f} = \frac{n_f - n_g}{n_f + n_g} \tag{10.85}$$

which yields

$$n_f^2 = n_a n_g, \quad n_f = \sqrt{n_g} \tag{10.86}$$

since $n_a = 1$ for air. For $n_g = 1.5$, this requires that $n_f = 1.225$. In practical cases, both surfaces of a glass plate or lens are optically coated, and it is possible to reduce the overall reflectance to a fraction of 1%.

Another application of thin films is the cool-beam PAR lamp. In this lamp, the vaporized aluminum normally present on the inside of the parabolic portion of the glass bulb is replaced by a multiple layer of metallic salts. The coating is spectrally selective, reflecting visible energy and transmitting infrared energy. The beam emerging from the front lens has only about one-third of the radiated power of a conventional PAR lamp of the same wattage.

The cool-beam lamp is one example of the use of so-called dichroic mirrors. Such mirrors are of two types, cold and hot. The cold type, as used in the cool-beam lamp, reflects visible radiation and transmits infrared. The hot type does the reverse. It is used in projection systems, in front of the lamp-reflector combination, to reflect the infrared radiation away from the film.

Dichroic lenses are used in colored PAR lamps. The interference film coating is placed on the inside of the cover lens and produces a given color by selective special transmission. Dichroic PAR lamps are available in blue, green, red, yellow, and amber colors.

EXAMPLE 10.14

In Fig. 10.42b, assume the film has a refractive index of 1.38 and the glass 1.50; further assume that the rays strike the glass at near-normal incidence. What film thickness is required for minimum reflectance at the center of the visible spectrum?

Solution. From Eq. (10.83) with $\cos \phi_f = 1$,

$$2t = \frac{\lambda}{2n_f} = \frac{550}{2 \times 1.38} = 199 \text{ nm}$$

Thus film thickness $t \approx 100$ nm.

10.11 DIFFRACTION

When electromagnetic waves pass through an aperture or pass the edge of an obstacle, they spread somewhat into the region that one would expect to be in shadow. This is the phenomenon of diffraction, the bending of light around edges of opaque objects.

An explanation of diffraction requires Huygens' principle and the concepts of interference. Historically, diffraction effects are divided into two classes: those in which the light source and the screen on which the diffraction pattern is observed are infinitely separated from the aperture producing the diffraction, and those where they are not. The first class is known as Fraunhofer* diffraction; the second as Fresnel diffraction. The former is much simpler to treat theoretically and can be simulated quite easily in the laboratory through the use of lenses that produce parallel rays, thus making the source and screen appear to lie at infinity.

Fraunhofer diffraction by a single slit, as shown in Fig. 10.43, will be considered first. In Fig. 10.43a, light from a monochromatic source S is converged by lens L_1 through a slit S_1 located at the focal point of lens L_2. Thus, the rays from L_2 passing through the slit S_2 are horizontal. They are converged by lens L_3 to the screen, which lies at the focal point of lens L_3.

The portion of Fig. 10.43a to the right of S_2 is shown expanded in Fig. 10.43b. Two very narrow wave trains of light are shown emerging from the top and center of S_2 and progressing at an angle α with the horizontal. They are converged by lens L_3 to point P on the screen, thus simulating a screen at infinity.

* Joseph Fraunhofer (1787–1826), a Bavarian glass grinder and specialist in experimental optics.

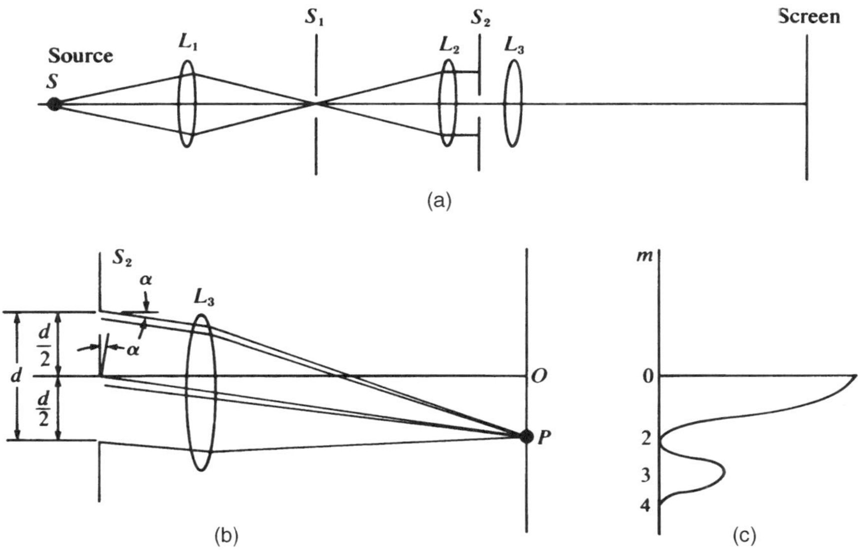

Figure 10.43 Fraunhofer diffraction.

The two wave trains depart from S_2 in phase, but the upper train travels an additional distance (d/2) sin α, in reaching the screen. If this path difference is λ/2 for a given source, annulment occurs at P. The same is true for the next two narrow wave trains immediately below the first two, and for all other consecutive pairs of wave trains. Thus, the light from each wave train in the upper half of slit S_2 is canceled by that in the lower half, and point P is dark. The condition is that

$$\frac{d}{2}\sin\alpha = \frac{\lambda}{2}, \quad \sin\alpha = \frac{\lambda}{d} \tag{10.87}$$

Suppose the slit is now considered divided into four segments, instead of two. By the same reasoning as before, segments 1 and 2 will annul, as will segments 3 and 4 if

$$\frac{d}{4}\sin\alpha = \frac{\lambda}{2}, \quad \sin\alpha = \frac{2\lambda}{d} \tag{10.88}$$

In general, annulment will occur if

$$\sin\alpha = \frac{m\lambda}{2d}, \quad \text{m an even integer} \tag{10.89}$$

The resulting light and dark band diffraction pattern is shown in Fig. 10.43c. The maximum constructive interference is for m = 0 (point O). The first annulment occurs for m = 2 (point P); the second for m = 4. For m = 3, constructive interface occurs, but the brightness is only one-third of that for m = 0. For this case, the slit is divided into three segments. The upper two will cancel for α such that Eq. (10.89) is satisfied for m = 3, leaving only the third segment to produce the bright band.

EXAMPLE 10.15

In Fig. 10.43b, assume the slit width is five wavelengths. What is the angular width of the central bright band of the diffraction pattern? What is the angle for the second null point?

Solution. From Eq. (10.89), with m = 2

$$\sin\alpha = \frac{2\lambda}{2(5\lambda)} = 0.2$$

$$\alpha = 11.5°$$

giving an angular width of 23°.

From Eq. (10.89), with m = 4

$$\sin\alpha = \frac{4\lambda}{2(5\lambda)} = 0.4$$

$$\alpha = 23.6°$$

Note that if the slit width were reduced to one wavelength, the central angle would increase to 180°, and no diffraction pattern would be observed. If the slit width were increased to 100 wavelengths, the central angle would decrease to about 1°, and the pattern would be very constrained. Generally, a slit width of 5 to 10 wavelengths is considered optimal for observing.

Single-slit diffraction suffers in that the boundaries of the light and dark bands are not sharply defined. Much better resolution can be obtained with a diffraction grating, first introduced by Fraunhofer, which is an assemblage of parallel slits, equal in width and separation. The diffraction grating is widely used in the quantitative analysis of spectra.

A diffraction grating is diagrammed in Fig. 10.44. Only five slits are shown, whereas the usual grating contains several thousand. Their spacing is d, which is usually of the order of 2 μm. A wave train is incident normally on the grating. The slits are

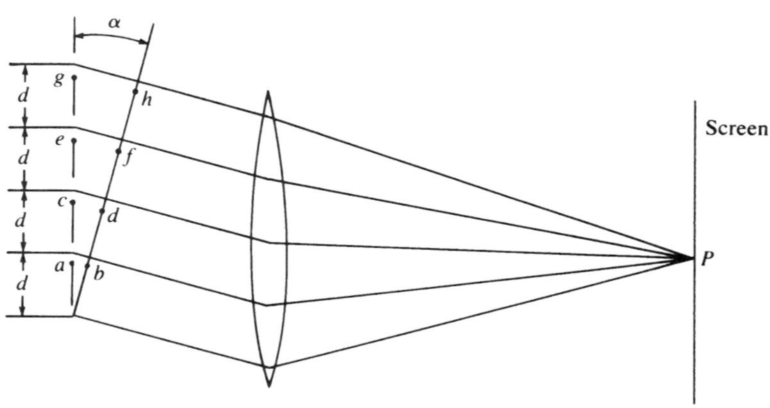

Figure 10.44 A diffraction grating.

assumed to be narrow enough so that the diffracted beam from each slit has sufficient angular breadth to interfere with all other diffracted beams. As in single-slit diffraction, the lens is inserted to make the diffraction pattern on the screen appear to be at infinity.

Consider first the light contained in very narrow elements at the lower extremity of each slit and leaving the slits at an angle α. Let α be such that the distance $ab = \lambda$. Then $cd = 2\lambda$, $ef = 3\lambda$, and $gh = 4\lambda$. The waves in these small elements reach point P in phase and thus constructively interfere. The same applies to all other corresponding elements in each slit.

If α is increased very slightly, near annulment occurs, assuming a large number of slits. Thus, the cutoff of the bright bands on the screen is very sharp. As α is increased further, eventually $ab = 2\lambda$, and a second bright band appears on the screen. The same is true for $ab = 3\lambda$, 4λ, etc. The angles for which maxima occur are obtained from

$$ab = d \sin\alpha = m\lambda \text{ (m an integer)}, \qquad \sin\alpha = \frac{m\lambda}{d} \tag{10.90}$$

If white light is incident on the grating, its spectral components are displayed on the screen. However, unlike a prism which produces one spectrum, a diffraction grating produces several on each side of the normal. These are called *first order, second order*, ... spectra, corresponding to $m = 1, 2, \ldots$. For $m = 0$ in Eq. (10.90), no spectrum is produced since the deviation (α) is zero for all wavelengths. The result is a white image on the screen.

For appreciable deviation, the spacing of the grating should be the same order of magnitude as the wavelength of the light. Commercial gratings customarily have 600 or 1200 lines per millimeter. Fraunhofer's first gratings were made by wrapping fine wire around two parallel screws. Gratings are now made by ruling the equidistant grooves on a glass or metal surface with a diamond stylus. Replicas can then be made on aluminized glass blanks for research and scientific use. Inexpensive 2 in × 2 in card-mounted diffraction gratings are also available for nonquantitative spectral analysis.

EXAMPLE 10.16

Find the angular width of the first-order diffraction pattern produced by a grating having 600 lines per millimeter when lighted with white light.

Solution. The grating spacing is 1.67 μm. It is assumed that white light extends from 400 to 700 nm. Then Eq. (10.90) yields

$$\sin\alpha = \frac{1(4\times10^{-7})}{1.67\times10^{-6}} = 0.239$$

$$\alpha = 13.9° \quad \text{for violet}$$

$$\sin\alpha = \frac{1(7\times10^{-7})}{1.67\times10^{-6}} = 0.149$$

$$\alpha = 24.8° \quad \text{for red}$$

Thus, the angular breadth is 10.9°.

10.12 POLARIZATION

In Sect. 1.3, it was pointed out that electromagnetic waves consist of electric and magnetic fields vibrating at 90° to each other in space, at 90° to the direction in space of the wave propagation, in time phase, and of the same frequency. Such waves are called *transverse*, in contrast to, for example, sound waves, which are *longitudinal* in that they vibrate along the direction of propagation. Interference and diffraction phenomena can be observed with both types of waves, but polarization effects can be observed only in transverse waves.

A radio antenna is a source of polarized waves. As shown in Fig. 10.45a, it provides an $\mathscr{E}$ field parallel and an H field perpendicular to the antenna, generated by oscillations, up and down, of electrons within the antenna. Such waves are called linearly polarized or plane polarized.

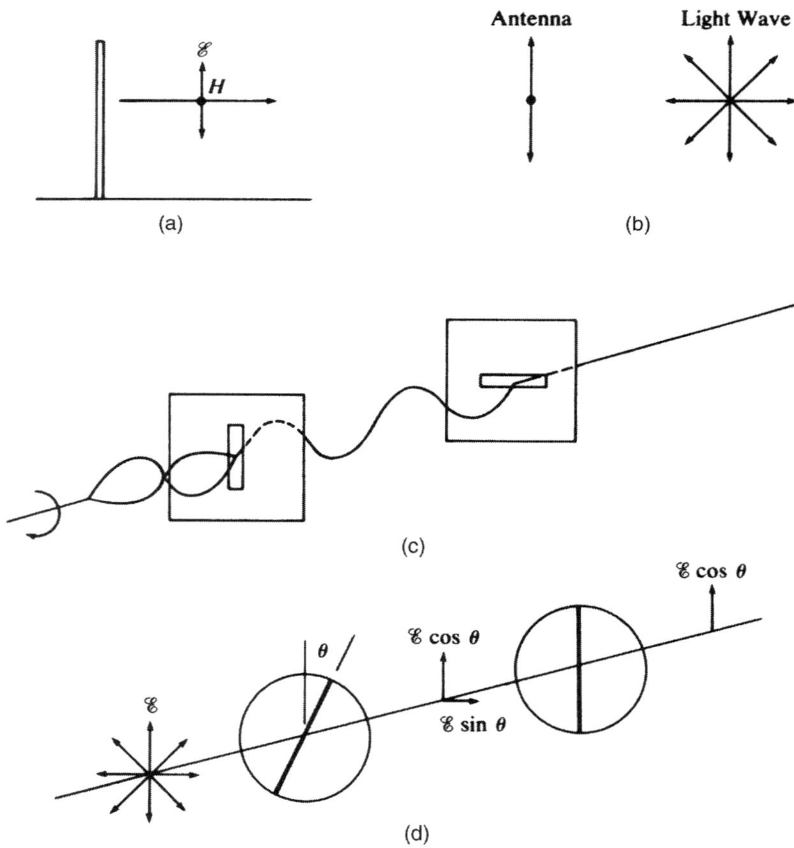

Figure 10.45 Pictorial representation of polarized light.

A light wave, unlike an antenna, normally has its $\mathscr{E}$ and H fields vibrating in all directions in the transverse plane. It can be thought of as being produced by an assemblage of antennas, oriented uniformly in all directions in the vertical plane in Fig. 10.45a. The resulting $\mathscr{E}$ field (which was shown by 0. Wiener in 1890 to be the field that produces the optical effects of polarized light) will also be oriented in all directions in this plane. The distinction between the $\mathcal{E}$ field of the antenna and the $\mathcal{E}$ fields of the light wave are noted diagrammatically in Fig. 10.45b. The light wave is seen to be a super-position of linearly polarized waves in all possible transverse directions.

To produce a polarized light wave requires some device to pass only those $\mathcal{E}$ vectors vibrating in a certain direction. Such a device is called a polarizer, and its effect is indicated in Fig. 10.45c. In this diagram, a piece of clothesline is threaded through two slits, one vertical and one horizontal, and tied to a post. A person at the left rotates the clothesline circularly as one would rotate a jump rope. The first polarizing slit (polarizer) removes all horizontal oscillations so that the clothesline is moving only in a vertical direction between the slits. This is linearly polarized wave motion. Now if a second polarizing slit (analyzer) is inserted horizontally, the clothesline will be motionless from there to the post.

If the slits in Fig. 10.45c are replaced by crystals, such as tourmaline, or plastics with imbedded crystals, such as Polaroid film, and if the rope twirler is replaced by a source of unpolarized light and the post by a photosensor, the diagram in Fig. 10.45d is obtained in which light is being polarized by directionally-selective absorption. Assume the polarizer is oriented at an angle θ from the vertical and produces a linearly polarized wave (single vector) of amplitude $\mathcal{E}$, which may be broken into a vertical component $\mathcal{E} \cos \theta$ and a horizontal component $\mathcal{E} \sin \theta$. The analyzer is oriented vertically so only the vertical component passes through. The transmission is maximum when $\theta = 0°$ and zero when $\theta = 90°$. Since light intensity is proportional to the square of the electric field intensity,

$$I = I_m \cos^2 \theta \tag{10.91}$$

where I_m is the intensity when polarizer and analyzer are aligned. This relation is called Malus' law.*

Suppose the source in Fig. 10.45d is partially polarized so that its $\mathcal{E}$ field amplitude is not the same in all directions. Place the sensor between the polarizer and analyzer and slowly rotate the polarizer, noting the reading of the sensor at each position. The degree of polarization of the source is given by

$$\text{Percent polarization} = \frac{I_{max} - I_{min}}{I_{max} + I_{min}}(100) \tag{10.92}$$

where I_{max} and I_{min} are the maximum and minimum intensities recorded by the sensor as the polarizer rotates through 360°.

There are four common methods of producing polarized light. We have already considered selective absorption. The other techniques are by reflection, refraction, and double refraction. A fifth (natural) technique, scattering, is present in the earth's atmosphere. Incident unpolarized sunlight, striking the atmosphere is scattered, with blue light being scattered more than red and becoming linearly polarized.

* Named for E.L. Malus, who discovered it experimentally in 1809.

Creating polarized light by reflection is perhaps the simplest method and was the first method historically, having been discovered by Malus in 1808. Malus observed that if a beam of unpolarized light was incident at a particular angle on a polished glass plate, the reflected beam was linearly polarized. The situation is diagrammed in Fig. 10.46a. Unpolarized light, which is indicated by arrows parallel and perpendicular to the plane of incidence between the ray and the normal to the surface, impinges on the glass plate at an angle ϕ. At one particular value of ϕ, given by

$$\tan \phi_\mu = \frac{n'}{n} \tag{10.93}$$

15% of the incident component perpendicular to the plane of incidence is reflected whereas none of the component parallel to this plane is reflected. Thus, the refracted ray contains 85% of the perpendicular component and all of the parallel component. The reflected ray is completely linearly polarized, but weak; the refracted ray is strong, but only partially polarized.

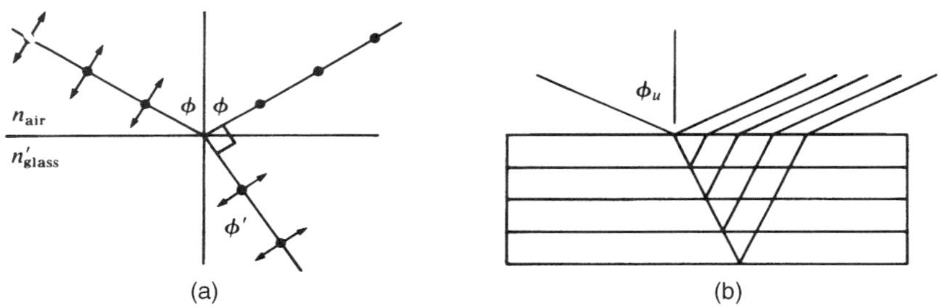

Figure 10.46 Polarization by reflection and refraction.

The relation in Eq. (10.93) is known as Brewster's law*, and the angle ϕ_μ is called the Brewster angle. It is about 57° for ordinary glass. Combining Brewster's law with Snell's law in Eq. (10.55) gives

$$\frac{\sin \phi_u}{\sin \phi'_u} = \frac{n'}{n} = \frac{\sin \phi_u}{\cos \phi_u} \tag{10.94}$$

$$\phi_u + \phi'_u = 90° \tag{10.95}$$

Thus, at the Brewster angle, the angles of incidence and refraction sum to 90°.

If the single glass surface is replaced by a pile of thin glass plates, as shown in Fig. 10.46b, the intensity of the reflected ray is increased, and the refracted ray becomes nearly completely polarized. What happens is that the refracted ray loses 15% of its perpendicular component at each refraction and, ultimately, becomes linearly polarized. The effect of 10 such plates is to reduce the intensity of the perpendicular transmitted component to about 4% of its original value.

* Named for Sir David Brewster, who discovered it experimentally in 1912.

The final technique for producing polarized light is by double refraction, also called birefringence. This requires that the medium be anisotropic, that is, that the velocity of a light wave within the medium be a function of the direction of travel. Common refractive materials, such as glass and plastic, are not doubly refracting, but certain crystals, notably calcite ($CaCO_3$) and quartz (SiO_2), exhibit this property.

Two sets of Huygens wavefronts, one spherical and the other ellipsoidal, emanate from each light wave surface in a doubly-refracting crystal. This results in two rays—an ordinary ray (O ray) from the spherical waves and an extraordinary ray (E ray) from the ellipsoidal waves, which propagate through the crystal. These two wavefronts are tangent to each other in only one direction in the crystal, called the *optic axis*.

Three orientations of E and O rays are shown in Fig. 10.47. In Fig. 10.47a, the incident ray is parallel to the optic axis, there is no double refraction, and the two wavefronts travel at equal velocities through the crystal. In Fig. 10.47b, the incident ray is perpendicular to the optic axis. Again, there is no double refraction, but the E ray travels at a higher velocity through the medium than the O ray. In Fig. 10.47c, which is the usual situation, the incident ray is inclined to the optic axis and both double refraction and differing velocities are present. If the crystal is rotated about the axis of the incident ray, the E ray rotates about the fixed O ray.

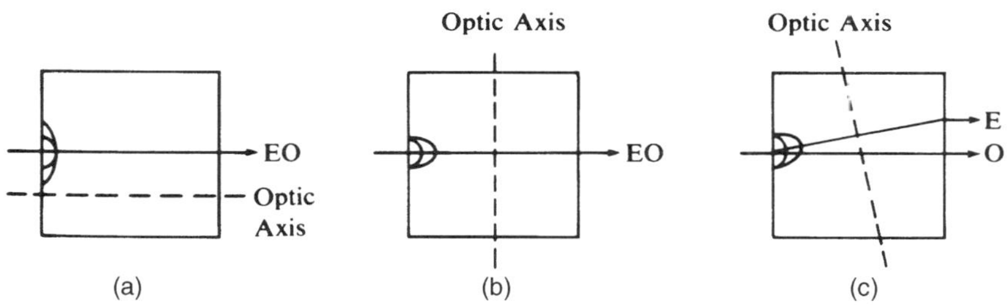

Figure 10.47 E and O rays in a crystal.

Snell's law holds for the O ray, but not for the E ray, since the velocity, and thus the index of refraction, for the latter is a function of direction. For example, in Fig. 10.47c, the incident ray is normal to the crystal face, but the E ray is not. For calcite n_O = 1.658 and n_E = 1.486, where the latter is customarily given for travel perpendicular to the optic axis, which is the direction of maximum velocity of the E ray. For quartz, the values are n_O = 1.544 and n_E = 1.553.

The most significant attribute of doubly-refracting crystals is that the E ray and the O ray are each linearly polarized, with the E ray oscillations at 90° in space to the O ray oscillations. Also the two rays emerge from the crystal out of time phase, with the amount of phase shift a function of path lengths within the crystal.

To explain this result, consider Fig. 10.48. In Fig. 10.48a, the two polarized rays are shown (they are not actually separated, but are shown that way to make the diagram clearer). The O ray is linearly polarized perpendicular to the page; the E ray in the plane of the page.

If the E and O rays are in time phase as they emerge from the crystal, the resultant ray is linearly polarized, as shown in Fig. 10.48b, and lies at an angle of 45°, if

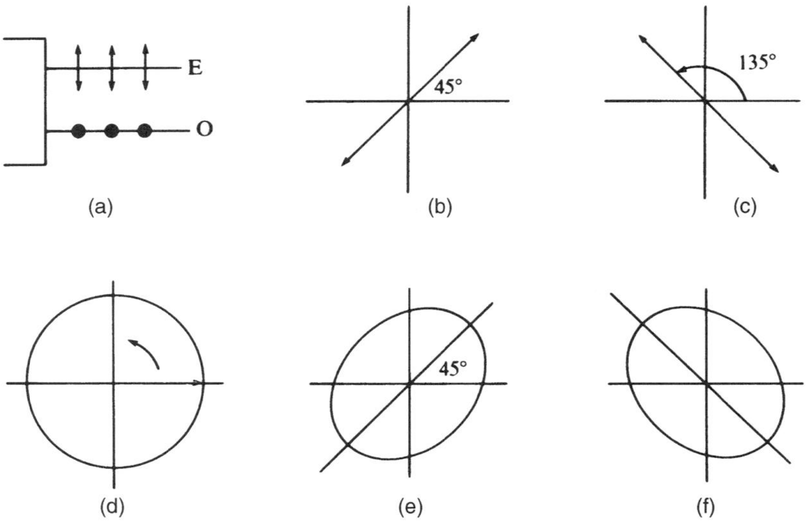

Figure 10.48 E and O ray polarization.

the E and O ray amplitudes are assumed equal. If the two rays emerge 180° apart in time phase, Fig. 10.48c results. Again the resultant is linearly polarized. Time shifts of 90° or 270° yield the result in Fig. 10.48d. This is circularly polarized light. The resultant vector does not oscillate along a given direction, as in Fig. 10.48b and c, but rather rotates in time, with constant amplitude, thus describing a circle locus. Figure 10.48e and f represent the situation where the time phase shift is 45° or 135°. Again the resultant vector rotates in space, but not with fixed amplitude; rather, it describes an elliptical locus, as shown.

For a given crystal, the pattern that is obtained depends on the frequency of the incident light, the indexes of refraction of the crystal, and the crystal's thickness. If the crystal provides a phase shift of 90°, it is called a *quarter-wave plate*; if 180°, a *half-wave plate*. Because phase shift is a function of frequency, these descriptions are valid only for a single color (wavelength) of light.

Assume the incident radiation is white light and that a crystal that is a half-wave plate for green light is inserted between the polarizer and analyzer in Fig. 10.45d. Linearly polarized white light emerges from the polarizer, strikes the plate, and is doubly refracted. The green component emerges from the plate rotated 90° in space and is linearly polarized. Other components emerge elliptically polarized. If the analyzer is oriented at 90° to the polarizer, the green component passes through unattenuated, whereas the other components, which are rotating elliptical vectors, are partially attenuated. If the analyzer is aligned with the polarizer, the green component is rejected, and the other components are again partially attenuated.

Ordinary cellophane tape is doubly refracting, and striking color effects can be observed by inserting various thicknesses of such tape between a polarizer and an analyzer. Another way of achieving color effects is through dichroism. Some doubly refracting materials, such as tourmaline, exhibit this property in which one of the linearly polarized rays is absorbed appreciably in passing through the crystal.

EXAMPLE 10.17

Derive an expression for the phase difference θ between the O and E rays after traversing a crystal plate whose optic axis is perpendicular to the incident ray. Use this expression to find the thickness of a quarter-wave calcite plate for 400-nm radiation.

Solution. The phase difference in radians is given by

$$\theta = 2\pi f\,(t_O - t_E)$$

where t_O and t_E are the times for the O and E rays to pass through the plate. Inserting expressions for times gives

$$\theta = 2\pi f\left(\frac{s}{v_O} - \frac{s}{v_E}\right)$$

where s is the plate thickness, and v_O and v_E are the velocities of the O and E rays in the plate. Using Snell's law

$$\theta = 2\pi fs\left(\frac{n_O}{nv} - \frac{n_E}{nv}\right)$$

But n = 1 for air and $\lambda = v/f$. Thus

$$\theta = \frac{2\pi s}{\lambda}(n_O - n_E) \tag{10.96}$$

For a quarter-wave plate, $\theta = \pi/2$ and, using the refractive indexes given in the text,

$$s = \frac{\lambda}{4(n_O - n_E)} = \frac{400}{4(1.658 - 1.486)} = 581 \text{ nm}$$

This section will conclude with a discussion of a few applications of polarized light. Perhaps the most well-known use is in sunglasses. It was noted earlier that when unpolarized light strikes a horizontal surface, the $\mathcal{E}$ field component parallel to the plane of incidence is attenuated in the reflected ray. Thus, the $\mathcal{E}$ field of the reflected ray is predominantly horizontal. In polarized sunglasses, the Polaroid film functions as an analyzer and is oriented vertically. Thus it eliminates the bulk of the reflected sunlight from ground surfaces. Sunglasses also absorb about 50% of the incident light, polarized or not; consequently, they serve as a filter as well as an analyzer.

As a second example, many years ago there was considerable interest in polarized automobile headlights. The scheme proposed is shown in Fig. 10.49. The headlight lenses and windshield of each automobile were to be of polarized material oriented

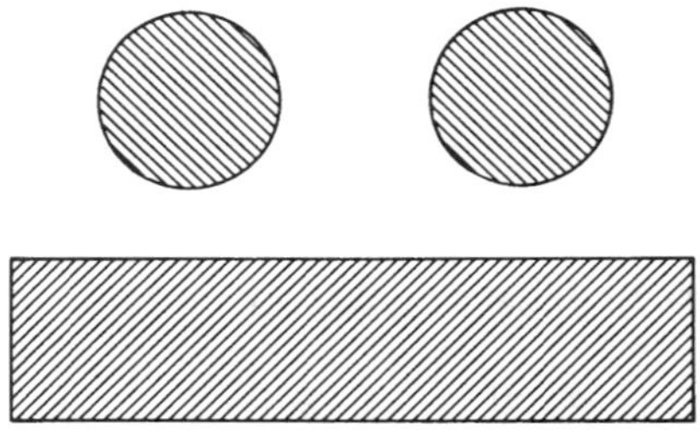

Figure 10.49 Proposal for polarized headlights.

at 45°. The driver would see the light from his own headlights, but not that from approaching headlights. What he would see is the reflection of the approaching car's headlights from the road, which would be polarized horizontally. The proposal was not adopted for several reasons: First, all cars would have had to convert to the new system at the same time, and this would have been costly and difficult logistically. Second, the intensity provided by the headlight filaments would have had to be increased by a factor of 3 to provide enough light on the road for the driver to see. This would have necessitated larger car generators. Last, a motorist approaching another at right angles at an intersection would have seen the full intensity of the other's headlights through the side window.

Another use of polarized light is in improving the CRF of a visual task. When unpolarized light from a ceiling-mounted luminaire reflects from a horizontal surface, particularly a visual task, the reflected light is predominantly horizontally-polarized. Such light is quite effective in producing veiling reflections, particularly at large viewing angles from the vertical. Multilayered polarized light panels that reduce this effect are available for use with fluorescent luminaires. The horizontal component of the unpolarized light emitted by the fluorescent lamps is reflected back to the source by the panel, whereas the vertical component passes through. Little absorption occurs because the reflected horizontal component is depolarized when it reflects from the inside of the luminaire. It then returns to the polarized panel, its vertical component passes through, and its horizontal component returns for a second reflection and depolarization. The result at the visual task is that the incoming light is largely vertically-polarized. Such light tends to be absorbed by the task and reradiated, rather than reflected. Thus, it does not result in the objectionable veiling reflections produced by horizontally-polarized light.* The effectiveness of this technique is very dependent on the geometry between the source, the task, and the observer, and thus is only moderately beneficial in general.

Polarized light is also used in industry, particularly in the detection of stress patterns in glass and plastic. These applications are based on photoelasticity, which is the phenomenon that certain materials, which are normally singly refracting, become doubly refracting when under stress. The material to be tested is placed between a crossed polarizer and analyzer. When the material is stressed, light and dark interference bands can be seen through the analyzer. The location of the bands, and their density, can be used to draw a stress diagram of the material.

* In polarized luminaries, the polarization is referred to as *circular*. Actually this is the geometric distribution of light about the nadir rather than the phase-shifted circular polarization discussed earlier.

10.13 FIBER OPTICS AND LIGHT GUIDES

In 1870, John Tyndall, a British physicist, demonstrated to the Royal Society that light could be transmitted along a curved path. Tyndall's apparatus consisted of an internally lighted water tank with a small hole in the side. The light emerging from the hole followed the curved stream of water downward. In 1880, William Wheeler patented a light distribution system consisting of pipes lined with reflective coating and fed by a central light source.

It was not until the middle of the twentieth century that light-channeling became feasible with practical optical devices. In 1966, Charles Kao and George Hackham, ITT scientists in England, originated the principle of information transmission through a cladded fiber. The light conductors are thin, flexible glass or plastic fibers, coated with glass or plastic cladding and often an outer protective sheath, and assembled in bundles. Their operation depends on the phenomenon of total internal reflection, discussed earlier in Sect. 10.7.

Two technological advances were instrumental in advancing the growth of fiber optics and remote-source illumination, as well as optical communication. Prior to the late 1960s, fibers contained so many impurities and surface defects that beam attenuation was as much as 99% in 10 m. In 1970, a high-purity, low-loss optical glass was developed by Corning Glass Works that permitted transmission of light up to 1 km without amplification. Current fibers absorb light at rate of about 1% per foot.

Because refractive index varies with wavelength, the light out of a fiber can appear quite different in color from the light in. Glass fibers have diameters of 50–150 µm and are manufactured in lengths of 10 km. They are usually assembled in bundles. Single-fiber cables are more efficient, but are difficult to align for splicing. Bundles are easier to work with, but have greater losses.

In the late 1970s, Dupont Chemicals introduced small-core plastic optical fibers. Lumenyte International introduced large-core plastic optical fibers in the mid 1980s. Small core plastic optical fibers (SCPOF) have diameters from 0.5 to 2.0 mm. They are typically bundled together to increase the active area of the fiber, thus increasing the light at the end fitting. Large core plastic optical fibers (LCPOF) range from 3.0 mm to 12.0 mm in diameter. Plastic fibers are less lossy than glass fibers, are more susceptible to UV and heat damage, and are somewhat less flexible in tight bends. They can yellow and become brittle over time.

The second major technological factor in fiber optic growth is the development of small semiconductor lasers and LEDs, some of which can be pulsed at rates of 1 billion per second. This permits modulation of a light wave, paving the way for optical and infrared communication.

Light is not actually bent as it travels through a fiber. Rather it follows a straight-line, zig-zag path, ricocheting off the sides of the rod. Because the index of refraction of the rod is greater than that of the surrounding medium (the cladding) and because most of the light strikes the sides of the rod at angles of incidence greater than the critical angle, nearly total internal reflection occurs, limited only by a smoothness of the rod's surface. In a straight rod, the angle of incidence is the same from one bounce to the next and is determined by the angle at which light enters the end of the rod. In a curved rod, the angles of incidence are greater for a given entry angle and thus the degree of curvature is limited by the critical angle of the glass. A rough guideline is that the radius of a bend should not be less than 10 times the diameter of a fiber.

The cladding around the fiber not only provides the appropriate index of refraction, but also protects the polished conducting surface and separates the conducting cylinders when they are bundled.

For total internal reflection to occur within a fiber, the light entering the end of the fiber must be confined within a certain acceptance cone half angle (θ_h). From Eq. 10.56, the critical angle for total internal reflection is

$$\sin \phi_c = \frac{n_2}{n_1} \tag{10.97}$$

From Fig. 10.50,

$$\theta_h = 90 - \phi_c \tag{10.98}$$

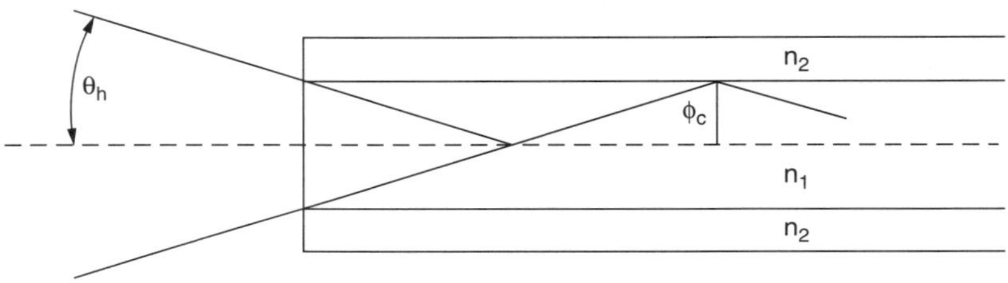

Figure 10.50 Internal reflection inside a fiber.

The numerical aperture (NA) is defined as

$$\mathrm{NA} = \sin \theta_h \tag{10.99}$$

which can be expressed as

$$\begin{aligned} \mathrm{NA} &= \sqrt{1 - \cos^2 \theta_h} \\ &= \sqrt{1 - \sin^2 \phi_c} \\ &= \sqrt{1 - n_2^2 / n_1^2} \end{aligned} \tag{10.100}$$

For common half angles of 30 to 40°, NA from Eq. (10.99) varies from 0.50 to 0.64. Then, for a core refractive index of 1.509, Eq. (10.100) gives a cladding refractive index range of 1.30 to 1.15.

There are two principal types of fibers in use. These are step-index type and the graded-index type, both diagrammed in Fig. 10.51. In the step-index type, shown in Fig. 10.51a, the index of refraction of the core material is constant across its diameter and is 1 to 5% greater than the index of refraction of the cladding. As noted earlier, the light rays propagate in straight zig-zag lines within the core. This type is available in glass cladding/glass core, plastic cladding/glass core, and plastic cladding/plastic core varieties, with the first variety the least lossy and the last the most lossy.

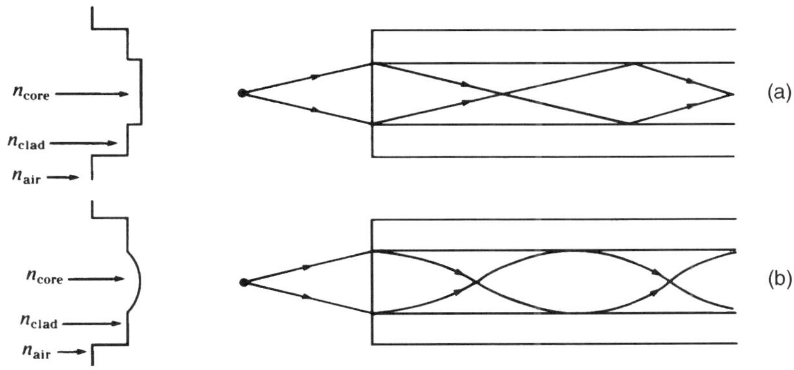

Figure 10.51 Principal types of fiber optics.

The other type of fiber, shown in Fig. 10.51b, is the graded-index type. In this fiber, the refractive index of the core is related parabolically to the core's radius. In this case, the rays follow a curved path within the core. As they move outward from the core axis, the index of refraction decreases, and they are bent back inward. In communications, the parabolic profile provides a greater bandwidth than the step profile and, thus, is preferable when optical pulses are to be transmitted.

Optical fibers emit light in two ways: at an end and along the side. With end emission, the rays zig-zag the length of the fiber and emerge in a cone angle comparable in size to the input angle. They may then be further optically controlled. For side emission, the surface of the cladding is abraded so that a portion of the light refracts through the cladding. There are losses in the fibers, but, in addition there are losses at the light source-fiber interface and at the fiber-output interface. Thus, delivered lumens per watt are only of the order of 20 lm/W.

A fiber optic lighting system consists of a light source, fiber optic cables, and output optics, with couplers between source and fiber and between fiber and output. The use of fiber optic lighting systems has increased dramatically in recent years. Their uniqueness is that they are "remote source" lighting, with the light source removed from the point of application. Thus, the hazards of electricity, lamp heat, and ballast noise are remote from the point of usage. Because one light source can feed many fibers, fewer sources are needed.

The light source may be a halogen lamp (MR16), a single ended metal halide lamp, a xenon lamp, or an LED (often a fan is needed to cool the lamp and the end of the fiber). The goal is to have as nearly a point source of light as possible. The lamp output is processed by reflector-lens optics and focused on the ends of the fibers, which are positioned relative to the source. Generally the reflector is elliptical, to permit focusing of the source rays on the fiber bundle. From 10 to 40% of source lumens are gathered by the fibers. Often the illuminator also includes UV/IR filtering to insure that there are none of these wavelengths at the output of the fiber. This makes fiber optics ideal for the lighting of artifacts. At the output, light can simply be allowed to spray from the ends of the fibers, a lens may be used to image the end of the fiber on a task providing highly-concentrated luminance, or diffusers or decorative fittings may be used.

Applications of fiber optics are myriad and increasing. Their first use was in the late 1950s when VanHeel, Hopkins, and Kapany developed a fiber optic probe, called an endoscope, to observe inside the body. Kapany invented a practical coated glass fiber and coined the term "fiber optics". Many improvements have been made in the device since. Basically it consists of two fiber optic bundles: one to deliver light into the body and the other to return an image to the physician. The device is used for examination, to detect ulcers, cancerous tissues, polyps, etc., and to enable the doctor to see to remove foreign objects and growths or collect tissue samples. Other medical applications of fiber optics include headlights for surgeons and dentists, and transducers for measuring temperature and blood pressure.

Perhaps the most important application of fiber optics is in optical communication. It is possible to transmit many thousands of conversations simultaneously using a single pair of glass fibers. Each voice signal is sampled 8000 times a second and the sampled signals are encoded in 8-bit numbers (256 signal levels), time multiplexed with the bits from other conversations, and sent over the fibers at the rate of 44.7×10^6 bits per second. At the receiving end, the signals are unscrambled and reconstructed. Fiber optic systems are free of the cross talk that plagues conventional cable transmission systems and are unaffected by random electrical disturbances such as lightning. Because they can be slipped into ducts alongside existing copper cables, they do not require space beyond what is used for conventional cables. Because of these advantages, telephone systems throughout the world now use fiber optics.

In the field of lighting, applications of fiber optics began slowly but have been mushrooming recently, as both lamp and luminaire manufacturers have entered the picture. Typical applications are:

- downlights (wall washing, horizontal illumination, localized lighting)
- display lights (jewelry, cosmetics, art work)
- linear lighting (display cases, edge lighting coves, architectural elements, contours)
- floor lighting (for pedestrian guidance)
- outdoor decorative lighting (archways, festoons, building outlines, towers)
- multipoint lighting (night sky simulation, clocks)
- hazardous locations (swimming pools, gangways to ships, explosion-prone areas).

Light guides (pipes) are very similar to fiber optics in their basic principles. The prism light guide was invented by Lorne Whitehead in 1978. The system consists of an electric light source (or daylight) at one end focused by a parabolic reflector, that feeds light in a small cone through an aperture to a long circular or square tube. Propagation through the tube is by total internal reflection, with sections cut along its length to extract light through diffusing or refracting lenses.

A reflecting mirror at the far end of the tube reflects light waves back for a second totally-reflecting tour. The source must be intense and concentrating; thus tungsten halogen, metal halide, xenon, and high pressure sodium are the sources of choice. The entire arrangement is shown in Fig. 10.52.

There are two types of plastic material used for the walls of most light guides. The first is a rigid acrylic, $^1/_4$-inch thick. The inner face of a sheet is flat; the outer face is ridged, with the sides of each ridge at 45° with the flat face to form a 45-45-90° prism for total internal reflection. The entire guide is eight inches in diameter, with a length from 8 to 40 feet.

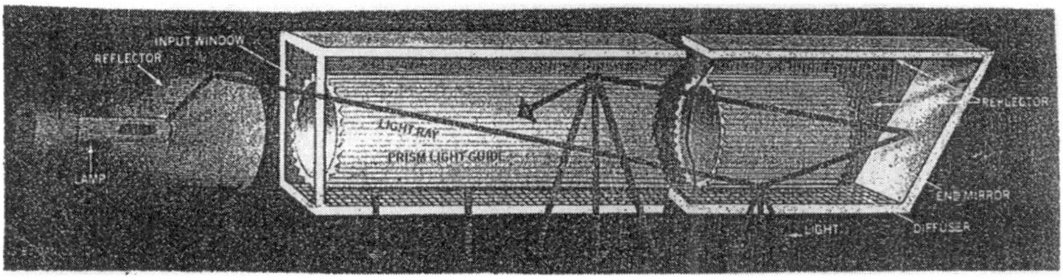

Figure 10.52 Prism light-guide system. *Source*: *Piping Light*, V. Gilmore, May, 1988. Courtesy of Popular Science.

The second type of material is flexible film, made by 3M Company. The film is 0.022-inch thick acrylic or 0.020-inch thick polycarbonate with precision 45-45-90° prism grooves on one side and a mirror-like finish on the flat side. Tubes are made by rolling the material with the smooth side inward and the grooves lengthwise.

Another type of light guide is a large-diameter metal tube with a specular silver interior coating having a reflectance of 95% or greater, used for straight runs only. A light source can be placed at each end to increase the effective length of lighted tube.

As in fiber optics, the angle of entry to the light guide is critical. For either the rigid or flexible guide material, a half angle of no more than 27.6° (a cone angle of 55.2°) is required. Assuming this constraint is met, one to ten percent of the light at each reflection will be transmitted (leaked) to illuminate the space. Next to no light will be absorbed.

If a narrow strip of diffuse reflecting material is placed inside the tube, light rays are scattered and pass through the tube walls. This enhances the natural "leakage".

Several of the applications of fiber optics can also be met by light guides. In hazardous areas, the light source can be housed in an adjacent room. With the source at the end, light guides can be placed over swimming pools and still be easy to service. They are used in rooms where electromagnetic radiation (the presence of metal) could cause problems. In a cold storage area, they can provide light without heat. They are very adept at piping sunlight, gathered by heliostats, deep into building spaces. Outside, they can provide linear accent illumination on the parapets of buildings.

REFERENCES

Baker, G., R. Heinisch, and I. Lewin: "New Techniques for Reflector Design and Photometry," *Journal of IES*, July 1977.

Blaker, W., and W. Rosenblum: *Optics*, Macmillan, New York, 1993.

Coaton, R., and A. Marsden: *Lamps and Lighting, Fourth Edition*, Arnold, 1997.

Committee on Light Control and Equipment Design of the IES: "IES Guide to Design of Light Control," Parts I-IV, *Illuminating Engineering*, November-December 1959, August 1967, August 1970.

Elmer, W.: *The Optical Design of Reflectors*, 2nd ed., John Wiley and Sons, New York, 1979.

Elmer, W.: "The Optics of Reflectors for Illumination," *IEEE Transactions on Industry*, Sept.-Oct., 1983.

Gilmore, V.: "Piping Light," *Popular Science*, May, 1988.

General Electric Company: "Light Measurement and Control," *Bulletin TP-118*, Nela Park, Cleveland, Ohio, 1971.

Halliday, D., and R. Resnick: *Physics, Parts 1 and 2*, Third Edition, John Wiley and Sons, New York, 1978.

Jenkins, F., and H. White: *Fundamentals of Physical Optics*, McGraw-Hill, New York, 1937.

Jolley, L., J. Waldram, and G. Wilson: *Theory and Design of Illuminating Engineering Equipment*, Chapman-Hall, London, 1931.

Kao, C.: *Optical Fiber Systems*, McGraw-Hill, New York, 1978.

Kleekamp, C., and Metcalf, B.: "Designer's Guide to Fiber Optics, Parts 1–4," *EDN*, January–March, 1978.

Magnum, S.: *Optical Fiber Illumination Systems*, General Electric Co., Nela Park, Cleveland, Ohio, 1995.

Myodo, O., and M. Karino: "A New Method for Computer-Aided Design of Luminaire Reflectors," *Journal of IES*, January 1982.

O'Shea, D.: *Elements of Modern Optical Design*, John Wiley and Sons, New York, 1985.

Rea, M., editor: *IESNA Lighting Handbook*, Ninth Edition, Illuminating Engineering Society of North America, New York, 2000.

Roberts, J., editor: *Advanced Lighting Guidelines*, New Buildings Institute, White Salmon, WA, 2001.

Waymouth, J., and R. Levin: *Designers Handbook: Light Source Applications*, GTE Products Corp., Danvers, Mass., 1980.

Wey, R.: "Fiberoptic Communications," 1980 *Laser Focus Buyer's Guide*, Newton, Mass., 1980.

Chapter 11

Exterior Lighting

11.1 INTRODUCTION

Although the same lighting fundamentals that have been utilized thus far in interior lighting apply to exterior lighting, their application is quite different for several reasons.

1. Except for the possibility of ground reflection or reflection from nearby structures, there is no reflected component to include in exterior lighting design. Also, the nighttime sky is dark, often resulting in high contrast.
2. Exterior lighting is often done for a variety of persons and/or tasks. For example, in sports lighting, light must be provided for the players, the referees, the spectators, and the TV cameras. Each of these viewers have different visual requirements.
3. The viewing of vertical, or perhaps oblique, surfaces is often of primary importance in exterior lighting. For example, it is more important for a tennis player to be able to see the side of the ball as it approaches than to be able to see its top.
4. There is no standard 25° viewing angle in exterior lighting. Viewing is often in all directions and thus the problem of providing glare-free seeing is increased.
5. Quite often, especially in sports lighting, the object being viewed is moving rapidly. This occurs occasionally indoors, particularly in industrial areas, but is absent from retail, school, and office lighting.
6. Illuminance levels are generally lower for outdoor activities. The reasons for this are that outdoor seeing tasks are not normally as difficult as those indoors and that many outdoor lighting installations cover large areas and are accomplished with luminaires mounted on poles whose height is much greater than the average 8-ft ceiling height in a room.
7. Very often exterior lighting conditions are such that vision is mesopic (<30 lx), involving both cones and rods.
8. Safety (of people) is often of paramount concern in exterior lighting. There are two major aspects. First, there must be enough light to perform the visual task safely. If an office worker makes an error because of insufficient lighting, it may be serious, but if a motorist is unable to see a pedestrian, it is much worse. Also there is personal safety. A person on a pathway, in a parking lot, or at a bus stop requires adequate horizontal and vertical illuminance, the latter to identify other persons nearby in a timely manner.
9. Security lighting is that part of exterior lighting designed to protect property by discouraging criminal activities through fear of detection and apprehension.

The goal here is to illuminate, either directly or by silhouette, any intruder in time to deter criminal action.

10. The issues of light trespass and light pollution are of concern. Light trespass is unwanted light from luminaires on adjacent property. It manifests itself as increased ambient illumination or distracting glare. Light pollution is the sky glow produced by outdoor lighting, caused by reflected light from the ground and structures, and from direct upward light from luminaires with improper cutoff.
11. A major goal of exterior lighting can be beautification. This includes landscape lighting, of parks, gardens and the like, and the floodlighting of buildings and monuments.
12. Exterior lighting is often provided for identification and attraction, through the lighting of signs, store fronts, and significant architectural features.

In the remainder of the chapter, the general concepts of exterior lighting will be covered, including the lamps and luminaires used and the calculation procedures employed. These concepts and procedures will be illustrated by focusing on four areas of exterior lighting, namely area floodlighting, building floodlighting, sports lighting, and roadway lighting.

11.2 FLOODLIGHT LUMINAIRES

Outdoor luminaires fall into two major categories, those that may be aimed and those that have fixed aiming. Floodlights, used extensively in sports lighting and the lighting of building facades, are in the first category; roadway and area luminaires for parking areas and walkways are in the second. The photometric characteristics of each type will be examined.

The National Electrical Manufacturers Association (NEMA) has defined four classes of floodlight luminaires (Rea, 2000). An enclosed heavy-duty (HD) floodlight is a rectangular weatherproof unit consisting of a metal housing and a separate removable reflector. It is provided with a hinged door and a cover glass whose light opening is equal (or greater) in diameter than that of the reflector. An enclosed general-purpose (GP) floodlight is also a weatherproof unit with a cover glass, but it is round and the housing forms the reflecting surface. A full cutoff shielded (S) floodlight is an enclosed unit that is mounted horizontally. It has negligible intensity upward. A reflector lamp (R) floodlight is simply a lamp holder housing a PAR or R lamp. Examples of floodlight luminaires are shown in Fig. 11.1.

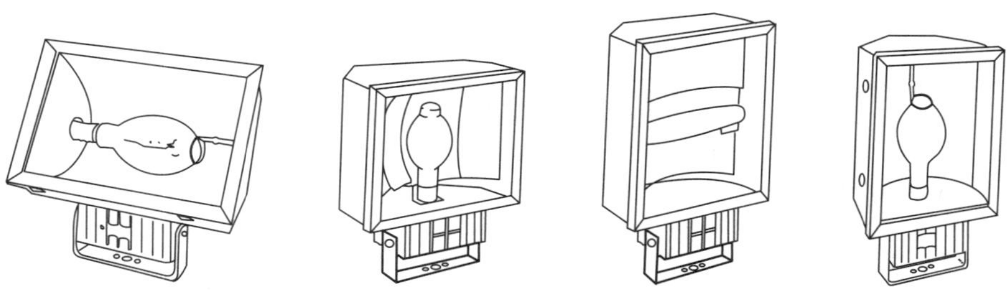

Figure 11.1 Examples of floodlights. *Source*: *IESNA Lighting Handbook*, 9th Edition, 2000. Courtesy of the Illuminating Engineering Society of North America.

There are several pieces of data that a floodlight manufacturer should provide to aid the user in choosing the proper luminaire. One of these is the field angle* of the unit. As was noted in Sect. 2.10, field angle is twice the angle from the beam axis at which the intensity is 10% of its maximum value. NEMA has established seven floodlight types based on their field angles. These are shown in Table 11.1. The field angle of some floodlighting units changes from one perpendicular plane to another. For such asymmetric units, two field angles are given. Thus, a floodlight with a horizontal field angle of 125° and a vertical field angle of 90° is designated as a Type 6 × 5 floodlight.

TABLE 11.1 NEMA FLOODLIGHT TYPES

Type	Field Angle (degrees)
1	$10 = \psi < 18$
2	$18 = \psi < 29$
3	$29 = \psi < 46$
4	$46 = \psi < 70$
5	$70 = \psi < 100$
6	$100 = \psi < 130$
7	$\psi \geq 130$

A manufacturer's photometric data sheet for a typical floodlight is shown in Fig. 11.2. General information about the unit is given in the lower right box, on a per 1000 lumen basis. It is common practice to present floodlight data in this relative manner. Then, for whatever lamp size is used, the data can simply be scaled by the actual number of lamp lumens divided by 1000. In Fig. 11.2, a 400-W high-pressure sodium lamp is specified. From Table 6.17, its initial lumen rating is 50,000 lm. Thus, a factor of 50 should be applied to the lumen and candela values in Fig. 11.2.

In the lower left section of Fig. 11.2, the lumen distribution of the floodlight is shown by angular zones. The three categories of photomatic coordinate systems were described in Sect. 8.6. Traditionally, floodlights are photometered on a Type B goniometer, illustrated in Fig. 8.13b. The photocell is located in a fixed position facing the floodlight, as shown. The floodlight is rotated about one axis with the other held constant, and vice-versa, which produces intensity traces on a sphere about the floodlight. In this procedure, horizontal angles are measured about the vertical axis and vertical angles about the horizontal axis.

If the floodlight is mounted directly over the 0,0 point with its axis perpendicular to the xy plane, as shown in Fig. 11.3a, the intensity traces project onto the plane as shown by the dashed lines in that figure. Note that each box subtends equal horizontal and vertical angular increments, in this case 10°, with the angles for the 20° to 30° box indicated in Fig. 11.3a. The actual horizontal area included within the two horizontal and two vertical angles is not square. Rather it has two straight sides, defined by the constant vertical angles, and two curved sides, defined by the constant horizontal angles. Thus, the grid in Fig. 11.2 is not the actual horizontal plane, but a transformation of it from the horizontal plane of Fig. 11.3a.

* Field angle is also called beam spread.

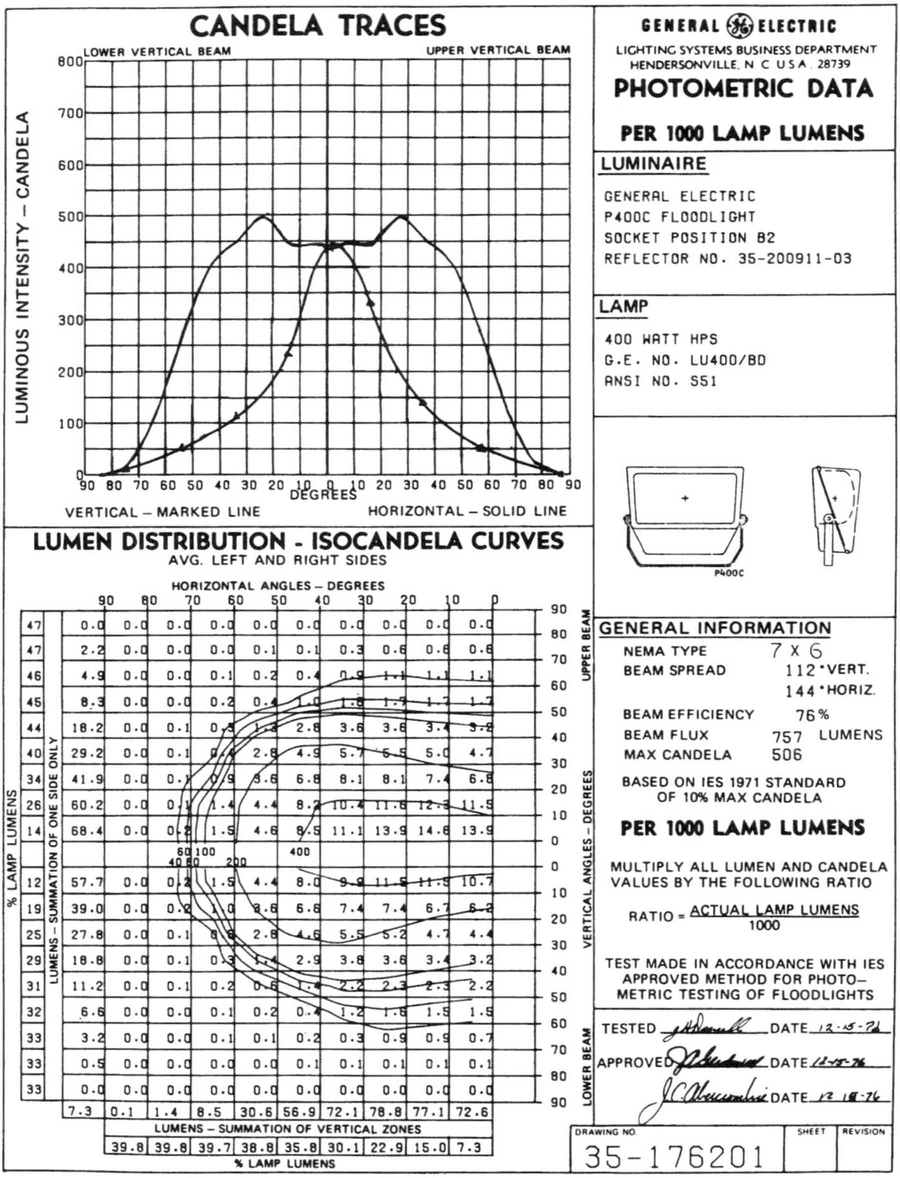

Figure 11.2 Typical floodlight photometric data sheet. *Source*: "Lighting System Product Application Guide," Hendersonville, N.C., 1980. Courtesy of General Electric Co.

The various angles involved in this discussion are shown in Fig. 11.3b. θ_V is the vertical angle, along the y axis, but the angle ϕ along the x axis is not called the horizontal angle. Rather, the horizontal angle is θ_H, which indeed is a horizontal angle from the point of view of the photometry, but is a lateral angle from the point of view of floodlighting geometry. The angle ψ is the angle of incidence at point B on the

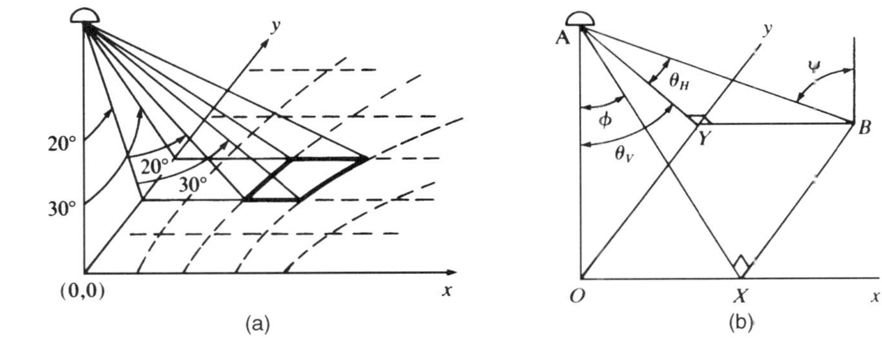

Figure 11.3 Typical floodlight angles and orientation.

horizontal plane. The several angles are related as follows:

$$\left.\begin{aligned}\cos\theta_v &= \frac{AO}{AY}\\ \cos\psi &= \frac{AO}{AB}\\ \cos\theta_H &= \frac{AY}{AB} = \frac{\cos\psi}{\cos\theta_v}\end{aligned}\right\} \tag{11.1}$$

The lumen data in each box in Fig. 11.2 are for one side of the floodlight only, it being assumed that the floodlight has left-right symmetry. In testing, the average between the two halves is taken. The summations of lumens for all horizontal zones on one side are given in the second column from the left; those for all vertical zones on one side in the second row from the bottom. Cumulative percentages of lamp lumens for both sides appear in the left-most column and for one side in the bottom most column.

The percent total efficiency of a floodlight is 100 times the ratio of total floodlight lumens to total lamp lumens. For the unit in Fig. 11.2, 47% of the lamp lumens are in the upper zones; 33% in the lower zones. Thus, the total luminaire efficiency is 80%. Alternatively, 39.8% of the lamp lumens are in the left zone, and doubling this gives about 80% efficiency.

Percent beam efficiency is less than percent total efficiency because only the lumens in the beam are included. In Fig. 11.2 the field angle is 112° × 144°. Thus the lumens out to ±56° vertically and ±72° horizontally are included in the beam (note that 10% of maximum intensity is 50.6 cd), giving a beam efficiency of 76%.

Shown superimposed on the lumen data in Fig. 11.2 are isocandela curves. These, together with the candela traces in the upper left section of Fig. 11.2, describe the intensity distribution pattern of the floodlight. The isocandela curves are generally drawn so that each successive curve has twice the intensity of the one before it, except near the periphery of the beam where more curves are given to improve the accuracy of illuminance calculations. Candela traces are usually drawn in the θ_H = constant and θ_v = constant planes in Fig. 11.3. For rotationally-symmetric units, only one candela trace need be drawn.

Because a floodlight is usually not aimed perpendicularly to a surface, it is desirable for the manufacturer to provide information in addition to that given in Fig. 11.2. Particularly useful are isoilluminance diagrams and charts such as shown in Fig. 11.4 for the floodlight in Fig. 11.2. In these diagrams, horizontal distances and aiming points are expressed in terms of mounting heights. This makes the shapes of the isoilluminance curves independent of mounting height and permits them to be labeled with identifying letters. Thus, for a given mounting height and lamp lumen output, the illuminance represented by each curve can be tabulated. This has been done for the 400-W sodium lamp for three mounting heights in Fig. 11.4.

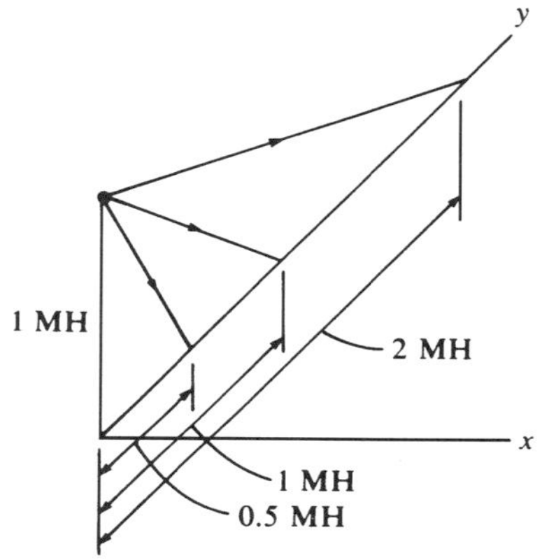

400 W Sodium
NEMA 7 × 6

Initial fc
Table

	Mounting Height		
	30	40	50
A	.18	.1	.06
B	.36	.2	.13
C	.89	.5	.32
D	1.8	1.0	.64
E	3.6	2.0	1.3
F	8.9	5.0	3.2

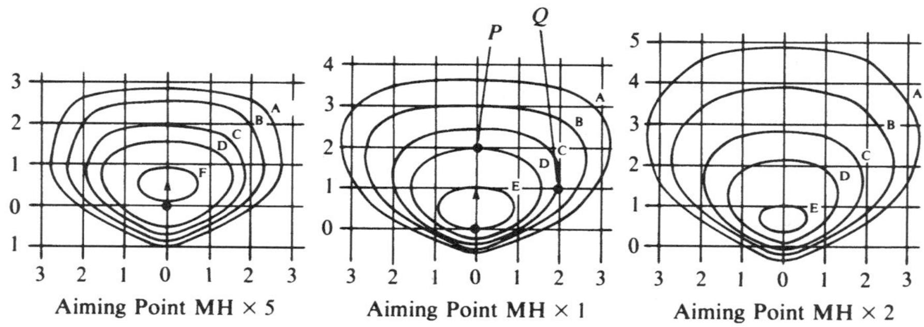

Figure 11.4 Isoilluminance diagrams. *Source*: GE "Lighting System Product Application Guide," Hendersonville, N.C., 1980. Courtesy of General Electric Co.

EXAMPLE 11.1

Calculate the illuminance at the labeled points P and Q in Fig. 11.4 for the floodlight in Fig. 11.2, with a mounting height of 30 feet and a 400-W high pressure sodium lamp.

Solution. The situation is shown in Fig. 11.5. The floodlight is aimed 1 MH directly in front of the pole. For the point P, 2 MH's in front of the pole, the vertical angle difference is 18.4° (63.4° – 45°). From the vertical candela trace in Fig. 11.2, I = 300 cd per 1000 lm. Thus I = 300 × 50 = 15,000 cd.

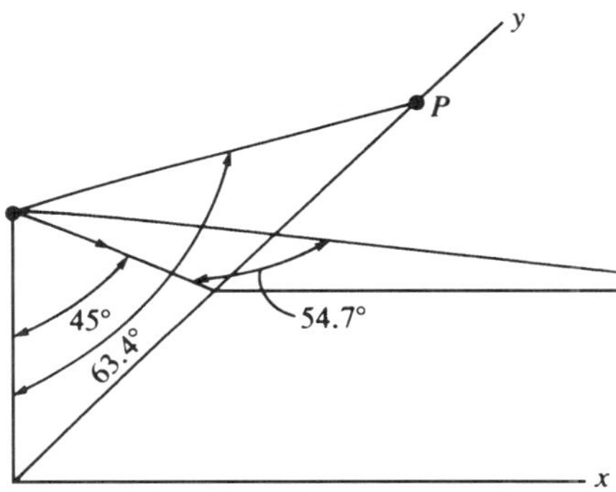

Figure 11.5

The form of the ISL in Eq. (2.6) is useful when luminaire mounting height is constant.

$$E = \frac{I \cos^3 \theta}{(MH)^2} \tag{11.2}$$

Inserting values gives

$$E = \frac{15{,}000 \cos^3 63.4}{30^2} = 1.5 \text{ fc}$$

For the point Q, 1 MH in front of the pole and 2 MH to the side, the angles θ_H and ψ in Fig. 11.3b are needed.

$$\cos \psi = \frac{1}{\sqrt{6}}, \quad \psi = 65.9°$$

$$\cos \theta_H = \frac{\sqrt{2}}{\sqrt{6}}, \quad \theta_H = 54.7°$$

From the horizontal candela trace in Fig. 11.2 for a horizontal angle of 54.7°, I = 50 × 290 = 14,500 cd. Then

$$E = \frac{14{,}500 \cos^3 65.9}{30^2} = 1.1 \text{ fc}$$

The illuminance values differ slightly from those indicated in Fig. 11.4. The isofootcandle diagrams are intended only for quick reference. For greater precision, the photometric data sheets, such as Fig. 11.2, should be used.

Isoilluminance diagrams are useful in estimating the coverage of a given floodlight and the effects of overlapping two or more floodlights. The curves are drawn so that a given curve provides approximately twice the illuminance of the curve immediately beyond it, for a given mounting height and lamp lumens. Thus curve C in Fig. 11.4 provides about twice the illuminance of curve B, etc. The centermost curve should be approximately the maximum illuminance for that aiming point. The outermost curve should be 1 to 2% of this value.

Another useful diagram is a set of utilization curves, shown in Fig. 11.6 for the floodlight in Fig. 11.2. These present the same summation of lumen data as was given in Fig. 11.2, but as a function of aiming angle. In Fig. 11.6, the ordinate is the cumulative percent of lamp lumens for a given aiming angle that fall in horizontal strips in front of and behind the pole, with the width of each strip expressed in mounting heights. An example will illustrate the calculation.

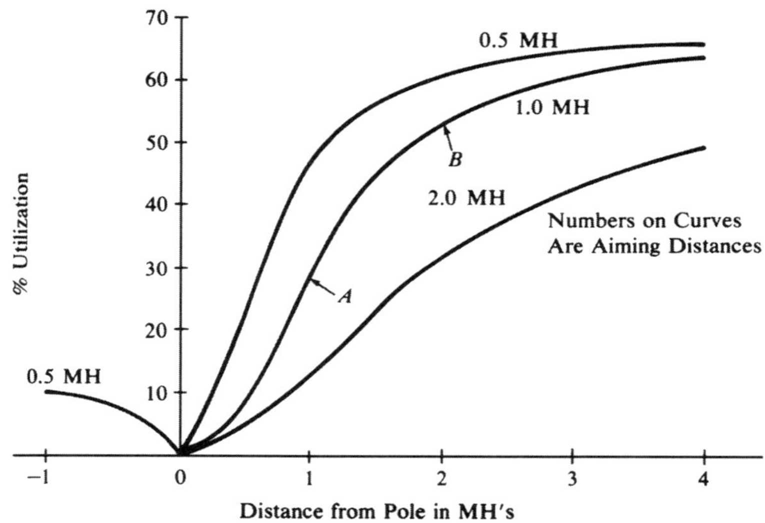

Figure 11.6 Utilization data for Fig. 11.2.

EXAMPLE 11.2

Verify the percent lamp lumen utilization at the labeled points in Fig. 11.6.

Solution. The aiming point is at 1 MH, so Fig. 11.5 may be used. It is desired to calculate the total lumens falling in a horizontal strip 45° wide immediately behind the aiming point and in one 18.4° wide immediately in front of the aiming point. From Fig. 11.2,

$$\phi_{0\text{ to }-45^\circ} = 2[57.7 + 39.0 + 27.8 + 18.8 + 0.5(11.2)]$$

$$= 297.8 \text{ lm}$$

$$\phi_{0\text{ to }-18.4^\circ} = 2[68.4 + 0.84(60.2)]$$

$$= 237.9 \text{ lm}$$

Since Fig. 11.2 is based on 1000 lamp lumens, the percent utilizations are

At A: 29.8%

At B: 53.6%(29.8 + 23.8)

11.3 CONSIDERATIONS IN FLOODLIGHTING DESIGN

There are several guidelines that are useful in floodlighting design.

1. The selection of the aiming point is a compromise between utilization and uniformity (Frier and Frier, 1980). Ideally, the utilization curve should appear as in Fig. 11.7a to optimize both, where k is the width of the area to be lighted in numbers of mounting heights.

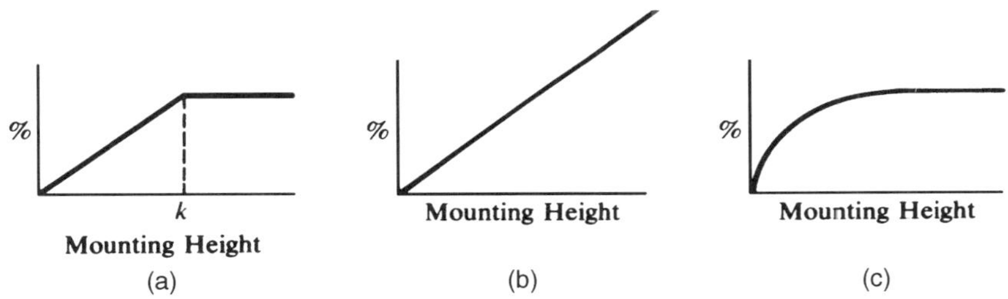

Figure 11.7 Utilization curves.

If the curve in Fig. 11.7a could be obtained, all of the floodlight flux would fall in the lighted area, the percent utilization would equal the total floodlight efficiency, and the flux distribution throughout the area would be uniform. Unfortunately, if the utilization curve is nearly linear within the lighted area, it is likely to continue rising outside that area, as in Fig. 11.7b. This wastes lumens. Or, if most of the lumens are contained within the area, it is likely that the utilization curve will have considerable curvature, providing nonuniform illumination, as in Fig. 11.7c.

Thus a compromise must be reached, and the aiming point chosen accordingly. For fixed mounting height, utilization increases as the aiming point moves toward the pole, but uniformity decreases. This is observed in Fig. 11.6. The utilization is greater for 0.5 MH aiming than for 2.0 MH, but the non-linearity of the curve is greater. The highest utilization (for a symmetrical vertical candela trace) occurs when the aiming line bisects the angle subtended at the floodlight by the near and far sides of the area to be lighted. For example, if the area to be lighted is 3 MH's wide with the pole on one of its boundaries ($\theta_v = 71.6°$ in Fig. 11.3b), the aiming point should be at 0.72 MH ($\theta_v = 35.8°$) for optimum utilization. However uniformity, as can be seen from Fig. 11.6, would be less than optimum. A goal should be a utilization of at least 50%, which occurs in Fig. 11.6 at about 2 MHs from the pole for a 1 MH aiming distance.

To explore the issue of uniformity further, recall that in Ex. 2.5 the aiming of a floodlight to obtain equal illuminances at the near and far sides of a rectangular area was discussd. Consider the two situations in Fig. 11.8. In Fig. 11.8a, a floodlight is

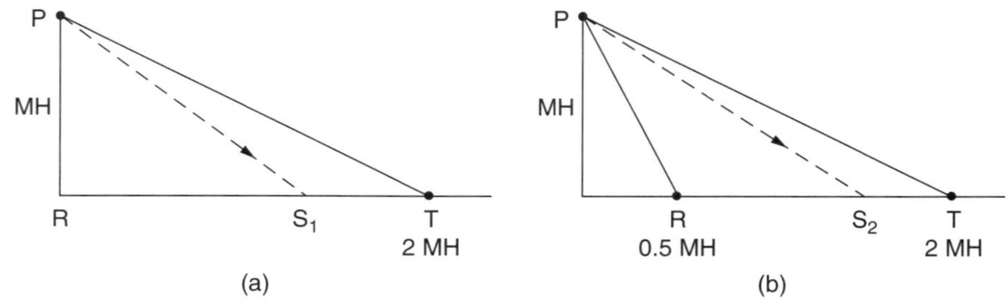

Figure 11.8 Aiming of floodlights.

mounted on one edge of a field 2 MHs wide and aimed at point S_1. In Fig. 11.8b, the aiming is at point S_2, the field is 1.5 MHs wide, and thus the pole setback is 0.5 MH. From Eq. (2.6), for the illuminances at R and T to be equal requires that

$$\left.\begin{aligned} \frac{I_R \cos^3 \beta_R}{(MH)^2} &= \frac{I_T \cos^3 \beta_T}{(MH)^2} \\ \frac{I_T}{I_R} &= \frac{\cos^3 \beta_R}{\cos^3 \beta_T} \end{aligned}\right\} \tag{11.3}$$

In Fig. 11.8a, $\beta_R = 0$ and $\beta_T = \tan^{-1} \frac{2}{1} = 63.43°$, giving $\frac{I_T}{I_R} = \frac{1}{0.0895} = 11.2$. In Fig. 11.8b, $\beta_R = \tan^{-1} \frac{0.5}{1} = 26.56°$ and $\beta_T = 63.43°$, yielding $\frac{I_T}{I_R} = \frac{0.7156}{0.0895} = 8.0$.

Thus, the intensity toward the far side must be 11 times the intensity toward the near side in Fig. 11.8a and eight times the intensity toward the near side in Fig. 11.8b. Generally a 10-times variation in intensity represents the limit of coverage of a floodlight beam. To determine the actual aiming points S in each case, the intensity distribution of the floodlight must be known and a trial and error solution must be done. As mentioned in Ex. 2.5, in general, the floodlight should be aimed ⅔ to ¾ of the distance from R toward T, as indicated by points S_1 and S_2 in Fig. 11.8, to achieve equal illuminances at R and T. This distance should not exceed 2 MHs.

2. The location of floodlighting poles varies with the application. For horizontal areas, such as parking lots, poles are generally located around the perimeter of the area to be lighted. For vertical areas, such as building fronts or signs, they are usually located on the ground some distance from the vertical surface.

For a nearly-square horizontal area, three configurations are most common, as shown in Fig. 11.9. In Fig. 11.9a floodlights are shown at the four corners of the area. This is probably the most effective layout for three reasons. First, most of the floodlight beam can be contained in the area being lighted. Second, the floodlight beams criss-cross, providing more uniform illumination and masking shadows produced by vertical objects. Last, probably only one floodlight per pole will be needed to achieve reasonable uniformity. The primary disadvantage is that the greatest number of poles is required.

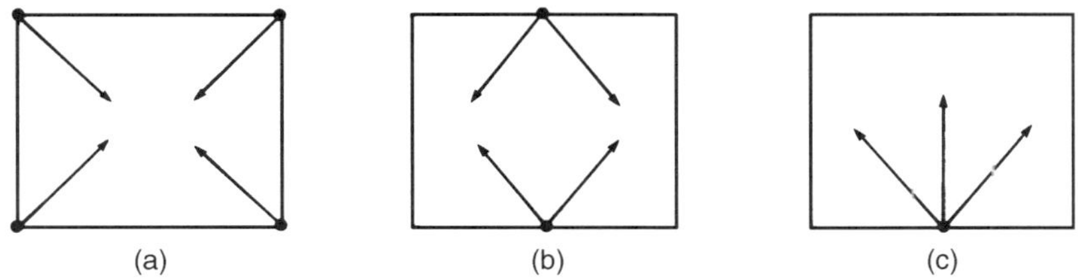

Figure 11.9 Floodlight configurations.

The layout in Fig. 11.9b is a compromise. Two floodlights will be needed on each pole. There is more spill light into surrounding areas, and uniformity is more difficult to achieve. The arrangement in Fig. 11.9c is the least satisfactory. It is particularly difficult to reach the far corners of the area to be lighted with this layout.

3. The evenness of illumination throughout an area is also an issue. Uniformity of illuminance is indicated by one of three ratios: average/minimum, maximum/minimum and maximum/average, or by a metric called coefficient of variation (CV), with the latter being used extensively in sports lighting. CV is defined by

$$CV = \frac{1}{E_{ave}} \sqrt{\frac{\Sigma_1^n (E_i - E_{ave})^2}{n}} \tag{11.4}$$

where E_{ave} is the average illuminance throughout the area
E_i is the ith illuminance
The radical quantity is the standard deviation.

In sports lighting, CV values should range up to 0.30 and values of E_{max}/E_{min} up to 4.0. In roadway lighting, E_{ave}/E_{min} should be held to 3 to 1 on major highways and 6 to 1 on local roads. For car dealerships, E_{max}/E_{min} should range from 5 to 1 to 10 to 1, and for parking lots, no greater than 15 to 1.

4. Coverage of a floodlighted area is dictated by pole height and pole spacing. Each floodlight can effectively light an area 2 MH wide. This determines pole height in Fig. 11.9. Thus the area in Fig. 11.9b can be 4 MH wide; that in Fig. 11.9c, 2 MH wide. In Fig. 11.9a the long dimension of the area should be no more than 4.5 MH or intermediate poles will be needed. Floodlight beams from adjacent floodlights should cross at their 25% of maximum intensity angles. The effective coverage horizontally is out to 45° each side of the aiming line. The 25% criterion means that 50% of maximum intensity will be provided at these beam intersections.

5. Finally, consider the types of lamps to use in floodlighting. Generally the choice is between tungsten halogen, metal halide, and high-pressure sodium. A comparison of these three source types is given in Table 11.2 (IES Sports and Recreational Areas Committee, 1988), where L, F, G and H refer to low, fair, good and high, respectively. For further details of specific lamps, the reader is referred to Chap. 6.

TABLE 11.2 LAMPS FOR EXTERIOR LIGHTING

Lamps	Lumens	lm/W	Life	Color	Control	Lumen Maintenance
TH	F	L	L	H	H	F
MH	H	G	G	G	G	G
HPS	H	H	H	F	G	H

11.4 RECOMMENDATIONS FOR AREA AND BUILDING FLOODLIGHTING

There are numerous outdoor areas requiring lighting, for both seeing and for safety and security. Parking lots need lighting for both driver and pedestrian visibility and safety, as well as to protect against vandalism and theft. Roadside rest areas are generally available 24 hours a day and must be lighted so as to be safe to enter and move about in at night. Sidewalks and walkways require not only horizontal illumination to see the path, but also vertical illumination to identify persons approaching. The latter task requires greater illuminance than the former. The same is true of pedestrian tunnels, but here much greater illuminance is required, both horizontal and vertical, because no other light, natural or electric, is present and because there are often areas where an aggressor could hide.

Service stations and car dealerships are two important outdoor retail activities. Unfortunately these operations are often overlit on the basis of the supposed need for high lighting levels for safety and security. In reality, the high levels are largely an advertising strategy. Very often, especially in gas station-convenience store lighting, there is excessive glare in driver and pedestrian eyes from the luminaires, particularly from those with side lenses under the canopy at the pump island.

Recommended illuminances for the outdoor area lighting areas just mentioned are given in Table 11.3. These are minimum average maintained footcandles unless otherwise specified.

The principles of building floodlighting are similar to those for area floodlighting, but there are differences.

a) There are two ways of aiming. If the floodlights are aimed directly at the surface, there are no shadows and details are obscured. If the surface is matte, the lighting is flat; if it is polished, a specular image of the luminaire may be observed if the luminaire is improperly located. The image may even appear to be within the building surface. If the floodlight grazes the surface, shadows are produced, details are enhanced and silhouetting can occur. However, one area or object may be hidden by another and edges of objects may be distorted.
b) As the distance of the floodlight from the building increases, shadows shorten and become softer but spill light can become a problem. Increasing luminaire size can also soften shadows. Increasing the number of luminaires and their locations can cause multiple shadows, which can be unpleasant.
c) At long viewing distances, the major architectural structure is what should be lighted; at short viewing distances, the light should enhance finish, texture, and shape.

TABLE 11.3 AREA ILLUMINANCE RECOMMENDATIONS

Area	Basic Level (fc)	Level for Security (fc)
Parking lots		
Horizontal	0.2	0.5
Vertical	0.1	0.25
Rest areas		
Entrances and exits	0.6–0.3	
Parking areas	0.6	
Activity areas	1.0	
Sidewalks		
Commercial	1.0 (Hor)	2.0 (Vert)
Intermediate	0.5 (Hor)	1.0 (Vert)
Residential	0.2 (Hor)	0.5 (Vert)
Walkways	0.5 (Hor)	0.5 (Vert)
Pedestrian tunnels	4.0 (Hor)	5.0 (Vert)
Service stations*		
Approaches	1.5–2	
Pump island	5–10	
Service areas	2–3	
Car dealerships*		
Driveways	1–3 (Max)	
Car rows	3–10 (Max)	

* The higher values are for competitive areas with light surrounds.

d) Building floodlights may be mounted on the ground, on poles or on rooftops. Ground mounting allows units to be hidden by shrubs, but exposes them to vandalism. Pole and roof-top mounting may limit flexibility and increase difficulty of servicing. Pole height should be ≥14 ft to minimize vandalism.

e) Building floodlighting, if ground-based, has the potential to create significant light pollution. Careful selection of field angle (NEMA type), along with appropriate shielding and proper aiming, are mandatory.

f) Recommended illuminance levels for floodlighting building exteriors are given in Table 11.4. These are average maintained fc. Note that the lighter the surface and the darker the surround, the less illuminance is required.

TABLE 11.4 ILLUMINANCE LEVELS FOR BUILDING FLOODLIGHTING

Surround	Surface	Vertical Illuminance (fc)
Bright	Light	50
Bright	Medium	70
Bright	Dark	100
Dark	Light	20
Dark	Medium	35
Dark	Dark	50

11.5 FLOODLIGHTING CALCULATIONS

There are two common methods for calculating the illuminance produced by a floodlight. One is the point-by-point method, using the inverse square law in the form of Eq. (11.2). Though tedious to perform manually, the method is easy to program on a computer and is as accurate and complete as the number of calculations the user wishes to make. Generally, the area to be lighted is divided into a square grid, like a checkerboard. The illuminance is calculated at the center of each square and at a few other points that may be deemed critical, for example, along the boundary of an area. If the resulting values are felt to be inadequate, the computer may be asked quite easily to recalculate illuminances for new mounting heights, lamp sizes, aiming points, and/or pole locations.

The second floodlight calculation procedure is the beam lumen method, which has certain similarities to the zonal cavity method of interior lighting design. In this method the average maintained illuminance is obtained from

$$E_m = \frac{n \times \phi_B \times CBU \times LLF}{A} \tag{11.5}$$

where
n = number of floodlights
ϕ_B = initial lamp beam lumens per floodlight (out to 10% of the average maximum candlepower)
CBU = coefficient of beam utilization
LLF = light loss factor
A = area to be lighted

The coefficient of beam utilization is the ratio of the lumens incident on the floodlit area to the total beam lumens. The determination of this parameter was very tedious (Murdoch, 1985), and the procedure has been replaced by the simplified lumen method.

To obtain CBU in the original method required that the boundary of the playing field be projected onto the lumen distribution grid of the floodlight, such as the grid in Fig. 11.2. The new method assumes that, because the area covered by the floodlight's horizontal distribution is large while that covered by the vertical distribution is small, there is negligible spill light along the horizontal axis of the beam off the ends of the area being lighted. This permits the area being lighted to be assigned all of the lumens in any rectangle bounded by two vertical (θ_v) angles on the grid in Fig. 11.2. Then Eq. (11.5) is modified to read

$$E_m = \frac{n \times \phi \times CU' \times AAF \times LLF}{A} \tag{11.6}$$

where
n = number of floodlights
ϕ = initial lamp lumens per floodlight
CU′ = preliminary coefficient of utilization
AAF = application adjustment factor
LLF = light loss factor
A = area to be lighted

TABLE 11.5 AAF FACTORS

	NEMA Horizontal Field Angle		
Beam Factor	Narrow (1, 2)	Medium (3, 4)	Wide (5, 6, 7)
≥ 4.5	0.95	0.85	0.80
3.0–4.4	0.90	0.80	0.75
2.0–2.9	0.85	0.75	0.70
≤ 1.9	0.75	0.65	0.55

The preliminary coefficient of utilization (CU′) is the ratio of lumens reaching A to total lamp lumens, assuming no spill light at the extremes of the beam horizontally. AAF is a factor applied to correct for any error in this assumption. Values of AAF, determined empirically, are presented in Table 11.5. The beam factor in that table relates to beam utilization at the corners of the area. It is given by

$$BF = \frac{W}{\sqrt{h^2 + s^2}} \tag{11.7}$$

where W = horizontal field width
h = luminaire mounting height
s = pole set back

An interpretation of BF can be obtained from Fig. 11.3b. Identifying OY as s, AO as h and YB as W, BF is simply tan θ_H, the lateral horizontal angle to a near corner of the area. However, this interpretation is flawed because W in Eq. (11.7) is defined as the full width of the area. If two BFs were defined, one to each near corner, Eq. (11.7) could be written as

$$BF = \tan\theta_{H1} + \tan\theta_{H2} \tag{11.8}$$

Note that values of AAF increase as horizontal field angle narrows and as beam factor increases, i.e., a narrow beam spread places more lumens on the area, as does a larger θ_H.

EXAMPLE 11.3

A playing field 360 ft × 160 ft is shown in Fig. 11.10. It is to be lighted by floodlights on four symmetrically-located poles, one of which is shown at point O, with a set back of 30 ft and a pole height of 50 ft. The floodlight is a NEMA Type 5 enclosed symmetric luminaire housing a lamp giving 33,300 initial lumens. There are five floodlights per pole. All lamps are aimed at a point (H) 60 ft in front of a given pole. Assuming a light loss factor of 0.75, find the average maintained horizontal illumination on the playing field.

Solution. The isocandela curves and lumen distribution by angular zones for the given floodlight are shown in Fig. 11.10. A three-dimensional view of the floodlight and field is shown in Fig. 11.11. To find CU′, the lumens incident on the field are

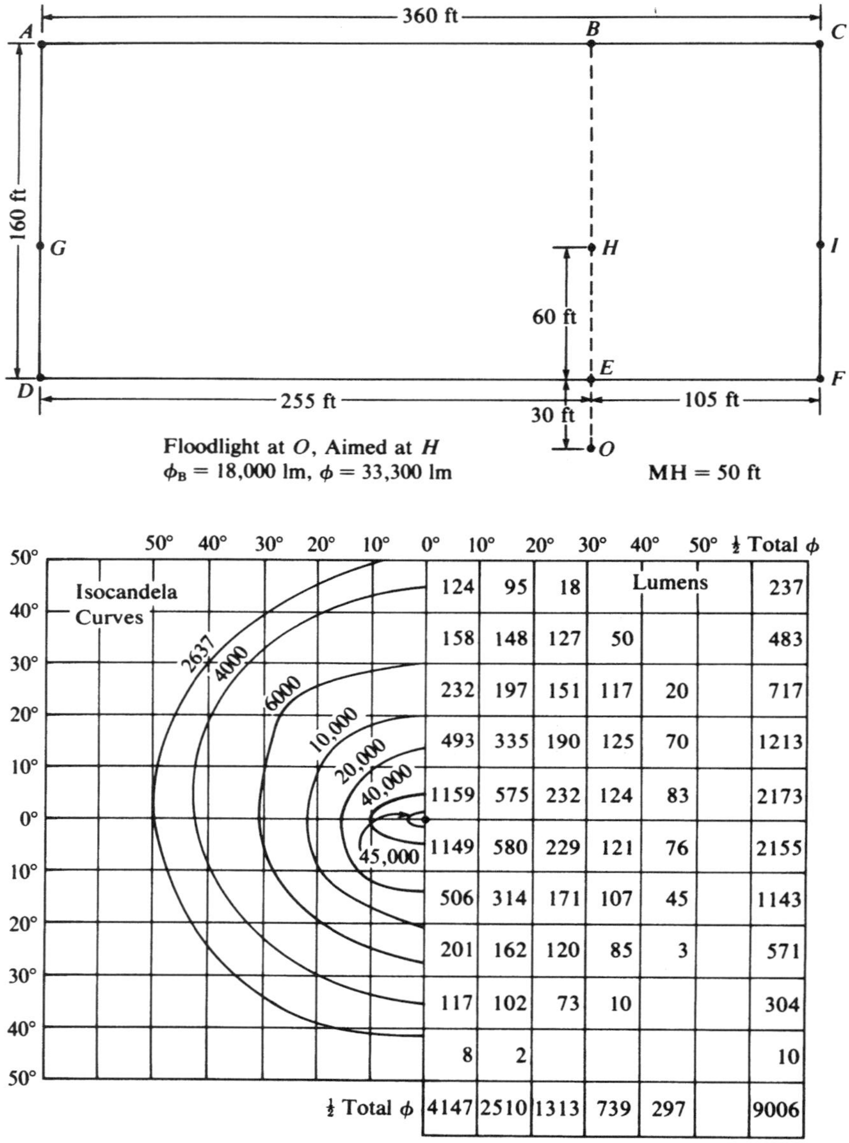

Vertical	0°–10°	10°–20°	20°–30°	30°–40°	40°–50°		½ Total ϕ
50°–40°	124	95	18				237
40°–30°	158	148	127	50			483
30°–20°	232	197	151	117	20		717
20°–10°	493	335	190	125	70		1213
10°–0°	1159	575	232	124	83		2173
0°–10°	1149	580	229	121	76		2155
10°–20°	506	314	171	107	45		1143
20°–30°	201	162	120	85	3		571
30°–40°	117	102	73	10			304
40°–50°	8	2					10
½ Total ϕ	4147	2510	1313	739	297		9006

Figure 11.10 Floodlighting design. ***Source*: *Recommended Practice for Sports Lighting*, IES Committee on Sports and Recreational Areas, 1969. Courtesy of the Illuminating Engineering Society of North America.**

needed. The bounding vertical angles are

$$\angle \text{EPH} = \tan^{-1}\frac{90}{50} - \tan^{-1}\frac{30}{50} = 60.94 - 30.96 = 30.0°$$

$$\angle \text{HPB} = \tan^{-1}\frac{190}{50} - \tan^{-1}\frac{90}{50} = 75.26 - 60.94 = 14.3°$$

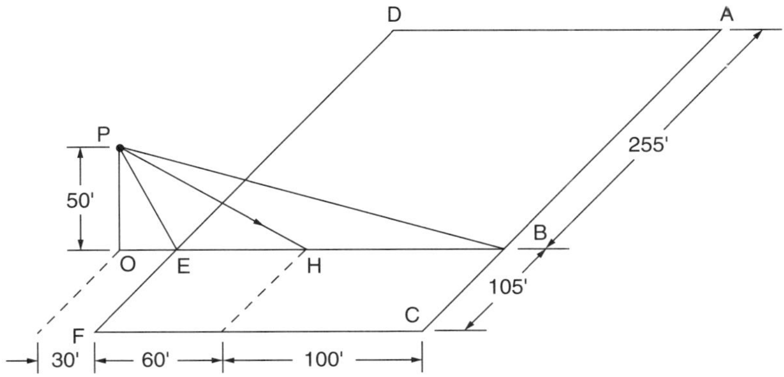

Figure 11.11 Pertinent to Ex. 11.3.

Thus, in Fig. 11.10, the numbers of lumens in the rectangular zones from 0 to 30.0° above the horizontal beam axis and from 0 to 14.3° below the horizontal axis are required. The summations are

Above: $2(2173 + 1213 + 717) = 8206 \text{ lm}$

Below: $2\left(2155 + \frac{4.3}{10} \times 1143\right) = 5293 \text{ lm}$

Thus, the preliminary coefficient of utilization is

$$\text{CU}' = \frac{8206 + 5293}{33,300} = 0.405$$

From Eq. (11.7), the beam factor is

$$\text{BF} = \frac{360}{\sqrt{50^2 + 30^2}} = 6.17$$

From Eq. (11.8), BF can also be expressed as

$$\text{BF} = \frac{105}{\sqrt{3400}} + \frac{255}{\sqrt{3400}} = 1.80 + 4.37 = 6.17$$

Now from Table 11.5

$$\text{AAF} = 0.80$$

Then Eq. (11.6) yields

$$E_m = \frac{20 \times 33{,}300 \times 0.405 \times 0.80 \times 0.75}{360 \times 160} = 2.8 \text{ fc}$$

11.6 OUTDOOR SPORTS LIGHTING PARAMETERS

Outdoor sports lighting is one of the most fascinating areas of floodlighting design, largely because of the variety and complexity of the visual tasks to be performed. There are three major constituencies that the lighting serves:

1. **The Participants.** The lighting must be sufficient, in quantity and quality, for the players and officials to perform their visual tasks at a level of skill and accuracy appropriate to the level of competition. The visual task often involves seeing a three-dimensional, fast-moving object, against a variety of backgrounds. There is often no "work plane," especially for aerial games. Both horizontal and vertical illuminance are important.

2. **The Spectators.** Lighting requirements must take into account the needs of spectators, particularly those spectators sitting in the farthest row of seats from the playing area. At this distance, the lighting levels needed are often higher than those required for the players, especially for older spectators, and, as for the players, vertical, as well as horizontal, illuminance is necessary.

3. **Television Requirements.** Television coverage of nighttime sports activities has increased dramatically since the development of the metal halide lamp, which made widespread color television coverage feasible. The needs of television are much like those of spectators—higher levels of both horizontal and vertical illuminance. The IESNA minimum illuminance recommendations for color television are not less than 75 fc of horizontal and vertical illuminance at any point on the playing surface. Light sources for TV broadcasting should have a CRI of at least 80.

Based on the needs of the three constituencies, there are four classes of playing facilities. Generally a close correlation exists between participant skill level and size of facility.

Class I—Competitive play in facilities for between 5,000 and 10,000 spectators.* This class applies to professional, semi-professional, and college sports.

Class II—Competitive play in facilities for up to 5000 spectators. This class is also for college sports, and for high school and amateur sports.

Class III—Competitive play with some spectator facilities. This class is for high schools, amateur leagues, and training facilities.

Class IV—Social and recreational play with no provision for spectators. This class includes the activities in Class III plus elementary schools and recreation fields.

* The lighting levels necessary for stadia with large spectator capacity are based more on the requirements of spectators and television than on participants' visual needs. See Sect. 11.8.

To define illuminance requirements, outdoor sports are divided into two groups: aerial and ground level. In both cases, the play may be multidirectional (baseball, soccer, field hockey, badminton) or unidirectional (golf, skiing, archery, skeet shooting).

Recommended illuminance levels for several outdoor sports are shown in Table 11.6. These are minimum recommended maintained average values on a horizontal plane 3 ft (0.9 m) above the ground, and are based on lamp lumens at 70% of rated life. Although it is recognized that vertical illuminance in sports lighting is often as significant, if not more so, than horizontal illuminance, it has been found that the vertical illuminances will be adequate if the horizontal illuminances in Table 11.6 are met and if the lighting equipment is located and aimed appropriately.

TABLE 11.6 RECOMMENDED ILLUMINANCES AND UNIFORMITIES

Sport	Class	Area	Hor. fc	CV	Max/Min
Baseball	I	Infield	150	0.13	1.7
		Outfield	100		
	II	Infield	100	0.21	2.5
		Outfield	70		
	III	Infield	50	0.25	3.0
		Outfield	30		
	IV	Infield	30	0.30	4.0
		Outfield	20		
Badminton			10	0.25	3.0
Golf		Tee box	5	0.35	5.7
		Fairway	3	0.25	3.0
		Green	5	0.25	3.0
Motor Racing	II	Track	30	0.21	2.5
		Finish line	75		
	III	Track	80	0.25	3.0
		Finish line	50		
Football, Lacrosse	I		100	0.13	1.7
and Field Hockey*	II		50	0.21	2.5
	III		30	0.25	3.0
	IV		20	0.30	4.0
Soccer	II		150	0.21	2.5
	III		100	0.25	3.0
	IV		50	0.30	4.0
Tennis	I		150	0.13	1.7
	II		100	0.21	2.5
	III		75	0.25	3.0
	IV		50	0.30	4.0
Horseshoes			5	0.30	4.0
Shuffleboard			5	0.25	3.0
Volleyball	III		30	0.25	3.0
	IV		20	0.30	4.0

* *Class* I applies only to football.

Source: IESNA Lighting Handbook, 9th Edition, 2000. Courtesy of the Illuminating Engineering Society of North America.

In outdoor sports lighting, it is important to have reasonable uniformity of illuminance on the playing area and throughout the space above the playing area. Sharp changes in illuminance from one part of the playing area to another through which a ball is traveling can make it appear as though the ball is accelerating or decelerating. This occurs when there is inadequate overlapping of floodlight beams.

Values of minimum acceptable uniformity are given in Table 11.6 in the terms of maximum values of both CV, defined earlier in Eq. (11.4), and the ratio of maximum to minimum illuminance. These are summarized for the four classes of play in Table 11.7.

TABLE 11.7 CV AND E_{MAX}/E_{MIN} (VALUES SHOULD NOT EXCEED THESE NUMBERS)

Class	CV	E_{max}/E_{min}
I	0.13	1.7
II	0.21	2.5
III	0.25	3.0
IV	0.30	4.0

EXAMPLE 11.4

To show the relationship between CV and E_{max}/E_{min}, consider a Class I case where there are eight illuminance readings given by 1.0 E_{min}, 1.1 E_{min}, 1.2 E_{min}, 1.3 E_{min}, 1.4 E_{min}, 1.5 E_{min}, 1.6 E_{min} and 1.7 E_{min}. Compute CV.

Solution. E_{ave} is 1.35 E_{min} and E_{max}/E_{min} is 1.7. From Eq. (11.4)

$$CV = \frac{1}{1.35}\sqrt{\frac{2(0.35^2 + 0.25^2 + 0.15^2 + 0.05^2)}{8}} = 0.17$$

If the readings were to be skewed toward E_{max} and E_{min} (1.00 E_{min}, 1.05 E_{min}, 1.10 E_{min}, 1.15 E_{min}, 1.55 E_{min}, 1.60 E_{min}, 1.65 E_{min} and 1.70 E_{min}), E_{ave} and E_{max}/E_{min} are unchanged, but CV becomes

$$CV = \frac{1}{1.35}\sqrt{\frac{2(0.35^2 + 0.30^2 + 0.25^2 + 0.15^2)}{8}} = 0.20$$

Thus, as anticipated, a larger CV means more illuminance values are farther from the average illuminance.

It is noted that a basic difference between CV and E_{max}/E_{min} is that the former does not rely on just two readings (E_{max} and E_{min}), but includes every illuminance reading in an array. It is thus the preferred metric to use.

As with interior lighting, there is a light loss factor for exterior lighting, and it is composed of non-recoverable and recoverable factors. These were delineated in Sect. 9.10. In the case of outdoor luminaires, the major recoverable factors are lamp lumen depreciation (LLD) and luminaire dirt depreciation (LDD). The lamp used in sports lighting is metal halide, with a typical LLD value of 0.82 at 40% life. LDD values are given in Table 11.8.

TABLE 11.8 LDD FACTORS FOR OUTDOOR LUMINAIRES

	TIME BETWEEN CLEANINGS (YEARS)		
Environment	**1**	**2**	**3**
Very clean	0.98	0.94	0.93
Clean	0.95	0.92	0.90
Moderate	0.92	0.87	0.84
Dirty	0.87	0.81	0.75
Very dirty	0.72	0.63	0.57

In sports lighting, there are two noncoverable factors: field factor (FF) and tilt factor (TF). Field factor consists of three components:

- Lamp lumen output variations from the value obtained during testing, upon which rated lumens are based.
- Ballast output variations from ballast to ballast.
- Luminaire intensity and flux distribution variations from luminaire to luminaire.

Experience indicates that there is at least a 15% loss due to these three factors; thus FF should not exceed 0.85.

Lamp tilt factor is a function of lamp position and, consequently, varies with aiming angle. It is defined as the ratio of lumens of an HID lamp in a given position to the lumens when the lamp is operated in the position used during testing. Generally an average TF is obtained for all the luminaires in a given installation. Its value can be as low as 0.82, but is typically 0.95.

Thus, in sports lighting, the overall light loss factor is

$$\mathrm{LLF} = \mathrm{LLD} \times \mathrm{LDD} \times \mathrm{FF} \times \mathrm{TF} \tag{11.9}$$

and is of the order of

$$\mathrm{LLF} = 0.82 \times 0.92 \times 0.85 \times 0.95 = 0.61$$

11.7 CONSIDERATIONS IN SPORTS LIGHTING DESIGN

There are several guidelines that are useful in outdoor sports lighting design. Recall that general guidelines for floodlighting design were given in Sect. 11.3. The guidelines in this section should be considered as supplementing those and applying particularly to sports lighting.

1. For multidirectional sports, light should be provided from enough directions at a given point to avoid harsh shadows. Some semi-directional lighting may be used to provide modeling. For unidirectional sports, such as archery and driving golf balls, much higher levels of vertical illuminance are needed in one direction. This often makes it possible to shield luminaires from the field of view.
2. Glare control is very important in outdoor sports. Not only will participants be adversely affected if glare is present, especially in aerial games, but the spectators will also suffer. In addition, glare can create light trespass and be annoying to inhabitants of buildings nearby, to pedestrians on adjacent walkways, and to drivers in vehicles on adjacent streets.

   As was discussed in Sect. 8.8, glare is a function of luminaire luminance, luminaire location, the visual size of the lighted portion of the luminaire, and the surrounding field luminance. Of particular importance is field luminance and its ratio to luminaire luminance, which is essentially the contrast between the luminaire and its background. In sports lighting, particularly in aerial games, the field is the dark sky, the worst possible background. Increasing surround luminances—on fences, the stands, and peripheral areas—can improve the situation.

   Other steps that can be taken are given below. For simplicity, they will be illustrated in terms of a rectangular playing area, but the conclusions are general.

   a) The NEMA field angles given in Table 11.1 should be consulted. As shown in Fig. 11.12a, if the field angle is too great, there is wasted light and increased glare in spectators', and possibly players', eyes. As the distance of the luminaire from the surface to be lighted increases, field angle must be decreased if the same playing surface area is to be lighted.
   b) A guideline for floodlighting design to minimize glare is that the angle between the horizontal playing surface and a line drawn through the lowest mounted floodlight to a point one-third the distance across the playing field should not be less than 30°. In equation form,

$$\mathrm{MH} = 0.577\left(\mathrm{SB} + \frac{\mathrm{W}}{3}\right) \tag{11.10}$$

   where SB is the setback and W is the field width. For a typical rectangular field of width 180 ft and setback of 50 ft, Eq. (11.10) gives a minimum mounting height of 63 ft.

   An alternative way of stating the criterion in Eq. (11.10) is to require that the mounting height be such that, for a given aiming angle, the upper edge of the field angle is a minimum of 10° below the horizontal. This is illustrated in Fig. 11.13, for a NEMA Type 4 beam of 60° field angle. For this case, the minimum aiming angle is 40° below the horizontal.

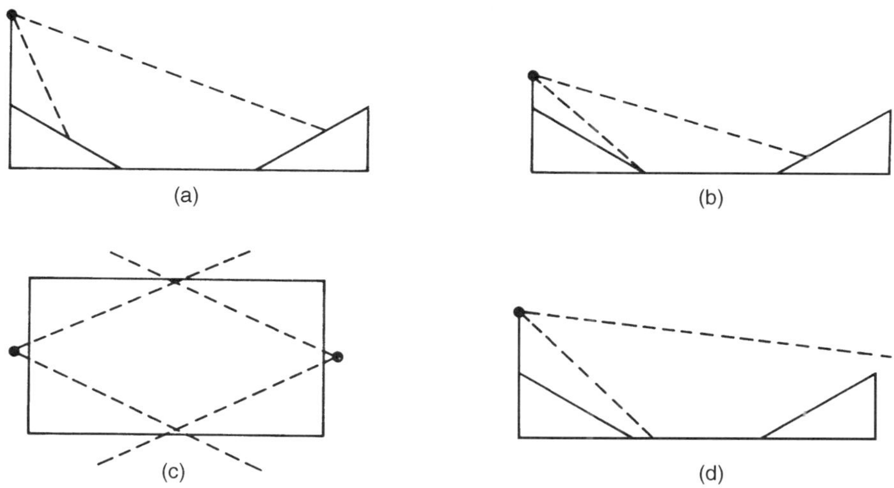

Figure 11.12 Factors affecting floodlight glare, (a) Field angle too great, (b) Mounting height too low, (c) Improper pole location, (d) Incorrect aiming.

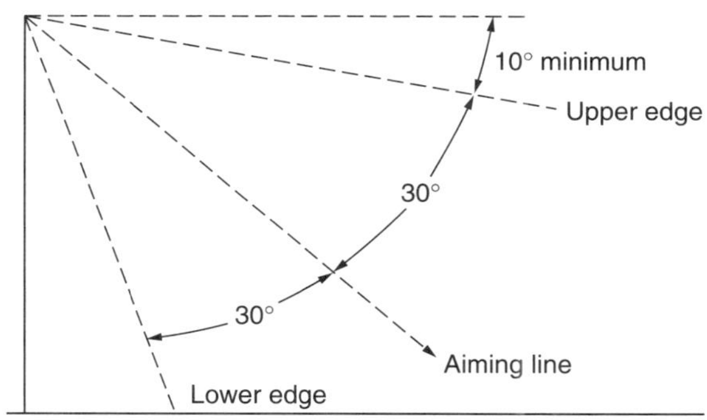

Figure 11.13 Field angle limitation—NEMA Type 4.

To compare the constraints in Eq. (11.10) and Fig. 11.13, consider again a field of width 180 ft and setback 50 ft. If the NEMA Type 4 floodlight in Fig. 11.13 were aimed 110 ft from the pole, the mounting height would need to be 92 ft (110 × tan 40). In general, Fig. 11.13 is a more demanding constraint than Eq. (11.10).

In conclusion, glare can occur if the mounting height is too low (Fig. 11.12b), if luminaires are in the normal line of sight of observers (Fig. 11.12c), or if the luminaire is aimed too high (Fig 11.12d).

c) In Sect. 11.3, luminaire spacing and mounting height were discussed in general in relation to aiming and coverage. In sports lighting, play may occur just as actively at the periphery of a playing area as at its center. Thus, it is important that there be adequate illuminance and uniform coverage in both regions.

Consider the rectangular playing area in Fig. 11.14, which could be used for football, soccer, lacrosse, or field hockey. Spacings and setbacks are given for typical four, six, and eight pole layouts. Note that as the setback distance is increased, fewer poles are required. Also note from Eq. (11.10), that as setback increases, mounting height must increase if glare is to be avoided.

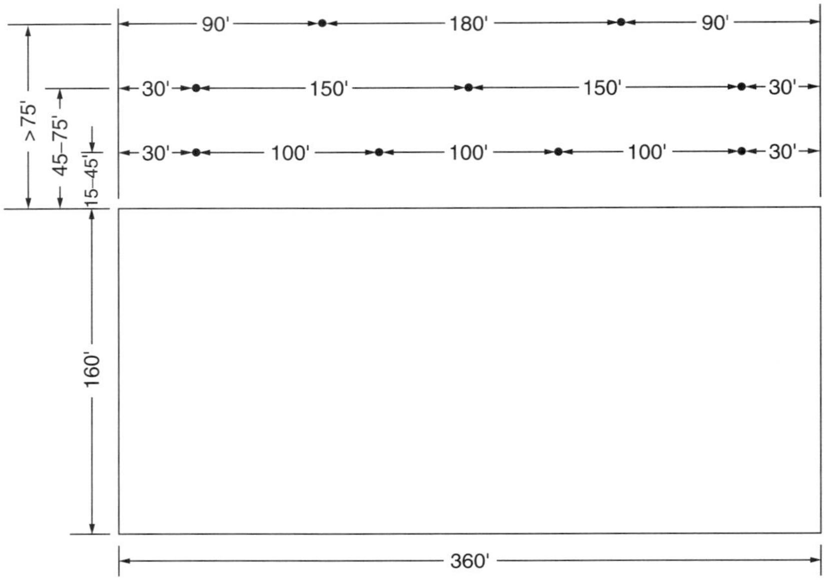

Figure 11.14 Spacings and setbacks for a rectangular playing field.

11.8 LARGE STADIA AND A NEIGHBORHOOD PLAYFIELD

The discussion of sports lighting will be concluded with two extremes: a review of the lighting of large baseball and football stadia and the design of the lighting for a community multi-sport recreational facility.

Sports played at night before thousands of spectators in large stadia have become a way of American life. The basic requirements for light sources for stadium lighting are that they permit good optical control, provide adequate color rendition, and operate economically. Prior to 1965, the only light source satisfying these three requirements well was incandescent. In particular, the 1000-W PS52 incandescent lamp was very widely used in outdoor sports lighting. An experiment in combined high-intensity discharge and incandescent lighting was tried in only one major U.S. stadium prior to 1965. This was Shea Stadium in New York, which was lighted with 953 clear 1000-W mercury lamps and 612 1500-W tungsten-halogen lamps in the early 1960s. Color rendering was satisfactory, but not good.

The advent and growth of color television sports broadcasting in the 1960s could not have occurred without two major developments. One was the introduction of the metal-halide lamp, particularly the 1000-W size, with its high efficacy, small source size, and good color-rendering properties. It made available 50% more lumens than its 1000-W mercury counterpart.

The second major development was that computer technology reached a point where it could be used to simplify the design of complex lighting systems. Computer technology provides the only economical (in money and time) method of precisely calculating aiming angles and light distributions.

The first major U.S. stadium to be relighted with 1000-W metal-halide lamps was the Municipal Stadium in Cleveland, Ohio, where 1318 incandescent units were replaced by 920 1000-W metal-halide units. The kilowatt load was decreased from 2306 to 1790 and the lumen output increased from 58 to 100 million.

It had long been customary, when using incandescent lamps for sports lighting, to operate them at 10% overvoltage. This gave 30% more lumens for 16% more watts, but reduced life to 20% of rated. This life amounted to about one playing season, after which the entire installation was relamped. The economics of this procedure were favorable. The same idea was used with mercury lamps. One-thousand watt lamps were operated on 1500-W ballasts, resulting in an increase of 50% in light output accompanied by a severe reduction in lamp life. The same idea was tried in the early 1000-W metal-halide installations. However, lamp life was too radically reduced. The 1000-W lamp was modified, but also a short-life 1500-W metal-halide lamp was introduced, and it has become the standard for outdoor stadium lighting. It currently gives up to 170,000 initial lumens and has a life of 3000 h. The first U.S. stadium to be relighted with the new lamp was Baltimore's Memorial Stadium, used for baseball. The 512 1500-W metal-halide units provided 500 fc (5500 lux) on the infield and 300 fc (3300 lux) on the outfield.

The lighting of Wrigley Field, home of the Chicago Cubs, is a very interesting sports lighting design, done by John P. Frier of the General Electric Co. Wrigley Field was the last major league baseball park to be lighted. Traditionalists were concerned that light poles would destroy the ambiance of the ivy-covered outfield walls. Thus, it was decreed that there would be no poles around the periphery of the outfield. Further, there was to be negligible light trespass on the adjacent neighborhood just outside the walls.

The solution was six luminaire towers on the roof, three along the right field line and three along the left field line. A narrow field angle was needed, so 546 NEMA Type 2 luminaires (18°–29°) were used, each containing a 1500-W metal halide lamp. Because there were no outfield poles, an elaborate criss-crossing computer-aiming program was developed. The result was an average maintained illuminance level of 370 fc (horizontal) and 200 fc (vertical) on the infield and 230 fc (horizontal) and 200 fc (vertical) on the outfield, with an average to minimum uniformity ratio of 1.4 on the infield and 2.6 on the outfield.

The neighborhood playfield, shown in Fig. 11.15*, is a multiuse area, typical of what is found in many small communities. It contains a soccer-football field 160 × 360 ft, a little league softball field with 60-ft basepaths, and a Babe Ruth baseball field with 90-ft basepaths. It is desired to light the area so that all of these sports may be played at night (not simultaneously!).

The first step in the design is to select appropriate illuminance levels. From Table 11.6, the caliber of activity is essentially Class IV. The levels chosen are 20 fc for the

---

* This example is adapted from a lighting design done by Donald L. Umbricht, Outdoor Lighting Specialist, GTE Lighting Products, Fall River, Mass, in the 1980s.

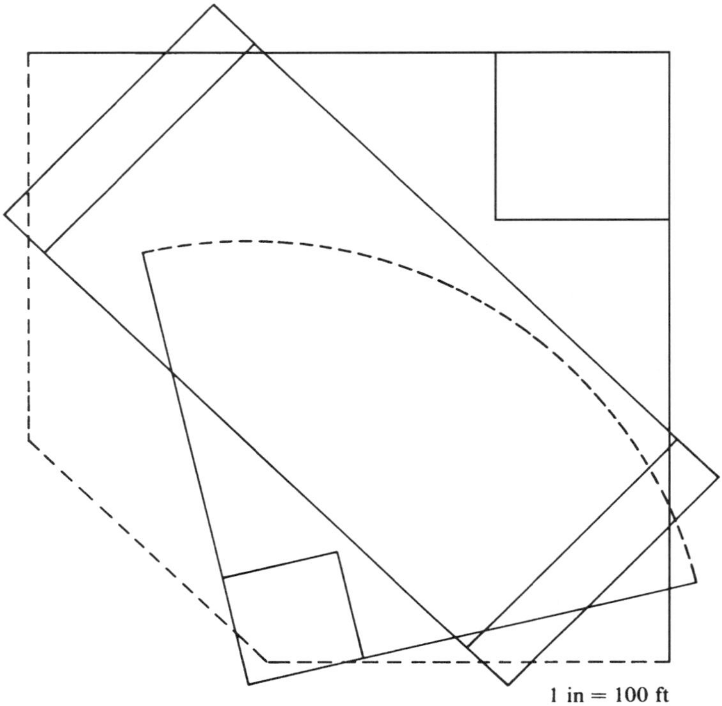

Figure 11.15 Neighborhood playfield area.

softball and baseball infields, 15 fc for the softball and baseball outfields, and 15 fc for the entire soccer field, average maintained.

The next step is to examine the standard layouts that have been developed for softball, baseball, and soccer fields (IES Sports and Recreational Areas Committee, 1969; Sylvania Outdoor Lighting, 1977). Two of these are reproduced in Fig. 11.16.

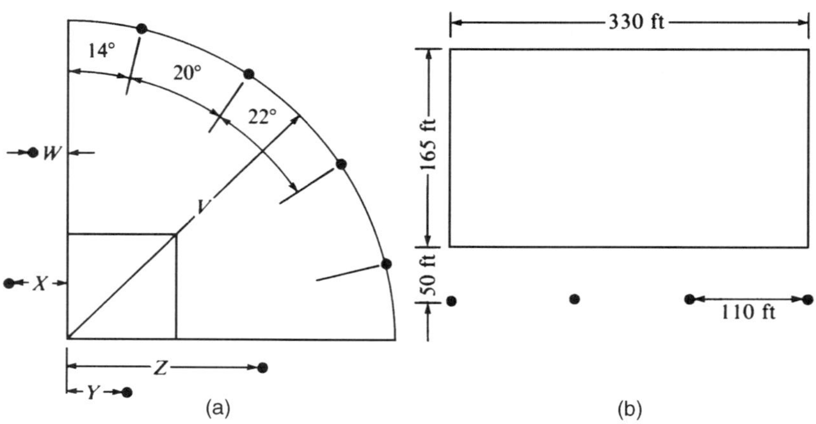

Figure 11.16 Standard baseball and soccer lighting layouts.

In the baseball layouts, eight poles are used, as indicated in Fig. 11.16a. For the 60-ft softball field, V = 185 ft, W = 20 to 30 ft, X = 30 to 50 ft, Y = 5 to 15 ft, and Z = 90 to 110 ft. For the 90-ft baseball field, these numbers are V = 400 ft, W = 30 to 60 ft, X = 40 to 80 ft, Y = 20 to 30 ft, and Z = 130 to 180 ft.

For the soccer layout, the constraint is setback distance, to insure that the poles do not interfere with baseball and softball. An 8-pole layout is chosen, with the pole locations as shown in Fig. 11.17. Setback distances of 112 ft and 75 ft were used for poles C2 and C3.

The choice of mounting height is next. For rectangular areas such as the soccer field, the guideline given in Eq. (11.10) yields

$$\text{MH} = 0.577\left(90 + \frac{160}{3}\right) = 82.7 \text{ ft}$$

The setback used in this calculation is an average setback of ≈90 ft for poles C2 and C3 in Fig. 11.17.

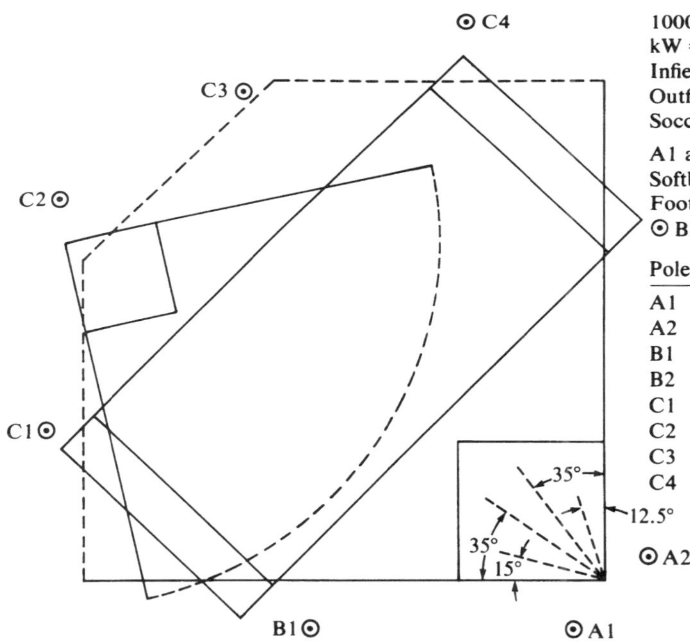

1000-W Clear MH
kW = 93.31, 75.95
Infields 20 fc
Outfields 15 fc
Soccer 15 fc

A1 and A2 Off for Softball and Football-Soccer

Pole	MH, ft	No.
A1	80	8
A2	80	8
B1	80	16
B2	80	16
C1	80	8
C2	80	12
C3	80	8
C4	80	10

Figure 11.17 Lighting layout for playfield.

For the softball and baseball fields, minimum mounting heights are given with the standard layouts (Sylvania Outdoor Lighting, 1977). The recommended minimum values are 40 to 50 ft for softball and 70 ft for baseball. Because the poles for softball are set back further than normal and because only one pole is behind the plate, the decision was to make all poles 80 ft high.

The next task is to choose a lamp and luminaire and to determine the number of luminaires needed per pole. A 1000-W metal-halide lamp was chosen (85 ballast watts), which (in the 1980s), provided 115,000 initial lumens, 92,000 mean lumens, and 10,000 h life.

TABLE 11.9 FLOODLIGHT SUMMARY FOR FIG. 11.17

Pole	Number of Units	NEMA Type	Kilowatts
A1	5	4	5.425
	3	6	3.255
A2	5	4	5.425
	3	6	3.255
B1	12	4	13.02
	4	6	4.34
B2	12	4	13.02
	4	6	4.34
C1	5	4	5.425
	3	6	3.255
C2	6	4	6.51
	6	6	6.51
C3	5	4	5.425
	3	6	3.255
C4	7	4	7.595
	3	6	3.255
	86		93.31

A general-purpose floodlight was chosen. The number of floodlights, their NEMA types, and the resulting kilowatt load are shown in Table 11.9. The total number of luminaires is 86, and they consume 93.31 kW, which is reduced to 75.95 kW when soccer or softball are played by turning off the units on poles A1 and A2.

11.9 ROADWAY AND AREA LUMINAIRES

Unlike floodlights, roadway and area luminaires are mounted horizontally and thus have fixed vertical aiming. The only flexibilities permitted are variation in mounting height and a rotation of the luminaire about its vertical axis, which is possible in some luminaires designed for installation on tall poles (high-mast luminaires).

Roadway luminaires have intensity distributions that are designed to light long narrow horizontal strips on one side of the luminaire, while minimizing the intensities on the other side of the luminaire. The intensity distributions up and down the narrow strip are generally the same. Any fixed-aimed luminaire that does not have this type of intensity distribution is called an area luminaire.

Roadways are classified in two ways, first according to vehicular traffic and second according to pedestrian activity. The former is affected by traffic speed, traffic volume, access points, and general complexity. There are five major categories:

1. Freeway

   A divided highway with no crossings at grade level. Such highways are subdivided into two types, A and B, with the former for urban areas and the latter for suburban/rural areas.

2. Expressway

   A divided highway with limited control of access and exchanges at major crossroads.

3. Major
   A roadway that is part of a system of principal through traffic in and out of a city.
4. Collector
   A roadway that connects major roadways and local streets within residential, commercial and industrial areas.
5. Local
   A street that provides direct access to residential, commercial, and industrial properties.

The second classification is concerned with vehicular/pedestrian interaction and is a strong function of pedestrian volume.

1. Commercial
   A municipal business area that is likely to have a heavy volume of nighttime pedestrians.
2. Intermediate
   Areas having retail stores, apartment complexes, and industrial buildings, with moderate nighttime pedestrian volume.
3. Residential
   Residential, or residential and small business, areas with few pedestrians at night.

There is a bit of jargon in connection with roadway and area luminaires that should be learned at the outset. Consider Fig. 11.18a. A roadway lighting unit is shown in a typical orientation with its pole set back from the curb and its luminaire overhanging the roadway. The point of zero vertical degrees is directly under the unit and labeled *nadir*. The line of zero lateral degrees extends directly in front of the luminaire, toward the street side. The line of 180 lateral degrees faces the house side, behind the luminaire.

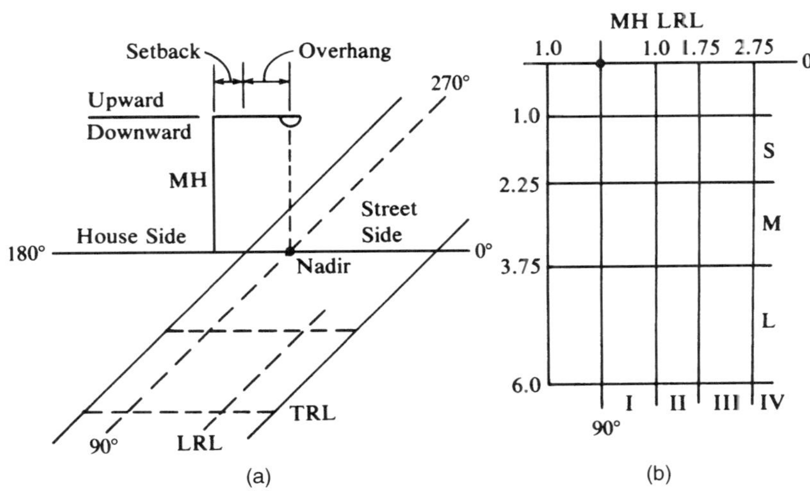

Figure 11.18 Roadway luminaire photometric nomenclature.

Transverse (TRL) and longitudinal (LRL) roadway lines are shown dashed in Fig. 11.18a, and are repeated on a more detailed grid pattern in Fig. 11.18b. These are given in terms of MH and have to do with three classifications of roadway luminaires established by the IESNA (IES Roadway Lighting Committee, 1993), based on a luminaire's vertical intensity distribution, lateral intensity distribution, and intensity distribution near 90° vertical angle.

The first is a vertical classification, determined by the location of the luminaire's maximum intensity. If this intensity falls between the 1.0 and 2.25 MH TRL lines, the vertical distribution is said to be *short* (S); between the 2.25 and 3.75 MH TRL lines, *medium* (M); and between the 3.75 and 6.0 MH TRL lines, *long* (L). The 1.0, 2.25, 3.75, and 6.0 MH TRL lines correspond to angles of incidence on the roadway of 45°, 66°, 75°, and 80°, respectively. Thus, if a luminaire has a medium distribution, its maximum intensity lies at a vertical angle between 66 and 75°. An example of such a luminaire is shown in Fig. 11.19.

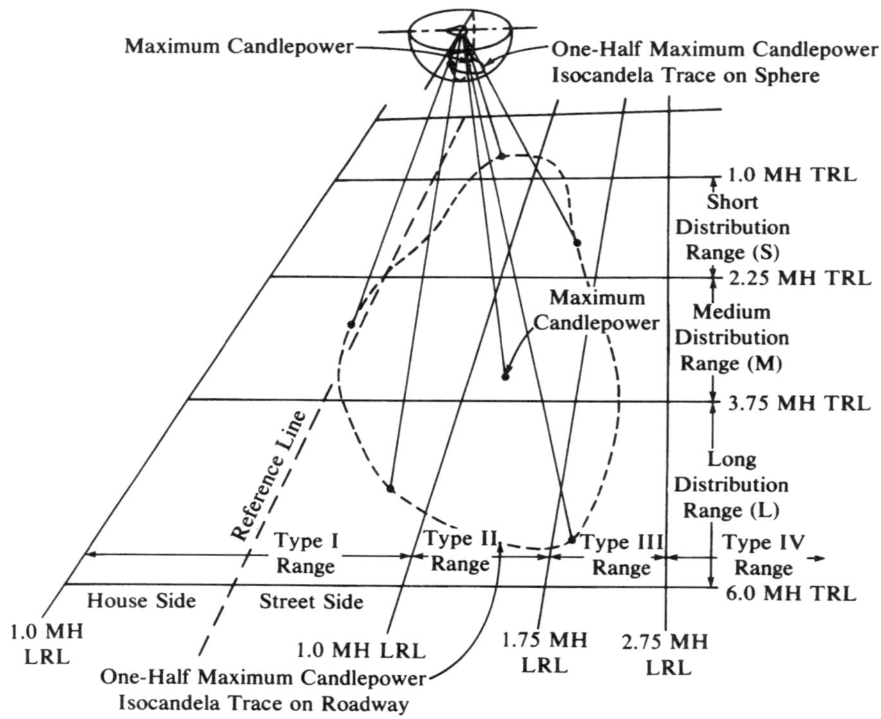

Figure 11.19 Roadway luminaire, medium, Type III. ***Source*: *IESNA Lighting Handbook*, 9th Edition, 2000. Courtesy of the Illuminating Engineering Society of North America.**

The second classification is a lateral classification and is a measure of the width of the beam. It is determined by the location on the roadway of an isocandela curve of one-half the maximum luminaire intensity. This is called the *one-half maximum candlepower isocandela trace* and four luminaire types are designated based on this trace, as shown in Fig. 11.18b.

Type I This type is for luminaires mounted near the center of the roadway, rather than near its edge. A luminaire is Type I if its one-half candela trace does not cross either of the 1.0 MH LRL lines while it is within the zone of maximum intensity (S, M, or L). A type I four-way luminaire has four such beams.

Type II This type and the two to follow are for luminaires mounted near the edge of a roadway. A luminaire is Type II if its one-half candela trace does not cross the 1.75 MH LRL line in the zone where the maximum intensity is located.

Type III A luminaire is Type III if its one-half candela trace crosses the 1.75 MH LRL line but does not cross the 2.75 MH LRL line in the zone where the maximum intensity is located. An example of a Type III luminaire is shown in Fig. 11.19.

Type IV A luminaire is Type IV if its one-half candela trace crosses the 2.75 MH LRL line in the zone where the maximum intensity is located.

There is a fifth intensity distribution (Type V) not shown in Fig. 11.18b. This is for luminaires mounted at the center of a four-way intersection and describes an intensity that is the same at all lateral angles.

The third classification of a roadway luminaire is based on its intensity distribution at large vertical angles and gives a measure of the potential for the luminaire to produce discomfort and/or disability glare. There are four luminaire cut off categorizations, which are given in Table 11.10.

TABLE 11.10 LUMINAIRE CUTOFF CATEGORIZATIONS

	max cd/1000 lm for $\theta > 90°$	max cd/1000 lm for $\theta = 90°$	max cd/1000 lm for $\theta = 80°$
Full Cutoff	0	0	100
Cutoff	—	25	100
Semicutoff	—	50	200
Noncutoff		No candela limitation	

Note: θ is the vertical angle above nadir.

With regard to Table 11.10, it should be noted that intensities in the 80–90° zone are likely to contribute to glare and light trespass on adjacent areas, whereas intensities above 90° will contribute to light pollution. In both instances light is wasted, resulting in higher energy costs per "useful lumen."

A typical photometric data sheet for a roadway luminaire is shown in Fig. 11.20. The luminaire is medium, Type III, semicutoff, and designed to house a high-pressure sodium lamp. In the general information box, the maximum intensity is given as 466 cd per 1000 lamp lumens and occurs at a vertical angle of 72.5°, placing it in the medium vertical distribution range. The intensities at vertical angles of 80° and 90° are 174 and 39 cd per 1000 lamp lumens, respectively, which is within the semicutoff range. The nadir candelas and footcandles, taken directly beneath the luminaire, are 111 and 0.123 per 1000 lm, respectively. The nadir footcandles are based on a 30-ft MH. Thus, the 0.123 is obtained by dividing 111 by 900.

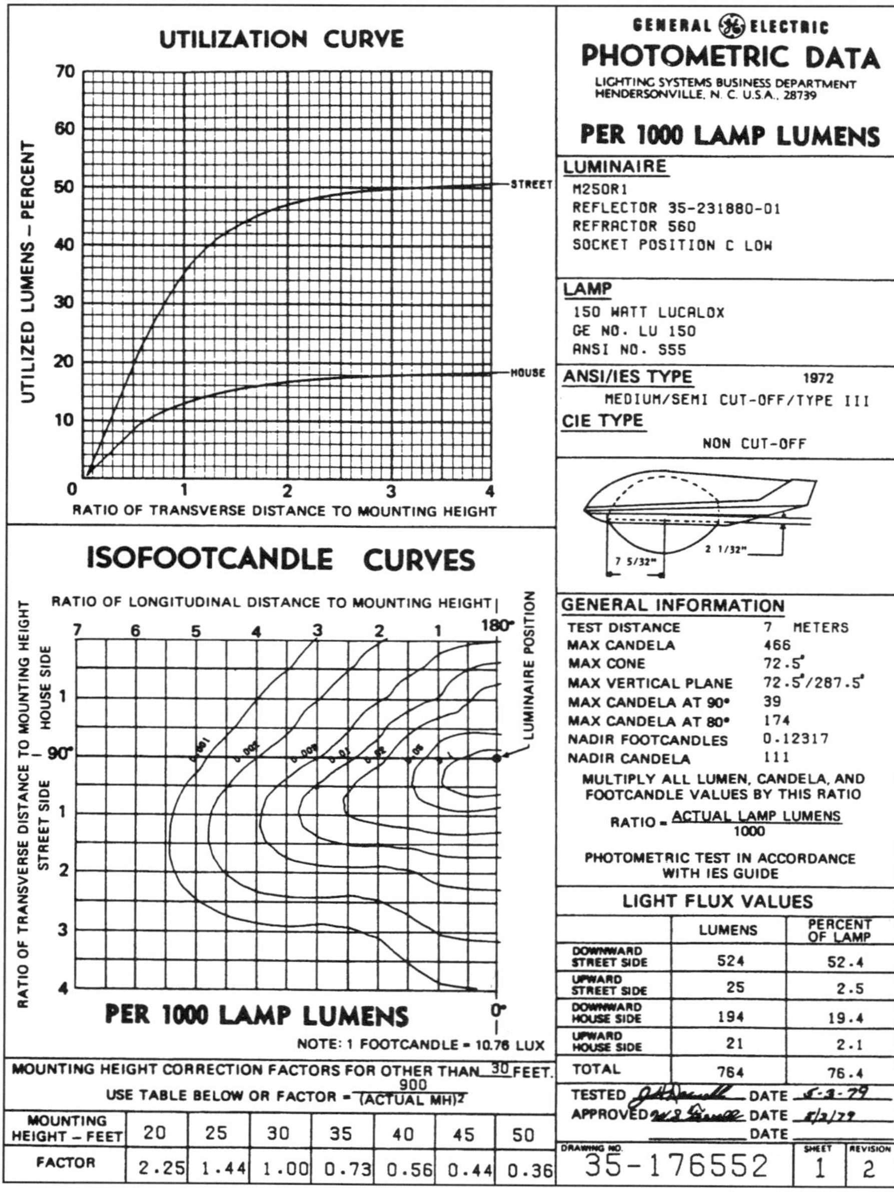

Figure 11.20 Roadway luminaire photometric data sheet. ***Source*****: "Lighting System Product Application Guide," Hendersonville, N.C., 1980. Courtesy of General Electric Co.**

In the light flux box, the lumen distribution of the luminaire is given on a per 1000 lamp lumen basis. 71.8% of the lamp lumens are directed downward, 52.4% to the street side, and 19.4% to the house side. 4.6% of the lamp lumens are directed upward. The efficiency of the luminaire is 76.4%.

The utilization curve for the luminaire appears in the left upper box. This is identical in form to the curves for a floodlight except that, instead of a curve for each aiming point, there is one curve for the street side and one for the house side. Again the issues of uniformity and utilization must be addressed. The discussion in Sect. 11.2 still applies except that, with a roadway luminaire, the aiming angle cannot be altered to improve either of these parameters. Rather, guidelines need to be set on the width of roadway that can be reasonably covered with a given luminaire.

From Fig. 11.20, on the street side, 35% of the lamp lumens impinge on the first MH of road width, 12% more on the second MH, 3% more on the third MH, and 1% more on the fourth MH. Saturation has certainly begun to occur by the 2-MH mark. Thus, as a general guideline, the mounting height of a roadway luminaire should be one-half the width of the street (minus the overhang if one exists). The only exception would be if a greater mounting height is needed to project lumens further up and down the street. This would improve uniformity while reducing utilization.

Isofootcandle curves on a per 1000 lamp lumen basis are shown in the lower left of Fig. 11.20. As in the case of floodlights, the values of illuminance are chosen such that a change of about 2:1 in illuminance is achieved between adjacent curves. A bulge in the illuminance curves occurs at a horizontal angle of 72.5° (also at the symmetrical angle of 287.5°). This is noted in the general information box under the listing "maximum vertical plane."

The isofootcandle values are based not only on 1000 lamp lumens, but also on 30 ft MH. Multiplying factors for other mounting heights are given at the bottom of Fig. 11.20.

11.10 ROADWAY LIGHTING DESIGN—ILLUMINANCE

There are three accepted IESNA metrics for quantifying roadway lighting. These are

- Illuminance—the average maintained horizontal illuminance in lx(fc) on the roadway surface and its uniformity (E_{avg}/E_{min}).
- Luminance—the average maintained road surface luminance in cd/m^2 (cd/ft^2) and its uniformity (L_{avg}/L_{min}).
- Small target visibility (STV)—the weighted average visibility level of an assemblage of targets on the roadway. Included are values of L_{avg} and the uniformity ratio L_{max}/L_{min}.

These procedures will be presented in this and the following two sections.

The average maintained horizontal illuminance from a roadway lighting installation is given by a formula very similar to that used in floodlighting design:

$$E_M = \frac{\phi_L(CU)(LLF)}{S(W)} \tag{11.11}$$

where ϕ_L = initial lamp lumens
CU = coefficient of utilization
LLF = light loss factor
W = width of the road (or area to be lighted)
S = spacing between luminaires

The coefficient of utilization is the ratio of the lumens falling on the area to be lighted to the total lamp lumens, both on an initial basis. CU values can be obtained directly from the utilization curve. For example, in Fig. 11.20 if the road width is 2 MH and there is no luminaire overhang, 47% of the lamp lumens reach the street and, thus, CU = 0.47.

Recommended illuminances and uniformity ratios for roadways are given in Table 11.11. In this table C, I, and R stand for the commercial, intermediate, and residential categorizations discussed earlier. R1 through R4 are roadway reflectance classifications and are described in Table 11.12. They are based on the specularity of the pavement and its lightness.

TABLE 11.11 RECOMMENDED AVERAGE MAINTAINED ROADWAY ILLUMINANCES AND UNIFORMITY RATIOS, ILLUMINANCES IN LUX (FOOTCANDLES)

Roadway Classification	Area Classification	Pavement Classification R1	R2, R3	R4	E_{avg}/E_{min}
Freeway	A	6(0.6)	9(0.8)	8(0.7)	3 to 1
	B	4(0.4)	6(0.6)	5(0.5)	
Expressway	C	10(0.9)	14(1.3)	13(1.2)	
	I	8(0.7)	12(1.1)	10(0.9)	3 to 1
	R	6(0.6)	9(0.8)	8(0.7)	
Major	C	12(1.1)	17(1.6)	15(1.4)	
	I	9(0.8)	13(1.2)	11(1.0)	3 to 1
	R	6(0.6)	9(0.8)	8(0.7)	
Collector	C	8(0.7)	12(1.1)	10(0.8)	
	I	6(0.6)	9(0.8)	8(0.7)	4 to 1
	R	4(0.4)	6(0.6)	5(0.5)	
Local	C	6(0.6)	9(0.8)	8(0.7)	
	I	5(0.5)	7(0.7)	6(0.6)	6 to 1
	R	3(0.3)	4(0.4)	4(0.4)	

Source: IESNA Lighting Handbook, 9th Edition, 2000. Courtesy of the Illuminating Engineering Society of North America.

TABLE 11.12 ROAD SURFACE CLASSIFICATIONS

Class	Description	Reflectance Characteristic
R1	• Portland cement, concrete road surface • Asphalt road surface with aggregate containing ≥15% artificial brightener	Mostly diffuse
R2	• Asphalt road surface with aggregate containing ≥60% gravel • Asphalt road surface with aggregate containing 10–15% artificial brightener	Mixed diffuse and specular
R3	• Asphalt road surface with dark aggregates providing rough texture	Slightly specular
R4	• Asphalt road surface with very smooth texture	Mostly specular

The usual situation is that the average maintained illuminance is prescribed by Table 11.11, and thus spacing and lamp size are the unknowns. To illustrate this, Eq. (11.11) can be rewritten as

$$S = \frac{CU(LLF)}{W(E_M)}(\phi_L) \qquad (11.12)$$

It is noted that if a larger lamp is used, spacing may be increased, thus saving poles and luminaires. There is a limit to this, however, for uniformity may deteriorate as spacing increases. It is desirable that the ratio of average to minimum illuminance not exceed 3:1, thus placing an upper bound on spacing. The IESNA (Rea, 2000) suggests that spacing be less than 4.5 MH for a short distribution luminaire, less than 7.5 MH for a medium distribution luminaire, and less than 12.0 MH for a long distribution luminaire. This will insure that the maximum intensity lines of two adjacent luminaires will at least meet on the roadway surface.

The spacing criteria apply for each of the three typical locations of pole placement along a roadway, namely, on one side, aligned on opposite sides, or staggered on opposite sides. Better uniformity can be achieved with units on both sides of the road, either aligned or staggered, but wiring costs are increased. If the road is wide, as in the case of major highways, poles on both sides are usually necessary to avoid excessive mounting heights.

For area lighting, S:MH ratios of 4.5:1 are recommended. To explain why this is so, refer to Fig. 11.21, in which a luminaire is shown mounted at the corner of a square area whose sides are each 4.5 MH long. The most difficult point to light in this situation is at the area's center (A), and thus the maximum intensity of the luminaire should be directed there. The vertical angle is 72.6°, which requires a luminaire with a Class M vertical distribution (66 to 75°). This is the class of luminaire generally used for area lighting. For the illuminance at A to equal that at C directly under the pole requires that the intensity toward A be 37 times as great as toward C. This is about the upper limit to be expected of a luminaire. At B, the vertical angle is 66.0°, and the intensity required is 15 times that to produce the same illuminance at C.

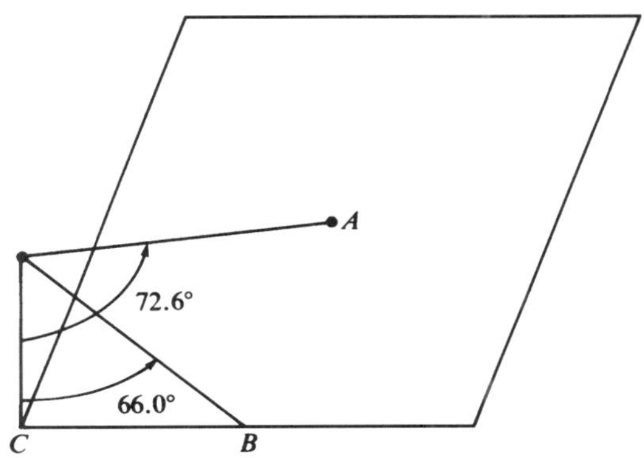

Figure 11.21 Pertaining to S:MH for area lighting. *Source*: Adapted from Industrial Lighting Systems, J.P. Frier and M.E.G. Frier, 1980. Courtesy of McGraw-Hill Inc.

EXAMPLE 11.5

For a given distance from a roadway or area lighting pole, what angle of incidence produces the maximum horizontal illuminance at a given point for a given intensity toward that point?

Solution. Using Eq. (2.6) and Fig. 2.5b, the horizontal illuminance at P is

$$E = \frac{I \cos^3 \beta}{h^2}$$

Noting that x = h tan β, E becomes

$$E = \frac{I \cos^3 \beta \tan^2 \beta}{x^2} = \frac{I \cos \beta \sin^2 \beta}{x^2}$$

Differentiating yields

$$\frac{dE}{d\beta} = \frac{I}{x^2}(-\sin^3 \beta + 2 \sin \beta \cos^2 \beta)$$

Equating to zero to find the maximum illuminance gives

$$-\sin^2 \beta + 2 \cos^2 \beta = 0$$

$$-\sin^2 \beta + 2(1 - \sin^2 \beta) = 0$$

$$\sin^2 \beta = 2/3$$

$$\beta = 54.74°$$

Inserting β into the illuminance equation yields

$$E = \frac{I}{x^2} \times 0.577 \times 0.667 = 0.385 \frac{I}{x^2}$$

Solving for I gives

$$I = 2.60\ E\ x^2 \tag{11.13}$$

For a given distance from the pole, Eq. (11.13) gives the lowest intensity to achieve a given light level, i.e., for 2 fc at 20 ft from the pole, the intensity would need to be 2080 cd and h would equal $\frac{x}{\tan \beta} = 14$ ft.

For roadway luminaires, the light loss factor is again composed of non-recoverable and recoverable factors. As discussed for sports lighting in Sect. 11.6, the major recoverable factors are LLD and LDD. Both metal halide and high-pressure sodium lamps are used in roadway and area lighting. Typical LLD values are 0.90 for high-pressure sodium at 50% life and, again, 0.82 for metal halide at 40% life.

In roadway lighting parlance, there is one non-recoverable light loss factor called equipment factor (EF). It includes some or all of the following:

- Voltage fluctuation
- Ballast and lamp variations
- Luminaire surface degradation
- Ambient temperature variation
- Changes in the physical surroundings.

The overall light loss factor is given by

$$\mathrm{LLF} = \mathrm{LLD} \times \mathrm{LDD} \times \mathrm{EF} \tag{11.14}$$

The IESNA offers no guidance on typical values of EF. Assuming a value of 0.90, typical light loss factors for roadway lighting are:

$$\mathrm{LLF} = 0.82 \times 0.92 \times 0.90 = 0.68 \text{ for metal halide}$$

$$\mathrm{LLF} = 0.90 \times 0.92 \times 0.90 = 0.75 \text{ for high pressure sodium.}$$

EXAMPLE 11.6

It is desired to light a major commercial artery 50 ft wide having an R4 pavement with the luminaire in Fig. 11.20 and a 30-ft. MH. The luminaires are to be mounted along one side of the street and are to overhang the curb by 5 ft.

Solution. From Table 11.11, the desired illuminance is 1.4 fc. It is 45 ft transversely from the luminaire to the far side of the street (1.5 MH). From the utilization curve in Fig. 11.20, the CU on the street side is 0.43. On the house side, the distance is $\frac{5}{30} = 0.17\,\mathrm{MH}$. The CU for this side is 0.03. Thus, the total CU is 0.46.

The LLF is composed of three parts. The assumption is a clean environment with units cleaned annually. From Table 11.8, LDD = 0.95. The luminaire in Fig. 11.20 uses a high-pressure sodium lamp. Assume for the moment that the lamp will be 250 W. Then from Table 6.17, LLD is 26,100/29,000 = 0.90. This is based on the lumens at 50% life, which, for the sodium lamp, is at about 12,000 h. Assume EF = 0.95. Thus, LLF = 0.81 from Eq. (11.14).

There is now sufficient data to use Eq. (11.12)

$$S = \frac{0.46(0.81)}{50(1.4)}(\phi_L) = 0.0053\phi_L$$

If the 250-W lamp is chosen,

$$S = 0.0053 \times 29{,}000 = 154 \text{ ft}$$

This is $\frac{154}{30} = 5.1$ MH, well within the spacing guideline for a medium vertical distribution (less than 7.5 MH).

As a cursory check on uniformity, the isofootcandle diagram in Fig. 11.20 is used in conjunction with the layout of the roadway in Fig. 11.22. The illuminance levels at three points, labeled A, B, and C in Fig. 11.22 are found and displayed in the chart in Table 11.13.

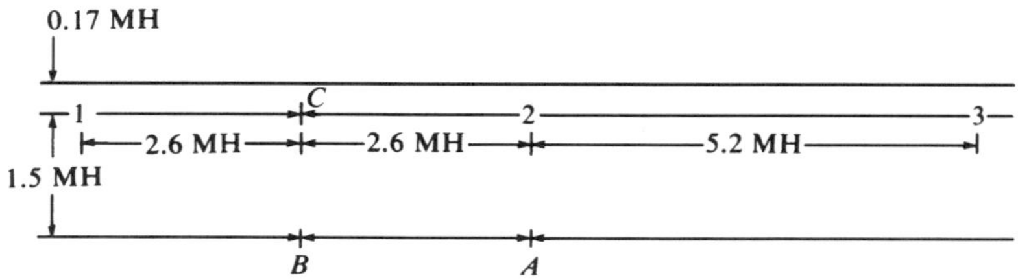

Figure 11.22 Checking roadway lighting uniformity.

TABLE 11.13

Luminaire	A	B	C
1	0.0013	0.01	0.009
2	0.015	0.01	0.009
3	0.0013	—	—
	0.018	0.02	0.018
$\times 29.0 \times 0.81$	0.42 fc	0.47 fc	0.42 fc

$$\frac{\text{Average}}{\text{Minimum}} = \frac{1.4}{0.42} = 3.3$$

The values for the A column are found as follows: For luminaire 1, proceed horizontally to the left 5.2 MH on the isofootcandle plot in Fig. 11.20. Then proceed vertically downward 1.5 MH. This locates a point between the 0.001 and 0.002 kilolumens isofootcandle curves. The value is estimated at 0.0013. The same value applies for luminaire 3. For luminaire 2, move vertically downward 1.5 MH, which locates the illuminance between the 0.01 and 0.02 isofootcandle curves. A value of 0.015 is estimated.

The isofootcandle values are for 1000 initial lamp lumens. They must be multiplied by the actual initial kilolumens times the light loss factor. The results are 0.42 fc, 0.47 fc, and 0.42 fc, respectively, for the A, B, and C locations. The ratio of average to minimum luminance is then found to be 3.3. This is slightly poorer than the guideline of 3:1. As a remedy, the spacing could be reduced slightly. Staggering the poles would not help in this situation because the illuminance at point A would be less with staggered poles. This is because, with poles 1 and 3 on the opposite side of the street in Fig. 11.22, their contribution to the illuminance at point A is reduced, since advantage is no longer taken of the bulge in the isofootcandle curves at 72.5° and 287.5° in Fig. 11.20.

A full assessment of the illuminances and uniformity on the roadway in Fig. 11.22 would require multi-point inverse square law calculations of illuminance by computer, using Eq. (2.6) multiplied by the total light loss factor.

11.11 AREA LIGHTING DESIGN—ILLUMINANCE

An example will be used to illustrate the general design procedure. An unusually-shaped hospital parking lot is shown in Fig. 11.23. Two $2^{1}/_{2}$-ft-high square pedestals spaced 178 ft apart are already present in the parking area, and the luminaire poles are to be placed on these. Since patient windows are on the front of the hospital building, it is desired to minimize the luminaire intensities in that direction.

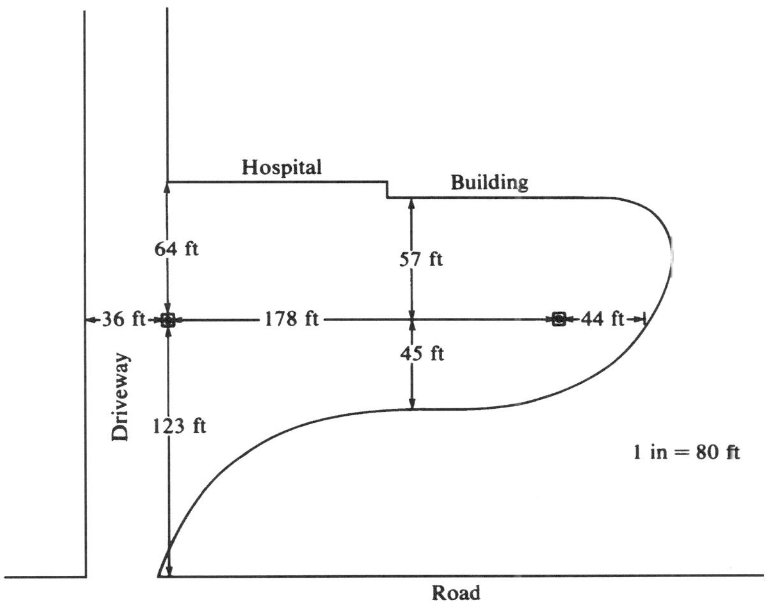

Figure 11.23 Hospital parking lot.

The roadway luminaire shown in Fig. 11.24 had been chosen previously by the hospital administrator; thus, the design was constrained to choosing a lamp, a mounting height, and an orientation of the luminaires on the poles. This is perhaps more often the case than not; the designer does not have complete freedom of choice, but rather is restricted by constraints of the area and by choices made by others prior to the designer arriving on the scene.

The first step is to select a tentative mounting height. Generally, as was discussed previously, spacing between luminaires should be less than 4.5 MH. Thus, with the 178 ft spacing predetermined, MH should be at least 40 ft. It was decided to use 40 ft poles that, with the $2^{1}/_{2}$ ft pedestal height, makes the mounting height 42.5 ft.

The next step is to select a lamp size. The IESNA recommended illuminance levels for open parking areas are given in Table 11.3. It was judged that the area being considered had a low level of activity and that the major issue was providing adequate horizontal illuminance for hospital employees and visitors to go safely to and from their cars. However, it was felt that some security should also be provided. A compromise was made to provide 0.5 fc, and a 400-W high-pressure sodium lamp was chosen tentatively.

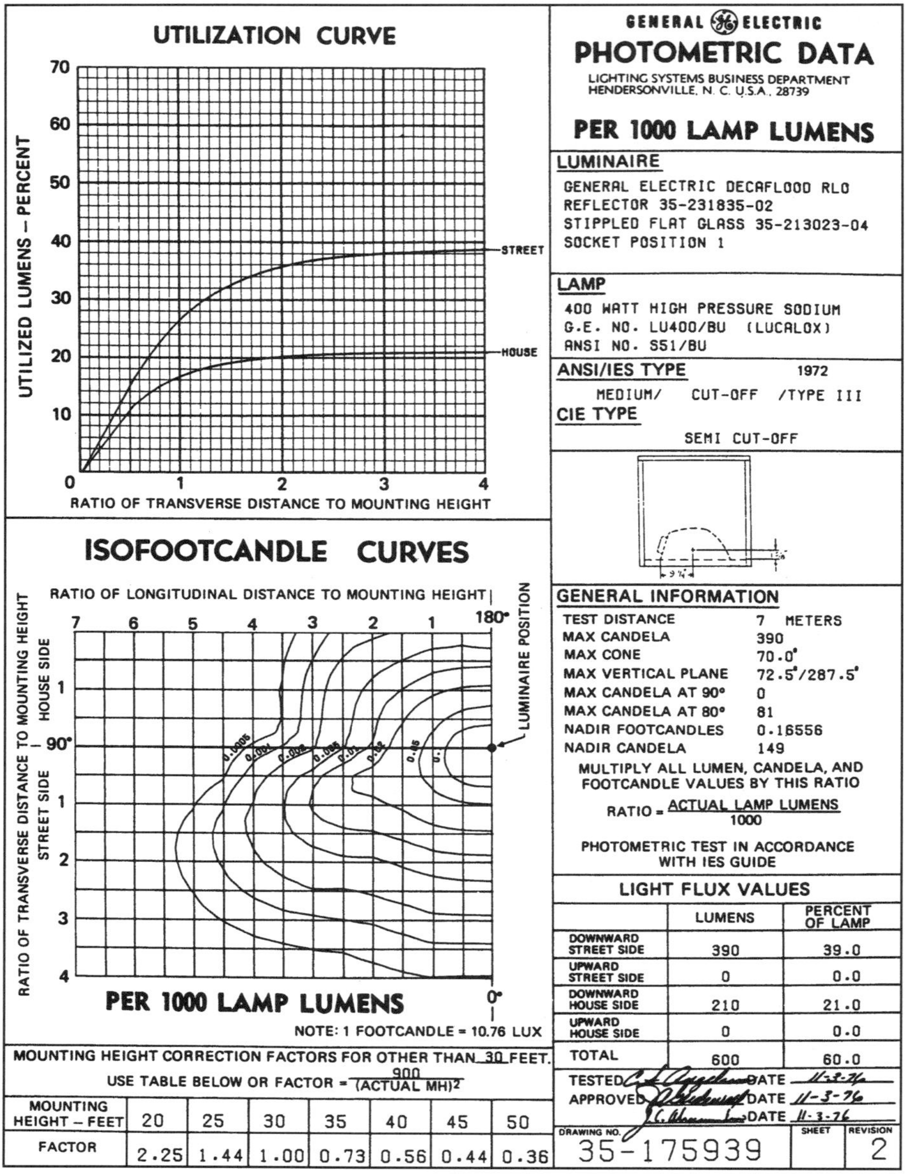

Figure 11.24 Luminaire for parking lot lighting. ***Source*****: "Lighting System Product Application Guide," Hendersonville, N.C., 1980. Courtesy of General Electric Co.**

From Table 6.17, the lamp emits 50,000 lm initially and 45,000 lumens (at 50% life). From Fig. 11.24, the initial illuminance directly under a pole, ignoring non-recoverable losses, is

$$E = \frac{149(50)}{(42.5)^2} = 4.1 \text{ fc}$$

In using the isofootcandle curves in Fig. 11.24, a multiplier of $50(30^2/42.5^2) = 25$ is required to adjust for mounting height and lamp lumens. Midway between the poles (2.25 MH), the initial illuminance from each luminaire is

$$E = 25(0.01) = 0.25 \text{ fc}$$

assuming the poles are mounted facing the street. Thus, it appears that the 400-W lamp will be sufficient to provide the desired illuminance.

In a similar fashion, the illuminances at various other locations produced by a single luminaire were calculated. These appear in Table 11.14. The parameter d is distance horizontally from the pole in feet. The data from Table 11.14 were used to prepare the template shown in Fig. 11.25. The template is drawn to the same scale as the layout of the parking area in Fig. 11.23.

TABLE 11.14 INITIAL HORIZONTAL ILLUMINANCES PRODUCED BY THE LUMINAIRE IN FIG. 11.24 (400-W HPS LAMP, 42.5 FT MH)

Location		Isofootcandles				
		2.5	1.25	0.5	0.25	0.125
Laterally (90°)	MH	0.80	1.2	1.75	2.10	2.5
	d	34	51	74	89	106
House side (180°)	MH	0.40	0.75	1.1	1.4	1.7
	d	17	32	47	60	72
Street side (0°)	MH	0.65	1	1.45	1.85	2.3
	d	28	43	62	79	98
45° Toward street	MH	0.78	1.13	1.77	2.26	2.83
	D	33	48	75	96	120
45° Toward house	MH	0.61	0.92	1.34	1.69	2.12
	d	21	39	57	72	90
17° Toward street	MH	—	—	2.4	3.05	3.5
	d	—	—	102	130	149
17° Toward house	MH	—	—	1.5	1.9	2.3
	d	—	—	64	81	98

Two of the templates shown in Fig. 11.25 should be made on transparent paper. These can then be overlaid on the parking area at the pole locations to determine the overall illuminance distribution. Because the curves have a bulge at 72.5° and 287.5°, it was decided to rotate the luminaire on the left pole 20° to the right and that on the right pole 20° to the left. This increases the illumination level between the poles and also on the driveway entering the lot. It has the disadvantages of spilling a bit more light onto the area adjacent to the right front of the lot. This turned out not to be a problem, other than the loss of utilization. The illuminances between the poles for both cases of luminaire orientation are given in Table 11.15. In that table, d is the horizontal distance from one pole toward the second pole, and MH is d given in mounting heights. It is seen that the uniformity ratio, taken as E_{max}/E_{min} is considerably improved when the luminaires are rotated by 20°.

The average illuminance throughout the area was not calculated, but it was judged that it would be adequate. In fact, it is possible that lamp size could have been

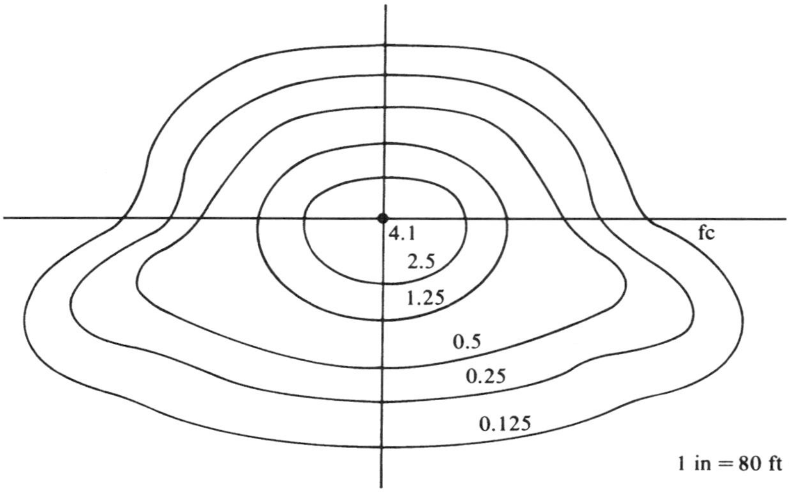

Figure 11.25 Isofootcandle curves (initial) for luminaire in Fig. 11.24. (400-W high-pressure sodium lamp, 42.5 ft MH.)

TABLE 11.15 ILLUMINANCES BETWEEN POLES

Luminaires Facing Street			Luminaires Rotated 20°		
MH, ft	d, ft	fc	MH, ft	d, ft	fc
0	0	4.1	0	0	4.1
0.8	34	2.5	0.85	36	2.8
1.2	51	1.3	1.30	55	1.6
1.75	74	0.6	2.30	98	1.3
2.1	89	0.5	3.25	138	2.8
2.5	106	0.6	3.70	157	3.3
3.0	127	1.3			
3.4	144	2.5			
$\frac{E_{max}}{E_{min}} = \frac{4.1}{0.5} = 8.2$			$\frac{E_{max}}{E_{min}} = \frac{4.1}{1.3} = 3.2$		

reduced to 250 W. However, this would have reduced all illuminance levels by 45%, and it was decided that this would produce a marginal installation.

Once again, contemporary computer programs can quickly perform the required multiple inverse square law iterations using Eq. (2.6). However, the visual aspects of the template approach are thought to be of value in understanding how the illuminance profile changes with different pole orientations.

11.12 ROADWAY LIGHTING DESIGN—LUMINANCE

For many years, roadway lighting design in North America was based on horizontal illuminance, as outlined in Sect. 11.10. However, it has been thought for some time that pavement luminance combined with the veiling luminance produced at the eye by stray light from roadway luminaires could provide better criteria than horizontal illuminance for estimating visibility of objects or features on or near a roadway.

In 1982, a new American Standard Practice for Roadway Lighting (IES Roadway Lighting Committee, 1983) was approved based on the luminance of roadway surfaces. However, because determination of pavement luminances and their subsequent use in roadway lighting design is a more difficult procedure than determining illuminances, and because use of illuminance criteria can result in adequate visibility if good design judgment is used, the Standard Practice has retained the illuminance method as an acceptable alternative procedure.

The measurement of roadway luminance is based on Figs. 11.26 and 11.27. Figure 11.26 depicts the geometry of the driver, the luminaire, and the roadway. The driver's eyes are located 1.45 m (about 5 ft) above the roadway with the line of sight 1° down from the horizontal. This places the driver 83 m (270 ft) from the point at which a luminance measurement is to be made.

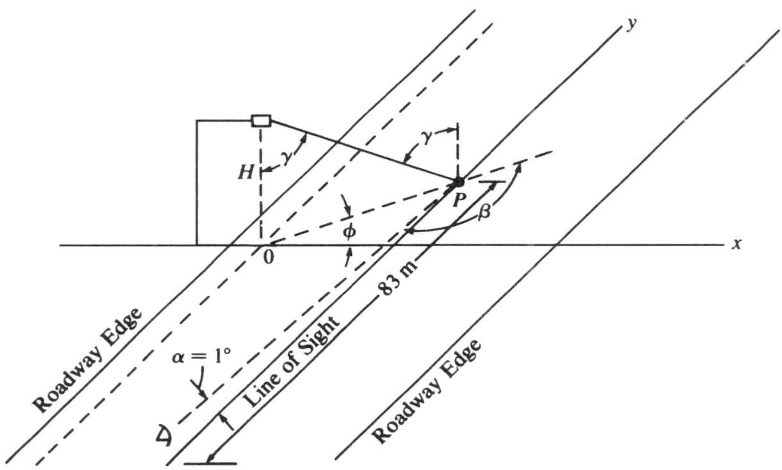

Figure 11.26 Roadway reflectance angles. ***Source*****:** ***IESNA Lighting Handbook*****, 9th Edition, 2000. Courtesy of the Illuminating Engineering Society of North America.**

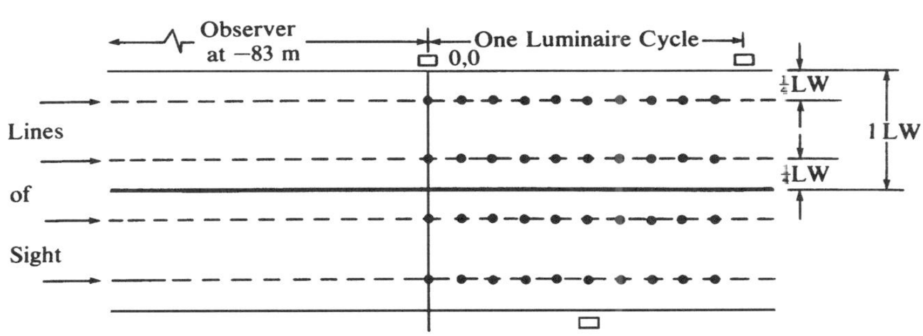

Figure 11.27 Roadway luminance geometry. ***Source*****:** ***IESNA Lighting Handbook*****, 9th Edition, 2000. Courtesy of the Illuminating Engineering Society of North America.**

As shown in Fig. 11.27, measurements are made along lines parallel to the edge of the roadway at the 1/4- and 3/4-lane-width (LW) points of each roadway lane. For each luminaire cycle, defined as the distance between luminaires on one side of the roadway, luminances are to be obtained at least 10 equally spaced points in each lane, with no more than 5 m between points. For each point, the observer is located 83 m to the left, as shown in Figs. 11.26 and 11.27. Figure 11.26 also shows the geometry for a typical luminaire, making a vertical angle γ and horizontal angle ϕ with the measurement point P. In calculating the luminance at P, at least one luminaire behind P and at least three luminaires ahead of P should be included.

The horizontal illuminance at P in Fig. 11.26 due to that single luminaire is

$$E_h = \frac{I(\gamma,\phi)\times\cos^3\gamma\times LLF}{H^2} \tag{11.15}$$

The luminance at point P in the direction of the observer is

$$L = \frac{I(\gamma,\phi)\times r(\beta,\gamma)\times LLF}{MF\times H^2} \tag{11.16}$$

where E_h = the horizontal illuminance provided by the single luminaire
L = the luminance at point P due to that luminaire in the direction of the observer
r = a reduced luminance coefficient (by definition). It is a measure of the reflectance of the pavement at point P in the β and γ angular directions for a specific angle α
MF = a multiplying factor used in r-tables to give large integer r values.

The total horizontal illuminance and the luminance are found by summing the contributions in Eqs. (11.15) and (11.16) from all luminaires.

Values of $r(\beta, \gamma)$ have been measured for the four standard road surfaces given in Table 11.12. One such table, for the R3 surface, is given in Table 11.16, for the case of $\alpha = 1°$, as indicated in Fig. 11.26. The metric r is a measure of bidirectional reflectance, defined as the ratio of the differential luminance of a ray of light from a point on a surface in a given direction to the differential illuminance incident on the surface at that point. From Eqs. (11.15) and (11.16),

$$r(\beta,\gamma) = \frac{L}{E_h}\times MF\times\cos^3\gamma \tag{11.17}$$

where L/E_h is the bidirectional reflectance and MF = 10,000. Thus $r(\beta, \gamma)$ is essentially bidirectional reflectance reduced by the cosine cubed of the incidence angle.

Assume now that the designer has determined the roadway and pavement classifications and has assessed the environment through which the roadway will pass. The next step is to consult Table 11.17. In that table, recommended luminance values and ratios are given for the various road and area classifications, where, as in Table 11.11, C, I, and R mean commercial, intermediate, and residential, respectively. In the first data column, the desired average maintained luminance of the roadway surface is shown. This is the average of the luminances calculated or measured at the points

TABLE 11.16 r-VALUES FOR SURFACE R3

tan γ \ β	0	2	5	10	15	20	25	30	35	40	45	60	75	90	120	150	180
0	294	294	294	294	294	294	294	294	294	294	294	294	294	294	294	294	294
0.25	326	326	321	321	317	312	308	308	303	298	294	280	271	262	253	244	240
0.5	344	344	339	339	326	317	308	298	289	276	262	235	217	204	199	199	194
0.75	357	353	353	339	321	303	285	267	244	222	204	176	158	149	149	136	140
1	362	362	352	326	276	249	226	204	181	158	140	118	104	100	100	100	100
1.25	357	357	348	298	244	208	176	154	136	118	104	83	73	70	74	77	78
1.5	353	348	326	267	217	176	145	117	100	86	78	72	60	57	60	60	62
1.75	359	335	303	231	172	127	104	89	79	70	62	51	45	44	46	45	47
2	326	321	280	190	136	100	82	71	62	54	48	39	34	34	35	36	38
2.5	289	280	222	127	86	65	54	44	38	34	25	23	22	23	24	24	25
3	253	235	163	85	53	38	31	25	23	20	18	15	15	14	15	16	17
3.5	217	194	122	60	35	25	22	19	16	15	13	9.9	9.0	9.0	11	12	13
4	190	163	90	43	26	20	16	14	12	9.9	9.0	7.4	7.0	7.1	8.3	9.0	9.9
5	145	109	60	24	16	12	9.0	8.2	7.7	6.8	6.1	4.3	3.2	3.3	4.3	6.5	7.1
6	113	77	36	15	11	9.0	8.0	6.5	5.1								
8	83	47	17	6.1	4.4	3.6	3.1										
10	65	32	9.0	3.3	2.4	2.0											
12	53	22	5.6	2.1	1.8												

Source: IESNA Lighting Handbook, 9th Edition, 2000. Courtesy of the Illuminating Engineering Society of North America.

TABLE 11.17 MAINTAINED LUMINANCE CRITERIA FOR ROADWAYS

Roadway Classification	Area Classification	L_{avg} (cd/m²)	$\frac{L_{avg}}{L_{min}}$	$\frac{L_{max}}{L_{avg}}$	$\frac{L_v}{L_{avg}}$
Freeway	C	0.6	3.5	6	0.3
	I	0.4	3.5	6	0.3
	R	—	—	—	—
Expressway	C	1.0	3	5	0.3
	I	0.8	3	5	0.3
	R	0.6	3.5	6	0.3
Major	C	1.2	3	5	0.3
	I	0.9	3	5	0.3
	R	0.6	3.5	6	0.3
Collector	C	0.8	3	5	0.4
	I	0.6	3.5	6	0.4
	R	0.4	4	8	0.4
Local	C	0.6	6	10	0.4
	I	0.5	6	10	0.4
	R	0.3	6	10	0.4

shown in Fig. 11.27. The second and third data columns are design guidelines so that reasonable uniformity will be achieved. The fourth column is a measure of disability glare. L_v is the veiling luminance at the observer's eyes in candelas per square meter and may be calculated from

$$L_v = \frac{10E_v}{\theta^2 + 1.5\theta} \tag{11.18}$$

where E_v = the vertical illuminance in the plane of the driver's pupil in lux
θ = the angle in degrees between the line of sight and the line from the observer to the luminaire. The line of sight is a horizontal line 1.45 m above the roadway, parallel to the curb and along one of the dashed lines in Fig. 11.27.

The luminance in Eq. (11.18) is for one luminaire. A summation over the n luminaires must be performed to obtain the overall veiling luminance. Then the ratio of L_v to L_{avg} can be calculated and compared with the guidelines in Table 11.17.

The next step in the design process is to obtain the recommended average maintained illuminance from Table 11.11. Then tentative selections of a lamp, a luminaire, and a mounting height are made. From the luminaire utilization curve, the coefficient of utilization is determined, and a light loss factor is calculated from the mean lumen data of the lamp and from Table 11.8.

With these quantities in hand, it is possible to calculate the pole spacing from Eq. (11.12). Then a tentative layout is drawn, tying down such factors as overhang, setback, stagger, etc. For each layout selected, horizontal illuminances should be calculated and compared with Table 11.11, and luminances should be calculated and compared with Table 11.17. The design should be iterated if the values or their ratios do not fulfill the criteria in these tables.

11.13 ROADWAY LIGHTING DESIGN—SMALL TARGET VISIBILITY

The third, and most recent, method of prescribing roadway lighting is through small target visibility (STV). In this procedure, the relative weighted average visibility level (RWVL) is determined in one direction of traffic flow for each of a group of targets spaced along the roadway.

The reason for including a new procedure is the recognition that the two existing procedures fail to provide illuminance and luminance patterns that aid in the recognition of low contrast objects. It is hoped that the utilization of the STV criterion will increase nighttime safety and driver performance through increased visibility.

The STV procedure pays considerable attention to the geometry of the lighting situation—source, driver, and object. If geometry is held constant, the luminance and illuminance levels on the roadway may be raised or lowered, but contrasts will likely stay the same. It is contrast that determines visibility, and geometry that determines contrast. STV addresses this issue directly.

The visibility of an object on the roadway depends on

- the luminance of the object
- the luminance of the object's immediate background
- the adaptation of the observer to the adjacent surroundings
- the veiling luminance at the eye from the luminaires
- the size of the object
- the time available for seeing
- the age of the observer.

It is important to realize how object contrast is manifested on a roadway. There are three possible contrast scenarios:

Alternating Contrast—

In this case, the luminaire intensity distribution is such as to provide high object luminance when pavement luminance is low, and vice-versa. Thus, the pavement luminance is made nonuniform intentionally, suggesting that luminaires be rather widely-spaced, and located opposite, rather than staggered. It is this model that is used to produce the highest STV.

All Positive Contrast—

In this case, the vertical object luminance is always greater than the horizontal pavement luminance, with the pavement luminance nearly uniform. This situation is used for one-way roads.

All Negative Contrast—

In this case, also predominantly for one-way roads, pavement luminance is again kept nearly uniform, and low levels of vertical illumination are provided on the object.

The targets used in the STV procedure are 18 cm × 18 cm flat, perfectly diffusing surfaces with a reflectance of 0.50. Target location is the same as for roadway luminance calculations, as shown in Figs. 11.26 and 11.27. Thus the first target is located

83 m from the observer, who is 60 years old with normal eyesight and whose eyes are 1.45 m above the roadway and looking straight ahead and downward at an angle of 1°. Subsequent targets are as shown in Fig. 11.27. Along each roadway lane line, targets should be located at 1/10 of the spacing between luminaires or 5 m, whichever is the lesser. Ten targets per roadway lane line should be observed.

For each target location, target luminance (L_t) is calculated at the center of the target as follows:

$$E_t = \frac{I(\gamma,\phi)\cos(90-\phi)\sin\gamma \times LLF}{d^2} \tag{11.19}$$

where d is the distance from the luminaire to the target. It is noted that

$$d \approx \frac{H-0.5T}{\cos\gamma} \tag{11.20}$$

where T is target height. Inserting Eq. (11.20) into Eq. (11.19) yields

$$E_t = \frac{I(\gamma,\phi)\cos(90-\phi)\sin\gamma\cos^2\gamma \times LLF}{(H-0.5T)^2} \tag{11.21}$$

The luminance of the target is its exitance divided by π.

$$L_t = \frac{I(\gamma,\phi)\cos(90-\phi)\sin\gamma\cos^2\gamma \times 0.5 \times LLF}{\pi(H-0.5T)^2} \tag{11.22}$$

Next the veiling luminance L_v is calculated from Eq. (11.18), and two background luminances are also calculated. L_{b1} is found at a point on the pavement next to the center of the bottom of the target. L_{b2} is found at a point on the pavement 11.77 m beyond the target, at the intersection on the pavement of a line from the observer through the center of the top of the target. Equation (11.16) is used to calculate these background luminances. Then

$$L_b = \frac{L_{b1} + L_{b2}}{2} \tag{11.23}$$

The next step is to determine adaptation luminance (L_a) of the surroundings, the log of L_a and the visual angle (VA) subtended by the target at the observer.

$$L_a = L_b + L_v \quad LL_a = \text{Log } L_a \tag{11.24}$$

$$\left.\begin{aligned} VA &= \tan^{-1}\frac{\text{target height}}{\text{distance to observer}} \\ &= 60\tan^{-1}\frac{0.18}{83} = 7.45 \text{ minutes} \end{aligned}\right\} \tag{11.25}$$

A quantity called "the sensitivity of the visual system," involving two parameters, F and L, is now determined. These depend on L_a, and there are three levels of L_a to consider, each producing a different F and L.

$$\left.\begin{aligned} & L_a \geq 0.6 \\ & F = [\mathrm{Log}(4.2841\, L_a^{0.1556}) + 0.1684\, L_a^{0.5867}]^2 \\ & L = (0.05946\, L_a^{0.466})^2 \end{aligned}\right\} \tag{11.26}$$

$$\left.\begin{aligned} & 0.00418 < L_a < 0.6 \\ & F = 10^{2(0.0866\, LL_a^2 + 0.3372\, LL_a - 0.072)} \\ & L = 10^{2(0.319\, LL_a - 1.256)} \end{aligned}\right\} \tag{11.27}$$

$$\left.\begin{aligned} & L_a \leq 0.00418 \\ & F = 10^{(0.346\, LL_a + 0.056)} \\ & L = 10^{(0.0454\, LL_a^2 + 1.055\, LL_a - 1.782)} \end{aligned}\right\} \tag{11.28}$$

With F and L in hand, a preliminary base visibility level appropriate for the adaptation luminance present can be calculated. This base level is called DL_1, given by

$$DL_1 = 2.6\left[\frac{\sqrt{F}}{VA} + \sqrt{L}\right] \tag{11.29}$$

DL_1 depends on the two sensitivity functions and the visual angle subtended by the object.

Succeeding steps in the procedure modify DL_1 as a function of time for seeing, observer age, and negative contrast. First a number of intermediate parameters are calculated:

$$\left.\begin{aligned} & B = \mathrm{Log}\ VA + 0.523 \\ & C = LL_a + 6 \\ & AA = 0.360 - \frac{0.0972\, B^2}{B^2 - 2.513\, B + 2.789} \\ & AL = 0.355 - \frac{0.1217\, C^2}{C^2 - 10.40\, C + 55.28} \\ & AZ = \frac{\sqrt{(AA)^2 + (AL)^2}}{2.1} \end{aligned}\right\} \tag{11.30}$$

First, DL is corrected for the length of time of observation (T) (usually 0.2 seconds)

$$DL_2 = DL_1 \frac{AZ + T}{T} \tag{11.31}$$

Next is a correction for the age of the observer (A) (usually assumed to be 60).

$$\left.\begin{aligned} &A \le 64, \quad FA = \frac{(A-19)^2}{2160} + 0.99 \\ &A > 64, \quad FA = \frac{(A-56.5)^2}{116.3} + 1.43 \\ &DL_3 = DL_2 \times FA \end{aligned}\right\} \tag{11.32}$$

The next step is to find the negative contrast adjustment factor (FCP). First, a quantity M is calculated, whose value depends on LL_a.

$$\left.\begin{aligned} &-2.4 < LL_a < -1, \quad M = 10^{-10^{-[0.075\,(LL_a+1)^2+0.245]}} \\ &LL_a \ge -1, \quad M = 10^{-10^{-[0.125\,(LL_a+1)^2+0.245]}} \end{aligned}\right\} \tag{11.33}$$

Then FCP is found from

$$\left.\begin{aligned} &FCP = 1 - \frac{M(VA)^{-0.6L_a}}{2.4\,DL_1\left(\dfrac{AZ+2}{2}\right)} \\ &\text{For } LL_a \le -2.4, \quad FCP = 0.5 \end{aligned}\right\} \tag{11.34}$$

If the target has negative contrast (target darker than background), the final correction to DL is

$$DL_4 = DL_3 \times FCP \tag{11.35}$$

VL and RWVL can now be calculated.

$$VL = \frac{L_t - L_b}{DL_4} \tag{11.36}$$

$$RWVL = 10^{-0.1\,VL} \tag{11.37}$$

VL in Eq. (11.36) is very similar to the Blackwell visibility levels in Sect. 3.14. As discussed in connection with Fig. 3.18, the VL levels are simply contrast multipliers. Thus, the ratio of $L_t - L_b$ to DL_4 in Eq. (11.36) is a measure of how much the contrast of the target differs from the base level. RWVL is employed because it has been found that larger VL values should be given lesser weight than smaller ones in determining the average VL.

After all RWVL values are calculated, an average RWVL is obtained

$$ARWVL = \frac{\sum_1^n RWVL}{n} \tag{11.38}$$

and STV is found from

$$STV = -10 \log (ARWVL) \tag{11.39}$$

Values of STV are displayed in Table 11.18.

TABLE 11.18 RECOMMENDED STV VALUES

Roadway Classification	Area Classification	STV	Pavement Luminance	
			L_{max}/L_{min}	L_{avg} (cd/m²)
Freeway	C	3.2	6.0	0.5
	I	2.6	6.0	0.4
	R	—	—	—
Expressway	C	3.8	6.0	0.5
	I	—	—	—
	R	—	—	—
Major	C	4.9	6.0	1.0
	I	4.0	6.0	0.8
	R	3.2	6.0	0.8
Collector	C	3.8	6.0	0.6
	I	3.2	6.0	0.5
	R	2.7	6.0	0.4
Local	C	2.7	10.0	0.5
	I	2.2	10.0	0.4
	R	1.6	10.0	0.3

REFERENCES

Armstrong, J.: "A New Measure of Uniformity of Lighting Installations," *JIES*, Summer, 1990.

Frier, J., and M. Frier: *Industrial Lighting Systems*, McGraw-Hill, New York, 1980.

General Electric Company: "Lighting System Product Application Guide," Hendersonville, N.C., 1980.

General Electric Company: "Building Floodlighting," *Lighting Application Bulletin*, Nela Park, Cleveland, Ohio, 1989.

IESNA Roadway Lighting Committee: *Recommended Lighting for Walkways and Class 1 Bikeways, DG-5,* Illuminating Engineering Society of North America, New York, 1994.

IESNA Roadway Lighting Committee: *Lighting for Parking Facilities, RP-20*, Illuminating Engineering Society of North America, New York, 1998.

IESNA Outdoor Environment Lighting Committee: *Lighting for Exterior Environments, RP-33*, Illuminating Engineering Society of North America, New York, 1999.

IESNA Roadway Lighting Committee: *Roadway Lighting, RP-8*, Illuminating Engineering Society of North America, New York, 1999.

IESNA Sports and Recreational Areas Committee: *Sports and Recreational Area Lighting, RP-6*, Illuminating Engineering Society of North America, New York, 2001.

Janoff, M.: "Subjective Ratings of Visibility and Alternative Measures of Roadway Lighting," *JIES*, 1989.

Janoff, M.: "The Relationship Between Visibility and Traffic Accidents," *JIES*, 1998.

Janoff, M.: "The Effect of Visibility on Driver Performance; A Dynamic Experiment," *JIES*, 1990.

Loch, C.: "Driver's View of the Roadway," *JIES*, January 1982.

Lowell, J., and L. Maloney: "Illuminance Selection Procedure for Sports Lighting," *Lighting Design and Application*, May 1983.

Murdoch, J.: *Illumination Engineering*, Macmillan, New York, 1985.

NEMA: "White Paper on Outdoor Lighting Issues and Quality Lighting," Luminaire Section, National Electric Manufacturers Association, Rosslyn, VA, 2000.

Rea, M., editor: *The IESNA Lighting Handbook, Ninth Edition*, Illuminating Engineering Society of North America, New York, 2000.

Roberts, J., editor: *Advanced Lighting Guidelines*, New Buildings Institute, White Salmon, WA, 2001.

Sylvania Lighting Center: "Outdoor Lighting," GTE Sylvania, Danvers, Mass. 1977.

Sylvania Outdoor Lighting: "Sports Lighting Made Easy," GTE Sylvania, Fall River, Mass., 1977.

Chapter 12

Daylighting

12.1 BACKGROUND AND INTRODUCTION

Prior to the 1973 Arab oil embargo, interest in daylighting in the United States was at a low level. With energy inexpensive, the electric lighting industry in this country was quick to point out the disadvantages of daylighting—glare, variability, difficulty of control, excessive illuminances. Although there was a modest amount of analytically-based work going on in daylighting, the major use of daylight was by architects who allowed it to spill into interior spaces to create visual effects, rather than to provide quality lighting for doing visual work.

All of that has of course changed in the intervening years. There has been a resurgence of interest in admitting daylight into interior spaces, not just to enhance those spaces psychologically and aesthetically, but to provide quality lighting at the appropriate quantitative level for good seeing. In fact, it has come to be realized that properly designed daylighting can often do a better job than electric lighting in minimizing veiling reflections and glare. However, it must unfortunately be acknowledged that the major reason for the renaissance in daylighting is economic, the energy cost reduction arising from being able to turn some or all of the electric lighting off in an interior space for a significant portion of the day. Be that as it may, daylighting today is an important aspect of interior lighting and gaining each year in interest and application among interior designers and architects.

What does daylighting provide? It gives the occupant of a space a connection with nature and the outdoors, a view in the distance. It provides a dynamic lighting environment (this can be a disadvantage if the luminance changes are too severe and rapid, as on a partly cloudy day with fast cloud motion). Daylighting can result in significant energy savings if the electric lighting is reduced or extinguished when illuminances reach certain levels. In the warm months there is a double savings—the electrical energy of the turned-off lights and the reduced air conditioning load to remove the heat generated by those lights. The spectral power of daylight is nearly uniform throughout the visible spectrum, providing relatively cool color temperatures and excellent color rendering. While there is an abundance of daylight, the challenge is to harness it effectively to produce quality lighting in a space.

This chapter will concentrate on the technical (quantitative) aspects of daylighting. However the architectural (qualitative) aspects will be discussed also, particularly as they interweave with the technical aspects. For additional excellent discussions of architectural daylighting, the reader is referred to the literature (Evans, 1981; Moore, 1985; Copping, 1987; Gordon, 1995; Michel, 1996; Loe, 1998; Tregenza, 1998).

The chapter's flow is to admit light from the sun and sky into interior spaces, guide it to points of usage, evaluate its quantity and quality there, and assess the potential for energy savings.

First, information on the luminances and illuminances to be expected from the sun and sky is assembled. Conventionally, two sky conditions have been considered, namely overcast and clear, and these will be addressed. A third sky condition, the partly cloudy sky, is of interest, but is difficult to quantify accurately.

With the sun and sky data in hand, the designer can predict the external illuminance—vertical, horizontal, or sloped, on the fenestration (window, skylight, etc.). This may require inclusion of not only direct radiation from the sun and sky, but also reflected radiation from the ground and adjacent structures, such as roofs and sides of buildings.

The third step in the process is to ascertain the transmission characteristics of the fenestration material. Generally, there are two types of transmittances to consider: direct transmittance of sunlight and diffuse transmittance of clear or overcast skylight. The transmittances of the fenestration for these two types of input may indeed be different, and the former is likely to depend on solar altitude.

Next, fenestration light controls are considered, including skylight wells, translucent glazing, overhangs, light shelves, and louvers. Each of these affects the distribution and quality of the lighting. Very often the designer will create a scale model of a space to determine the effects of these controls.

The fifth step is to process the luminous flux that enters the interior space. Three procedures will be presented: the lumen method for skylighting; the lumen method for sidelighting; and the daylight factor method. The first of these permits calculation of the average horizontal illuminance on the work plane and is very similar to the lumen method of electric lighting design presented in Chap. 9. The second procedure yields the horizontal illuminance at five selected work-plane reference points on a line perpendicular to the window wall at its center. The third procedure permits calculation of the illuminance at any work-plane point within the space as a fraction of the horizontal illuminance outdoors at the same instant of time from an unobstructed overcast sky (in the absence of direct sunlight).

Finally, a sample energy analysis is presented to ascertain the potential cost savings from reducing the level of electric lighting, or turning it off, during a portion of the day.

12.2 THE SUN

The sun is our source of daylight. Although a relatively small star, it nevertheless has a huge reservoir of energy. Its visible surface is 864,000 miles in diameter (diameter of the photosphere). It has a volume 1.3×10^6 as great as the earth and its mass is 0.3×10^6 that of the earth. It is 92.9 million miles from earth and subtends an angle of 2.53°. The sun's average surface temperature is 5750 K. It emits radiant energy at the rate of 6.25 kW/cm^2. Its relatively high temperature gives it a luminous efficacy of 94.2 lm/W.

The earth receives only one part in 10^9 of the total energy radiated by the sun. This amounts to 1350 W/m^2 outside the earth's atmosphere and 1200 W/m^2 at sea level on a clear day, giving a yearly energy intake of over 10^{18} kWh.

The solar radiation on a horizontal surface at sea level on a clear day with the sun at zenith is shown in Fig. 12.1. About 40% of this received power is in the visible region of the spectrum and 55% is in the infrared. The peak of the sunlight radiation curve outside the atmosphere is at 475 nm. The peak is shifted slightly to about 500 nm for zenith solar radiation at sea level. Curves for late afternoon and early morning would show considerably greater shift. This, of course, is due to selective absorption of solar radiation by the earth's atmosphere and accounts for the orange appearance of the sun early and late in the day. An additional effect is that the angle of incidence of the sun's rays is much greater early and late in the day than it is at noon. This reduces the incident power density at all wavelengths.

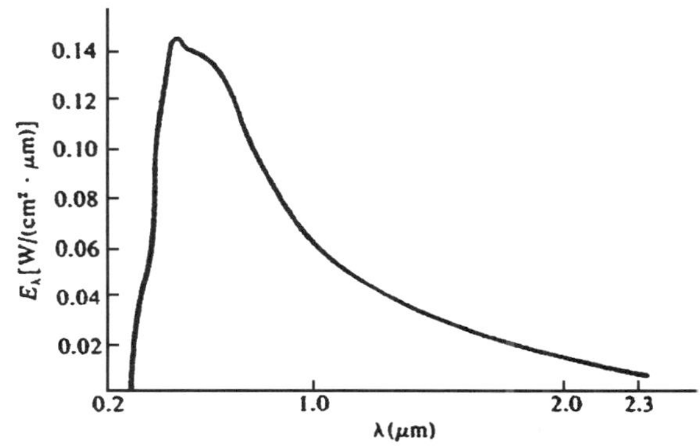

Figure 12.1 Horizontal solar radiation at sea level on a clear day with the sun at zenith. *Source*: Adapted from P. Moon, *The Scientific Basis of Illuminating Engineering*, 1936. Courtesy of McGraw-Hill Inc. and Dover Publications Inc.

EXAMPLE 12.1

Calculate the illuminance at sea level from the sun at zenith. Repeat for a full moon at zenith.

Solution. The luminance of the sun at zenith at sea level is (Rea, 2000)

$$L_s = 1600 \times 10^6 \text{ cd/m}^2 \ (149 \times 10^6 \text{ cd/ft}^2)$$

Its diameter is 8.64×10^5 mi (45.62×10^8 ft), and its distance from earth is 92.9×10^6 mi (49.05×10^{10} ft). Thus:

$$\text{Projected area} = \frac{\pi \times (45.62)^2 \times 10^{16}}{4} = 16.35 \times 10^{18} \text{ ft}^2$$

$$\text{Intensity} = 149 \times 10^6 \times 16.35 \times 10^{18} = 24.35 \times 10^{26} \text{ cd}$$

$$E = \frac{24.35 \times 10^{26}}{(49.05)^2 \times 10^{20}} = 10{,}100 \text{ fc}$$

The moon has a diameter of 2160 mi (11.4×10^6 ft) and is 239,000 mi (12.6×10^8 ft) from earth. The luminance of a full moon at zenith at sea level is (Rea, 2000)

$$L_m = 2500 \text{ cd/m}^2 \text{ (232 cd/ft}^2)$$

$$\text{Projected area} = \frac{\pi \times (11.4)^2 \times 10^{12}}{4} = 10.21 \times 10^{13} \text{ ft}^2$$

$$E = \frac{23.69 \times 10^{15}}{(12.6)^2 \times 10^{16}} = 0.015 \text{ fc}$$

12.3 SOLAR ILLUMINANCE

The earth receives roughly one part in a billion of emitted solar power. The average irradiance (radiant power density) on a surface perpendicular to the sun's rays and just outside the earth's atmosphere is

$$E_s = \int_o^{\infty} E_{s\lambda} d_{\lambda} = 1353 \text{ W/m}^2 \text{ (125.7 W/ft}^2) \tag{12.1}$$

where $E_{s\lambda}$ is solar spectral irradiance in watts per unit of surface area per micron, and λ is in microns. Values of the integral are taken from IES RP-16-96. E_s is known as the "solar constant".

The average illuminance on a surface perpendicular to the sun's rays and lying just outside the earth's atmosphere can be calculated from

$$E_{sc} = k \int_{0.38}^{0.78} E_{s\lambda} V(\lambda) d\lambda = 127.5 \text{ klx (11,850 fc)} \tag{12.2}$$

where $V(\lambda)$ is the spectral luminous efficiency of the eye, and k is the maximum luminous efficacy (683 lm/W). E_{sc} is called the "solar illuminance constant". Note the similarity of Eq. (12.2) to Eq. (4.21).

Dividing the result in Eq. (12.2) by that in Eq. (12.1) gives a solar efficacy outside the earth's atmosphere of 94.2 lm/W. This amounts to 0.1 fc for each $\mu W/cm^2$.

The quantity E_{sc} is an average illuminance. It must be adjusted for a given date to account for the slightly elliptical shape of the earth's orbit around the sun. The actual illuminance on any day of the year is given by

$$E_{xt} = E_{sc} \left[1 + 0.034 \cos \frac{360}{365} (J - 2) \right] \tag{12.3}$$

where J is the Julian day number (1-365).

E_{xt} is the solar illuminance outside the earth's atmosphere on day J on a surface perpendicular to the sun's rays. Note that the maximum variation of E_{xt} from E_{sc} is ±3.4%. To obtain the solar illuminance E_{dn} at sea level on the same day on a similarly-oriented surface requires accounting for attenuation through the earth's atmosphere. Atmospheric attenuation is a function of the composition of the atmosphere and the length of the path traversed by the sun's rays. Thus, E_{dn} can be expressed as

$$E_{dn} = E_{xt}\varepsilon^{-cm} \tag{12.4}$$

where c is the optical atmospheric extinction coefficient, and m is the relative optical air mass. This latter quantity is the ratio of the mass of atmosphere in the actual earth-sun path to the mass when the sun is directly overhead at sea level, and is expressed by

$$m = \frac{1}{\sin a_t} \tag{12.5}$$

where a_t is the solar altitude, the angle of elevation of the sun above the horizon.

The atmospheric extinction coefficient has been measured at several locations throughout the world with a variety of results. Consensus values appear to be c = 0.21 for a clear sky and c = 0.80 for a partly cloudy sky. Thus, when the sun is directly overhead, from Eq. (12.4) it can be expected that 11,850 × 0.81 = 9600 fc will be obtained from the sun at sea level on a clear day and 5300 fc on a partly cloudy sky.

Solar illuminance on a perpendicular surface is seldom of interest. Rather it is of more interest to know the illuminances provided by sunlight on horizontal and vertical surfaces (skylights and windows). To obtain these latter illuminances requires that the sun's location in the sky be determined for a given time, date, and location on the earth in terms of its altitude (vertical) angle a_t and its azimuth (horizontal) angle a_s.

These two angles, and others of relevance, are shown in Fig. 12.2. In that figure, the sun is located in the southeast, and a vertical window faces (arbitrarily) east. The angles are named in Table 12.1. a_t and a_i have been explained. a_s is the angle between south and the projection of the sun on the horizontal plane. South is zero azimuth, west of south is positive, and east of south is negative. North is ±180° azimuth. a_p is called the profile angle. It is the angle in the plane perpendicular to the window between the horizontal and the sun's projection on the plane. It is very useful in determining the shadowing effect of overhangs. a_e is the angle between the normal to the window and south, in this case −90°. a_z is a_s-a_e, which is the angle between the solar azimuth and the window azimuth.

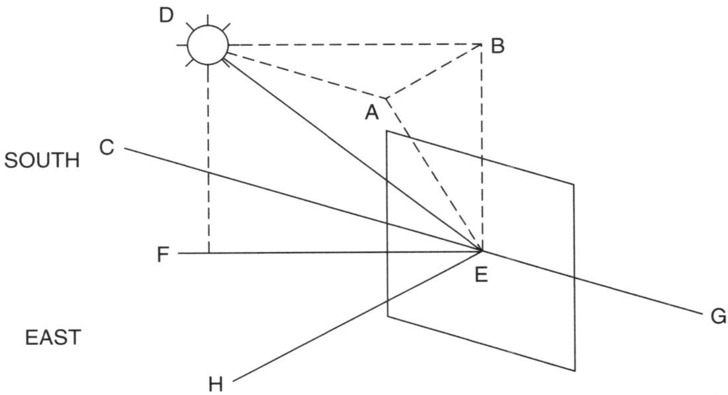

Figure 12.2 Daylighting angles.

TABLE 12.1 DAYLIGHTING ANGLES

Angle	Name	Symbol
DEF	solar altitude	a_t
DEH	angle of incidence	a_i
CEF	solar azimuth	a_s
AEH	profile angle	a_p
CEH	window azimuth	a_e
FEH	solar-window azimuth	$a_z = a_s - a_e$

Relations between the various angles are, from Fig. 12.2

$$\cos a_i = \cos a_t \cos a_z \tag{12.6}$$

$$\tan a_p = \frac{\tan a_t}{\cos a_z} \tag{12.7}$$

The horizontal illuminance and the illuminance on a vertical surface are given by:

$$E_h = E_{dn} \sin a_t \tag{12.8}$$

$$E_v = E_{dn} \cos a_i = E_{dn} \cos a_t \cos a_z \tag{12.9}$$

EXAMPLE 12.2

In Fig. 12.2, let $a_t = 45°$, $a_e = -90°$, and $a_s = -30°$. Find a_z, a_i, and a_p. Find E_{dn} and then find E_h and E_v.

Solution.

$$a_z = a_s - a_e = 60°, \quad a_i = \cos^{-1}(\cos 45 \cos 60) = 69.3°, \quad a_p = \tan^{-1}\left(\frac{\tan 45}{\cos 60}\right) = 63.4°.$$

From Eq. (12.4) and neglecting the correction in Eq. (12.3),

$$E_{dn} = 11{,}850 e^{-\frac{.21}{\sin 45}} = 8805 \text{ fc}$$

From Eqs. (12.8) and (12.9),

$$E_h = 8805 \sin 45 = 6225 \text{ fc}$$

$$E_v = 8805 \cos 69.3 = 3110 \text{ fc}$$

Solar altitude and azimuth can be expressed in terms of the local latitude L, the solar declination δ, and the local solar time ST. The angles are shown in Fig. 12.3, where a celestial sphere has been placed around and centered at an observer's point on the earth's surface. The celestial equator is in the plane of the earth's equator, and the celestial poles are at the ends of the earth's axis of rotation. The local latitude L is shown as the angle

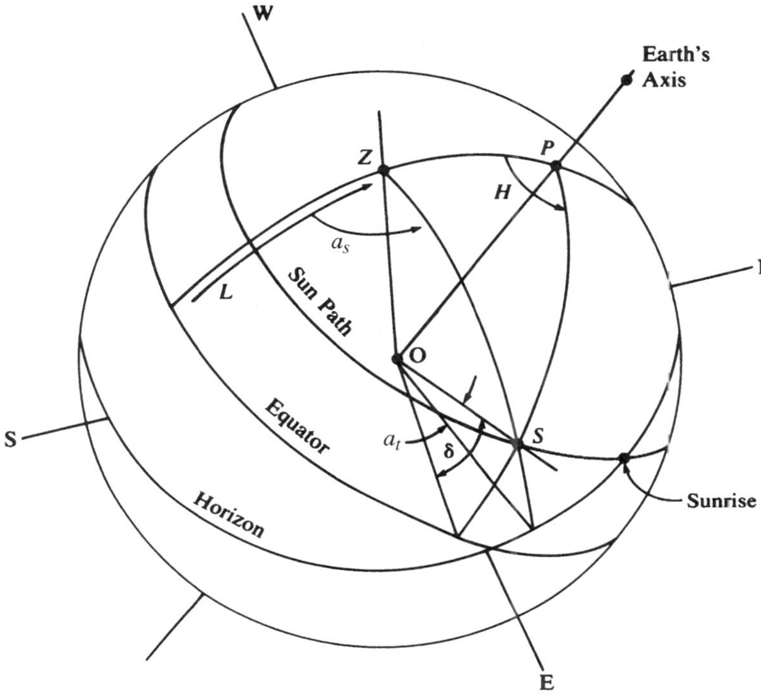

Figure 12.3 Pertaining to solar altitude and azimuth.

between the plane of the equator and the observer's zenith. It is also the angle between the earth's axis of rotation and the observer's north horizon line. Solar declination δ is the angular distance of the sun north of the celestial equator, the complement of the angle between the earth-sun line and the earth's axis of rotation. This angle arises because the earth's axis is tilted at an angle of 23.45° with respect to the normal to its orbital plane. Each year in northern latitudes, δ is +23.45° around June 21, −23.45° around December 21, and 0° around March 21 and September 21. On any date, δ may be obtained approximately from

$$\delta = 23.45 \sin \frac{360}{368}(\mathrm{J} - 81) \tag{12.10}$$

For example, on March 21, J is 80 and Eq. (12.10) gives a declination of −0.4°.

Because an observer's location is generally not at the center of a time zone, and because the earth's orbit around the sun is slightly elliptical and thus its orbital velocity varies slightly throughout the year, solar time, as would be measured by a sun dial, varies slightly from local clock time. There are two corrections to make in going from the observer's clock time to solar time. The first is a longitude time correction for the observer's location and is given by

$$\Delta \mathrm{T} = \frac{1}{15}(\mathrm{SM} - \mathrm{L}) \tag{12.11}$$

where SM is the longitude of the center of the time zone (standard meridian), L is the longitude of the observer (both in degrees), and ΔT is the time difference in hours between the center of the time zone and the observer's location. There are 24 times zones, so each time zone is 15° wide.

The second correction, which addresses the variation due to the elliptical orbit, is given by the equation of time, which is

$$ET = 0.164 \sin \frac{720}{365.25}(J - 81.6) - 0.127 \sin \frac{360}{365.25}(J - 2.5) \quad (12.12)$$

where ET is in hours. The equation of time yields a maximum variation of about ±17 min (±0.265h) during the course of a year, with the sun running behind local time by about 17 min in February and ahead of local time by about 17 min in October and November.

With the two corrections defined,

$$ST = TZ + ET + \Delta T \quad (12.13)$$

where ST is solar time and TZ is time zone time. To equate ST to a solar-time angle H in degrees, ST is multiplied by 15, since 1 hr of time corresponds to 15 degrees of angle. Thus

$$H = 15\ ST \quad (12.14)$$

EXAMPLE 12.3

Find the solar declination δ and the solar-time angle H at 8:30 a.m. Central Standard time on October 21 at 32 degrees north latitude and 94 degrees west longitude.

Solution. Tables 12.1 and 12.2 will be of assistance in solving this and similar problems. From Table 12.1, October 21 is Julian day 294. Thus, from Eq. (12.10)

$$\delta = 23.45 \sin \frac{360(213)}{368} = 23.45 \sin 208° = -11.1°$$

TABLE 12.1

Month	Days in month	J
January	31	i
February	28	31 + i
March	31	59 + i
April	30	90 + i
May	31	120 + i
June	30	151 + i
July	31	181 + i
August	31	212 + i
September	30	243 + i
October	31	273 + i
November	30	304 + i
December	31	334 + i

TABLE 12.2

Zone	Standard Meridian West of Greenwich
Atlantic	60°
Eastern	75°
Central	90°
Mountain	105°
Pacific	120°
Yukon	135°
Alaska-Hawaii	150°
Bering	165°

From Table 12.2, the middle of the Central time zone is at 90 degrees west longitude. From Eq. (12.11)

$$\Delta T = \frac{90-95}{15} = -0.333 \text{ hr}$$

From Eq. (12.12)

$$\begin{aligned} ET &= 0.164 \times 0.854 - 0.127 \times -0.955 \\ &= 0.258 \text{ hr} \end{aligned}$$

Then, from Eq. (12.13),

$$ST = 8.50 + 0.26 - 0.33 = 8.45 \text{ am (8:27 am).}$$

and from Eq. (12.14),

$$H = 15 \times 8.45 = 126.7°$$

With L, δ, and H known, Fig. 12.3 can be utilized to obtain expressions for solar altitude a_t and azimuth a_s. The desired relationships can be obtained by applying the cosine and sine laws of spherical trigonometry to the spherical triangle ZPS in Fig. 12.3. The resulting equations are

$$\sin a_t = \sin L \sin \delta - \cos L \cos \delta \cos H \qquad (12.15)$$

$$\tan a_s = \frac{\cos \delta \sin H}{\cos L \sin \delta + \sin L \cos \delta \cos H} \qquad (12.16)$$

For four particular situations, solar altitude may be calculated very easily to give guidance in design. These are at noon on the two solstices and the two equinoxes.

The appropriate equations are:

$$\left.\begin{array}{ll} \text{June 21} & a_t = 90 - (L - 23.45) \\ \text{March 21 and} & \\ \text{September 21} & a_t = 90 - L \\ \text{December 21} & a_t = 90 - (L + 23.45) \end{array}\right\} \qquad (12.17)$$

EXAMPLE 12.4

Find the solar altitude and azimuth for the date, time, and location in Ex. 12.3.

Solution. Applying Eqs. (12.15) and (12.16) directly gives

$$\sin a_t = \sin 32 \sin(-11.1) - \cos 32 \cos(-11.1) \cos 126.7$$

$$= 0.530 \times (-0.193) - 0.848 \times 0.981 \times (-0.598) = 0.395$$

$$a_t = 23.30°$$

$$\tan a_s = \frac{\cos(-11.1) \sin 126.7}{\cos 32 \sin(-11.1) + \sin 32 \cos(-11.1) \cos 126.7}$$

$$= \frac{0.981 \times 0.802}{0.848 \times (-0.193) + 0.530 \times 0.981 \times (-0.598)} = -1.658$$

$$a_s = -58.90 \text{ (east of south)}$$

With a means now available of calculating solar altitude and azimuth, the original problem of determining the illuminances on horizontal and vertical surfaces produced by the sun can be addressed. The equations to use are Eq. (12.8) and (12.9), in conjunction with Fig. 12.4.

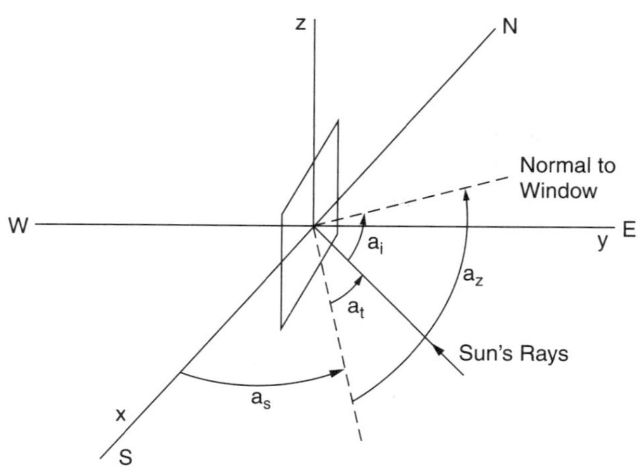

Figure 12.4 Angles involved in calculating E_h and E_v.

EXAMPLE 12.5

For the date, time, and location of Example 12.3, find the solar illuminance on a horizontal skylight and on a vertical window facing east.

Solution. From Eqs. (12.3) and (12.4), the perpendicular solar illuminance is

$$E_{dn} = 11{,}850\left[1 + 0.034\cos\frac{360\times 292}{365}\right]e^{-0.210/\sin 23.3}$$

$$= 11{,}850(1.011)(0.588) = 7040 \text{ fc}$$

The illuminance on the skylight is, from Eq. (12.8),

$$E_h = 7040 \sin 23.3 = 2780 \text{ fc}$$

For the vertical window, a_z is 90 – 79.4 = 10.6°, and Eq. (12.9) yields

$$E_v = 7040 \cos 23.3 \cos 10.6 = 6360 \text{ fc}$$

E_{dn} from Eq. (12.4), is plotted in Fig. 12.5. Also plotted there are E_h from Eq. (12.8) and E_v from Eq. (12.9), the latter for various values of a_z, the horizontal angle between the sun's location and the normal to the vertical surface.

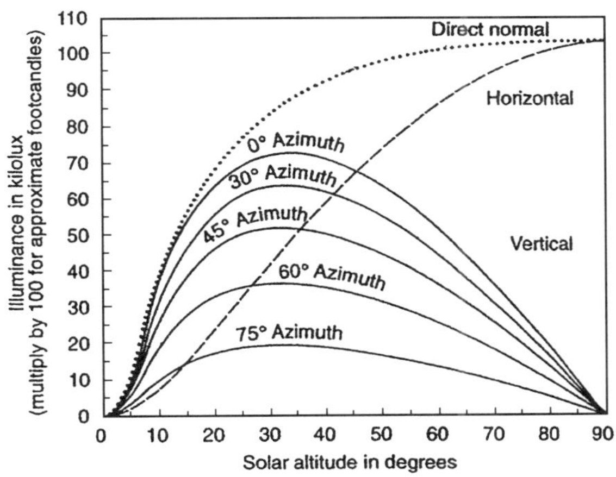

Figure 12.5 Illuminance produced by the sun under clear sky conditions. *Source: IESNA Lighting Handbook*, 9th edition, 2000. Courtesy of the Illuminating Engineering Society of North America.

Values of solar altitude for selected dates, times, and latitudes are given in Table 12.3 to facilitate use of Fig. 12.5.

The curves for vertical illuminance in Fig. 12.5 rise and then fall. To explore this, return to Eq. (12.9), and consider the case when $a_z = 0$. Then

$$E_v = E_{dn}\cos a_t = E_{xt}\varepsilon^{-\frac{c}{\sin a_t}}\cos a_t \tag{12.18}$$

TABLE 12.3 SOLAR ALTITUDE IN DEGREES VERSUS LATITUDE, DATE, AND TIME

		A.M.	6	7	8	9	10	11	Noon
Latitude	Date	P.M.	6	5	4	3	2	1	Noon
30°N	June 21		12	24	37	50	63	75	83
	Mar.-Sept. 21			13	26	38	49	57	60
	Dec. 21				12	21	29	35	37
34°N	June 21		13	25	37	50	62	74	79
	Mar.-Sept. 21			12	25	36	46	53	56
	Dec. 21				9	18	26	31	33
38°N	June 21		14	26	37	49	61	71	75
	Mar.-Sept. 21			12	23	34	43	50	52
	Dec. 21				7	16	23	27	28
42°N	June 21		16	26	38	49	60	68	71
	Mar.-Sept. 21			11	22	32	40	46	48
	Dec. 21				4	13	19	23	25
46°N	June 21		17	27	37	48	57	65	67
	Mar.-Sept. 21			10	20	30	37	42	44
	Dec. 21				2	10	15	20	21

Taking the derivative of Eq. (12.18) and setting it equal to zero yields

$$c \cos^2 a_t = \sin^3 a_t \tag{12.19}$$

For c = 0.21, Eq. 12.19 gives $a_t = 32.1°$, which is the value of a_t at the peak of the 0° azimuth curve in Fig. 12.5.

12.4 SKY LUMINANCE AND ILLUMINANCE

Sky conditions are classified by the IESNA in one of two ways:

Sky Ratio (SR): This method is used by the National Institute of Standards and Technology (NIST). It defines sky ratio as the ratio of horizontal sky irradiance to global horizontal irradiance. The defined conditions are:

clear—SR ≤ 0.3
partly cloudy—0.3 < SR < 0.8
overcast—0.8 ≤ SR

This ratio approaches 1 when a_t approaches 0 and thus is inaccurate at low solar altitudes.

Sky Cover (SC): This method is used by the National Oceanic and Atmospheric Administration (NOAA). It estimates the amount of cloud cover in tenths by visual observation. The definitions are:

clear—0.0 to 0.3
partly cloudy—0.4 to 0.7
overcast—0.8 to 1.0

Thus, a sky cover of 0.5 means that half the sky is covered by clouds.

It should be noted that these definitions do not address the uniformity of a cloud cover, its ceiling, its density, or its thickness. Of particular note is a sky with 100% cloud cover, which can vary from dense overcast to thin haze. Also of concern is whether the sun is obscured. Both an overcast sky and a well-placed cloud in a clear sky can cover the sun. The sky categorizations in the next three sections are based on the sky cover model.

The horizontal illuminance produced by an unobstructed sky, excluding direct sunlight, is given by

$$E_h = A + B \sin^C a_t \tag{12.20}$$

This general equation is the synthesis of the work of many researchers (see Gillette, 1984). It includes the gathering of sky luminance data for clear and overcast conditions, originally by H. Kimball and I. Hand at the U.S. Weather Bureau from 1919 to 1922, the empirical work of P. Moon and D. Spencer (Moon, 1942) for the overcast sky, and the analysis of clear skies by R. Kittler (Kittler, 1967). Experimentally-derived values for A, B, and C are given in Table 12.4.

TABLE 12.4 CONSTANTS FOR EQ. (12.20)

Sky Condition	A Klx (fc)	B Klx (fc)	C
Overcast	0.30 (28)	21.0 (1950)	1.0
Partly cloudy	0.30 (28)	45.0 (4180)	1.0
Clear	0.80 (74)	15.5 (1440)	0.5

In Eq. (12.20), A is the horizontal illuminance to be expected at sunrise and sunset. The second term in the equation shows clearly that solar altitude is the primary driving function in determining the value of E_h. Other effects, such as atmospheric turbidity, local atmospheric pressure, and cloud type and amount are generally neglected.

Each of the three sky conditions will be discussed in further detail.

12.5 OVERCAST SKY

For the overcast sky, Moon and Spencer (Moon, 1942) developed an empirical equation for sky luminance, which was adopted by the CIE in 1955 and has been universally used ever since to represent such skies.

$$L = \frac{L_z}{3}(1 + 2\cos\theta) \tag{12.21}$$

where L is the sky luminance at a point P in cd/m², L_z is the luminance at zenith in cd/m², given by

$$L_z = 123 + 8600 \sin a_t, \tag{12.22}$$

and θ is the angle from the zenith to the point P. The situation is diagrammed in Fig. 12.6. To find the horizontal illuminance at point 0, Eq. (2.21), repeated here, is used:

$$dE_{ho} = L \cos\theta \, d\omega \tag{12.23}$$

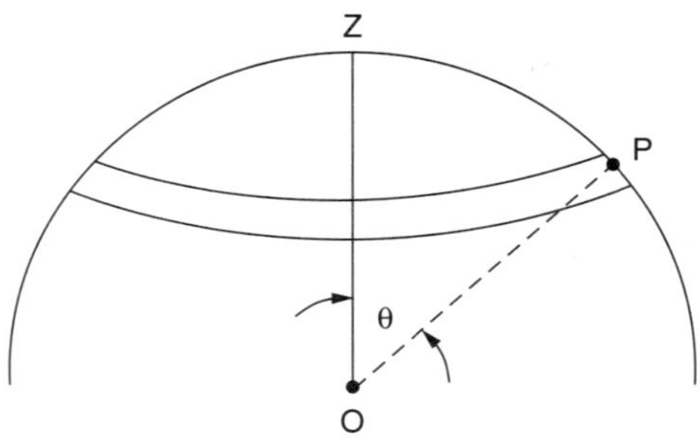

Figure 12.6 Overcast sky differential segment.

where

$$d\omega = 2\pi \sin\theta \, d\theta \tag{12.24}$$

and the "o" stands for overcast.

The integration of Eq. (12.23) to obtain E_h for L in Eq. (12.21) was carried out in Ex. 2.11, giving

$$E_{ho} = \frac{7\pi}{9} L_z = \frac{7\pi}{9}(123 + 8600 \sin a_t) \tag{12.25}$$

$$\left.\begin{aligned} E_{ho} &= 0.30 + 21 \sin a_t \text{ Klx} \\ &= 28 + 1950 \sin a_t \text{ fc} \end{aligned}\right\} \tag{12.26}$$

It is noted from Table 12.4 that the coefficients for the overcast condition match those in Eqs. (12.26), showing that for the overcast sky case, the horizontal illuminance can indeed be derived from the sky luminance.

Solving Eq. (12.25) for L_z yields

$$L_z \equiv ZL \times E_{ho} = \frac{9}{7\pi} E_{ho} = 0.409 \, E_{ho}. \tag{12.27}$$

ZL in Eq. (12.27) is called a zenith luminance factor and is calculated at the same solar altitude as E_{ho}. It is single-valued for the overcast sky.

There are several points to note about the standard overcast sky:

a) Luminance varies with polar angle (θ), but not with azimuth angle.
b) Zenith luminance is three times horizon luminance. Thus a given area of skylight receives more lumens from an overcast sky than does an equal-size window.
c) The vertical illuminance in any direction is roughly 40% of the horizontal illuminance.

$$E_{vo} = 0.396 \, E_{ho} \approx 0.4 \, E_{ho} \tag{12.28}$$

This result can be derived by an integration similar to that in Ex. 2.11.

d) The standard overcast sky is most representative of conditions in northern Europe and is likely not to apply as well in the U.S. to hazy desert skies or sky conditions on the coasts.

e) The maximum horizontal and vertical illuminances are 1978 and 791 fc, respectively.

The illuminances on horizontal and vertical surfaces from a standard overcast sky as a function of solar altitude are shown in Fig. 12.7.

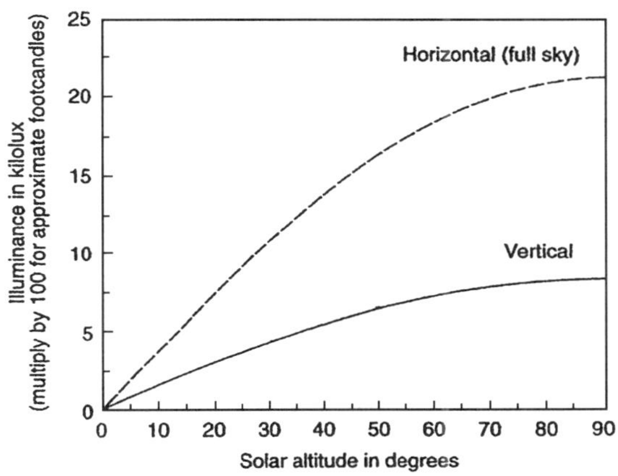

Figure 12.7 Illuminances from a standard overcast sky. ***Source*: *IESNA Lighting Handbook*, 9th edition, 2000. Courtesy of the Illuminating Engineering Society of North America.**

EXAMPLE 12.6

The astrodome in Fig. 12.8 is lighted by the CIE standard overcast sky. Assume the dome transmittance is 90% and the solar altitude is 30°. What is the horizontal illuminance at P?

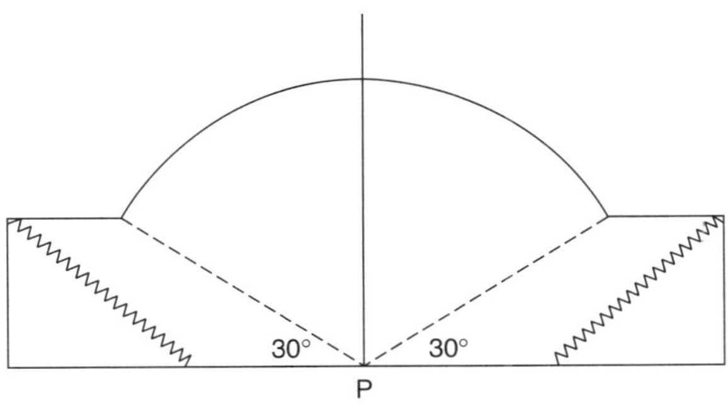

Figure 12.8

Solution. From Eqs. (12.23) and (12.24)

$$dE_{ho} = 2\pi \times 0.9L \cos\theta \sin\theta \, d\theta$$

Then, with the aid of Eq. (12.21),

$$E_{ho} = \int_{o}^{\pi/3} \frac{2\pi \times 0.9L_z}{3} (1 + 2\cos\theta) \cos\theta \sin\theta \, d\theta$$

$$= \frac{1.8\pi L_z}{3} \left[\frac{\sin^2\theta}{2} - \frac{2\cos^3\theta}{3} \right]_{o}^{\pi/3}$$

$$= \frac{1.8\pi L_z}{3} \left(\frac{3}{8} - \frac{1}{12} + \frac{2}{3} \right) = 1.81\ L_z$$

For a solar altitude of 30°

$$L_z = 123 + 8600 \times 0.5 = 4423\ \text{cd/m}^2$$

Thus

$$E_{ho} = 1.81 \times 4423$$

$$= 8000\,\text{lx} = 740\,\text{fc}$$

12.6 CLEAR AND PARTLY CLOUDY SKIES

For clear and partly cloudy skies, luminance is a function of several angles, and there is no easy integration to find E_h. The clear sky luminance equation was developed by R. Kittler (Kittler, 1967) and adopted by the CIE in 1973.

$$L = 3.650\ L_z \ \frac{(0.91 + 10e^{-3\gamma} + 0.45\ \cos^2\gamma)\ (1 - e^{-0.32\sec\varsigma})}{(0.91 + 10e^{-z_o} + 0.45\ \cos^2 Z_o)} \qquad (12.29)$$

where (all angles in radians)

- L is the sky luminance at point P in Fig. 12.9
- L_z is the zenith sky luminance
- γ is the angle between sun line to the observer point and the line from that point to P
- ζ is the angle between the zenith line and the line from the observer to point P
- α is the azimuth angle from the sun to point P
- $Z_o = \frac{\pi}{2} - a_t$

  From the geometry in Fig. 12.9, γ is given by

$$\cos\gamma = \cos Z_o\ \cos\zeta + \sin Z_o\ \sin\zeta\ \cos\alpha \qquad (12.30)$$

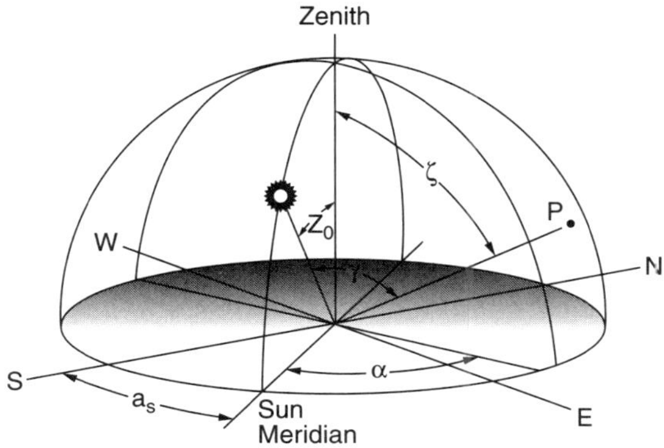

Figure 12.9 Sky angles in connection with Eq. (12.29). ***Source*****: *IESNA Lighting Handbook*, 9th edition, 2000. Courtesy of the Illuminating Engineering Society of North America.**

The equation for the luminance of the partly cloudy sky is

$$L = 1.816 L_z \frac{(0.526 + 5e^{-1.5\gamma})(1 - e^{-0.80 \sec \zeta})}{0.526 + 5e^{-1.5Z_o}} \tag{12.31}$$

As in Eq. (12.22), L_z for the clear and partly cloudy sky cases may be expressed in terms of horizontal illuminance and a zenith luminance factor

$$L_z = ZL \times E_{hc}, \quad L_z = ZL \times E_{hp} \tag{12.32}$$

where the "c" stands for clear and the "p" for partly cloudy. For these cases, however, there is no single ZL, but a family of them as a function of solar altitude, as shown in Table 12.5.

TABLE 12.5 ZL VALUES FOR CLEAR AND PARTLY CLOUDY SKIES

a_t in degrees	ZL (clear)	ZL (partly cloudy)
90	1.034	0.637
80	0.664	0.501
70	0.445	0.413
60	0.314	0.343
50	0.234	0.292
40	0.185	0.255
30	0.156	0.230
20	0.142	0.214
10	0.139	0.205
0	0.144	0.201

For a given horizontal illuminance, calculated from Eq. (12.20), the ZL value from Table 12.5 at the given solar altitude can be used with Eq. (12.29) or Eq. (12.31) to find the appropriate sky luminance.

As in the standard overcast sky case, there are several points to note about the standard clear sky.

a) Sky luminance varies with both polar and azimuth angles.
b) The portion of a clear sky surrounding the sun has the highest luminance, as much as 12 times the zenith luminance.
c) The portion of the sky 90° in vertical angle opposite the sun has the lowest luminance and is the bluest.
d) Horizon luminance is three times greater than zenith luminance and can occasionally be six times greater at the horizon point opposite the sun.
e) Horizontal illuminance is driven by the square root of the sine of the solar altitude.
f) The maximum horizontal illuminance is 1514 fc. (It is 4208 fc for the partly cloudy sky).

Illuminances on vertical and horizontal surfaces from standard clear skies are given in Fig. 12.10.

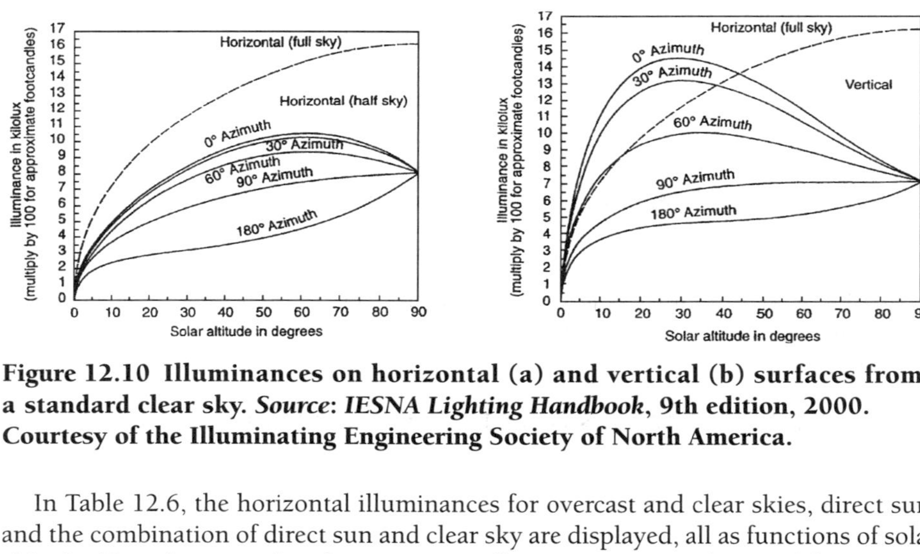

Figure 12.10 Illuminances on horizontal (a) and vertical (b) surfaces from a standard clear sky. *Source*: *IESNA Lighting Handbook*, 9th edition, 2000. Courtesy of the Illuminating Engineering Society of North America.

In Table 12.6, the horizontal illuminances for overcast and clear skies, direct sun, and the combination of direct sun and clear sky are displayed, all as functions of solar altitude. Note that on a clear day, E_{hs} essentially swamps out E_{hc} (5 times) for $a_t > 47°$. Note also that $E_{ho} = E_{hc}$ at $a_t = 36°$ and exceeds it for all $a_t > 36°$.

EXAMPLE 12.7

For the situation in Ex. 12.3, find the horizontal illuminance due to an overcast sky and a clear sky plus direct sun.

Solution. From Ex. 12.4, the solar altitude and azimuth are 23.3° and −58.9°, respectively. From Ex. 12.5, the horizontal illuminance due to the sun is 2780 fc.

From Eq. (12.20) for the overcast sky,

$$E_{ho} = 28 + 1950 \sin 23.3 = 800 \text{ fc}$$

TABLE 12.6 HORIZONTAL ILLUMINANCES—SKY AND SUN—FC

a_t	E_{ho}	E_{hc}	E_{hs}	$E_{hc} + E_{hs}$
0	28	74	0	74
10	367	674	617	1291
20	695	916	2193	3109
30	1003	1092	3893	4985
40	1281	1228	5494	6722
50	1522	1334	6900	8234
60	1717	1414	8053	9467
70	1860	1465	8905	10370
80	1948	1503	9429	10930
90	1978	1514	9606	11120

For the clear sky, Eq. (12.20) gives

$$E_{hc} = 74 + 1440 \sin^{0.5} 23.3 = 980 \text{ fc}$$

Thus the horizontal illuminance for the clear sky plus direct sun is 3760 fc.

12.7 REFLECTED GROUND LIGHT

Sunlight and skylight that are reflected from ground surfaces to windows or from roofs to clerestories or monitors can be a significant source of daylighting. To develop a measure of this effect, consider the situation in Fig. 12.11a. A strip of ground G receives light from the sky and sun and a portion of it is reflected to a point on a window W. What is needed is the relationship for the illuminance at W produced by a luminance at G. This problem is considered in general for a variety of area sources in Chap. 13. For now, an equation for the illuminance provided by a rectangular plane source to a point on a perpendicular surface is required. The diagram in Fig. 12.11b applies.

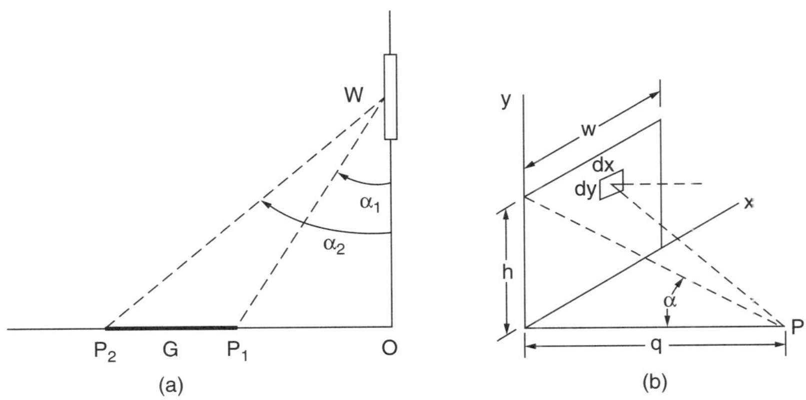

Figure 12.11 Pertaining to ground light.

The plane is assumed to be perfectly diffusing with a constant luminance L. It lies in the xy plane, with dimensions w and h. The point P is a distance q perpendicularly outward from a corner of the plane. A small differential element of dimensions dx and dy is identified. The illuminance in a horizontal plane at P from this element is, from Eq. (2.30),

$$dE = \frac{L\,q\,y}{(q^2 + x^2 + y^2)^2}\,dx\,dy \tag{12.33}$$

which, when integrated, yields

$$\left.\begin{aligned} E &= \int_o^h \int_o^w \frac{L\,q\,y}{(q^2 + x^2 + y^2)^2}\,dx\,dy \\ &= \frac{L}{2}\left(\tan^{-1}\frac{w}{q} - \frac{q}{\sqrt{h^2 + q^2}}\tan^{-1}\frac{w}{\sqrt{h^2 + q^2}}\right) \end{aligned}\right\} \tag{12.34}$$

The angles in Eq. (12.34) are in radians.

EXAMPLE 12.8

Find the illuminance at point P in a horizontal plane produced by the rectangular source shown in Fig. 12.12. The source is perfectly diffusing and has a luminance of 2 cd/in^2.

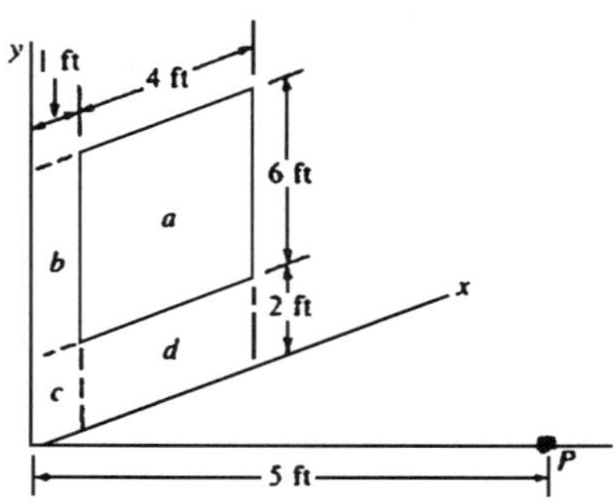

Figure 12.12

Solution. In Fig. 12.12, four areas are labeled. What is wanted is the illuminance produced by area a. To get it, the principle of superposition is used to find the illuminance produced by area abcd, from which is subtracted out the illuminance produced by areas b, c, and d. Specifically,

$$E_a = E_{abcd} - E_{bc} - E_{cd} + E_c \tag{12.35}$$

The reason for the $+E_c$ is that area c has been removed twice by subtracting E_{bc} and E_{cd}. The data needed are tabulated.

Area	h	w	Eq. (12.34)
abcd	8	5	75.9
bc	8	1	20.4
cd	2	5	13.0
c	2	1	3.9

Then, from Eq. (12.35), $E = 75.9 - 20.4 - 13.0 + 3.9 = 46$ fc.

In dealing with reflected ground or roof light, rectangular strips that are long compared with their width are of concern. Letting w approach ∞ in the second Eq. (12.34) and doubling the result to let the strip extend from −∞ to +∞ gives

$$\left.\begin{aligned} E &= L\left(\frac{\pi}{2} - \frac{q}{\sqrt{h^2+q^2}} \times \frac{\pi}{2}\right) \\ &= \frac{\pi L}{2}(1-\cos\alpha) \end{aligned}\right\} \tag{12.36}$$

where α is as identified in Fig. 12.11b.

Now reconsider the ground-window geometry in Fig. 12.11a. The illuminance E_{wg} on the window from ground strip G of luminance L_g is, from Eq. (12.36),

$$\begin{aligned} E_{wg} &= \frac{\pi L_g}{2}[1-\cos\alpha_2 - (1-\cos\alpha_1)] \\ &= \frac{\pi L_g}{2}(\cos\alpha_1 - \cos\alpha_2) \end{aligned} \tag{12.37}$$

Let the illuminance on the ground surface be E_g and the reflectance of the ground surface be ρ_g. Then, assuming perfect diffusion,

$$L_g = \frac{M_g}{\pi} = \frac{\rho_g E_g}{\pi} \tag{12.38}$$

and

$$E_{wg} = \frac{\rho_g E_g}{2}(\cos\alpha_1 - \cos\alpha_2) \tag{12.39}$$

A listing of reflectances of typical ground surfaces is given in Table 12.7. The percentages are average values.

TABLE 12.7 GROUND SURFACE REFLECTANCES

Material	Reflectance (%)
grass (dark green)	6
asphalt (dirt free)	7
earth	7
gravel	13
sandstone	18
vegetation	25
brick (buff)	44
brick (red)	30
concrete	40
cement	27
limestone	45
snow (new)	75

EXAMPLE 12.9

In Fig. 12.11a, assume that there is a concrete road parallel to the building and extending from 8 ft to 28 ft from the base of the building. The rest of the ground cover is dark green grass. If the illuminance on the ground surfaces is 2000 fc, what is the illuminance on the center of the window 8 ft above the ground?

Solution. The illuminance is the summation of three illuminances obtained from Eq. (12.39).

$$E_{wg} = 1000[0.06(1 - \cos 45) + 0.40(\cos 45 - \cos 74) + 0.06(\cos 74 - 0)]$$

$$= 1000(0.018 + 0.173 + 0.017) = 208 \text{ fc}$$

12.8 FENESTRATIONS

A fenestration is an aperture in a building for the admission of daylight. It is filled with glazing materials and optical controls. There are a variety of architectural forms for admitting daylight into buildings. These fall into two general classes: sidelights and toplights. Several of these are diagrammed in Fig. 12.13.

Sidelights are windows, clerestories, or sawtooths, which receive light in a vertical plane. The amount of light flux admitted depends on the area of the opening, the glazing material, and any optical control elements (overhangs, light shelves, blinds).

The distribution and penetration of the light depend on the location of the sidelight and, again, the optical controls. In general, significant daylight penetration from a window is about 2 times the distance from the floor to the top of the window (window head height). A light shelf can increase this distance. For these penetrations to occur, it is important that the sky view in the upper half of the window not be obstructed.

As shown in Fig. 12.13, a unilateral design produces an exponential fall off of illumination as the space is penetrated (Fig. 12.13a). Uniformity is significantly improved with a bilateral design (Fig. 12.13b) or with a clerestory (Fig. 12.13c and d).

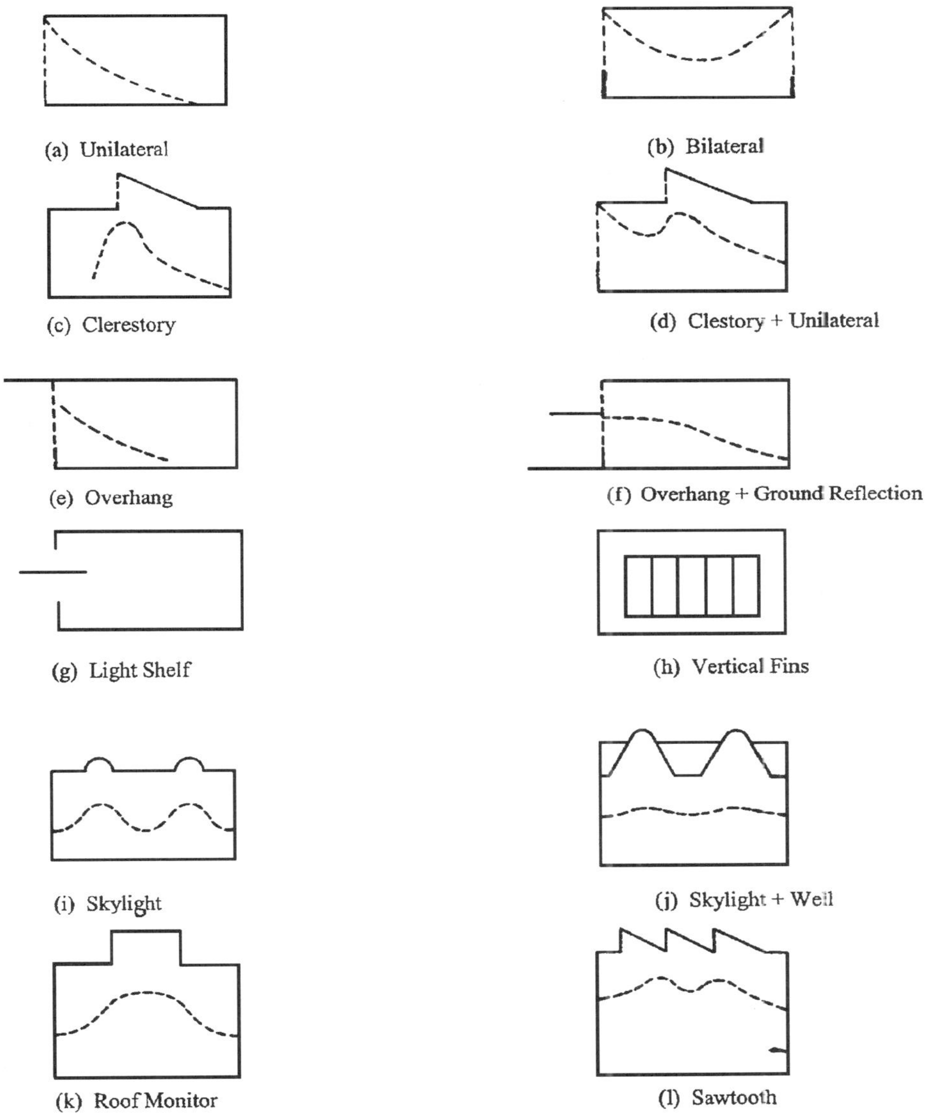

Figure 12.13 Types of fenestrations (dashed lines represent illuminance distributions).

Direct sunlight on windows is generally to be avoided, because of the glare it produces. Sometimes windows may be oriented so as to minimize this problem. Also translucent or tinted glazing materials may be used. The latter, however, can severely reduce the amount of daylight admitted.

Another solution is to use overhangs (Fig. 12.13e) to shield the room from direct sunlight. These also reduce the illumination from the sky and the penetration of the daylight into the space. A slightly more complicated procedure (Fig. 12.13f) is to use the overhang as a reflective surface to redirect some of the light deeper into the room. Also reflections from ground surfaces help to replace the light near the window that is lost due to the overhang.

Light shelves (Fig. 12.13g) are useful for shading and redirection of light. They provide more uniform illumination—less near the window and more deeper into the space. Vertical fins or louvers (Fig. 12.13h) are especially useful for east and west exposures to eliminate direct sunlight at low solar altitudes.

In toplighting, the double-domed skylight (Fig. 12.13i) is the most popular unit. It should be combined with a light well (Fig. 12.13j) to provide uniform horizontal illuminance. The interior of the well should be painted a matte-finish white, and the well angle with the vertical should be about 30°.

A roof monitor (Fig. 12.13k) is another method of toplighting, but does not usually achieve the uniformity of a skylighting installation. It can be used in high-ceiling industrial settings, as can the sawtooth arrangement (Fig. 12.13l). The latter is quite popular in warehouse and storage areas.

12.9 SKYLIGHTS

Most office buildings are multistory, and only the top floor can have skylights, so why are skylights important?

- 70% of U.S. commercial buildings are single story and 62% of commercial building floor area is directly under a roof (only 25% is easily lit by sidelighting).
- A given skylight area produces 2 to 5 times the horizontal illuminance as the same window area (because the light hits the work plane nearly normally).
- Domed skylights are efficient light gatherers at all solar altitudes.
- Skylights afford greater uniformity of horizontal illuminance throughout the day.
- Skylights are not in the direct field of view. Thus glare can be minimized.
- Skylights are independent of building orientation.
- Properly designed skylights are very energy efficient.

The aforementioned points caused the Department of Energy in 1977 to state that: "Skylights are the most energy-efficient and cost-effective passive solar application". It is ironic to note that a steam-powered electric plant fed from an on-shore oil well has an overall yield of around 8%. Thus 1250 equivalent watts in the well are required to light a 100 watt light bulb. In contrast, 100 square feet of north-facing window, with the sun at 20° solar altitude, harvests 40,000 lumens or 400 vertical footcandles. For a south facing window, multiply by 15. Thus there is an excess of daylight available. The problem is to bring it into building spaces so as to provide quality lighting.

There are three major types of skylights—residential, commercial-industrial, and architectural. Most of the first two types are domed and square or rectangular. The third type takes many forms—barrel vault, pyramid, lean-to, cluster, and daylight sandwich panels. Daylight sandwich panels are often used also in window walls.

Skylights have two transmittances: a diffuse transmittance (T_d) for the sky, which is a single value, because the sky is assumed to be diffuse, and a direct transmittance (T_D)

for the sun, which is a function of the angle of incidence. T_D is essentially the same as T_d for $a_t > 40°$, but falls rapidly for lesser solar altitudes.

Transmittances are given by the manufacturer as flat-sheet transmittances. Values for transparent and translucent plastic are given in Table 12.8.

TABLE 12.8 FLAT-SHEET TRANSMITTANCES

	1/8"	3/16"	1/4"
Clear	0.92	0.92	0.92
Light Translucent	0.72	0.60	0.53
Medium Translucent	0.53	0.42	0.36
Dense Translucent	0.32	0.22	0.17

Generally, two flat sheets are domed and then placed together in a sandwich, providing an energy-efficient fenestration. The width-to-height ratio for doming is generally 4 to 1. Doming alters the transmission of a flat sheet as follows:

$$T_{DM} = 1.25\ T_{FS}(1.18 - 0.416\ T_{FS}) \qquad (12.40)$$

For transparent glazing,

$$T_{DM} = 1.25 \times 0.92(1.18 - 0.416 \times 0.92) = 1.00\ T_{FS} = 0.92$$

Thus there is no change. For translucent grazing, there is a change. For example, for $^3/_{16}$-in medium translucent glazing,

$$T_{DM} = 1.25 \times 0.42(1.18 - 0.416 \times 0.42) = 1.26 \times T_{FS} = 0.53$$

This is a 26% increase and is caused by the doming process, which thins a sheet at its center. An additional beneficial effect of doming is that light gathering is increased at low solar altitudes, because of a greater incidence angle of the sun's rays.

When two sheets are placed together, the transmittance is not the product of the two transmittances, because of multiple reflections between the two sheets. A formula known as the Pierson equation gives the overall transmittance.

$$T = \frac{T_1 T_2}{1 - R_1 R_2} \qquad (12.41)$$

where T_1 and T_2 are the transmittances of the individual sheets, and R_1 and R_2 are the reflectances. For example, suppose two $^3/_{16}$-in medium translucent sheets are paired. The value of T is

$$T = \frac{0.42 \times 0.42}{1 - 0.58 \times 0.58} = 0.27$$

The product of the transmittances is 0.18. Thus, the transmittance has increased by 50%. Usually a clear and a translucent sheet are paired, and the increase in transmittance is about 15%.

Occasionally triple-glazing is used. In this case, the overall transmittance is given by

$$T = \frac{\dfrac{T_1T_2}{1-R_1R_2} \times T_3}{1-\left(1-\dfrac{T_1T_2}{1-R_1R_2}\right)R_3} = \frac{T_1\,T_2\,T_3}{(1-R_3)(1-R_1R_2)+T_1T_2R_3} \tag{12.42}$$

An alternative top-lighting assembly is the daylight sandwich panel. Daylight sandwich panels consist of two reinforced translucent fiberglass sheets bonded to an I beam frame of interlocked extruded aluminum (AL) or thermally-broken (TB) composite and filled with translucent fiberglass batts. Panels can be curved or flat.

Flat panels are normally $2^3/_4$ in thick. Standard panel size is 12 in by 24 in, and panels are assembled in grids 4 to 5 ft wide and 3 to 20 ft long. The exterior face sheet is $^5/_{64}$ in thick; the interior sheet $^3/_{64}$ in thick. Sheets are available in light translucent crystal (c) and medium translucent white (w).

The panels are highly-insulating and diffuse light very well. Their U-factors (a measure of total energy flow through in BTU/hr-ft^2-°F) range from 0.53 to 0.10 and light transmittances from 0.50 to 0.05 for 2-$^3/_4$ in. thick panels. A summary of these factors is given in Table 12.9.

TABLE 12.9 PARAMETERS FOR DAYLIGHT SANDWICH PANELS

U Factor		Light Transmittance		
AL	TB	c/c	c/w	w/w
0.53	0.53	0.50	0.35	0.20
0.29	0.23	0.30	0.20	0.15
0.26	0.18	0.20	0.15	0.10
0.22	0.14	0.15	0.12	0.08
0.18	0.10	0.10	0.08	0.05

Daylight sandwich panels are used in a myriad of applications in ceiling and wall systems, and provide quality diffuse lighting.

- as the upper portion of a split window
- as an entire window wall
- spanning an atrium
- as a barrel vault sky dome
- as the wall of a cylindrical rotunda
- as a walkway cover

There are a number of "rules of thumb" that are useful in skylighting design.

- To optimize both energy conservation and lighting effectiveness, skylights should provide the desired illuminance for approximately one-half the operating year. To achieve this requires that skylight openings represent 5–8% of the ceiling area.

- A skylight should have a well (not required of daylight sandwich panels), if uniform illumination is to be achieved. The skylight width and well height should be essentially equal. Well walls should be painted matte white.
- A 90° well traps too much light and produces a roller-coaster light distribution. A 45° well uses too much ceiling area (a 4 ft × 4 ft skylight with a 4-ft well height has a 12 ft × 12 ft ceiling opening). Research has shown (Serres, 1991) that a well splayed at 30° with the normal to the skylight is optimum.
- If no well is present, the spacing depends on the glazing. The spacing criterion is the ratio of the center-to-center skylight spacing to the ceiling height above the floor. This ratio should not exceed 0.5 for a clear skylight, 0.8 for light translucent, 1.2 for medium translucent, and 1.5 for dense translucent. If a 30° well painted matte white is present, the recommended ratio is 1.5, independent of the glazing.
- Light floor and ceiling finishes improve day light uniformity and reduce the contrast between the skylight opening and the adjacent ceiling area, thus reducing glare.
- Skylights can be integrated with ceiling architecture to block direct sunlight from the space below. This is particularly important in museums, where exposure to direct sunlight can permanently damage art work. Examples of such architectural designs are shown in Fig. 12.14a and b.

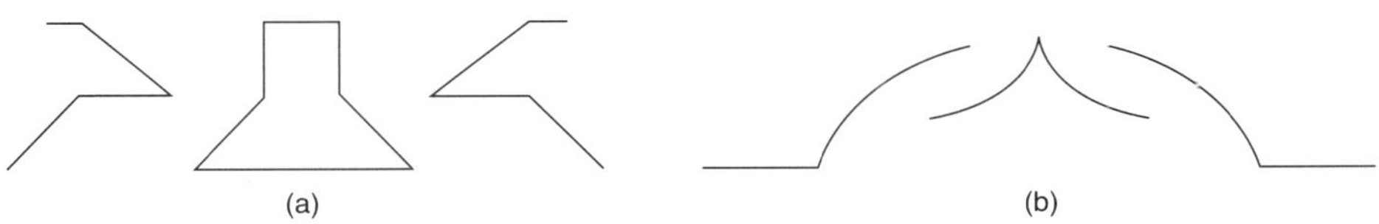

Figure 12.14 Skylights that obscure direct sunlight.

12.10 SIDELIGHTS

Sidelights were introduced in Fig. 12.13. This section concentrates on vertical windows and their various optical controls, as illustrated in Fig. 12.15.

- Vertical Window. The simplest sidelighting solution is a double-glazed vertical window with clear glass and no controls. This is usually satisfactory for a north facing window, unless there is a light-finished south-facing building directly opposite.
- Overhangs. An overhang is shown in Fig. 12.15a. Its purpose is to prevent direct sun penetration. Its effect is to widen the room and thus reduce daylight penetration distance.

  The size of an overhang depends on site latitude. In Boston (43° N latitude), to completely shield a 4-ft high vertical window facing the sun at noon on March or September 21 requires an overhang of 3.7 ft. In Mexico City (26° N latitude) under the same conditions only a 1.8-ft overhang is needed. On December 21, these widths are 9.4 ft and 4.6 ft. Thus, overhangs are more architecturally-practical at lower latitudes.

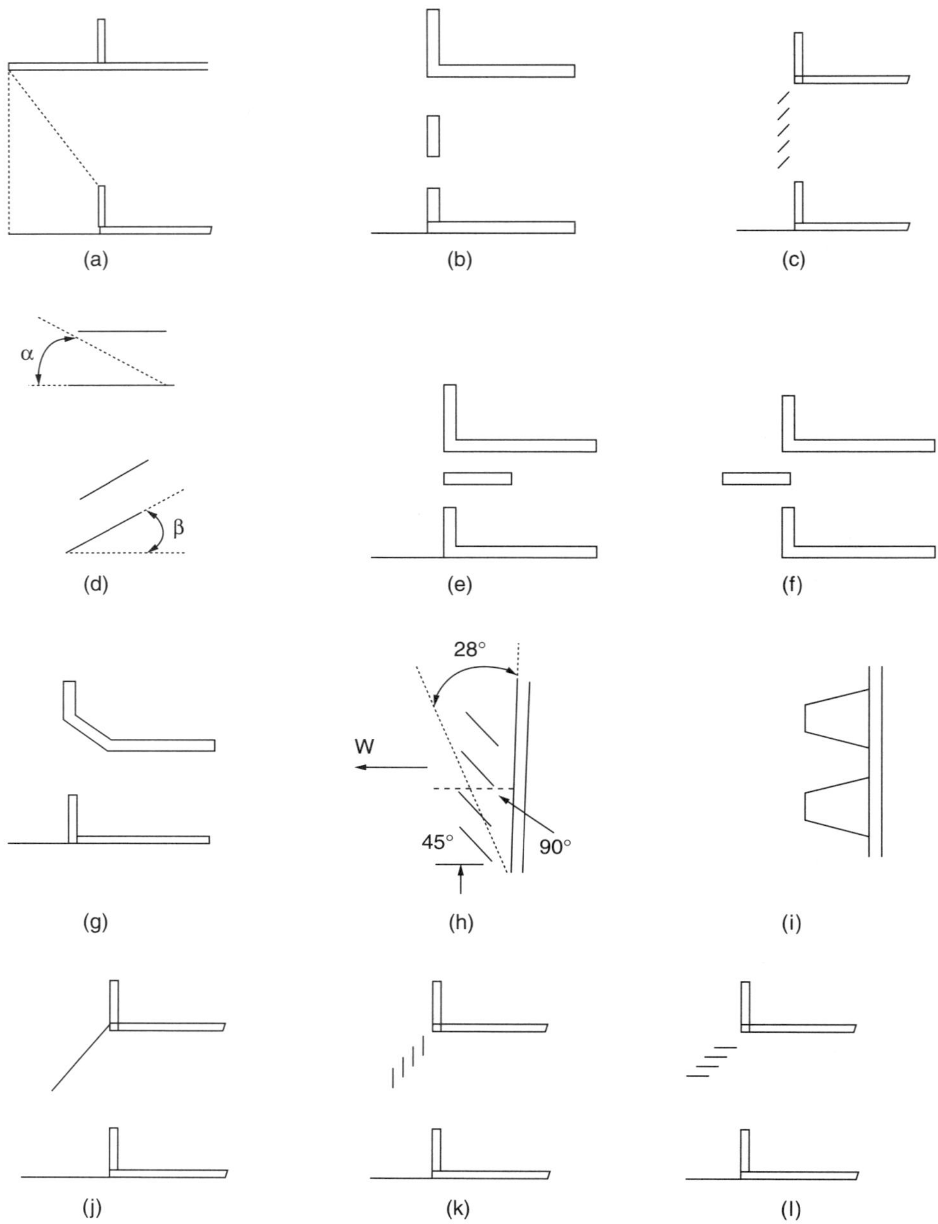

Figure 12.15 Sidelighting configurations.

If ground reflection is present, the underside of an overhang should have a light-colored matte finish. Overhangs are relatively ineffective on east and west facing windows because of lower sun angles.

- Split Window. A split window (Fig. 12.15b) can take several forms. The lower part can be clear glazed to allow a view out. The upper part can be translucent, using, for example, a daylight sandwich panel.

- Horizontal Blinds. These are shown in Fig. 12.15c and can be used on both the simple window and the split window. In the split window case, two sets can be used, giving separate control over light levels near to the window and deeper into the space. Blinds control direct sunlight, reduce glare from the sky, and, if oriented properly, redirect light to the ceiling.

  There are two angles to consider in connection with blinds (Fig. 12.15d). The first is the cut off angle (α), which is determined by the lowest solar altitude during the working day. It is customarily 20°. The second is the slope angle (β), which is set to optimize transmission. Its usual value is 45°.
- Light Shelves. The purpose of a light shelf is to improve daylight uniformity. One should provide whatever daylighting level is required deep in a space, and then use a light shelf to reduce the illuminance near the window to improve uniformity. The lighting level deep in the space will change very little.

  There are two types of light shelves—internal and external (Fig. 12.15e and f), or a combination of the two may be used. Their use means that the window head can be higher than in the simple window case because the shelf intercepts direct sunlight. Internal shelf depth should equal the height of the upper window; external shelf depth should equal the height of the shelf above the work plane. Room ceiling height should be at least 9 ft so that the shelf can be placed above eye level.
- Light Scoop. A sloped ceiling near a window wall (Fig. 12.15g) raises the window head without raising the remaining ceiling height. If the window is a single unit, diffuse daylight penetration will increase. If the window is split, it allows extra vertical height for the light shelf.
- Vertical Louvers. Vertical louvers or building fins (Fig. 12.15h and i—plan views) are particularly useful in controlling early morning and late afternoon sunlight. Their use is generally in east or west facing facades. In Fig. 12.15h, the window is facing west, and the louvers are at a 45° angle with the normal to the window. This provides a cut off angle of 90° with the window and results in 28° of north sky daylight being blocked, a disadvantage of the system. Vertical fins on the building produce essentially the same blocking effect of north daylight.
- Awnings. Three types of awnings are shown in Fig. 12.15j, k, and l. An awning, such as can be wound up and down by a shop owner each day, is shown in Fig. 12.15j. If it is opaque, it simply hides the sun and reduces the daylight incident on the window. If it is translucent, it replaces the sun as a glare source, and because it is so near, it is a large area glare source and will cause discomfort. Large areas of daylight sandwich panels near an observer produce a similar effect.

  The vertical and horizontal slat awnings in Fig. 12.15k and l redirect the light incident on them. In Fig. 12.15k, there will be more light near the window and less penetration. The opposite is true in Fig. 12.15l.

As with skylighting, there are several "rules of thumb" for sidelighting design.

- Daylight Penetration. As a very rough guide for a simple window, there will be significant daylighting from 0 to 2 window head heights into a space, and partial daylighting in the region from 2 to 4 window head heights.

- Building Configuration. Floor layout preference is narrow (no more than 60 to 70 ft wide) rectangular, with the long sides of the building facing north and south.
- Window Orientation. The rankings, in order of preference, are south, north, east, and west facing. South windows need overhangs or blinds. East and west windows require vertical fins or louvers. North windows usually do not need controls.
- Miscellaneous. Orient partitions and VDT screens perpendicular to windows. Separately switch outer rows of lights. Design overhangs to shade in summer and admit solar radiation in winter. Splay walls around windows and finish the splayed areas with light matte paint.

An example of the effect of room geometry on daylight penetration is shown in Fig. 12.16. All three floor plans have the same area, 10,000 ft^2. The first space is square, the second is rectangular, and the third has an atrium space. Assume windows are 7.5 ft high and thus full daylight penetration is up to two window height (15 ft) and partial penetration from two to four window heights (30 ft). The square area has 51% of its area fully daylit, 33% partially daylit, and 16% not at all daylit. With the rectangle, 59% is fully daylit and 41% partially so. For the atrium situation, 100% of the space is fully daylit. The example is idealized, but it does show the effect of the geometrical factors in admitting the maximum amount of daylight into a space.

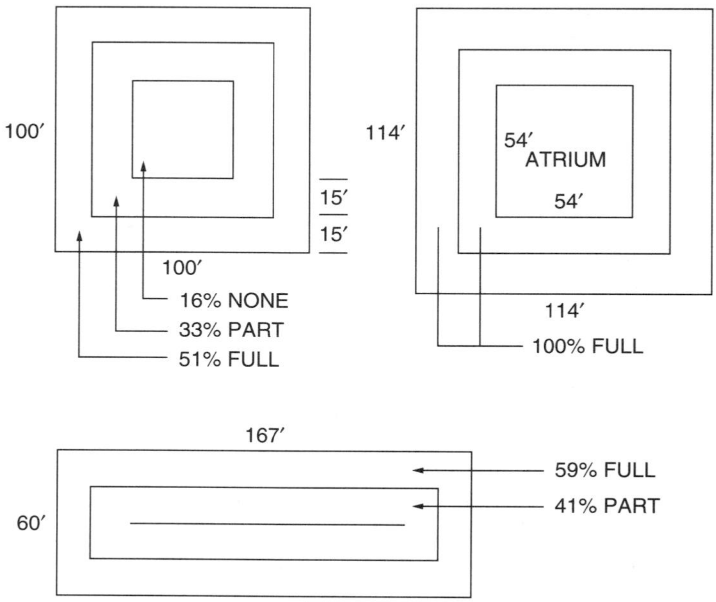

Figure 12.16 Daylight penetration.

12.11 GLAZING MATERIALS

All daylighting apertures in buildings have glass or plastic glazing materials, often film coated. The different glasses and plastics used will now be considered.

Clear Glass

Clear sheets of window glass are formed by the float glass manufacturing process diagrammed in Fig. 12.17. Raw batch materials are carefully mixed and then melted in a furnace tank that may hold up to 1000 tons. A ribbon of molten glass from the tank is floated on rollers over a surface of liquid tin. The sheet is polished as it proceeds, insuring that the two surfaces are flat and parallel. The sheet then passes through a lehr, or annealing oven, which can be as long as 200 ft. Following the lehr is a cutting table, where the continuous sheet is cut into large sheets (which are later cut to window size). Following cutting, additional processing can occur: tempering, laminating, tinting, coating, fritting, and patterning.

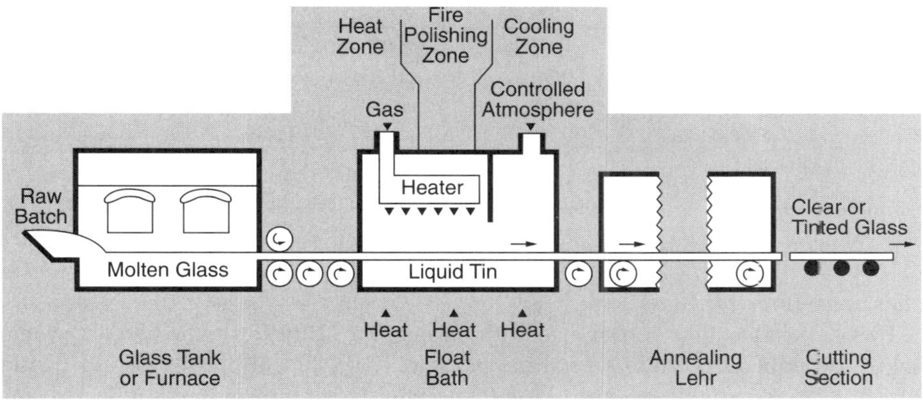

Figure 12.17 Float glass-making process. *Source*: Selecting the Right Glass, 1987. Courtesy of the Pittsburg Plate Glass Co.

The process for making plate glass for mirrors, display windows, and windshields is somewhat different. Rather than floating, the molten glass is fed through two water-cooled rollers that press the glass into a continuous sheet of very even thickness, such that the resulting glass is essentially distortion free.

Clear glazing can be single, double, or triple. Double glazing is the norm for energy efficiency, but triple glazing, with one of the layers a film, is becoming more common.

The National Fenestration Rating Council (NFRC) has established four metrics for rating glazing units:

- U Factor (the reciprocal of R factor)

  This measures the total energy flow (radiation, conduction, and convection) through a window assembly. The units are BTU/hr-ft^2-°F, and the measurement is done at the center of the glass (COG) with 70°F inside, 0°F outside, and a 15 mile-per-hour wind outside.
- Solar Heat Gain Coefficient (SHGC)

  This metric replaces a previous metric called shading coefficient (SC)* and measures the fraction of solar heat gain that is transmitted through a window.

* Multiply SHGC by 1.15 to obtain SC.

- Visible Light Transmittance
  The fraction τ of the energy within the visible spectrum that is transmitted through the glazing.
- Emissivity (see Sect. 4.7)
  The ability of a body to absorb and radiate energy. $\varepsilon = 1.0$ for a perfect absorber; 0 for a perfect reflector.

Values of these metrics for single, double, and triple clear glazing are given in the first three rows in Table 12.10.

TABLE 12.10 GLAZING SPECIFICATIONS FOR CLEAR GLASS

Type	U Factor	SHGC	τ	ε
Single-glaze	1.12	0.82	0.88	0.84
Double-glaze	0.50	0.70	0.78	0.84
Triple-glaze	0.35	0.70	0.75	0.84
Clear/low ε	0.34	0.65	0.75	0.30
Clear/low ε_{uv}*	0.29	0.37	0.70	1.10
Clear/low ε/low ε	0.17	0.56	0.64	0.10

* A low ε coating that blocks ultraviolet wavelengths as well.

Translucent (Frosted) Glass

This glass has a milky appearance, with the degree of diffusion a function of the density of the imbedded white crystalline particles. Translucent glasses do not distort color and are essentially non-absorbing. Thus, what is not transmitted is reflected, and their temperature does not rise.

If the transmittance of translucent glazing is kept below 50%, "hot spots" from sunlight are unlikely to be present. However, if the glazing is in the direct field of view it can be a source of glare.

Tinted (Gray and Bronze) Glasses

Tinted glasses reduce daylight transmission. Gray glasses can make an occupant of an interior space feel as though the exterior sky is forever overcast. For that reason, transmittance should be at least 40%. Some gray glasses are spectrally-neutral; others have relatively lower transmittances for wavelengths above 550 nm, making the grayness appear cool.

Bronze glasses both reduce transmittance and distort color, producing an enhancement of warmer colors, both for the view out and the view within.

Fritted Glass

These glasses have an opaque ceramic pattern superimposed on clear glass, which provides approximately 50% transmittance and 50% reflectance. Such glasses do not distort color and are useful in architectural skylights to admit sunlight and skylight during the day and reflect electric light at night.

Prismatic Glass (Glass Block)

Prismatic glass blocks are highly insulating but often provide a somewhat distorted view through. They are available in 8 to 12-in squares, usually 4 in thick. The blocks can be clear, giving high light transmittance, patterned, or prismatically-lensed. They are exceptionally rugged and can soften glare and reduce heat gain.

Plastic Glazing

Plastic glazing has many of the characteristics of glass, but it is lighter and scratches more easily. The fiberglass sheets used in daylight sandwich panels have already been discussed. Skylights are generally made of acrylic, which supports the weight of snow, ice, and wind loads well. Acrylic is reasonably resistant to ultraviolet radiation and to heat, but unfortunately, is a material that continues to burn once ignited, so it must be used where flame-retardant material is not required. As noted earlier, skylights are usually double-domed. They are mounted at the factory in a double-skin extruded aluminum curb with fiberglass insulation. A dry gasket sealing system forms the two domes together and to the curb, insuring a fully waterproof unit.

In the late 1970s, the Lawrence Berkeley Laboratory began developing a window coating that was transparent to light, but reflected heat. In 1981, the first low-emissivity (low ε) window was produced, and by the mid 1990s, about one-third of residential and one-fifth of commercial window sales were low ε coated units.

The principle of low ε coatings is shown in Fig. 12.18a. Solar energy is incident from the left and passes through the left pane with little attenuation. The low ε coating is applied to the inner face of the right pane. It is spectrally selective, allowing the visible and short wavelength infrared radiation to pass, but rejecting long wavelength infrared radiation. Inside the room, the light and short wavelength infrared are absorbed and converted to long wavelength infrared, which is reflected from the low-ε film and thus does not escape the space. Newer low ε coatings are even more spectrally selective and block UV radiation, as well as more of the long-wavelength infrared. Characteristics of typical low ε glasses are shown in the last three rows of Table 12.10.

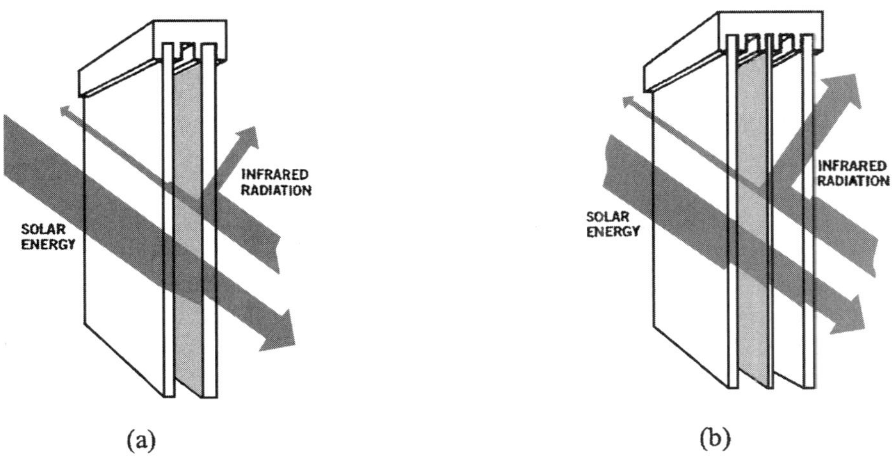

Figure 12.18 Low-emissivity coatings. *Source*: Superwindows, V. Gilmore, March, 1986. Courtesy of Popular Science.

The window in Fig. 12.18b reduces heat transfer even more than the one in Fig. 12.18a. It contains a coated polyester film layer midway between the inner and outer panes.

In newer versions there are two films and three spaces, all filled with an inert gas (argon or krypton). Optimum space thickness is about $^1/_2$ in.

Low ε coatings are applied in two ways: (a) a sheet of glass is coated with a layer of silver amid two layers of zinc oxide by a sputtering process involving argon gas atoms, (b) a coating of tin oxide is baked into a glass sheet, creating what is know as pyrolytic glass. The latter is tougher, but has higher ε values than the former.

The recent literature (Erwine, 1996, and Roberts, 2001) discusses warm and cool windows, distinguishing between the situations that exist during heating and cooling seasons. In heating-dominated buildings with normal double glazing, conduction/convection accounts for 70% of the heat lost, and radiation 30%. Low ε coatings affect radiation. Thus the emphasis in warm buildings should be to reduce the U factor. Single panes have a U factor of about 1; double panes about $^1/_2$. The newest super insulating windows, with multiple glass layers, one or more low ε coatings and gas fillings, can achieve U factors of about $^1/_8$.

In cool windows, the goal is to keep the solar heat out. This is the situation in many commercial buildings, particularly in warmer climates. 45% of the sun's energy is light. It is desired to admit this energy while rejecting the rest. Until recently, the only solution was tinted glass, which reduced interior light levels, often requiring additional electric lighting. Now spectrally-selective coatings are available that transmit light while blocking UV and IR. They are usually applied to the inner face of the outer pane.

Coolness factor is defined as the ratio of visible transmittance to the solar heat gain coefficient. Thus

$$CF = \frac{\tau}{SHGC} \tag{12.43}$$

Selective blue-green tinted coatings have CFs of 1.3, and spectrally-selective coatings have CFs as high as 1.6. These compare with a CF of 1.1 for standard double glazing.

12.12 DAYLIGHT MODELS

Daylight models can be very effective in providing the designer with a detailed view of how sunlight and skylight will be distributed in an interior space. With such models, a variety of sidelighting and toplighting options can be explored quickly and at modest cost.

The scale of a daylight model is most often 24 to 1 ($^1/_2$ in = 1 ft). Thus, for example, a 60 ft × 60 ft museum area with a 24 ft ceiling height would be represented by a 30 in × 30 in × 12 in model. The material of choice for modeling is $^3/_{16}$-in thick white foam core, available in 4 ft × 4 ft sheets and easily cut by an exacto knife. The model can be assembled using duct tape and pins. The outside surfaces should be covered with aluminum foil because foam core does transmit light, which will affect illuminance readings during testing.

Desired wall reflectances can be replicated in the model by lining wall surfaces with construction paper (matte) or glossy card (specular). Wall color is usually not an issue and thus most models use shades of gray. Major partitions and other internal structures should be modeled accurately and to scale.

Overhangs, light shelves, skylight wells, and the like should be modeled as accurately as possible, keeping in mind, for example, that cardboard of thickness $^1/_{16}$ in represents $1^1/_2$ in thickness in the actual fenestration. It is also possible to create small skylight domes (Parent, 1989). A 4 ft × 4 ft skylight with a 1-ft high dome becomes a 2 in × 2 in × $^1/_2$ in unit when modeled.

Scale models permit an easy assessment of the relationship between building mass and site geometry. External ground reflectances should be represented accurately, and exterior objects, such as trees, should be proportionally scaled.

If all glazing is clear, fenestration openings can be left uncovered and a correction factor of 0.92 applied to all interior illuminance readings. For translucent and tinted glazing, thin plastic material should be used in the model, with the proper transmittance and light distribution.

It is difficult to model electric lighting in a daylight model. Modeling lamps and luminaires at $^1/_{24}$ actual size is not feasible. Tiny incandescent (grain of wheat) lamps and very small diameter discharge sources may be used to give a qualitative notion of the effects of electric lighting.

The easiest and most inexpensive way to test a daylight model is with a mirrored daylight chamber to simulate the sky and a theater spotlight to simulate the sun. A test chamber 5 ft × 5 ft × $2^1/_2$ ft can be rather easily constructed. Its floor should be matte gray, and its sides should be mirrored. Its ceiling should be translucent sheet with rows of closely-spaced fluorescent lamps above. Such a chamber replicates an overcast sky condition quite well. To replicate a clear sky accurately is not practical, except at a large commercial or governmental testing facility.

Measurements should be made with a multicell illuminance meter. Usually measurements are used to obtain Daylight Factors (see Sect. 12.15). This requires measuring the horizontal illuminance of the unobstructed sky external to the model and taking a number of horizontal illuminance readings in a grid pattern inside the model. A small porthole is cut into the model to admit the light cell wires. A second porthole should be cut for visual observation. It is important that each light cell be as small as possible (even a cell as small as $^1/_2$ in in diameter translates into one 12 in in diameter in the actual space).

Testing of a model under direct sunlight can be done in the lab with a theater spotlight on an adjustable-height pole. The model is placed on a table and the "sun" located at selected altitude and azimuth angles. Internal illuminances may be measured, but usually it is more a matter of observing glare trouble spots and patterns of light and shadow within the model, for various "sun" positions, and evaluating configurations of light shelves, overhangs, and the like.

An alternative to having a daylight test chamber and an artificial sun is simply to place the model on a card table and test it outdoors. In this case, some type of sundial is useful to establish solar time. The advantage of outdoor testing is that the model may be examined under a variety of sun and sky conditions. The disadvantage is that sky conditions and solar position change, the former sometimes rapidly, making it difficult to take measurements.

12.13 LUMEN METHOD OF SKYLIGHTING

This section is the first of three devoted to calculation procedures for obtaining horizontal illuminance on the work plane in daylit interiors. In it, a method is developed for calculating the average work-plane illuminance provided by diffusing skylights mounted at or slightly above roof level. The procedure is denoted as the lumen method of skylighting (Rea, 2000) and is very similar to the lumen method of interior lighting presented in Chap. 9, with the ceiling-mounted luminaires being replaced by skylights.

The basic formula, which should be compared with Eq. (9.2), is

$$E_i = E_h \frac{A_t}{A_w} \tau (CU)(LLF) \tag{12.44}$$

where E_i = average incident maintained illuminance on the work plane from skylights
E_h = horizontal illuminance from sky and sun on the exterior of the skylighting elements
τ = net transmittance of skylight and well
A_t = gross projected horizontal area of the skylighting elements
A_w = area of the work plane
CU = coefficient of utilization
LLF = light loss factor

The calculation procedure may be divided into four steps:

a) Determination of the horizontal illuminance E_h on the exterior of the skylighting elements
b) Determination of the net transmittances, direct and diffuse, of the skylighting elements
c) Determination of the coefficient of utilization and light loss factor
d) Determination of
   a) the average maintained horizontal illuminance E_i on the work plane (if the number and area of the skylighting elements are specified)

   or

   b) the number and area of the skylighting elements (if the average maintained horizontal illuminance is specified).

Step 1 Incident Horizontal Illuminance

The procedures outlined in Sects. 12.3–12.6 are used to find the total horizontal illuminance on the exterior of the skylights. Generally, the analysis is done for both overcast sky and clear sky plus direct sun conditions. Then, an initial design judgment is made as to how many skylights to use, how large each skylight should be, and whether or not controls, such as louvers or lenses, should be installed at the bottoms of the skylight well openings.

Step 2 Net Skylight Transmittances

As was discussed in Sect. 12.9, there are two transmittances to consider when dealing with skylights. One is the direct transmittance (T_D), which is used whenever direct sunlight impinges on the skylight and which varies with the solar altitude, decreasing rapidly for flat sheets at low values of this angle. The second is the diffuse transmittance (T_d), which is used with clear and overcast sky luminances and is presumed not to vary with solar altitude. Occasionally manufacturers provide transmittance data showing the variation of T_D with angle, but more often, they simply provide a single number for each flat sheet, as presented in Table 12.8. From these numbers, the transmittances for domed sheets are calculated from Eq. (12.40) and for double-glazing from Eq. (12.41).

It remains to include the effect of any light well present between the skylight dome and the ceiling plane of the room. The reflections within such a well will reduce the overall transmittance of the skylighting system. First a well cavity ratio is calculated from

$$\mathrm{WCR} = \frac{5h(w+1)}{wl} \tag{12.45}$$

where h, w, and l are well height, width, and length, respectively. Second, well efficiency (N_w) is defined as the fraction of the luminous flux emerging from the skylight dome that enters the room from the well opening. Well efficiency, as a function of well cavity ratio and well-wall reflectance, is displayed in Fig. 12.19. Third, because a skylight has a non light-transmitting frame to hold the dome in place, it is necessary to include the ratio of net to gross skylight area (R_a). Finally, if any diffusers, lenses, louvers, or other optical controls are present, their transmittances (T_c) must be included. Then the net transmittance of the skylight-well system can be found from

$$\tau_d = T_d \times N_w \times R_a \times T_c \tag{12.46a}$$

$$\tau_D = T_D \times N_w \times R_a \times T_c \tag{12.46b}$$

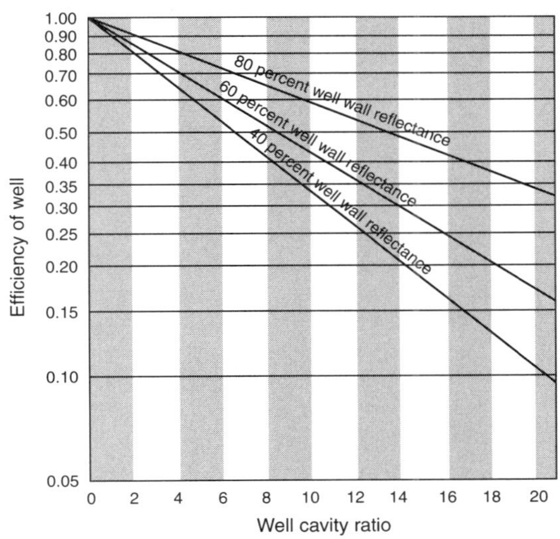

Figure 12.19 Efficiency of well versus well index. ***Source: IESNA Lighting Handbook*, 9th edition, 2000. Courtesy of the Illuminating Engineering Society of North America.**

Step 3 Coefficient of Utilization and Light Loss Factor

In this step, the fraction of the luminous flux incident on the skylight that reaches the work plane is sought. This is made up of two parts: the net transmittance (τ) determined in Step 2; and the room coefficient of utilization (CU), which is the fraction of the luminous flux entering the room which reaches the work plane. This coefficient is determined in much the same manner as was CU in the lumen method. First the room cavity ratio is found from Eq. (9.3), repeated here.

$$RCR = \frac{5h_c(l + w)}{l \times w} \tag{12.47}$$

where h_c is the ceiling height (bottom of the well) above the work plane.

With the room cavity ratio determined, CU for a 20% floor reflectance may be obtained from Table 12.11. For other floor reflectances, Table 9.8 may be used to obtain the appropriate multiplying factor, as in the lumen method.

TABLE 12.11 COEFFICIENTS OF UTILIZATION FOR SKYLIGHTS (based on 20% floor reflectance)

Ceiling Reflectance (Percent)	RCR	Wall Reflectance (Percent) 50	30	10
80	0	1.19	1.19	1.19
	1	1.05	1.00	0.97
	2	0.93	0.86	0.81
	3	0.83	0.76	0.70
	4	0.75	0.67	0.60
	5	0.67	0.59	0.53
	6	0.62	0.53	0.47
	7	0.57	0.49	0.43
	8	0.54	0.47	0.41
	9	0.53	0.46	0.41
	10	0.52	0.45	0.40
50	0	1.11	1.11	1.11
	1	0.98	0.95	0.92
	2	0.87	0.83	0.78
	3	0.79	0.73	0.68
	4	0.71	0.64	0.59
	5	0.64	0.57	0.52
	6	0.59	0.52	0.47
	7	0.55	0.48	0.43
	8	0.52	0.46	0.41
	9	0.51	0.45	0.40
	10	0.50	0.44	0.40

Source: *IESNA Lighting Handbook*, 9th edition, 2000. Courtesy of the Illuminating Engineering Society of North America.

The product of τ and CU gives the fraction of the lumens incident on the skylight that reach the work plane when all surfaces are clean. There are two light losses: room surface dirt depreciation (RSDD) and skylight dirt depreciation (SDD). RSDD may be obtained using Eq. (9.12). Little data is available on SDD factors. As a rough guide, 0.75 will be used for office areas and 0.65 for industrial areas, based on cleaning once annually.

Step 4 Illuminance or Number and Size of Skylights

Let the number of skylights be N, and the gross area of each skylight be A. Then the gross skylight area is

$$A_t = N \times A \tag{12.48}$$

On an overcast day, the average maintained horizontal work-plane illuminance for a particular date, time, and latitude is

$$E_{io} = E_{ho} \frac{A_t}{A_w} \tau_d (CU)(LLF) \tag{12.49}$$

where the o denotes overcast and the d denotes net diffuse transmittance.

For a day with clear sky and direct sun, the illuminance equation becomes

$$E_{ics} = E_{hc} \frac{A_t}{A_w} \tau_d (CU)(LLF) + E_{hs} \frac{A_t}{A_w} \tau_D (CU)(LLF) \tag{12.50}$$

where cs denotes clear sky plus direct sun, c denotes clear sky, s denotes direct sun, and d and D denote net diffuse and direct transmittances. If only one transmittance is given by the manufacturer, Eq. (12.50) becomes

$$E_{ics} = (E_{hc} + E_{hs}) \frac{A_t}{A_w} \tau (CU)(LLF) \tag{12.51}$$

EXAMPLE 12.10

Consider an office area 40 × 30 ft with a 10 ft ceiling. Assume reflectances of 75% ceiling, 50% walls, and 20% floor.

The room is to be lighted by six 3 × 3 ft double-domed transparent over translucent skylights mounted on 9-in high, 90° light wells whose wall reflectance is 60%. Each skylight has a 1.5-in wide retaining frame to hold the dome in place.

The transmittances of the flat sheets of plastic from which the dome is made are given as

	T_d	T_D
Transparent	0.79	0.92
Translucent	0.46	0.45

The depreciation factors are RSDD = 0.95 and SDD = 0.75.

Find the average horizontal illuminance on the work plane at 10 a.m. on March and September 21 at 42° north latitude for both overcast sky and clear sky plus direct sun conditions.

Solution.

Step 1 From Table 12.3, the solar altitude is 40°. Then the exterior horizontal illuminances are, from Table 12.6,

Solar Altitude	E_{ho}(fc)	E_{hc}(fc)	E_{hs}(fc)
40°	1281	1228	5494

Step 2 When the flat sheets are domed, their transmittances [from Eq. (12.40)] become

	T_d	T_D
Transparent	0.84	0.92
Translucent	0.57	0.56

Combining them into a double dome yields [from Eq. (12.41)] $T_d = 0.51$, $T_D = 0.53$. Each well is $3 \times 3 \times 0.75$ ft, giving a well cavity ratio from Eq. (12.45) of 2.5 and, from Fig. 12.19, a well efficiency of 0.80. With a 1.5-in frame, the ratio of net to gross skylight area is 0.84. No louvers or lenses are present so the net transmittances are, from Eq. (12.46),

$$\tau_d = 0.51(0.80)(0.84) = 0.34, \qquad \tau_D = 0.53(0.80)(0.84) = 0.36$$

Step 3 From Eq. (12.47), the room cavity ratio, assuming a $2^1/_2$-ft work plane, is 2.19. Then double interpolation in Table 12.11 yields a CU of 0.88. The light loss factor is

$$\text{LLF} = 0.95(0.75) = 0.71$$

Step 4 The average horizontal illuminances on the work plane are, from Eqs. (12.50) and (12.51),

$$E_{io} = 1281\left(\frac{54}{1200}\right)(0.34)(0.88)(0.71) = 12 \text{ fc}$$

$$E_{ics} = 1228\left(\frac{54}{1200}\right)(0.34)(0.88)(0.71)$$

$$+5494\left(\frac{54}{1200}\right)(0.36)(0.88)(0.71)$$

$$= 12 + 56 = 68 \text{ fc}$$

EXAMPLE 12.11

For the office area in Example 12.10, it is desired to have a minimum of 25 fc on the work plane on June 21 at 10 a.m. at 42° north latitude. How many skylights of the type described in Example 12.10 are required?

Solution. For the minimum illuminance, an overcast sky is assumed. From Table 12.3, the solar altitude is 60°. Then, from Table 12.6, the horizontal illuminance on the skylights is 1717 fc. Only the diffuse transmittance is needed and the other factors in Example 12.10 are unchanged. Thus, from Eq. (12.49)

$$25 = 1717\left[\frac{N(9)}{1200}\right](0.34)(0.88)(0.71)$$

$$N = 9\,\text{skylights}$$

These could be arranged as shown in Fig. 12.20. The general guideline for spacing skylights so as to provide uniform work-plane illuminance is not to exceed 1.5 times the ceiling height above the work plane ($1.5 \times 7.5 \approx 11$ ft) for a well splayed 30°. The suggested layout does not quite meet this criterion, and the well is splayed 0°, suggesting a lesser spacing. The portion of the roof area covered by skylights is $\frac{81}{1200} = 0.07$, which is within a second general guideline of 4 to 8%.

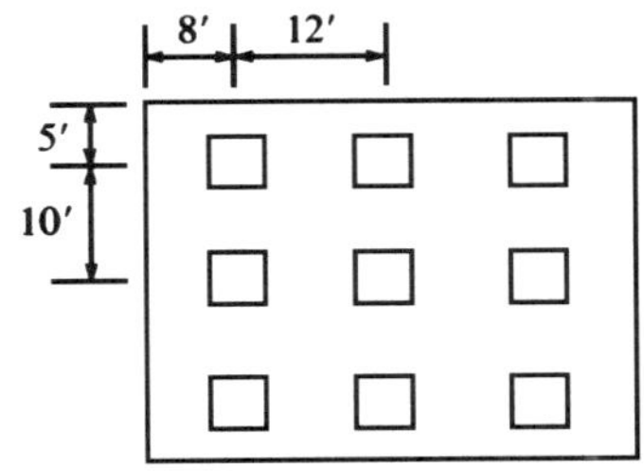

Figure 12.20 Skylight layout.

12.14 LUMEN METHOD OF SIDELIGHTING

The lumen method of sidelighting (Rea, 2000) is a procedure for obtaining the horizontal illuminance at five prescribed points on the work plane produced by daylight entering the space through a vertical window wall. It takes into account not only the skylight impinging on the window, but also the sunlight and skylight reflected onto the window from a uniform ground surface. It excludes direct sunlight on the window.

The method includes simple window controls, such as shades, drapes, and films, which reduce the transmittance of the window. It does not accommodate window blinds, overhangs, light shelves, and the like. There are more elaborate procedures to handle these cases (Mistrick, 1994).

Whereas in toplighting it is only necessary to know the horizontal illuminance provided externally by the sun and full sky, in sidelighting it is required to know both the vertical and horizontal external illuminances, to deal with skylight on the window and direct sunlight and skylight on ground surfaces.

The model for the lumen method of sidelighting is shown in Fig. 12.21. It presupposes a rectangular room of width W and depth D with a horizontal ceiling. The floor cavity extends from the bottom of the windows to the floor and has $\rho_{fc} = 30\%$. The ceiling cavity extends from the top of the windows to the ceiling and has $\rho_{cc} = 70\%$. Wall reflectance is $\rho_f = 50\%$. Room cavity height is labeled H.

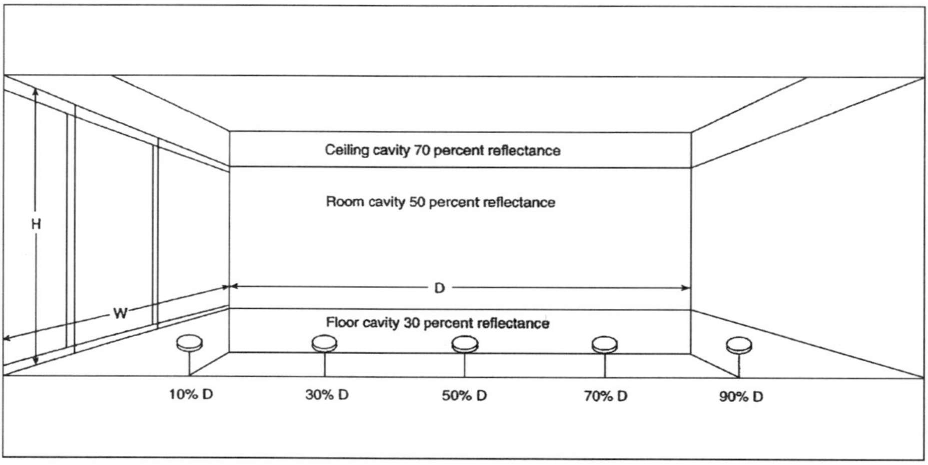

Figure 12.21 Model for the lumen method of sidelighting. ***Source*****:** ***IESNA Handbook*****, 9th edition, 2000. Courtesy of the Illuminating Engineering Society of North America.**

The window extends the entire length W of one wall. The five measurement points are at the top of the floor cavity and on a line perpendicular to the window wall at its center, as shown in Fig. 12.21.

The maintained horizontal illuminance E_i from daylighting at one of the points on the work plane in Fig. 12.21 is given by

$$E_i = E_v \tau(CU)(LLF) \qquad (12.52)$$

where E_v is the total vertical illuminance on the window from the sky and the ground, τ is the net transmittance of the window wall, CU is the coefficient of utilization, and LLF is the light loss factor. Application of Eq. (12.52) involves superposition of the illuminances from the sky and ground.

The net transmittance τ is the product of three factors

$$\tau = T \times R_a \times T_c \qquad (12.53)$$

where T is the transmittance of the glazing, R_a is the ratio of net to gross window area (allowing for a window frame and mullions), and T_c is the transmittance of any simple window controls present. Typical glazing transmittances are given in Table 12.12.

The light loss factor includes window dirt depreciation (WDD) and room surface dirt depreciation (RSDD). Average values of WDD are 0.83 for an office and 0.71 for a factory.

The exterior illuminance E_v is the total vertical illuminance on the window, excluding direct sunlight. It consists of the illuminance from the sky, which can be obtained from Figs. 12.7 and 12.10b, and the illuminance from the ground, which is found from Eq. (12.39) by allowing α_1 to be 0° and α_2 to be 90°. Thus

$$E_{wg} = \frac{\rho_g E_g}{2} \qquad (12.54)$$

where E_g is E_{ho} (Fig. 12.7) or $E_{hc} + E_{hs}$ (Figs. 12.5 and 12.10a).

TABLE 12.12 TYPICAL GLASS TRANSMITTANCES

Glass	Thickness (in)	τ (%)
Clear	1/8	89
Clear	3/16	88
Clear	1/4	87
Clear	5/16	86
Grey	1/8	61
Grey	3/16	51
Grey	1/4	44
Grey	5/16	35
Bronze	1/8	68
Bronze	3/16	59
Bronze	1/4	52
Bronze	5/16	44
Thermopane	1/8	80
Thermopane	3/16	79
Thermopane	1/4	77

Coefficients of utilization for each of the five points in Fig. 12.21 are given in Tables 12.13, a through e, for the sky component and in Table 12.13f for the ground component. There are two cases to consider. If the window has transparent glazing, the coefficients of utilization depend on the ratio of the vertical to horizontal illuminances at the center of the window from the half sky seen by the window. CUs are given for five values of this ratio in Tables 12.13, a through e.

If the glazing is translucent (not image-preserving), or shades or drapes are used, the vertical illuminances on the window from the sky and ground are summed. The CU for the sky component is obtained from Table 12.13b, for an E_v/E_h ratio of 1.0 based on a uniform sky. In both the transparent and translucent cases, Table 12.13f is used to obtain the CU for the ground component.

The components in the last two paragraphs are summarized in the following equations.

Transparent case

$$E_i = (E_{vs} \times CU_s + E_{vg} \times CU_g)\ \tau\ (LLF) \tag{12.55}$$

Translucent case

$$E_i = 0.5\ (E_{vs} + E_{vg})\ (CU_s + CU_g)\ \tau\ (LLF) \tag{12.56}$$

where the subscripts "s" and "g" denote sky and ground.

EXAMPLE 12.12

Carry out the lumen method of sidelighting for the following situation:

40°N latitude, September 21, noon

Room:	40 ft × 30 ft × 12 ft
	$\rho_{cc} = 70\%$, $\rho_w = 50\%$, $\rho_{fc} = 30\%$
Sky:	Overcast
Ground:	Concrete

TABLE 12.13 COEFFICIENTS OF UTILIZATION FOR WINDOWS WITHOUT BLINDS

a) $E_v/E_h = 0.75$

Room Depth/ Window Height	Percent D	Window Width/Window Height .5	1	2	3	4	6	8	Infinite
	10	.824	.864	.870	.873	.875	.879	.880	.883
	30	.547	.711	.777	.789	.793	.798	.799	.801
1	50	.355	.526	.635	.659	.666	.669	.670	.672
	70	.243	.386	.505	.538	.548	.544	.545	.547
	90	.185	.304	.418	.451	.464	.444	.446	.447
	10	.667	.781	.809	.812	.813	.815	.816	.824
	30	.269	.416	.519	.544	.551	.556	.557	.563
2	50	.122	.204	.287	.319	.331	.339	.341	.345
	70	.068	.116	.173	.201	.214	.223	.226	.229
	90	.050	.084	.127	.151	.164	.167	.171	.172
	10	.522	.681	.739	.746	.747	.749	.747	.766
	30	.139	.232	.320	.350	.360	.366	.364	.373
3	50	.053	.092	.139	.163	.174	.183	.182	.187
	70	.031	.053	.081	.097	.106	.116	.116	.119
	90	.025	.041	.061	.074	.082	.089	.090	.092
	10	.405	.576	.658	.670	.673	.675	.674	.707
	30	.075	.134	.197	.224	.235	.243	.243	.255
4	50	.028	.050	.078	.094	.104	.112	.114	.119
	70	.018	.031	.048	.059	.065	.073	.074	.078
	90	.016	.026	.040	.048	.053	.059	.061	.064
	10	.242	.392	.494	.516	.521	.524	.523	.588
	30	.027	.054	.086	.102	.111	.119	.120	.135
6	50	.011	.023	.036	.044	.049	.055	.056	.063
	70	.009	.018	.027	.032	.035	.040	.041	.046
	90	.008	.016	.023	.028	.031	.034	.035	.040
	10	.147	.257	.352	.380	.387	.391	.392	.482
	30	.012	.026	.043	.054	.060	.067	.070	.086
8	50	.006	.013	.021	.026	.029	.033	.035	.043
	70	.005	.011	.017	.021	.023	.026	.027	.034
	90	.004	.010	.015	.019	.021	.023	.025	.030
	10	.092	.168	.248	.275	.284	.290	.291	.395
	30	.006	.014	.026	.032	.036	.041	.044	.059
10	50	.003	.008	.014	.017	.019	.022	.024	.032
	70	.003	.007	.012	.014	.016	.018	.019	.026
	90	.003	.006	.011	.013	.015	.016	.017	.024

b) $E_v/E_h = 1.0$

Room Depth/ Window Height	Percent D	Window Width/Window Height .5	1	2	3	4	6	8	Infinite
	10	.671	.704	.711	.715	.717	.726	.726	.728
	30	.458	.595	.654	.668	.672	.682	.683	.685
1	50	.313	.462	.563	.589	.598	.607	.608	.610
	70	.227	.362	.478	.515	.527	.530	.532	.534
	90	.186	.306	.424	.465	.481	.468	.471	.472
	10	.545	.636	.658	.660	.661	.665	.666	.672
	30	.239	.367	.459	.484	.491	.499	.501	.506
2	50	.121	.203	.286	.320	.335	.348	.351	.355
	70	.074	.128	.192	.226	.243	.259	.264	.267
	90	.058	.101	.156	.188	.207	.215	.221	.223
	10	.431	.561	.607	.613	.614	.616	.615	.631
	30	.133	.223	.306	.337	.348	.357	.357	.366
3	50	.058	.103	.155	.183	.197	.211	.213	.218
	70	.037	.064	.098	.119	.132	.147	.150	.154
	90	.030	.051	.079	.098	.110	.122	.126	.129
	10	.339	.482	.549	.560	.563	.566	.565	.593
	30	.078	.139	.204	.234	.247	.258	.260	.272
4	50	.033	.060	.094	.114	.126	.139	.143	.150
	70	.022	.039	.061	.074	.083	.095	.099	.104
	90	.019	.032	.050	.061	.070	.080	.084	.089
	10	.211	.343	.433	.453	.458	.461	.461	.518
	30	.033	.065	.103	.123	.135	.145	.148	.167
6	50	.015	.029	.047	.057	.064	.073	.077	.086
	70	.011	.021	.033	.040	.045	.051	.054	.060
	90	.010	.019	.028	.034	.038	.044	.046	.052
	10	.135	.238	.326	.353	.362	.366	.367	.452
	30	.016	.034	.058	.072	.080	.090	.094	.116
8	50	.008	.017	.027	.034	.039	.045	.048	.059
	70	.006	.013	.021	.026	.028	.032	.035	.043
	90	.005	.012	.019	.023	.025	.029	.031	.038
	10	.090	.165	.244	.272	.283	.290	.291	.395
	30	.009	.020	.036	.045	.052	.060	.064	.087
10	50	.005	.010	.019	.023	.026	.030	.033	.044
	70	.004	.009	.015	.018	.020	.023	.025	.033
	90	.003	.008	.014	.016	.018	.020	.022	.030

c) $E_v/E_h = 1.25$

Room Depth/ Window Height	Percent D	Window Width/Window Height .5	1	2	3	4	6	8	Infinite
	10	.578	.607	.614	.619	.621	.633	.634	.635
	30	.405	.525	.580	.594	.599	.612	.614	.615
1	50	.287	.423	.519	.547	.556	.569	.571	.573
	70	.218	.347	.461	.501	.515	.522	.525	.526
	90	.186	.307	.428	.473	.491	.483	.486	.487
	10	.472	.549	.566	.569	.570	.574	.575	.581
	30	.221	.337	.422	.447	.456	.465	.467	.472
2	50	.120	.202	.285	.321	.337	.353	.357	.361
	70	.078	.136	.204	.242	.261	.281	.287	.290
	90	.064	.112	.174	.211	.233	.244	.251	.253
	10	.377	.488	.527	.533	.534	.536	.536	.549
	30	.130	.217	.298	.329	.341	.352	.353	.362
3	50	.062	.110	.165	.195	.211	.228	.231	.237
	70	.040	.070	.109	.132	.147	.166	.171	.175
	90	.033	.057	.090	.112	.127	.142	.148	.152
	10	.300	.424	.484	.494	.497	.499	.499	.524
	30	.080	.143	.209	.240	.255	.267	.269	.283
4	50	.036	.066	.104	.126	.140	.156	.160	.168
	70	.024	.043	.068	.083	.094	.109	.115	.120
	90	.021	.036	.056	.070	.080	.092	.099	.103
	10	.193	.314	.395	.415	.420	.423	.423	.476
	30	.036	.071	.113	.136	.149	.161	.165	.186
6	50	.017	.033	.053	.065	.074	.084	.089	.100
	70	.012	.024	.037	.045	.050	.058	.061	.069
	90	.011	.021	.031	.038	.043	.049	.053	.060
	10	.128	.226	.310	.337	.346	.351	.352	.433
	30	.019	.039	.066	.082	.092	.104	.109	.134
8	50	.009	.019	.031	.040	.045	.052	.056	.069
	70	.007	.015	.023	.029	.032	.037	.040	.049
	90	.006	.013	.021	.025	.028	.032	.035	.043
	10	.088	.164	.241	.270	.282	.290	.291	.396
	30	.011	.024	.043	.054	.062	.071	.076	.103
10	50	.005	.012	.022	.026	.030	.035	.038	.052
	70	.004	.010	.017	.020	.023	.026	.028	.038
	90	.004	.009	.016	.018	.020	.023	.025	.034

d) $E_v/E_h = 1.50$

Room Depth/ Window Height	Percent D	.5	1	2	3	4	6	8	Infinite
		Window Width/Window Height							
1	10	.503	.528	.536	.541	.544	.557	.558	.559
	30	.359	.464	.514	.528	.534	.549	.550	.552
	50	.261	.384	.471	.499	.508	.524	.526	.527
	70	.204	.325	.432	.470	.485	.497	.499	.500
	90	.179	.295	.412	.456	.475	.474	.477	.478
2	10	.412	.477	.490	.492	.493	.498	.499	.505
	30	.201	.304	.379	.402	.410	.422	.424	.429
	50	.115	.192	.269	.304	.320	.339	.343	.347
	70	.078	.136	.204	.241	.261	.286	.292	.295
	90	.066	.117	.183	.221	.246	.262	.271	.273
3	10	.331	.426	.458	.461	.462	.465	.465	.477
	30	.121	.202	.275	.304	.316	.327	.329	.337
	50	.062	.109	.164	.193	.209	.228	.232	.238
	70	.041	.073	.114	.138	.154	.176	.183	.188
	90	.035	.062	.099	.123	.141	.159	.169	.173
4	10	.265	.372	.422	.430	.433	.435	.435	.456
	30	.077	.137	.199	.229	.243	.256	.259	.272
	50	.037	.069	.107	.130	.144	.161	.167	.175
	70	.026	.046	.073	.089	.101	.119	.126	.132
	90	.022	.039	.063	.078	.090	.106	.114	.120
6	10	.173	.281	.351	.368	.373	.375	.375	.422
	30	.037	.073	.115	.137	.151	.164	.168	.189
	50	.018	.036	.058	.071	.080	.092	.098	.110
	70	.013	.026	.040	.049	.056	.064	.069	.078
	90	.012	.023	.035	.043	.048	.057	.062	.070
8	10	.117	.207	.282	.305	.314	.319	.320	.393
	30	.020	.042	.071	.087	.098	.111	.116	.143
	50	.010	.021	.035	.044	.050	.058	.063	.078
	70	.007	.016	.026	.032	.036	.041	.045	.055
	90	.076	.014	.023	.028	.031	.036	.040	.049
10	10	.082	.153	.224	.250	.262	.269	.271	.368
	30	.012	.026	.047	.059	.068	.078	.084	.114
	50	.006	.014	.024	.030	.034	.040	.044	.060
	70	.005	.011	.019	.022	.025	.029	.032	.043
	90	.004	.010	.017	.020	.023	.026	.028	.038

e) $E_v/E_h = 1.75$

Room Depth/ Window Height	Percent D	.5	1	2	3	4	6	8	Infinite
		Window Width/Window Height							
1	10	.435	.457	.465	.471	.474	.486	.488	.489
	30	.317	.407	.452	.466	.471	.486	.488	.489
	50	.234	.343	.422	.447	.456	.472	.475	.476
	70	.187	.297	.395	.430	.445	.458	.461	.462
	90	.168	.276	.384	.426	.444	.447	.450	.451
2	10	.357	.412	.422	.424	.424	.430	.431	.436
	30	.180	.271	.335	.356	.363	.375	.378	.381
	50	.106	.177	.246	.278	.293	.313	.318	.321
	70	.074	.130	.194	.229	.249	.274	.282	.284
	90	.065	.116	.181	.219	.244	.264	.273	.276
3	10	.288	.369	.394	.397	.397	.400	.401	.411
	30	.110	.183	.247	.272	.282	.294	.296	.304
	50	.058	.104	.154	.181	.196	.215	.221	.226
	70	.040	.072	.112	.136	.152	.176	.184	.188
	90	.035	.063	.101	.126	.144	.166	.177	.182
4	10	.232	.324	.365	.371	.373	.375	.375	.394
	30	.071	.127	.183	.209	.222	.235	.238	.250
	50	.036	.067	.104	.125	.139	.157	.163	.171
	70	.025	.046	.072	.089	.101	.119	.127	.134
	90	.022	.041	.065	.082	.095	.114	.124	.130
6	10	.153	.247	.307	.320	.324	.326	.327	.367
	30	.035	.070	.109	.130	.143	.155	.160	.180
	50	.018	.036	.058	.071	.080	.091	.098	.110
	70	.013	.026	.041	.051	.058	.067	.073	.082
	90	.012	.023	.037	.046	.052	.062	.069	.078
8	10	.104	.184	.249	.269	.276	.281	.282	.346
	30	.020	.042	.070	.086	.096	.109	.115	.141
	50	.010	.022	.036	.046	.052	.060	.066	.081
	70	.008	.017	.027	.033	.038	.044	.048	.059
	90	.007	.015	.024	.030	.034	.040	.044	.054
10	10	.074	.138	.201	.223	.233	.240	.242	.328
	30	.012	.027	.048	.059	.067	.078	.084	.114
	50	.006	.014	.026	.032	.036	.043	.047	.064
	70	.005	.011	.020	.024	.027	.031	.034	.046
	90	.004	.010	.018	.022	.024	.028	.031	.042

f) Ground component

Room Infinite Depth/ Window Height	Percent D	.5	1	2	3	4	6	8	
		Window Width/Window Height							
1	10	.105	.137	.177	.197	.207	.208	.210	.211
	30	.116	.157	.203	.225	.235	.241	.243	.244
	50	.110	.165	.217	.241	.252	.267	.269	.270
	70	.101	.162	.217	.243	.253	.283	.285	.286
	90	.091	.146	.199	.230	.239	.290	.292	.293
2	10	.095	.124	.160	.178	.186	.186	.189	.191
	30	.082	.132	.179	.201	.212	.219	.222	.225
	50	.062	.113	.165	.189	.202	.214	.218	.220
	70	.051	.093	.141	.165	.179	.194	.198	.200
	90	.045	.079	.118	.140	.153	.179	.183	.185
3	10	.088	.120	.157	.175	.183	.185	.163	.167
	30	.059	.107	.154	.176	.187	.198	.193	.198
	50	.039	.074	.114	.134	.146	.157	.166	.170
	70	.031	.055	.085	.101	.111	.122	.127	.130
	90	.028	.047	.070	.083	.092	.107	.113	.115
4	10	.073	.113	.154	.174	.183	.187	.176	.184
	30	.040	.082	.127	.148	.159	.170	.177	.185
	50	.025	.049	.078	.094	.103	.113	.117	.123
	70	.020	.036	.054	.065	.071	.079	.083	.087
	90	.019	.032	.046	.054	.060	.069	.073	.076
6	10	.056	.106	.143	.164	.175	.184	.173	.194
	30	.021	.050	.081	.098	.107	.117	.123	.138
	50	.013	.027	.041	.049	.054	.060	.064	.072
	70	.011	.021	.029	.033	.035	.039	.041	.046
	90	.011	.020	.026	.030	.032	.035	.037	.042
8	10	.036	.082	.122	.143	.156	.166	.170	.208
	30	.011	.029	.050	.062	.070	.078	.082	.101
	50	.007	.016	.024	.028	.031	.035	.038	.046
	70	.000	.013	.010	.020	.021	.023	.026	.030
	90	.006	.013	.017	.019	.020	.022	.023	.028
10	10	.024	.061	.109	.120	.131	.144	.147	.200
	30	.006	.017	.034	.040	.046	.053	.056	.076
	50	.004	.010	.016	.018	.020	.023	.024	.033
	70	.004	.009	.013	.014	.015	.016	.016	.022
	90	.004	.009	.013	.013	.014	.015	.016	.021

Source: IESNA Lighting Handbook, 9th edition, 2000. Courtesy of the Illuminating Engineering Society of North America.

Note: Percent D is the relative distance from the window to the opposite wall.

Window:	W = 40 ft, H = 7 ft, Sill height = 3 ft, 3/16" grey glass
Frame:	20% of opening area
Light Loss:	WDD = 0.83, RSDD = 0.93

Solution. From Table 12.3, the solar altitude is 50°. Then from Eq. (12.26), $E_h = 1520$ fc and, from Eq. (12.28), $E_v = 609$ fc. From Table 12.7, the ground surface reflectance is 0.40 and, from Eq. (12.54),

$$E_{wg} = \frac{0.40 \times 1520}{2} = 304 \text{ fc}$$

From Table 12.12, the glass transmittance is 51%. The value of R_a is 0.8. There are no window controls. Thus, from Eq. (12.53),

$$\tau = 0.51 \times 0.80 = 0.41$$

The light loss factor is

$$LLF = 0.83 \times 0.93 = 0.77$$

The vertical half-sky illuminance is 609 fc, as calculated earlier from Eq. (12.28). The horizontal half-sky illuminance is $\frac{1520}{2} = 760$ fc, half the value given by Eq. (12.26). Thus the ratio $E_v/E_h = 609/760 = 0.80$. In Table 12.13, the room depth to window height ratio is D/H = 30/7 = 4.3. The room width to window height ratio is W/H = 40/7 = 5.7. Approximating these values as 4 and 6, respectively, and using interpolation between Tables 12.13a and b gives the CU values in Table 12.14. The ground component of CU is obtained from Table 12.13f and is also entered in Table 12.14. Then, using Eq. (12.55), the maintained illuminances can be calculated. These also appear in Table 12.14.

TABLE 12.14 VALUES FOR EXAMPLE 12.12

Percent D	CU_s	CU_g	E_i fc
10	0.653	0.187	144
30	0.246	0.170	64
50	0.117	0.113	33
70	0.077	0.079	22
80	0.063	0.069	19

EXAMPLE 12.13

Repeat Example 12.12 for a clear sky with the window facing east.

Solution. The solar altitude is still 50° and the solar azimuth is 0° (noon). Thus there is no direct sun on the window and the lumen method of sidelighting applies. From Eq. (12.20), $E_h = 1330$ fc; from Table 12.6, $E_{hs} = 6900$ fc; and from Fig. 12.10b, for a solar-window azimuth of 90°, $E_v = 700$ fc. From Eq. (12.54),

$$E_{wg} = \frac{0.40(1330 + 6900)}{2} = 1650 \text{ fc}$$

The values of τ and LLF remain the same.

The vertical half-sky illuminance is 700 fc and the horizontal half-sky illuminance, from Fig. 12.10a, is also 700 fc. This gives an $E_v/E_h = 1.0$, calling for Fig. 12.13b. CU_g values are unchanged. The values of CU_s and the illuminances are shown in Table 12.15.

TABLE 12.15 VALUES FOR EXAMPLE 12.13

Percent D	CU_s	CU_g	E_i fc
10	.566	.187	222
30	.258	.170	146
50	.139	.113	90
70	.095	.079	62
90	.080	.069	54

12.15 DAYLIGHT FACTOR METHOD

The procedures described in the previous two sections are popular in the United States for obtaining horizontal work-plane illuminance in daylit interiors. In Europe, where cloudy skies predominate, another procedure, called the daylight factor method, is widely used.

Daylight factor (DF) is defined as the ratio of the illuminance at a point on a plane produced by the luminous flux received directly or indirectly at that point from a sky of a given luminance distribution to the illuminance on a horizontal plane produced by an unobstructed hemisphere of this same sky.

In the above definition, direct sunlight is excluded. Thus the basic daylight factor method, as it has been applied abroad, has been limited to overcast sky cases, in particular to the uniform overcast sky and to the CIE overcast sky. Work in this country has sought to extend the method to include clear skies as well, but the complications are severe. In this section, treatment will be limited to overcast skies.

The three ways in which daylight may reach a point on a horizontal plane within a room are shown in Fig. 12.22. The sky component (SC) is that portion of the daylight

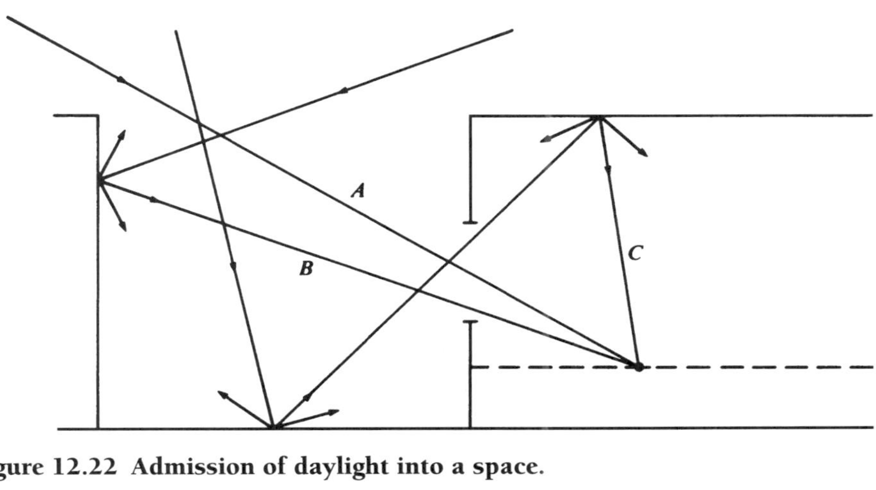

Figure 12.22 Admission of daylight into a space.

factor due to daylight received directly at the point from the sky. It is indicated by ray A in Fig. 12.22. The externally reflected component (ERC) is that portion of the daylight factor due to daylight received directly at the point from external reflecting surfaces. Ray B in Fig. 12.22 depicts this situation. The internally reflected component (IRC) is that portion of the daylight factor due to daylight that reaches the point from internal reflecting surfaces. A typical situation of this type is shown by ray C in Fig. 12.22. Note that the IRC includes light that enters the space directly and then is reflected to the point, as well as light that is reflected into the space from external surfaces and then is further reflected before reaching the point.

The daylight factor is the sum of its three components, given by

$$DF = SC + ERC + IRC \tag{12.57}$$

Each of these components will be examined in greater detail and procedures for calculating them will be developed.

Sky Component

The situation for the sky component is very similar to that for reflected ground light in Sect. 12.7, where the sky replaces the ground. The pertinent equation is Eq. (12.34). First, four angles are defined in Fig. 12.23 and using Eqs. (12.58).

$$\left.\begin{aligned} \sin\gamma &= \frac{h}{\sqrt{h^2+q^2}} \\ \sin\beta &= \frac{w}{\sqrt{w^2+q^2}} \\ \sin\gamma_1 &= \frac{h}{\sqrt{w^2+h^2+q^2}} \\ \sin\beta_1 &= \frac{w}{\sqrt{w^2+h^2+q^2}} \end{aligned}\right\} \quad 12.58$$

Inserting Eqs. (12.58) into Eq. (12.34) gives

$$E_w = \frac{L}{2}(\beta - \beta_1 \cos\gamma) \tag{12.59}$$

where “w” denotes window.

Consider the uniform sky case, where L is constant. From Eq. (2.39), the horizontal illuminance produced by such an unobstructed sky is

$$E_h = \pi L \tag{12.60}$$

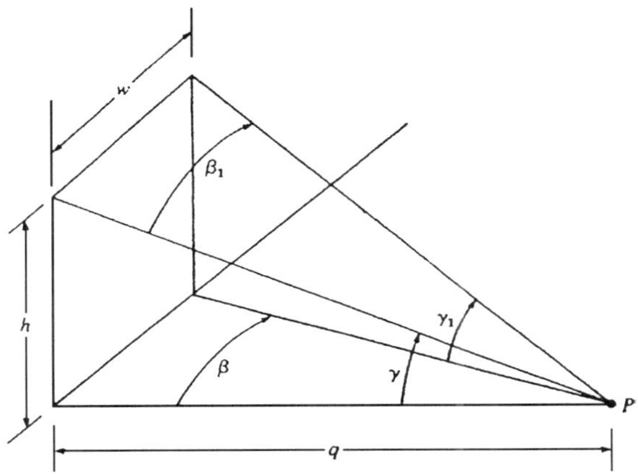

Figure 12.23 Pertaining to sky component.

Assume the window glazing has a transmittance τ. Then the sky component may be written as

$$SC_{uni} = \frac{E_w \tau}{E_h} = \frac{\tau}{2\pi}(\beta - \beta_1 \cos\gamma) \tag{12.61}$$

For the CIE overcast sky case, for which the sky luminance is given by Eqs. (12.21) and (12.22), a derivation similar to that leading to Eq. (12.61) yields

$$SC_{CIE} = \frac{\tau}{7\pi}\left[\frac{3}{2}(\beta - \beta_1 \cos\gamma) + 2\sin^{-1}(\sin\beta \sin\gamma) - \sin 2\gamma \sin\beta_1\right] \tag{12.62}$$

The Building Research Station at Watford in England has prepared tables of SC values (in %) for both the uniform and CIE overcast skies, as functions of h/q and w/q (Hopkinson et al., 1966). These are reproduced as Tables 12.17 and 12.18, based on the geometry in Fig. 12.24 (where w is window width). The tables include an allowance for the transmittance of a single sheet of clear glass (about 85%). They are based on the diagram in Fig. 12.23, with the point P on a line perpendicular to the lower left corner of the window. For other locations of point P, the method of superposition, illustrated in Example 12.8, should be used to obtain the net sky component, except that, if a portion of the window lies below the work plane, that portion should be neglected in sky component calculations.

It remains to consider the effect of external obstructions (buildings, etc.) on the sky component. Consider the situation depicted in Fig. 12.24. A building of height H located a distance D from the window partially obscures the sky from the reference point P. The angle of obstruction is defined by

$$\alpha = \tan^{-1}\frac{H - 2.5}{D + q} \tag{12.63}$$

TABLE 12.17 SKY COMPONENTS (%)—UNIFORM OVERCAST SKY

h/q \ w/q	0.1	0.2	0.3	0.4	0.5	0.6	0.8	1.0	1.5	2.0	3.0	6.0	α°
∞	1.3	2.6	3.9	5.0	6.1	7.0	8.8	10.1	12.3	13.6	14.7	15.3	90
5.0	1.3	2.6	3.8	5.0	6.0	7.0	8.7	10.0	12.0	13.1	14.0	14.6	79
3.0	1.2	2.5	3.6	4.8	5.7	6.7	8.2	9.4	11.4	12.4	13.1	13.5	72
2.0	1.1	2.3	3.2	4.2	5.1	5.9	7.2	8.3	9.9	10.7	11.3	11.5	63
1.8	1.0	2.2	3.0	4.0	4.8	5.6	6.8	7.8	9.4	10.1	10.6	10.8	61
1.6	0.97	2.0	2.9	3.8	4.5	5.2	6.4	7.3	8.6	9.3	9.8	10.0	58
1.4	0.91	1.9	2.7	3.5	4.2	4.8	5.9	6.7	7.8	8.5	8.8	9.0	54
1.2	0.82	1.7	2.4	3.1	3.8	4.3	5.2	5.9	6.9	7.4	7.7	7.8	50
1.0	0.72	1.5	2.1	2.7	3.2	3.7	4.5	5.0	5.8	6.2	6.3	6.5	45
0.9	0.65	1.4	1.9	2.4	2.9	3.4	4.0	4.5	5.1	5.5	5.6	5.7	42
0.8	0.57	1.2	1.6	2.1	2.6	3.0	3.6	4.0	4.5	4.7	4.8	4.9	39
0.7	0.50	1.0	1.4	1.8	2.2	2.5	3.0	3.3	3.7	3.9	4.0	4.0	35
0.6	0.40	0.83	1.1	1.5	1.8	2.1	2.4	2.7	2.9	3.1	3.2	3.2	31
0.5	0.30	0.63	0.86	1.2	1.4	1.6	1.8	2.0	2.1	2.3	2.3	2.4	27
0.4	0.21	0.43	0.59	0.80	0.94	1.1	1.2	1.4	1.5	1.6	1.6	1.7	22
0.3	0.13	0.25	0.33	0.46	0.54	0.64	0.73	0.82	0.87	0.92	0.94	0.96	17
0.2	0.05	0.11	0.16	0.22	0.27	0.31	0.36	0.40	0.44	0.46	0.47	0.48	11
0.1	0.01	0.02	0.04	0.07	0.09	0.11	0.13	0.14	0.15	0.16	0.17	0.18	6

Source: R.G. Hopkinson, P. Petherbridge, and J. Longmore, "Simplified Daylight Tables." National Building Studies Report #26, HMSO, London, 1958; reprinted in R.G. Hopkinson, P. Petherbridge, and J. Longmore, *Daylighting*, Heineman, London, 1966.

TABLE 12.18 SKY COMPONENTS (%)—CIE OVERCAST SKY

h/q \ w/q	0.1	0.2	0.3	0.4	0.5	0.6	0.8	1.0	1.5	2.0	3.0	6.0	α°
∞	1.3	2.5	3.7	4.9	5.9	6.9	8.4	9.6	11.9	13.0	14.2	14.9	90
5.0	1.2	2.4	3.7	4.8	5.9	6.8	8.3	9.4	11.4	12.7	13.7	14.1	79
3.0	1.2	2.3	3.5	4.5	5.5	6.4	7.8	8.7	10.4	11.7	12.4	12.6	72
2.0	1.0	2.0	3.1	4.0	4.8	5.6	6.7	7.5	8.9	9.7	10.0	10.2	63
1.8	0.97	1.9	2.9	3.8	4.6	5.3	6.3	7.1	8.3	9.0	9.3	9.5	61
1.6	0.90	1.8	2.7	3.5	4.2	4.9	5.8	6.5	7.6	8.2	8.5	8.6	58
1.4	0.82	1.6	2.4	3.2	3.8	4.4	5.2	5.9	6.8	7.3	7.5	7.6	54
1.2	0.71	1.4	2.1	2.7	3.3	3.8	4.5	5.0	5.8	6.1	6.2	6.3	50
1.0	0.57	1.1	1.7	2.2	2.6	3.0	3.6	4.0	4.5	4.7	4.8	5.0	45
0.9	0.50	0.99	1.5	1.9	2.2	2.6	3.1	3.4	3.8	4.0	4.1	4.2	42
0.8	0.42	0.83	1.2	1.6	1.9	2.2	2.6	2.9	3.2	3.3	3.4	3.4	39
0.7	0.33	0.68	0.97	1.3	1.5	1.7	2.1	2.3	2.5	2.6	2.7	2.8	35
0.6	0.24	0.53	0.74	0.98	1.2	1.3	1.6	1.8	1.9	2.0	2.1	2.1	31
0.5	0.16	0.39	0.52	0.70	0.82	0.97	1.1	1.3	1.4	1.5	1.5	1.5	27
0.4	0.10	0.25	0.34	0.45	0.54	0.62	0.75	0.89	0.95	0.96	0.97	0.98	22
0.3	0.06	0.14	0.18	0.26	0.30	0.34	0.42	0.47	0.50	0.51	0.52	0.53	17
0.2	0.03	0.06	0.09	0.11	0.12	0.14	0.20	0.21	0.22	0.23	0.23	0.24	11
0.1	0.01	0.02	0.02	0.03	0.03	0.04	0.05	0.05	0.06	0.07	0.07	0.08	6

Source: R.G. Hopkinson, P. Petherbridge, and J. Longmore, "Simplified Daylight Tables." National Building Studies Report #26, HMSO, London, 1958; reprinted in R.G. Hopkinson, P. Petherbridge, and J. Longmore, *Daylighting*, Heineman, London, 1966.

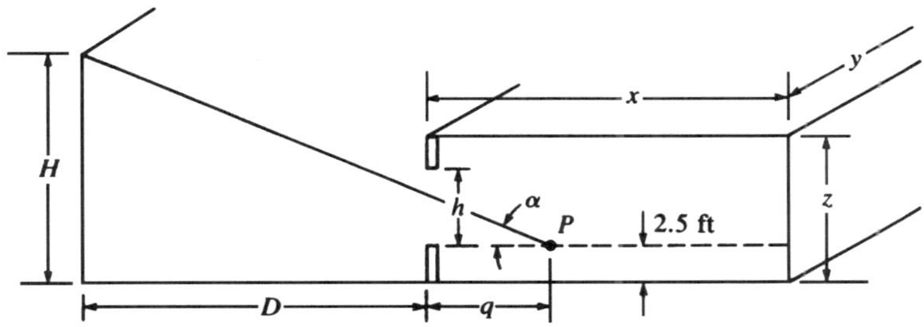

Figure 12.24 Geometry applicable to Tables 12.17 and 12.18.

Obstruction angles are included in Tables 12.17 and 12.18. For a given w/q, one enters either table at the appropriate value of α and obtains the sky component to be subtracted because of the obstruction. Alternatively, one may compute the (H − 2.5)/(D + q) ratio and enter the tables directly to obtain the amount to be subtracted.

EXAMPLE 12.14

In Fig. 12.24, let H = 27.5 ft, D = 40 ft, h = 7 ft and q = 10 ft. If the window is 16 ft wide and point P is on a line perpendicular to the bottom center of the window, find the net sky component if the glazing is a single clear glass sheet and the sky is CIE overcast.

Solution.

$$\tan\alpha = \frac{H-2.5}{D+q} = 0.5$$

$$\frac{h}{q} = 0.7, \quad \frac{w}{q} = 0.8$$

From Table 12.18, the sky component for the unobstructed window is 2(2.1) = 4.2%. For the obstruction, α = 27° and the sky component is 2(1.1) = 2.2%. Thus the net sky component is

$$SC = 4.2 - 2.2 = 2.0\%$$

Externally Reflected Component

The externally reflected component is small and, as long as there are overcast skies and surfaces of medium to low reflectances, approximations may be used in its calculation. For a uniform overcast sky, it is customary to take the luminance of the obstruction as 0.10 of the sky luminance. For the CIE overcast sky, the average luminance may be taken as $\frac{7}{9}L_{zo} = 0.778\ L_{zo}$, from Eq. (12.25). Very roughly, the near-horizon average sky luminance is one-half of this value. Thus, for the CIE overcast sky, it is customary to take the luminance of obstructions as 0.20 of the near-horizon average sky luminance if the obstruction angle is less than 20°. Otherwise a value of 0.10 is used.

With these luminance modifications, the externally reflected component can be obtained by using Tables 12.17 and 12.18 and the procedure explained previously for obtaining the sky component due to an obstruction. For the uniform sky, we multiply the sky component for the obstruction by 0.10; for the CIE sky by 0.10 or 0.20, depending on α.

EXAMPLE 12.15

Find the externally reflected component for the situation in Example 12.14.

Solution. The sky is CIE overcast. From Example 12.14, the sky component for the obstruction was 2.2%. The obstruction angle is greater than 20°, so the externally reflected component is

$$\text{ERC} = 0.1\ (2.2) = 0.22\%$$

Note that the ERC is indeed relatively small, only 11% of the net sky component.

Internally Reflected Component

The procedure for obtaining the internally reflected component is based on the theory of the integrating sphere, presented in Sect. 2.10, and the split-flux principle (Hopkinson et al., 1966). As shown in Fig. 12.25, the split-flux principle divides the flux entering the room through the window into two components by inserting a horizontal plane at the middle of the window. One component of flux is that entering the room directly from the sky or by reflection from obstructions above the dividing plane. The other component is produced entirely by reflections, from ground surfaces and from obstructions below the dividing plane.

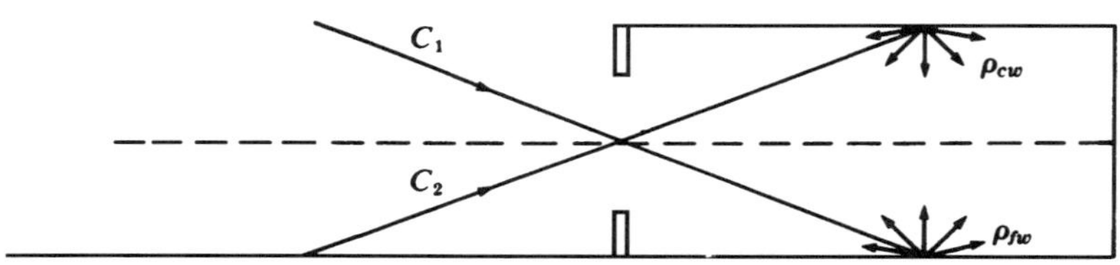

Figure 12.25

The indirect illuminance at any point on the inside of an integrating sphere is shown in Eq. (2.50) to be

$$E = \frac{\rho\phi}{4\pi r^2(1-\rho)} \tag{12.64}$$

where ρ is the sphere reflectance, r is its radius, and ϕ is the flux entering the space. It follows that $4\pi r^2$ is the surface area of the enclosure and $\rho\phi$ is the flux resulting from the first reflection from that surface area.

The split-flux procedure assumes that the result in Eq. (12.64), which is for a perfectly diffusing sphere with a single value of reflectance, may be applied to a rectangular room with a variety of reflectances. Thus, in the denominator of Eq. (12.64), $4\pi r^2$ is replaced by A, the sum of the areas of the ceiling, floor, and walls, including the window, and ρ is replaced by ρ_{av}, the weighted-average reflectance of the room surfaces, again including the window with an assumed reflectance of 15%. The numerator of Eq. (12.64) is altered to express the first reflected flux ϕ_r as the sum of two fluxes:

$$\phi_r = \tau E_{w1} A_w \rho_{fw} + \tau E_{w2} A_w \rho_{cw} \tag{12.65}$$

where E_{w1} and E_{w2} are the illuminances on the outside of the window due to the fluxes above and below, respectively, the dividing plane in Fig. 12.25 and ρ_{fw} and ρ_{cw} are the weighted-average reflectances of the lower and upper portions, respectively, of the room, each excluding the window wall.

The average internally reflected component is the ratio of the illuminance in the modified form of Eq. (12.64) to the illuminance E_H from the unobstructed sky hemisphere. Thus it can be written (IRC in %)

$$\begin{aligned} \mathrm{IRC} &= \frac{\tau E_{w1} A_w \rho_{fw} + \tau E_{w2} A_w \rho_{cw}}{E_H A(1-\rho_{av})} \\ &= \frac{\tau A_w}{A(1-\rho_{av})}\left(\frac{\rho_{fw} E_{w1}}{E_H} + \frac{\rho_{cw} E_{w2}}{E_H}\right) \\ &= \frac{\tau A_w}{A(1-\rho_{av})}(C\rho_{fw} + 0.05\rho_{cw}) \end{aligned} \tag{12.66}$$

The final form of Eq. (12.66) requires explanation. In the first term inside the parentheses, E_{w1}/E_H has been replaced by C. Values of this constant have been calculated for various obstruction angles and are tabulated in Table 12.19 for the CIE overcast sky, with the luminance of the obstruction assumed to be 0.10 of the average sky luminance and with the obstruction assumed to extend to infinity laterally.

TABLE 12.19

$\alpha°$	C(%)
0	39
10	35
20	31
30	25
40	20
50	14
60	10
70	7
80	5

In the second term in parentheses, the ratio E_{w2}/E_H is replaced by 0.05. The assumption is made that E_{w2} is produced by a perfectly diffusing ground surface whose reflectance is 10%. From Eq. (12.37), with $\alpha_1 = 0°$, $\alpha_2 = 90°$ and $\rho = 0.1$, the value of E_{w2} is seen to be 0.05 E_H.

EXAMPLE 12.16

For the situation in Example 12.14, assume the room has dimensions x = 20 ft, y = 30 ft, and z = 11 ft (Fig. 12.24). Further assume that $\rho_c = 80\%$, $\rho_w = 50\%$, and $\rho_f = 30\%$. Find the internally reflected component and the overall daylight factor.

Solution.

$$A_w = 7(16) = 112 \text{ ft}^2$$

$$A = 2(30)(20) + 2(30)(11) + 2(20)(11) = 2300 \text{ ft}^2$$

The dividing plane is located 6 ft above the floor. Denoting the areas above and below this plane by A_a and A_b,

$$A_a = 30(20) + 30(5) + 2(20)(5) = 950 \text{ ft}^2$$

$$A_b = 30(20) + 30(6) + 2(20)(6) = 1020 \text{ ft}^2$$

The weighted-average reflectances are

$$\rho_{av} = \frac{0.8(600) + 0.3(600) + 0.5(988) + 0.15(112)}{2300} = 0.51$$

$$\rho_{cw} = \frac{0.8(600) + 0.5(350)}{950} = 0.69$$

$$\rho_{fw} = \frac{0.3(600) + 0.5(420)}{1020} = 0.38$$

where $\tau = 85\%$ and thus window reflectance is 15%.

From Table 12.19 for the given obstruction angle of 27°, interpolation yields C = 27%. Now Eq. (12.66) gives

$$\text{IRC} = \frac{0.85(112)}{2300(1-0.51)}[0.27(0.38) + 0.05(0.69)] = 0.012$$

Thus the internally reflected component is 1.2% and, from Eq. (12.57) the daylight factor is

$$\text{DF} = 2.0 + 0.2 + 1.2 = 3.4\%$$

As an alternative to Eq. (12.66), Table 12.20 may be used to obtain the minimum IRC (points farthest from the window) for a standardized room of floor area 36 m^2 (387 ft^2), ceiling height 3m (9.8 ft), ceiling reflectance ρ_c = 70% and obstruction angle $\alpha = 20°$.

TABLE 12.20 MINIMUM VALUES OF IRC (%)

Window Area as Percentage of Floor Area	Floor Reflectance (%) 10				20				40			
	Wall Reflectance (%)											
	20 (%)	40 (%)	60 (%)	80 (%)	20 (%)	40 (%)	60 (%)	80 (%)	20 (%)	40 (%)	60 (%)	80 (%)
2	—	—	0.1	0.2	—	0.1	0.1	0.2	—	0.1	0.2	0.2
5	0.1	0.1	0.2	0.4	0.1	0.2	0.3	0.5	0.1	0.2	0.4	0.6
7	0.1	0.2	0.3	0.5	0.1	0.2	0.4	0.6	0.2	0.3	0.6	0.8
10	0.1	0.2	0.4	0.7	0.2	0.3	0.6	0.9	0.3	0.5	0.8	1.2
15	0.2	0.4	0.6	1.0	0.2	0.5	0.8	1.3	0.4	0.7	1.1	1.7
20	0.2	0.5	0.8	1.4	0.3	0.6	1.1	1.7	0.5	0.9	1.5	2.3
25	0.3	0.6	1.0	1.7	0.4	0.8	1.3	2.0	0.6	1.1	1.8	2.8
30	0.3	0.7	1.2	2.0	0.5	0.9	1.5	2.4	0.8	1.3	2.1	3.3
35	0.4	0.8	1.4	2.3	0.5	1.0	1.8	2.8	0.9	1.5	2.4	3.8
40	0.5	0.9	1.6	2.6	0.6	1.2	2.0	3.1	1.0	1.7	2.7	4.2
45	0.5	1.0	1.8	2.9	0.7	1.3	2.2	3.4	1.2	1.9	3.0	4.6
50	0.6	1.1	1.9	3.1	0.8	1.4	2.3	3.7	1.3	2.1	3.2	4.9

Correction factors for different floor areas, different ceiling reflectances and to convert from minimum to average IRC are given in Table 12.21.

EXAMPLE 12.17

For the values in Examples 12.14 to 12.16, find the average IRC using Tables 12.20 and 12.21.

Solution. The window area as a percentage of the floor area is

$$\frac{WA}{FA} = \frac{7\times16}{20\times30}\times100 = 18.7\%$$

The floor reflectance is 30% and the wall reflectance is 50%. Thus a triple interpolation in Table 12.20 is needed. The result is IRC = 0.96%. The actual floor area is 600 ft^2. The wall reflectance is 50%. From Table 12.19a, a correction factor of 1.03 is obtained. The ceiling reflectance is 80%, giving a multiplier of 1.1 from Table 12.21b. A 1.35 multiplier is needed from Table 12.21c to convert from minimum to average IRC. The product of the multipliers is 1.53, giving an average IRC of 1.47%.

This is higher than the more accurate value given by Eq. (12.66). No correction factors are given for ceiling height and obstruction angle. The latter is greater (27°) than that for Table 12.80 (20°). In addition, the point P in Fig. 12.24 is 10 ft from the window in a 20-ft deep room. The IRC at this point may or may not equal the average IRC. Thus Eq. (12.66) is to be preferred in IRC calculations.

TABLE 12.21 IRC CONVERSION FACTORS

(a) Different Floor Areas

Floor Area m^2 (ft^2)	Wall Reflectance (%) 20	40	60	80
	Conversion Factor			
1 (10.8)	0.6	0.7	0.8	0.9
36 (388)	1.0	1.0	1.0	1.0
100 (1080)	1.4	1.2	1.0	0.9

(b) Different Ceiling Reflectances

Ceiling Reflectance (%)	Conversion Factor
40	0.7
50	0.8
60	0.9
70	1.0
80	1.1

(c) From Minimum to Average IRC

Wall Reflectance (%)	Conversion Factor
20	1.8
40	1.4
60	1.3
80	1.2

Very often the average daylight factor on a horizontal reference plane, usually the work plane, is desired. This can be obtained from the following empirical equation:

$$ADF = \frac{\tau A_g \theta}{A_s(1-\rho^2)}\% \qquad (12.67)$$

where: τ is the decimal transmittance of the glazing
A_g is the net glazing area
A_s is the total interior surface area, including windows, in the same units as A_g
ρ is the area-weighted average reflectance of all interior room surfaces, including windows
θ is the angle in degrees in the vertical plane subtended by the portion of the sky that is visible from the center of the window.

Experience has shown that when ADF is 5% or greater, an interior space appears well-lit. When the ADF is 3% or less, the interior space appears dimly-lit.

The ADF is a suitable metric for shallow side-lighted rooms and for rooms lighted by skylights. It is not a valid metric for deep side-lighted rooms because of the effect of the exponential decay of daylighting as distance from the windows increases. Limiting room depth is 3 to 4 window heights above the work plane.

EXAMPLE 12.18

For the situation in Examples 12.14 to 12.17, find the average daylight factor.

Solution. The glazing transmittance τ is assumed to be 85%. The net glazing area is $A_g = 7 \times 16 = 112$ ft^2. The total interior surface area is $A_s = 2(20 \times 30 + 20 \times 11 + 30 \times 11) = 2300$ ft^2. The weighted average reflectance ρ of all interior room surfaces is 0.51 from Example 12.16. The angle θ is $90\text{-}\alpha = 90–27 = 63°$. Inserting values into Eq. (12.67) yields

$$\text{ADF} = \frac{0.85 \times 112 \times 63}{2300(1 - 0.51^2)} = 3.5\%$$

12.16 DAYLIGHTING ENERGY ANALYSIS

In this section a procedure is presented for calculating the lighting energy that can potentially be saved by turning off some or all of the electric lighting in a room that has skylights. It is not a procedure for determining the total energy balance within a space, although it could be extended to do so by including heating and air conditioning considerations and the thermal losses of skylights. Rather it is a procedure for determining the lighting energy tradeoffs between electric lighting and skylighting based on the ability of each to provide average horizontal footcandles on a work plane.

To illustrate how the procedure works, consider a certain skylighting installation that provides 30 fc on the work plane at a specified latitude, date, time, and sky condition. Further assume that the electric lighting provides 50 fc. Then, if the electric lighting can be switched on and off in parts, it is possible to turn off 60 percent of the electric lighting under the assumed conditions. If the entire electric lighting installation, including ballasting, consumes P watts, then 0.6P watts can be saved. Since each watt represents 3.413 Btu/h, the saving is 2.048P Btu/h. If, on another day, the skylighting produces 60 fc, then all of the electric lighting can be turned off and the saving is 3.413P Btu/h. In this latter case, there is more illumination in the room space than required, but credit can only be taken for the first 50 fc.

It is relatively straightforward, though computationally laborious, to find the number of Btu's saved per year. In the procedure to be presented here, the simplest situation will be considered, namely turning none of the electric lighting off until the work-plane illuminance provided by skylighting exceeds the design illuminance, and then turning it all off. Thus either 3.413P Btu/h will be saved or nothing will be saved. The procedure will be presented in a step-by-step manner, and an example worked to illustrate the necessary calculations.

Step 1 Use the lumen method described in Chap. 9 to find the number of watts (including ballast watts) of electric lighting required to produce the desired average work-plane illuminance level within the given room.

Step 2 Use the lumen method of skylighting described in Sect. 12.13 to find the external horizontal illuminance on the skylights required to produce the desired average work-plane illuminance. This is done for both the overcast sky and the clear sky plus direct sun cases.

Step 3 From Table 12.6, determine the solar altitude at which the illuminance determined in Step 2 is achieved, again for both the overcast sky and clear sky plus direct sun cases.

Step 4 For the given latitude, determine the number of hours on a given day that the solar altitude exceeds the value obtained in Step 2 for both sky cases. Generally this is done on the 21st day of each month using data such as that presented in Table 12.22 (for 40° north latitude as an example). Such data may be calculated from Eq. (12.15).

TABLE 12.22 SOLAR ALTITUDE (IN DEGREES) VERSUS DATE AND TIME FOR 40°N LATITUDE

	Solar time (A.M.)							
Date	5	6	7	8	9	10	11	12
Jan. 21				8.1	16.8	23.8	28.4	30.0
Feb. 21			4.3	14.8	24.3	32.1	37.3	39.2
Mar. 21			11.4	22.5	32.8	41.6	47.7	50.0
Apr. 21		7.4	18.9	30.3	41.3	51.2	58.7	61.6
May 21	1.9	12.7	24.0	35.4	46.8	57.5	66.2	70.0
June 21	4.2	14.8	26.0	37.4	48.8	59.8	69.2	73.5
July 21	2.3	13.1	24.3	35.8	47.2	57.9	66.7	70.6
Aug. 21		7.9	19.3	30.7	41.8	51.7	59.3	62.3
Sept. 21			11.4	22.5	32.8	41.6	47.7	50.0
Oct. 21			4.5	15.0	24.5	32.4	37.6	39.5
Nov. 21				8.2	17.0	24.0	28.6	30.2
Dec. 21				5.5	14.0	20.7	25.0	26.6

Step 5 For the location of interest, determine the percentages of clear and overcast days each month. This data, by city, is available from the U.S. Weather Bureau. In Table 12.23, the percentages are shown for New York City (40°N latitude).

From this data and the data of Step 4, determine the total number of hours each month, for both sky cases, that the work-plane illuminance provided by skylighting exceeds the prescribed value. Sum these figures to determine the number of hours per year that this is the case. Multiply the number of hours by 3.413P to determine the number of Btu's per year saved by turning off the electric lighting.

EXAMPLE 12.19

Assume the installation in Example 12.10 is changed from 6 to 12 skylights with all other factors remaining the same, and it is located in New York City. Further assume that an average maintained horizontal illuminance of 30 fc is desired. The electric

TABLE 12.23 PERCENTAGE OF CLEAR AND OVERCAST DAYS NEW YORK CITY (40°N LATITUDE)

Period	Days	% Clear	% Overcast
Dec. 22–Jan. 21	31	52	48
Jan. 22–Feb. 21	31	57	43
Feb. 22–Mar. 21	28	59	41
Mar. 22–Apr. 21	31	60	40
Apr. 22–May 21	30	61	39
May 22–June 21	31	62	38
June 22–July 21	30	64	36
July 22–Aug. 21	31	63	37
Aug. 22–Sept. 21	31	63	37
Sept. 22–Oct. 21	30	62	38
Oct. 22–Nov. 21	31	57	43
Nov. 22–Dec. 21	30	53	47

lighting is to be provided by luminaire 7 in Table 9.7, with two F032T8 lamps and electronic ballasts. An LLF = 0.74 is assumed. Find the number of Btu's per year that may be saved if the electric lighting is turned off whenever the illuminance level from skylighting exceeds 30 fc.

Solution.

Step 1 Proceeding through the steps of the lumen method procedure in Sects. 9.5 to 9.8 shows that 15 two-lamp fluorescent luminaires are needed, each consuming 62 W (from Table 6.15). Thus the total wattage consumption is 930 W.

Step 2 With little error, the value of $T_{Dn} = 0.36$ in Example 12.10 maybe used for both the direct sun and clear sky transmittances. With this simplification, the horizontal illuminances on the work plane for 12 skylights from Step 4 in Example 12.10 are

$$E_{io} = 0.0191\ E_{ho}$$

$$E_{ics} = 0.0202\ (E_{hc} + E_{hs})$$

Thus, to achieve 30 fc,

$$E_{ho} = 52.4\ E_{io} = 1570 \text{ fc}$$

$$E_{hc} + E_{hs} = 49.5\ E_{ics} = 1485 \text{ fc}$$

Step 3 From Table 12.6, with interpolation, the solar altitudes at which the illuminance levels in Step 2 are reached are

$$h = 52° \quad \text{(overcast sky)}$$

$$h = 11° \quad \text{(clear sky plus direct sun)}$$

Step 4 It is now necessary to use Table 12.22 to determine the average number of hours during a given month that the solar altitude exceeds the values in Step 3. This information is tabulated below, where H_s is the number of hours per day at the start of the period, H_e is the number of hours per day at the end of the period, and H_{av} is the average number of hours per day during the period.

	Overcast (52°)			Clear (11°)		
Period	H_s	H_e	H_{av}	H_s	H_e	H_{av}
Dec. 22–Jan. 21	0	0	0	6.6	7.3	6.9
Jan. 22–Feb. 21	0	0	0	7.3	8.7	8.0
Feb. 22–Mar. 21	0	0	0	8.7	10.1	8.9
Mar. 22–Apr. 21	0	3.8	1.9	10.1	11.4	10.7
Apr. 22–May 21	3.8	4.8	4.3	11.4	12.3	11.8
May 22–June 21	4.8	5.4	5.1	12.3	12.6	12.4
June 22–July 21	5.4	5.1	5.2	12.6	12.4	12.5
July 22–Aug. 21	5.1	3.9	4.5	12.4	11.5	11.9
Aug. 22–Sep. 21	3.9	0	1.9	1.5	10.1	10.8
Sept. 22–Oct. 21	0	0	0	10.1	8.8	9.4
Oct. 22–Nov. 21	0	0	0	8.8	7.4	8.1
Nov. 22–Dec. 21	0	0	0	7.4	6.6	7.0

Step 5 The numbers of clear and overcast days are determined from Table 12.23. From these and the average hours per day that the solar altitude exceeds the required amount, the total number of hours per year that the electric lighting can be extinguished may be found. The tabulation appears below.

	Overcast (52°)			Clear (11°)		
Period	H_{av}	Days	Hours	H_{av}	Days	Hours
Dec. 22–Jan. 21	0	14.9	0	6.9	16.1	111.1
Jan. 22–Feb. 21	0	13.3	0	8.0	17.7	141.6
Feb. 22–Mar. 21	0	11.5	0	8.9	16.5	146.9
Mar. 22–Apr. 21	1.9	12.4	23.6	10.7	18.6	199.0
Apr. 22–May 21	4.3	11.7	50.3	11.8	18.3	215.9
May 22–June 21	5.1	11.8	60.2	12.4	19.2	238.1
June 22–July 21	5.2	10.8	56.2	12.5	19.2	240.0
July 22–Aug. 21	4.5	11.5	51.8	11.9	19.5	232.1
Aug. 22–Sep. 21	1.9	11.5	21.9	10.8	19.5	210.6
Sept. 22–Oct. 21	0	11.4	0	9.4	18.6	174.8
Oct. 22–Nov. 21	0	13.3	0	8.1	17.7	143.4
Nov. 22–Dec. 21	0	14.1	0	7.0	15.9	111.3
			264			2165

The total hours are 2429. Thus the potential Btu per year savings are 2429 (930) (3.413) = 7.7×10^6 Btu.

REFERENCES

Architectural Aluminum Manufacturers Association: "Voluntary Standard Procedure for Calculating Skylight Annual Energy Balance," AAMA Publication 1602, Chicago, Ill., 1977.

Coaton, J., and A. Marsden: *Lamps and Lighting*—Fourth Edition, John Wiley and Sons, New York, 1997.

Copping, B. (editor): *Window Design*, The Chartered Institution of Building Services Engineers, London, 1987.

Dickinson, W., and P. Cheremisinoff: *Solar Energy Technology Handbook*. Part A, Dekker, New York, 1980.

Dietz, P., J. Murdoch, J. Pokoski, and J. Boyle: "A Skylight Energy Balance Analysis Procedure," *JIES*, October 1981.

Duffie, J., and W. Beckman: *Solar Energy Thermal Processes*, Wiley, New York, 1974.

Erwine, B.: "A View to the Future," *Lighting Design & Application*, February 1996.

Evans, B.: *Daylight in Architecture*, McGraw-Hill, New York, 1981.

Gillette, G.: *A Daylight Model for Building Energy Simulation*, NBS Building Science Series 152, Washington, 1983.

Gillette, G., and T. Kusuda: "A Daylighting Computation Procedure for Use in DOE-2," *JIES*, October 1982.

Gillette, G., W. Pierpoint, and S. Treado: "A General Illuminance Model for Daylight Availability," *JIES*, July 1984.

Gilmore, V.: "Superwindows," *Popular Science*, March 1986.

Gordon, G., and J. Nuckolls: *Interior Lighting for Designers*, Third Edition, John Wiley and Sons, New York, 1995.

Heshong, L., and D. Mahone, F. Rubinstein, and J. McHugh: *Skylighting Guidelines*, The Heschong-Mahone Group, San Francisco, CA, 1998.

Hopkinson, R.G., P. Petherbridge, and J. Longmore: *Daylighting*, Heinemann, London, 1966.

IESNA Calculation Procedures Committee: *Recommended Practice for the Lumen Method of Daylight Calculation Procedures, RP-23-89,* Illuminating Engineering Society of North America, New York.

IESNA Calculation Procedures Committee: *Recommended Practice for the Calculation of Daylight Availability, RP-21-84* (reaffirmed in 1994), Illuminating Engineering Society of North America, New York.

IESNA Daylighting Committee: *Recommended Practice of Daylighting, RP-5-99*, Illuminating Engineering Society of North America, New York.

IESNA Nomenclature Committee: *Nomenclature and Definition for Illuminating Engineering, RP-16-96*, Illuminating Engineering Society of North America, New York.

International Commission on Illumination: "Daylight:International Recommendations for the Calculation of Natural Light," *CIE Report #16*, Paris, 1970.

International Commission on Illumination: "Standardization of Luminance Distribution of Clear Skies," *CIE Report #22*, Paris, 1973.

Kimball, H., and I. Hand: "Daylight Illumination on Horizontal, Vertical and Sloping Surfaces,"*Transactions of the Illuminating Engineering Society*, vol. XVIII, May 1923.

Kittler, R.: "Standardization of Outdoor Conditions for the Calculation of Daylight Factor with Clear Skies," CIE Intercessional Conference on Sunlight in Buildings, Amsterdam, 1967.

Kittler, R.: "A Universal Calculations Method for Simple Predetermination of Natural Radiation on Building Surfaces and Solar Collectors," *Building and Environment*, vol. 16, no. 3, Great Britain, 1981.

Libbey-Owens-Ford Co.: "How to Predict Interior Daylight Illumination," Toledo, Ohio, 1976.

Linforth, E.: "Efficiency of Domed Acrylic Skylights," *Illuminating Engineering*, October 1958.

Loe, D., and K. Mansfield: *Daylighting Design in Architecture*, BRESCU, England, 1998.

Lynes, J.: *Principles of Natural Lighting*, Applied Science Publishers, London, 1968.

Michel, L.: *Light: The Shape of Space*, John Wiley and Sons, New York, 1996.

Mistrick, R., and S. Chen: "A Lumen Method of Sidelighting Based on Profile Angle," *JIES*, 1994.

Moon, P., and D. Spencer: "Illumination from a Non-Uniform Sky," *Illuminating Engineering*, December, 1942.

Moore, F.: *Concepts and Practice of Architectural Daylighting*, Van Nostrand Reinhold, New York, 1985.

Murdoch, J., T. Oliver, and G. Reed: "Luminance and Illuminance Characteristics of Translucent Daylighting Sandwich Panels," *JIES*, 1991.

Murdoch, J., and R. Levin: "A Portable Sky Luminance Meter,"*Energy and Buildings*, 1984.

Murdoch, J.: "A Procedure for Calculating the Potential Savings in Lighting Energy from the Use of Skylights," *JIES*, July 1977.

Parent, M., and J. Murdoch: "Skylight Dome-Well System Analysis from Intensity Distribution Data," *Lighting Research and Technology*, December 1989.

Pierson, O.L.: *Acrylics for the Architectural Control of Solar Energy*, Rohm and Haas, Philadelphia, 1962.

Rea, M., editor: *The IESNA Lighting Handbook*—Ninth Edition, Illuminating Engineering Society of North America, New York, 2000.

Roberts, J., editor: *Advanced Lighting Guidelines*, New Buildings Institute, White Salmon, WA, 2001.

Selkowitz, S. (editor): *Skylight Handbook: Design Guidelines*, American Architectural Manufacturers Association, Palatine, IL, 1987.

Serres, A., and J. Murdoch: "Intensity Distribution of Skylight/Well Systems that have Prismatic Lenses," *JIES*, Summer, 1991.

Tregenza, P., and D. Loe: *The Design of Lighting*, E&FN Spon, London, 1998.

Chapter 13

Lighting Calculations

13.1 THE SCOPE OF THE CHAPTER

The illuminances provided by area sources to receiving points or areas on planes have been considered only briefly thus far (Sects. 2.5-2.8 and Sects. 12.5-12.7). In this chapter, such sources will be examined in detail, and the concept of the transfer of luminous flux from surfaces to points and from surfaces to surfaces will be developed.

The transfer of luminous flux from a source to a receiver can be classified in five ways:

a) point source to a point on a receiving plane
b) point source to a receiving area
c) constant luminance area source to a point on a receiving plane
d) constant luminance area source to a receiving area
e) area source whose luminance varies with angle and/or with location to either a point on a receiving plane or to a receiving area.

Type (a) is the basic inverse square law formulation in Eq. (1.27) of a point source whose intensity is known providing illuminance at a point on a receiving plane a given distance away from the source.

Type (b) is typified by the procedure that was developed in Sect. 2.9 to obtain the total luminous flux emitted by a lamp or luminaire. It involved knowing the intensity distribution curve of the source, then calculating by inverse square law the lumens in differential zonal areas and summing. This method was also used to obtain the average illuminance on a surface from a point source.

Type (c) was first illustrated by finding the illuminance at a point on a ground plane from a constant luminance sky in Sect. 2.8. It was further demonstrated in connection with reflected ground light in Sect. 12.7. It is Type (c) that beckons the use of configuration factors, which will be developed for a variety of source shapes in the next several sections of this chapter.

Type (d) has been explored only slightly in connection with the development of the lumen method in Chap. 9. The lumen method is based on determining flux transfer from one surface to another in a room. The connecting metric is form factor. Later in this chapter, form factors will be compiled for several source and receiver shapes and form factor algebra will be developed to describe the flux transfer between a variety of surfaces. Then the basics of the lumen method will be developed and a number of applications of form factors to flux transfer will be considered.

Type (e) will not be treated here. It is a complex topic, but is well-covered in the literature (DiLaura, 1996-1997; Rea, 2000).

13.2 STRIP, TUBE AND RECTANGULAR SOURCES

In computing the illuminance at a point produced by a source of irregular shape or by many sources of regular shape but separated from one another, it is useful to use the "building-block" approach. The way this method works is as follows: A few standard shapes of sources are defined, and formulas for the illuminances they produce are worked out. The formulas are expressed with distances, angles, and luminances as variables, so that the user can insert the values applicable to a particular lighting situation. The principle of superposition is then used, and the illuminances are summed that are produced by whatever combination and number of standard source shapes are required to match the installation under consideration.

These ideas of modeling a physical system, of using several simple entities to replicate a complex system, and of employing the principle of superposition to obtain the aggregate answer are powerful ideas that are widely used in physical science and engineering. The principle of superposition states that the response of a system to several stimuli is the sum of the responses of the system to the individual stimuli applied separately. The only constraint on the system is that it must be linear, that is, if a stimulus S produces a response R, then a stimulus kS must produce a response kR, where k is any number. This constraint assures that a system will not saturate under the increased stimulus, which would cause it to give no greater response than it did before (this happens in sound systems that are not very high fidelity—turning up the volume control too far produces little change in volume and introduces distortion).

Strip Source

Consider Fig. 13.1 which shows a strip source of constant luminance L and width W in the yz plane. The illuminance at point P opposite the middle of the strip is desired, on a plane parallel to the yz plane. The procedure is to find the illuminance at P produced by the small segment of width W and length dy, and then integrate to find the contributions of all such segments.

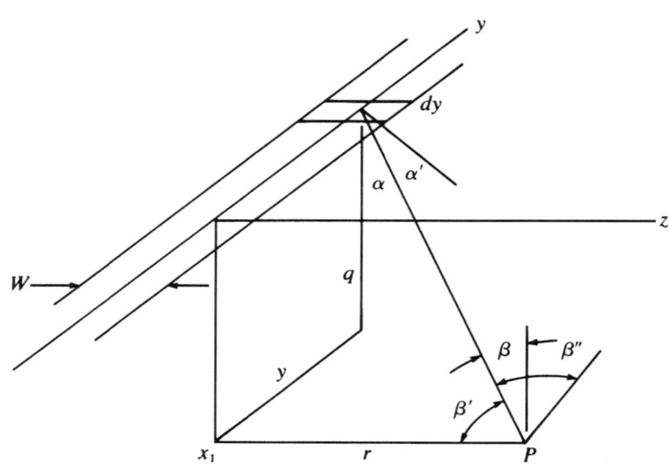

Figure 13.1 Strip source.

Eq. (2.28) is used. The distance from the source segment to P is $\sqrt{y^2+q^2+r^2}$. The angles α and β are equal, and their cosines are $q/\sqrt{y^2+q^2+r^2}$. Inserting these quantities into Eq. (2.28) yields

$$dE_H = \frac{LWq^2\,dy}{(y^2+q^2+r^2)^2} \tag{13.1}$$

which, when integrated between the limits of y = o and h, where h is the length of the strip, gives

$$E_H = \frac{LWq^2}{2(q^2+r^2)}\left(\frac{h}{q^2+r^2+h^2}+\frac{1}{\sqrt{q^2+r^2}}\tan^{-1}\frac{h}{\sqrt{q^2+r^2}}\right) \tag{13.2}$$

where the subscript H denotes that the illuminance is on a horizontal plane in Fig. 13.1. If the strip extended from –h to h (point P opposite the center of the strip) Eq. (13.2) would be multiplied by 2.

If h becomes infinite, indicating that the strip source is very long compared with the distance from it to the point P, the first term inside the parentheses in Eq. (13.2) vanishes, and the second term becomes $\pi/(2\sqrt{q^2+r^2})$. The result is

$$E_H = \frac{\pi LWq^2}{4(q^2+r^2)^{3/2}} \tag{13.3}$$

If h becomes very small compared with the distance to P, Eq. (13.2) becomes

$$E_H = \frac{LWq^2}{2(q^2+r^2)}\left(\frac{h}{q^2+r^2}+\frac{h}{q^2+r^2}\right) = \frac{LWq^2h}{(q^2+r)^2} \tag{13.4}$$

where the approximation has been used that for small values of an angle the tangent of the angle equals the angle in radians.

The illuminances produced at points on planes other than those parallel to the plane of the strip are often of interest. An example is the illuminance produced on a wall by a lighting fixture on a ceiling. In Fig. (13.1), to find the vertical illuminance at P on a plane parallel to the xy plane requires only that β be replaced by β'. Thus, in Eq. (13.1), the cosine term is changed from $q/\sqrt{y^2+q^2+r^2}$ to $r/\sqrt{y^2+q^2+r^2}$. Since q and r are constant, there is no change n the integration and the result is

$$E_{V_{\parallel}} = E_H\frac{r}{q} \tag{13.5}$$

where $V_{\parallel}$ means the illuminance at P on a vertical plane parallel to the strip.

To find the vertical illuminance at P on a plane parallel to the xz plane, β becomes β″ in Fig. 13.1 and the cosine term in Eq. (13.1) must be changed to $y/\sqrt{y^2+q^2+r^2}$. Since y is a variable, this does change the integration. Equation (13.1) becomes

$$dE = \frac{LW\,qy\,dy}{(y^2+q^2+r^2)^2} \tag{13.6}$$

which gives

$$E_{V_\perp} = \frac{LWq}{2}\int_0^h \frac{2y\,dy}{(y^2+q^2+r^2)^2}$$

$$= \frac{LWq}{2}\left(\frac{-1}{y^2+q^2+r^2}\right)\Bigg|_0^h = \frac{LW\,qh^2}{2(q^2+r^2)(q^2+r^2+h^2)} \tag{13.7}$$

$V_\perp$ is used to indicate that the illuminance at P is on a vertical plane perpendicular to the strip.

EXAMPLE 13.1

Ten 8-ft long single-lamp fluorescent fixtures are mounted end-to-end on the ceiling of a long corridor in Fig. 13.2. It is desired to find the illuminance at point P at the midpoint of the corridor (between fixtures 5 and 6) and 6 ft out from directly beneath the fixtures. The fixtures have a width of 6 in, negligible thickness, and a constant luminance of 1 cd/in^2.

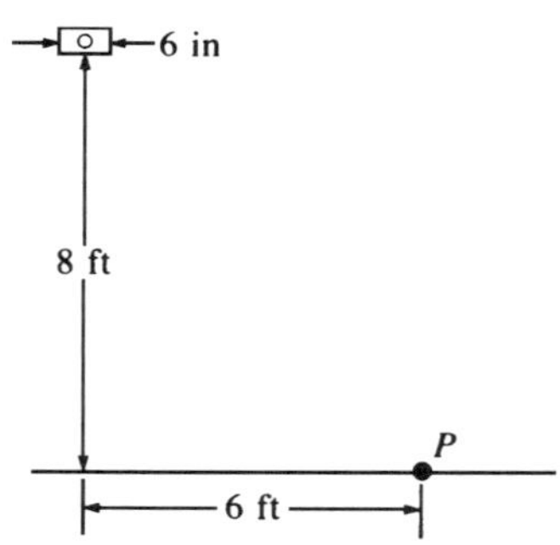

Figure 13.2

Solution. Even though the array of fixtures is not infinite in length, it is long enough (40 ft each side of the midpoint) compared with the minimum distance from the fixtures to point P (10 ft) that twice Eq. (13.3) can be used with reasonable accuracy. Inserting values gives

$$E_H = \frac{\pi(1)(144)\left(\frac{6}{12}\right)(64)}{2(1000)} = 7.24\text{ fc}$$

where the 4 in the denominator of Eq. (13.3) has been replaced by 2 because the strip runs from –h to h. If twice the exact relationship in Eq. (13.2) is used, the answer is

$$E_H = \frac{1(144)\left(\frac{6}{12}\right)(64)}{100}\left(\frac{40}{1700}+\frac{1}{10}\tan^{-1}\frac{40}{10}\right) = 7.19 \text{ fc}$$

Thus the error from using Eq. (13.3) is less than 1%.

In Table 13.1, the illuminances obtained from strip sources (Fig. 13.2) of different lengths have been tabulated. It is seen that the illuminance does not differ appreciably from the value for an infinite strip until the number of units is reduced to four, where an error of about 8% results. At this point the length of the strip each side of the midpoint is 16 ft, giving a ratio of 1.6:1 of this length to the minimum distance to point P (10 ft). It can be reasonably concluded that if the ratio of strip half-length to the minimum distance from fixture to point of measurement is 2:1 or greater, there will be no serious error introduced if Eq. (13.3), rather than Eq. (13.2), is used to compute E_H.

TABLE 13.1

Number of 8-ft units	Value of h (feet)	Footcandles from twice Eq. (13.2)	% Error
∞	∞	7.24	0
10	40	7.19	0.7
8	32	7.17	1.0
6	24	7.06	2.5
4	16	6.73	7.6
2	8	5.36	35.1

Tube Source

The only difference between a tube source and a strip source is that the angle α changes. The tube source always presents a projected width of W towards point P in Fig. 13.1, whereas a strip source does not. Thus, for a tube source, α changes to α' in Fig. 13.1, and $\cos\alpha$ changes from $q/\sqrt{y^2+q^2+r^2}$ to $\sqrt{q^2+r^2}/\sqrt{y^2+q^2+r^2}$ in Eqs. (13.1) and (13.6). Since q and r are not variables, none of the integrations change and, the following general result is obtained

$$E_{tube} = \sqrt{1+\frac{r^2}{q^2}}\,E_{strip} \tag{13.8}$$

When $r = 0$, that is, when the point P is directly under the source, $E_{tube} = E_{strip}$. Otherwise $E_{tube} > E_{strip}$.

Rectangular Plane

This source shape was considered in Sect. 12.7, and an expression was derived for the illuminance produced at a point on a plane perpendicular to the source [Eq. (12.34)]. It remains to consider a point on a plane parallel to the source plane.

The situation is shown in Fig. 13.3. The plane source has a luminance L, dimensions h and w, and lies in the xy plane. The point P lies in a parallel plane a distance q outward from a corner of the plane. A small differential element of dimensions dx and dy is selected for consideration. The illuminance at P produced by this element is, from Eq. (2.28),

$$dE = \frac{Lq^2 dx\,dy}{(q^2 + x^2 + y^2)^2} \tag{13.9}$$

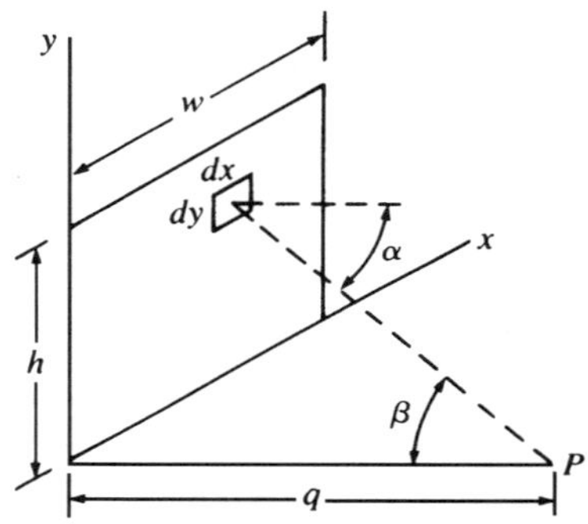

Figure 13.3 Rectangular plane.

Integration must be done twice to obtain E. The result is

$$E_{\|} = \frac{L}{2}\left(\frac{h}{\sqrt{h^2+q^2}}\tan^{-1}\frac{w}{\sqrt{h^2+q^2}} + \frac{w}{\sqrt{w^2+q^2}}\tan^{-1}\frac{h}{\sqrt{w^2+q^2}}\right) \tag{13.10}$$

where the symbol $\|$ indicates that the illuminance is on a plane parallel to the source plane.

For completeness, Eq. (12.34) is repeated,

$$E_{\perp} = \frac{L}{2}\left(\tan^{-1}\frac{w}{q} - \frac{q}{\sqrt{h^2+q^2}}\tan^{-1}\frac{w}{\sqrt{h^2+q^2}}\right) \tag{13.11}$$

where the symbol $\perp$ indicates that the plane of P is perpendicular to the plane of source. Note that the angles in Eqs. (13.10) and (13.11) must be entered in radians.

EXAMPLE 13.2

A perfectly diffusing panel 30 ft × 3 ft has a luminance of 1.2 cd/in^2. Find the illuminance at a point P on a desktop, 6 ft below the bottom of the panel and 10 ft out from its center.

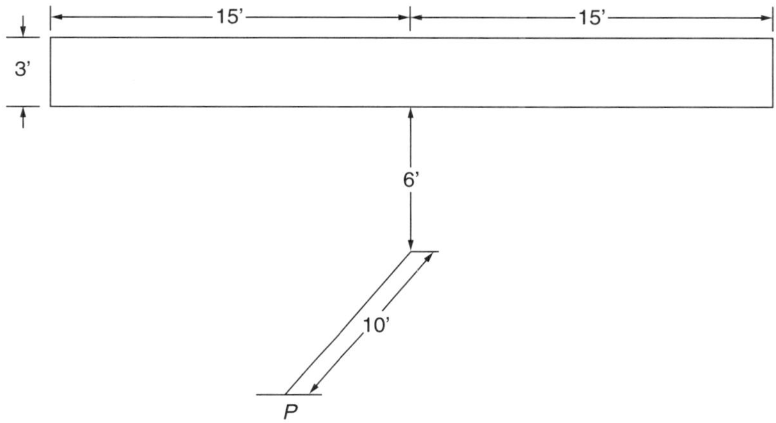

Figure 13.4 Luminous panel.

Solution. Superposition is used to find the illuminance at P as the illuminance produced by a 9 ft × 30 ft panel minus that produced by a 6 ft × 30 ft panel. From Eq. (13.11)

$$E_{\perp} = 2 \times \frac{1.2 \times 144}{2}\left(\tan^{-1}\frac{15}{10} - \frac{10}{\sqrt{9^2+10^2}}\tan^{-1}\frac{15}{\sqrt{9^2+10^2}}\right.$$

$$\left. - \tan^{-1}\frac{15}{10} + \frac{10}{\sqrt{6^2+10^2}}\tan^{-1}\frac{15}{\sqrt{6^2+10^2}}\right)$$

$$= 142.8\,(-0.743 \times 0.840 + 0.857 \times 0.910)$$

$$= 142.8 \times 0.156 = 22 \text{ fc}$$

13.3 FLUX TRANSFER, CONFIGURATION FACTORS AND FORM FACTORS

The starting point in this development is the basic definition of luminance given in Eq. (2.13) and repeated here.

$$L = \frac{d^2\Phi}{d\omega\, dA \cos\theta} \tag{13.12}$$

This equation describes the luminance in a given direction as the luminous flux per unit of solid angle and per unit of projected area in that direction.

Consider luminous flux radiating from differential elements of two surfaces, as shown in Fig. 13.5. The cone defined by $d\omega_1$ intersects surface 2 in dA_2, and the cone defined by $d\omega_2$ intersects surface 1 in dA_1.

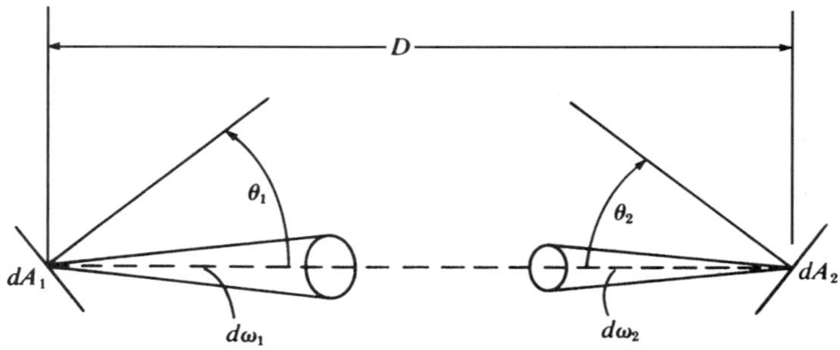

Figure 13.5 Flux transfer.

From Eq. (13.12),

$$L_1 = \frac{d^2\Phi_{12}}{d\omega_1\ dA_1\ \cos\theta_1} \tag{13.13}$$

$$L_2 = \frac{d^2\Phi_{21}}{d\omega_2\ dA_2\ \cos\theta_2} \tag{13.14}$$

where Φ_{12} refers to the flux from dA_1 to dA_2, and Φ_{21} to the flux from dA_2 to dA_1. L_1 is the luminance of dA_1 in the direction of dA_2, and L_2 is the luminance of dA_2 in the direction of dA_1.

Each of these equations may be revised to yield

$$d^2\Phi_{12} = L_1\cos\theta_1 dA_1\left(\frac{dA_2\cos\theta_2}{D^2}\right) \tag{13.15}$$

$$d^2\Phi_{21} = L_2\cos\theta_2\, dA_2\left(\frac{dA_1\cos\theta_1}{D^2}\right) \tag{13.16}$$

Consider now that surface 1 is a finite surface of area A_1, rather than a differential surface of area dA_1. From Eqs. (13.15) and (13.16)

$$d\Phi_{12} = L_1\int_{A_1}\frac{\cos\theta_1\cos\theta_2\, dA_2}{D^2}\,dA_1 \tag{13.17}$$

$$d\Phi_{21} = L_2\int_{A_1}\frac{\cos\theta_1\cos\theta_2\, dA_2}{D^2}\,dA_1 \tag{13.18}$$

where it is assumed that the luminances L_1 and L_2 are constant (the surfaces A_1 and dA_2 are Lambertian). From Eqs. (13.17) and (13.18),

$$\frac{d\Phi_{12}}{d\Phi_{21}} = \frac{L_1}{L_2} = \frac{M_1}{M_2} = \frac{M_1}{d\Phi_2/dA_2} \tag{13.19}$$

where M_1 and M_2 are the luminous exitances of A_1 and dA_2, respectively, and $d\Phi_2$ is the total flux emitted by dA_2. Rearranging Eq. (13.19) yields

$$\frac{d\Phi_{21}}{d\Phi_2} = \frac{d\Phi_{12}}{M_1 dA_2} = \frac{E_2}{M_1} \equiv c_{21} \tag{13.20}$$

Equation (13.20) is the definition of configuration factor. The configuration factor c_{21} is the fraction of the flux emitted by dA_2 that is received directly by A_1. It is obtained as the ratio of the illuminance at dA_2 produced by the flux received directly from A_1 due to the luminous exitance of A_1. This is often called the *wrong-way law*, in the sense that the configuration factor from 2 to 1 is defined in terms of the illuminance at a point on 2 produced by the luminous exitance of an area at 1, but it is the direction of flux transfer that determines the subscripts.

The derivation is now repeated assuming both 1 and 2 are finite areas, giving

$$\Phi_{12} = L_1 \int_1 \int_2 \frac{\cos\theta_1 \cos\theta_2}{D^2} dA_1 \, dA_2 \tag{13.21}$$

$$\Phi_{21} = L_2 \int_1 \int_2 \frac{\cos\theta_1 \cos\theta_2}{D^2} dA_1 \, dA_2 \tag{13.22}$$

These yield

$$\frac{\Phi_{12}}{\Phi_{21}} = \frac{L_1}{L_2} = \frac{M_1}{M_2} = \frac{M_1}{\Phi_2 / A_2} \tag{13.23}$$

from which is obtained

$$\frac{\Phi_{21}}{\Phi_2} = \frac{\Phi_{12}}{M_1 A_2} = \frac{\overline{E}_2}{M_1} \equiv f_{21} \tag{13.24}$$

Equation (13.24) defines form factor. The form factor f_{21} is the fraction of the flux emitted by area A_2 that is received directly by area A_1. It is obtained as the ratio of the average illuminance on A_2 produced by the flux received directly from A_1 due to the luminous excitance of A_1 (again, a wrong-way law).

A reciprocity relationship can be developed for form factors by noting that the average illuminance at A_2 can also be expressed as

$$\overline{E}_2 = \frac{\Phi_{12}}{A_2} = \frac{\Phi_1 f_{12}}{A_2} = \frac{M_1 A_1 f_{12}}{A_2} \tag{13.25}$$

Then, from Eqs. (13.24) and (13.25),

$$A_2 f_{21} = A_1 f_{12} \tag{13.26}$$

Configuration factors relate to illuminance at a point. Form factors relate to average illuminance on a plane. Configuration factors and their applications will be considered in the next three sections, and form factors will be returned to later in the chapter.

13.4 MORE ON CONFIGURATION FACTORS

The configuration factor c_{21}, as defined in Sect. 13.3, is the ratio of the iluminance at point 2 produced by the flux received directly from surface 1 to the luminous exitance of surface 1. From Eq. (13.20) the illuminance at point 2 is

$$E_2 = c_{21}M_1 \tag{13.27}$$

In Sects. 2.12 and 13.2, formulas for the illuminances at a point produced by disc, strip, tube, and rectangular sources were dervied. To obtain configuration factors for these geometries, the illuminance in each of these equations must be divided by the luminous exitance of the source.

Infinite Horizontal Plane

For the perfectly diffusing infinite horizontal plane shown in Fig. 13.6, the configuration factor to the point P beneath it is obtained from Eqs. (2.36) and (2.63) as

$$E = \pi L = M \tag{13.28}$$

and thus

$$c = 1 \tag{13.29}$$

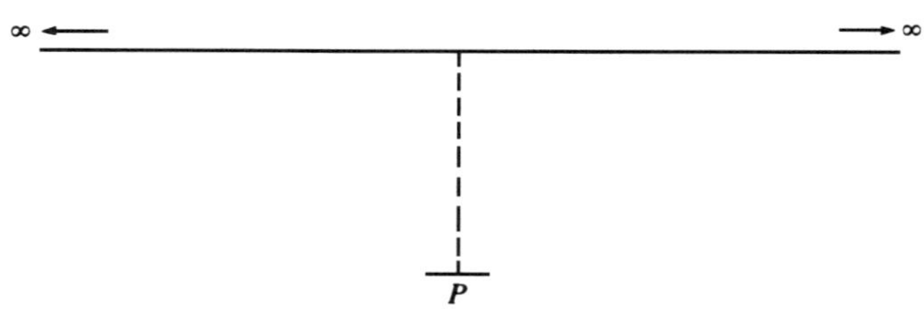

Figure 13.6 Infinite horizontal plane.

EXAMPLE 13.3

Assume the wall and ceiling surfaces of the rectangular box in Fig. 13.7a are perfectly diffusing and have the same luminance. Find the configuration factor at a point P anywhere on the floor of the box.

Solution. The configuration for this geometry is the same as for the infinite plane, because, with the walls perfectly diffusing, it is impossible for the observer at P to distinguish the vertical walls from their projection on an infinite horizontal plane, as shown in Fig. 13.7b. It is as though the walls were rotated 90° about their intersection lines with the ceiling and then extended to infinity. Thus, the configuration factor for the equal-luminance, perfectly diffusing rectangular box is c = 1, irrespective of the location of P on the floor of the box.

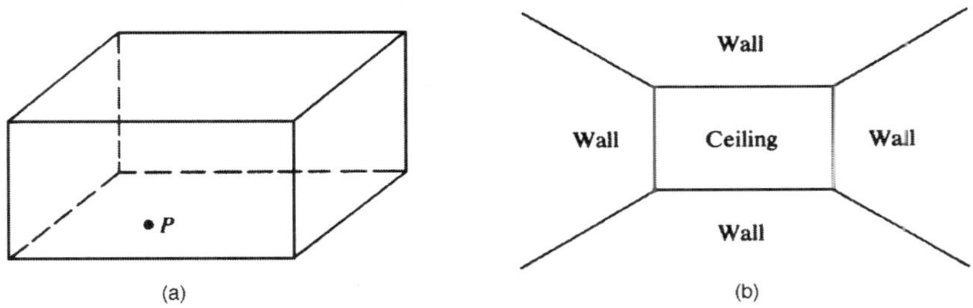

Figure 13.7

Rectangle

The illuminance produced by a perfectly diffusing rectangular source at a point P on a line perpendicular to one corner of the source is given by Eq. (13.10) if P is in a plane parallel to the source and by Eq. (13.11) if P is in a plane perpendicular to the source. The configuration factor in each case is obtained by dividing the illuminance by $M = \pi L$.

These two configuration factors may be placed in an alternative useful form by making the following identifications in Fig. 13.8:

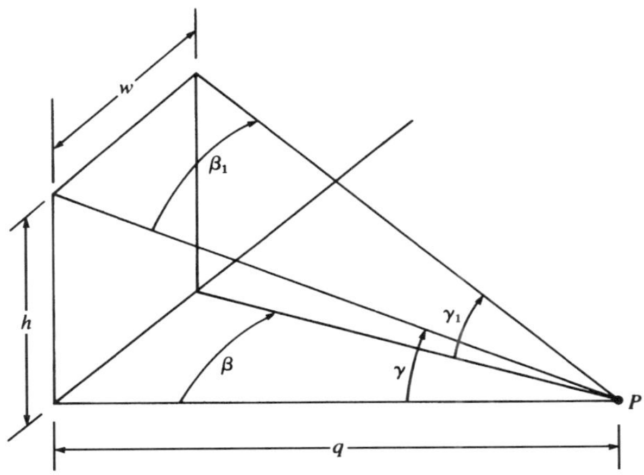

Figure 13.8

$$\left.\begin{aligned} \sin\gamma &= \frac{h}{\sqrt{h^2+q^2}} \\ \sin\beta &= \frac{w}{\sqrt{w^2+q^2}} \\ \sin\gamma_1 &= \frac{h}{\sqrt{w^2+h^2+q^2}} \\ \sin\beta_1 &= \frac{w}{\sqrt{w^2+h^2+q^2}} \end{aligned}\right\} \qquad (13.30)$$

Inserting these into Eqs. (13.10) and (13.11) yields

$$c_{\|} = \frac{1}{2\pi}(\beta_1 \sin\gamma + \gamma_1 \sin\beta) \tag{13.31}$$

$$c_{\perp} = \frac{1}{2\pi}(\beta - \beta_1 \cos\gamma) \tag{13.32}$$

If the configuration factor is desired for a point on a perpendicular line that does not originate at a corner of the source, the technique of superposition can be used, as was done in Example 13.2.

EXAMPLE 13.4

Assume the box in Fig. 13.7 is a cube and that point P is at the center of the floor. Further assume that the luminous exitance of the walls is M_w and that of the ceiling is $M_c > M_w$. Find the horizontal illuminance at point P produced by these exitances.

Solution. The illuminance may be obtained as the superposition of the illuminances produced by an infinite plane of luminance exitance M_w and a square of luminous exitance $M_c - M_w$. The square is composed of four of the areas in Fig. 13.8, all equal. From Eqs. (13.30),

$$\sin\gamma = \sin\beta = \frac{0.5}{\sqrt{1.25}} = 0.447$$

$$\sin\gamma_1 = \sin\beta_1 = \frac{0.5}{\sqrt{1.50}} = 0.408$$

$$\gamma_1 = \beta_1 = 0.421 \text{ rad}$$

Then, from Eq. (13.31), multiplied by 4, the configuration factor for the square ceiling is

$$c_{\|} = \frac{2}{\pi}[0.421(0.447) + 0.421(0.447)] = 0.240$$

The configuration factor for the infinite plane is 1.0. Thus

$$E_P = M_w(1) + 0.240(M_c - M_w)$$
$$= 0.240 M_c + 0.760 M_w$$

Circular Disc

The illuminance produced at P in Fig. 13.9 by a perfectly diffusing disc source is given in Eq. (2.58). The configuration factor is

$$c = \frac{R^2}{R^2 + D^2} = \frac{1 - \cos\theta}{2} \tag{13.33}$$

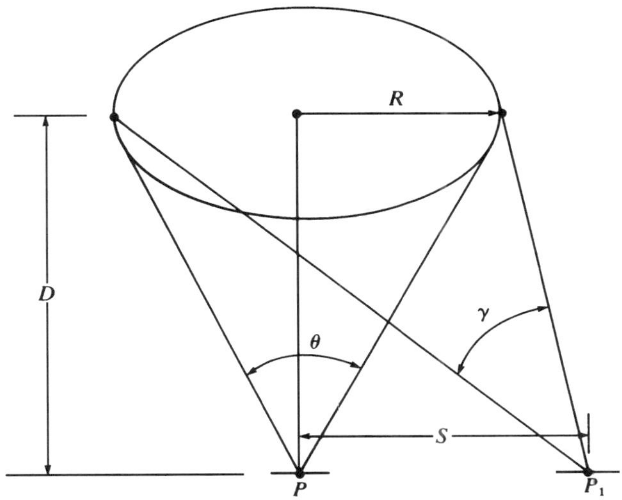

Figure 13.9

If the point is displaced a distance S horizontally from the normal to the disc's center to a point P_1, c becomes

$$c = \frac{1 - \cos\gamma}{2} = \frac{1}{2}\left[1 - \frac{D^2 + S^2 - R^2}{\sqrt{(D^2 + S^2 + R^2)^2 - 4R^2S^2}}\right] \tag{13.34}$$

If the disk in Fig. 13.9 becomes a perfectly diffusing circular dome with the same θ and γ angles, there will be no change in the configuration factors in Eqs. (13.33) and (13.34), assuming $S \le R$ in Eq. (13.34).

EXAMPLE 13.5

A cylindrical ceiling is lighted by fluorescent lamps in a cove around the periphery, as shown in Fig. 13.10a. The luminance of the ceiling of the cylinder is 40 cd/ft^2; that of its wall, 10 cd/ft^2. Find the illuminance at point P.

Solution. The geometry is detailed in Fig. 13.10b.

$$\alpha_1 = \tan^{-1}\frac{10}{14} = 35.54°$$

$$\alpha_2 = \tan^{-1}\frac{10}{12} = 39.81°$$

From Eq. (13.33)

$$c_{pc} = \frac{1 - \cos 2\alpha_1}{2} = 0.338$$

$$c_{p,cw} = \frac{1 - \cos 2\alpha_2}{2} = 0.410$$

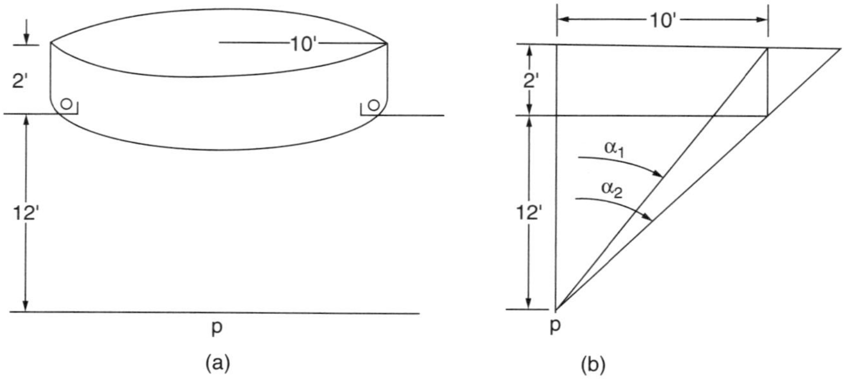

Figure 13.10 Cylindrical rotunda.

where "pc" means from point p to the ceiling of the cylinder and "ρ, cw" means from point p to the ceiling and walls of the cylinder.

From Eq. (13.27), and using superposition,

$$
\begin{aligned}
E_p &= c_{p,cw}\, M_w + c_{pc}\,(M_c - M_w) \\
&= \pi[c_{p,cw} L_w + c_{pc}\,(L_c - L_w) \\
&= \pi(0.410 \times 10 + 0.338 \times 30) \\
&= 45 \text{ fc}
\end{aligned}
$$

Strip and Tube Sources

The illuminances produced in various planes at P in Fig. 13.11 by a perfectly diffusing strip source are given by Eqs. (13.2), (13.5), and (13.7). It is noted from Fig. 13.11 that

$$
\left.
\begin{aligned}
\tan\beta &= \frac{h}{\sqrt{q^2 + r^2}} \\
\sin^2\beta &= \frac{h^2}{q^2 + r^2 + h^2} \\
\sin^2\gamma &= \frac{q^2}{q^2 + r^2}
\end{aligned}
\right\} \tag{13.35}
$$

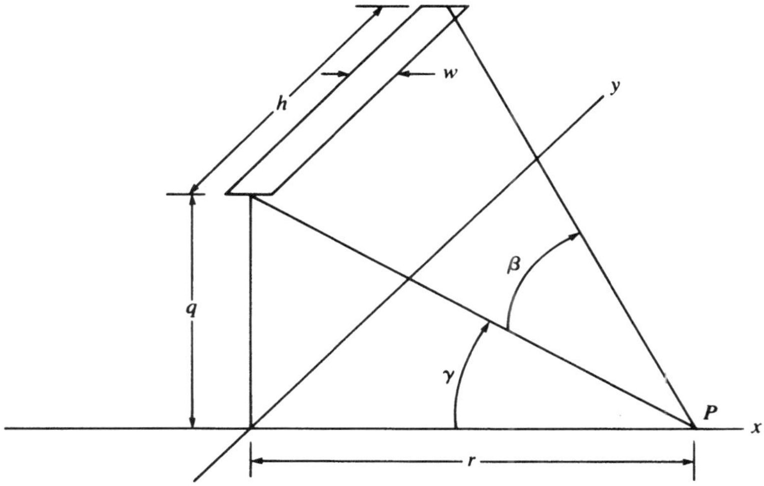

Figure 13.11

Inserting Eqs. (13.35) into the three strip-source equations yields

$$c_H = \frac{W}{2\pi h}\sin^2\gamma(\sin^2\beta + \beta\tan\beta) \tag{13.36}$$

$$c_{V_\parallel} = \frac{r}{q}c_H \tag{13.37}$$

$$c_{V_\perp} = \frac{W}{2\pi q}\sin^2\gamma\sin^2\beta \tag{13.38}$$

For tube sources, the only change is that noted in Eq. (13.8). Thus

$$c_{tube} = \frac{1}{\sin\gamma}c_{strip} \tag{13.39}$$

Sloped Planes

A plane of dimensions m by f and sloped at angle ϕ is shown in Fig. 13.12. The configuration factor from the plane to point P in a horizontal plane a distance a from a corner of the intersection of the two planes can be shown to be

$$c = \frac{1}{2\pi}\left[\tan^{-1}\frac{m}{a} + \frac{f\cos\phi - a}{A}\tan^{-1}\frac{m}{A} + \frac{m\cos\phi}{B}\left(\tan^{-1}\frac{f - a\cos\phi}{B} + \tan^{-1}\frac{a\cos\phi}{B}\right)\right] \tag{13.40}$$

where

$$A = \sqrt{f^2 + a - 2af\cos\phi}, \quad B = \sqrt{m^2 + a^2\sin^2\phi} \tag{13.41}$$

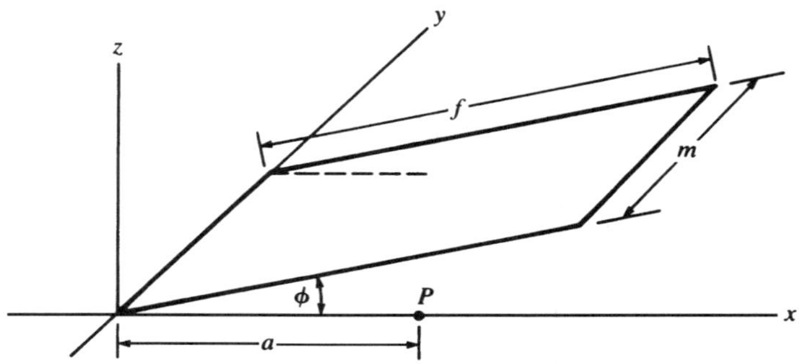

Figure 13.12

Letting $m \to \infty$ in Eq. (13.40) and doubling the answer gives the configuration factor for a sloped plane of infinite extent in and out of the page. The result is

$$c = \frac{1}{2}\left(1 + \frac{f\cos\phi - a}{\sqrt{f^2 + a^2 - 2af\cos\phi}}\right) \tag{13.42}$$

Now letting $f \to \infty$ in Eq. (13.42) gives the configuration factor for a sloped plane that is infinite in both directions. For this case,

$$c = \frac{1}{2}(1 + \cos\phi) \tag{13.43}$$

EXAMPLE 13.6

The perfectly diffusing skylight in Fig. 13.13 is 5 ft in width, of infinite extent in and out of the page, and has a luminance of 200 cd/ft^2. Find the horizontal illuminance at point P.

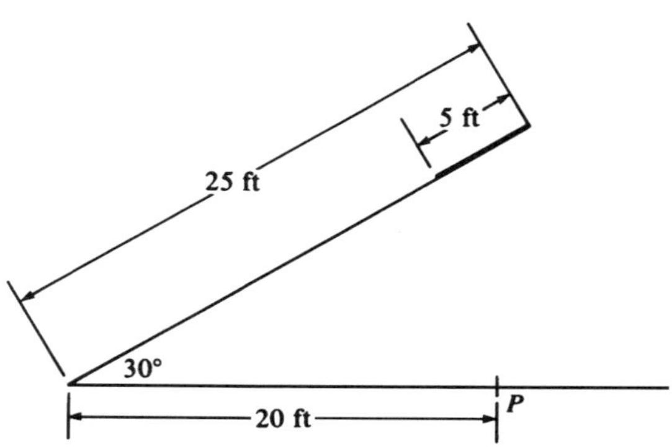

Figure 13.13

Solution. Using superposition and Eq. (13.42), the configuration factor is

$$c = \frac{1}{2}\left[\frac{25(0.866)-20}{\sqrt{625+400-1000(0.866)}} - \frac{20(0.866)-20}{\sqrt{400+400-800(0.866)}}\right]$$

$$= 0.194$$

Then $E_p = 200\pi(0.194) = 122$ fc.

The procedure in this section up to now has been to derive configuration factors in rectangular coordinates and then transform them to angular coordinates. It is reasonable to inquire if some of the configuration factors could be derived directly in angular coordinates. Such a development is possible (Levin, 1971) and will now be presented.

A perfectly diffusing area source S of arbitrary shape, but prescribed angular boundaries (α_1, α_2, β_1, β_2) is shown in Fig. 13.14. It lights a point P. The differential solid angle from a small element dA of the source to the point P is

$$d\omega = \cos\beta\, d\alpha\, d\beta \tag{13.44}$$

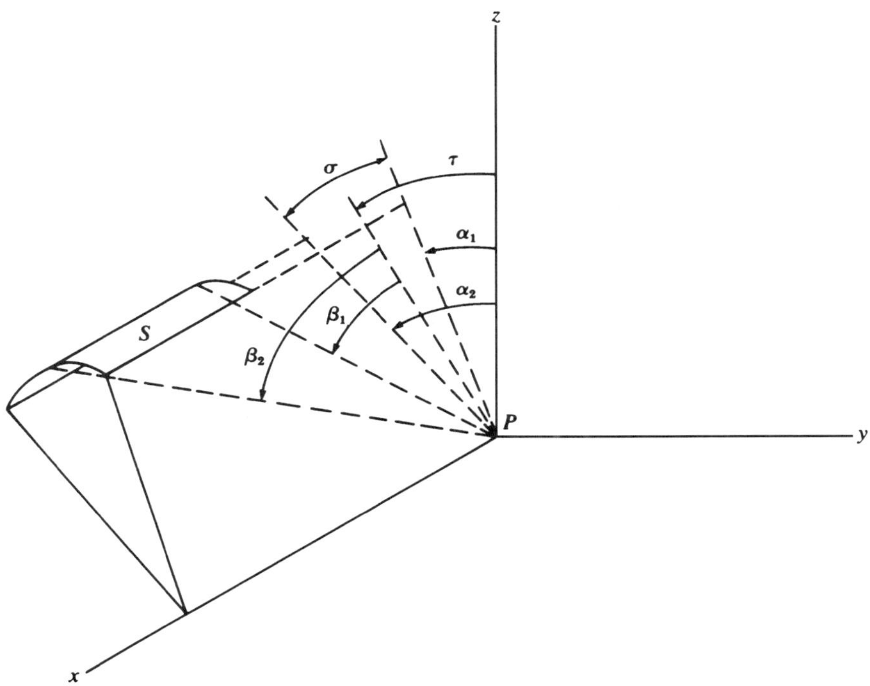

Figure 13.14 (*Source:* R.E. Levin, "Illumination Due to Area Sources Expressed in Angular Coordinates." *Journal of the Illuminating Engineering Society of North America*, October, 1971. Courtesy of the Illuminating Engineering Society of North America.)

The line from the source element to P makes an angle γ with the normal to the horizontal plane given by

$$\cos\gamma = \cos\alpha\cos\beta \tag{13.45}$$

Then, from Eq. (2.30), the horizontal illuminance at P is

$$E_P = \int_\beta\int_\alpha L\cos\alpha\cos^2\beta\, d\alpha\, d\beta \tag{13.46}$$

Performing the double integration gives

$$E_p = L(\sin\alpha_2 - \sin\alpha_1)\,[f(\beta_2) - f(\beta_1)] \tag{13.47}$$

where

$$f(\beta) = \frac{1}{2}\beta + \frac{1}{4}\sin 2\beta \tag{13.48}$$

Because the source is perfectly diffusing, the result in Eq. (13.47) is independent of source shape within the angular boundaries. However, it does assume that β_1 and β_2 are independent of α. This means that the ends of S must be circular arcs in planes normal to the x axis. Deviations from this requirement are not serious when the source is reasonably narrow.

It is often useful to express the result in Eq. (13.47) in terms of the angles σ and τ in Fig. 13.14, rather than α_1 and α_2. Noting that

$$\alpha_1 = \tau - \frac{\sigma}{2}, \qquad \alpha_2 = \tau + \frac{\sigma}{2} \tag{13.49}$$

Eq. (13.47) can be rewritten as

$$E_P = 2L\cos\tau\,\sin\frac{\sigma}{2}\,[f(\beta_2) - f(\beta_1)] \tag{13.50}$$

The result in Eq. (13.50) can be used to obtain the horizontal illuminance at point P in Fig. 13.11 produced by the given strip source. The result should verify the configuration factor in Eq. (13.36). From Fig. 13.11,

$$\beta_1 = 0, \qquad \beta_2 = \beta \tag{13.51a}$$

$$\cos\tau = \sin\gamma \tag{13.51b}$$

$$\sin\frac{\sigma}{2} \simeq \frac{\sigma}{2} = \frac{W\sin\gamma}{2\sqrt{q^2 + r^2}} \tag{13.51c}$$

Equation (13.51c) results from first observing that, with W small compared with the other distances present, the angle σ is small and thus the angle and its sine are approximately equal. Then σ is found by dividing projected arc length by radius

Inserting Eqs. (13.51) into Eq. (13.50) yields

$$E_P = 2L\sin\gamma \frac{W\sin\gamma}{2\sqrt{q^2+r^2}}\left(\frac{1}{2}\beta + \frac{1}{4}\sin 2\beta\right)$$

$$= \frac{LW\sin^2\gamma}{2h}\left(\frac{h}{\sqrt{q^2+r^2}}\beta + \frac{h}{2\sqrt{q^2+r^2}}\sin 2\beta\right)$$

$$= \frac{LW\sin^2\gamma}{2h}(\beta\tan\beta + \sin^2\beta) \qquad (13.52)$$

Dividing by πL yields c_H.

13.5 CONFIGURATION FACTOR METHOD: DIRECT COMPONENT

The configuration factor method permits the calculation of the direct component of the horizontal illuminance at a point on the work plane from a perfectly diffusing source located on the ceiling.

Consider Fig. 13.15. It shows a row of five 4 × 4 ft fluorescent luminaires, whose intensity is assumed to be a cosine distribution, mounted 8 ft above a work plane. The horizontal illuminance on the work plane at point P is desired.

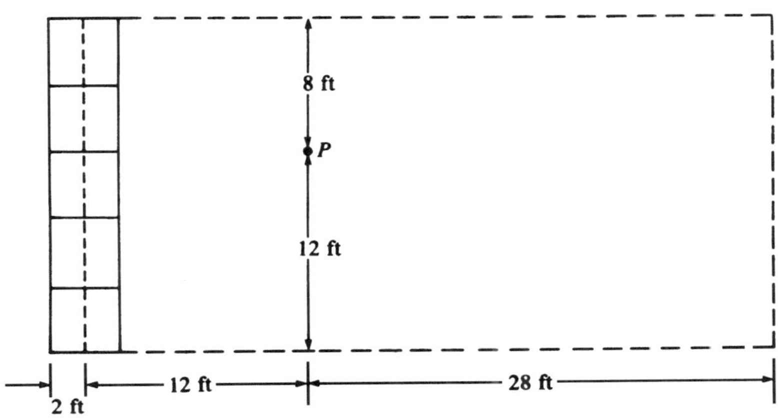

Figure 13.15

The first step in the procedure is to create a room cavity, indicated by the dashed lines in Fig. 13.15, whose width w is the length of the luminaire row and whose length l is twice that amount. The room cavity is always drawn so that its left wall contains the center line of the luminaires. The height of the cavity is the mounting height of the luminaires above the work plane. For the situation in Fig. 13.15,

$$\mathrm{RCR} = \frac{5(8)(60)}{800} = 3.0$$

Next, the grid in Fig. 13.16 is created and the coordinates of the point P are located. The grid is 10×20 squares, so P is located

Along l: $\frac{12}{40}(20) = 6$ squares in from the left

Along w: $\frac{8}{20}(10) = 4$ squares down from the top

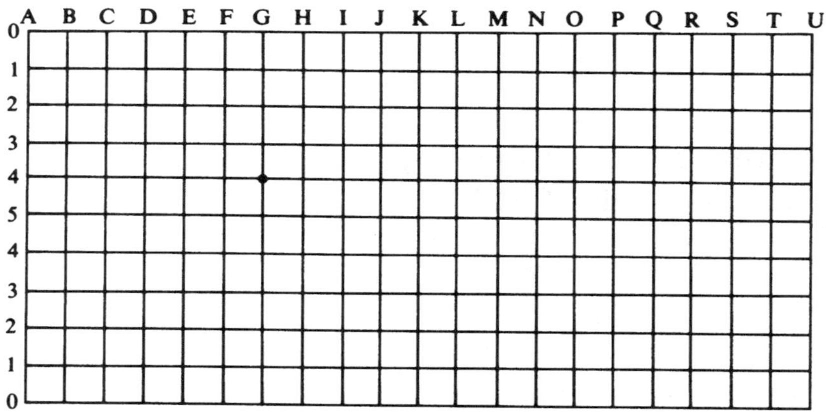

Figure 13.16

Since "6" corresponds to "G" in Fig. 13.16, P lies at coordinates (G, 4). Note that the vertical numbering of squares need only be from 0 to 5 because of symmetry.

A set of normalized configuration factors (c_n) for a strip source is given in Table 13.2. These were calculated from Eq. (13.36) for rooms whose length-to-width ratio is $l/w = 2$ and for strips whose width to length ratio is $W/w = 0.1$. For the given problem, it is found from Table 13.2 that $c_n = 0.0144$. The luminaire width-to-length ratio is $\frac{4}{20} = 0.2$; thus $c = 2(0.0144) = 0.0288$.

Now assume that each fluorescent luminaire in Fig. 13.15 has an intensity $I = I_0 \cos\theta$, where θ is the angle with the normal to the luminaire. From Sect. 2.11, each luminaire emits πI_0 lumens and thus has a luminous exitance of $\pi I_0/16$ lumens per square foot (a constant luminance of $I_0/16$ candelas per square foot). Assume that $I_0 = 2400$ cd is given. Then the horizontal luminance at point P is

$$E_P = \frac{\pi}{16}(2400)(0.0288) = 13.6\,\mathrm{fc}$$

TABLE 13.2 NORMALIZED CONFIGURATION FACTORS (STRIP SOURCE, l/w = 2, W/w = 0.1)

RCR	A0	B0	C0	D0	E0	F0	G0	I0	K0	P0	U0
1	0.1873	0.0958	0.0318	0.0124	0.0058	0.0031	0.0018	0.0007	0.0004	0.0001	0
2	0.0931	0.0763	0.0473	0.0269	0.0154	0.0092	0.0058	0.0026	0.0013	0.0003	0.0001
3	0.0611	0.0557	0.0434	0.0307	0.0209	0.0141	0.0097	0.0048	0.0025	0.0007	0.0003
4	0.0446	0.0423	0.0363	0.0290	0.0221	0.0164	0.0121	0.0067	0.0038	0.0011	0.0004
5	0.0345	0.0333	0.0300	0.0256	0.0210	0.0167	0.0131	0.0079	0.0048	0.0016	0.0006
6	0.0275	0.0268	0.0249	0.0222	0.0190	0.0159	0.0131	0.0086	0.0056	0.0020	0.0008
7	0.0225	0.0221	0.0209	0.0191	0.0169	0.0147	0.0125	0.0088	0.0060	0.0024	0.0010
8	0.0187	0.0184	0.0176	0.0164	0.0149	0.0133	0.0116	0.0086	0.0062	0.0027	0.0012
9	0.0157	0.0155	0.0150	0.0142	0.0131	0.0119	0.0106	0.0082	0.0062	0.0029	0.0014
10	0.0134	0.0133	0.0129	0.0123	0.0115	0.0106	0.0097	0.0077	0.0060	0.0030	0.0015

RCR	A1	B1	C1	D1	E1	F1	G1	I1	K1	P1	U1
1	0.3124	0.1557	0.0470	0.0169	0.0075	0.0038	0.0021	0.0008	0.0004	0.0001	0
2	0.1338	0.1079	0.0644	0.0350	0.0194	0.0112	0.0069	0.0029	0.0014	0.0004	0.0001
3	0.0797	0.0722	0.0553	0.0383	0.0254	0.0168	0.0113	0.0054	0.0028	0.0007	0.0003
4	0.0549	0.0519	0.0442	0.0348	0.0261	0.0191	0.0139	0.0074	0.0042	0.0012	0.0004
5	0.0408	0.0393	0.0352	0.0298	0.0242	0.0191	0.0148	0.0088	0.0053	0.0017	0.0006
6	0.0316	0.0308	0.0285	0.0252	0.0215	0.0179	0.0146	0.0094	0.0060	0.0021	0.0009
7	0.0252	0.0247	0.0233	0.0212	0.0188	0.0162	0.0137	0.0095	0.0064	0.0025	0.0011
8	0.0206	0.0203	0.0194	0.0180	0.0163	0.0144	0.0126	0.0092	0.0066	0.0028	0.0013
9	0.0171	0.0169	0.0163	0.0153	0.0141	0.0128	0.0114	0.0088	0.0065	0.0030	0.0014
10	0.0144	0.0143	0.0138	0.0132	0.0123	0.0113	0.0103	0.0082	0.0063	0.0031	0.0016

RCR	A2	B2	C2	D2	E2	F2	G2	I2	K2	P2	U2
1	0.3595	0.1792	0.0558	0.0202	0.0088	0.0044	0.0025	0.0009	0.0004	0.0001	0
2	0.1595	0.1287	0.0768	0.0414	0.0226	0.0129	0.0078	0.0032	0.0016	0.0004	0.0001
3	0.0943	0.0852	0.0649	0.0446	0.0293	0.0191	0.0126	0.0059	0.0030	0.0008	0.0003
4	0.0636	0.0600	0.0508	0.0397	0.0296	0.0214	0.0154	0.0081	0.0045	0.0013	0.0005
5	0.0462	0.0444	0.0398	0.0335	0.0270	0.0211	0.0162	0.0095	0.0056	0.0017	0.0007
6	0.0351	0.0342	0.0316	0.0278	0.0236	0.0195	0.0158	0.0101	0.0064	0.0022	0.0009
7	0.0276	0.0270	0.0255	0.0231	0.0204	0.0175	0.0147	0.0101	0.0068	0.0026	0.0011
8	0.0222	0.0219	0.0209	0.0193	0.0175	0.0154	0.0134	0.0098	0.0069	0.0029	0.0013
9	0.0183	0.0180	0.0174	0.0163	0.0150	0.0136	0.0121	0.0092	0.0068	0.0031	0.0015
10	0.0152	0.0151	0.0146	0.0139	0.0130	0.0119	0.0108	0.0086	0.0066	0.0032	0.0016

RCR	A3	B3	C3	D3	E3	F3	G3	I3	K3	P3	U3
1	0.3689	0.1864	0.0597	0.0220	0.0097	0.0048	0.0027	0.0010	0.0005	0.0001	0
2	0.1719	0.1393	0.0839	0.0456	0.0249	0.0142	0.0085	0.0035	0.0016	0.0004	0.0001
3	0.1037	0.0938	0.0714	0.0490	0.0321	0.0209	0.0137	0.0063	0.0032	0.0008	0.0003
4	0.0699	0.0659	0.0557	0.0434	0.0322	0.0232	0.0166	0.0086	0.0047	0.0013	0.0005
5	0.0503	0.0484	0.0432	0.0363	0.0291	0.0227	0.0174	0.0100	0.0059	0.0018	0.0007
6	0.0378	0.0368	0.0339	0.0298	0.0253	0.0209	0.0168	0.0106	0.0067	0.0023	0.0009
7	0.0294	0.0288	0.0271	0.0246	0.0216	0.0185	0.0155	0.0106	0.0071	0.0026	0.0011
8	0.0235	0.0231	0.0220	0.0204	0.0184	0.0162	0.0140	0.0102	0.0072	0.0030	0.0013
9	0.0191	0.0189	0.0182	0.0171	0.0157	0.0142	0.0126	0.0095	0.0070	0.0032	0.0015
10	0.0159	0.0157	0.0152	0.0145	0.0135	0.0124	0.0112	0.0088	0.0068	0.0033	0.0016

RCR	A4	B4	C4	D4	E4	F4	G4	I4	K4	P4	U4
1	0.3716	0.1888	0.0612	0.0229	0.0101	0.0051	0.0028	0.0010	0.0005	0.0001	0
2	0.1771	0.1440	0.0873	0.0477	0.0262	0.0149	0.0089	0.0036	0.0017	0.0004	0.0001
3	0.1086	0.0983	0.0751	0.0515	0.0338	0.0219	0.0144	0.0066	0.0033	0.0008	0.0003
4	0.0735	0.0693	0.0586	0.0457	0.0338	0.0243	0.0174	0.0089	0.0048	0.0013	0.0005
5	0.0528	0.0508	0.0453	0.0380	0.0304	0.0237	0.0181	0.0104	0.0061	0.0018	0.0007
6	0.0395	0.0384	0.0354	0.0311	0.0263	0.0216	0.0174	0.0109	0.0068	0.0023	0.0009
7	0.0306	0.0299	0.0281	0.0255	0.0224	0.0191	0.0160	0.0109	0.0072	0.0027	0.0011
8	0.0243	0.0239	0.0228	0.0210	0.0190	0.0167	0.0144	0.0104	0.0073	0.0030	0.0013
9	0.0197	0.0194	0.0187	0.0176	0.0161	0.0145	0.0129	0.0098	0.0072	0.0032	0.0015
10	0.0163	0.0161	0.0156	0.0148	0.0138	0.0126	0.0114	0.0090	0.0069	0.0033	0.0016

RCR	A5	B5	C5	D5	E5	F5	G5	I5	K5	P5	U5
1	0.3722	0.1893	0.0617	0.0232	0.0103	0.0052	0.0029	0.0011	0.0005	0.0001	0
2	0.1786	0.1453	0.0883	0.0484	0.0266	0.0152	0.0091	0.0037	0.0017	0.0004	0.0001
3	0.1101	0.0997	0.0762	0.0524	0.0343	0.0223	0.0146	0.0067	0.0033	0.0008	0.0003
4	0.0747	0.0704	0.0596	0.0464	0.0343	0.0247	0.0176	0.0091	0.0049	0.0013	0.0005
5	0.0536	0.0516	0.0460	0.0386	0.0309	0.0240	0.0183	0.0105	0.0061	0.0018	0.0007
6	0.0401	0.0390	0.0359	0.0315	0.0266	0.0219	0.0176	0.0111	0.0069	0.0023	0.0009
7	0.0310	0.0303	0.0285	0.0258	0.0226	0.0193	0.0162	0.0110	0.0073	0.0027	0.0011
8	0.0245	0.0241	0.0230	0.0213	0.0191	0.0169	0.0146	0.0105	0.0074	0.0030	0.0013
9	0.0199	0.0196	0.0189	0.0177	0.0163	0.0147	0.0130	0.0098	0.0072	0.0032	0.0015
10	0.0164	0.0162	0.0157	0.0149	0.0139	0.0127	0.0115	0.0091	0.0069	0.0034	0.0016

EXAMPLE 13.7

A row of five 2 × 4 ft fluorescent luminaires, each of whose intensity is $I = 2000 \cos \theta$ candelas, is mounted 7 ft above the work plane, as shown in Fig. 13.17. Find the horizontal illuminance at point P.

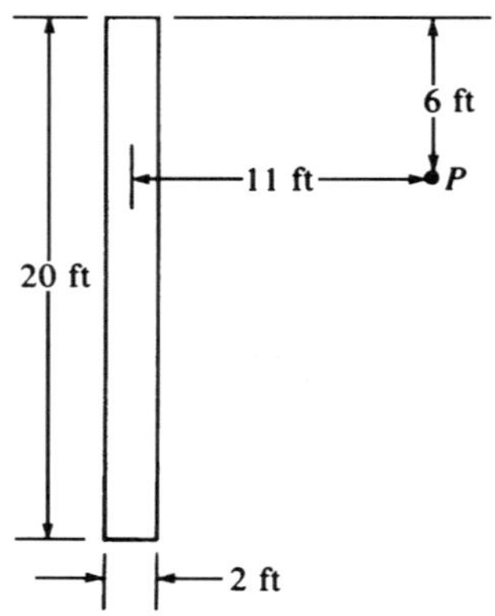

Figure 13.17

Solution. The luminaire row length is 20 ft. Thus the room cavity ratio is

$$\text{RCR} = \frac{5(7)(60)}{20(40)} = 2.625$$

Point P is located

Along l: $$\frac{11}{40}(20) = 5.5 \text{ squares in}$$

Along w: $$\frac{6}{20}(10) = 3 \text{ squares down}$$

Thus P lies between F and G on the grid at F – G, 3.

Interpolation is required to obtain c_n from Table 13.2:

RCR	F3	F – G, 3	G3
2	0.0142		0.0085
2.625	0.0184	0.0151	0.0118
3	0.0209		0.0137

$$c_n = 0.0151$$

The luminaire ratio is $W/w = \frac{2}{20} = 0.1$ giving $c = 1(0.0151) = 0.0151$.

Each luminaire emits 2000π lumens and thus has a luminous exitance of $2000\pi/8 = 250\pi$ lumens per square foot. Thus

$$E_P = 250\pi(0.0151) = 12 \text{ fc}$$

In addition to its basic limitation to perfectly diffusing sources, the configuration factor method using strip sources is limited by the accuracy of Eq. (13.36). This equation is valid only as long as the width W is small compared with the distance from source to receiver. For applications involving rows of fluorescent luminaires on ceilings, the method is generally quite accurate for luminaires up to 4 ft in width. For example, the exact illuminance at P in Fig. 13.15 is 14.0 fc, which represents less than a 3% error. For luminaires of greater width, such as luminous ceiling panels, another configuration factor should be used, namely that in Eq. (13.31) for a point on a plane to a parallel rectangular source. That procedure will now be developed.

A rectangular luminous ceiling is shown in Fig. 13.18. It is assumed to be perfectly diffusing with a luminous exitance M = 300 lumens per square foot. The horizontal illuminance at point P, on a work plane 10 ft below the ceiling, is desired.

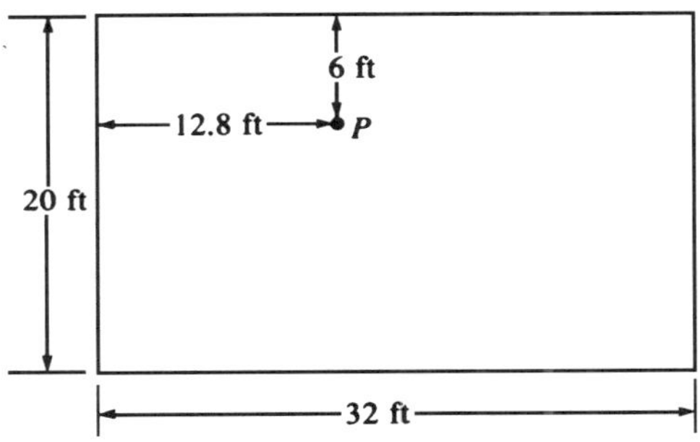

Figure 13.18

As in the previous procedure, the first step is to obtain the room cavity ratio.

$$\text{RCR} = \frac{5(10)(52)}{20(32)} = 4.0625$$

Next a grid is created as shown in Fig. 13.19, where each block represents 10% of the room length and width. The lettering and numbering of the coordinates indicate the symmetry that exists with a rectangular element.

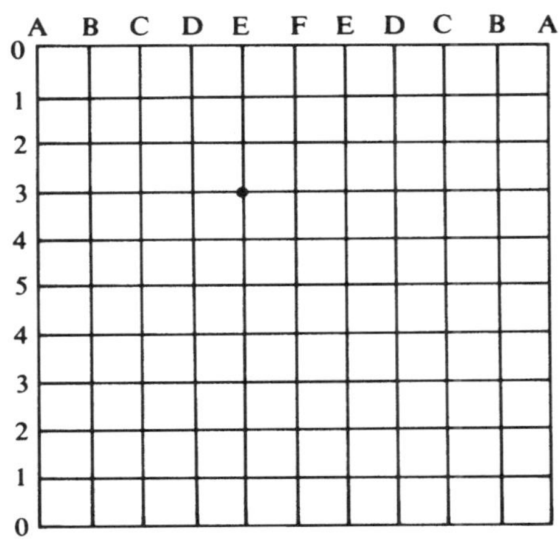

Figure 13.19

The location of point P is shown on the grid in Fig. 13.19. Its coordinates are determined from

Along l: $$\frac{12.8}{32}(10) = 4(E)$$

Along w: $$\frac{6}{20}(10) = 3$$

A set of configuration factors for a square ceiling element to a point on a horizontal work plane below is given in Table 13.3. For the given coordinates and room cavity ratio,

RCR	E3
4	0.6096
4.0625	0.6034
5	0.5012

$$c = 0.6034$$

Thus $$E_P = 300(0.6034) = 180 \text{ fc}$$

In this, and the two examples to follow, it is justifiable to use four significant figures in the computations, but not in the ultimate illuminance, because the configuration factors in Table 13.3 are for square rooms whereas the rooms in the examples are rectangular.

EXAMPLE 13.8

Assume the ceiling indicated by the dashed lines in Fig. 13.15 is uniformly luminous with a luminous exitance of 200 lm/ft^2. Find the horizontal illuminance at point P.

Solution. The room cavity ratio is 3.0, as before. P is located

Along l: $$\frac{12}{40}(20) = 3 \text{ (D)}$$

Along w: $$\frac{8}{20}(10) = 4$$

From Table 13.3, c = 0.7255. Then $E_P = 200(0.7255) = 150$ fc.

EXAMPLE 13.9

Use the configuration factor method for rectangular sources to find an approximate value for the illuminance at P in Example 13.7.

TABLE 13.3 CONFIGURATION FACTORS SQUARE ROOM (l/w = 1), CEILING TO POINT ON THE FLOOR

RCR	A0	B0, A1	C0, A2	D0, A3	E0, A4	F0, A5	B1
1	0.2480	0.4244	0.4708	0.4837	0.4882	0.4893	0.7393
2	0.2421	0.3526	0.4159	0.4448	0.4571	0.4605	0.5227
3	0.2329	0.3092	0.3654	0.3989	0.4158	0.4208	0.4159
4	0.2212	0.2774	0.3231	0.3543	0.3718	0.3773	0.3513
5	0.2078	0.2507	0.2872	0.3138	0.3296	0.3348	0.3048
6	0.1935	0.2270	0.2559	0.2777	0.2911	0.2955	0.2678
7	0.1790	0.2054	0.2284	0.2459	0.2569	0.2606	0.2369
8	0.1648	0.1859	0.2041	0.2181	0.2269	0.2299	0.2105
9	0.1512	0.1681	0.1827	0.1939	0.2009	0.2033	0.1875
10	0.1385	0.1522	0.1639	0.1728	0.1785	0.1804	0.1676
RCR	**C1, B2**	**D1, B3**	**E1, B4**	**F1, B5**	**C2**	**D2, C3**	**E2, C4**
1	0.8144	0.8333	0.8394	0.8410	0.9010	0.9235	0.9308
2	0.6160	0.6563	0.6728	0.6773	0.7269	0.7752	0.7949
3	0.4928	0.5375	0.5595	0.5660	0.5848	0.6381	0.6641
4	0.4108	0.4808	0.4729	0.4799	0.4813	0.5286	0.5546
5	0.3505	0.3836	0.4031	0.4095	0.4039	0.4426	0.4653
6	0.3031	0.3295	0.3457	0.3511	0.3437	0.3742	0.3928
7	0.2642	0.2851	0.2981	0.3024	0.2953	0.3191	0.3338
8	0.2318	0.2481	0.2584	0.2619	0.2557	0.2742	0.2857
9	0.2043	0.2172	0.2252	0.2280	0.2229	0.2372	0.2462
10	0.1809	0.1910	0.1974	0.1996	0.1954	0.2066	0.2137
RCR	**F2, C5**	**D3**	**E3, D4**	**F3, D5**	**E4**	**F4, E5**	**F5**
1	0.9327	0.9478	0.9558	0.9587	0.9641	0.9662	0.9683
2	0.8002	0.8279	0.8495	0.8554	0.8721	0.8783	0.8847
3	0.6719	0.6967	0.7255	0.7340	0.7557	0.7647	0.7739
4	0.5628	0.5808	0.6096	0.6186	0.6399	0.5494	0.6591
5	0.4727	0.4852	0.5102	0.5184	0.5367	0.5453	0.5541
6	0.3990	0.4077	0.4281	0.4349	0.4496	0.4568	0.4642
7	0.3388	0.3450	0.3611	0.3666	0.3781	0.3838	0.3896
8	0.2896	0.2942	0.3067	0.3110	0.3199	0.3243	0.3289
9	0.2493	0.2527	0.2624	0.2657	0.2726	0.2760	0.2795
10	0.2161	0.2187	0.2262	0.2288	0.2431	0.2367	0.2395

Solution. To solve this problem using rectangular sources, create two nearly square room cavities, as shown in Fig. 13.20, whose widths are w = 20 ft and whose lengths are l = w ± W/2 = 19 ft and 21 ft. The room cavity ratios are

$$RCR_{19} = \frac{5(7)(39)}{20(19)} = 3.592, \qquad RCR_{21} = \frac{5(7)(41)}{20(21)} = 3.417$$

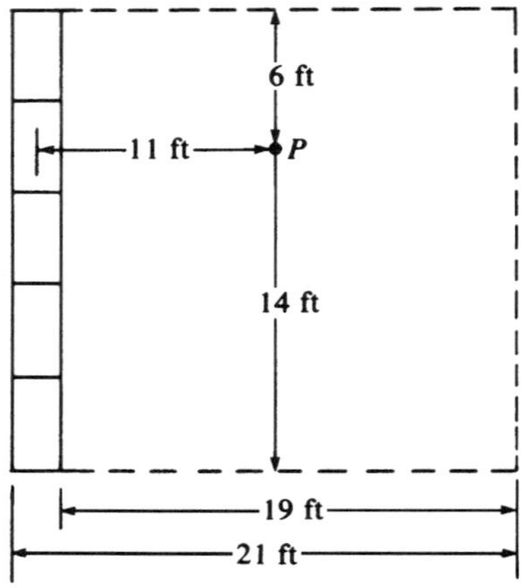

Figure 13.20

The locations of P are obtained from

19-ft room: $10\left(\dfrac{10}{19}\right) = 5.263, \quad 10\left(\dfrac{6}{20}\right) = 3$

21-ft room: $10\left(\dfrac{12}{21}\right) = 5.714, \quad 10\left(\dfrac{6}{20}\right) = 3$

giving coordinates in the range F – E, 3 for both rooms.

Then Table 13.3 yields

		F3		E3
19 ft room	3	0.7340		0.7255
	3.592	0.6657	0.6634	0.6569
	4	0.6186		0.6096
21 ft room	3	0.7340		0.7255
	3.417	0.6859	0.6797	0.6772
	4	0.6186		0.6096

Thus the net configuration factor is c = 0.6797 – 0.6634 = 0.0163, giving an illuminance of

$$E_P = 250\pi(0.0163) = 13 \text{ fc}$$

This answer is slightly in error because Table 13.3 was derived for square elements, whereas those in Fig. 13.20 are slightly rectangular.

Slight errors in the configuration factors for the 19 ft and 21 ft rooms can cause a larger error in their difference. Thus, as mentioned earlier, strip source configuration factors should be used for luminaires 4 ft or less in width.

It may seem surprising that configuration factors for square ceiling elements are used in rectangular room problems. To explore this, replace the room in Fig. 13.18 by a square room 20 × 20 ft. For the same RCR, its height would be

$$h = \frac{lw(RCR)}{5(l + w)} = \frac{w(RCR)}{10} = \frac{20(4.0625)}{10} = 8.125 \text{ ft}$$

The relative location of the point P′ in the square room would be 8 ft in from one edge and 6 ft in from the other. What is being said is that the configuration factor for the point P in Fig. 13.18 is very nearly the same as for the point P′ in Fig. 13.21. For this to remain true requires that the length of the room be not excessively greater than its width (less than 3:1).

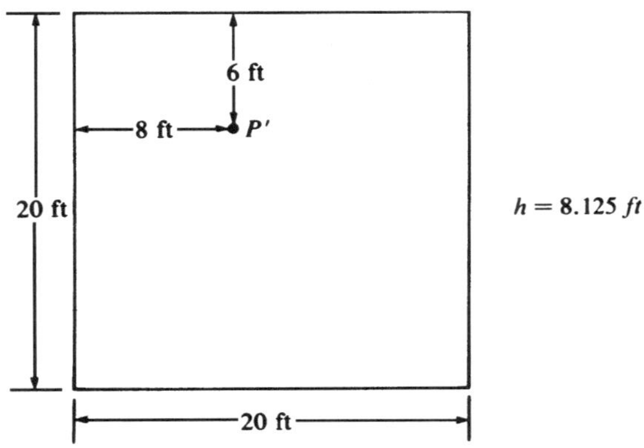

Figure 13.21

13.6 INFINITE PLANE METHOD: REFLECTED COMPONENT

Determination of the horizontal illuminance at a point on the work plane from reflected flux also involves the use of configuration factors and, in addition, the use of the infinite plane, discussed earlier in Example 13.4. The method assumes that the walls, because they are perfectly diffusing, can be replaced by their infinite projections on the ceiling plane. Then, assuming the walls all have the same luminous exitance, the total reflected illuminance at the point on the work plane is the illuminance produced by the infinite plane of the walls, plus the illuminance produced by a finite ceiling plane, whose luminous exitance is the difference between its actual luminous exitance and that of the walls. In equation form,

$$E_P = 1.0M_w + c(M_c - M_w) \tag{13.53}$$

where c is the configuration factor from the point to the ceiling plane, and M_w and M_c are given by Eqs. (9.14) and (9.15), respectively. Inserting these latter equations into Eq. (13.53) yields

$$E_P = \frac{\Phi_i}{A}[WEC + c(CEC - WEC)] \tag{13.54}$$

To illustrate the procedure, consider a factory room 20 × 30 ft with luminaires mounted 8 ft above the work plane. Assume the room reflectances are 70% for the ceiling, 50% for the walls, and 20% for the floor. Assume further that the direct fluorescent luminaire in Table 9.13 is used, that there are 12 of them, and that each houses two F032 fluorescent lamps. The initial horizontal illuminance due to reflected flux on a desk located 2 ft from the long wall at its midpoint is desired.

The calculations follow:

$$\phi_I = 12(2)\ (2950) = 70{,}800 \text{ lm (from Table 6.10)}$$

$$A = 20(30) = 600 \text{ ft}^2$$

Desk location (Fig. 13.19): Fl

$$\text{RCR} = \frac{5(8)(50)}{600} = 3.333$$

$$c = 0.537 \qquad \text{(Table 13.3)}$$

$$\text{WEC} = 0.132 \qquad \text{(Table 9.13)}$$

$$\text{CEC} = 0.079 \qquad \text{(Table 9.13)}$$

$$M_w = \frac{70{,}800 \times 0.132}{600} = 15.6 \text{ lm/ft}^2 \quad \text{[Eq. (9.14)]}$$

$$M_c = \frac{70{,}800 \times 0.079}{600} = 9.3 \text{ lm/ft}^2 \quad \text{[Eq. (9.15)]}$$

$$E_P = 15.6 + 0.537(9.3 - 15.6) = 12 \text{fc}$$

The procedure just developed does not require that the luminaire be perfectly diffusing, but it does assume that the walls are perfectly diffusing and have the same luminous exitance. There is a procedure for perfectly diffusing luminous walls that do not have the same luminous exitance (see Balogh, 1966; O'Brien and Balogh, 1967). This procedure uses Eq. (13.32) to develop a table of configuration factors, similar to Table 13.3, for the case of a point lying in a plane perpendicular to the luminous element. Then a method very similar to that in Sect. 13.5 for rectangular ceiling elements is used to find the illuminance at the point on the work plane.

13.7 ANGULAR COORDINATE DIRECT ILLUMINANCE COMPONENT (DIC) METHOD

The DIC method (IES, 1974; Jones et al., 1969; Kaufman, 1984) is an additional procedure for determining the direct component of illuminance at a point. Its particular virtue is that it does not require that the sources be Lambertian. It is most applicable to continuous rows of fluorescent luminaires, although it may also be used with individual luminaires that are separated.

Inherent in the method is locating the point P with respect to the luminaire in Fig. 13.22a. This location is specified in terms of the mounting height H, the longitudinal angle α, and the lateral angle β, where

$$\alpha = \tan^{-1}\frac{L}{H}, \qquad \beta = \tan^{-1}\frac{R}{H} \tag{13.55}$$

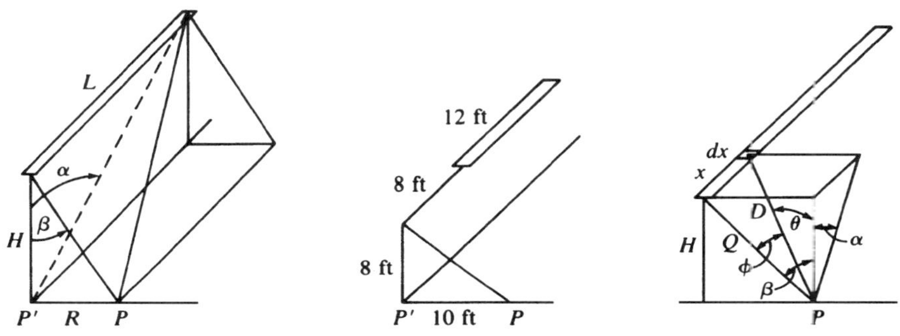

Figure 13.22 Angular coordinates for DIC method.

Note that the plane containing β lies at the end of the luminaire row and is perpendicular to it.

Data for use in calculating the direct component of horizontal illuminance must be prepared by the manufacturer. An example for a prismatic wrap-around luminaire (Jones et al., 1969) is shown in Table 13.4. The numbers are horizontal footcandles on the work plane at point P for selected displacements R and luminaire row lengths L, as expressed by β and α, with H set at 6 ft. For other mounting heights, the values listed must be multiplied by 6/H. Since the point P does not normally lie directly out from the end of a luminaire row, it is often necessary to use superposition to find the net illuminance at P.

TABLE 13.4 DIRECT HORIZONTAL ILLUMINANCE COMPONENTS FOR PRISMATIC WRAP-AROUND FLUORESCENT LUMINAIRE (fc)

α \ β	5	15	25	35	45	55	65
0–10	9.4	8.9	7.8	5.8	3.1	0.9	0.2
0–20	18.4	17.3	15.2	11.4	6.1	1.8	0.4
0–30	25.9	24.6	22.0	16.5	8.8	2.7	0.6
0–40	31.7	30.1	27.1	20.8	11.0	3.6	0.8
0–50	35.0	33.2	29.7	22.8	12.3	4.4	1.0
0–60	36.1	34.3	30.7	23.7	13.0	4.9	1.2
0–70	36.5	34.7	31.1	24.1	13.4	5.2	1.4
0–80	36.7	34.9	31.3	24.3	13.6	5.4	1.5
0–90	36.7	34.9	31.3	24.3	13.6	5.4	1.5

Source: J.R. Jones, R.C. Levere, N. Ivanicki, and P. Chesebrough, "Angular Coordinate Systems and Computing Illumination at a Point." *Illuminating Engineering*, April 1969.

EXAMPLE 13.10

For the situation in Fig. 13.22b, determine the horizontal illuminance at point P produced by the luminaire in Table 13.4 using the DIC method.

Solution. The problem must be solved by finding the illuminance produced by a 20-ft luminaire and subtracting that produced by an 8-ft luminaire. From Eq. (13.55), the necessary angles are

$$\begin{aligned} L &= 20\text{ ft}, & L &= 8\text{ ft}, \\ H &= 8\text{ ft}, & H &= 8\text{ ft}, \\ R &= 10\text{ ft}, & R &= 10\text{ ft}, \\ \alpha &= 68.2^\circ, & \alpha &= 45.0^\circ, \\ \beta &= 51.3^\circ, & \beta &= 51.3^\circ. \end{aligned}$$

The required interpolations in Table 13.4 are

	45	51.3	55
60	13.0		4.9
68.2		8.18	
70	13.4		5.2
40	11.0		3.6
45.0		6.83	
50	12.3		4.4

Then

$$E_P = \frac{6}{8}(8.18 - 6.83) = 1.0 \text{ fc}$$

In order to develop data such as that in Table 13.4 for a given luminaire, it is first necessary to calculate the luminaire intensities at the given α and β angles from published intensity values in the parallel, perpendicular, and 45° planes. This is done using an intensity interpolation technique, such as will be discussed in the next section. In this particular instance, a piecewise cosine intensity interpolation was used (Jones et al., 1969), with the α angles taken at the midpoints of the zones in Table 13.4 (at $\alpha = 5°, 15°, ..., 85°$). The resulting intensity data for the luminaire in Table 13.4 are given in Table 13.5.

The next step is to obtain a set of zone multipliers, each of which gives the illuminance at point P per 1000 cd provided by an element of a luminaire at a given β and included within a given α range. These multipliers for horizontal illuminance are given in Table 13.6.

The zone multipliers may be derived with the aid of Fig. 13.22c. The differential horizontal illuminance at P produced by the luminaire element dx is

$$dE = \frac{dI \cos^3 \theta}{H^2} \tag{13.56}$$

TABLE 13.5 INTENSITY DATA FOR LUMINAIRE IN TABLE 13.4

Polar angle	Parallel	Perpendicular	45°
5	1300	1310	1310
15	1250	1370	1310
25	1150	1445	1330
35	980	1455	1310
45	640	1220	1150
55	275	660	450
65	135	290	240
75	85	200	180
85	40	145	125

α \ β	5	15	25	35	45	55	65	75	85
5	1310	1365	1440	1453	1213	658	290	200	145
15	1260	1320	1400	1390	1150	640	286	199	145
25	1155	1210	1320	1282	1022	600	280	196	145
35	982	1000	1045	1105	810	530	272	194	144
45	638	625	595	546	475	420	260	191	144
55	270	268	262	260	254	250	227	188	143
65	135	137	140	146	152	177	200	178	142
75	85	86	87	88	90	100	112	150	140
85	40	80	40	41	41	43	48	51	114

Source: J.R. Jones, R.C. Levere, N. Ivanicki, and P. Chesebrough, "Angular Coordinate System and Computing Illumination at a Point," *Illuminating Engineering*, April 1969.

TABLE 13.6 ZONE MULTIPLIERS

α \ β	5	15	25	35	45	55	65	75	85
0–10	7.144	6.519	5.395	3.993	2.576	1.379	0.533	0.127	0.0049
10–20	6.934	6.360	5.315	3.988	2.613	1.421	0.578	0.135	0.0052
20–30	6.515	6.033	5.141	3.961	2.663	1.507	0.632	0.151	0.0059
30–40	5.904	5.545	4.847	3.880	2.753	1.633	0.723	0.180	0.0072
40–50	5.113	4.875	4.400	3.691	2.790	1.787	0.858	0.230	0.0096
50–60	4.163	4.033	3.760	3.324	2.708	1.918	1.044	0.318	0.0145
60–70	3.093	2.038	2.917	2.712	2.391	1.910	1.234	0.471	0.0264
70–80	1.942	1.925	1.893	1.833	1.722	1.549	1.221	0.675	0.0657
80–90	0.842	0.842	0.837	0.830	0.823	0.790	0.749	0.578	0.1596

Source: J.R. Jones, R.C. Levere, N. Ivanicki, and P. Chesebrough, "Angular Coordinate System and Computing Illumination at a Point," *Illuminating Engineering*, April 1969.

The differential intensity is

$$dI = LW \cos\phi \, dx \tag{13.57}$$

where L is the luminance of the element in the direction of P and W cos ϕ dx is the projected area of the element toward P. Combining gives

$$dE = \frac{LW \cos\phi \cos^3\theta dx}{H^2} \tag{13.58}$$

It is noted that

$$\tan\phi = \frac{x}{Q}, \qquad \cos\beta = \frac{H}{Q} \tag{13.59}$$

giving

$$x = \frac{H\tan\phi}{\cos\beta}, \qquad dx = \frac{H\sec^2\phi\, d\phi}{\cos\beta} \tag{13.60}$$

It is further noted that

$$\cos\theta = \frac{H}{D}, \qquad \cos\phi = \frac{Q}{D} \tag{13.61}$$

and thus

$$\cos\theta = \cos\phi\,\cos\beta \tag{13.62}$$

Inserting Eqs. (13.60) and (13.62) into Eq. (13.58) gives

$$dE = \frac{LW\cos^2\beta}{H}\cos^2\phi\, d\phi \tag{13.63}$$

and integrating yields

$$E = \frac{LW\cos^2\beta}{2H}\left(\phi + \sin\phi\cos\phi\right)\Big|_{\phi_1}^{\phi_2} \tag{13.64}$$

The intensity of a 4 ft element located at

$$\phi_m = \frac{\phi_1 + \phi_2}{2} \tag{13.65}$$

is

$$I = 4LW\cos\phi_m \tag{13.66}$$

Inserting Eq. (13.66) into Eq. (13.64) gives

$$E = \frac{I\cos^2\beta}{8H\cos\phi_m}\left(\phi + \sin\phi\cos\phi\right)\Big|_{\phi_1}^{\phi_2} \tag{13.67}$$

The zone multipliers are calculated for $I = 1000$ cd and $H = 6$ ft. Inserting these numbers yields

$$ZM = \frac{1000\cos^2\beta}{48\cos\phi_m}\left(\phi + \sin\phi\,\cos\phi\right)\Big|_{\phi_1}^{\phi_2} \tag{13.68}$$

EXAMPLE 13.11

Calculate the zone multiplier for $\beta = 25°$ and $\alpha = 0$ to $10°$ and use it to find the horizontal component of illuminance produced by the luminaire in Table 13.4. Repeat for $\alpha = 0$ to $20°$.

Solution. The first step is to find the ϕ angles for each α. From Fig. 13.22c and Eq. (13.59),

$$\tan\alpha = \frac{x}{H} = \frac{\tan\phi}{\cos\beta} \tag{13.69}$$

$\alpha = 0°$: $\tan\phi = 0$, $\phi = 0°$, $\sin\phi = 0$, $\cos\phi = 1$

$\alpha = 10°$: $\tan\phi = 0.1763(0.9063) = 0.1598$, $\phi = 9.0780° = 0.1584$ rad. $\sin\phi = 0.1578$, $\cos\phi = 0.9875$

$$\phi_m = \frac{9.0780 + 0}{2} = 4.5390°, \qquad \cos\phi_m = 0.9969$$

Inserting these values into Eq. (13.68) gives

$$ZM = \frac{1000(0.8214)}{48(0.9969)}[0.1584 + 0.1578(0.9875)] = 5.395$$

This is the zone multiplier appearing in Table 13.6 for $\beta = 25°$ and $\alpha = 0$ to $10°$. Multiplying by the intensity value in thousands for $\beta = 25°$ and $\alpha = 5°$ from Table 13.5 gives

$$E = 5.395(1.440) = 7.8 \text{ fc}$$

which appears in Table 13.4.

For $\alpha = 0$ to $20°$, we must first calculate the zone multiplier for $\beta = 10$ to $20°$. The necessary quantities are

$\alpha = 20°$: $\phi = 18.256° = 0.3186$ rad, $\sin\phi = 0.3133$, $\cos\phi = 0.9497$.

$\alpha = 10°$: $\phi = 9.078° = 0.1584$ rad, $\sin\phi = 0.1578$, $\cos\phi = 0.9875$.

$$\phi_m = \frac{9.078 + 18.256}{2} = 13.667°$$

$$ZM = 5.315$$

Then

$$E = 7.8 + 5.315(1.400) = 7.8 + 7.4 = 15.2 \text{ fc}$$

13.8 INVERSE SQUARE LAW APPROXIMATIONS

Throughout the book, interpolation has been employed, sometimes doubly and triply, to obtain values of parameters. Linear (first degree) interpolation has consistently been used, even though it was an approximation. The possibility of error with linear interpolation is especially great when dealing with rapidly-changing intensity distribution curves.

In this section, second and third degree interpolation techniques will be examined and inverse square law approximations will be revisited (the latter were first examined in Sect. 2.12). The goal is to develop guidelines for minimizing errors in calculations of illuminance at a point resulting from both the size of luminaire elements chosen and the values of their (interpolated) intensities.

In recent years, computer procedures have been developed that calculate the illuminance at a point from an array of luminaires. In these procedures, the luminaires are discretized into elements, such that the maximum dimension of each element is small in comparison with the distance from the center of the element to the task. This constraint permits use of the inverse-square law (ISL) to calculate the illuminance at the task as the summation of the illuminances from the luminaire elements.

The horizontal illuminance E on a visual task produced by a point source of light can be calculated by the ISL as

$$E = \frac{I(\theta, \phi)\cos\theta}{d^2} \tag{13.70}$$

where $I(\theta, \phi)$ is the intensity of the source in the direction of the task, θ and ϕ are the polar and azimuth angles describing the location of the source with respect to the task, and d is the distance from source to task. In interior spaces the sources are customarily all at the same mounting height h above the task and it is more useful to have Eq. (13.70) in the form

$$E = \frac{I(\theta,\phi)\cos^3\theta}{h^2} \tag{13.71}$$

The geometry to be used in this section is shown in Fig. 13.23. The task is located at the origin of an xy coordinate system such that +y is always the viewing direction. A typical luminaire element is shown, with its center at x_e, y_e, h. The location of the center may also be described in terms of h, the polar angle θ, and the azimuth angle ϕ, as shown. θ can vary from 0 to 90° whereas ϕ can vary from 0 to ±180°. A luminaire angle ψ is also shown. ψ is 0° when the luminaire is viewed lengthwise and is 90° for crosswise viewing. ϕ and ψ are not necessarily the same angles. ϕ is the azimuth angle at the task with respect to the viewing direction whereas ψ is the azimuth angle at the luminaire with respect to the luminaire axis.

Consider that the element shown in Fig. 13.23 has an area A_e and is part of a larger luminaire of area A_l. The approximate illuminance on the task from this element is, from Eq. (13.71),

$$E_{ISL} = \frac{A_e}{A_l}\left(\frac{I_{mid}(\theta,\phi)\cos^3\theta_{mid}}{h^2}\right) \tag{13.72}$$

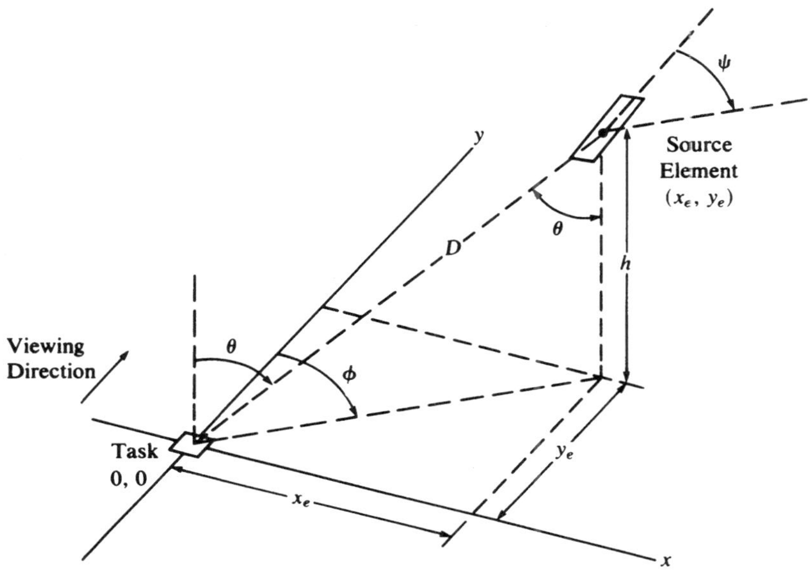

Figure 13.23

where A_e/A_l is the fraction of the luminaire intensity attributable to the element and "mid" denotes that I, θ, and φ are with reference to the element's center. The subscript "ISL" denotes that E is an approximate illuminance obtained from the inverse-square law. The total illuminance on the task is a summation of the illuminances from all the luminaire elements.

In using Eq. (13.72), two problems arise. The first involves the determination of I_{mid} for each source element. The second has to do with deciding how many source elements are needed.

Luminaire manufacturers generally present intensity distribution data in 5 or 10° increments in polar angle θ for three values of luminaire azimuth angle ψ, namely parallel (0°), perpendicular (90°), and at 45° to the luminaire axis. It is thus necessary to develop interpolation procedures for obtaining intensity values at other θ and ψ angles.

A second-degree polynomial will be used for ψ interpolations (since only three values of intensity are generally available for each θ), and a third-degree polynomial for those in θ. For the ψ interpolation,

$$I(\psi) = a_0 + a_1\psi + a_2\psi^2 \tag{13.73}$$

Substituting ψ = 0, 45, and 90 into Eq. (13.73) gives three equations with three unknowns, yielding

$$\left.\begin{aligned} a_0 &= I_0 \\ a_1 &= \frac{-3I_0 + 4I_{45} - I_{90}}{90} \\ a_2 &= \frac{I_0 - 2I_{45} + I_{90}}{4050} \end{aligned}\right\} \tag{13.74}$$

For each value of θ, the interpolation passes a parabola through the intensity values at $\psi = 0$, 45, and 90°, resulting in three coefficients that are then used to calculate $I(\psi)$ for the given value of ψ. This interpolation is done prior to the θ interpolation, and is repeated four times for each intensity determination, yielding intensity values labeled I_1, I_2, I_3, and I_4 at four consecutive θ angles, each Δθ apart, with two on each side of the given θ value. The procedure is illustrated in Table 13.7, where it is desired to obtain the intensity at $\psi = 56.9°$ and $\theta = 43.9°$. Here $\Delta\theta = 5°$ and it is necessary to perform the ψ interpolation at $\theta = 35$, 40, 45, and 50°. The resulting intensity values are $I_1 = 947$ cd, $I_2 = 857$ cd, $I_3 = 699$ cd, and $I_4 = 505$ cd.

TABLE 13.7 ILLUSTRATING DOUBLE INTERPOLATION TO OBTAIN I(ψ, θ) IN CANDELAS

θ	I_0	I_{45}	$I_{56.9}$	I_{90}
35°	290	860	(I_1)947	1050
40°	260	780	(I_2)857	940
43.9°			(I)738	
45°	220	650	(I_3)669	690
50°	180	490	(I_4)505	400

First interpolation: ψ = 56.9°

θ	a_0	a_1	a_2	I cd
35°	290	16.89	−0.09383	$I_1 = 947$
40°	260	15.56	−0.08889	$I_2 = 857$
45°	220	13.89	−0.09630	$I_3 = 699$
50°	180	11.33	−0.09877	$I_4 = 505$

Second interpolation: θ = 43.9°, θ′ = 8.9°

θ	A_0	A_1	A_2	A_3	I cd
43.9°	947	−9.067	−2.000	0.0427	738

For the θ interpolation there are two cases, namely $\Delta\theta = 5°$ and $\Delta\theta = 10°$. The pertinent equations are:

$$\Delta\theta = 5°: \qquad I(\theta') = A_0 + A_1\theta' + A_2(\theta')^2 + A_3(\theta')^3 \tag{13.74}$$

Letting $\theta' = 0$, 5, 10, and 15 in Eq. (13.74) gives four equations with four unknowns, yielding

$$\left.\begin{aligned} A_0 &= I_1 \\ A_1 &= \frac{-11I_1 + 18I_2 - 9I_3 + 2I_4}{30} \\ A_2 &= \frac{2I_1 - 5I_2 + 4I_3 - I_4}{50} \\ A_3 &= \frac{-I_1 + 3I_2 - 3I_3 + I_4}{750} \end{aligned}\right\} \tag{13.75}$$

$$\Delta\theta = 10°: \qquad I(\theta') = A_0' + A_1'\theta' + A_2'(\theta')^2 + A_3'(\theta')^3 \tag{13.76}$$

where

$$\begin{aligned} A_0' &= A_0, \quad A_1' = \frac{1}{2}A_1 \\ A_2' &= \frac{1}{4}A_2, \quad A_3' = \frac{1}{8}A_3 \end{aligned} \tag{13.77}$$

The use of θ′ requires explanation. To avoid large numbers in the calculation of $I(\theta')$, it is desirable to shift the $\theta = 0°$ axis to the value of θ at I_1. Thus, in the example, $\theta' = 0$ corresponds to $\theta = 35°$. This yields values of θ′ between 5 and 10° for $\Delta\theta = 5°$ and between 10 and 20° for $\Delta\theta = 10°$, in Eqs. (13.74) and (13.76), respectively. For the example in Table 13.7, $\theta' = 8.9°(43.9 - 35)$ for $\theta = 43.9°$.

It was mentioned earlier that certain cautions must be used with polynomial interpolations. To examine these constraints for a parabolic interpolation, recast Eq. (13.73) into a standard form for a parabola and obtain

$$I(\psi) - \left(a_0 - \frac{a_1^2}{4a_2}\right) = a_2\left(\psi + \frac{a_1}{2a_2}\right)^2 \tag{13.78}$$

For the values of θ and intensity in Table 13.7, Eq. (13.78) becomes

$$\left.\begin{aligned} \theta = 35°: &\quad I(\psi) - 1050 = -0.09383(\psi - 90°)^2 \\ \theta = 40°: &\quad I(\psi) - 941 = -0.08889(\psi - 87.5°)^2 \\ \theta = 45°: &\quad I(\psi) - 721 = -0.09630(\psi - 72.1°)^2 \\ \theta = 50°: &\quad I(\psi) - 505 = -0.09877(\psi - 57.4°)^2 \end{aligned}\right\} \tag{13.79}$$

Equations (13.79) are plotted in Fig. 13.24 and the twelve intensity values used in their calculation are marked, as are the four calculated intensities. Note that each curve has one point of zero slope [at the values in parentheses in Eqs. (13.79)] and that the curves do not have zero slope at $\psi = 0°$ or at $\psi = 90°$ (except by coincidence for the $\theta = 35°$ curve at $\psi = 90°$). This poses a problem because most luminaires have a symmetry which requires that the slopes of the intensity distribution curves be zero at $\psi = 0°$ and $\psi = 90°$. The parabolic interpolation violates this constraint. Also, the parabolic interpolation permits an intensity value at a given ψ which is greater than the three values of intensity used to determine it. This may or may not be correct. The parabolic interpolation is quite good for luminaires possessing near-rotational symmetry but is likely to be in error for batwing luminaires, corridor luminaires, and luminaires designed to provide relatively uniform illuminance for rectangular areas.

An alternative to the parabolic interpolation on ψ has been suggested (Levin, 1984). The desire is to have an interpolation that (a) passes through three points, (b) has zero slope at its end points, and (c) possesses a derivative with no more than one zero in the interval. Such a function is

$$y = ax^4 + bx^3 + cx^2 \tag{13.80}$$

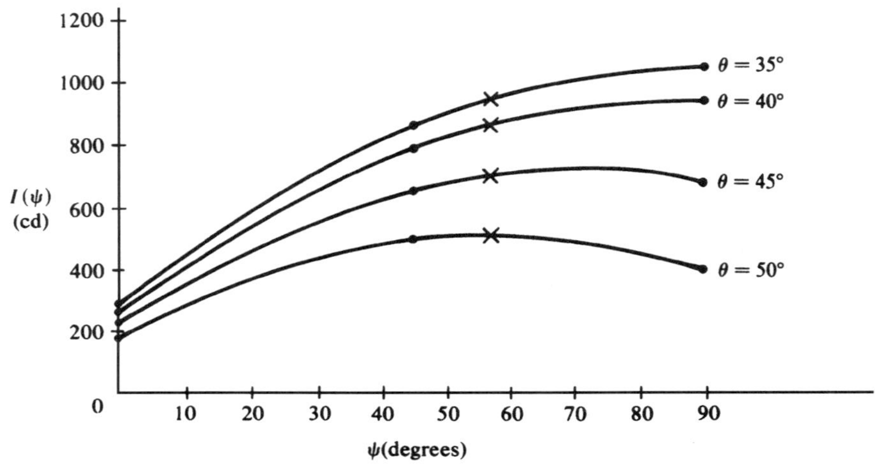

Figure 13.24

To utilize this equation for the ψ interpolation, it is normalized so that the range $0 < \psi < 90$ becomes $0 < x < 1$ and $y(0) = I(0°)$, $y(\frac{1}{2}) = I(45°) - I(0°) \equiv A$ and $y(1) = I(90°) - I(0°) \equiv B$. Then

$$A = \frac{1}{16}a + \frac{1}{8}b + \frac{1}{4}c \tag{13.81a}$$

$$B = a + b + c \tag{13.81b}$$

$$0 = 4a + 3b + 2c \tag{13.81c}$$

where Eq. (13.81c) arises from requiring the derivative to be zero at $x = 1$. Solution of Eqs. (13.81) yields

$$\left.\begin{aligned} a &= 16A - 8B \\ b &= -32A + 14B \\ c &= 16A - 5B \end{aligned}\right\} \tag{13.82}$$

and inserting these results into Eq. (13.80) gives

$$y = (16A - 8B)x^4 - (32A - 14B)x^3 + (16A - 5B)x^2 \tag{13.83}$$

For the data in Table 13.7 at $\theta = 35°$,

$$A = 860 - 290 = 570, \qquad B = 1050 - 290 = 760$$

$$x = \frac{56.9}{90} = 0.632$$

$$a = 3040, \quad b = -7600, \qquad c = 5320$$

$$y = 692$$

$$I_1 = 982 \text{ cd}$$

For the other values of θ,

$$I_2 = 882 \text{ cd} \qquad \text{for } \theta = 40°$$
$$I_3 = 715 \text{ cd} \qquad \text{for } \theta = 45°$$
$$I_4 = 506 \text{ cd} \qquad \text{for } \theta = 50°$$

Then, performing the θ interpolation indicated by Eqs. (13.74) and (13.75) gives $I = 765$ cd. This compares with $I = 738$ using the parabolic interpolation. The difference in this case is slight (less than 3%).

The discussion will turn now to the number of source elements required, which of course depends on the level of accuracy desired. From Eq. (13.72), this accuracy depends on how I and θ vary from one portion of the luminaire element to another. Much has been written in the past about requiring that the distance from source to task be at least five times the maximum source dimension for the inverse-square law to yield accurate answers. This is true if it is desired to have an error in E_{ISL} less than ±1%; but if a ±5% error is tolerable, then this distance can often be equal to the maximum source dimension if the intensity is reasonably constant over the source element, and the source is reasonably Lambertian (Murdoch, 1981).

Table 13.8 has been prepared, using the symbology in Fig. 13.23, to display the variables that must be considered in selecting element lengths for rows of fluorescent luminaires. The table is based on $h = 7$ ft (Fig. 13.23) and can be used for rows parallel or perpendicular to the viewing direction. Dimensions for the latter are shown in parentheses.

In selecting element lengths, four factors have been considered.

1. *Luminaire length.* Element lengths were chosen to fit a standard 4-ft luminaire, or integral multiples thereof. For example, in Table 13.8 there is no element with $y_1(x_1) = 2$ ft and $y_2(x_2) = 6$ ft, where $y_1(x_1)$ and $y_2(x_2)$ are the coordinates of the ends of the element (see Fig. 13.23).
2. *Variation in polar angle θ.* A row directly over the task was considered first. For such a row, $x_e(y_e) = 0$ and there is no variation with azimuth angle ϕ, except in switching from $\phi = 0°$ directly in front of the task to $\phi = 180°$ directly behind it for $x_e = 0$. Element lengths in this row were chosen such that the change in θ would be limited to approximately 10° in traversing each element length. Without knowing how the intensity of a given luminaire will vary with θ, it was felt that $\Delta\theta \sim 10°$ was the maximum change in θ that should be allowed. Also this choice is consistent with contemporary luminaire intensity distribution data, which is provided by the manufacturer in no more than 10° increments in θ. Values of Δθ for the element lengths chosen are given in Table 13.8.
3. *Variation in azimuth angle ϕ.* Intensity distribution data are normally provided in no less than three azimuth planes, each separated by 45° in ϕ. Again because the actual intensity distribution is not known, it was decided to limit variations in ϕ across an element to approximately half this value, or 22.5°, although in a very few cases, where variations in Δθ were small, a Δϕ greater than 22.5° was allowed. Values of Δϕ for the element lengths chosen are also given in Table 13.8.
4. *% Error.* As a general guideline, element lengths were chosen to keep % error below 1% for elements within 8 ft horizontally of the task. For $h = 7$ ft, this means less than 1% error in E_{ISL} for all θ less than about 50°. Beyond 50°, it was felt that, in general, luminaire intensity would be lower and a greater % error in E_{ISL} (up to as much as 8%) could be tolerated.

TABLE 13.8 ELEMENT LENGTHS FOR ROWS PARALLEL (PERPENDICULAR) TO VIEWING DIRECTION

x_e, ft* (y_e, ft)	y_1, ft* (x_1, ft)	y_2, ft* (x_2, ft)	Δθ°	Δφ°
0	0	1.2	10	0
	1.2	2.5	10	0
	2.5	4	10	0
	4	6	11	0
	6	8	8	0
	8	12	11	0
	12	20	11	0
2	0	1.2	3	31
	1.2	2.5	6	20
	2.5	4	8	12
	4	6	9	8
	6	8	8	4
	8	12	10	4
	12	20	11	4
4	0	1.7	2	23
	1.7	4	7	22
	4	6	7	11
	6	8	6	7
	8	12	9	9
	12	20	10	7
6	0	1.7	2	16
	1.7	4	4	18
	4	8	9	19
	8	12	7	10
	12	20	9	10
8	0	4	2	27
	4	12	8	30
	12	20	8	12
10	0	4	2	22
	4	12	9	28
	12	20	7	13
12	0	8	4	34
	8	20	9	25
14	0	20	11	55

*All x and y values can be either + or −.

EXAMPLE 13.12

To illustrate the use of Table 13.8, select element lengths for a row of fluorescent luminaires parallel to the viewing direction and displaced 5 ft horizontally from the task ($x_e = 5$ ft). The row runs from 8 ft behind the task to 16 ft in front of the task($y = -8$ ft to $+ 16$ ft).

Solution. With $x_e = 5$ ft, the $x_e = 4$ column will be used, rather than the $x_e = 6$ column, to be conservative. The first element lies between −8 ft and −6 ft, the second between −6 ft and −4 ft, ... the second last between 8 ft and 12 ft, and the last element between +12 ft and +16 ft (it would lie between +12 and +20 ft if the row went that far). 10 elements are needed to represent the row.

Table 13.8 may be used also in the selection of element sizes for arrays of square luminaires. For a 2 × 2 ft luminaire, Table 13.8 shows that, for x_e (y_e) less than 4 ft, four 1-ft square elements should be used, outside a 4-ft radius circle, the luminaire need not be subdivided. A 3 × 3 ft luminaire should be divided into four 1.5-ft square elements within a 6-ft radius circle and taken as a single element outside this circle.

13.9 MORE ON FORM FACTORS

The remainder of the chapter will be devoted to form factors and their application. Recall from Eq. (13.24) that a form factor is a measure of the luminous flux transfer from one surface to another. Specifically, f_{21} is the fraction of the flux emitted by area A_2 that is received directly by area A_1. It is calculated by dividing the average illuminance on A_2 caused by the flux received directly from A_1 by the luminous exitance of A_1.

Form factors form the basis for the lumen method for obtaining average illuminance in a space. They can be used in calculating the transfer of flux through a skylight well or a light pipe, and in a variety of other applications that will be explored in the following sections.

The derivation of form factors for most geometries is straight forward theoretically but often leads to multiple integrals with difficult boundary conditions. In this section, the form factor will be derived for the relatively simple case of parallel circular disks and the form factors for several other geometries will be given without proof.

Two such (perfectly diffusing) disks of radius a and b separated by a distance h are shown in Fig. 13.25. Disk 1 will be treated as the source, with a luminance L, and disk 2 as the receiver, and the form factor f_{12} from disk 1 to disk 2 will be derived.

From the development in Eqs. (13.21) to (13.24),

$$f_{12} = \frac{\phi_{12}}{\phi_1} \tag{13.84}$$

the fraction of the flux emitted by surface 1 that is received directly by surface 2.

In Fig. 13.25, a perfectly diffusing sphere of radius r is circumscribed around the two disks. It is noted that all flux impinging on either of the disks also impinges on the segment of the sphere intercepted by that disk. Thus the problem changes to finding the flux transfer between two sphere segments.

In Sect. 2.10, the integrating sphere was discussed and it was noted that the illuminances at all points on the sphere were the same and that the contributions to the illuminance at a point from all other points were the same. Equation (2.47) expressed this as

$$dE = \frac{LdA}{4r^2} \tag{13.85}$$

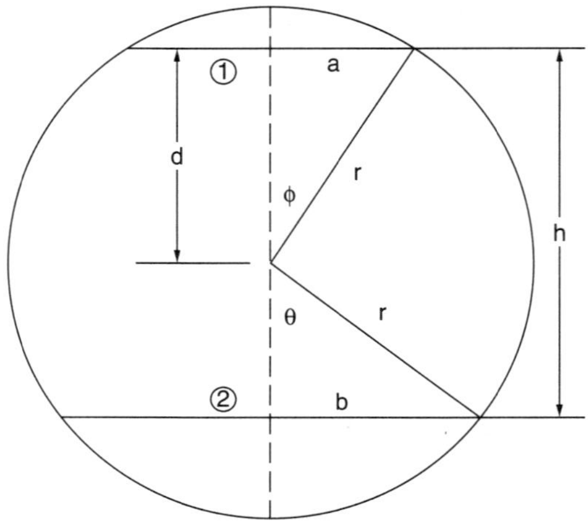

Figure 13.25 Sphere method for circular disks.

The two sphere segments in Fig. 13.25 have areas

$$\left.\begin{aligned} A_1 &= 2\pi r^2 (1-\cos\phi) \\ A_2 &= 2\pi r^2 (1-\cos\theta) \end{aligned}\right\} \tag{13.86}$$

The illuminance at any point on the sphere from the source (A_1) is

$$E = \frac{LA_1}{4r^2} = \frac{\pi L}{2}(1-\cos\phi) \tag{13.87}$$

The flux from source to receiver is

$$\begin{aligned} \phi_{12} = EA_2 &= \frac{\pi L}{2}(1-\cos\phi) 2\pi r^2 (1-\cos\theta) \\ &= \pi^2 L r^2 (1-\cos\phi)(1-\cos\theta) \end{aligned} \tag{13.88}$$

The total flux from the source is

$$\begin{aligned} \varphi_T = E(A_s - A_1) &= \frac{\pi L}{2}(1-\cos\phi)[4\pi r^2 - 2\pi r^2 (1-\cos\phi)] \\ &= \pi^2 L r^2 (1-\cos\phi)(1+\cos\phi) \end{aligned} \tag{13.89}$$

Then

$$f_{12} = \frac{\phi_{12}}{\phi_T} = \frac{1-\cos\theta}{1+\cos\phi} \tag{13.90}$$

Usually a, b, and h are given in Fig. 13.25. To find θ and ϕ, note that

$$r^2 = a^2 + d^2 = b^2 + (h - d)^2 \tag{13.91}$$

giving

$$d = \frac{h^2 - a^2 + b^2}{2h} \tag{13.92}$$

Then

$$\tan\phi = \frac{a}{d}, \qquad \tan\theta = \frac{b}{h-d} \tag{13.93}$$

EXAMPLE 13.13

In Fig. 13.26, a perfectly diffusing circular disk source of luminance 2 cd/in^2 and diameter 5 ft lights a 15 ft diameter floor area. Find the average illuminance on this floor surface.

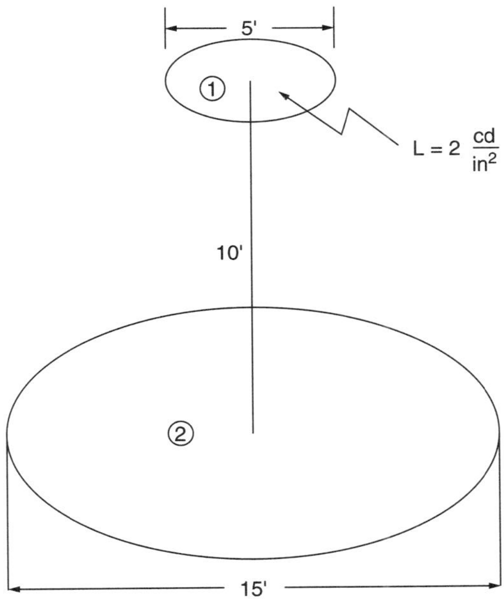

Figure 13.26

Solution. To circumscribe a sphere around the two disks, its center must first be found. Using the notation in Fig. 13.25,

$$a = 2.5', \qquad b = 7.5', \qquad h = 10'$$

$$\text{d from Eq. (13.92)} = \frac{100 - 6.25 + 56.25}{20} = 7.5'$$

From Eqs. (13.93),

$$\phi = \tan^{-1}\frac{2.5}{7.5} = 18.43°, \qquad \cos\phi = 0.949$$

$$\theta = \tan^{-1}\frac{7.5}{2.5} = 71.57°, \qquad \cos\theta = 0.316$$

From Eq. (13.90),

$$f_{12} = \frac{1-0.316}{1+0.949} = 0.351$$

The illuminous exitance of the source is

$$M_1 = \pi L = \pi \times 2 \times 144 = 905 \text{ lm/ft}^2$$

The flux emitted by the source is

$$\phi_1 = M_1 A_1 = 905 \times \pi \times 6.25 = 17{,}800 \text{ lm}$$

Thus

$$\phi_{12} = 0.351 \times 17800 = 6250 \text{ lm}$$

and

$$E_{av} = \frac{\phi_{12}}{A_2} = \frac{6250}{\pi \times 56.25} = 35 \text{ fc}$$

The next shapes to be considered are the two concentric parallel rectangular surfaces shown in Fig. 13.27. The form factor for this configuration is given without proof as

$$f = \frac{2}{\pi st}\left[\frac{1}{2}\ln\frac{(1+s^2)(1+t^2)}{1+s^2+t^2} + t\sqrt{1+s^2}\tan^{-1}\frac{t}{\sqrt{1+s^2}} + s\sqrt{1+t^2}\tan^{-1}\frac{s}{\sqrt{1+t^2}} - t\tan^{-1}t - s\tan^{-1}s\right] \tag{13.94}$$

where

$$s = \frac{w}{h}, \qquad t = \frac{\ell}{h} \tag{13.95}$$

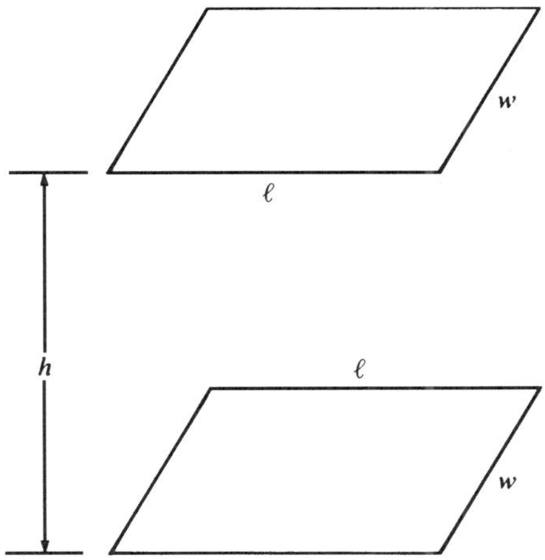

Figure 13.27 Concentric parallel rectangles.

To find the form factor for two concentric parallel infinitely long strips, let ℓ approach infinity. Then t approaches infinity in Eq. (13.94), giving

$$f = 0 + \frac{2}{\pi s}\sqrt{1+s^2} \times \frac{\pi}{2} + 0 - \frac{2}{\pi s} \times \frac{\pi}{2} - 0$$

$$= \frac{\sqrt{1+s^2} - 1}{s} \tag{13.96}$$

EXAMPLE 13.14

In Fig. 13.28, the source is a long ceiling strip 5 ft wide with a luminance of 0.5 cd/in^2. Find the average illuminance on a parallel and concentric floor strip 10 ft away.

Solution. Assume that ℓ is very large and w = 5 ft. Thus $s = \frac{5}{10} = 0.5$ and

$$f = \frac{\sqrt{1+0.25} - 1}{0.5} = 0.236$$

The exitance of the source is

$$M = 0.5 \times 144 \times \pi = 226 \text{ lm/ft}^2$$

The flux from the source is

$$\phi_{\text{source}} = 226 \times 5 \times \ell \times 0.236 = 266.7\ell \text{ lm}$$

Thus

$$E_{\text{av}} = \frac{226.7\ell}{5\ell} = 53 \text{ fc}$$

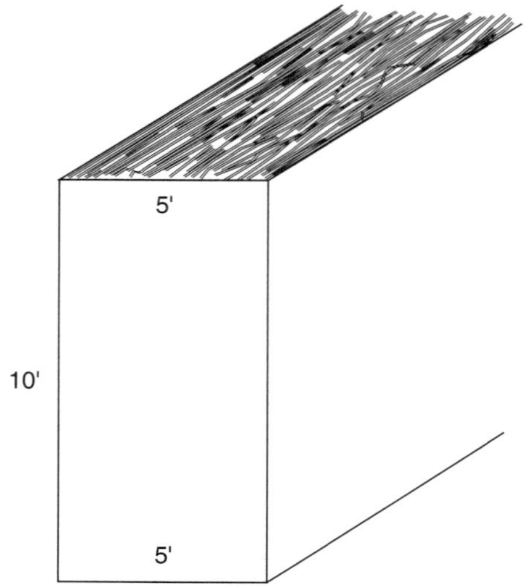

Figure 13.28

Consider now the basic rectangular parallelepiped room cavity shown in Fig. 13.29. Its "ceiling" is the luminaire plane, and its "floor" is the work plane. Let the walls be numbered 1, the ceiling 2, and the floor 3. Then the form factor in Eq. (13.84) becomes $f_{23} = f_{32}$, because the areas of the floor and ceiling are the same and thus so are the form factors, from Eq. (13.26).

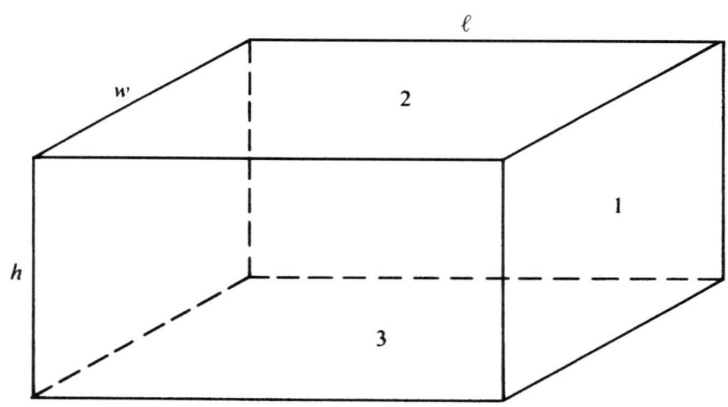

Figure 13.29 Basic room cavity.

All other form factors for the space in Fig. 13.29 may be expressed in terms of f_{23}. Thus

$$f_{21} = f_{31} = 1 - f_{23} \tag{13.97}$$

$$f_{12} = f_{13} = \frac{A_2}{A_1} f_{21} = \frac{A_2}{A_1} (1 - f_{23}) \tag{13.98}$$

$$f_{11} = 1 - 2f_{12} = 1 - 2 \frac{A_2}{A_1} (1 - f_{23}) \tag{13.99}$$

A word of justification for each of these equations is required. Equation (13.97) is true because the sum of the form factors for a given surface must be unity. Equation (13.98) arises because of the reciprocity relationships between the walls and the ceiling and floor. Equation (13.99) addresses the situation that the walls radiate to themselves, as well as to the floor and ceiling. Again, in that equation, the sum of the form factors for a given surface must be unity.

As shown by Eq. (13.95), the form factors for a rectangular space depend on the width:height and length:height ratios. Using Eq. (9.3), Eq. (13.95) may be modified to

$$s = \frac{5}{CR}\left(1 + \frac{w}{\ell}\right), \qquad t = \frac{5}{CR}\left(1 + \frac{\ell}{w}\right) \tag{13.100}$$

Thus f in Eq. (13.94), and all the other form factors for a rectangular parallepiped, may be expressed in terms of cavity ratio and length:width ratio. The effect of ℓ/w on f_{23} is small numerically. Thus the CU tables have been standardized for $\ell/w = 1.6$. The form factors for this length:width ratio are listed in Table 13.9. To generate the table, f_{23} is calculated from Eq. (13.94). Then Eqs. (13.97) to (13.99) are used to calculate the other form factors, after inserting A_2/A_1, obtained by modifying Eq. (9.22) to read

$$CR = 2.5\frac{A_1}{A_2}, \qquad \frac{A_2}{A_1} = \frac{2.5}{CR} \tag{13.101}$$

where A_1 is the wall area and A_2 is the work plane (ceiling) area.

TABLE 13.9

CR	f_{11}	f_{12}, f_{13}	f_{21}, f_{31}	f_{23}, f_{32}
0	0.000	0.500	0.000	1.000
1	0.133	0.434	0.173	0.827
2	0.224	0.388	0.311	0.689
3	0.298	0.351	0.421	0.579
4	0.361	0.320	0.511	0.489
5	0.415	0.292	0.585	0.415
6	0.463	0.269	0.645	0.355
7	0.504	0.248	0.694	0.306
8	0.540	0.230	0.735	0.265
9	0.573	0.214	0.769	0.231
10	0.601	0.199	0.798	0.202

Equation (13.94) is complex analytically. The following approximation has been developed, which is accurate within 0.4% for CRs of 20 or less:

$$f_{23} = 0.026 + 0.503e^{-0.270CR} + 0.470e^{-0.119CR} \tag{13.102}$$

EXAMPLE 13.15

A conical light well is shown in Fig. 13.30 with $f_{23} = 0.183^*$. Find f_{21}, f_{32}, f_{31}, f_{12}, f_{13} and f_{11}, where the surface numbering matches that in Fig. 13.29.

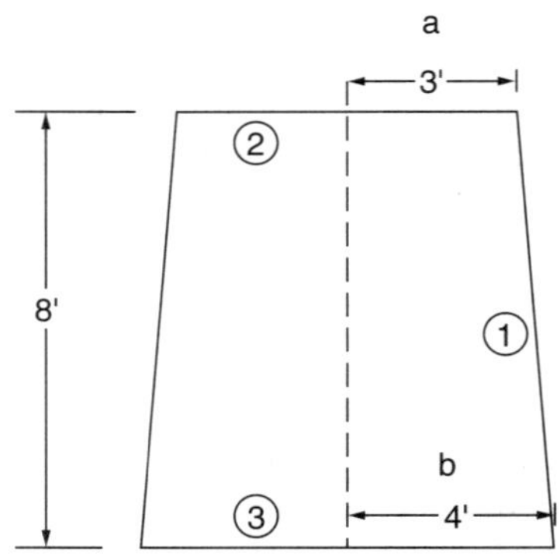

Figure 13.30

Solution.

$$f_{21} = 1 - f_{23} = 0.817$$

$$A_2 = 9\pi \text{ ft}^2, \qquad A_3 = 16\pi \text{ ft}^2$$

$$f_{32} = \frac{A_2}{A_3} f_{23} = \frac{9}{16} \times 0.183 = 0.103$$

$$f_{31} = 1 - f_{32} = 0.897$$

$$A_1^{**} = \pi(b^2 - a^2)\sqrt{1 + \left(\frac{h}{b-a}\right)^2} = \pi(16-9)\sqrt{1+\left(\frac{8}{4-3}\right)^2} = 56.44\ \pi \text{ ft}^2$$

$$f_{12} = \frac{A_2}{A_1} f_{21} = \frac{9\pi}{56.44\pi} \times 0.817 = 0.130$$

$$f_{13} = \frac{A_3}{A_1} f_{31} = \frac{16\pi}{56.44\pi} \times 0.897 = 0.254$$

$$f_{11} = 1 - f_{12} - f_{13} = 1 - 0.130 - 0.254 = 0.616$$

---

* This may be derived from Eq. (13.90).

** The area of the curved surface of a right cone whose height is z and base is of radius r is

$$A = \pi r^2 \sqrt{1 + \left(\frac{z}{r}\right)^2} \tag{13.103}$$

13.10 FORM FACTOR ALGEBRA AND RELATIONSHIPS

a) Flux Reciprocity

If $M_1 = M_2$, then $\phi_{12} = \phi_{21}$, irrespective of the sizes and orientations of A_1 and A_2. Thus the flux directly impinging on receiver surface 2 from source surface 1 is the same as the flux that would be received by surface 1 if surface 2 were the source with the same exitance.

b) Form Factor Reciprocity

This was developed in Eq. (13.26) and states in general that

$$f_{ij}A_i = f_{ji}A_j \tag{13.104}$$

where i and j represent the ith and jth surfaces in an array.

c) Form Factors for Offset Rectangles

Consider the geometry in Fig. 13.31. It is desired to find f_{14}. Since form factors are geometric values, they cannot depend on values of exitance. Therefore, in Fig. 13.31, let $M_1 = M_2$. Define ϕ_{12-34} as the flux from 12 to 34 and let $\phi_{(12)}$ be the total flux from 12. Then

$$f_{12-34} = \frac{\phi_{12-34}}{\phi_{(12)}} = \frac{\phi_{13} + \phi_{14} + \phi_{23} + \phi_{24}}{\phi_{(12)}}$$

$$= \frac{\phi_{13}}{\phi_1} \times \frac{\phi_1}{\phi_{(12)}} + \frac{\phi_{14}}{\phi_1} \times \frac{\phi_1}{\phi_{(12)}} + \frac{\phi_{23}}{\phi_2} \times \frac{\phi_2}{\phi_{(12)}} + \frac{\phi_{24}}{\phi_2} \times \frac{\phi_2}{\phi_{(12)}} \tag{13.105}$$

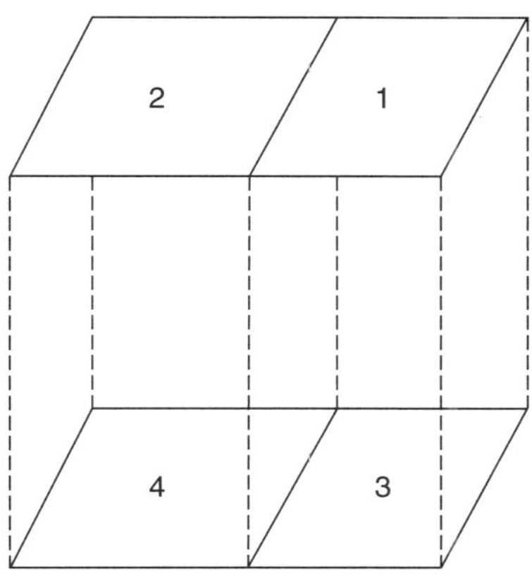

Figure 13.31 Offset rectangles.

Using the assumption of equal exitances yields

$$\frac{\phi_1}{\phi_{(12)}} = \frac{M_1A_1}{M_1A_1 + M_2A_2} = \frac{M_1A_1}{M_1(A_1 + A_2)} = \frac{A_1}{A_1 + A_2} \tag{13.106}$$

and

$$\frac{\phi_2}{\phi_{(12)}} = \frac{A_2}{A_1 + A_2} \tag{13.107}$$

Substituting Eqs. (13.106) and (13.107) into Eq. (13.105) yields

$$f_{12-34} = f_{13}\frac{A_1}{A_1 + A_2} + f_{14}\frac{A_1}{A_1 + A_2} + f_{23}\frac{A_2}{A_1 + A_2} + f_{24}\frac{A_2}{A_1 + A_2} \tag{13.108}$$

f_{23} needs to be replaced in Eq. (13.108).

$$\left.\begin{aligned} A_2 f_{23} &= A_3 f_{32}, \\ f_{23} &= \frac{A_3}{A_2} f_{32} = \frac{A_1}{A_2} f_{14} \end{aligned}\right\} \tag{13.109}$$

Inserting gives

$$f_{12-34} = f_{13}\frac{A_1}{A_1 + A_2} + f_{24}\frac{A_2}{A_1 + A_2} + f_{14}\left(\frac{A_1}{A_1 + A_2} + \frac{A_1}{A_2} \times \frac{A_2}{A_1 + A_2}\right) \tag{13.110}$$

and solving for f_{14} yields

$$f_{14} = \frac{f_{12-34}(A_1 + A_2) - f_{13}A_1 - f_{24}A_2}{2A_1} \tag{13.111}$$

EXAMPLE 13.16

A very long rectangular tube is shown in Fig. 13.32. It is desired to find f_{63}, f_{64}, f_{14}, and f_{23}.

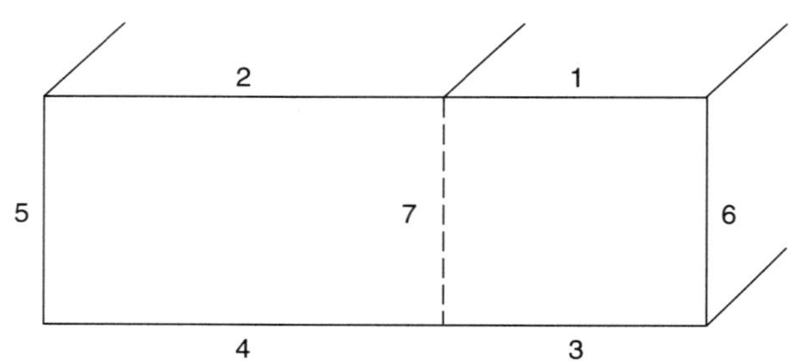

Figure 13.32

Solution. Create a dashed plane surface 7. Find f_{13}, f_{24}, f_{12-34}, f_{56}, and f_{67} from Eq. (13.96). Then, since the form factors must sum to 1,

$$f_{63} = f_{61} = \frac{1 - f_{67}}{2}$$

$$f_{64} = f_{62} = \frac{1 - f_{65} - 2f_{63}}{2} = \frac{f_{67} - f_{65}}{2}$$

Finally, f_{14} may be obtained from Eq. (13.11), and f_{23} from the second Eq. (13.109).

d) Form Factors for Unequal Size Rectangles

An extension of Eq. (13.111) for the case of a rectangle to a larger rectangle is shown in Fig. 13.33. Using superposition, the form factor is given by

$$f_{1-436} = \frac{f_{12-34}(A_1 + A_2) + f_{15-36}(A_1 + A_5) - f_{24}A_2 - f_{56}A_5}{2A_1} \tag{13.112}$$

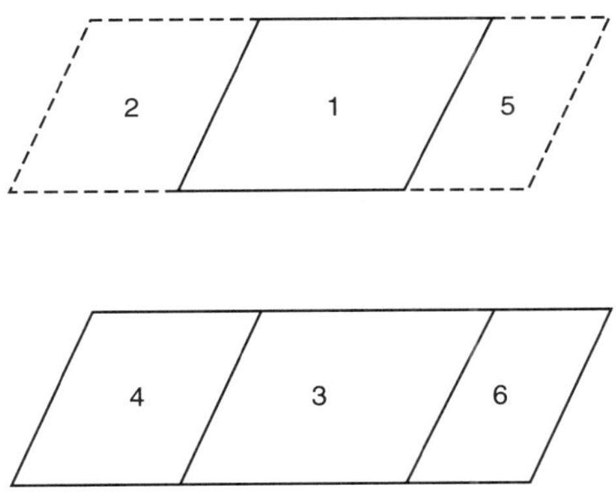

Figure 13.33 Unequal size rectangles.

EXAMPLE 13.17

A long corridor 7 ft wide and 8 ft high has a long row of 1 ft × 4 ft translucent-lensed luminaires centered as shown in Fig. 13.34. The (constant) luminance is 250 cd/ft^2. Neglecting reflections, find the average illuminances on the floor and walls.

Solution. It is assumed that t in the second Eq. (13.95) is infinite. From the first Eq. (13.95),

$$s = \frac{1}{8} = 0.125 \qquad \text{for sections 1 and 3}$$

$$s = \frac{3}{8} = 0.375 \qquad \text{for sections 2 and 4}$$

$$s = \frac{4}{8} = 0.500 \qquad \text{for sections 12 and 34}$$

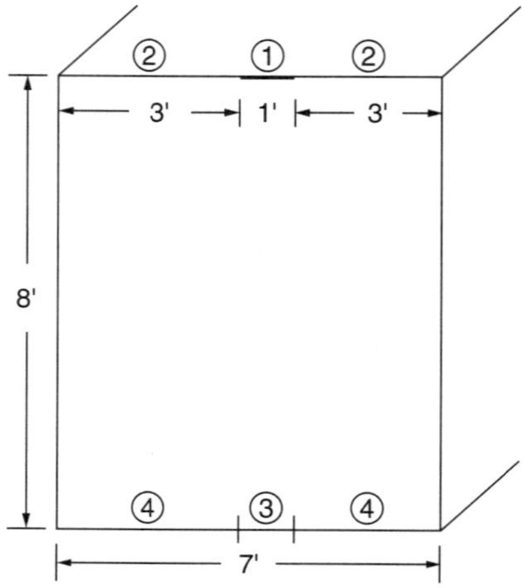

Figure 13.34

From Eq. (13.96)

$$f_{13} = \frac{\sqrt{1+0.0156}-1}{0.125} = 0.062$$

$$f_{24} = \frac{\sqrt{1+0.1406}-1}{0.375} = 0.181$$

$$f_{12-34} = \frac{\sqrt{1+0.25}-1}{0.5} = 0.236$$

f_{1-434}, is also needed, and from Eq. (13.112) is given by

$$f_{1-434} = \frac{f_{12-34}\,(A_1 + A_2) - f_{24}A_2}{A_1}$$

$$f_{1-434} = \frac{0.236 \times 4 - 0.181 \times 3}{1} = 0.401 \qquad (13.113)$$

Then the form factor from the source to the walls is

$$f_{1w} = 1 - 0.401 = 0.599$$

The source exitance is $250\,\pi = 785$ lm/ft^2. Thus the flux emitted by the source is

$$\phi_s = 785 \times 1 \times \ell = 785\ell \text{ lm}$$

and to the walls and floor is

$$\phi_f = 785\ell \times 0.401 = 315\ell \text{ lm}$$

$$\phi_w = 785\ell \times 0.599 = 470\ell \text{ lm}$$

Finally, the illuminances are

$$E_f = \frac{315\ell}{7\ell} = 45 \text{ fc}$$

$$E_w = \frac{470\ell}{16\ell} = 29 \text{ fc}$$

e) Form Factors for Perpendicular Walls
The situation is diagrammed in Fig. 13.35 and f_{15} is desired. From Eqs. (13.94) and (13.26), $f_{23} = f_{32}$, $f_{14} = f_{41}$ and $f_{56} = f_{65}$ can be found. Then

$$\left.\begin{aligned} 2f_{15} &= 1 - f_{14} - 2f_{12} \\ 2f_{21} &= 1 - f_{23} - 2f_{25} \\ 2f_{52} &= 1 - f_{56} - 2f_{51} \end{aligned}\right\} \quad (13.114)$$

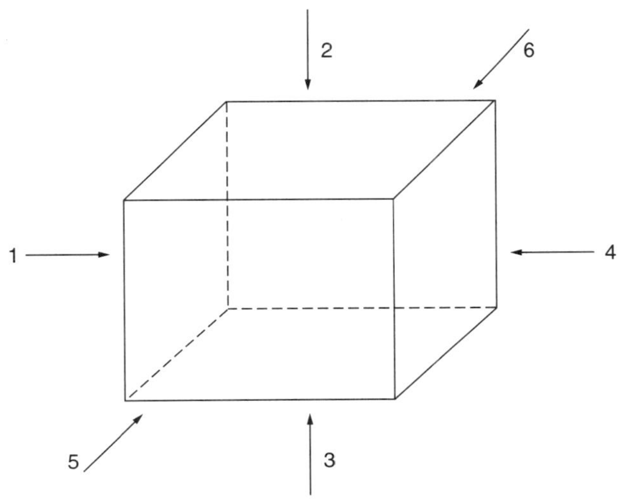

Figure 13.35 Perpendicular walls.

Substituting the second and third of these equations into the first yields

$$2f_{15} = 1 - f_{14} - 2\,\frac{A_2}{A_1}\left(\frac{1 - f_{23} - 2f_{25}}{2}\right)$$

$$= 1 - f_{14} - \frac{A_2}{A_1}(1 - f_{23}) + 2\frac{A_2}{A_1} f_{25}$$

$$= 1 - f_{14} - \frac{A_2}{A_1}(1 - f_{23}) + 2\frac{A_2}{A_1} \times \frac{A_5}{A_2}\left(\frac{1 - f_{56} - 2f_{51}}{2}\right)$$

$$= 1 - f_{14} - \frac{A_2}{A_1}(1 - f_{23}) + \frac{A_5}{A_1}(1 - f_{56}) - 2f_{15}$$

Thus

$$f_{15} = \frac{1 - f_{14} - \frac{A_2}{A_1}(1 - f_{23}) + \frac{A_5}{A_1}(1 - f_{56})}{4} = f_{16} \tag{13.115}$$

It follows that

$$f_{12} = f_{13} = \frac{1 - f_{14} - 2f_{15}}{2} \tag{13.116}$$

and

$$f_{25} = f_{26} = \frac{1 - f_{23} - 2f_{21}}{2} = \frac{1 - f_{23} - 2\frac{A_1}{A_2}f_{12}}{2} \tag{13.117}$$

EXAMPLE 13.18

In Fig. 13.35, let the enclosure be a rectangular solid of width 1 ft, height 1.25 ft and depth 4 ft. Find the form factors for the six surfaces.

Solution. For f_{23}, w = 1 ft, l = 4 ft, h = 1.25′. From Eq. (13.95), s = 0.8 and t = 3.2. Inserting these values into Eq. (13.94) gives $f_{23} = 0.281 = f_{32}$.

For f_{14}, w = 1.25 ft, l = 4 ft, h = 1 ft. Thus s = 1.25 and t = 4. This gives $f_{14} = 0.400 = f_{41}$.

Similarly, for f_{56}, w = 1 ft, l = 1.25 ft, h = 4 ft and s = 0.25, t = 0.3125, giving $f_{56} = 0.024 = f_{65}$.

From Eq. (13.115),

$$f_{15} = \frac{1 - 0.400 - \frac{4}{5}(1 - 0.281) + \frac{1.25}{1}(1 - 0.024)}{4}$$

$$= 0.067 = f_{16}$$

Then Eq. (13.116) yields

$$f_{12} = f_{13} = \frac{1 - 0.400 - 0.067}{2} = 0.233$$

and Eq. (13.117) gives

$$f_{25} = f_{26} = \frac{1 - 0.281 - 2 \times \frac{5}{4} \times 0.233}{2} = 0.068$$

As a check,

$$f_{14} + 2f_{12} + 2f_{15} = 0.400 + 0.466 + 0.134 = 1.0$$

13.11 ONE-BOUNCE AND TWO-BOUNCE FLUX TRANSFER ANALYSIS

Before the general case of multiple-bounce flux transfer within a six-sided enclosure is addressed, single- and double-bounce flux transfer will be examined for three situations: a skylight, a window, and a cavity, as shown in Fig. 13.36a, b, and c, respectively.

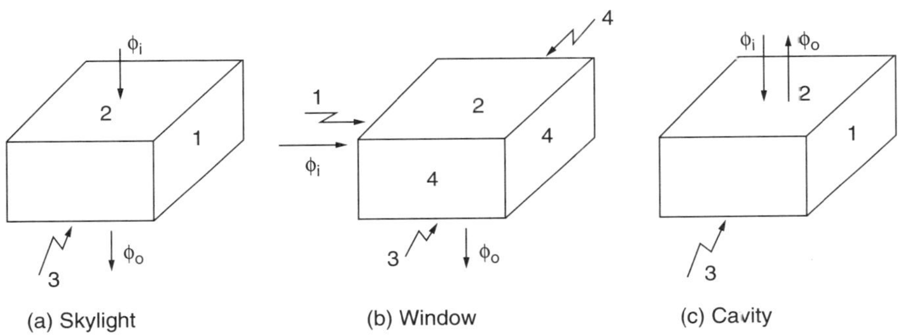

Figure 13.36 Single-and double bounce flux transfer cases.

In one-bounce flux transfer analysis, the initial flux entering the space from <u>outside</u> by transmission is permitted one bounce before exiting the space. If the initial flux is produced from <u>inside</u> the space by reflection, that first reflection is not counted as a bounce. Applying these rules to the three situations in Fig. 13.36 yields

<u>Skylight</u>

ϕ_i is into face 2

ϕ_o is out of face 3

$$\phi_o = \phi_i(f_{23} + f_{21}\rho_1 f_{13}) \tag{13.118}$$

<u>Window</u>

ϕ_i is into face 1

ϕ_o is out of face 3

$$\phi_o = \phi_i(f_{13} + f_{12}\rho_2 f_{23} + f_{14}\rho_4 f_{43}) \tag{13.119}$$

where the three non-window walls are considered as one surface.

<u>Cavity</u>

ϕ_i is into face 2

ϕ_o is out of face 2

$$\phi_o = \phi_i(f_{21}\, \rho_1 f_{12} + f_{23}\rho_3 f_{32}) \tag{13.120}$$

If a second bounce is allowed in Fig. 13.36, the following results are obtained:

Skylight

$$\phi_o \text{ added} = \phi_i(f_{21}\rho_1 f_{11}\rho_1 f_{13} + f_{21}\rho_1 f_{12}\rho_2 f_{23}) \tag{13.121}$$

Window

$$\begin{aligned}\phi_o \text{ added} = \phi_i(&f_{12}\rho_2 f_{24}\rho_4 f_{43} + f_{12}\rho_2 f_{21}\rho_1 f_{13} \\ &+ f_{14}\rho_4 f_{41}\rho_1 f_{13} + f_{14}\rho_4 f_{44}\rho_4 f_{43} \\ &+ f_{14}\rho_4 f_{42}\rho_2 f_{23})\end{aligned} \tag{13.122}$$

Cavity

$$\begin{aligned}\phi_o \text{ added} = \phi_i(&f_{21}\rho_1 f_{13}\rho_3 f_{32} + f_{21}\rho_1 f_{11}\rho_1 f_{12} \\ &+ f_{23}\rho_3 f_{31}\rho_1 f_{12})\end{aligned} \tag{13.123}$$

EXAMPLE 13.19

For the floor cavity in Fig. 13.37, with ρ_w = 50% and ρ_f = 20%, find the cavity reflectance.

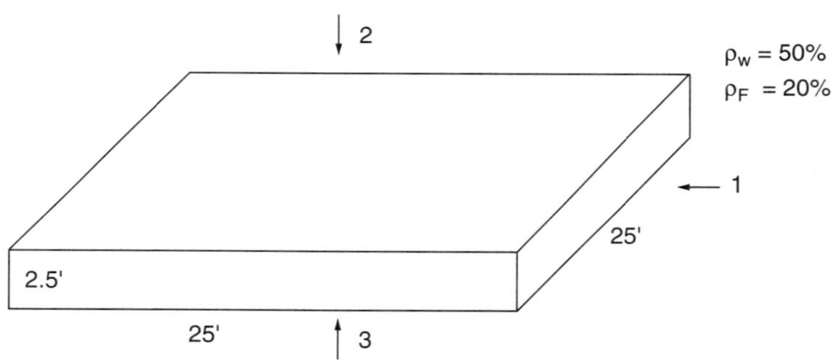

Figure 13.37

Solution. The situation is that of Fig. 13.36c. From Eq. (9.3), the cavity ratio is

$$CR = \frac{5 \times 2.5 \times 50}{625} = 1.0$$

From Table 13.9,

$$f_{23} = f_{32} = 0.827, \qquad f_{21} = f_{31} = 0.173,$$

$$f_{12} = f_{13} = 0.434 \quad \text{and} \quad f_{11} = 0.133$$

Using Eq. (13.120) for one-bounce analysis gives

$$\phi_o = \phi_i\,(0.173 \times 0.5 \times 0.434 + 0.827 \times 0.2 \times 0.827)$$

$$\rho_{fc} = \frac{\phi_o}{\phi_i} = 0.174(17.4\%)$$

Using Eq. (13.123) for two-bounce analysis gives additional output flux of

$$\begin{aligned}\phi_o \text{ added} = \phi_i\,(&0.173 \times 0.5 \times 0.434 \times 0.2 \times 0.827\\ &+ 0.173 \times 0.5 \times 0.133 \times 0.5 \times 0.434\\ &+ 0.827 \times 0.2 \times 0.173 \times 0.5 \times 0.434)\end{aligned}$$

$$\frac{\phi_o \text{ added}}{\phi_i} = 0.015\ (1.5\%)$$

Thus $\rho_{fc} = 0.174 + 0.015 = 0.189(18.9\%)$
The percent error of one-bounce versus two-bounce is

$$\frac{18.9 - 17.4}{18.9} = 8\%$$

Table 9.5 gives an exact value of 19% for ρ_{fc}. Thus two-bounce analysis gives excellent accuracy and one-bounce analysis gives acceptable accuracy.

EXAMPLE 13.20

A cylindrical recess in a ceiling is shown, with a ceiling reflectance of 70% and a well reflectance of 50%. It is lighted by a wide-beam floodlight of negligible dimensions. The lamp delivers 2000 lumens to the side of the cylinder and 4000 lumens to its top. Using one-bounce flux transfer analysis, find the lumen output from the ceiling recess and its efficiency as a luminaire. Find the luminous exitances of the top and side.

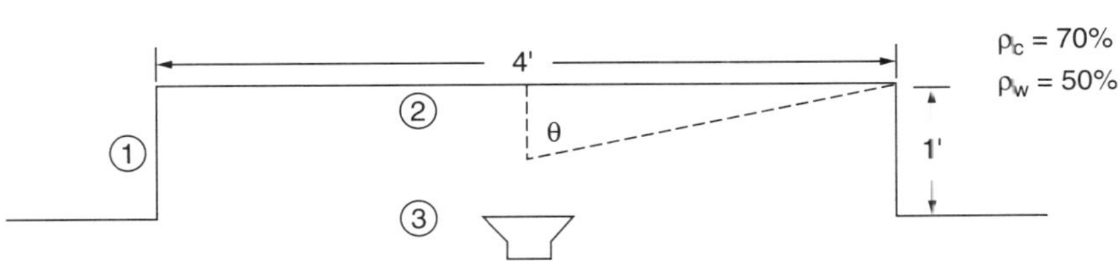

Figure 13.38

Solution. The angle θ is obtained from

$$\theta = \tan^{-1}\frac{2}{0.5} = 76°, \qquad \cos\theta = 0.2425$$

The form factors are found from Eqs. (13.26), (13.90), and (13.97).

$$f_{23} = f_{32} = \frac{1-\cos\theta}{1+\cos\theta} = \frac{1-0.2425}{1+0.2425} = 0.610$$

$$f_{21} = f_{31} = 1 - f_{23} = 0.390$$

$$f_{12} = f_{13} = \frac{A_2}{A_1} f_{21} = \frac{\pi r^2}{2\pi r \ell} f_{21} = \frac{4\pi}{4\pi} f_{21} = 0.390, \quad \text{where } r = 2 \text{ ft and } \ell = 1 \text{ ft.}$$

$$f_{11} = 1 - 2f_{12} = 0.220$$

The initial flux reflected into the recess is

$$\begin{aligned}\phi_{in} \equiv \phi_c + \phi_w &= 4000 \times 0.7 + 2000 \times 0.5 \\ &= 2800 + 1000 = 3800 \text{ lm}\end{aligned}$$

From one-bounce flux transfer analysis, the flux out of the enclosure is

$$\begin{aligned}\phi_{out} &= \phi_c(f_{23} + f_{21}\rho_1 f_{13}) \\ &\quad + \phi_w(f_{13} + f_{12}\rho_2 f_{23} + f_{11}\rho_1 f_{13}) \\ &= 2800(0.610 + 0.390 \times 0.50 \times 0.610) \\ &\quad + 1000(0.390 + 0.390 \times 0.70 \times 0.610 + 0.220 \times 0.50 \times 0.390) \\ &= 2800 \times 0.686 + 1000 \times 0.599 \\ &= 1921 + 599 = 2520 \text{ lm}\end{aligned}$$

Thus the luminaire efficiency is

$$\eta = \frac{2520}{6000} \times 100 = 42\%$$

To obtain the exitances, the total flux reflected from the ceiling and walls (ϕ_{ct} and ϕ_{wt}) is needed

$$\begin{aligned}\phi_{ct} &= (4000 + 2000 \times 0.50 \times 0.390)0.7 \\ &= 2800 + 273 = 3073 \text{ lm} \\ \phi_{wt} &= (2000 + 4000 \times 0.70 \times 0.390 + 2000 \times 0.50 \times 0.220) \\ &= 1000 + 546 + 110 = 1656 \text{ lm}\end{aligned}$$

Then the exitances are

$$M_c = \frac{\phi_{ct}}{A_c} = \frac{3073}{4\pi} = 245 \text{ lm/ft}^2$$

$$M_w = \frac{\phi_{wt}}{A_w} = \frac{1656}{4\pi} = 132 \text{ lm/ft}^2$$

Note that the output flux can be represented by

$$\begin{aligned}\phi_{out} &= \phi_{ct} \times f_{23} + \phi_{wt} \times f_{13} \\ &= 3073 \times 0.610 + 1656 \times 0.340 \\ &= 1875 + 645 = 2520 \text{ lm}\end{aligned}$$

13.12 MULTIPLE-BOUNCE FLUX TRANSFER ANALYSIS

Multiple-bounce flux transfer analysis will now be developed, by considering the general case first and then returning to a six-sided space so as to derive the basis for coefficients of utilization.

Consider a multisurfaced space with all surfaces perfectly diffusing. Assume the dimensions of the space and the reflectances of its surfaces are known. Consider two surfaces, m and n, within the space. Establish the following definitions:

M_{on} = initial luminous exitance of surface n
M_n = final luminous exitance of surface n
ρ_n = reflectance of surface n
E_n = illuminance on surface n due to the other surfaces
f_{mn} = fraction of flux emitted by surface m that falls on surface n

"Initial" means flux leaving the surface prior to reflections, and "final" means the total flux ultimately leaving the surface, including reflected flux. Thus the final luminous exitance of surface n is its initial exitance, due to sources, plus the exitances caused by fluxes from all other surfaces, due to reflections, and can be expressed as

$$M_n = M_{on} + \rho_n E_n \tag{13.124}$$

Also

$$E_{n(m)} = \frac{M_m A_m f_{mn}}{A_n} = M_m f_{nm} \tag{13.125}$$

where $E_{n(m)}$ is the contribution of surface m to the illuminance E_n. Using Eqs. (13.124) and (13.125),

$$M_n = M_{on} + \rho_n \sum_m M_m f_{nm} \tag{13.126}$$

For the room cavity in Fig. 13.29, Eq. (13.126) yields the following three equations:

$$\left.\begin{aligned} M_1 &= M_{o1} + \rho_1(M_1 f_{11} + M_2 f_{12} + M_3 f_{13}) \\ M_2 &= M_{o2} + \rho_2(M_1 f_{21} + M_3 f_{23}) \\ M_3 &= M_{o3} + \rho_3(M_1 f_{31} + M_2 f_{32}) \end{aligned}\right\} \tag{13.127}$$

These may be rearranged and placed in matrix form to yield

$$\begin{bmatrix} 1 - \rho_1 f_{11} & -\rho_1 f_{12} & -\rho_1 f_{13} \\ -\rho_2 f_{21} & 1 & -\rho_2 f_{23} \\ -\rho_3 f_{31} & -\rho_3 f_{32} & 1 \end{bmatrix} \begin{bmatrix} M_1 \\ M_2 \\ M_3 \end{bmatrix} = \begin{bmatrix} M_{o1} \\ M_{o2} \\ M_{o3} \end{bmatrix} \tag{13.128}$$

EXAMPLE 13.21

Use multiple-bounce flux transfer analysis to find the floor cavity reflectance ρ_{fc} in Fig. 13.39. Wall reflectance is 50%, and floor reflectance is 20%.

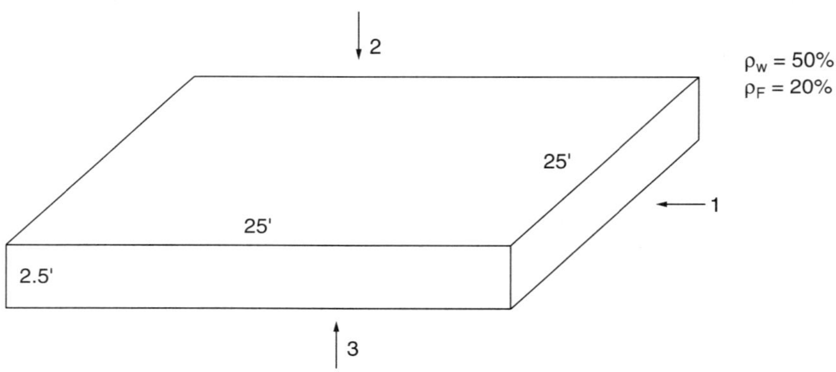

Figure 13.39

Solution. The cavity is the same as in Ex. 13.19 and Fig. 13.37. The initial input exitance is $M_{o2} = \phi_i/A_2$. From Ex. 13.19,

$$f_{23} = f_{32} = 0.827, \quad f_{21} = f_{31} = 0.173,$$
$$f_{12} = f_{13} = 0.434 \quad \text{and} \quad f_{11} = 0.133$$

From Eq. (13.127),

$$M_1 = 0 + \rho_1(M_1 f_{11} + M_2 f_{12} + M_3 f_{13})$$
$$M_2 = M_{o2} + 0$$
$$M_3 = 0 + \rho_3(M_1 f_{31} + M_2 f_{32})$$

where it is noted that $M_{o1} = M_{o3} = \rho_2 = 0$. Combining and simplifying yields

$$(1-\rho_1 f_{11})M_1 = \rho_1 f_{12} M_{o2} + \rho_1 f_{13} M_3$$
$$M_3 = \rho_3 f_{31} M_1 + \rho_3 f_{32} M_{o2}$$

and inserting numbers yields

$$0.9335M_1 = 0.2170M_{o2} + 0.2170M_3$$
$$M_3 = 0.0346M_1 + 0.1654M_{o2}$$

Solving for M_1 and M_3 gives

$$M_1 = 0.2731M_{o2}, \qquad M_3 = 0.1748M_{o2}$$

The flux leaving the cavity is

$$\begin{aligned}\phi_o &= M_1 A_1 f_{12} + M_3 A_3 f_{32} \\ &= 0.273M_{o2} \times 250 \times 0.434 + 0.175M_{o2} \times 625 \times 0.827 \\ &= 29.62M_{o2} + 90.45M_{o2} = 120.1M_{o2}\end{aligned}$$

Then

$$M_o = \frac{\phi_o}{A_2} = \frac{120.1M_{o2}}{625} = 0.19M_{o2}$$

giving

$$\rho_{fc} = \frac{M_o}{M_{o2}} = 0.19$$

It is noted that two-bounce flux transfer (in Ex. 13.19) gave the same answer, indicating the adequacy of two-bounce analysis in problems involving rectangular parallelepipeds.

13.13 CALCULATING CU, WEC AND CEC

The general result in Eq. (13.128) will now be applied to derive the coefficients of utilization and the exitance coefficients used with the lumen method in Chap. 9. All of the form factors in Eq. (13.128) may be expressed in terms of $f_{23} \equiv f$ and CR, using

Eqs. (9.3) and (13.97) through (13.99). This yields

$$\begin{bmatrix} 1-\rho_1\left[1-\frac{5}{CR}(1-f)\right] & -\rho_1\frac{2.5}{CR}(1-f) & -\rho_1\frac{2.5}{CR}(1-f) \\ -\rho_2(1-f) & 1 & -\rho_2 f \\ -\rho_3(1-f) & -\rho_3 f & 1 \end{bmatrix}\begin{bmatrix} M_1 \\ M_2 \\ M_3 \end{bmatrix} = \begin{bmatrix} M_{o1} \\ M_{o2} \\ M_{o3} \end{bmatrix} \tag{13.129}$$

The inversion of the matrix in Eq. (13.129) is straightforward, but algebraically complicated. The result is

$$\begin{bmatrix} M_{11} & M_{12} & M_{13} \\ M_{21} & M_{22} & M_{23} \\ M_{31} & M_{32} & M_{33} \end{bmatrix}\begin{bmatrix} M_{o1} \\ M_{o2} \\ M_{o3} \end{bmatrix} = \begin{bmatrix} M_1 \\ M_2 \\ M_3 \end{bmatrix} \tag{13.130}$$

where

$$\left.\begin{aligned}
M_{11} &= \frac{1}{1-\rho_1}\left\{1-\left(\frac{\rho_1}{1-\rho_1}\right)\left(\frac{2.5}{CR}\right)\left[\frac{C_1(C_2+C_3)}{C_0}\right]\right\} \\
M_{12} &= \left(\frac{1}{1-\rho_2}\right)\left(\frac{\rho_1}{1-\rho_1}\right)\left(\frac{2.5}{CR}\right)\left(\frac{C_1C_2}{C_0}\right) \\
M_{13} &= \left(\frac{1}{1-\rho_3}\right)\left(\frac{\rho_1}{1-\rho_1}\right)\left(\frac{2.5}{CR}\right)\left(\frac{C_1C_3}{C_0}\right) \\
M_{21} &= \left(\frac{1}{1-\rho_1}\right)\left(\frac{\rho_2}{1-\rho_2}\right)\left(\frac{C_1C_2}{C_0}\right) \\
M_{22} &= \frac{1}{1-\rho_2}\left\{1-\left(\frac{\rho_2}{1-\rho_2}\right)\left[\frac{C_2(C_1+C_3)}{C_0}\right]\right\} \\
M_{23} &= \left(\frac{1}{1-\rho_3}\right)\left(\frac{\rho_2}{1-\rho_2}\right)\left(\frac{C_2C_3}{C_0}\right) \\
M_{31} &= \left(\frac{1}{1-\rho_1}\right)\left(\frac{\rho_3}{1-\rho_3}\right)\left(\frac{C_1C_3}{C_0}\right) \\
M_{32} &= \left(\frac{1}{1-\rho_2}\right)\left(\frac{\rho_3}{1-\rho_3}\right)\left(\frac{C_2C_3}{C_0}\right) \\
M_{33} &= \frac{1}{1-\rho_3}\left\{1-\left(\frac{\rho_3}{1-\rho_3}\right)\left[\frac{C_3(C_1+C_2}{C_0}\right]\right\}
\end{aligned}\right\} \tag{13.131}$$

where

$$\left.\begin{aligned} C_1 &= \frac{(1-\rho_1)(1-f^2)}{(2.5/CR)\rho_1(1-f^2)+f(1-\rho_1)} \\ C_2 &= \frac{(1-\rho_2)(1+f)}{1+\rho_2 f} \\ C_3 &= \frac{(1-\rho_3)(1+f)}{1+\rho_3 f} \\ C_0 &= C_1 + C_2 + C_3 \end{aligned}\right\} \qquad (13.132)$$

Now, place luminaires on the ceiling of the room cavity in Fig. 13.29. For each luminaire, eighteen 10° flux zones will be defined, in the manner of Sect. 2.9. Let n be the zone index, where $1 \le n \le 18$, with zone 1 adjacent to the nadir and zone 18 adjacent to the zenith. The lumens in each zone are given by Eq. (2.43), and summations can be written as

$$\Phi_D = \sum_{n=1}^{9} \Phi_n \qquad (13.133)$$

$$\Phi_U = \sum_{n=10}^{18} \Phi_n \qquad (13.134)$$

where Φ_n is the lumens in zone n, Φ_D is the total downward lumens, and Φ_U is the total upward lumens.

It is desired to relate Φ_D and Φ_U to the total lamp lumens Φ_T. This is done by writing

$$\eta_D = \frac{\Phi_D}{\Phi_T} \qquad (13.135)$$

$$\eta_U = \frac{\Phi_U}{\Phi_T} \qquad (13.136)$$

Thus η_D is the fraction of the total lamp lumens that goes directly to the walls and floor in Fig. 13.29, and η_U is the fraction of the total lamp lumens that travels directly to the ceiling.

The downward lumens need to be separated into two parts, those traveling directly to the walls and those going directly to the floor in Fig. 13.29. This requires determination of a parameter called the direct ratio (D_m), the fraction of the downward luminaire flux that reaches the work plane directly. This ratio can be written as

$$D_m = \frac{1}{\Phi_D} \sum_{n=1}^{9} k_{mn} \Phi_n \qquad (13.137)$$

where k_{mn} is called a zonal multiplier and is a function of both cavity ratic, denoted by m, and zone n. k_{mn} is the fraction of the downward flux in zone n that reaches the work plane directly for a given cavity ratio m.

In addition to being functions of cavity ratio and zone, the zonal multipliers also depend somewhat on the spacing-to-mounting height ratio (S:MH) of the luminaires. Formerly, coefficient of utilization tables (Table 9.7) were based on an S:MH of 0.4 for fluorescent luminaires and 0.7 for incandescent and HID luminaires. Zonal multipliers for these S:MH ratios were calculated and presented in tabular form.

However, it has been shown (Levin, 1977) that k_{mn} is not a significant function of S:MH. Rather it depends more on luminaire location. An approximating equation for k_{mn} has been developed,

$$k_{nm} = e^{-Am^B} \tag{13.138}$$

where A and B are constants for each flux zone and are given in Table 13.10.

TABLE 13.10 ZONE CONSTANTS

n	A	B
1	0	0
2	0.041	0.98
3	0.070	1.05
4	0.100	1.12
5	0.136	1.16
6	0.190	1.25
7	0.315	1.25
8	0.640	1.25
9	2.10	0.80

With the direct ratio in hand, the coefficient of utilization can be calculated. From Eq. (13.130),

$$M_3 = M_{31}M_{o1} + M_{32}M_{o2} + M_{33}M_{o3} \tag{13.139}$$

M_{31}, M_{32}, and M_{33} are obtainable from Eqs (13.131) and (13.132). It remains to determine M_{o1}, M_{o2}, and M_{o3}. The fluxes reaching the cavity walls, ceiling, and floor are

$$\Phi_1 = (1 - D_m)\eta_D\Phi_T \tag{13.140a}$$

$$\Phi_2 = \eta_U\Phi_T \tag{13.140b}$$

$$\Phi_3 = D_m\eta_D\Phi_T \tag{13.140c}$$

Thus the initial exitances are

$$\left.\begin{aligned} M_{o1} &= \frac{\rho_1(1-D_m)\eta_D\Phi_T}{A_1} \\ M_{o2} &= \frac{\rho_2\eta_U\Phi_T}{A_2} \\ M_{o3} &= \frac{\rho_3 D_m\eta_D\Phi_T}{A_3} \end{aligned}\right\} \tag{13.141}$$

The coefficient of utilization is

$$\mathrm{CU} = \frac{M_3 A_3}{\rho_3 \Phi_T} \tag{13.142}$$

Using Eqs. (13.131), (13.139), and (13.141), Eq. (13.142) becomes

$$\mathrm{CU} = \frac{2.5\rho_1 \eta_D (1 - D_m) C_1 C_3}{CR(1-\rho_1)(1-\rho_3)C_o} + \frac{\rho_2 \eta_U C_2 C_3}{(1-\rho_2)(1-\rho_3)C_0} + \frac{D_m \eta_D}{1-\rho_3}\left[1 - \frac{\rho_3 C_3 (C_1 + C_2)}{(1-\rho_3)C_0}\right] \tag{13.143}$$

The wall and ceiling exitance coefficients are considered next. Equation (13.130) yields

$$M_1 = M_{11} M_{o1} + M_{12} M_{o2} + M_{13} M_{o3} \tag{13.144}$$

$$M_2 = M_{21} M_{o1} + M_{22} M_{o2} + M_{23} M_{o3} \tag{13.145}$$

From Eqs. (9.14) and (9.15), ignoring light loss,

$$\mathrm{WEC} = \frac{M_1 A_3}{\Phi_T} \tag{13.146}$$

$$\mathrm{CEC} = \frac{M_2 A_3}{\Phi_T} \tag{13.147}$$

Then Eqs. (13.131), (13.141), (13.144), and (13.145) give

$$\mathrm{WEC} = \frac{2.5}{CR}\left\{\frac{\rho_1(1-D_m)\eta_D}{1-\rho_1}\left[1 - \frac{2.5\rho_1 C_1 (C_2 + C_3)}{CR(1-\rho_1)C_0}\right] + \frac{\rho_1 \rho_2 \eta_U C_1 C_2}{(1-\rho_1)(1-\rho_2)C_0} + \frac{\rho_1 \rho_3 D_m \eta_D C_1 C_3}{(1-\rho_1)(1-\rho_3)C_0}\right\} \tag{13.148}$$

$$\mathrm{CEC} = \frac{2.5\rho_1 \rho_2 (1-D_m)\eta_D C_1 C_2}{CR(1-\rho_1)(1-\rho_2)C_0} + \frac{\rho_2 \eta_U}{1-\rho_2}\left[1 - \frac{\rho_2 C_2 (C_1 + C_3)}{(1-\rho_2)C_0}\right] + \frac{\rho_2 \rho_3 D_m \eta_D C_2 C_3}{(1-\rho_2)(1-\rho_3)C_0} \tag{13.149}$$

13.14 RADIATION COEFFICIENTS AND EFFECTIVE CAVITY REFLECTANCE

In addition to the quantities already calculated, there are three others that are of use in interior lighting computations. These are the wall direct radiation coefficient (WDRC), the wall reflected radiation coefficient (WRRC), and the effective cavity reflectance (ρ_{eff}).

The wall luminous exitance M_1 in Eq. (13.144) is the result of two wall illuminances, one resulting from flux going directly from the luminaires to the walls and the second due to flux reflected from the room surfaces to the walls. These two illuminances may be expressed as

$$E_{1D} = \frac{\Phi_T \mathrm{WDRC}}{A_3} \tag{13.150}$$

$$E_{1R} = \frac{\Phi_T \mathrm{WRRC}}{A_3} \tag{13.151}$$

where the D and R denote direct and reflected, respectively. It is also noted that

$$E_1 = \frac{M_1}{\rho_1} = E_{1D} + E_{1R} \tag{13.152}$$

Now, Eqs. (13.146), (13.150), and (13.151) may be inserted into Eq. (13.152) to yield

$$\frac{\mathrm{WEC}}{\rho_1} = \mathrm{WDRC} + \mathrm{WRRC} \tag{13.153}$$

If WDRC can be determined, then Eq. (13.153) may be used to find WRRC. The flux going directly to the walls is given by Eq. (13.140a). Thus

$$E_{1D} = \frac{(1 - D_m)\eta_D \Phi_T}{A_1} \tag{13.154}$$

Equating the right sides of Eqs. (13.150) and (13.154) and employing Eq. (9.22) gives

$$\mathrm{WDRC} = (1 - D_m)\eta_D \frac{A_3}{A_1} = \frac{2.5}{\mathrm{CR}} \eta_D (1 - D_m) \tag{13.155}$$

In Sect. 9.7, it was pointed out that the analytical expression from which Table 9.5 was developed was complex. With the concepts of flux transfer and form factor established, that expression can now be derived.

Figure 13.40 depicts a cavity. Surfaces 1 represent the walls of the cavity, surface 2 the base, and the dashed surface 3 the cavity opening. These surfaces have areas A_1, A_2, and A_3 and reflectances ρ_1, ρ_2, and 0, respectively. Also, the initial exitances M_{o1} and M_{o2} are zero, as there are no sources of light within the cavity.

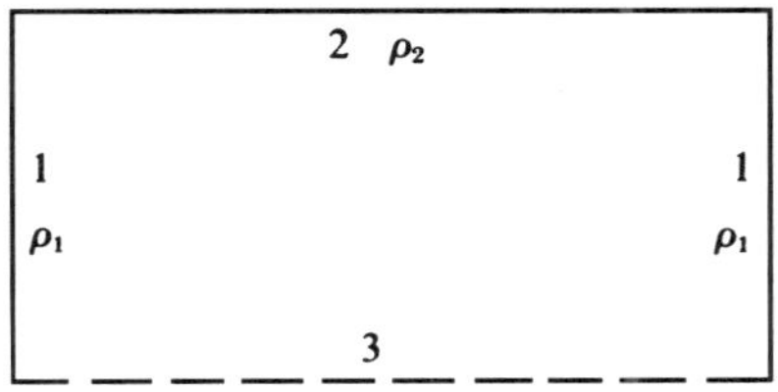

Figure 13.40 Depicting a cavity.

The flux into the cavity is $M_{o3}A_3$; that out of the cavity, using Eq. (13.26), is $(M_1f_{31} + M_2f_{32})A_3$ where M_1 and M_2 are the final exitances, after interreflections, of the walls and the base of the cavity. Thus the effective cavity reflectance is

$$\rho_{eff} = \frac{M_1f_{31} + M_2f_{32}}{M_{o3}} \tag{13.156}$$

M_1 and M_2 must now be found under the given constraints. From Eq. (13.128),

$$\begin{bmatrix} 1-\rho_1f_{11} & -\rho_1f_{12} & -\rho_1f_{13} \\ -\rho_2f_{21} & 1 & -\rho_1f_{23} \\ 0 & 0 & 1 \end{bmatrix} \begin{bmatrix} M_1 \\ M_2 \\ M_3 \end{bmatrix} = \begin{bmatrix} 0 \\ 0 \\ M_{o3} \end{bmatrix} \tag{13.157}$$

This yields

$$\left.\begin{aligned} (1-\rho_1f_{11})M_1 - \rho_1f_{12}M_2 &= \rho_1f_{13}M_{o3} \\ -\rho_2f_{21}M_1 + M_2 &= \rho_2f_{23}M_{o3} \end{aligned}\right\} \tag{13.158}$$

which gives

$$\left.\begin{aligned} M_1 &= \frac{(\rho_1f_{13} + \rho_1\rho_2f_{12}f_{23})M_{o3}}{1-\rho_1f_{11} - \rho_1\rho_2f_{12}f_{21}} \\ M_2 &= \frac{[(1-\rho_1f_{11})\rho_2f_{23} + \rho_1\rho_2f_{21}f_{13}]M_{o3}}{1-\rho_1f_{11} - \rho_1\rho_2f_{12}f_{21}} \end{aligned}\right\} \tag{13.159}$$

Inserting these equations into Eq. (13.156) yields

$$\rho_{eff} = \frac{f_{31}(\rho_1 f_{13} + \rho_1\rho_2 f_{12} f_{13}) + f_{32}(1-\rho_1 f_{11})\rho_2 f_{23} + \rho_1\rho_2 f_{21} f_{13}}{1-\rho_1 f_{11} - \rho_1\rho_2 f_{12} f_{21}} \qquad (13.160)$$

The result in Eq. (13.160) can be simplified by using Eqs. (13.97) to (13.99) to express all of the form factors in terms of $f = f_{23}$. Letting $A_2/A_1 = A$, the result is

$$\rho_{eff} = \frac{\rho_1\rho_2 f[2A(1-f)-f] + \rho_1 A(1-f)^2 + \rho_2 f^2}{1-\rho_1[1-2A(1-f)] - \rho_1\rho_2 A(1-f)^2} \qquad (13.161)$$

13.15 NEAR FIELD PHOTOMETRY

Conventional photometry, first presented in Sect. 2.12, is based on the assumption that the inverse square law holds. As has been seen, this requires that the distance from the source to the receiver be five times the maximum source dimension, if an accuracy of the order of 1% is to be achieved.

There are several cases where so-called "far-field" photometry is invalid, such as with indirect lighting of a ceiling plane, wall washing from a wall bracket or track light, and with various task lighting luminaires. In such cases, "near-field" photometry is required (IESNA Testing Procedures Committee, 2000).

Both far- and near-field photometry depend on measurements of illuminance, from which intensities are derived. The governing equation is the inverse square law, solved for intensity.

$$I = \frac{ED^2}{\cos\theta} \qquad (13.162)$$

In most far-field photometry, the light cell face is perpendicular to a line from source to cell and thus $\theta = 0$ and $\cos\theta = 1$. Also, the distance D is held constant throughout the measurements. These conditions do not pertain to near-field photometry, i.e., both θ and D are variables as illustrated in Fig. 13.41. In that figure, illuminances are measured

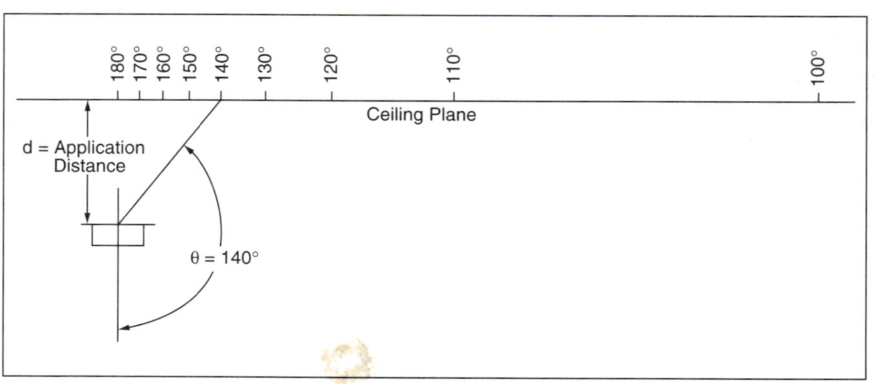

Figure 13.41 Near-field illuminance measurements. *Source*: *IESNA LM-70*, 2000. Courtesy of the Illuminating Engineering Society of North America.

on the ceiling plane at the intersection points with circles created by the polar angles from 180° to at least 100° (for a given application distance d, the distance from the 180° point to the 100° point on the ceiling is 5.7d, which may exceed the ceiling dimensions). If the luminaire is unsymmetrical, illuminance readings are taken at several azimuth angles for each polar angle, and averaged. Thus data points are located by rotating the luminaire to the desired azimuthal angle and sliding the light cell outward to the desired polar angle.

After illuminance, distance, and angle of incidence are measured at each designated point, "equivalent" intensities are calculated from Eq. (13.162). These equivalent intensities are valid only for that specific geometry and application distance d.

Two photometric testing procedures have been developed for obtaining equivalent intensity values. The first is a "linear procedure" and is illustrated in Fig. 13.42. In this method, the light cell is mounted facing downward on a horizontal track. The vertical distance of the track from the source is adjustable, so that illuminance readings can be taken at the various polar angles for several application distances.

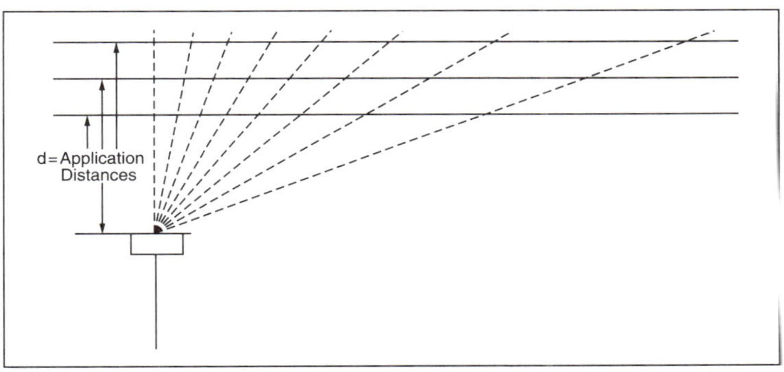

Figure 13.42 Linear procedure. ***Source*****:** ***IESNA LM-70*****, 2000. Courtesy of the Illuminating Engineering Society of North America.**

The second method is called the "scanning procedure", and is illustrated in Fig. 13.43. In this procedure the light cell always faces down (in the horizontal plane), but is located on radial polar angle lines at a series of points equidistant from the source center. In this case the radius from the source center, rather than the application distance is the variable, and is varied in steps ranging generally from 8 in to 60 in.

In both methods, interpolations may be necessary to determine intermediate equivalent intensities. Generally, linear interpolation is not sufficiently accurate, and techniques such as discussed in Sect. 13.8 are required. For example, in the linear procedure, it has been determined (Bradford, 1992) that a quadratic interpolation scheme, in which a second-degree polynomial is passed through three measured values, yields the most accurate results, with the caveat that the changes in application distance during measurements are held to 6 in, rather than 12 in.

In both testing procedures, it is necessary to blend near-field and far-field data. Thus far-field measurements of intensities are required to scale properly the near-field measurements. The calibration factor is the ratio of the flux resulting from application distance measurements to that from far field measurements in a cone from 180° to the lowest polar angle utilized (100°). The accuracy of application distance measurements compared with far-field measurements is 5% for 100°.

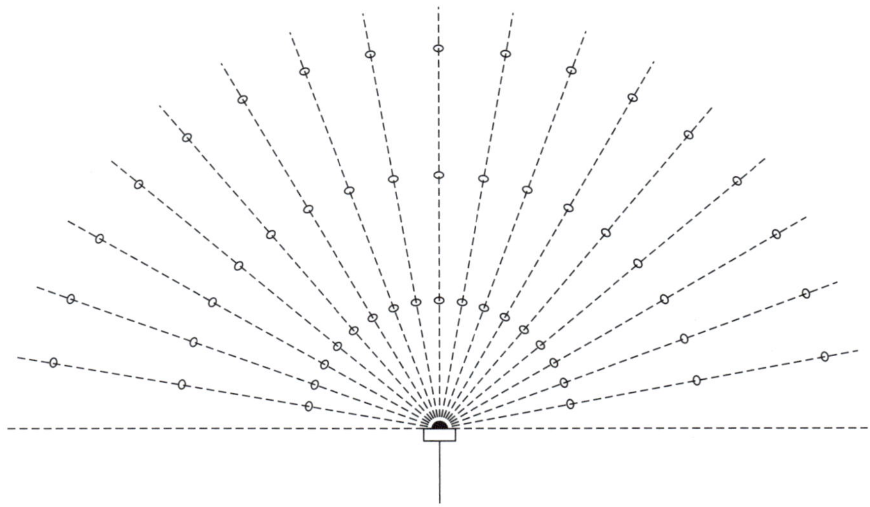

Figure 13.43 Scanning procedure. *Source*: *IESNA LM-70*, 2000. Courtesy of the Illuminating Engineering Society of North America.

REFERENCES

Ashdown, I.: *Radiosity*, John Wiley and Sons, New York, 1994.

Balogh, E.: "Infinite Plane Luminance Difference Technique for Computing Illumination," *Illuminating Engineering*, April 1966.

Bradford, R., and S. Stannard: "Refinements to Application Distance Photometry," *JIES*, Summer, 1992.

DiLaura, D.: "On the Simplification of Radiative Transfer Calculations," *JIES*, October 1982.

DiLaura, D.: "On the Development of a Recursive Method for the Solution of Radiative Transfer Problems," *JIES*, Summer, 1992.

DiLaura, D., and S. Santoro, "Non-Diffuse Radiative Transfer 1-5," *JIES*, 1996–1997.

IESNA Design Practice Committee: "The Determination of Illumination at a Point in Interior Spaces, *JIES*, January 1974.

IESNA Testing Procedures Committee: *IESNA Approved Guide to Near-Field Photometry*, LM-70, Illuminating Engineering Society of North America, New York, 2000.

Jones, J., R. Levere, N. Ivanicki, and P. Chesebrough: "Angular Coordinate System and Computing Illumination at a Point," *Illuminating Engineering*, April 1969.

Kaufman, J.: *IES Lighting Handbook, Reference Volume*, Illuminating Engineering Society, New York, 1984.

Levin, R.: "Cavities, Coefficients and Direct Ratios," *JIES*, August 1977.

Levin, R.: "Illumination Due to Area Sources Expressed in Angular Coordinates," *JIES*, October 1971.

Levin, R.: "Interpolation of Intensity Distributions," Annual Conference of the Illuminating Engineering Society, St. Louis, August 1984.

Moon, P.: *The Scientific Basis for Illuminating Engineering*, McGraw-Hill, New York, 1936.

Murdoch, J.: "Extension of the Configuration Factor Method to Strip Sources," *JIES*, April 1984.

Murdoch, J.: "Inverse Square Law Approximation of Illuminance," *JIES*, January 1981.

Ngai, P., J. Zhang and F. Zhang: "Near-Field Photometry: Measurement and Application for Fluorescent Luminaires," *JIES*, Summer, 1992.

Ngai, P., J. Zhang and Y. Quan: "On Computation using Far-Field and Near-Field Photometry, *JIES*, Summer, 1993.

O'Brien, P.: "Lighting Calculations for Thirty-Five Thousand Rooms," *Illuminating Engineering*, April 1960.

O'Brien, P.: "Numerical Analysis for Lighting Design," *Illuminating Engineering*, April 1965.

O'Brien, P.: "Effective Reflectance of Room Cavities with Specular and Diffuse Surfaces," *Illuminating Engineering*, April 1966.

O'Brien, P. and E. Balogh: "Configuration Factors for Computing Illumination Within Interiors," *Illuminating Engineering*, April 1967.

Rea, M., editor: *IESNA Lighting Handbook,* Ninth Edition, Illuminating Engineering Society of North America, New York, 2000.

Stannard, S., and J. Brass: "Application Distance Photometry," *JIES*, Winter, 1990.

Chapter 14

Lighting Economics

14.1 INTRODUCTION

Lighting economics is basic economic analysis applied to lighting. The principles of basic economics are covered in a myriad of textbooks. In this chapter they will be utilized to perform economic analyses of lighting systems.

The typical cost breakdown for a lighting system is shown in the pie chart in Fig. 14.1. This chart shows that operational costs (energy plus maintenance) represent two-thirds of the total cost over the life of a lighting system. Thus it is important to design lighting systems that are energy and maintenance efficient, and it is necessary to develop economic analysis methods to insure that this is the case.

Lighting economic analysis can be divided into two categories, called first-level and second-level analysis. The basic distinction between these is that the latter considers the time value of money. First-level methods are compelling because they are simple to apply and yield quick "back of the envelope" answers to lighting costs and paybacks. However, they are prone to error, particularly because they do not deal with future costs, nor do they consider system lifetime. Two such methods will be considered: cost of light and simple payback. Then Life Cycle Cost/Benefit Analysis (LCCBA) will be presented as the second-level method.

14.2 COST OF LIGHT METHOD

The cost of light method is based on finding the cost to produce one million lumen hours of light. This cost can be expressed as

$$C = \frac{10^6}{\phi}\left(\frac{X+Y}{L} + \frac{RP}{1000}\right) \tag{14.1}$$

where C = cost in cents of 10^6 lumen hours
ϕ = average lamp lumens
X = lamp cost in cents
Y = labor cost to replace a lamp in cents
L = rated lamp life in hours
P = average input watts per lamp
R = cost of energy in cents/kilowatt hour

Equation (14.1) may be thought through as follows: The number of hours of burning to reach 10^6 lumen hours is $10^6/\phi$. Thus the number of lamps required is $10^6/\phi L$. The initial cost is the number of lamps times the lamp and labor cost per lamp, $\frac{10^6}{\phi L}(X+Y)$. The energy cost is the product of the hours of burning times the energy rate times the power per lamp, giving $\frac{10^6 RP}{1000\,\phi}$ (1000 is to convert watts to kilowatts).

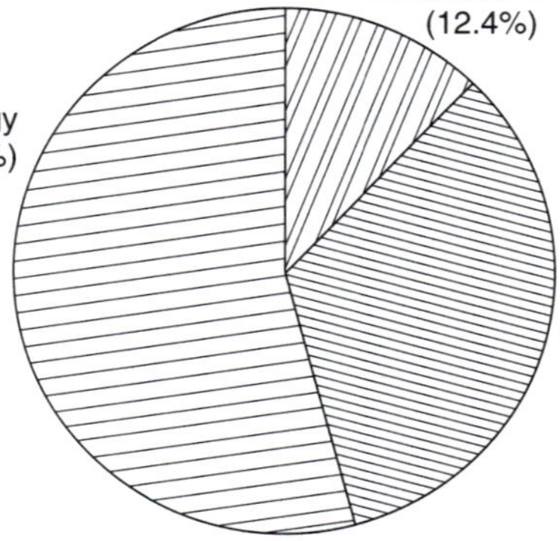

Figure 14.1 Lighting system costs in a typical building.

EXAMPLE 14.1

Find the cost of producing a million lumen hours for a 100-watt incandescent lamp emitting 1750 initial lumens and having a life of 750 hours. Assume lamp cost is 55¢, labor cost is $2, and energy is 10¢ per kWh. Find the annual operating cost if the lamp is burned 8 hours per day, five days per week.

Solution. Assume that average watts are 0.96 of initial watts and that average lumens are 0.91 of initial lumens. From (Eq. 14.1)

$$C = \frac{10^6}{1750 \times 0.91}\left(\frac{55+200}{750} + \frac{10 \times 0.96 \times 100}{1000}\right)$$

$$= 628(0.34 + 0.96) = 816¢ = \$8.16$$

Note that the energy cost ($6.03) is 2.8 times the initial cost ($2.14), a typical result.

The total hours of burning annually are 8 × 5 × 52 = 2080 hours. Thus the annual operating cost is

$$A = \frac{2080}{628} \times 6.03 = \$19.97$$

EXAMPLE 14.2

A store owner uses 20 incandescent sign lamps rated at 120 V, 750 h, and 100 W on a separately-metered electric circuit. He operates these lamps 4 hours each night for 30 days and receives a power bill of $23.76. He notes that his energy rate is 9¢/kWh. What is the store owner's voltage?

Solution. Let P be the initial power to the store owner's lamps. Then

$$\frac{120P \times 20 \times 9}{10^5} = 0.216P = \$23.76$$

giving

$$P = 110\ \text{W}$$

From Eq. (6.3)

$$\frac{P}{P_o} = \left(\frac{V}{V_o}\right)^{1.54} = 1.10$$

giving

$$\frac{V}{V_o} = 1.06, \quad V = 1.06 \times 120 = 128\ \text{V}$$

14.3 SIMPLE PAYBACK METHOD

The second method of first-level analysis is simple payback. It is generally used to compare a proposed retrofitting with an existing installation. It tells the investor the amount of time to pay off an initial expenditure and is obtained by equating the initial cost in lamps, equipment, and labor of the proposed system to the decrease in annual operating cost. In equation form,

$$X + Y + Z = \frac{\Delta P \times R \times H \times T}{1000} \tag{14.2}$$

where X, Y, and R are as in Eq. (14.1), Z is equipment cost in cents (ballasts, luminaires, wiring), ΔP is the saving in average power, H is the number of burning hours per year, and T is the payback period in years. Solving for T yields

$$T = \frac{1000(X + Y + Z)}{\Delta P \times R \times H} \tag{14.3}$$

Generally a payback period of 1-2 years is quite acceptable, 2-4 years is marginal, and greater than four years is unacceptable.

It should be noted that simple payback answers the question "How soon will I recover my investment?" It fails to consider whether the new system will continue to perform as well after the payback time is over as it has up to that time and whether the two systems have different expected lifetimes.

EXAMPLE 14.3

It is proposed to replace 100 75-watt incandescent lamps in a downlighting installation with 67-watt energy savers. The 75-watt lamp emits 1210 lumens; the 67 watt lamp 1130 lumens. Each lamp has 750 hours of life. If energy costs 10¢ per kWh, the energy-saving lamps cost 60¢ each, the installation cost is $1.00 per lamp, and the lamps operate 2000 hours per year, find the payback time.

Solution. Inserting the data into Eq. (14.3) gives

$$T = \frac{1000(60+100)\,100}{800 \times 10 \times 2000} = 1.0 \text{ yrs}$$

This looks like a good bargain, but observe that the lighting level will be lower with the energy-saving lamps. The same lighting level requires

$$\frac{1210}{1130} \times 100 = 107 \text{ lamps}$$

Now $\Delta P = 100 \times 75 - 107 \times 67 = 331$ W and

$$T = \frac{1000(60+100)\,107}{331 \times 10 \times 2000} = 2.6 \text{ yrs}$$

In this calculation, the need for seven new luminaires and their installation has been ignored. Assume a cost of $15 per luminaire, including wiring and labor. Now

$$T = \frac{1000(160 \times 107 + 105 \times 100)}{331 \times 10 \times 2000} = 4.2 \text{ yrs}$$

The lesson from Ex. 14.3 is that one should be very careful using simple payback analysis, particularly in establishing the criteria that will be used in making the assessment.

EXAMPLE 14.4

A restaurant replaces 90 60-watt incandescent wall lights with 2-lamp compact fluorescent units, each lamp and its ballast consuming 10 watts. Each 60-watt lamp emits 870 lumens and has a life of 750 hours; each CFL emits 400 lumens and lasts 10,000 hours. Costs are:

60-W lamp	—	64¢
CFL	—	$4.77
ballast and adaptor	—	$8.95
hardware and labor per unit	—	$6.25
energy cost	—	9¢/kWh

If lamps are burned 12 hours per day, find the simple payback time in years.

Solution. The burning hours per year are $H = 12 \times 365 = 4380$ h. The watts saved are $\Delta P = 60 - 2 \times 10 = 40$ W per unit or 3600 W total. The initial costs are

$$X + Y + Z = (2 \times 4.77 + 2 \times 8.95 + 6.25)90$$

$$= 33.69 \times 90 = \$3032.10$$

Now from Eq. (14.3)

$$T = \frac{1000 \times 3032.10 \times 100}{3600 \times 9 \times 4380} = 2.1 \text{ years}$$

As in the first solution in Ex. 14.2, this solution is based on unequal illuminances. For equal illuminances, there must be $\frac{870}{800} \times 90 = 98$ CFL luminaires. This changes the power saved to

$$\Delta P = 60 \times 90 - 2 \times 10 \times 98 = 3440 \text{ W}$$

The initial costs increase to

$$X = \frac{98}{90} \times 3032.10 = \$3301.62$$

Thus the payback time lengthens to

$$T = \frac{1000 \times 3301.62 \times 100}{3440 \times 9 \times 4380} = 2.4 \text{ years}$$

Whether equal illuminance needs to be a criterion in this case is open to question. Probably the restaurant owner would choose to go with the slightly lower lumen total provided by the CFLs and use the same number of luminaires.

14.4 COSTS

What are the costs associated with the economic analysis of lighting systems? They fall into three major categories: initial costs, operating costs, and fixed charges. Some of these costs have been identified in the previous two sections; they will be discussed more fully here.

Initial costs are those costs that are paid "up front", to obtain and install the lighting system and related equipment. They include:

- Lamps
- Luminaires (and ballasts)
- Electrical controls
- Optical controls
- Wiring and conduit
- Labor
- Air conditioning

The cost of the first four items is often the most significant part of initial cost, as much as 70%. It is directly related to the number and quality of the luminaires. Their number is a direct function of the desired average maintained illuminance level. Thus this level should be selected thoughtfully to provide both adequate visibility and an economical system. The quality of the four items determines the quality of the lighting installation, for example, minimization of glare, lighting uniformity, a balance between vertical and horizontal illuminances, luminance ratios, and amenity. It is also true that a higher grade luminaire can be easier to install, through plug-in wiring for example, and thus reduce labor costs.

It should go without saying that energy-saving lamps and ballasts should be used. These may add slightly to initial costs, but will yield significant savings in operating costs. Also, electronic ballasts are lighter and smaller than their magnetic counterparts, decreasing luminaire weight, and thus making installation easier.

Wiring costs consist of the branch-circuit wiring, the feeders to the luminaires, and the required electrical controls (switching). Sometimes these are calculated on an average cost per connected kilovolt ampere ($/kVA). kVA is related to the kilowatts to the luminaires by

$$\mathrm{kVA} = \frac{\mathrm{kW}}{\mathrm{PF}} \tag{14.4}$$

where PF is luminaire power factor and kW is the total luminaire power. Because volt-amperes are the product of voltage and current, a higher supply voltage for the same power will result in a lower current, which, in turn, will require a smaller wire size or fewer branch circuits, thus reducing wiring (and labor) costs.

Installation (labor) cost is also composed of two parts: wiring cost and luminaire cost. These labor costs are directly related to the number of luminaires and less directly to luminaire quality. The electrical labor rate can vary dramatically from area to area and is also influenced by Federal, state, and local building codes.

There may be tangential initial costs caused by the installation of a lighting system. These result from the heat energy produced by the lamps. If, for example, the new lighting system produces more heat than the previous system, additional air conditioning may be required, and this will be an initial cost. It is also quite possible that the new lighting system will be less demanding on air conditioning, requiring no initial cost increment (and saving energy). To determine the increments or decrements in air conditioning involves many variables that are difficult to quantify. Thus, unless these costs are felt to be significant, they should not be included in cost analyses.

There are two operating costs: energy costs and maintenance charges. As Fig. 14.1 shows, energy costs are the most significant factor in lighting system costs, representing over 50% of the total. The four ingredients are power to the luminaires, hours of operation, energy rate, and demand charge. A 20% increase, say, in energy rate means at least a 10% increase in overall costs. Likewise a 20% decrease in number of luminaires or hours of operation represents a 10% decrease in overall costs.

Demand charge is the charge imposed by the electric company for users who wish to consume energy during peak demand periods. Demand charges are separate from energy charges. They are assessed monthly on the user's peak demand for power. The rate is given in dollars per peak kilowatt per month.

Maintenance costs occur as a lighting system ages. They involve

- replacing burned-out lamps
- replacing ballasts

- replacing optical controls (lenses, baffles, louvers)
- cleaning luminaires

Sect. 9.10 indicates that four recoverable factors cause the lumen output of a lighting installation to decrease over time. Two of these, burnouts and lumen depreciation, are directly attributable to the lamps. Fig. 14.2 shows the effect of these factors. The plot in Fig. 14.2a is a lamp mortality curve. The life rating of a lamp on this curve is when there are 50% survivors. The plot in Fig. 14.26b is a lumen depreciation curve. At 100% life, each lamp in this example is providing 80% of initial lumens, on average. The plot in Fig. 14.2c is a plot of lumens times survivors and gives the lumen output of the remaining lamps. At 100% life, the lumen output of all the lamps is 40% of its initial value.

Examine the situation at 70% life, marked by ticks in Fig. 14.2. At this point, there are 90% survivors, and they are providing, on average, 85% of initial lumens per lamp. Thus the total lumen output is 76.5% of its initial value. This is the point in lamp operation where it is customary to group relamp the entire installation. At this point, over 70% of the lumen hours available from the installation have already been used.

Group relamping can reduce the operating cost of a lighting system, as well as maintain illuminance levels. It saves labor costs per unit, because it is cheaper to replace all lamps at once, rather than one at a time. A rough guideline to justify group relamping is if the cost difference between spot replacing and group replacing one lamp exceeds the cost of one new lamp (discounted for bulk purchase).

The third cost category is fixed charges. These consist of equipment depreciation charges, local, state and Federal taxes, insurance coverage, and interest paid on loans. These costs are called "fixed" because they are independent of the amount of time that a lighting system is used.

14.5 TIME VALUE OF MONEY

Before discussing Life Cycle Cost/Benefit Analysis (LCCBA), which is the preferred second-level analysis method for the economic judging of lighting systems. the interrelationship of time and money will be explored. There are several mathematical relations that need to be developed.

Analysis of the time value of money allows the comparison of dollars now vs. dollars in the future. The following are defined:

P = present value in dollars
F = future value in dollars
i = annual interest rate as a decimal
n = number of years between P and F

Consider investing \$1500 now at an annual interest rate of 8%. After one year, $F_1 = 1500 + 0.08 \times 1500 = \$1620.00$. After two years, $F_2 = 1620 + 0.08 \times 1620 = 1500\ (1 + 2 \times 0.08 + 0.08^2) = 1500\ (1 + 0.08)^2$. Extrapolating this result gives the general relation

$$F = P\ (1 + i)^n \tag{14.5}$$

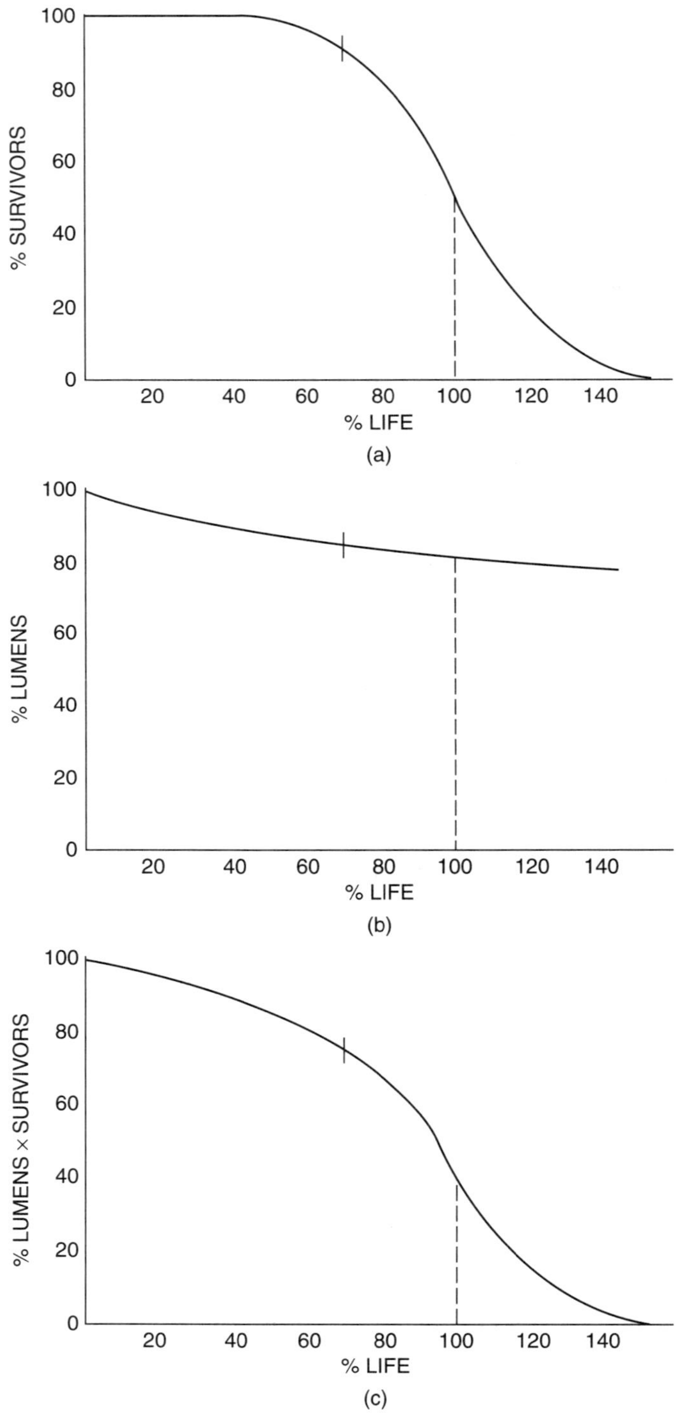

Figure 14.2 Lumens, life, and survivors.

EXAMPLE 14.5

The Manhattos Indians sold the island of Manhattan to the Dutch in 1626 for \$24. If they could have deposited the money in a bank at 5% annual interest, what would their investment be worth now (2002)?

Solution. Using Eq. (14.5)

$$F = 24(1 + 0.05)^{376} = \$2{,}225{,}295{,}500$$

The inverse of Eq. (14.5) is often of interest. It allows us to determine the present value of a future sum of money, discounted n years back in time at an interest rate i.

$$P = F\frac{1}{(1+i)^n} \tag{14.6}$$

EXAMPLE 14.6

What interest rate will permit a U.S. savings bond to be sold at \$100 and mature to \$200 in 10 years?

Solution. From Eq. (14.6).

$$100 = 200(1+i)^{-10}$$

$$(1+i)^{10} = 2, \quad 1+i = 1.072, \quad i = 7.2\%$$

Consider next the situation where the future value is a sum of annual payments over time and its present value must be found. Let A = the amount of the annual payment in dollars. The present value of n future payments is, from Eq. (14.6),

$$P = \frac{A}{1+i} + \frac{A}{(1+i)^2} + \frac{A}{(1+i)^3} + \cdots + \frac{A}{(1+i)^{n-1}} + \frac{A}{(1+i)^n} \tag{14.7}$$

Multiplying both sides by (1 + i) gives

$$P(1+i) = A + \frac{A}{1+i} + \frac{A}{(1+i)^2} + \cdots + \frac{A}{(1+i)^{n-2}} + \frac{A}{(i+1)^{n-1}} \tag{14.8}$$

Subtracting Eq. (14.7) from Eq. (14.8) yields

$$P(1+i-1) = A - \frac{A}{(1+i)^n} \tag{14.9}$$

and solving for P and A give

$$P = A\frac{(1+i)^n - 1}{i(1+i)^n} \tag{14.10}$$

$$A = P\frac{i(1+i)^n}{(1+i)^n - 1} \tag{14.11}$$

The factor in Eq. (14.10) is called the uniform present worth factor, and P is the present value of a stream of n annual payments. The factor in Eq. (14.11) is called uniform capital recovery factor, and A is the annual payment over n years equivalent to paying P dollars today.

EXAMPLE 14.7

A home is purchased for $200,000, 30% is placed down, and a 30 year mortgage at a 6% annual interest rate is taken. What is the monthly payment?

Solution. Find A, using Eq. (14.11). The monthly interest rate is $\frac{6}{12} = 0.5\%$ and n is $30 \times 12 = 360$.

$$A = 140{,}000\ \frac{0.005\ (1.005)^{360}}{1.005^{360} - 1}$$

$$= 140{,}000 \times 0.0059955 = \$839.37$$

EXAMPLE 14.8

A person has just won the lottery and will receive $20,000 quarterly for the next 20 years. Assuming an interest rate of 6% what is the present value of the prize?

Solution. P must be found from Eq. (14.10), with i = 1.5% and n = 80.

$$P = 20{,}000\ \frac{(1+0.015)^{80} - 1}{0.015\ (1+0.015)^{80}}$$

$$= 20{,}000 \times 46.407323 = \$928{,}146.47$$

The future value of an annual stream of payments can be found by combining Eqs. (14.5) and (14.10).

$$F = A\frac{(1+i)^n - 1}{i(1+i)^n} \times (1+i)^n = A\frac{(1+i)^n - 1}{i} \tag{14.12}$$

Alternatively, the annual payment to create a future value is given by

$$A = F\frac{i}{(1+i)^n - 1} \tag{14.13}$$

The factor in Eq. (14.13) is called a uniform sinking fund factor.

Annual costs may escalate over the time period of interest. Let r = the rate of escalation as a decimal. Thus

$$A_n = A(1 + r)^n \tag{14.14}$$

where A is now the amount of the initial annual payment in dollars. The present value of the kth payment is, from Eq. (14.6),

$$P_k = A\,\frac{(1+r)^k}{(1+i)^k} \tag{14.15}$$

and the present value of all the payments is

$$P = \sum_{k=1}^{n} A\,\frac{(1+r)^k}{(1+i)^k} \tag{14.16}$$

Expanding Eq. (14.16) to obtain a closed form representation gives

$$P = A\,\frac{1+r}{1+i} + A\left(\frac{1+r}{1+i}\right)^2 + A\left(\frac{1+r}{1+i}\right)^3 + \cdots + A\left(\frac{1+r}{1+i}\right)^{n-1} + A\left(\frac{1+r}{1+i}\right)^n \tag{14.17}$$

Multiplying both sides by $\frac{1+i}{1+r}$ yields

$$P\left(\frac{1+i}{1+r}\right) = A + A\,\frac{(1+r)}{(1+i)} + A\left(\frac{1+r}{1+i}\right)^2 + \cdots + A\left(\frac{1+r}{1+i}\right)^{n-2} + A\left(\frac{1+r}{1+i}\right)^{n-1} \tag{14.18}$$

Now subtracting Eq. (14.17) from Eq. (14.18) gives

$$\left.\begin{aligned} P\left(\frac{1+i}{1+r}\right) - P &= A - A\left(\frac{1+r}{1+i}\right)^n \\ P\left(\frac{i-r}{1+r}\right) &= A\left[1-\left(\frac{1+r}{1+i}\right)^n\right] \\ P &= A\,\frac{1+r}{i-r}\left[1-\left(\frac{1+r}{1+i}\right)^n\right] \end{aligned}\right\} \tag{14.19}$$

EXAMPLE 14.9

A company's annual electricity bill is currently \$3000. The energy rate is expected to escalate 4% annually for the next five years. What is the present worth of the energy costs over this period if the interest rate is 6%?

Solution. From Eq. (14.19)

$$P = A\,\frac{1.04}{0.02}\left[1-\left(\frac{1.04}{1.06}\right)^{5}\right] = 4.724\,A$$

Thus

$$P = 3000 \times 4.724 = \$14{,}172$$

14.6 LIFE CYCLE COST BENEFIT ANALYSIS (LCCBA)

Life cycle cost benefit analysis is used to compare the economics of two or more proposed lighting systems. It includes the following costs: hardware, wiring, labor, taxes, insurance, energy, and maintenance. It is generally done in tabular form in two sections—initial costs and operating costs—and it considers the time value of money, either by obtaining the present value of all future expenses or by calculating the equivalent annual cost of ownership.

Define the following quantities for a lighting system:

IC = initial cost
AOC = annual operating cost
SV = salvage value at the end of life

To obtain the present value of these quantities, it is noted that IC is already in the present, SV is at the end of life and may be brought to the present by Eq. (14.6), and that AOC occurs annually and may be brought to the present by Eq. (14.10). Thus

$$P = IC + SV\frac{1}{(1+i)^n} + AOC\,\frac{(1+i)^n - 1}{i(1+i)^n} \tag{14.20}$$

To obtain the equivalent annual cost, it is noted that AOC is already an annual expense, SV is a future value that can be converted to an annual cost through Eq. (14.13), and IC is an initial value that is expressed as an annual cost through Eq. (14.11). Thus

$$A = AOC + SV\,\frac{i}{(1+i)^n - 1} + IC\,\frac{i(1+i)^n}{(1+i)^n - 1} \tag{14.21}$$

EXAMPLE 14.10

It is estimated that the relamping of a lighting installation in 6 years will cost \$200 and that annual cleaning costs will be \$100 per year during that period of time. If the interest rate is 5%, what is the total present value of these expenses?

Solution. Eqs. (14.6) and (14.10) are required.

$$P = 200\,\frac{1}{(1.05)^6} = \$149.24$$

$$P = 100\,\frac{(1.05)^6 - 1}{0.05(1.05)^6} = \$507.57$$

Thus the total present value is $656.81.

EXAMPLE 14.11

A lighting system costs $2000 initially and has annual energy and maintenance charges of $250. There are no fixed charges or salvage costs. Find the present value and the annual equivalent cost for this installation if its life is 8 years and the interest rate is 7%.

Solution. From Eqs (14.20) and (14.21)

$$P = 2000 + 250\,\frac{(1.07)^8 - 1}{0.07(1.07)^8}$$

$$= 2000 + 250 \times 5.9713 = \$3492.83$$

$$A = 250 + 2000\,\frac{0.07(1.07)^8}{(1.07)^8 - 1}$$

$$= 250 + 2000 \times 0.16747 = \$584.94$$

Without considering the time value of money, the initial total value of the installation would be $\$2000 + 8 \times 250 = \$4000$ and the average annual outlay in initial dollars would be $4000/8 = \$500$.

EXAMPLE 14.12

Consider two lighting systems: (1) 34-watt fluorescent lamps in recessed three-lamp troffers with prismatic lenses, (2) 175-watt metal halide lamp in a single-lamp recessed downlight with a glass refractor. Data for the two systems are given in Table 14.1.

Find the present value and the equivalent annual cost for each system, assuming an interest rate of 6%.

Solution. The assemblage of costs is presented in Table 14.2, divided between initial cost, annual operating cost (salvage value is listed as an initial cost, negative because money is to be received), and fixed cost, assumed to be 15% of total initial cost, excluding lamps.

Eqs. (14.20) and (14.21) will be used to calculate the present value (worth) and equivalent annual cost of each lighting system.

TABLE 14.1

	System 1	System 2
Watts per luminaire, including ballast	108	210
Initial lumens per lamp	2750	14,400
Number of luminaires	10	6
Burning hours per year	3000	3000
Lamp life, hours@ 10 hrs/start	26,000	10,000
Energy rate per kWh	8¢	8¢
Lamp replacement method	Spot	Spot
Replacement labor cost per lamp	$2.50	$2.50
Lamp cost	$2.00	$20.00
Luminaire cost	$60	$130
Wiring cost per luminaire	$20	$20
Labor cost per luminaire	$25	$20
Cleaning cost per luminaire	$2	$1
Cleanings per year	1	1
Salvage value	$100	$200
System life, years	10	10

TABLE 14.2

INITIAL COST	System 1	System 2
Total kilowatts	1080	1260
Total luminaire cost	$600	$780
Total wiring cost	$200	$120
Total labor cost	$250	$120
Total lamp cost	$60	$120
Total initial cost	$1110	$1140
Salvage value	–$100	–$100
OPERATING COST		
Annual lamp replacements	3.5	1.8
Total replacement cost	$15.75	$40.50
Annual cleaning cost	$20.00	$6.00
Total annual maintenance cost	$35.75	$46.50
Annual energy cost	$259.20	$302.40
Total annual operating cost	$294.95	$348.90
FIXED COST		
Total initial cost, less lamps	$1050	$1020
Annual fixed costs	$157.50	$153.00
Total annual cost	$425.45	$501.90

System 1

$$P = 1110 - 100 \frac{1}{(1+0.06)^{10}} + 452.45 \frac{(1+0.06)^{10} - 1}{0.06(1+0.06)^{10}}$$

$$= 1110 - 100 \times 0.5584 + 452.45 \times 7.3601$$

$$= 1110 - 55.84 + 3330.07 = \$4384.23$$

$$A = 452.45 - 100\frac{0.06}{(1+0.06)^{10}-1} + 1110\frac{0.06(1+0.06)^{10}}{(1+0.06)^{10}-1}$$

$$= 452.45 - 100 \times 0.0759 + 1110 \times 0.1359$$

$$= 452.45 - 7.59 + 150.85 = \$595.71$$

System 2

$$P = 1140 - 100 \times 0.0759 + 501.90 \times 7.3601$$

$$= 1140 - 7.59 + 3594.03 = \$4726.44$$

$$A = 501.90 - 100 \times 0.0759 + 1140 \times 0.1359$$

$$= 501.90 - 7.59 + 154.93 = \$649.24$$

It is seen that System 1 is a slightly better economical choice than System 2, but small changes in any of the variables could lead to a different conclusion.

14.7 ENVIRONMENTAL COSTS

In the previous analyses, environmental costs have not been considered. There are several ways in which lighting can impact the environment, and all of these can have economic consequences for the producer and/or the consumer.

Three environmental "costs" of lighting will be examined, with the word "cost" implying both economic and societal consequences. Then LEED and the concept of green design will be introduced.

Air-Borne Emissions

Every kilowatt hour generated by a fossil-fuel power plant results in the emission to the atmosphere of 1.5 pounds of carbon dioxide, 5.8 grams of sulfur dioxide, and 2.5 grams of nitrous oxides. These emissions effect air quality, induce acid rain, and cause global warming.

Since nearly 20% of the nation's electricity is used for lighting, energy-efficient lighting designs can reduce significantly these airborne contaminants. For example a 10% reduction in electricity demand would lower annual emissions by an estimated 232 million tons for carbon dioxide, 1.1 million tons for sulfur dioxide, and 0.8 million tons for nitrous oxides (Roberts, 2001).

The Federal Energy Policy Act (EPACT) was signed into law on October 24, 1992. Although only a small portion of the Act concerned lighting, it has had a major effect on lighting energy consumption. The Act outlawed several inefficient lamps, particularly incandescent reflector lamps and the widely-used 40-watt T12 cool white fluorescent lamps. The former have been replaced by much more efficient halogen and compact fluorescent lamps and the latter by 34-watt energy-saving T12 lamps first, and later by 32-watt T8 lamps. A summary of EPACT requirements for lamps is given in Table 14.3.

TABLE 14.3 EPACT LAMP REGULATIONS

a) Incandescent Reflector Lamps

Lamp Watts	Minimum Average Lamp Efficacy (lm/W)
40–50	10.5
51–66	11.0
67–85	12.5
86–115	14.0
116–155	14.5
156–205	15.0

b) Linear and U-shaped Fluorescent Lamps

Lamp Type	Nominal Lamp Watts	Minimum CRI	Minimum Average Lamp Efficacy (lm/W)
4-foot minimum bi-pin	>35 W	69	75.0
	≤35 W	45	75.0
2-foot U-shaped	>35 W	69	68.0
	≤35 W	45	64.0
8-foot slimline	>65 W	69	80.0
	≤65 W	45	64.0
8-foot high output	>100 W	69	80.0
	≤100 W	45	80.0

Prior to EPACT, the Federal government passed the National Appliance Energy Conservation Amendments (NAECA) in 1988. These included regulations on ballasts by requiring minimum ballast efficacy factors, as given by Eq. (6.10). These requirements outlawed standard magnetic ballasts, while accepting energy-saving magnetic and electronic ballasts. A summary of NAECA regulations appears in Table 14.4.

TABLE 14.4 NAECA BALLAST REGULATIONS

Ballast-Lamp System	Nominal Lamp Watts	Ballast Efficacy Factor
1 F40T12	40	1.805
2 F40T12	80	1.06/1.05
2 F96T12	150	0.570
2 F96T12HO	220	0.390

Note: BEFs generally apply to ballasts designed for an input of 120 V or 277 V. However, for two-lamp F40T12, the two BEFs listed are for 120/277 V.

In addition to emitting the three airborne gases, fossil-fuel power plants also send mercury into the atmosphere. Worldwide (1994 data) estimated 16,000 tons of mercury are released to the atmosphere each year from all sources, natural and man-made. The breakdown (NEMA, 1996), shows that fossil-fuel plants contribute 3000 tons per year, equivalent to 0.04 milligrams for each kilowatt hour sold, natural sources 7000 tons, and other man-made sources 6000 tons. Lamps account for 0.2% of the total mercury in the environment.

Ballast Toxicity

Enactment of the Toxic Substance Control Act (TSCA) in 1976 outlawed polychlorinated biphenyls (PCBs) in equipment. By 1979, manufacturers eliminated PCBs from the capacitors in their ballasts. Such ballasts are labeled "No PCB".

This chemical is very toxic, and ballasts containing PCBs, some of which still exist in fluorescent luminaires, must be disposed of with care. In addition to being an environmental cost, disposal of ballasts containing PCBs is also an economic cost that needs to be included in economic analyses. Modern magnetic and electronic ballasts do not contain hazardous materials and do not require special disposal.

Ballasts containing PCBs are classified in two ways for disposal purposes. Intact ballasts, with no PCB leakage, may be disposed of in a municipal landfill, but this is not recommended because they will eventually leak. Rather they should be placed in sealed 55-gallon drums and placed in a hazardous-waste landfill. Leaking ballasts should be disposed of by high temperature incineration of either the entire ballast or, preferably, of the ballast's capacitor, with the rest of the ballast recycled.

Lamp Toxicity

Lamps contain lead and mercury. Lead is found in solders in most incandescent and HID lamps, but lead-free solders are becoming common. Lead is also found in glass, but the emergence of T8, and now T5, fluorescent lamps has reduced the amount of glass required per lamp, over T12, thus reducing the amount of lead.

Mercury is absolutely necessary for the operation of fluorescent and HID light sources. However, the amount of mercury per lamp can be, and has been, significantly reduced. For example, the average 4-ft fluorescent lamp contained 48 mg of mercury in 1985. By 1994, that number had been reduced to 23 mg, and today it averages 12 mg and is as low as 5 mg.

In 1990, the Environmental Protection Agency (EPA) introduced its Toxicity Characteristic Leaching Procedure (TCLP) test that classifies a fluorescent lamp as hazardous if its mercury content is greater than 6 mg. If a lamp is hazardous, it must be disposed of in a prescribed manner.

Since 2000, lamps have been included in the Federal Universal Waste Rule (including batteries, pesticides, and thermostats), which encourages recycling, but has numerous exemptions for small quantity users and for lamps passing the TCLP test.

Recycling, by Federal Law is the responsibility of the user, and thus must be included as a cost in any economic analysis. Recycling is not self-supporting. Spent lamps are not redeemable; recycled glass often contains residual mercury; and recycled aluminum end caps and mercury have minimal value. Also, there are many exceptions to the Universal Waste Rule, as well as strict EPA guidelines for handlers, transporters, and recyclers in applying the rule. Thus only about 15% of lamps are recycled currently at a cost of about 50¢ per lamp.

The alternative way of disposing of a lamp is the traditional method of placing it in a landfill. Historically, lamps have comprised only about 4% of all mercury deposited in landfills. Many studies of the effects of mercury in landfills have been done. Their conclusion is that 99.99% of the mercury in a landfill is safely contained, thus making landfills a viable option for lamp disposal. It is ironic that there is greater concern with mercury in the ground rather than in the air, yet, for a single fluorescent lamp, a fossil-fuel plant will emit 2 to 3 times the amount of mercury to the atmosphere as compared with the quantity of mercury that will escape to the atmosphere from the disposal of a spent lamp.

Green Design

In recent years, there has been increasing interest in "green" buildings. The US Green Buildings Council (USGBC) defines green design as "design and construction practices that significantly reduce or eliminate the negative impact of buildings on the environment and occupants".

A rating system, called LEED (Leadership in Energy and Environmental Design) has been developed by the USGBC to evaluate building compliance with green principles. Four levels of certification are provided, called Certified, Silver, Gold and Platinum, based on a scale that allocates points in five categories for all aspects of building design.

Building owners can earn points for lighting in two areas — energy performance and system controllability. Points are given for meeting or exceeding requirements of energy codes, for using occupancy and photosensors, for employing building lighting on-off and dimming controls, for providing individual occupant lighting control at workstations, and for controlling daylight. The latter is particularly important, as many LEED projects are designed to integrate daylighting with the architecture so as to optimize daylight penetration.

REFERENCES

Barker, W.: "Life-Cycle Costing", *Plant Engineering*, May 1978.

Belcher, M.: "Lighting Cost Analysis: Is There a Better Way", *LD+A*, June 1989.

English, C., J. Smith and W. Daiber: *Energy Management and Economics, ED-100.9*, IESNA, New York, 1999.

English, C.: *Lighting Economics, ED-150.9*, IESNA, New York, 1993.

Fetters, J.: *Handbook of Lighting Surveys and Audits,* CRC Press, Boca Raton, FL, 1998.

Helms, R., and M. Belcher: *Lighting for Energy-Efficient Luminous Environments*, Prentice-Hall, Englewood Cliffs, N.J., 1991.

IESNA Lighting Economics Committee: *Recommended Practice for the Economic Analysis of Lighting, RP-31-96*, IESNA, New York, 1996.

Lindsey, J.: *Applied Illumination Engineering*, Fairmount Press, Inc., Lilburn, GA, 1991.

Maniccia, D.: "Lighting Controls 'Leed' Green to Gold", *LD+A*, January, 2003.

National Electrical Manufacturers Association: *Fluorescent Lamps and the Environment*, NEMA, Rosslyn, VA, 1996.

Osram/Sylvania: *A Guide to the Energy Policy Act of 1992*, OS Inc, Danvers, MA, 1994.

Rea, M., editor: *IESNA Lighting Handbook*, Ninth Edition, IESNA, New York, 2000.

Roberts, J., editor: *Advanced Lighting Guidelines*, New Buildings Institute, White Salmon, WA, 2001.

Serres, A.: *Lighting Economics*, University of New Hampshire internal publication, Durham, NH, 1987.

Walitsky, P.: "Shedding Light on Fluorescent Waste", *ECON*, January 1996.

Problems

CHAPTER 1

1.1 An electric field intensity is given by

$$\mathscr{E} = 100 \sin 2\pi \left(5\times10^5 t - \frac{s}{600}\right) \text{V/m (s in meters, t in seconds)}$$

(a) What are the period and wavelength?
(b) Sketch $\mathscr{E}$ versus s for a fixed t and $\mathscr{E}$ versus t for a fixed s.

1.2 An electron is at rest at point a in Fig. P1.2. It accelerates toward point c. Find the following:
(a) Work done, KE gained, and PE lost as the electron moves from a to b; from a to c
(b) Force on the electron and acceleration of the electron
(c) Velocity of the electron at points b and c and the time to reach each point

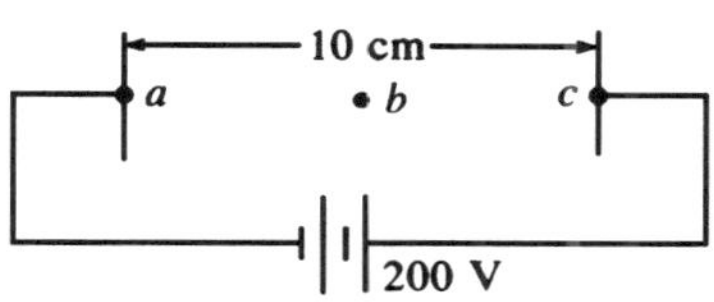

Figure P1.2

1.3 (a) Given that 1 J = 9.48×10^{-4} Btu, how many kilowatts per square meter does 1 Btu per hour per square foot represent?
(b) A sodium lamp emits a single spectral line at 0.59×10^{-6} m. Express this wavelength in microns, nanometers, and Angstroms.

1.4 An electron with a KE of 2.0×10^{-18} J collides with a hydrogen atom. What is the highest energy level to which the electron in the hydrogen atom can be raised? What wavelength radiation will be emitted when the electron reverts to its normal state from this level?

1.5 If the iris of a human eye contracts from 8 mm in diameter to 2 mm in diameter, what is the change in the illuminance on the retina?

1.6 The mesopic range of vision extends from a background luminance of 0.01 cd/m^2 to 3.0 cd/m^2. How many log units does this represent?

1.7 A perfectly diffusing spherical glass globe is 14 inches in diameter, has a luminance of 3 cd/in^2, and has a transmittance of 60%. It absorbs 40%. What is the intensity (assumed uniform) of the incandescent lamp at its center?

1.8 An incandescent lamp bulb, assumed spherical, has a uniform intensity of 130 cd and a diameter of $2\frac{5}{8}$ in. Find the luminance and luminous exitance of the bulb, assuming perfect diffusion.

1.9 The illuminance at B from the square perfectly diffusing luminous wall panel 10 ft^2 in area in Fig. P1.9 is 1.0 fc. The panel's transmittance is 45%. Find

a) The intensity of the panel when viewed by an observer at A (far away).
b) The luminance of the right side of the panel.
c) The illuminance on the panel.
d) The luminous flux incident on the panel.

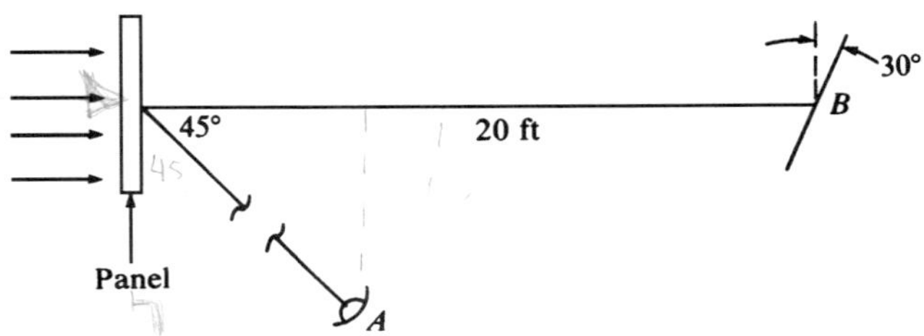

Figure P1.9

1.10 A 3 ft × 3 ft square by 6 in high luminaire has a milky white bottom panel that is perfectly diffusing with $\tau = 50\%$. The luminaire sides and top are opaque. The luminaire contains four fluorescent lamps, each emitting 2000 lm. 20% of the lamp lumens are absorbed and do not reach the bottom panel. Find the following:

a) Illuminance on the inside of the bottom panel.
b) Luminous exitance of the bottom panel.
c) Luminance of the bottom panel at 30° from the vertical.

1.11 A long straight street is lighted by incandescent lamps on poles 20 ft high and spaced 50 ft apart. Each lamp has an intensity of 2000 cd, uniform in every direction. Find the illuminance on the street midway between any two poles. Find the illuminance on a small vertical sign 5 ft off the ground at the same location.

1.12 A circular floodlight 4 in. in diameter is aimed at point P. It has an intensity of 5000 cd toward P and 2500 cd toward Q. Find the illuminances at P and Q. Find the luminances of the floodlight seen at P and Q.

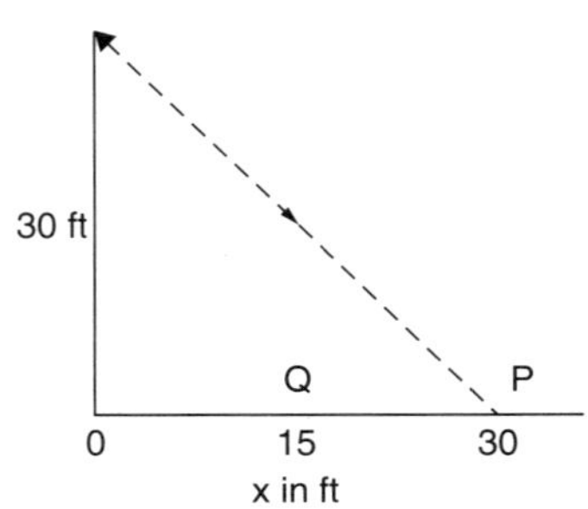

Figure P1.12

1.13 An 8 in diameter translucent ($\tau = 60\%$) disk lighting fixture in Fig. P1.13 is perfectly diffusing with a luminance of $400\ \frac{\text{cd}}{\text{ft}^2}$.

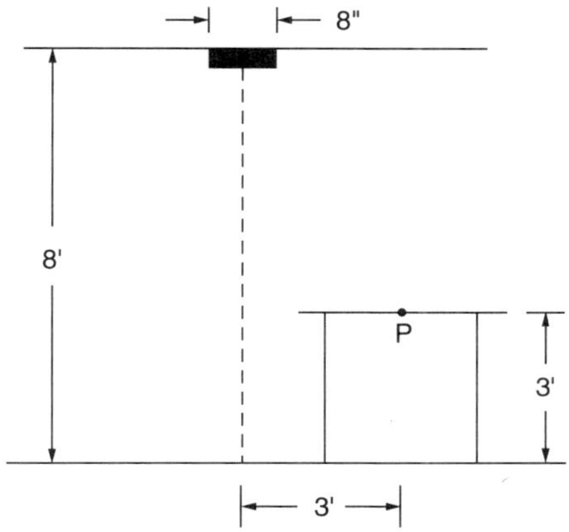

Figure P1.13

a) How many lumens does the fixture emit?
b) What is the illuminance on the inside of the disk?
c) What is the illuminance on the table top at point P?

CHAPTER 2

2.1 A 100-W PAR 36 spot lamp is aimed at the middle of a mannequin M in Fig. P2.1. Find the following:
a) The illuminance on the mannequin's middle, as seen by customer A.
b) The luminance in the eye's of customer B.

I_{cd}	$\alpha°$
9200	0
8000	5
4600	10
1800	15
700	20
0	25

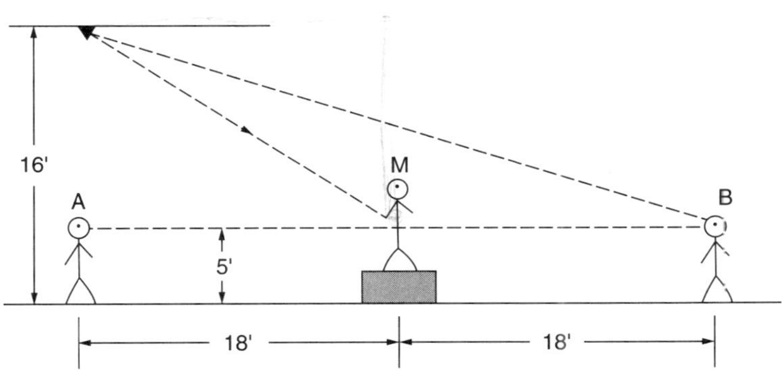

Figure P2.1

2.2 A 150-W floodlamp has the following data:

Degrees from lamp axis	Intensity in candelas
0	3500
5	3210
10	2820
15	2040
20	1360
30	290
40	240
50	190
60	100

One of these lamps is mounted on a pole 20 ft high and aimed at a point on the ground 6 ft from the pole. Find the illuminance at a second point on the ground 8 ft from the first point and 10 ft from the pole (a 6-8-10 triangle).

2.3 A shuffleboard runway is to be lighted by three PAR lamps on a pole aimed as shown in Fig. P2.3. Derive a general expression for the luminance at point P in terms of θ and the intensities. Find the illuminance at point P when $\theta = 65°$.

Angle, degrees	Intensity, cd
0	16,000
10	14,000
20	10,000
30	2,000
40	500
50	0

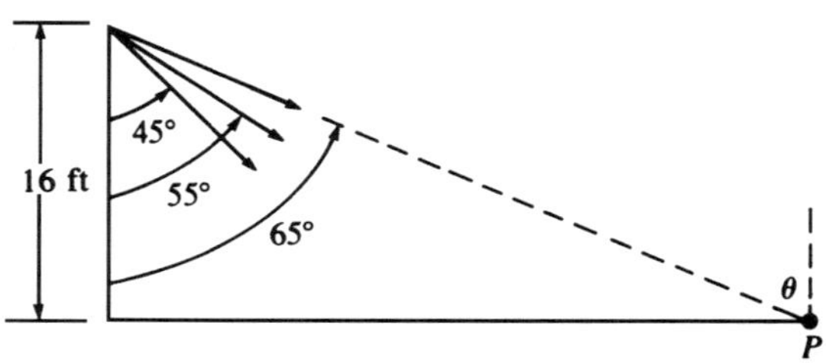

Figure P2.3

2.4 In a basement recreation room with an 8-ft ceiling, a remodeling contractor has specified 150-W recessed incandescent units to light the ping-pong table. The table is 9 ft × 5 ft and is 3 ft high. Three 120-V, 150-W lighting units are to be used—one centered over each end of the table and one over the middle of the table, as shown in Fig. P2.4. Two types of lenses are available for the lighting units—a prismatic clear glass lens and an opal diffusing lens. Find the illuminances on the table center line at the points shown. Which lens would you choose?

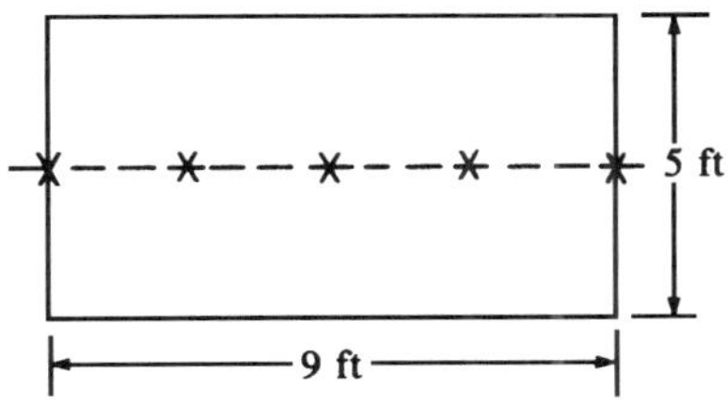

Figure P2.4

Prismatic lens		Opal lens	
Polar angle, degrees	Candelas	Polar Angle, degrees	Candelas
85	7.0	85	44.7
75	41.2	75	86.5
65	67.5	65	135
55	112	55	179
45	248	45	222
35	457	35	259
25	635	25	289
15	755	15	309
5	805	5	320
0	810	0	321

2.5 A 300-W PAR 56 wide flood light has the following intensity data:

$\alpha°$	I_{cd}
0	12,000
5	11,500
10	10,500
15	7,500
20	4,500
25	1,500

At what angle α is the lamp's luminance 50% of its value at $\alpha = 0$?

2.6 The candlepower distribution curve of a hypothetical batwing lamp-reflector unit is shown in Fig. P2.6. How many lumens does the unit emit?

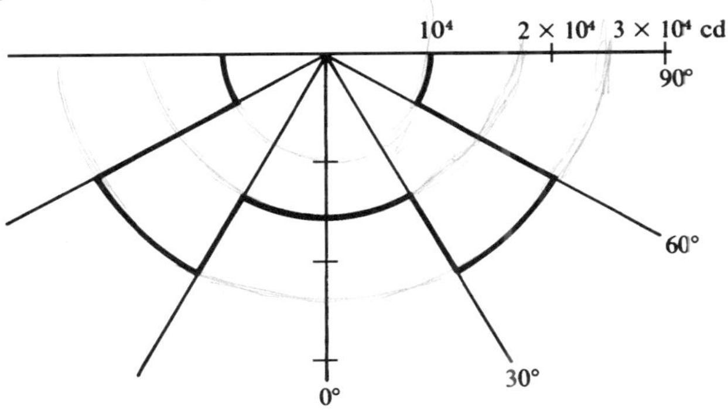

Figure P2.6

2.7 A floodlight lights a circular pond in Fig. P2.7. What is the average horizontal illuminance on the pond? What is the ratio of maximum to average illuminance on the pond? What is the ratio of average to minimum illuminance on the pond?

I_{cd}	$\alpha°$
2400	0
2280	5
2040	10
1760	15
1000	20
400	25
0	30

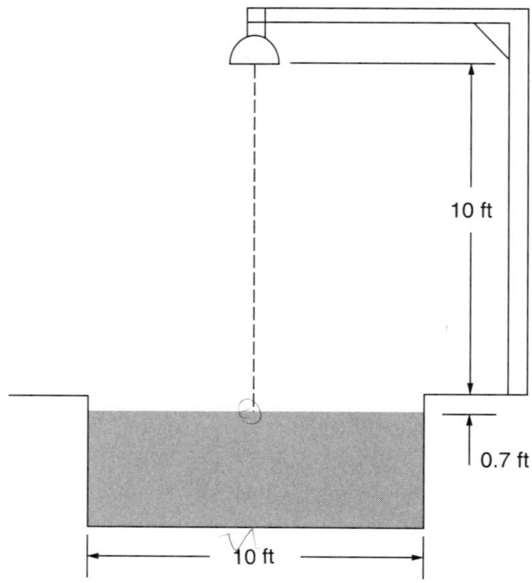

Figure P2.7

2.8 Intensity data are given for a 2 ft × 2 ft parabolic troffer containing two U-shaped fluorescent lamps each emitting 1750 lumens.

a) Calculate the lumens emitted by the fixture in the 0–30° zone.

b) What luminance is seen by the viewer in the chair in Fig. P2.8 looking up at the fixture?

c) What is the efficiency of the luminaire?

$\alpha°$	I_{cd}	$\alpha°$	I_{cd}
0	1575	50	609
5	1561	55	458
10	1560	60	177
15	1507	65	79
20	1445	70	46
25	1349	75	27
30	1259	80	13
35	1158	85	5
40	947	90	0
45	771		

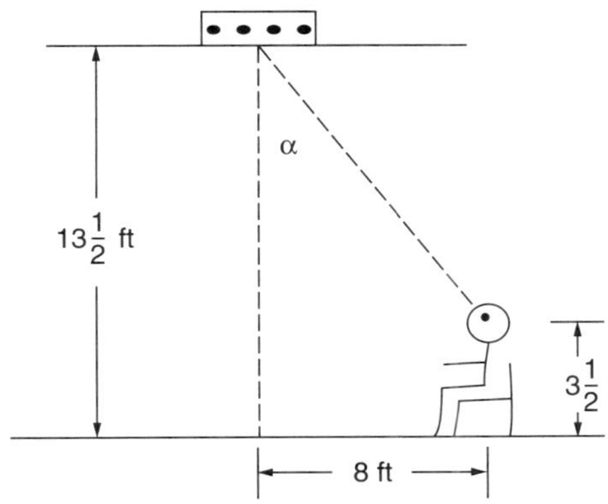

Figure P2.8

2.9 A company has installed a large array of 400-W high-pressure sodium lamp and reflector units in its greenhouse to provide electric light for rose growing. The installation consists of 225 lamps and reflectors mounted in a grid arrangement, as shown in Fig. P2.9, an average of 5 ft above the buds of the plants. Each lamp has an initial output of 47,000 lm. The candlepower distribution data for the lamp-reflector combination is given below.

Polar angle, degrees	Initial I, cd	Polar angle, degrees	Initial I, cd
0	4500	50	9000
5	4650	55	8980
10	4800	60	8400
15	5200	65	6900
20	5700	70	4900
25	6300	75	2700
30	6950	80	1400
35	7530	85	750
40	8100	90	400
45	8650		

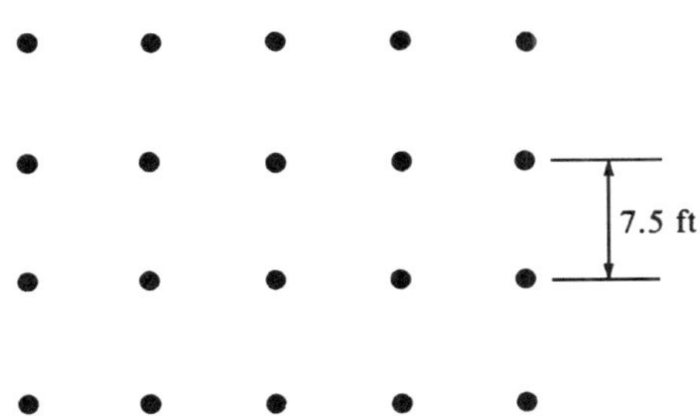

Figure P2.9

a) What is the initial horizontal illuminance on a rose bud directly beneath a lamp in the center of the array?
b) What is the efficiency of the reflector?
c) At what distance horizontally from directly under a lamp-reflector unit is the horizontal illuminance on a bud due to that single unit down to 10% of its maximum value?

2.10 An incandescent lamp of 1000 cd uniform intensity is mounted in a small reflector whose reflectance is 80%. All of the lumens in the 90 to 180° zone are intercepted by the reflector. Of those lumens reflected, 60% enter the 0 to 20° zone. What percent of the total lamp lumens end up in the 0 to 20° zone? What is the apparent candlepower of the lamp-reflector combination in the 0 to 20° zone? Assume the viewer is at least 10 times the reflector diameter away from the source.

2.11 An 8 square inch flat disk source has an intensity given by:

$$I = 5\cos^2\alpha \text{ cd}, \quad \alpha \leq 90°$$
$$I = 0 \text{ cd}, \quad \alpha > 90°$$

where α is the angle with the perpendicular to the disk. Find the following:
a) An expression for the luminance of the disk as a function of α.
b) The lumens leaving the disk.
c) The luminous exitance of the disk.

Repeat the problem for

$$I = 5\cos\alpha \text{ cd}, \quad \alpha \leq 90°$$
$$I = 0 \text{ cd}, \quad \alpha > 90°$$

2.12 A circular fluorescent lamp lights a point P in a horizontal plane 4 ft directly below the lamp's center, as shown in Fig. P2.12. The lamp is a 1-in diameter tube and has a luminance of 4.4 cd/in^2. Find the approximate illuminance at point P and the apparent intensity of the lamp in the direction of P. In working the problem, assume the inverse-square law holds and thus that the lamp is located on a circular locus $11\frac{1}{2}$ in from the lamp's center.

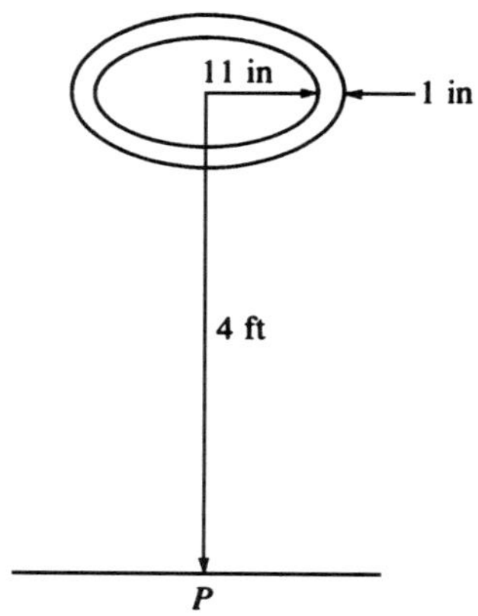

Figure P2.12

2.13 A standard cool white fluorescent lamp has the following data: 40 W, 48 in long, $1\frac{1}{2}$ in in diameter, L = 4.2 cd/in^2. The lamp is mounted 5 ft above a desk. What is the approximate illuminance at a point on the desk directly under the lamp's center? In working the problem, break the lamp into four 1-ft long pieces, assume each is a point source, find the approximate illuminance produced by each using the inverse-square law, and sum the results.

Repeat the problem with two 2-ft pieces and with a one 4-ft piece. Compare results.

2.14 Intensity data are given for the completely indirect batwing lighting fixture in Fig. P2.14, suspended 2.9 ft from the ceiling of a 16-ft diameter circular room. Ceiling reflectance is 70%. Find the average luminous exitance of the ceiling.

$\alpha°$	I_{cd}
90	0
100	0
110	200
120	400
130	700
140	900
150	1000
160	800
170	600
180	500

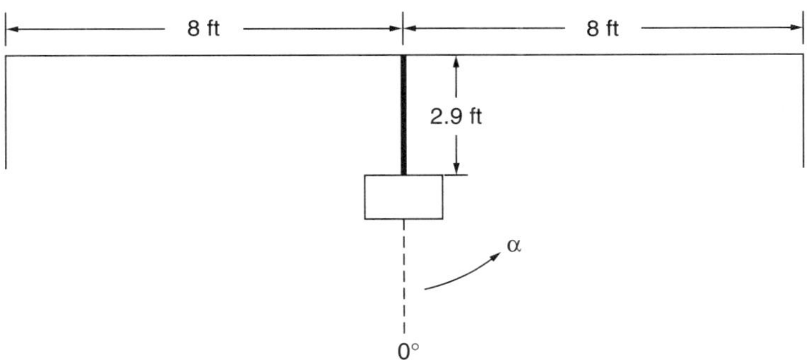

Figure P2.14

2.15 A circular park in the center of the city is surrounded by buildings, as shown in Fig. P2.15. The sky is overcast, and its luminance is approximated by:

$$L = 1900 \text{ cd/m}^2 \ (0 < \theta < 30°) \quad \text{and} \quad L = 1700 \text{ cd/m}^2 \ (30 < \theta < 60°),$$

where θ is the angle measured from the zenith. Find the horizontal illuminance at P. Ignore lumens reflected from the buildings.

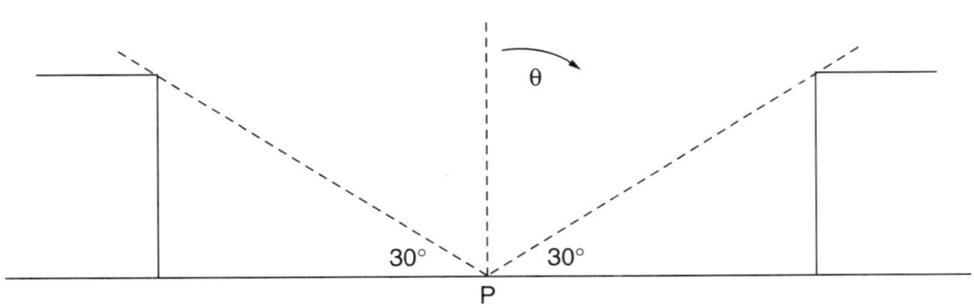

Figure P2.15

2.16 The spherical dome in Fig. P2.16 is perfectly diffusing, has a luminance of 0.2 cd/in^2 and lights the lobby floor below.

a) What is the solid angle in steradians subtended by the dome at the point P?

b) What is the horizontal illuminance at point P?

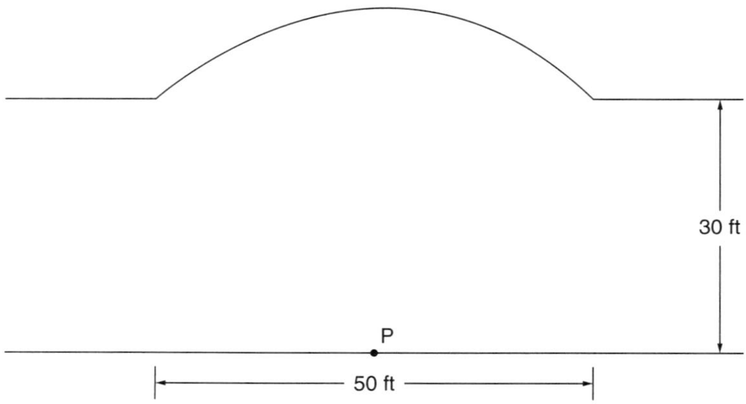

Figure P2.16

2.17 The circular open rotunda in Fig. P2.17 is 50 ft. in diameter and 25 ft high. Its dome-shaped ceiling is 20 ft high and is supported by eight pillars. The inside of the dome is perfectly diffusing and is lighted uniformly by hidden cove lighting. Its luminance is L = 0.10 cd/in^2. Find the horizontal illuminance at the center of the rotunda floor.

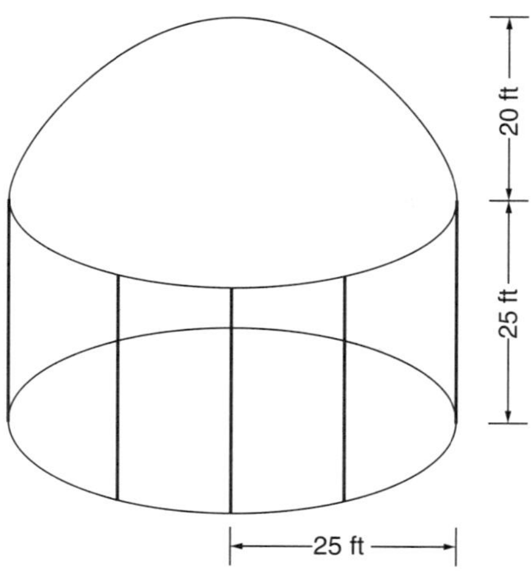

Figure P2.17

2.18 At what height h should the uniform point source of I candelas in Fig. P2.18 be mounted to produce the maximum illuminance on the tilted plane at point P? Find E_{max} if I = 500 cd.

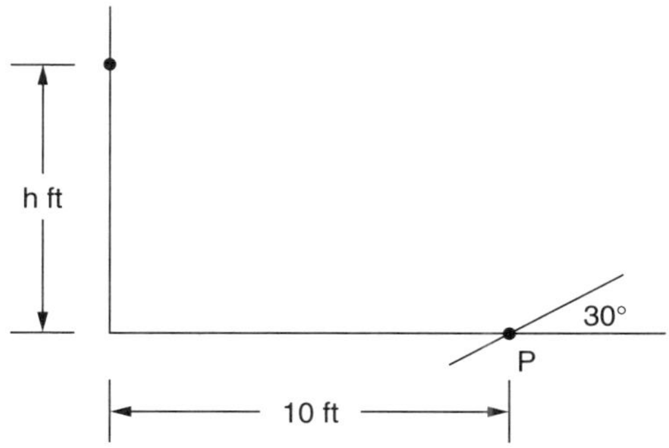

Figure P2.18

2.19 The horizontal footcandles produced on a gym floor by a luminaire containing a 400-watt high pressure sodium lamp are given below as a function of x, the distance on the floor from a point directly under the luminaire. If the luminaire mounting height is 22.5 ft, what is the total lumen output of the luminaire.

x (ft)	E_H (fc)	x (ft)	E_H (fc)
0	31	15	14
3	30	18	10
6	27	21	7
9	23	24	3.5
12	18	27	0

2.20 A new WILDCAT DINER is opening in Durham, NH. Its sign is to be lighted by two 150-watt PAR 38 spot lamps mounted on two foot high poles six feet out from the corners of the building, as shown in Fig. P2.20. Lamp A is aimed at point A, lamp B at point B.

a) Find the illuminances on the sign at points A and C.

b) A customer is sitting at a table at point D and looking toward the lamp at point A through clear window glass ($\tau = 90\%$). What is the luminance in the customer's eyes produced by the lamp at point A?

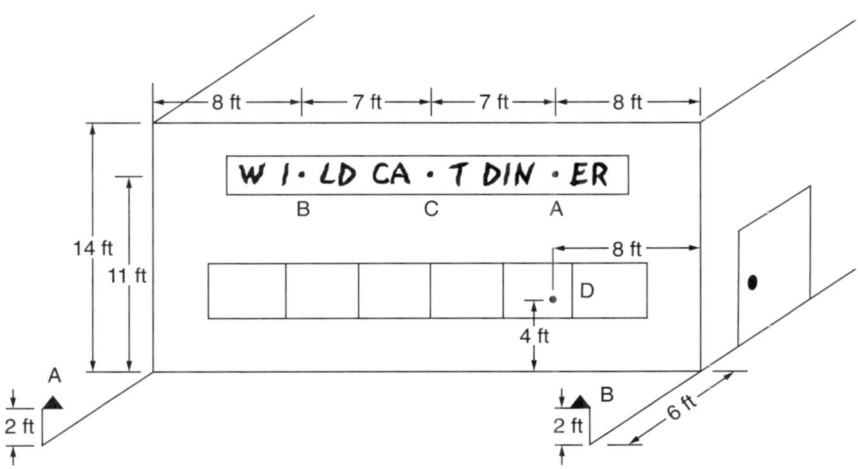

Figure P2.20

I_{cd}	$\alpha°$
11,000	0
8,300	5
4,400	10
1,500	15
700	20
500	25
400	30

CHAPTER 3

3.1 What is the threshold size of the black object in Fig. 3.9 on a white background of luminance $L_d = 2$ lm/ft^2? How large is the object if it is viewed from 14 in away?

3.2 What is the minimum perceptible contrast of a 4-min luminous disk at a background luminance of 10 lm/ft^2? Compare the values obtained for MPC from Figs. 3.11 and 3.17. Explain the difference.

3.3 A white dot 0.02 in in diameter is just seen from 69 in away when its luminance is 1 lm/ft^2. Find its contrast and the luminance of its background if the time for viewing is 1 s, using the expanded version of Fig. 3.17 in Fig. P3.3.

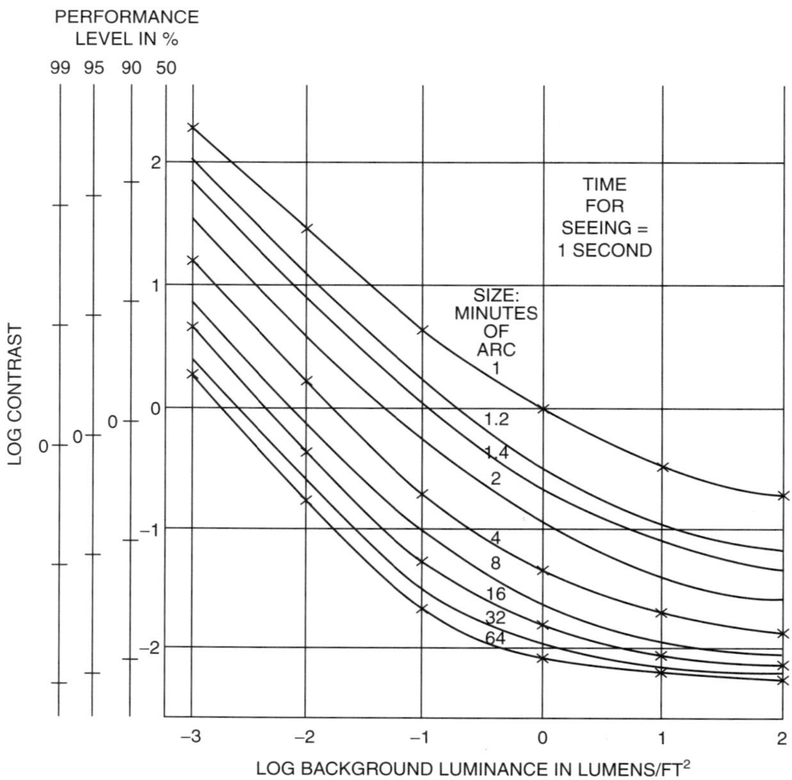

Figure P3.3

3.4 Two single line-spectrum light sources of the same wattage are observed under photopic lighting conditions. One has a wavelength of 500 nm; the other 650 nm. What is the ratio of their luminances?

3.5 a) One point of book type is 1/72 in measured from the top of the longest ascending letter to the bottom of the longest descending letter. Calculate the visual angle in minutes of arc made by 12-point type at 15 in viewing distance.

b) Another measure of type size is the width of a letter, called stroke width. For 12-point type, it is 1/38 in. Recalculate the visual angle in minutes of arc for this definition.

3.6 Calculate the visual angle in micro-steradians and in minutes of arc of a 3/64-in diameter drilled hole at a viewing distance of 15 in.

3.7 The reflectances of an overexposed copy of bookprint are ρ_d (task detail) = 0.10 and ρ_b (task background) = 0.65. What is the contrast? What ρ_b is required for a contrast of 0.88 with the same ρ_d? What is the maximum contrast that can be achieved with $\rho_d = 0.10$?

3.8 An L_b of 20 lm/ft² produces VL1 for a certain visual task in a given lighting environment. Using the expanded version of Fig. 3.19 in Fig. P3.8, what change in contrast is required to produce VL8 in the same lighting environment if L_b is reduced to 10 lm/ft²?

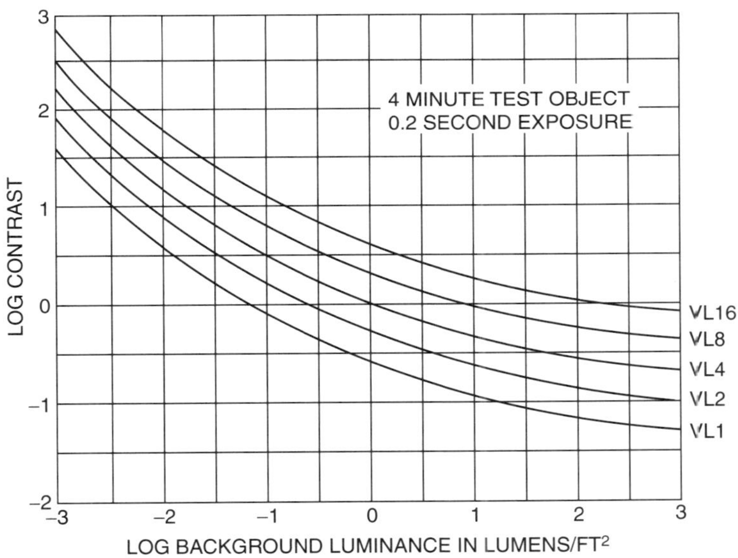

Figure P3.8

3.9 A task has $\tilde{C}_0 = 0.8$ and $\rho_b = 0.75$. Using Fig. P3.8, find the background luminance in lm/ft² and the illuminance in fc to give a visibility of VL8 in a diffuse hemisphere. The same task is now placed in a real lighting environment and its CRF is measured as 0.85. Find the background luminance and illuminance to give a visibility of VL8 in the real lighting environment.

3.10 A lighting system provides 110 fc on a visual task. L_b at the 25° viewing angle is 95 lm/ft². CRF has been measured as 0.8. Using Fig. P3.8, find the illuminance in the hemisphere that will provide the same visibility.

3.11 A car approaches a beige dog on a highway at dusk, as shown in Fig. P3.11. Immediately past the dog, the road slopes at 5°. The vertical illuminance on the dog from the sky is 0.1 fc. The horizontal illuminance on the road from the sky is 0.2 fc. The dog has a reflectance of 50%, the road 20%, both in the direction of the car. The car's headlamps have a combined intensity of 90,000 cd in the direction of the dog. Using Fig. P3.3 and assuming the dog is a rectangle 2 ft high and 3 ft wide, can it be seen with certainty by the driver?

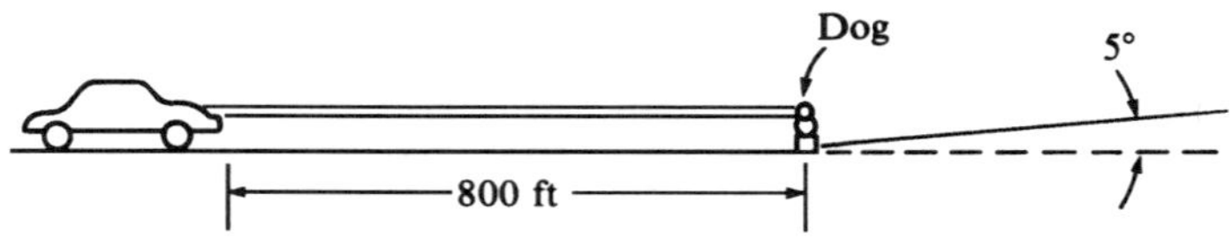

Figure P3.11

3.12 An automobile collides with the locomotive of a freight train at a grade crossing at dusk as shown in Fig. P3.12 (an actual case near Goffstown, NH).

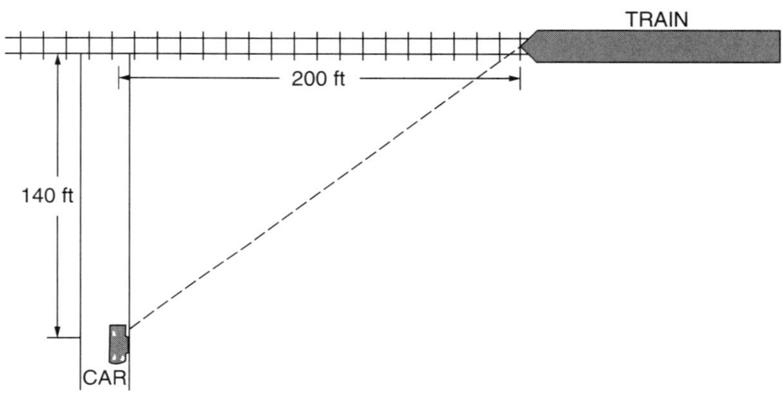

Figure P3.12

The car is 140 ft from the crossing. If the driver applies his brakes now, he can stop just before the crossing. When the car is 140 ft from the crossing the train is 200 ft from the crossing. Data:

Vertical illuminance = 0.12 fc
ρ locomotive = 14%
ρ foliage = 10%
Locomotive height = 14 ft

Using Fig. P3.3, determine whether the driver can see the train with 99% certainty in time to stop just short of the crossing. Assume the car windshield transmittance is 100%.

3.13 An automobile license plate with light ($\rho_t = 75\%$) $\frac{1}{2}$-inch wide letters and numbers on a dark ($\rho_b = 25\%$) background is lighted to 1 fc by the license plate light. At what distance away can a policeman on foot read the letters and numbers with certainty? What does this distance become if the policeman is in a patrol car with a dirty windshield ($\tau = 70\%$), the license plate light bulbs are dirty ($\tau = 50\%$), and the license plate is dirty ($\rho_b = 40\%$ and $\rho_t = 60\%$)?

3.14 a) A researcher conducting Weston's experiment with Landolt C rings identifies 24 rings correctly in one minute, misses 3 rings and takes $\frac{1}{2}$ second to cancel a ring. Calculate the researcher's visual performance in rings/minute.

b) In Fig. 3.15, there are 100 Landolt C rings. Weston's experiment is to be performed by the reader, first under a normal level of lighting and second under a low level. In the first test, count those rings with the gap facing northwest; in the second test count those with the gap facing southeast. Note the elapsed time for the 100 rings in each test. Assume $\frac{1}{2}$ second to mark an identified ring. Calculate visual performance.

CHAPTER 4

4.1 The spectral power density curve of an incandescent radiator in milliwatts per square centimeter per 10 nanometers is shown. Find the following:
(a) Total milliwatts per square centimeter in the 500 to 510 nm range
(b) Total milliwatts per square centimeter in the 600 to 640 nm range
(c) Total milliwatts per square centimeter in the 400 to 700 nm range

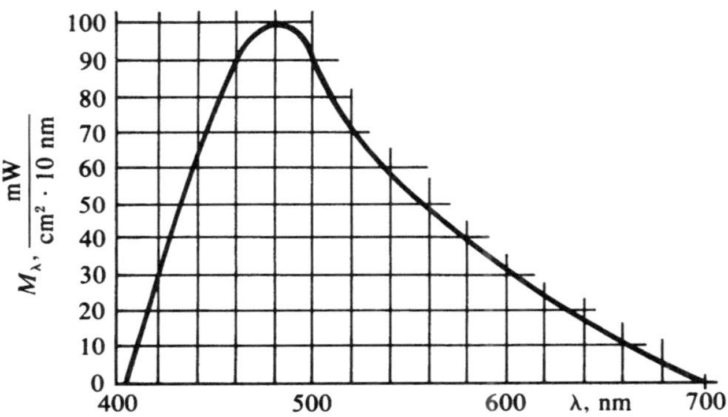

Figure P4.1

4.2 The filament of an incandescent lamp is 0.006 cm in diameter and 60 cm. long. It consumes 100 W. Assuming the filament may be considered a blackbody radiator, what is its operating temperature in kelvins? At what wavelength does the maximum value of M_λ occur? Is this filament an efficient light source? Explain.

4.3 Assume the straight-line approximation shown in Fig. P4.3 represents the spectral power density of a blackbody radiator. Determine the absolute value in W/(cm^2 • μm) of $M_{\lambda,max}$. At what temperature is the blackbody operating?

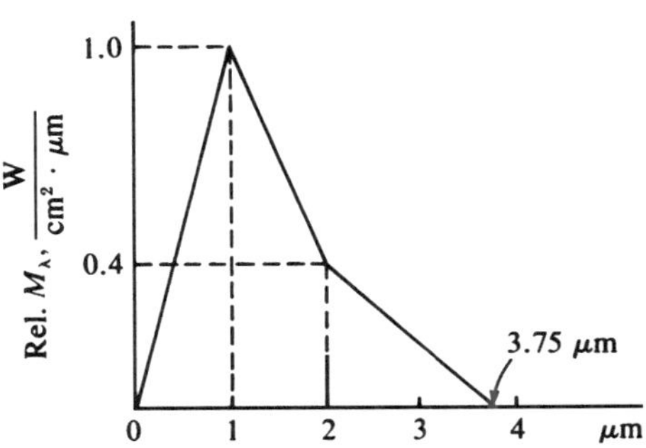

Figure P4.3

4.4 Relative values of M_λ in W/(cm^2 • μm) in the visible spectrum are given below for a blackbody radiator at 2800 K. What percent of the total power density is in the visible spectrum at this temperature?

Relative		Relative		Relative	
λ, μm	M_λ	λ, μm	M_λ	λ, μm	M_λ
0.38	0.075	0.51	0.531	0.64	1.311
0.39	0.093	0.52	0.585	0.65	1.373
0.40	0.113	0.53	0.640	0.66	1.433
0.41	0.137	0.54	0.697	0.67	1.492
0.42	0.163	0.55	0.756	0.68	1.550
0.43	0.193	0.56	0.816	0.69	1.608
0.44	0.225	0.57	0.876	0.70	1.663
0.45	0.261	0.58	0.938	0.71	1.718
0.46	0.299	0.59	1.000	0.72	1.770
0.47	0.340	0.60	1.062	0.73	1.822
0.48	0.384	0.61	1.125	0.74	1.871
0.49	0.431	0.62	1.188	0.75	1.919
0.50	0.480	0.63	1.250	0.76	1.965

The absolute value of M_λ at 0.59 μm is 88.39 W/(cm² • μm).

4.5 Planck's equation for a blackbody radiator may be plotted on log-log paper to facilitate calculations. Taking the log of Eq. (4.4) yields

$$\log M_\lambda = \log [C_1 \ (e^{C_2/\lambda T} - 1)^{-1}] - 5 \log \lambda$$

For values of λ and T such that λT is constant, the equation has the form $y = mx + b$. Thus each line of slope −5 on the log-log plot is a line of constant λT.

Assume two temperatures, T_1 and T_2, and two values of λT, k_1 and k_2, as shown in Fig. P4.5. Show that

$$\log M_3 - \log M_1 = \log M_4 - \log M_2$$

and

$$\log \lambda_1 - \log \lambda_3 = \log \lambda_2 - \log \lambda_4$$

This proves that a single log-log curve can be used in blackbody radiator calculations and that the coordinate axes can be shifted to accommodate various temperatures.

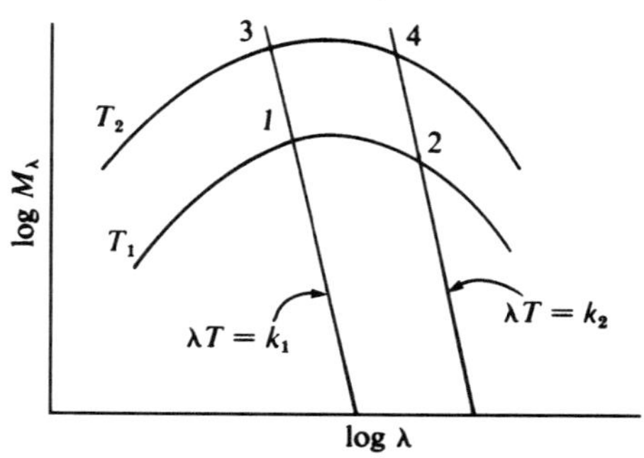

Figure P4.5

4.6 A tungsten filament has the following data: 75 W, 120 V, 1150 initial lumens, $\ell = 16$ in, $d = 0.0018$ in, 18.5% losses. Its effective emissivity versus temperature is shown below. Find its operating temperature (by trial and error).

T, K	$\bar{\varepsilon}$
2000	0.254
2100	0.263
2200	0.286
2300	0.296
2400	0.308
2500	0.316
2600	0.328
2700	0.332
2800	0.336
2900	0.342
3000	0.348

4.7 A coiled tungsten filament with $\bar{\varepsilon} = 0.336$ operates at 2800 K inside a gas-filled incandescent lamp with 18% losses. The filament occupies a cylindrical space 2.0 mm in diameter and 2.1 cm long. How much power does it consume?

4.8 A cylindrical gray body radiator operating at 6 V and 3000 K emits 200 W/cm^2 of surface area and consumes 600 W. At this temperature its resistivity is 50 μ ohm-cm and its losses are 20%. Find the following:

a) diameter and length
b) emissivity

4.9 A tungsten filament radiates 100 W. Its length is 30 cm, and its diameter is 0.40 mm. Its weighted average emissivity is 0.28. At what temperature does it operate? If it were a black body of the same dimensions and radiating 100 W, at what wavelength would its radiated spectral power be maximum?

4.10 A coiled tungsten filament operating in a gas-filled incandescent lamp has the following values:

$$\ell_f = 90 \text{ cm}, \quad d_f = 0.08 \text{ mm}, \qquad \ell_c = 1.2 \text{ mm}, \quad d_c = 0.8 \text{ mm}$$

$$D = 19.35 \text{ gm/cm}^2, \quad L = 1000 \text{ hr}, \quad \text{losses} = 18\%.$$

Find the following:

a) filament operating temperature
b) input watts
c) output lumens

4.11 A tungsten filament has the following values:

coil length = 3.0 cm
coil diameter = 2.5 mm
filament length = 50 cm
filament diameter = 0.08 mm
density = 19.35 gm/cm^3
resistivity = 92×10^{-6} Ω-cm

The filament is operated at a temperature of 3000 K in a gas filled incandescent lamp with 20% losses. Find:

a) weighted average emissivity
b) power supplied to the filament
c) life of the filament
d) lamp voltage

4.12 Find the luminance of the perfectly diffusing surface for which ρ_λ is the spectral reflectance, E_λ is the incident spectral radiant power density, and $V(\lambda)$ is an approximation to the spectral luminous efficiency curve of the eye (as shown in Fig. P4.12).

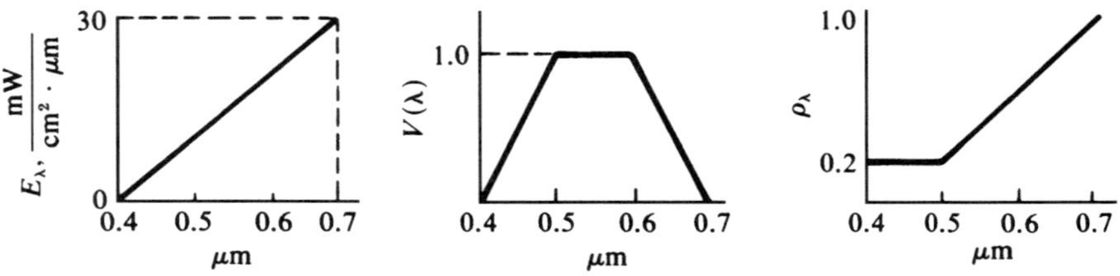

Figure P4.12

4.13 A gaseous discharge lamp emits two visible lines, one at 500 nm, the other at 600 nm. If 30 W of power goes into these lines and they give a combined output of 8900 lm, find the lumens produced by each line.

4.14 A color-corrected mercury lamp is a mercury lamp to which a phosphor has been added to the inside of the outer bulb to augment the orange and red portions of the spectrum. Assume a 400-W color-corrected mercury lamp has its four visible lines present with the given wattage values and further assume that the spectral power distribution of the phosphor may be represented as shown in Fig. P4.14. Calculate the efficacy of the lamp.

λ, nm	P_i, W
404.7	80
435.8	40
546.1	10
578.0	20

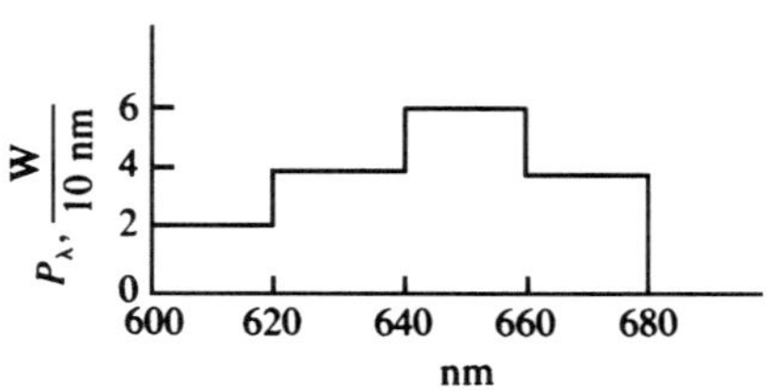

Figure P4.14

4.15 Given are two single line-spectrum uniform point sources at 500 nm and 650 nm with equal radiant intensities of 50 W/sr. The sources are mounted on poles 8 ft above the ground and are separated by 12 ft. At what point on the ground on a line connecting the poles will the horizontal illuminance provided by the sources be equal? What is the total horizontal illuminance at this point?

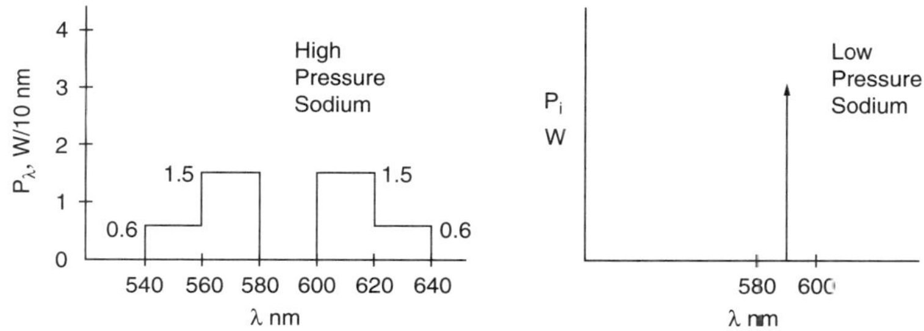

Figure P4.16

4.16 The spectral power distributions of a high-pressure and low-pressure sodium lamp are given in Fig. P4.16.
a) Find the lumen output of the high pressure lamp.
b) Find P_i such that the lumen output of the two lamps is the same.

4.17 A triphosphor fluorescent lamp has three 20-nm wide bands centered at 440, 550, and 620 nm. The lamp consumes 32 watts, 30% of which appears in the three bands. The lamp is designed so that the lumens in the three bands have the relationship: $4\ \Phi_{440} = \Phi_{550} = \Phi_{620}$. What is the lumen output of the lamp?

4.18 A vacuum incandescent lamp with an uncoiled filament operates at a temperature of 2570 K. Use the curves in Fig. 4.11 to find
a) Life of the filament
b) Watts consumed by the filament
c) Lumens emitted by the filament
d) Lumens per watt efficacy of the filament
e) Operating voltage of the filament

The filament is 22.2 in. long and 0.0012 in in diameter. Assume losses represent 6% of the input power. Tungsten has a density of 19.35 gm/cm^3.

4.19 Repeat Prob. 4.18 for a gas-filled incandescent lamp with a coiled filament operating at 2870 K. The filament has an uncoiled length of 22.8 in, a coiled length of $\frac{15}{16}$ in, an uncoiled diameter of 0.0025 in and a coiled diameter of $\frac{1}{30}$ in. Losses represent 18% of input power.

CHAPTER 5

5.1 Find the maximum speed at which electrons will be ejected from a cesium surface whose work function is 1.8 eV when that surface is irradiated by monochromatic sodium light at 589 nm. Would the maximum speed increase or decrease if the radiation were changed to a mercury line at 253.7 nm?

5.2 An electron at rest leaves the filament of a fluorescent lamp and is accelerated to a velocity of 1.35×10^6 m/s, at which time it collides with a mercury atom, exciting the latter.
a) What is the wavelength of the emitted photon?
b) What is the kinetic energy of the departing electron?
c) What is the wavelength of the electron just prior to collision?
d) What is the mass of the emitted photon?

5.3 Argon is used in fluorescent lamps to facilitate starting. Assume an electron leaves the cathode of a 48-in fluorescent lamp with zero initial velocity. Assume a lamp voltage of 120 V dc and a positive arc column of length 46 in with a gradient of 2 V/in. Assume negligible voltage drop at the anode.

a) To what velocity must the electron be accelerated if it is to be just able to ionize an argon atom and help initiate the arc?
b) What is the least distance from the filament at which the electron has sufficient energy to ionize an argon atom?
c) How many 253.7-nm exciting collisions could the electron make with the mercury atom if it had the energy of part (a)?

5.4 A quantum of energy at 253.7 nm impinges on a phosphor and results in the emission of a quantum of energy at 579 nm. What percent of the input energy is lost in this conversion? Could a quantum of energy at 700 nm produce the 579-nm photon? Explain.

5.5 What is the angular momentum in joule-seconds of the $n = 4$, $\ell = 3$ energy level in an atom?

5.6 A neon atom in its first excited state collides with a mercury atom in its rest state. A 253.7-nm photon is observed. Immediately thereafter, the neon atom collides with another mercury atom in its rest state, ionizing the latter. What is the maximum velocity of the electron that is freed by the second collision?

5.7 The voltage distribution versus distance from one electrode of a 4-ft fluorescent lamp is given by

$$v = 15x, \qquad 0 < x < 1$$

$$v = 1.8x + 13.2, \qquad 1 < x < 48$$

where x is the distance from the electrode in inches.

a) Find the distance from the electrode at which an electron, starting from rest at the electrode, will have the highest probability of ionizing a mercury atom. What is the electron's velocity at that point?
b) Assume that, due to prior collisions, an electron arrives at $x = 1$ in with zero velocity. How much further must it travel to have the highest probability of producing a 185 nm mercury photon?
c) What is the maximum number of excitation collisions immediately yielding radiation that the electron in (b) can make from $x = 1$ in to $x = 48$ in?

5.8 A gaseous discharge lamp emits 33.5 mW and 10^{17} photons per second at a single wavelength. What is that wavelength?

5.9 Assume an electron with a mass M_1 and traveling with a velocity v_{1i} collides with an atom of mercury of mass M_2 which is standing still. If the collision is elastic, find expressions for v_{1f}, the final velocity of the electron, and v_{2f}, the final velocity of the atom, in terms of v_{1i}, assuming that $M_2 \gg M_1$ and that the collision is "head on," rather than glancing.

5.10 A gas discharge tube is shown in Fig. P5.10. Electrons leave the left electrode with a velocity of 0.6×10^6 m/s. Mercury atoms are located at $x = 4.5$, 12, 21, and 25 cm. The voltage along the tube is given by

$$V = 3x, \qquad 0 < x < 5 \text{ (Crooke's dark space)}$$

$$V = 0.4x + 13, \qquad 5 < x < 10 \text{ (negative glow)}$$

$$V = -0.2x + 19, \qquad 10 < x < 15 \text{ (Faraday dark space)}$$

The remainder of the tube is positive column. Collision rules are

1. If an electron has enough energy to ionize a mercury atom, ionization will occur.
2. If an electron has enough energy to excite a mercury atom, it will excite it to the highest possible level.

Describe what happens at each collision. If a photon is emitted, find its wavelength.

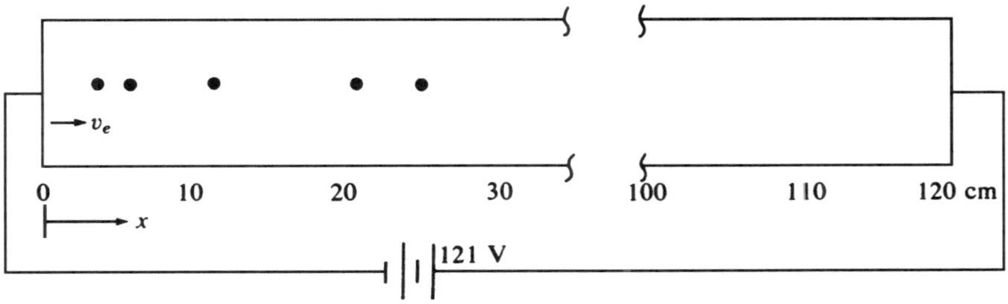

Figure P5.10

5.11 Four electrons, each with an initial KE of 0.5 ev, leave the cathode of a fluorescent lamp, whose voltage vs distance curve is shown in Fig. P5.11.

a) The first electron collides with a mercury atom 1.30 inches from the cathode. Describe completely what is most likely to happen—energy exchanged?, excitation?, ionization?, photon emitted?, wavelength?

b) The second electron collides with a mercury atom 7.25 inches from the cathode Repeat (a).

c) The third electron collides with an argon atom 2.75 inches from the cathode. Repeat (a).

d) The fourth electron collides with a mercury atom in its lowest metastable state at 0.75 inches from the cathode. Repeat (a)

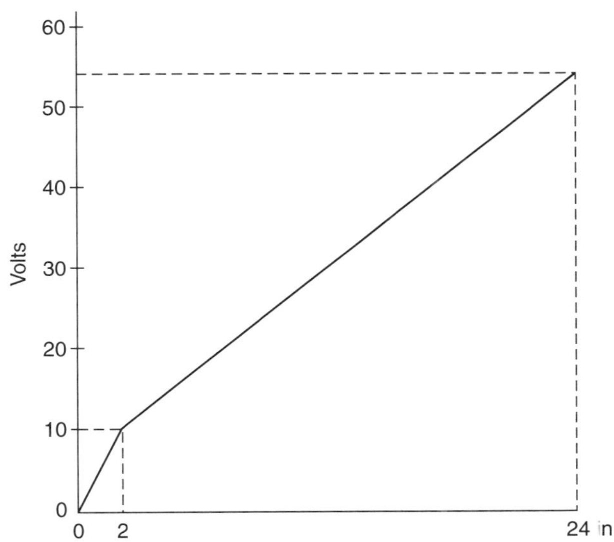

Figure P5.11

5.12 What is the range of quantum ratios for fluorescent lamps in the visible spectrum (from 400 to 700 nm)?

5.13 Solar radiation falls on the atmosphere outside the earth's surface at the rate of 1353 w/m^2. How many photons/cm^2 • min does this represent, assuming an average solar wavelength of 550 nm?

5.14 A gas in a discharge tube has $\alpha = 100$ electrons per meter and $\beta = 2 \times 10^{-4}$ secondary electrons per positive ion. What electrode spacing is required for a self-maintained discharge?

5.15 Assume a gas atom has a diameter of 2.5×10^{-10} m. In a container of these atoms at p = 10^{-2} mm of mercury and T = 273 K, find
a) The density of the gas atoms
b) The mean free path of an electron

CHAPTER 6

6.1 A log-log plot of lumens versus volts for a 75-W, 1150-initial lumen, 120-V incandescent lamp is shown in Fig. P6.1. Find the initial lumens emitted if the lamp is operated at 130 V.

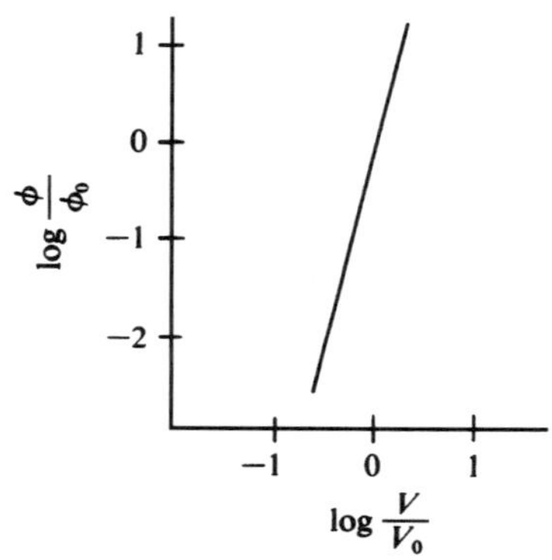

Figure P6.1

6.2 At what voltages should a 120-V incandescent lamp rated at 1000 h life and 1500 initial lumens be operated to give
a) 1500 h life
b) 1950 initial lumens

6.3 The owner of an auditorium feels that his 40 100-W incandescent downlights are producing too much light. He decides to purchase dimming equipment to reduce his lighting level by 20%. Calculate the new values of power, life, lumens per watt, and correlated color temperature that this change will cause.

6.4 A fluorescent lamp in series with a magnetic ballast across a 120-V, 60 Hz supply draws 0.5A. The lamp and ballast consume 30 W. What is the power factor of the circuit? What will the current from the source become if a capacitor is placed in parallel with the lamp and ballast to make the power factor 0.9? What is the capacitor current? What capacitor current would be required to achieve unity power factor?

6.5 An F40T12 fluorescent lamp has the following data:

cathode fall distance = 1.5 in
cathode fall gradient = 10 V/in
positive column gradient = 2 V/in
loading = 0.145 W/in^2

Find lamp voltage, lamp current, and lamp power factor.

6.6 In creating the 4-ft T8 lamp, the following design goals were to be met:

Loading = 0.18 W/in^2
Lamp PF = 0.90
positive column gradient = 2.5 V/in
cathode fall = 16 volts
cathode fall distance = 1 in

a) Find lamp volts, amps, and watts. Compare with actual ratings of 135 V, 265 mA, and 32 W.
b) The lamp is connected in series with a reactor ballast to an autotransformer. Requirements are that the transformer secondary voltage = 1.8 times the lamp voltage and that the ballast consumes 5 W. Find the inductance and resistance of the ballast.

6.7 The rectangular 60 Hz voltage wave of height 35.30 V in Fig. P6.3 can be represented by the following Fourier series:

$$v = 45\sin \omega t + \frac{45}{3}\sin 3\,\omega t + \frac{45}{5}\sin 5\,\omega t + \frac{45}{7}\sin 7\,\omega t + \frac{45}{9}\sin 9\omega t \text{ volts}$$

where $\omega = 2\pi f = 377$ r / s.

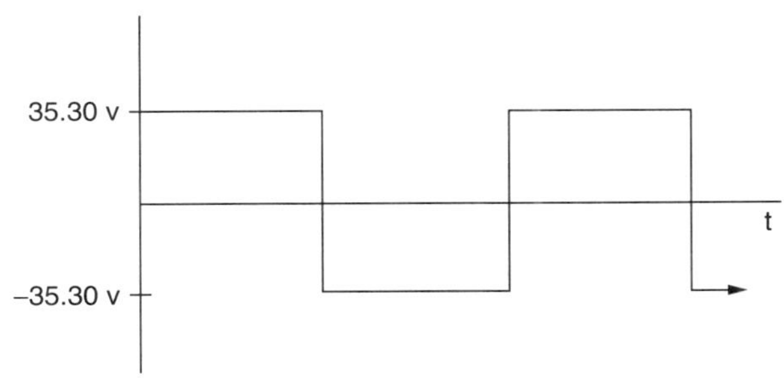

Figure P6.3

Find the THD produced by this rectangular wave. What is the RMS value of the wave?

6.8 The relation between the instantaneous values of voltage across and current through a T12 fluorescent lamp is given by

$$v = \frac{2.37\ell}{0.56 + i} + 12.5$$

where ℓ is tube length in inches, v is instantaneous lamp voltage in volts, and i is instantaneous lamp current in amperes. If the starting voltage is twice the operating voltage for 0.1 A maximum instantaneous lamp current, what is the maximum length lamp that can be started from a 120-V RMS source? If ℓ is 0.75 of this maximum length, what value of series resistance must be used to limit the maximum instantaneous lamp current to 1 A?

6.9 A fluorescent lamp has the following specifications:

Lamp watts = 26.0
Lamp amperes = 0.49
Lamp volts = 60

Two of these lamps are operated in a lead-lag parallel circuit directly from a 116 V, 60 Hz supply. Each ballast consumes 6 W and has the same inductance. The overall circuit operates at unity power factor. Find L and C.

6.10 An 8 ft T12 fluorescent lamp has the following data:

Lamp current = 1.5 A
Loading = 0.422 W/in^2
Positive column gradient = 1.656 V/in
$V_{start} = 2\,(V_{run})$

Two of these lamps are to operate at 60 Hz in a lead-lag circuit powered from an autotransformer. RMS readings on the primary side of the autotransformer are

$$V = 240\text{ V},\quad I = 2.104\text{ A},\quad P = 505\text{ W}$$

Find the inductance and resistance of each inductor, the capacitance of the lead-circuit capacitor, and the autotransformer turns ratio. Assume the inductors are identical.

CHAPTER 7

7.1 (a) The CIE spectral primaries are to be combined so as to match 6 W of pure spectral radiation at 600 nm. Find the number of watts and lumens of each primary required.
(b) Find R, G, and B for 6 W of pure spectral radiation at 600 nm. Find the luminous flux and chromaticity coordinates of this radiation.
(c) The same primaries are combined to match 6 W/nm of equal energy white. How many lumens of each primary are required? How many units?

7.2 Find the tristimulus values and the chromaticity coordinates of a line spectrum source having the following spectral data:

λ, nm	M_i, W/cm^2
490	1.70
550	3.20
680	7.60

Describe the color of the source.

7.3 It is desired to color match equal energy white using two pure spectral colors, one of which is 1 W at 480 nm. What must the other spectral color be and how many watts of it are required?

7.4 Relative spectral radiances from a gold fluorescent lamp are given in the table below. Find the tristimulus values and the chromaticity coordinates of the overall radiance.

λ, μm	Rel. L_λ	λ, μm	Rel. L_λ
0.51	1	0.61	21
0.53	4	0.63	11
0.55	12	0.65	6
0.57	19	0.67	2
0.59	24	0.69	1

In what color region of the CIE diagram does the lamp lie? What spectral color does the lamp most nearly match?

7.5 A color sample is illuminated by the source in Prob. 7.2. The spectral reflectances of the sample are:

λ, nm	ρ_i
490	0.10
550	0.15
680	0.40

Find the chromaticity coordinates of the lighted sample and its dominant wavelength and purity.

7.6 Compute the chromaticity coordinates of the color that results from the combination of the following colors:

	x	y	L, cd/ft²
1.	0.272	0.494	12.0
2.	0.329	0.218	9.5

What is the luminance of the resultant color?

7.7 Three colored fluorescent lamps have the following specifications:

Lamp	x	y	lm/W
A	0.600	0.300	10
B	0.500	0.400	20
C	0.200	0.300	30

All lamps have the same wattage. Find the chromaticity coordinates of the resultant color produced by four of lamp A, two of lamp B, and three of lamp C.

7.8 The colors from two sources are to be mixed. What is necessary in order that the chromaticity coordinates of the resultant color lie at the midpoint of the line connecting the chromaticity coordinates of the individual sources?

7.9 An area is lighted with luminaires, each of which contains one 400-W metal-halide lamp and one 400-W high-pressure sodium lamp. What is the chromaticity of the resulting illumination?

	HPS	MH
CCT(K)	2100	4500
x	0.512	0.362
y	0.420	0.371
ϕ, lm	50,000	34,000

7.10 A conference room is lighted by four fluorescent luminaires, each containing two cool white fluorescent lamps, and by 10 incandescent downlights, each containing one tungsten halogen lamp (see Table 7.4). Each fluorescent lamp emits 3150 lm; each tungsten halogen lamp 1900 lm. Find the chromaticity coordinates of the combination.

7.11 Illuminant C lights a color sample such that the reflected light has $X = 77.9$, $Y = 65.5$, $Z = 34.1$. Find the dominant wavelength and excitation purity of the reflected light and the reflectance of the sample.

7.12 A lamp has chromaticity coordinates of $x = 0.40$ and $y = 0.40$. What is its correlated color temperature?

7.13 Two lamps have the following chromaticity coordinates

Lamp A	$x = 0.50$,	$y = 0.40$
Lamp B	$x = 0.20$,	$y = 0.30$

What two spectral colors, when combined in different amounts, can reproduce the chromaticity coordinates of either lamp?

7.14 A metal halide lamp has an indium line at 450 nm, a thallium line at 535 nm and a sodium line at 590 nm, with relative power densities of 1:3:2. It lights the human face whose spectral reflectance is given in Fig. P7.14. Find the chromaticity coordinates of the source and of the complexion.

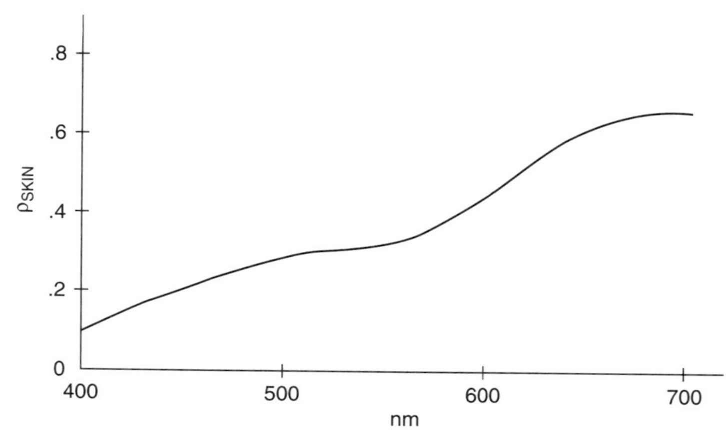

Figure P7.14

7.15 An object has the following values when lighted by illuminant C:

$$Y = 70.1, \quad x = 0.455, \quad y = 0.484$$

Convert these to L^*, a^*, b^* coordinates and calculate h^* and c^*. Locate the color on Fig. 7.16 and determine its Munsell notation from Table 7.7.

7.16 Two identical spaces are lighted by (a) metal halide downlights ($S/P = 1.80$) and tungsten halogen downlights ($S/P = 1.50$). Assume each source has the same lumens and CRI. Find the ratio of visually effective lumens for $k = 0.73$ and $k = 1.0$.

CHAPTER 8

8.1 For the luminaire in Fig. P8.1:

a) To which of the six luminaire flux distribution classifications does it be_ong?

b) What is its efficiency? How many lumens does the lamp emit?

c) How many lumens does the luminaire deliver into the 0–60° cone directly downward?

d) What is the average luminance of the luminaire when viewed crosswise at a 45° vertical angle?

ACME NUMBER
3151-4R

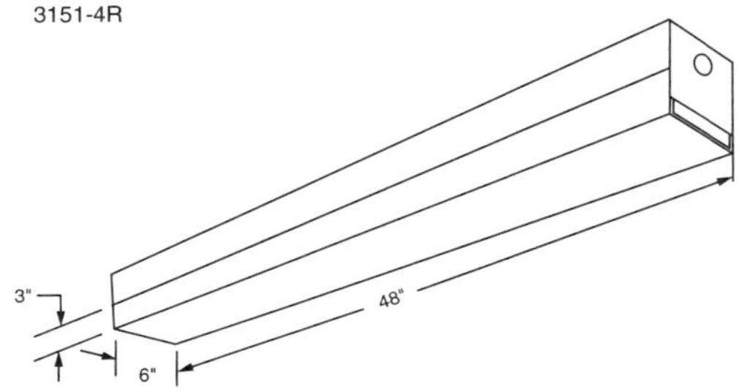

Also available in 2', 3' and tandem-wired (8') lengths

- Slim, low-profile housing - diffuser assembly hinges or latches from either side on four steel torsion springs
- Choice of prismatic acrylic diffuser or flat white opal acrylic - high-impact version optional
- Diffuser ends injection-molded to match diffuser composition
- End plates welded to channel for clean, finished appearance
- Channel cover secured by quarter-turn fasteners for easy access

PHOTOMETRICS CA 140 A*

COEFFICIENTS OF UTILIZATION ZONAL CAVITY

Pic	20%																	
Pcc	80%				70%				50%			30%						
Pw	70%	50%	30%	10%	70%	50%	30%	10%	50%	30%	10%	50%	30%	10%				
1	81	77	74	71	78	74	71	69	69	67	65	65	63	61				
2	74	67	62	57	71	65	60	56	60	56	53	56	53	50				
3	67	59	53	48	64	57	51	47	53	49	45	50	46	43				
4	62	52	45	40	59	51	44	39	47	42	38	44	40	36				
5	56	46	39	34	54	45	38	33	42	36	32	39	34	31				
6	52	41	34	29	49	40	33	29	37	32	28	35	30	27				
7	47	37	30	25	45	36	29	25	34	28	24	32	27	23				
8	44	33	26	22	42	32	26	21	30	25	21	28	23	20				
9	40	30	23	19	39	29	23	18	27	22	18	26	21	17				
10	37	27	21	16	36	26	20	16	25	19	16	23	18	15				

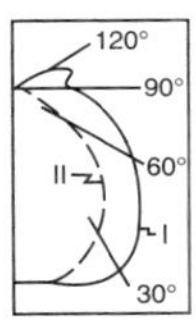

CANDLEPOWER

ANGLE	ALONG	22.5°	45.1°	57.5°	ACROSS
0	586	586	586	586	586
5	589	588	592	593	594
15	584	592	609	628	628
25	525	554	594	636	645
35	463	518	577	620	628
45	347	422	485	513	514
55	205	298	388	446	437
65	116	197	291	345	367
75	64	110	185	263	288
85	20	63	96	159	188
90	7	38	77	126	159
95	6	44	79	124	157
105	6	37	84	166	188
115	3	42	85	113	119
125	1	19	38	56	60
135	0	7	16	24	25
145	0	5	12	18	18
155	0	2	6	10	10
165	0	0	1	1	1
175	0	0	0	0	0
180	0	0	0	0	0

For photometrics on other configurations, see Technical Data section of Lithonia representative

Full report available. Request LRL 182-98

SPACING CRITERION: 1.2 × MOUNTING HEIGHT

* STANDARD BALLAST
F40T12/CW LAMP
(3150 LUMENS)

ZONAL LUMEN SUMMARY

ZONAL	LUMENS	%LAMP	%FIXTURE
0-30	681	21.6	28.0
30-60	995	31.6	40.9
60-90	442	14.0	18.1
90-100	317	10.1	13.0
TOTAL	2435	77.3	100.0

Figure P8.1 (Courtesy Lithonia Lighting).

8.2 For the luminaire in Fig. 8.14, calculate the average luminance crosswise at 65° vertical angle and compare with the value given in the table.

8.3 Consider two 2 ft × 4 ft recessed 2-lamp louvered fluorescent luminaires with premium magnetic ballasts.

Lamp	Lum. Eff.	BF	P_{in} W	ϕ lm
F40T12	74%	0.88	84	3150
F32T8	80%	0.93	70	2850

Calculate LER and C. Do the LER values meet the benchmark values of Table 8 3?

8.4 A lensed luminaire with luminous dimensions 11 in × 22 in is located at point s in Fig. 8.7. Its intensities in the directions of points A and O are 900 cd and 600 cd, respectively. The dimensions are H = 6 ft, X = 10 ft, Y = 8 ft. What luminances are seen by observers at A and O?

8.5 For the louvered luminaire in Fig. P8.5, find the lamp cut off angle, the louver cut off angle, and the shielding angle. Lamps are one inch in diameter and centered in each 5-in segment.

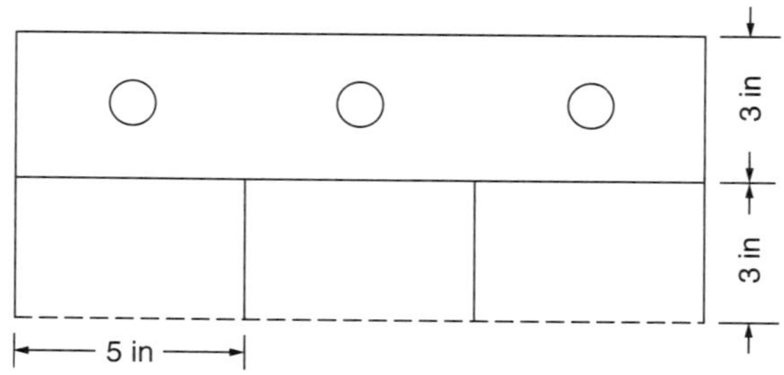

Figure P8.5

8.6 The luminaire in Fig. 8.14 lights an 18 ft × 30 ft room whose ceiling is 6 ft above desk height. What is the minimum number of luminaires required so that the spacing criteria are not exceeded, assuming a spacing of one-half the criteria from luminaire to wall? If 55% of the luminaire lumen output reaches the plane of the desks, what is the average horizontal illuminance?

8.7 For the luminaires in Figs. 8.14, 8.15, 8.16, and P8.1, answer the following questions:
a) Which is the most efficient? What is there in its design that makes this so?
b) Which provides the highest percentage of lumens in the 0–30° zone? The 0–60° zone? What is there in its design that governs this?
c) Which requires the least number of luminaires, for given ceiling dimensions, to not exceed the spacing criteria.
d) Which has the lowest average luminance at 55° vertical angle with crosswise viewing? At 85°? Which luminaires satisfy the IESNA criteria in Table 8.4?
e) Which luminaire is best for placing light on the walls? To answer this question, consider the percent of lamp lumens in the 60–90° zone.

8.8 Two glare sources are 31° up from and 45° to the left and right of the line of sight of an observer. Each source has a luminance of 3085 cd/m^2 and subtends a solid angle of 0.015 steradian at the observer. The field luminance is 137 cd/m^2. Find DGR and VCP.

8.9 Six 4-ft long fluorescent luminaires, assumed perfectly diffusing, are mounted in a row on the ceiling of a room as shown in Fig. P8.9. Each luminaire has a lighted area of 4 ft^2 and a luminance of 1720 cd/m^2. The observer is at point A, directly under the end of the first luminaire. The luminance of wall B is 240 cd/m^2. Calculate the M of each luminaire, the DGR, and the VCP.

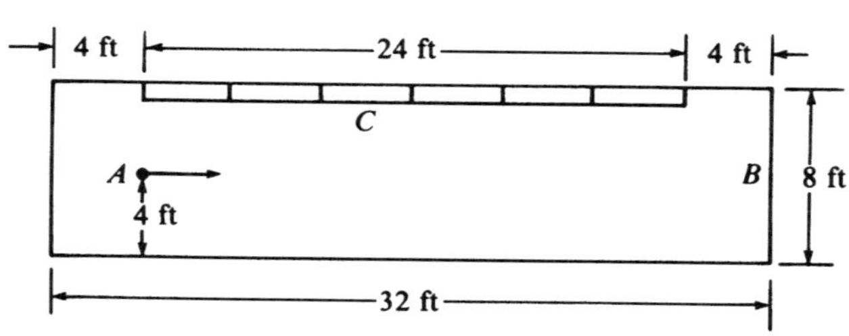

Figure P8.9

8.10 Note the M of source C from Prob. 8.9. Now split source C into two sources of 2 ft^2 each and calculate the DGR. Repeat for source C split into four sources of 1 ft^2 each. Compare the results. Is the DGR calculation procedure consistent?

8.11 A 2 ft by 2 ft luminaire is centered on the ceiling of a room 22 ft directly ahead of and 4 ft up from an observer's line of sight. The luminaire intensity toward the observer is 266 cd and the background luminance is 240 cd/m^2. Find M, DGR, and VCP. What intensity is required to place the luminaire on the boundary between comfort and discomfort?

8.12 For the luminaire in Prob. 8.11, calculate the UGR. Assume $\omega_F = 5$ sr

CHAPTER 9

Note: Assume work-plane height is 30 in unless otherwise stated.

9.1 A school room has the following data:

Dimensions: 28 × 32 × 10 ft
Reflectances: 80%, 50%, 30%
Illuminance level: 50 fc average maintained
Luminaire: #7
Lamp: F32T8
LLF (total): 0.85

Find the number of luminaires needed and the watts per square foot required.

9.2 A room 40 × 30 × 10 ft has the following reflectances:

Ceiling	70%
Walls	50%
Floor	30%
Windows	5%
Chalkboards	15%

Windows occupy 10% of the wall area, chalkboards 5%. The room is lighted with F32T8 U lamps ($\phi_{ini} = 2700$ lm) in luminaire #6. Assuming a recoverable light loss factor of 0.80, how many lamps are needed to provide 40 average maintained footcandles?

9.3 The science classroom in Fig. P9.3 is 32 ft × 40 ft × 9 $\frac{1}{2}$ ft. It has two chalkboards (CB) and one tack board (TB), each 3 $\frac{1}{2}$ ft high. $\rho_{CB} = 30\%$ and $\rho_{TB} = 40\%$. The windows (W) are 5 ft high with $\rho_w = 10\%$.

a) Choose an appropriate illuminance level
b) Assign reasonable room surface reflectances
c) Choose a luminaire and lamp
d) Determine a reasonable light loss factor
e) Find CU and the number of luminaires needed.
f) Calculate the watts per ft^2 required
g) Sketch a lighting layout.

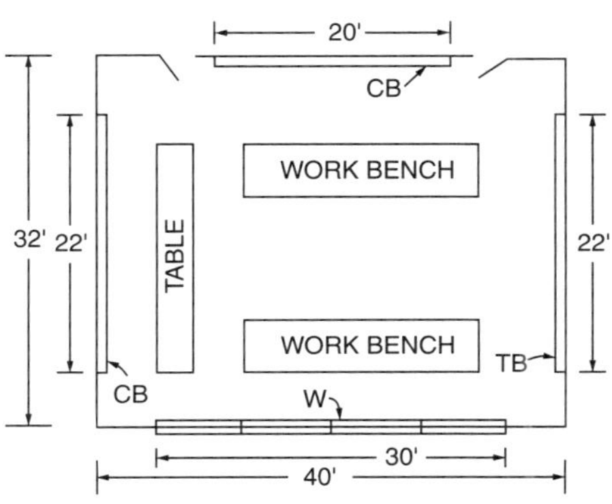

Figure P9.3

9.4 An elementary classroom is shown in Fig. P9.4. It is 20 ft × 30 ft and has a 10 ft ceiling. The room reflectances are ceiling 80%, walls 50%, and floor 30%. The tack board and blackboard are each 60 ft^2. Windows are 4 ft × 5 ft. Tack board reflectance is 30%; chalkboard reflectance is 15%; window reflectance is 10%.

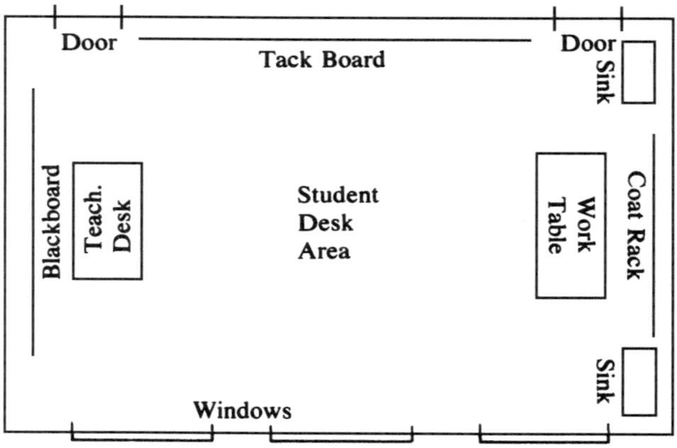

Figure P9.4

Design a lighting installation to provide 35 average maintained footcandles using luminaire #9 suspended 2 ft. Let the recoverable light loss factor be 0.80.

9.5 A lobby area is shown in Fig. P9.5. The entrance is on the far left. The 40 ft wall is entirely glass. The lobby is to be lighted to an average maintained level of 20 fc using F32T8 U lamps (ϕ_{ini} = 2700 lm) and luminaire #8. A recoverable light loss factor of 0.80 is to be assumed. Calculate the number of luminaires required and sketch a possible lighting layout. Assume the work plane is the floor.

Ceiling height	10 ft
Ceiling reflectance	70%
Wall reflectance	50%
Floor reflectance	30%
Glass reflectance	5%

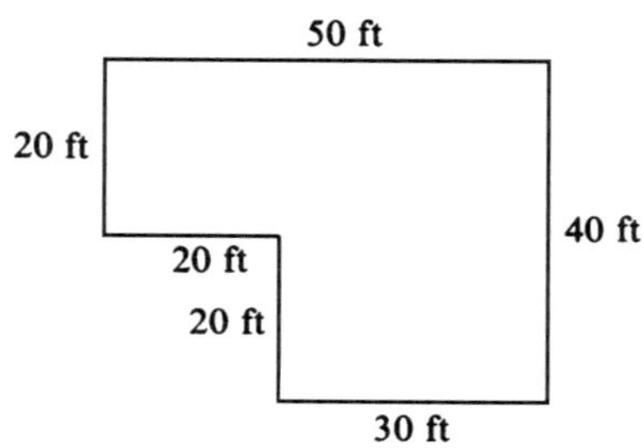

Figure P9.5

9.6 Assume a very long and wide room with $\rho_c = 80\%$, $\rho_w = 50\%$ and $\rho_f = 20\%$ is lighted with luminaire #10. Calculate the CU and verify its value in Table 9.7. Repeat for luminaire #9 and luminaire #5. From these results, comment on the CU's for long wide rooms for direct, direct-indirect and indirect luminaires.

9.7 A 20 ft × 30 ft × 10 ft room is lighted by luminaire #8. Lamps are replaced when there are 5% burnouts. The luminaire is Category IV and cleaning is done every two years. The room atmosphere is clean. Calculate two recoverable light loss factors, based on the old and the new luminaire dirt depreciation data.

9.8 A room with two ceiling heights is shown in Fig. P9.8. It is lighted by luminaire #6, surface-mounted to the 10-ft ceiling and suspended 5 ft from the 15-ft ceiling. Each luminaire contains two F32T8 U lamps, emitting 2700 initial lumens. Room reflectances are $\rho_c = 70\%$, $\rho_w = 50\%$ and $\rho_f = 30\%$. Assume an overall light loss factor of 0.70.

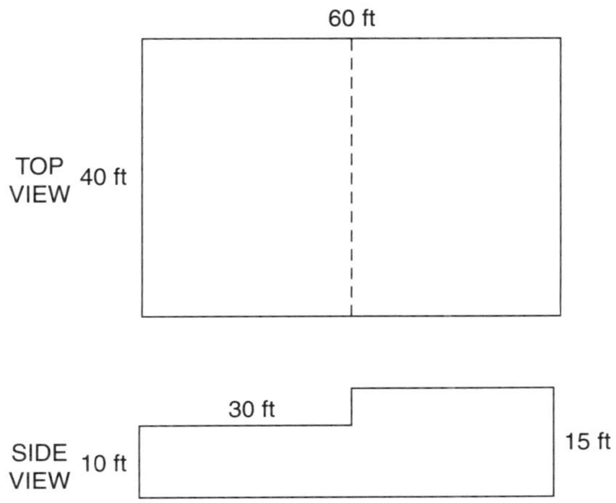

Figure P9.8

The space is an open office area. The work done is cross checking computer printouts of bank ID numbers against bank payments and includes typing and bookkeeping.

a) Select an appropriate average maintained illuminance level.
b) Determine the number of luminaires required.
c) Sketch a layout, showing all spacings.

9.9 A small rotunda is 25 ft in diameter with 12-ft high walls. The top is a hemispherical dome. Reflectances are: walls 30%, dome 70%, floor 20%. The indirect luminaire in Table 9.13 is suspended from the center of the dome to a point 12 ft above the floor. The luminaire houses a 400-W pulse start metal halide lamp.

Calculate the average initial illuminance on the floor.

9.10 A plan view of a 9-ft high room is shown in Fig P9.10. Point X is on a desk top 30 in above the floor. The luminaires (shown by squares) are 2 ft × 2 ft. Luminaire intensity distribution data is given. Find the illuminance on the desk at point X.

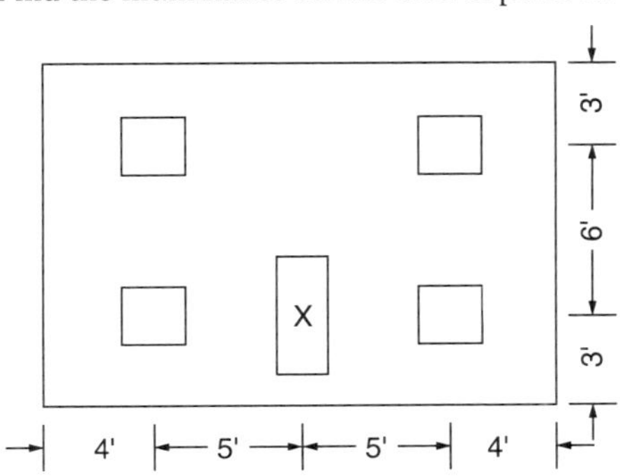

Figure P9.10

Angle	0°	45°	90°
5	1160.	1160.	1160.
15	1135.	1130.	1135.
25	1050.	1060.	1060.
35	940.	935.	940.
45	780.	790.	785.
55	625.	620.	625.
65	430.	442.	425.
75	252.	257.	247.
85	100.	112.	99.
95	52.	58.	52.
105	49.	56.	49.
115	44.	51.	45.
125	39.	45.	39.
135	33.	36.	34.
145	26.	28.	27.
155	17.	20.	19.
165	12.	13.	13.
175	9.	10.	10.

9.11 The old mill building in Fig P9.11 has been converted into a discount house. It is to be lighted to an initial average illuminance of 50 fc using the batwing luminaire in Table 9.13 and 60-W high-output fluorescent lamps. The luminaires are mounted on the undersides of the roof trusses. How many are needed?

$$\rho_c = 70\%$$

$$\rho_w = 30\%$$

$$\rho_f = 20\%$$

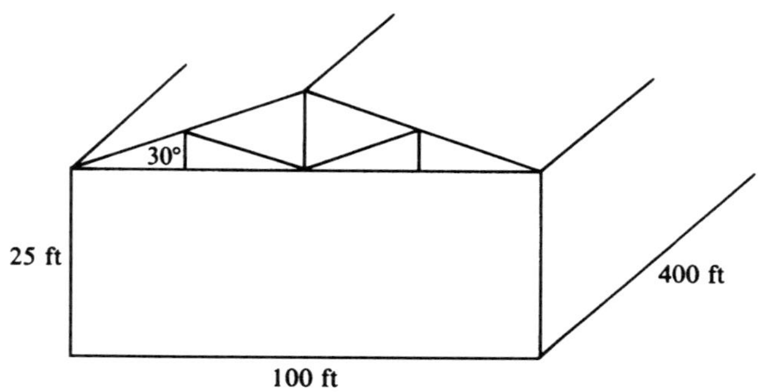

Figure P9.11

9.12 The room shown in Fig. P9.12 has the following data:

Ceiling reflectance	70%
Reflectance of two long walls and beams	55%
Reflectance of two short walls	35%
Floor reflectance	30%

A row of eight F32T8 two-lamp luminaires (#7) is surface mounted on the bottom of each beam. Find the average initial illuminance.

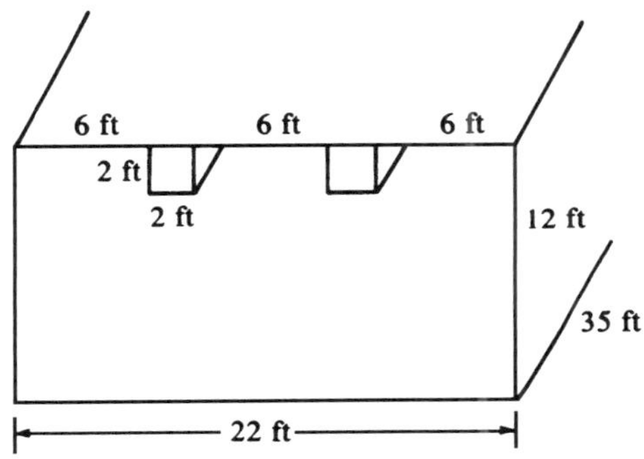

Figure P9.12

9.13 The room shown in Fig. P9.13 is 12 ft deep and has reflectances of

Ceiling	80%
Walls	40%
Floor	30%

It is lighted by six indirect fluorescent luminaires (#9) each containing two F32T8 fluorescent lamps (three rows, two luminaires per row). The luminaires are suspended from the ceiling as shown. Find the average initial illuminance on the work plane.

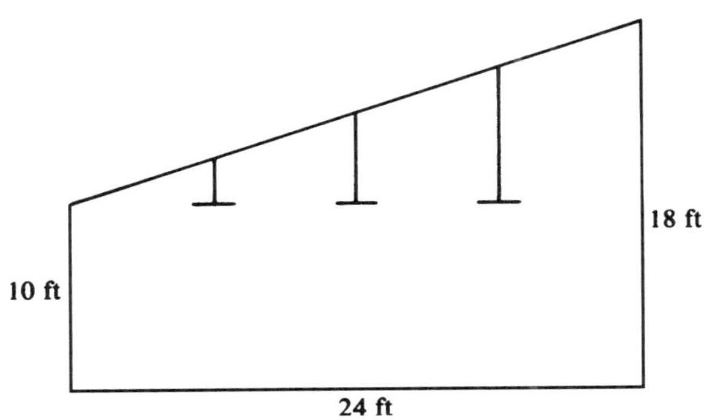

Figure P9.13

9.14 The room shown in Fig. P9.14 is 30 × 60 × 111/2 ft. It is lighted with six rows of 14 luminaires each. Each luminaire has one 34WT12 fluorescent lamp. The ceiling is composed of 72 4 ft × 4 ft coffers, each 18 in deep, surrounded by 12-in wide beams. Reflectances are: beams and coffers 80%, walls 50%, and floor 20%. Tables of coefficients of utilization and exitance coefficients are given for the luminaire. Find the following:

(a) Average initial footcandles on the work plane
(b) Average initial footcandles on the walls
(c) Floor cavity luminous exitance
(d) Ceiling cavity luminous exitance

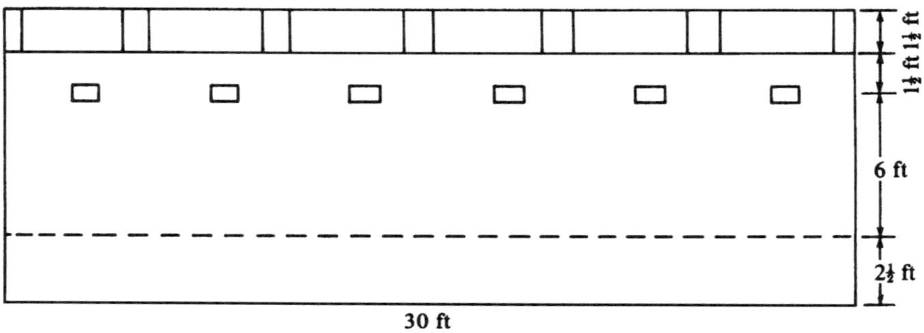

Figure P9.14

ρ_{cc}		70			50	
ρ_w	50	30	10	50	30	10
RCR			% CU for 20% ρ_{fc}			
1	61	59	57	46	44	43
2	54	50	47	41	38	36
3	47	43	39	36	33	30
4	42	37	34	32	29	26
5	37	32	29	28	25	23
			% Wall exitance coefficient			
1	18	10	3	13	8	2
2	17	9	3	12	7	2
3	15	8	2	11	6	2
4	14	7	2	10	6	2
5	13	7	2	10	5	1
			% Ceiling exitance coefficient			
1	56	55	54	39	38	37
2	56	54	52	38	37	37
3	55	53	51	38	37	36
4	55	52	50	38	37	35
5	55	52	50	38	36	35

9.15 In the partitioned space shown in Fig. P9.15, ceiling height is 11 $\frac{1}{2}$ ft and partition height is 6 ft. Reflectances are: ceiling, 80%; walls 50%; floor, 30%. There are two partition reflectances, 50% and 30%. The luminaire is #6 in Table 9.7 and the lamp is a F32T8 U lamp, giving 2700 initial lumens. Desks are 30 in high.

a) Find the CU for each partition reflectance.
b) For each CU find the number of luminaires to give 30 initial average footcandles on the desks in the cubicles. Do a layout.
c) How many luminaires are "visible" to the point × on a desk at the center of a cubicle?

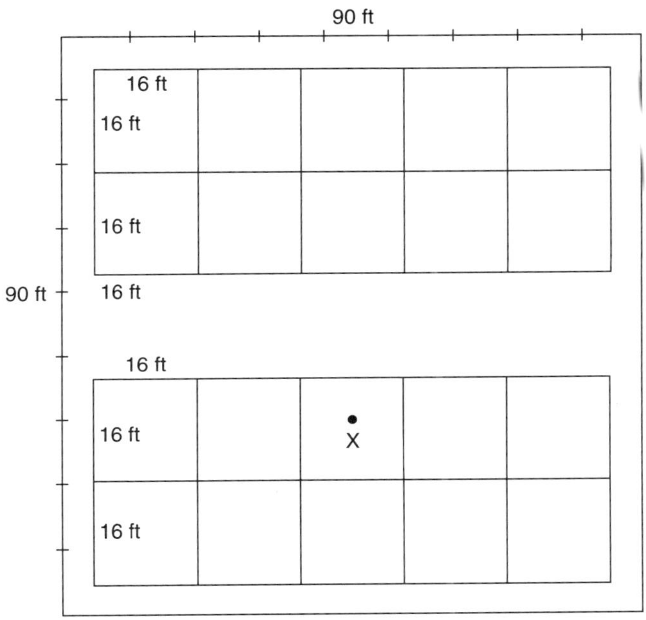

Figure P9.15

9.16 The cross-section of a long semicircular pedestrian walkway is shown in Fig. P9.16. It is to be lighted on each side by a continuous row of single-lamp fluorescent luminaires in a recessed cove 12 ft above the floor. The luminaires are 20% efficient and their reflectors are oriented so that no light is directed downward. Each lamp provides 2800 initial lumens. The entire walkway surface has 50% reflectance. Find the initial average illuminance on the floor.

Note: CUs for indirect luminaires are in direct ratio to their efficiencies.

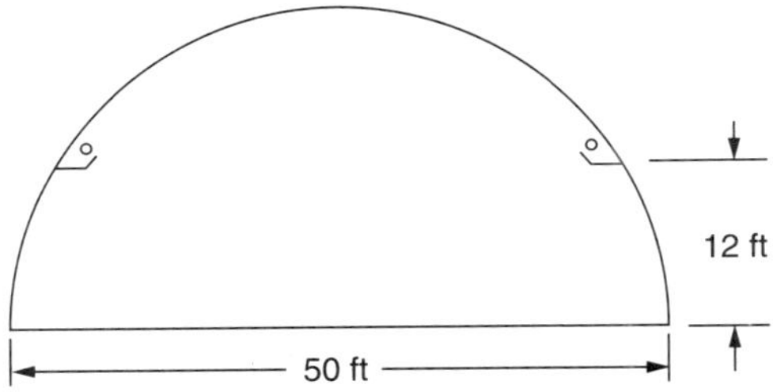

Figure P9.16

9.17 A small auditorium 12 ft high with a 5 ft high blackboard covering the diagonal wall is shown in Fig. P9.17. Reflectances are: ceiling 70%; walls 50%; floor 20%; blackboard 10%. It is to be lighted with luminaire #2 containing a 32-W compact fluorescent lamp. Desired initial average illuminance is 30 fc. Find the number of luminaires required and do a layout.

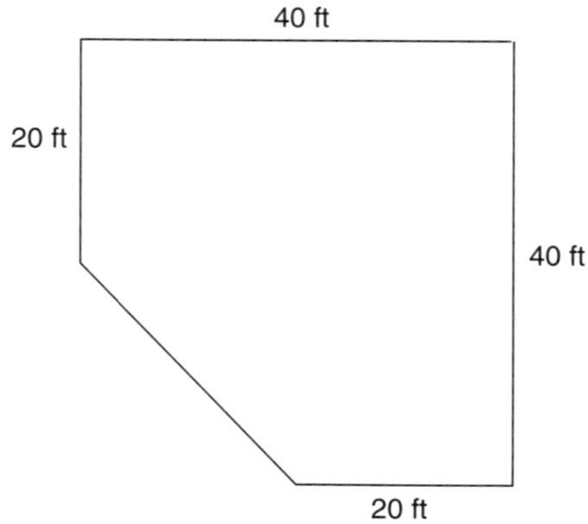

Figure P9.17

CHAPTER 10

10.1 The prism shown in Fig. P10.1 has a refractive index of $\sqrt{2}$. What is the angle between the two light rays as they emerge from the prism? What would this angle be if the prism were immersed in water?

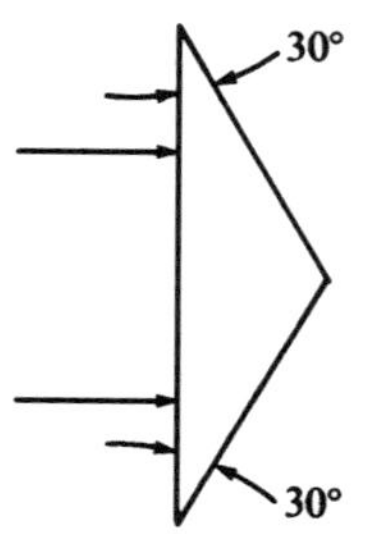

Figure P10.1

10.2 A concave mirror is to form the image of the filament of a PAR lamp on a screen 4 m from the mirror. The filament is 5 mm high and the desired image height is 40 cm. Find
a) Radius of curvature of the mirror.
b) Location of the filament.

10.3 A solid glass hemisphere is placed with its flat face down on a tabletop. The hemisphere has a radius of 10 cm and a refractive index of 1.5. It is lighted from above by a cylindrical beam of light 1 cm in diameter centered on the hemisphere axis. What is the diameter of the circle of light on the table?

10.4 The focal length of a convex lens is to be measured. An object is imaged by the lens on a screen 12 cm from the lens. When the lens is moved 2 cm further from the object, the screen must be moved 2 cm closer to the object to obtain an image. What is the focal length of the lens?

10.5 (a) The near point of an eye is 100 cm in front of the cornea. What type and focal length of lens should be used so that a person can see clearly an object 25 cm in front of the cornea? Draw a ray diagram.

(b) The far point of an eye is 1 m in front of the cornea. What type and focal length of lens should be used so that a person can see clearly an object at infinity? Draw a ray diagram.

10.6 The spotlight system in Fig. P10.6 consists of a circular reflector, two identical plano-convex lenses, a diaphragm, and a double-convex lens. The source is a 400-W tungsten-halogen lamp emitting 7500 lm. Assume it is a uniform point source. The reflector has a reflectance of 90%. Each lens has a transmittance of 92% and an index of refraction of 1.85. The curved lens surfaces are circular and the lenses may be assumed thin. Find the following:

a) Lumens leaving the double-convex lens
b) Radius of each plano-convex lens
c) Diameter of the diaphragm if it is to just allow all of the lumens from the second plano-convex lens to pass through.
d) Radius of the double-convex lens.

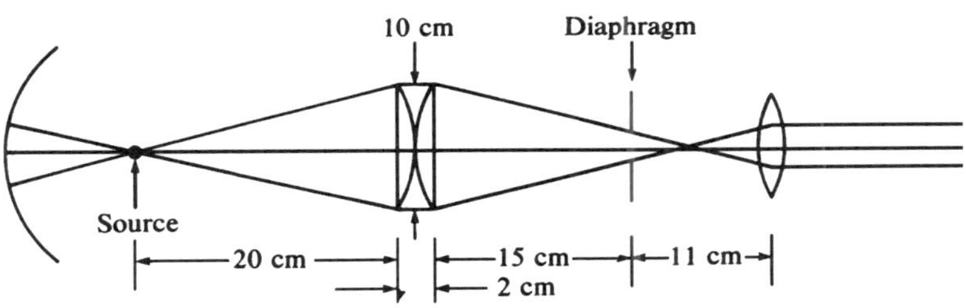

Figure P10.6

10.7 The reflector in Fig. P10.7 consists of a parabola and an ellipse. Its reflectance is 80%. The lamp, assumed to be a uniform point source, has an intensity of 100 cd and is located at the focal point of the parabola and at one of the focal points of the ellipse. Find the average illuminance on the circular sign.

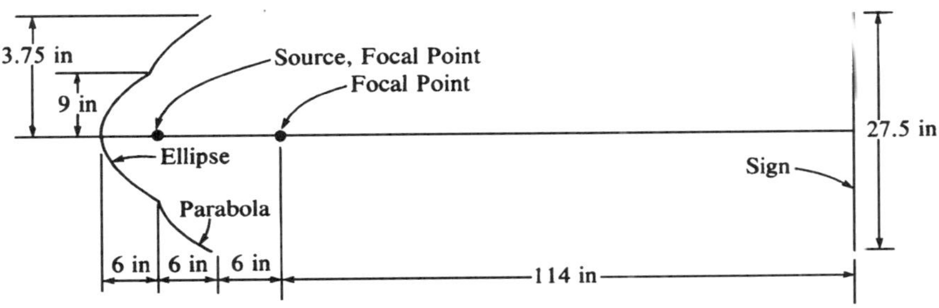

Figure P10.7

10.8 A surgical spotlight is shown in Fig. P10.8. It consists of an elliptical reflector, with a 10° hole in its center, and a small parabolic reflector. Lumens from the source (assumed point and uniform) reflect from both reflectors before becoming a concentrated parallel beam. Find the following:

a) Lumens in the parallel beam
b) Illuminance on the circular plate

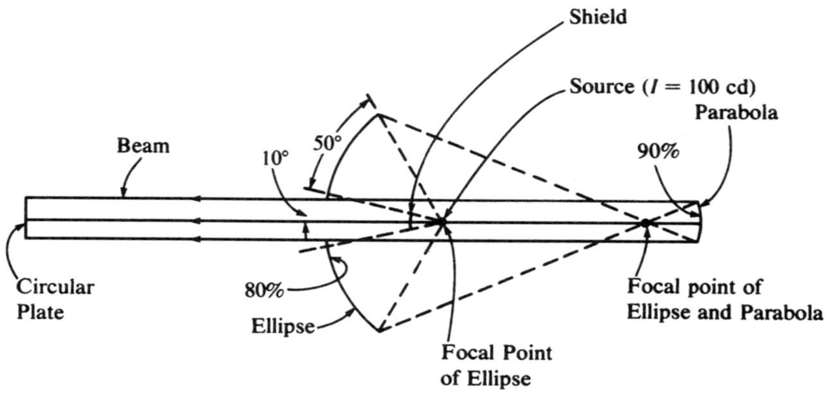

Figure P10.8

10.9 (a) A parabolic reflector is to be designed to fit into a cylindrical enclosure 5 in deep and 4 in in radius. Find the focal length of the parabola and accurately draw its shape.

(b) An elliptical reflector is to be designed to fit into the same enclosure, such that exactly half of the complete ellipse is inside the enclosure. Find the two focal points of the ellipse and accurately draw its shape.

10.10 The lamp whose intensity values are given in Table 10.1 is to be used to design a reflector with a 70° cutoff angle such that the lamp-reflector combination has the following intensity distribution:

Zone, degrees	Lamp reflector I_{mid}, cd
0–10	720
10–20	780
20–30	840
30–40	900
40–50	580
50–60	380
60–70	240

Prepare a table similar to Table 10.1 for this data. Determine the required ρ of the reflector. Assume the pattern in Fig. 10.21a is to be used. Prepare graphs similar to those in Fig. 10.22. Then develop the reflector contour, as in Fig. 10.23.

10.11 An 8-ft long fluorescent luminaire is to be designed with a nearly-parabolic specular cylindrical reflector (ρ = 85%). The luminaire opening is 18 in and is centered 36 in above the ground. The underside of the top of the luminaire is black. The lamp is an 8-ft T12 high output fluorescent rated at 15,000 initial lumens. The horizontal illuminance at P, 50 ft away, is to be maximized and as much of the distance from 0 to P as possible is to be lighted, but there is to be no light projected above the horizontal.

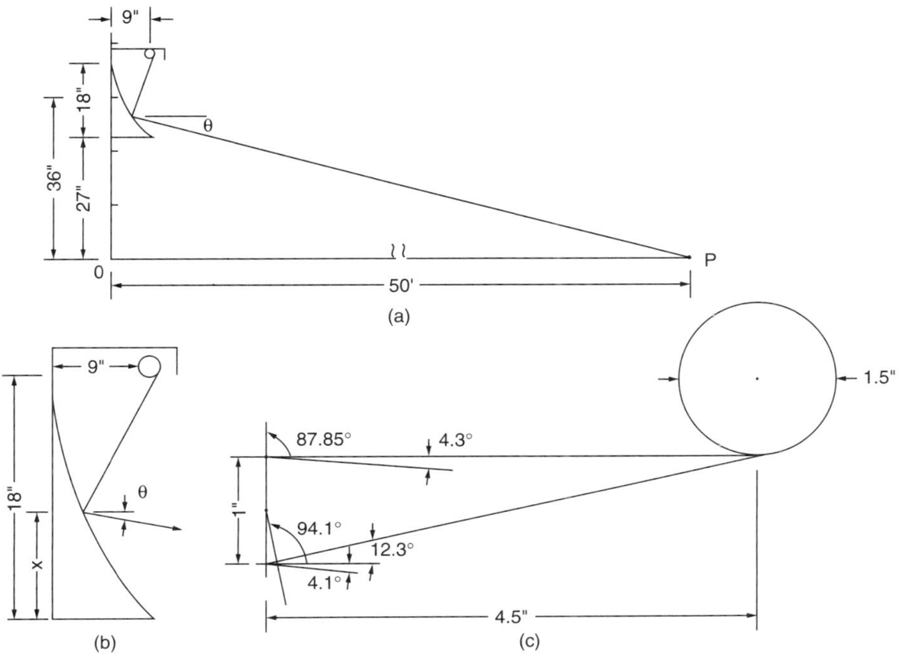

Figure P10.11

An enlargement of the lamp-reflector is shown in Fig. P10.11b. A useful starting point is to plot θ vs x, where θ is the angle with the horizontal of a ray from the right edge of the lamp to the reflector to P. Then any ray from the left edge of the lamp to the same point on the reflector will hit the ground between O and P.

Use the θ vs x graph to develop the reflector contour as a series of straight line segments, the first two of which are shown in Fig. P10.11c. Then find the luminance of the luminaire opening, the horizontal illuminance at P from a long row of luminaires and the width of the lighted band between O and P, excluding direct light from the lamp.

10.12 A wall-washing luminaire is recessed in a ceiling near a wall. The luminaire is 4 ft long and contains one 2800-lumen fluorescent lamp and a specular reflector ($\rho = 0.85$). It has a 48-in by 12-in opening. Assuming that the lamp is a Lambertian cylinder and that the reflector aperture is fully flashed, what is the maximum intensity that can be achieved?

10.13 A two-sided translucent sign is 20 ft high, 48 ft long and $1\frac{1}{2}$ ft thick. The sides have 35% transmittance and 40% reflectance. The top, bottom, and ends are opaque. The sign will be lighted internally by horizontal rows of 2800-lumen T8 fluorescent lamps. Calculate the number of lamps to provide a sign luminance of 40 cd/ft^2.

10.14 Two slits are spaced 0.3 mm apart and 50 cm from a screen. Find the distance on the screen between the second and third dark lines of the interference pattern when the slits are illuminated by 600 nm radiation.

10.15 What is the thinnest film of refractive index 1.20 in which constructive interference of violet light (400 nm) can occur by reflection? What is the thinnest film for destructive interference? What index of refraction should the glass have if the destructive interference is to produce near-annulment?

10.16 What wavelength of light will produce an intensity maximum at an angle of 20° when passed through a diffraction grating having 6000 lines per centimeter?

10.17 Radiation from a mercury lamp is to be passed through a diffraction grating. How many lines per centimeter should the grating have so that the first- and second-order visible spectra of mercury are separated by 5°? The visible mercury lines are 405, 454, 546, and 579 nm.

10.18 A beam of linearly polarized light at 589 nm is incident on a calcite crystal at right angles to the optic axis. Find the wavelengths of the ordinary and extraordinary rays in the crystal.

10.19 The phase difference δ between the E and O rays after traversing a crystal plate between the polarizer and analyzer is given by

$$\delta = \frac{2\pi}{\lambda}(n_o - n_e)t$$

where λ is the wavelength in air and t is the crystal thickness. What is the minimum thickness of a quarter-wave calcite plate for 400-nm radiation? Repeat for 500 nm, 600 nm, and 700 nm.

10.20 A beam of linearly polarized light strikes a calcite crystal at an angle of 60° with the optic axis. What are the ratios of the amplitudes of the two refracted beams and of their intensities?

CHAPTER 11

11.1 What aiming angle of a floodlight, with respect to the vertical, will produce the maximum horizontal illuminance at a given distance from its pole base? Assume the maximum candela point of the floodlight is along the aiming line.

11.2 For an aiming distance of 0.5 MH, verify the values of % utilization given in Fig. 11.6 for distances from the pole of 0.5 MH, 1 MH, and 2 MH.

11.3 The floodlight in Fig. 11.2 is aimed at a point 1.5 MH horizontally from the base of the pole.
(a) What percent of the total lamp lumens lie in the horizontal strip from the pole out to 1 MH?
(b) If the mounting height is 40 ft, find the initial horizontal illuminance at a point 40 ft out from the pole in the direction of the aiming line and 23 ft to the right of the plane of the aiming line and the pole.

11.4 A portion of a lumen distribution diagram for a floodlight is shown in Fig. P11.4. For a 50 ft mounting height, sketch and dimension the zone on the ground containing the 500 lm, assuming the floodlight is aimed straight down.

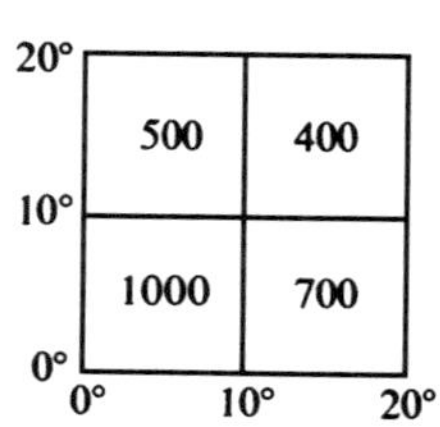

Figure P11.4

11.5 In the floodlighting design in Ex. 11.3, assume the mounting height is increased to 70 ft but that the aiming angle is unchanged.
(a) Find the new pole location.
(b) Calculate the new bounding vertical angles and preliminary coefficient of utilization.
(c) Determine BF, AAF and the average maintained illuminance on the field. Would you recommend the change to 70 ft?

11.6 Two of the floodlights whose data appear in Fig. 11.10 are used to light a large billboard, as shown in Fig. P11.6. The floodlights are located at points P and Q and aimed at points O and R on the billboard. The lines FP and QG are horizontal and perpendicular to the billboard. Find the average illuminance on the billboard.

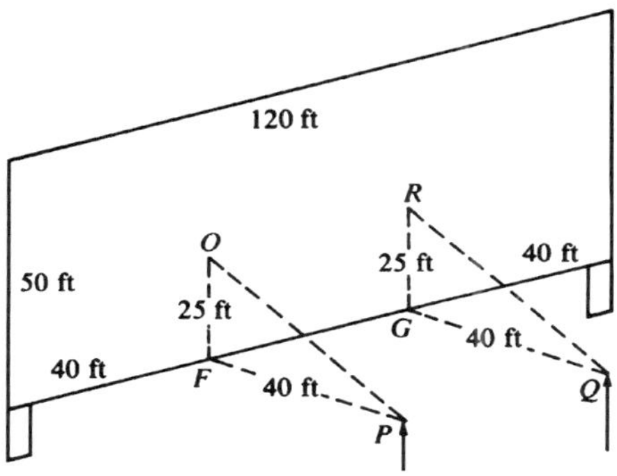

Figure P11.6

11.7 It is desired to light a 30-ft wide road to an average maintained illuminance of 1.4 fc using the luminaire in Fig. 11.20. The mounting height is 30 ft and the luminaires are mounted directly over the curb. The lamp is a 250-W high-pressure sodium. Assume the light loss factor is 0.75. Find the maximum spacing between luminaires.

11.8 The luminaire in Fig. 11.24, housing a 250-W high-pressure sodium lamp, is mounted atop a 40-ft pole at A in Fig. P11.8. What is the initial horizontal illuminance at B?

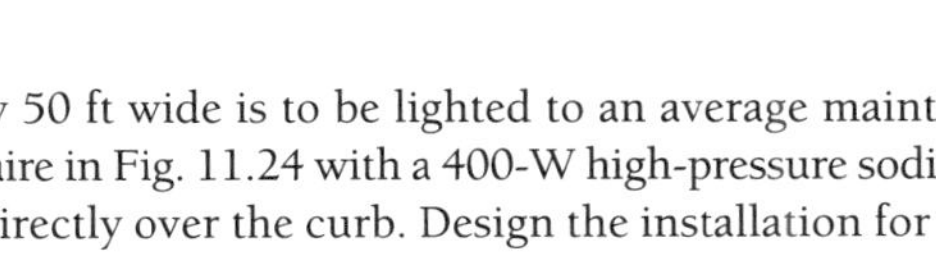

Figure P11.8

11.9 A commercial artery 50 ft wide is to be lighted to an average maintained illuminance of 2 fc using the luminaire in Fig. 11.24 with a 400-W high-pressure sodium lamp. Luminaires are to be mounted directly over the curb. Design the installation for $E_{ave}/E_{min} \leq 3$ and justify that this criterion is achieved.

11.10 A parking lot 240 × 480 ft is to be lighted to an average maintained horizontal illuminance of 1 fc using the luminaire in Fig. 11.24. Two luminaires are to be mounted back-to-back on each pole. Select mounting height, the number and location of poles, and the wattage size of high-pressure sodium lamps. Calculate CU and LLF and verify that 1 fc is achieved.

11.11 Ten illuminance readings in fc are taken on a small playing field: 40, 38, 22, 37, 21, 24, 29, 28, 31, 30. Find:

$$CV,\ E_{max}/E_{min},\ E_{ave}/E_{min},\ E_{max}/E_{ave}$$

11.12 The luminaire in Fig. 11.24 houses a 250-W pulse start metal halide lamp. Pole height is 40 ft and the pole is mounted at the curb of a commercial street.

a) What is the initial horizontal illuminance on the street side 80 ft laterally and 50 ft outward from the curb?
b) What is the initial horizontal illuminance on the curb directly under the luminaire?
c) Assume the street is 60 ft wide and the sidewalk 20 ft wide. What is the percent utilization?
d) What is the overall luminaire efficiency?
e) Within what range does the luminaire's field angle lie?

11.13 A shipyard unloading dock adjacent to the water is shown. It is to be lighted by the floodlight in Fig. 11.2. No poles can be located on the loading area, along the water, or along the left boundary (///).

Design the installation to provide an average maintained illuminance of 5 fc. Include:

- pole location and number
- number of floodlights
- lamp, type and wattage
- mounting height
- aiming points

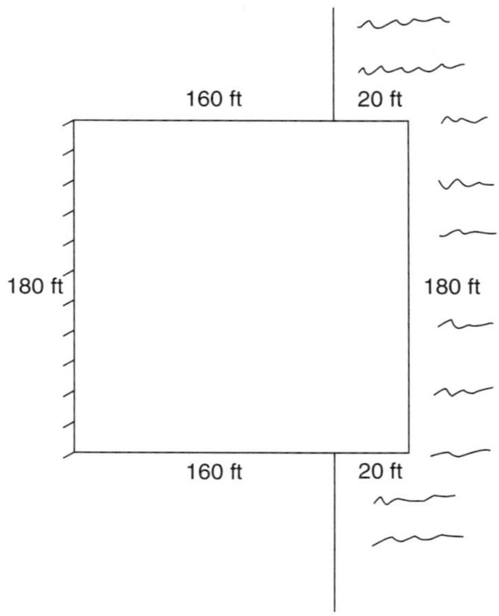

Figure P11.13

11.14 An 800-ft × 400-ft parking lot is shown in Fig. P11.14. It holds 1040 cars. The parking bays are 60 ft wide. Each parking space has an area of 10 ft × 20 ft = 200 ft^2. Driving lanes are 20 ft wide except in the center of the lot, where they are 40 ft wide. The lot is outdoors, not in a parking garage.

Constraints on the lighting design are that poles cannot be located in parking or driving lanes. The luminaire to be used is that in Fig. 11.24. Design the lighting:

- Select an illuminance level
- Select a uniformity criterion
- Select pole locations and mounting height
- Determine the number of poles
- Select a lamp wattage
- Determine the number of luminaires and their orientation.

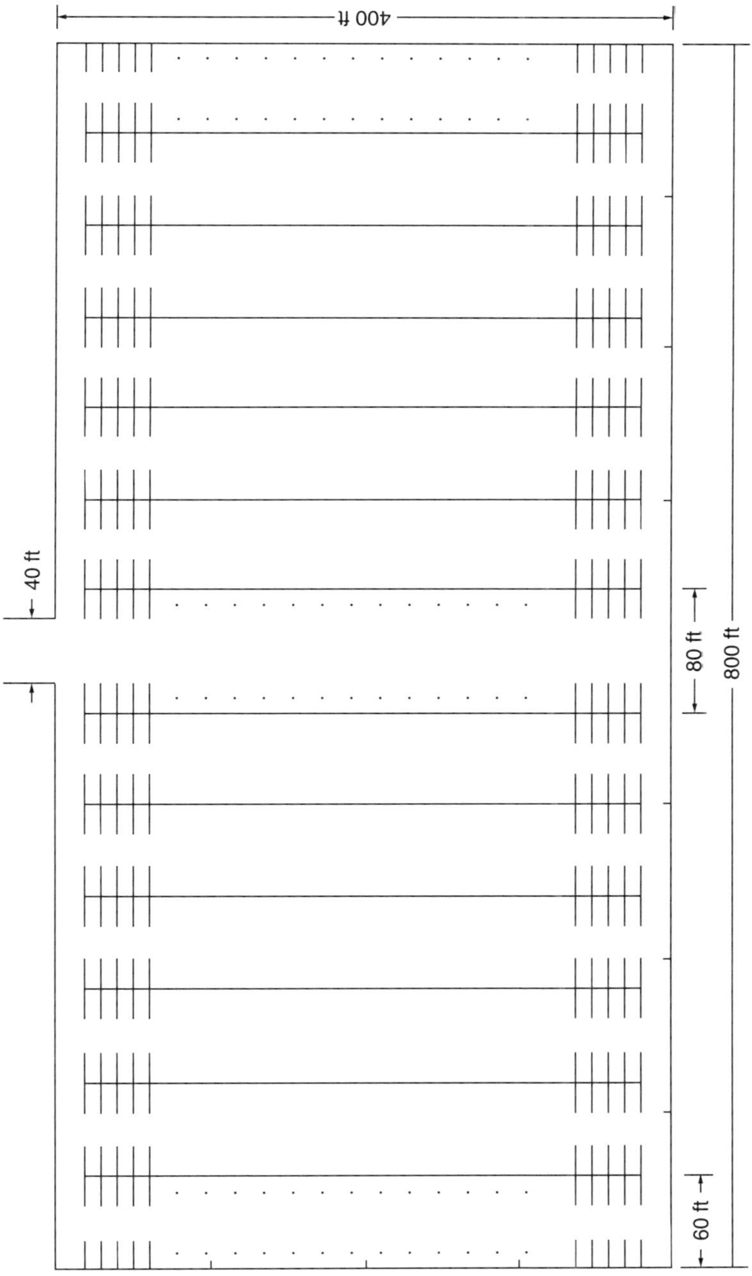

Figure P11.14

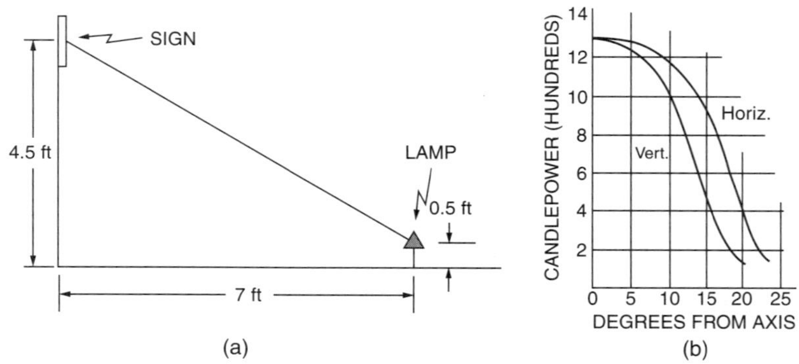

Figure P11.15

11.15 A white sign with black letters is shown in Fig. P11.15a. It is located at the entrance to a housing development. The letters on the sign have a reflectance of 10%; the background 70%.

The sign is to be lighted by 50PAR36 WFL lamps staked in the ground as shown. Intensity distribution data are given in Fig. P11.15b.

A design level of 15 fc on the sign is desired. Determine the number of lamps required and their spacing to achieve reasonable uniformity and minimum spill light.

CHAPTER 12

12.1 Values of solar irradiation in W/(cm^2 • μm) on a horizontal surface for a clear day with the sun at zenith are given in the table below. Find the following:

a) kW/m^2 incident on the surface

b) fc on the surface

c) luminous efficacy of the sun under these conditions.

λ	E_λ	λ	E_λ	λ	E_λ
0.30	0	0.65	0.132	1.00	0.065
0.35	0.042	0.70	0.125	1.20	0.045
0.40	0.080	0.75	0.108	1.40	0.032
0.45	0.132	0.80	0.096	1.60	0.024
0.50	0.147	0.85	0.085	2.00	0.012
0.55	0.139	0.90	0.075	2.20	0.007
0.60	0.139	0.95	0.068	2.40	0.006

12.2 The surface temperature of the sun is 5700 K and the sun's diameter is 1.4×10^9 m. What is the radiation from the sun in watts, assuming the sun is a black body?

12.3 Let the solar altitude be 50° and the solar azimuth −30°. For a window facing southeast find:

- solar-window azimuth angle
- angle of incidence
- profile angle.

12.4 For the solar altitude and azimuth in Prob. 12.3, find E_{dn}, E_h, and E_v.

12.5 Find the solar altitude and azimuth angles at 9:30 am EST on April 21 at 80° W longitude and 40° N latitude.

12.6 For the conditions in Prob. 12.5, find the horizontal illuminance and the illuminances on vertical surfaces facing east and south produced by the sun on a clear day.

12.7 For a solar altitude of 43°, calculate the horizontal and vertical illuminances caused by a CIE overcast sky. Repeat for the horizontal illuminance produced by a CIE clear sky.

12.8 A classroom is shown in Fig. P12.8. It has three windows with dimensions and spacings given. The windows have a transmittance of 85%. Assume an overcast sky with an average luminance of $1000/\pi$ cd/ft^2. Find the direct component of horizontal illuminance at point P, 10 ft perpendicularly from the center window at sill height.

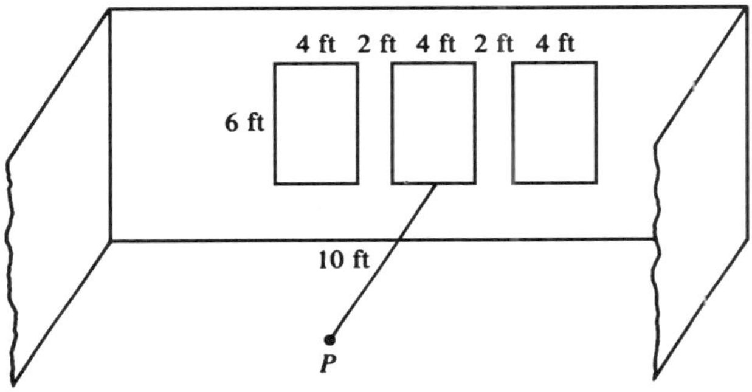

Figure P12.8

12.9 It is desired to calculate the horizontal illuminance at point P on the floor of the room in Fig. P12.9 due to light coming in through the window.

The illuminance on the outside of the window is made up of three components:

- Direct sunlight
- Clear skylight
- Reflected ground light.

Assume it is noon at 46° north latitude on June 21. The direct illuminance from the sun on a plane perpendicular to the sun is 9400 fc. The clear sky illuminance on the window is 1400 fc. The ground area around the window is grass with a reflectance of 6%. The window faces south and has a transmittance of 85%.

Find the following:

(a) Luminance of the inside of the window in candelas per square inch
(b) Horizontal illuminance at point P on the floor in footcandles.

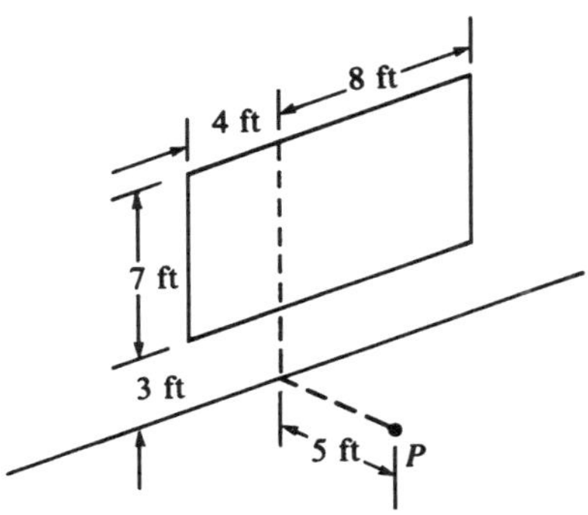

Figure P12.9

12.10 A circular horizontal field is completely surrounded by trees, whose tops are at an angle of 20° with the horizontal at point P. The sky is CIE overcast and the sun is at an altitude of 30°. Find the horizontal illuminance at P. Neglect reflections from the trees.

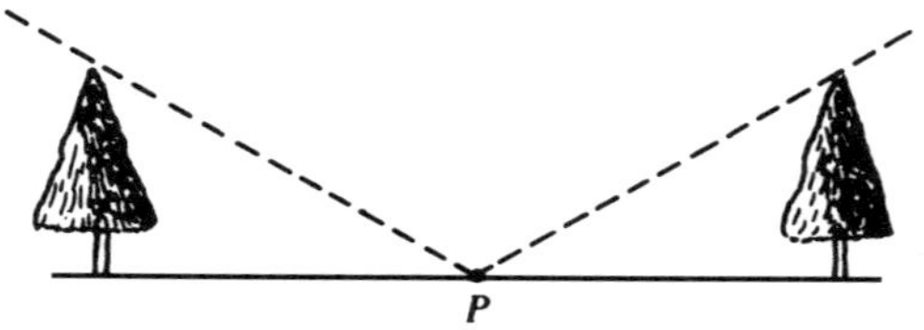

Figure P12.10

12.11 An astrodome with a perfectly diffusing plastic skylight having a transmittance of 90% is shown in Fig. P12.11. It is an overcast day and the horizontal illuminance on the skylight is given by $E_H = 400 \cos\theta$ fc.

Assume second base is at the center of the astrodome. Find the horizontal illuminance on second base and the vertical illuminance on a runner standing on second base. Neglect reflections.

What would these illuminances become if the horizontal illuminance on the dome were constant at 400 fc?

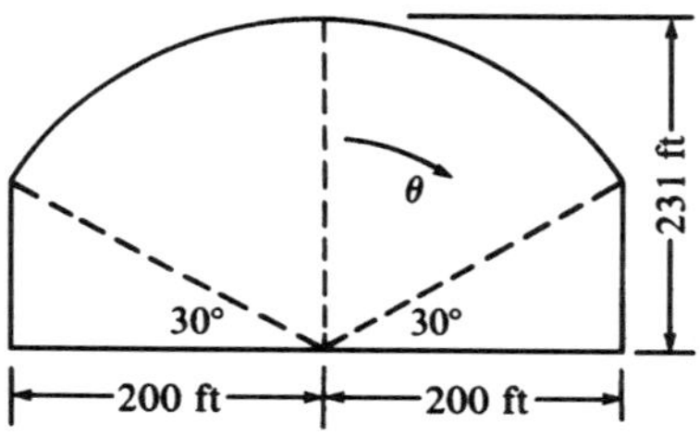

Figure P12.11

12.12 The luminance distribution for an overcast sky at a solar altitude of 40° is shown below, as a function of the polar angle θ, where a relative L = 1.0 corresponds to 1850/π cd/ft². Calculate the horizontal illuminance on the ground. What is the luminance of a uniform sky that would product the same illuminance?

Δθ, degrees	Relative L
0–10	1.00
10–40	0.95
40–50	0.85
50–60	0.75
60–70	0.65
70–80	0.55
80–85	0.45
85–90	0.40

12.13 A wall has a 30 ft by 8 ft window with sill height of 3 ft. Nearest the window is a grass strip 10 ft wide, beyond which is a gravel playground. The site is at 42° north latitude and the window faces south. It is noon on September 21 on a clear day. What is the total illuminance on the center of the window (7 ft above the ground)?

12.14 A large office is 100 × 40 ft with a 10-ft ceiling. Its reflectances are: ceiling 80%, walls 50%, and floor 20%. It is to be lighted by 4 × 4 ft double-domed (transparent over translucent) skylights whose transmittances are:

	τ_d	τ_D
Transparent	0.75	0.90
Translucent	0.40	0.50

Each skylight has a 15-in high curb, a 3-in wide retaining frame, and a 60% well reflectance. Depreciation factors are RSDD = 0.95, SDD = 0.80. The sky is CIE overcast.

Find the number of skylights to provide an average maintained horizontal illuminance of 30 fc at 11 a.m. at 42° north latitude on March and September 21. What average maintained horizontal footcandles will the installation provide if the sky is CIE clear?

12.15 A warehouse lighted by skylights is shown in Fig. P12.15. There are 10 translucent flat skylights on each side of the roof. Their transmittance is 40%, and their underside reflectance is 45%. The exterior illuminance on the skylights on the right side of the roof is 5000 fc, that on the left side 2500 fc. SDD = 0.75 and RSDD = 0.95. Room surface reflectances are ceiling 70%, walls 30%, and floor 20%. Find the average maintained horizontal illuminance on the work plane.

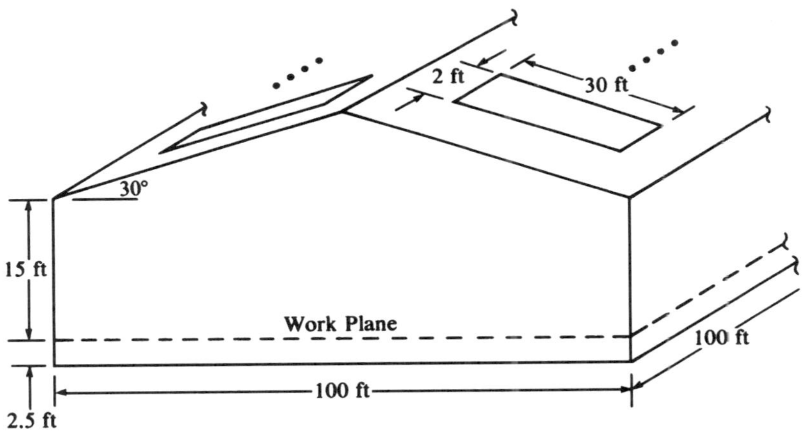

Figure P12.15

12.16 Using the lumen method of sidelighting, find the horizontal illuminances under the following constraints:

Clear day, noon, Sept. 21, 46° north latitude

Room:	40 × 30 × 12 ft
	$\rho_{cc} = 70\%$, $\rho_w = 50\%$, $\rho_{fc} = 30\%$
Office area:	WDD = 0.83, RSDD = 0.95
Ground surface:	vegetation
Windows:	9 ft × 30 ft
	$\frac{1}{4}$-in clear glass
	Sill height: 3 ft
	Direction: Facing east
	Controls: None
	Frame: 20% opening

12.17 In Fig. 12.24, the following values apply: H = 20 ft, D = 50 ft, q = 8 ft, h = 6 ft, x = 20 ft, y = 40 ft, z = 10 ft. The window is of clear glass with a width of 16 ft. Room reflectances are ceiling 70%, walls 50%, and floor 30%. Find the overall daylight factor at P produced by the CIE overcast sky.

12.18 For the situation in Prob. 12.17, find the average daylight factor.

12.19 For a daylight factor of 3 to 5% on a CIE overcast day with the sun at 30° solar altitude, what range of interior illuminances will be achieved? For the same range of interior illuminances, what daylight factor range would be required on a clear day, including direct sun?

CHAPTER 13

13.1 The diagram in Fig. P13.1 shows a proposed bowling alley lighting installation (bowler at left, pins at right). The circles represent 4-ft, $1\frac{1}{2}$-in diameter, 40-W fluorescent lamps, centered over the alley. Each lamp has a luminance of 4.2 cd/in^2. Neglecting ceiling and floor reflectances, find the horizontal illuminance at point P.

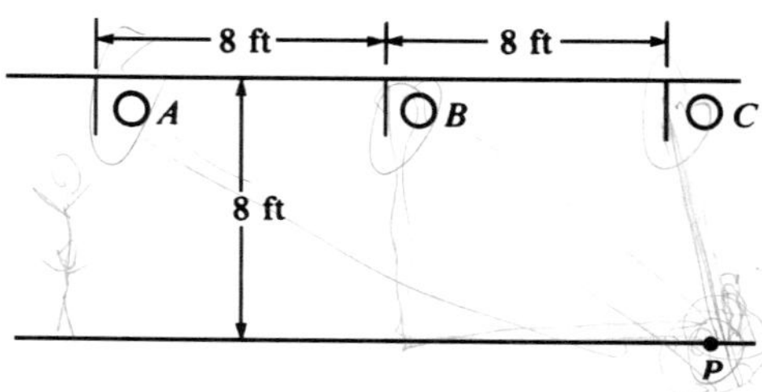

Figure P13.1

13.2 A room has three rows of fluorescent lamps on its ceiling. The rows are each 32 ft long and are separated by 8 ft. Find the illuminance on a desk top 6 ft below the ceiling and located

(a) Beneath the middle of the middle row
(b) Beneath one end of the middle row
(c) Beneath the middle of one outside row

Assume that the lamps are 1 1/2 in in diameter with a luminance of 4 cd/in^2.

13.3 A top view of two rectangular luminaires is shown in Fig. P13.3. They are perfectly diffusing with $L = 630/\pi$ cd/ft^2. Find the horizontal illuminance at P, 10 ft below the luminaires.

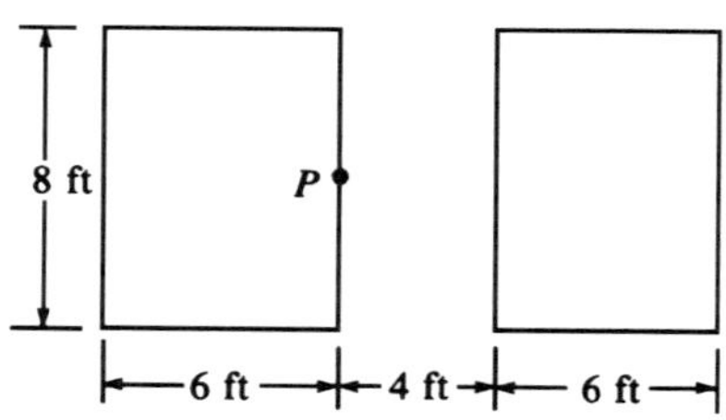

Figure P13.3

Compare your answer with that obtained by considering each rectangle as a point source located at its center.

13.4 A room 40 × 20-ft with a 10-ft ceiling height has a wall exitance of 30 lm/ft^2 and a ceiling exitance of 40 lm/ft^2. All surfaces are perfectly diffusing. Find the horizontal illuminance at the center of the floor.

13.5 Two library book stacks are shown in Fig. P13.5. Each is 10 ft long and 6 ft high and they are separated by 6 ft. The books on the shelves are lighted by a 10-ft long ceiling-mounted strip luminaire 6 in wide and centered between the shelves. The luminaire is perfectly diffusing with L = 4 cd/in^2. Find the vertical and horizontal illuminances at P.

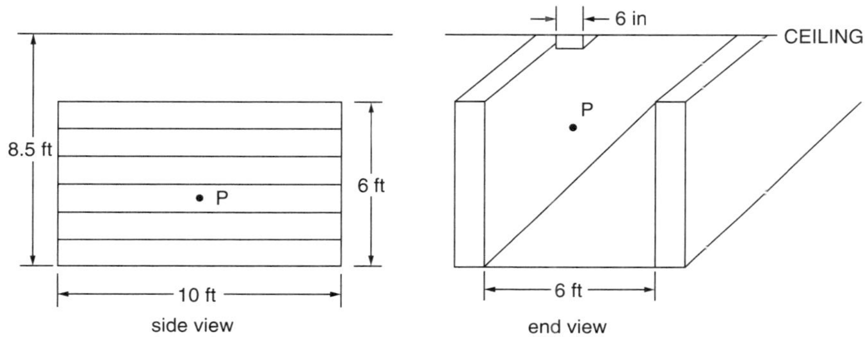

Figure P13.5

13.6 A luminaire consisting of two lighted rectangular panels is shown in Fig. P13.6. Assume that $\ell \gg 4$ ft and that the panels are perfectly diffusing. Find c_{21} (from panels to P) and find the horizontal illuminance at P if L = 0.5 cd/in^2.

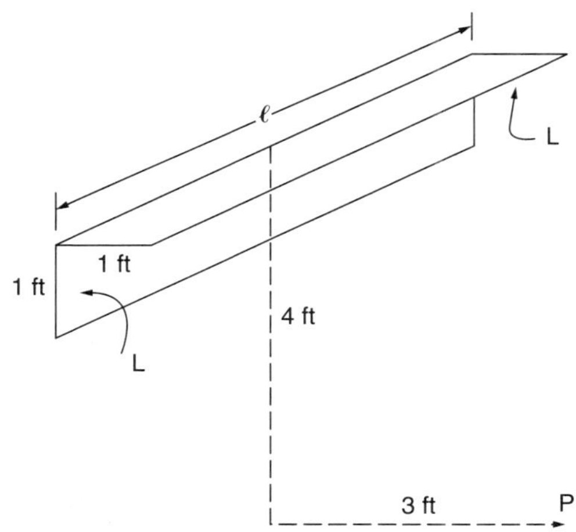

Figure P13.6

13.7 Using Eq. (13.31), show that the configuration factor in Fig. P13.7 at point P on a horizontal plane caused by a perfectly diffusing surface A that is generated by an infinitely long line moving parallel to itself between the limits θ and ϕ is given by $c = \frac{1}{2}(\cos\theta - \cos\phi)$.

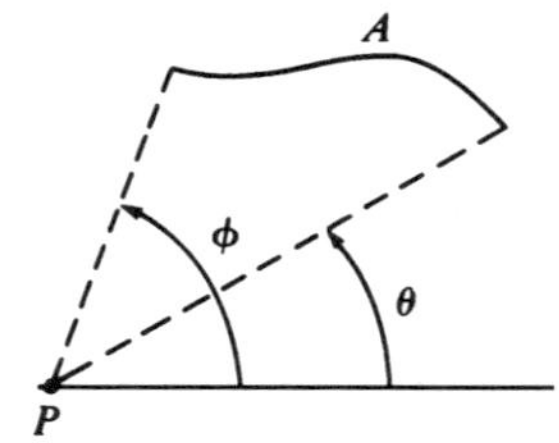

Figure P13.7

13.8 Show that the configuration factor in Fig. P13.18 at point P on a horizontal plane caused by a perfectly diffusing circular disc source of radius r is given by Eq. (13.34), repeated here

$$c=\frac{1-\cos\gamma}{2}=\frac{1}{2}\left[1-\frac{D^2+S^2-R^2}{\sqrt{(D^2+S^2+R^2)^2-4R^2S^2}}\right]$$

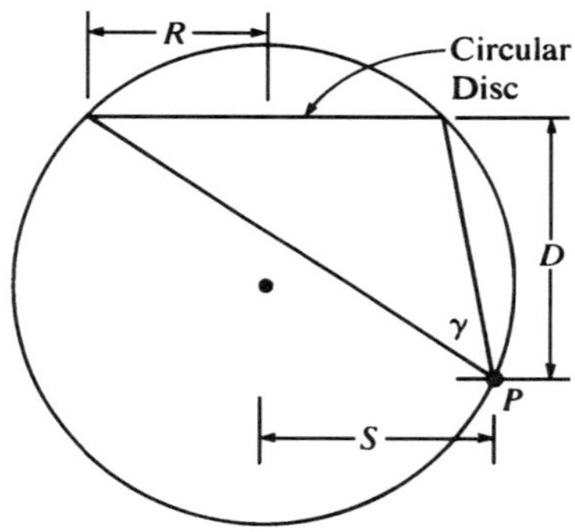

Figure P13.8

13.9 A perfectly diffusing skylight is shown in Fig. P13.9.

$$L_{sky}=1200\,\frac{cd}{ft^2},\quad \tau=50\%$$

Find the configuration factor from the skylight to P and the illuminance at P.

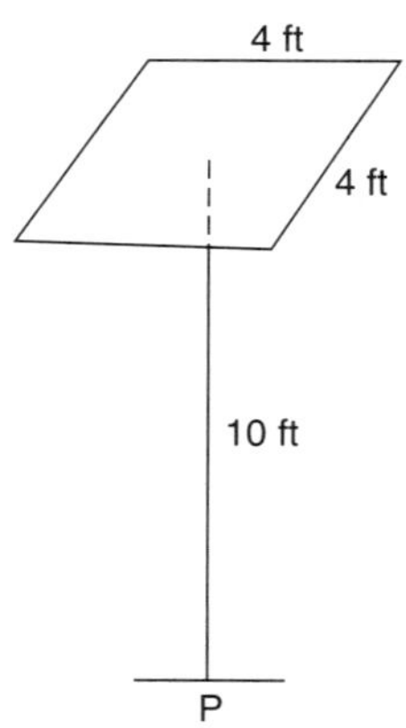

Figure P13.9

13.10 Use Eq. (13.31) to find an exact value for the configuration factor for the situation in Fig. 13.15. What percent error results from using the configuration factor method for strip sources in finding this configuration factor?

13.11 Let the point P in Fig. 13.17 be located such that the 6 ft and 11 ft distances become 8 ft and 16 ft, respectively. What must the maximum intensity I_0 become so that the horizontal illuminance at P is unchanged?

13.12 The room in Fig. P13.12 is 32 × 20 ft and has an 8-ft ceiling height. It is lighted by two perfectly diffusing 12 ft × 8 ft ceiling elements with a luminous exitance of 300 lm/ft^2. Use the configuration factor method for rectangular sources to find the horizontal illuminance at point P on the floor.

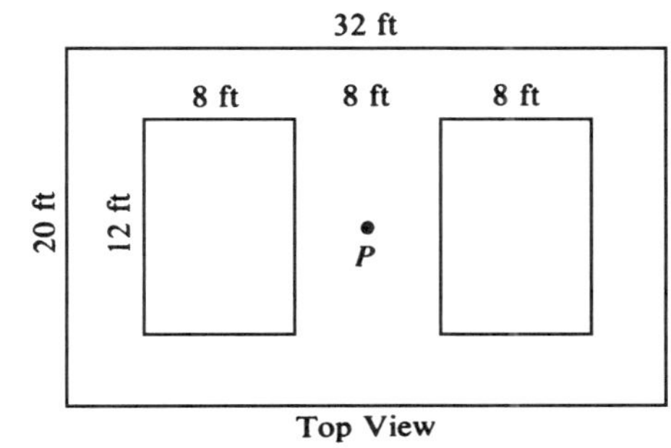

Figure P13.12

13.13 A room 36 ft × 24 ft with a 10-ft ceiling height is lighted by three rows of fluorescent luminaires, with eight luminaires per row. The direct luminaire in Table 9.13 is used and houses two 2800-lumen fluorescent lamps. Room reflectances are $\rho_c = 80\%$, $\rho_w = 50\%$, and $\rho_f = 20\%$. Calculate the reflected component of initial horizontal illuminance on the work plane at a point 10.8 ft from a short wall and 9.6 ft from a long wall.

13.14 Calculate the zone multiplier for $\alpha = 30$ to $40°$ and $\beta = 25°$ and verify its value in Table 13.6. Use the result to calculate the horizontal component of illuminance produced by an element of the luminaire in this zone whose intensity data is given in Table 13.5.

13.15 Using the data in Table 13.7, find the intensity at $\psi = 30°$ and $\theta = 40°$.

(a) Using the parabolic interpolation procedure.

(b) Using the fourth-power equation procedure.

Compare the results and comment.

13.16 Three rows of the prismatic wraparound luminaires of Table 13.4 are shown in Fig. P13.16. Find the horizontal illuminance at point P.

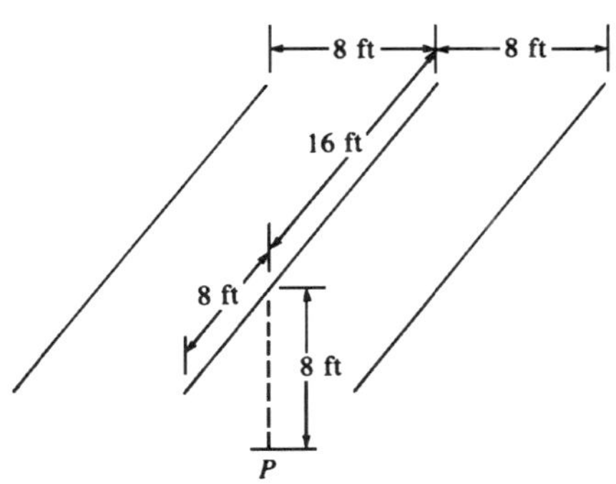

Figure P13.16

13.17 Intensity distribution data in candelas are given for a batwing fluorescent unit, where I_{90° represents intensity values in a vertical plane perpendicular to the fixture axis, I_{0° parallel to the fixture axis, and I_{45° in between the two.

Four of these 4-ft units are mounted as shown in Fig. P13.17 on a ceiling which is 7 ft above the work plane.

Subdivide the lighting units into appropriate lengths and calculate the horizontal illuminance on the task. Use linear interpolation to obtain intensity values.

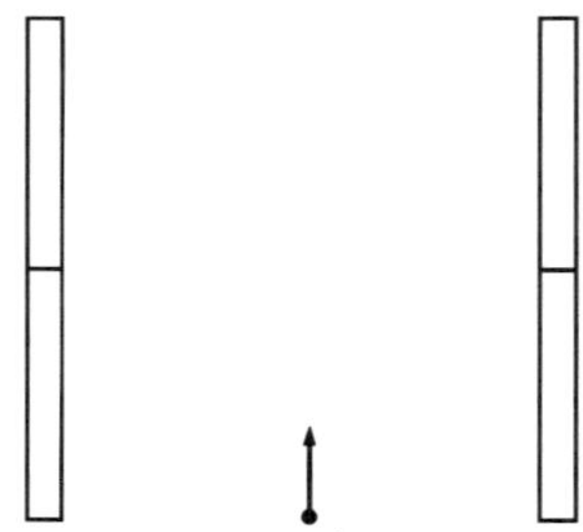

Figure P13.17

$\theta°$	I_{90°	I_{45°	I_{0°
0	349	348	348
5	390	368	349
10	533	437	351
15	730	551	351
20	941	699	354
25	1047	808	326
30	1088	872	307
35	1054	859	285
40	935	778	260
45	685	648	217
50	402	493	179
55	305	344	159
60	276	254	146
65	250	193	127
70	223	158	106
75	185	128	78
80	141	92	48
85	32	32	13

13.18 A short, vertical, 70-W, 5800-lm, high-pressure sodium lamp is located at the center of a right circular cylinder as shown in Fig. P13.18.

(a) Assuming the lamp is a Lambertain cylinder, find its intensity at $\theta = 90°$.

(b) Find the luminous fluxes to the top and walls of the cylinder. Use these to find the initial luminous exitances of these surfaces.

(c) Find the various form factors.

(d) Calculate the final luminous exitances from the top and sides of the cylinder.

(e) Find the configuration factors to point P and the horizontal illuminance at point P.

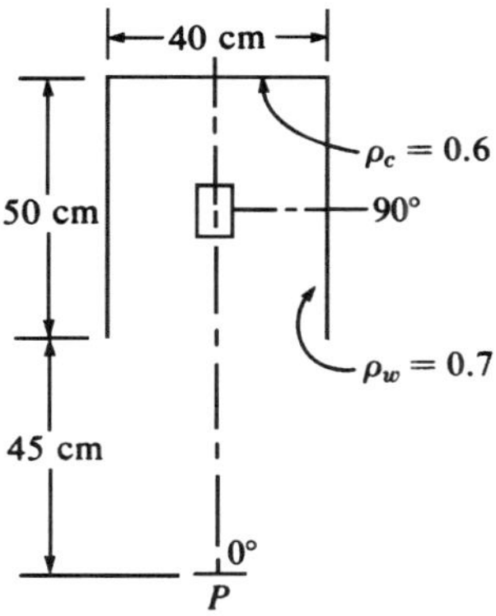

Figure P13.18

13.19 The form factor for two identical parallel rectangles is given in Eq. (13.94). If $\phi = \theta$, Eq. (13.90) gives the form factor for two identical circular disks. Let $\ell = w$ to make the rectangles become squares and let $b = \ell/\sqrt{\pi}$ to make the disks and squares have the same area. For equal separation distances, show that the two form factors are essentially the same if $\alpha = h/b \geq 1$.

13.20 For the ceiling recess in Ex. 13.20, calculate the final luminous exitances and the output flux by multiple bounce flux transfer analysis.

13.21 For the long rectangular tube in Ex. 13.16 let the portion numbered 2 be 9 ft wide, the portion numbered 1 be 6 ft wide and the height be 6 ft. Find f_{14} and f_{64}.

13.22 A rectangular fluorescent luminaire is mounted on the wall as shown in Fig. P13.22. The lamp is F40 T12 (3200 lumens). The luminaire is very long in and out of the page. The top and bottom are open. The right side is translucent plastic ($\tau = 60\%$, $\rho = 40\%$). The left side is opaque white ($\rho = 80\%$).

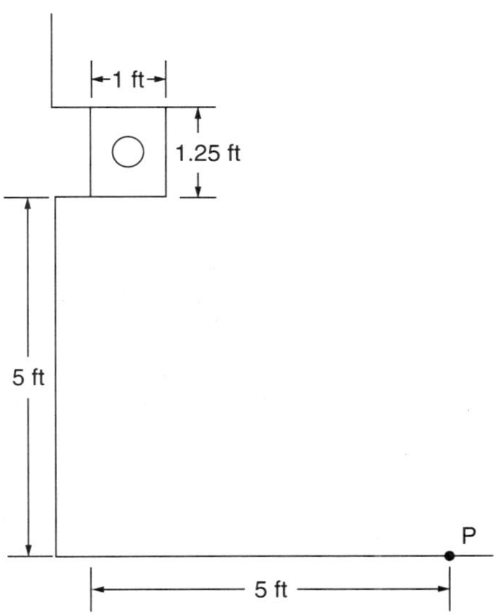

Figure P13.22

a) Find the form factors
b) Find the initial internal luminous exitances of the left and right sides.
c) Find the external luminous exitance of the right side by
   - one-bounce analysis
   - two-bounce analysis
   - multiple-bounce analysis
d) Find the horizontal illuminance at P through configuration factors. Assume the lamps are not visible from P.

13.23 A 12-ft diameter, 6-ft high cylindrical roof monitor is shown in Fig. P13.23.

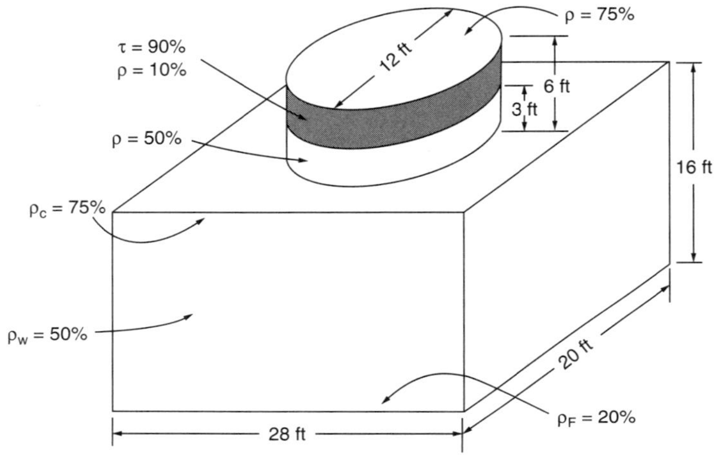

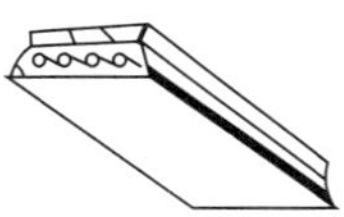

Typical Intensity Distribution and Per Cent Lamp Lumens		ρ_{cc} →	80			70			50			30			10			0
		ρ_w →	50	30	10	50	30	10	50	30	10	50	30	10	50	30	10	0
Maint. Cat.	SC	RCR ↓	Coefficients of Utilization for 20 Per Cent Effective Floor Cavity Reflectance (ρ_{FC} = 20)															
V	1.2	0	.69	.69	.69	.67	.67	.67	.64	.64	.64	.61	.61	.61	.59	.59	.59	.58
		1	.60	.58	.55	.59	.57	.55	.56	.55	.53	.54	.53	.51	.52	.51	.50	.49
0% ↑		2	.52	.49	.45	.51	.48	.45	.49	.46	.44	.47	.45	.43	.46	.44	.42	.40
		3	.46	.41	.38	.45	.41	.37	.43	.40	.37	.42	.39	.36	.40	.38	.35	.34
$57\frac{1}{2}$% ↓		4	.41	.36	.32	.40	.35	.32	.39	.34	.31	.37	.34	.31	.36	.33	.30	.29
		5	.36	.31	.28	.36	.31	.27	.35	.30	.27	.33	.30	.27	.32	.29	.26	.25
		6	.33	.28	.24	.32	.27	.24	.31	.27	.24	.30	.26	.23	.29	.26	.23	.22
		7	.30	.25	.21	.29	.25	.21	.28	.24	.21	.28	.24	.21	.27	.23	.21	.20
		8	.27	.22	.19	.27	.22	.19	.26	.22	.19	.25	.21	.19	.25	.21	.19	.17
		9	.25	.20	.17	.25	.20	.17	.24	.20	.17	.23	.20	.17	.23	.19	.17	.16
		10	.23	.18	.15	.23	.18	.15	.22	.18	.15	.22	.18	.15	.21	.18	.15	.14

Figure P13.23

The upper half of the cylinder-wall is clear glass; the lower half opaque. The vertical illuminance on the exterior of the monitor is 500 fc.

a) Find the luminous flux entering the large room from the cylinder by both one-bounce and multiple-bounce flux transfer analysis.
b) Find the average illuminance on the workplane, using the CU table given for a perfectly diffusing luminaire.

CHAPTER 14

14.1 Find the cost of producing one million lumen hours of light by an incandescent lamp under the following conditions:

Line voltage = 121 V
Rated lamp voltage = 115 V
Energy cost = 8¢/kWh
Lamp cost = 55¢
Labor cost = $2.00
Rated initial lamp power = 100 Watts
Rated lamp life = 750 h
Rated initial luminous flux = 1630 lm
Lumen depreciation factor = 0.91
Wattage depreciation factor = 0.96

14.2 The morality and lumen depreciation curves for a large installation of experimental incandescent lamps are shown in Fig. P14.2. What life rating should the manufacturer place on these lamps? What percentage of the total lumen hours of the installation occurs before 100% life is reached? At 70% life, the customary time for group replacement, what is the total lumen output in % of initial?

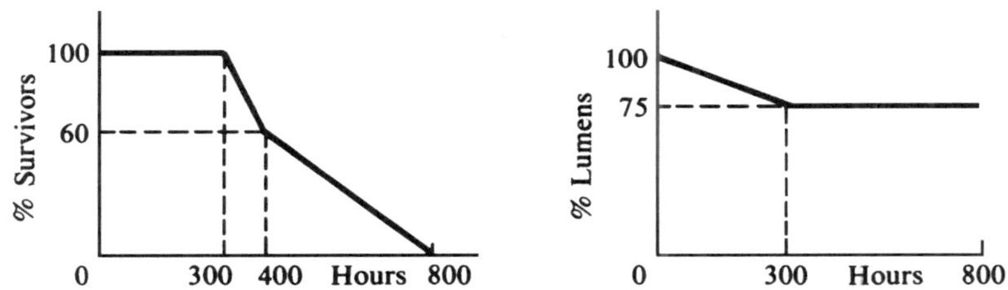

Figure P14.2

14.3 In an outdoor tennis court complex, 20 750-W, 17,040-lumen incandescent lamps, one to a pole, are to be replaced by 250-W, 22,000-lumen metal halide lamps (plus 50-W ballasts). Costs are:

Lamp	$45
Ballast	$40
Luminaire	$100
Labor	$20
Removal	$100 per pole
Energy rate	12¢/kWh

Lamps are burned an average of 4 hrs/day each day for 22 weeks.

a) For the same illuminance level, what is the simple payback time?

b) For the same number of poles and luminaires, what is the simple payback time?

What is the annual energy cost saving for each solution?

14.4 A motel replaces 1000 75-W incandescent lamps (1210 lm) with 18-W compact fluorescent lamps (1250 lm). Costs for each socket are:

Lamp	$8.00
Hardware	$3.00
Labor	$4.00

If energy costs 12¢/kWh and lamps are burned 2000 hrs/year, find the simple payback time.

14.5 A parking lot is lighted by 24 1200-W tungsten halogen lamps (29,000 lumens and 2000 hours life). There are six poles and 4 lamps per pole. The lot is to be relighted to approximately twice the illuminance with 1000-W metal halide lamps (115,000 lumens and 10,000 hours life). The data are

Burning hours	—	6 per night
Energy cost	—	12¢/kWh
Replacement luminaire and ballast	—	$660 each
Ballast watts	—	100 W
Lamp cost		$80 each
Installation cost	—	$175 per luminaire
Pole cost, installed	—	$850 each

Find simple payback time in years and simple life cycle cost in dollars.

14.6 A 200,000 ft^2 government field office was relighted.

Old lighting

3600 fluorescent luminaires each containing four 40-W T12 lamps and one 20-W magnetic ballast.

New lighting

Same luminaires relamped with two 32-W T8 lamps and one 7-W electronic ballast. Each luminaire was cleaned, and a silvered-film reflector inserted.

Costs

T8 lamp – $5, electronic ballast – $20, reflector – $15, hardware (each luminaire) – $2, labor per luminaire – $40, energy –10¢/kWh, lamp recycling cost – 50¢

Other data

Hours of operation – 7 am to 10 pm
Days of operation – 250 per year
Utility rebate – $50,000

Find simple payback time in years.

14.7 A college gymnasium was relighted with T5 fluorescent lamps and electronic ballasts, replacing T12 high output fluorescent lamps and magnetic ballasts. New luminaires were installed, considerable rewiring was done and a lighting control system was added. The data are (energy cost is 8¢/kWh):

PRERETROFIT

36 4-lamp luminaires, each containing two 33-W ballasts. 24 luminaires were 8 ft long with 110-watt lamps; 12 luminaires were 4 ft long with 60-watt lamps. Annual hours of burning were 3950 hr.

POSTRETROFIT

40 4-ft luminaires, each containing 3 55-W lamps, with a 7-W ballast for two of the lamps and a 6-W ballast for the third lamp. There are three levels of control, with one, two, and three lamps per luminaire on. The expected use is three lamps per luminaire on for 1350 hours (games) and one lamp per luminaire on for 2600 hours (practices), giving the same total usage of 3950 hours annually.

INITIAL COSTS

Luminaires (including ballasts)	\$207.80 each
Lamps	\$9.23 each
Controls	\$2300.00 total
Labor	\$9500.00 total

Find: Annual saving in energy cost and simple payback time.

14.8 A manufacturing firm is lighted with 7500 F40T12 fluorescent lamps. These are to be replaced by 34-W energy-saving T12 lamps. Lamps cost \$2.50 each and are burned 4000 hours per year. Lamp life is 20,000 hours and energy costs 10¢/kWh. What is the energy saving per year and over the life of the lamps? What is the total saving over the life of the lamps?

14.9 A father agrees to relight his son's home three years from now at an estimated cost of \$2000. How much must be placed in the bank now to cover this cost if the interest rate is 6%?

14.10 A lighting system costs \$1000 initially and \$200 for three years. What is the present value of these costs if the interest rate is 8%?

14.11 Two lighting systems are to be compared based on simple payback and discounted payback. The comparative costs are:

	A	B
Initial cost	**\$10,000**	**\$10,500**
Year 1 Operating cost	\$500	\$300
Year 2 Operating cost	\$600	\$400
Year 3 Operating cost	\$700	\$500
Year 4 Operating cost	\$800	\$600

Note that System B has a higher initial cost than System A but lower annual operating costs.

a) Determine the simple payback time in years of the extra investment initially in System B.

b) Determine the discounted payback time in years of the extra investment. This is defined as the time at which the present value of the future operating costs equals the initial cost. Assume an interest rate of 7%.

14.12 A company incurs annual lighting maintenance costs of \$1500. These costs are anticipated to escalate 5% per year for the next five years and then hold steady for the next five years. Find the present value of these costs if the interest rate is 7%.

14.13 A lighting installation has a salvage value of \$2000 after five years of service. What annual payment is equivalent to the salvage value if the interest rate is 7%?

14.14 The initial cost of a lighting system is \$125,000. Find the annual cost of this system over its 10 year life span, based on 6% interest.

14.15 A lighting system is to be designed for a gymnasium on a college campus. The choices are a metal halide system or a VHO fluorescent system. Determine, on an economic basis only, which system is best, by comparing them on a present value basis.

General Information

Room	60 ft × 30 ft × 35 ft
Illuminance	30 fc maintained
Energy rate	7¢/kWh
System life	10 years
Interest rate	7%
Burning hours per year	3100 hr

Group lamp replacement at 70% life

System A

195-W VHO fluorescent lamp
2-lamp luminaire
CU = 0.270, LLF = 0.544
Ballast (2-lamp) 15-W
14 luminaires
14,000 initial lamp lumens
Lamp life (10 hrs/start) – 15,000 hr
Lamp replacement interval – 10,500 hr

System B

400-W clear metal halide lamp
High-bay luminaire
CU = 0.345, LLF = 0.557
Ballast (1-lamp) 60-W
8 luminaires
35,500 initial lamp lumens
Lamp life (10 hrs/start) – 20,000 hr
Lamp replacement interval – 14,000 hr

Costs

	A	B
Luminaire	$80	$150
Wiring (per luminaire)	$25	$40
Labor (per luminaire)	$40	$35
Replacement lamp cost	$8	$40
Labor cost per lamp	$4	$10
Cleaning cost per luminaire	$12	$12
Cleanings per year	1	1
Salvage value	$150	$100

ANSWERS TO SELECTED PROBLEMS

1.2 (a) $W_{ab} = 1.6 \times 10^{-17}$ J, $W_{ac} = 3.2 \times 10^{-17}$ J
(b) $F = 3.2 \times 10^{-6}$ N, $a = 0.35 \times 10^{15}$ m/s^2
(c) $v_b = 0.59 \times 10^7$ m/s, $t = 0.017$ μs, $v_c = 0.84 \times 10^7$ m/s, $t = 0.024$ μs

1.4 $n = 3$, $\lambda = 103$ nm

1.7 I = 770 cd

1.8 L = 24.0 cd/in^2, M = 75.5 lm/in^2

1.9 (a) 327 cd (b) 0.321 cd/m^2 (c) 322 fc (d) 3220 lm

1.10 (a) 800 fc (b) 400 lm/ft^2 (c) 127 cd/ft^2

1.12 $E_P = 2$ fc, $E_Q = 2$ fc, $L_P = 398$ cd/in^2, $L_Q = 210$ cd/in^2

2.1 17.6 fc, 139 cd/in^2

2.2 E = 2.3 fc

2.3 E = 12 fc

2.5 (a) 509 cd (b) 502 fc (c) 0.7 fc

2.6 $\phi = 96{,}950$ lm

2.7 10.3 fc, 2 to 1, 4 to 1

2.8 1201 lm, 322 cd/ft^2, 87%

2.9 (a) E = 560 fc (b) 74% (c) 11 ft

2.10 27%, 9000 cd

2.13 11.02 fc, 11.18 fc, 12.10 fc

2.14 9.3 lm/ft^2

2.15 390 fc

2.18 3.4 fc

2.20 $E_A = E_B = 6$ fc, $E_c = 9$ fc

3.1 2.7×10^{-3} in

3.3 C = 1.75, $L_b = 0.36$ fL

3.4 3 to 1

3.7 $\rho_b = 83\%$

3.9 18 fL, 24 fc, 52.5 fL, 70 fc

3.11 Yes

3.14 25.9 rings/min

4.3 $M_{\lambda\,max} = 252\ W/cm^2 \bullet \mu m$

4.4 9.2%

4.6 T = 2725 K

4.7 190 W

4.8 $\ell = 8.21$ cm, d = 0.093 cm, $\varepsilon = 0.435$

4.10 T = 3000 K, P = 58.9 W, $\phi = 1330$ lm

4.12 $L = 0.34\ cd/cm^2$

4.14 58 lm/W

4.15 9.3 ft from 500-nm source, 96 fc

4.16 $\phi = 4014$ lm, $P_i = 7.8$ W

4.18 201 h, 45 W, 584 lm, 12.9 lm/W, 166 V

5.2 (a) 253.7 nm (b) 4.95×10^{-20} J (c) 0.538 nm (d) 8.7×10^{-36} kg

5.4 56.2%

5.5 3.16×10^{-34} J-s

5.6 $v_{max} = 6.8 \times 10^5$ m/s

5.7 4.21 in, 2.70×10^6 m/s, 3.72 in, 18

5.8 593 nm

5.10 ionize, excite (noλ), elastic, excite (185 nm), elastic

5.13 2.25×10^{19} photons/cm^2 • min

5.14 8.5 cm

5.15 $n = 3.53 \times 10^{20}/m^3$, $\lambda_e = 5.77$ cm

6.1 1580 lm

6.4 $I_s = 0.277A$, $I_c = 0.312A$

6.5 108 V, 480 mA, PF = 0.77

6.6 (a) 133.5 V, 0.257A, 30.8 W
(b) 1.4 h, 76 Ω

6.7 43%, 35.3 V

6.9 L = 0.36 h, C = 6.9 μfd

6.10 L = 0.366 h, R = 16.7 Ω, C = 6.48 μfd, $N_1/N_2 = 1.46$

7.1 (a) red: 492 W, 1370 lm; green: 1.50 W, 1000 lm; blue: 0 W, 0 lm
(b) R = 0.106, G = 0.019, B = 0
(c) red: 6 units, 77,200 lm; green: 6 units, 354,000 lm; blue: 6 units, 4340 lm

7.2 X = 1.7970, Y = 3.6668, Z = 0.8186, x = 0.286, y = 0.584

7.3 580 nm, 0.77 W

7.4 X = 75.003, Y = 66.339, Z = 0.504, x = 0.529, y = 0.468

7.5 x = 0.354, y = 0.563, 571 nm, 41%

7.6 x = 0.309, y = 0.317, L_t = 21.5 cd/ft^2

7.9 x = 0.447, y = 0.399

7.10 x = 0.394, y = 0.391

7.11 590 nm, 49%, 0.655

7.14 Source: x = 0.33, y = 0.46
Face: x = 0.38, y = 0.49

7.15 h* = 86.3°, C* = 93.6, 5Y9/13

8.2 4140 cd/m^2

8.4 L_A = 6.2 cd/in^2, L_O = 5.8 cd/in^2

8.6 43 fc

8.8 VCP = 85%

8.9 VCP = 78%

8.11 VCP = 90%

8.12 UGR = 8.45

9.1 16 luminaires, 1.1 W/ft^2

9.2 42 lamps

9.4 8 luminaires

9.5 37 luminaires

9.7 LLF (old) = 0.65, LLF (new) = 0.74

9.8 52 luminaires

9.9 18 fc

9.10 30 fc

9.11 460 luminaires

9.12 59 fc

9.13 25 fc

9.15 257 luminaires, 241 luminaires

9.16 8 fc

9.17 33 lamps

10.1 30°, 4°

10.3 2/3 cm

10.5 **(a)** f = 33 cm **(b)** f = −100 cm

10.6 **(a)** 167 lm **(b)** 17 cm **(c)** 2.5 cm **(d)** 10.2 cm

10.8 **(a)** 220 lm **(b)** 5950 fc

10.10 82%

10.12 3790 cd

10.13 140 lamps

10.15 167 nm, 1.44

10.17 3650 lines/cm

10.18 355 nm, 396 nm

10.20 $\sqrt{3}$, 3

11.1 54.8°

11.3 (a) 19.2% (b) 2.6 fc

11.5 CU′ = 0.328, E = 2.1 fc

11.6 CU′ = 0.432, E = 2.7 fc

11.7 S = 181 ft

11.8 0.3 fc

11.9 MH = 35 ft, S = 133 ft

11.11 CV = 0.21

11.13 1000 W HPS

12.1 1.02 kW/m^2, 9290 fc

12.2 3.67×10^{26} W

12.5 $a_t = 43.1°$, $a_s = -64.6°$

12.8 31 fc

12.9 (a) 10.1 cd/in^2 (b) 650 fc

12.10 930 fc

12.11 210 fc, 49.6 fc, 270 fc, 70.4 fc

12.12 1510 fc

12.13 4540 fc

12.14 26 skylights, 195 fc

12.15 113 fc

12.17 5.3%

12.18 2.8%

13.1 6.6 fc

13.2 31 fc, 16 fc, 27 fc

13.3 114 fc

13.5 19 fc, 38 fc

13.6 29 fc

13.9 0.0484, 29 fc

13.10 c = 0.0297

13.11 5500 cd

13.13 27 fc

13.14 5.1 fc

13.17 31 fc

13.18 (a) 588 cd (b) 0.165 lm/cm^2, 0.569 lm/cm^2 (d) 0.810 lm/cm^2, 1.225 lm/cm^2 (e) 172 fc

13.20 147 lm/cm^2, 263 lm/cm^2

13.21 $f_{14} = 0.238$, $f_{64} = 0.11$

13.22 **(b)** 146 lm/ft^2, 73 lm/ft^2
(d) 14.7 fc

13.23 $\phi_{out} = 212{,}000$ lm, E = 22 fc

14.1 $8.42

14.3 **(a)** 4.9 yrs, **(b)** 6.2 yrs

14.5 2.7 yrs, $26,820

14.6 1.84 yrs

14.7 $3721, 5.7 yrs

14.10 $1515.42

14.11 2.50 yrs, 2.85 yrs

14.12 $12,684.09

14.14 $16,983.49

Index

ILLUMINATING ENGINEERING SOCIETY OF NORTH AMERICA

Joseph B. Murdoch, Author of *Illuminating Engineering: From Edison's Lamp to the LED,* is a long standing member of the Illuminating Engineering Society of North America (IESNA). The information below offers some brief descriptions of the mission and services of the IESNA.

Purpose

The IESNA is the recognized technical authority on illumination. IESNA was founded in 1906, and for almost 100 years, its objective has been to communicate information on all aspects of good lighting practice to its members and the lighting community through a variety of programs, publications, and services.

Membership

Membership in the IESNA provides numerous opportunities for education, career advancement, and networking. With more than 9,000 members, the Society represents the interests of all professionals whose work involves lighting, offering diversity and exchange of information between various sectors of the community.

Local Section

The IESNA has over 100 sections throughout the United States, Canada and Mexico, each offering numerous educational courses, seminars and special events to a broad geographical base of members.

Standing Committees

Over 70 technical, research, and design and application committees study and report on all aspects of lighting. These committees, composed of representatives from every segment of the lighting industry, set consensus standards and recommended practices.

Publications

Keep current with the progress of the industry and the latest innovative technologies. Presently, over 100 documents on lighting are published in categories ranging from design guides and guidelines, to recommended practices. Members receive substantial discounts on all IESNA publications.

IESNA Lighting Handbook

The indispensable reference for industry professionals. Included in this one comprehensive volume is everything you need to know about lighting—explanations of concepts, techniques, applications, procedures and systems.

IESNA Educational Programs

The IESNA offers many educational programs that are given by local sections and corporations. They include "Lighting Education Fundamentals" (ED-100) and an updated version of "Intermediate Level Lighting" (ED-150). And, newly created online courses including "Discover Lighting," a basic, introduction to the world of lighting, and

"Security Lighting for People, Property, and Public Spaces," provide web based learning opportunities.

Lighting Design + Application (LD+A)

LD+A is a unique monthly magazine, popular with both experienced professionals and novices with an interest in design and lighting applications. Each monthly issue features a unique application topic, technical articles, product development, current issues, IESNA events, and employment opportunities. Subscription is free with membership.

IESNA Annual Conference

Hundreds of professionals gather at the yearly IESNA Annual Conference to update their technical knowledge, recognize achievements, attend luncheons where Society honors and the IIDA Awards are presented. And the highlight social event, the President's Banquet, presents a wonderful opportunity to enjoy the company of friends and associates in the lighting industry.

Lightfair International

LIGHTFAIR INTERNATIONAL, cosponsored by the IESNA, is the largest architectural lighting trade show in the United States, with over 19,000 annual attendees and more than 700 exhibitors. It sets the pace for future industry developments, as manufacturers display their newest and most advanced products.

IESNA Annual Street and Area Lighting Conference

Dedicated to the continuing education of street and area lighting personnel from utilities, municipalities, public works, transportation departments, manufacturers and distributors, the IESNA ANNUAL STREET AND AREA LIGHTING CONFERENCE is the only one of its kind. The conference program includes exhibits of the industry's leading manufacturers and three days of educational programming on outdoor lighting topics.

Awards Program

Every year the IESNA membership nominates outstanding professionals in the field to receive the IESNA Medal, the Louis B. Marks Award, the Distinguished Service Award, and IESNA Fellow status.

International Illumation Design Awards

The IIDA Program is one of the most important and visible programs in the IESNA. It recognizes the fruition of all the Society's research, development, and education. Its purpose is not only to reward excellence in design, but also to educate by example.

This is just a small sample of the values and benefits of membership in the IESNA. For more information, write: IESNA, 120 Wall Street, 17th Floor, New York, NY 10005, or call 212-248-5000, ext. 117

To order additional copies of this book or other titles from Visions Communications, please use the following information:

Visions Communications
200 East 10th Street, #714
New York, NY 10003

212-529-4029 (ph & fax)
917-734-1523 (mobile)
www.visionsbooks.com
info@visionsbooks.com

For further information about submitting a manuscript, please call 212-529-4029.